Lecture Notes in Computational Vision and Biomechanics

Volume 30

The research related to the analysis of living structures (Biomechanics) has been a source of recent research in several distinct areas of science, for example, Mathematics, Mechanical Engineering, Physics, Informatics, Medicine and Sport. However, for its successful achievement, numerous research topics should be considered, such as image processing and analysis, geometric and numerical modelling, biomechanics, experimental analysis, mechanobiology and enhanced visualization, and their application to real cases must be developed and more investigation is needed. Additionally, enhanced hardware solutions and less invasive devices are demanded.

On the other hand, Image Analysis (Computational Vision) is used for the extraction of high level information from static images or dynamic image sequences. Examples of applications involving image analysis can be the study of motion of structures from image sequences, shape reconstruction from images, and medical diagnosis. As a multidisciplinary area, Computational Vision considers techniques and methods from other disciplines, such as Artificial Intelligence, Signal Processing, Mathematics, Physics and Informatics. Despite the many research projects in this area, more robust and efficient methods of Computational Imaging are still demanded in many application domains in Medicine, and their validation in real scenarios is matter of urgency.

These two important and predominant branches of Science are increasingly considered to be strongly connected and related. Hence, the main goal of the LNCV&B book series consists of the provision of a comprehensive forum for discussion on the current state-of-the-art in these fields by emphasizing their connection. The book series covers (but is not limited to):

- Applications of Computational Vision and Biomechanics
- Biometrics and Biomedical Pattern Analysis
- Cellular Imaging and Cellular Mechanics
- Clinical Biomechanics
- Computational Bioimaging and Visualization
- Computational Biology in Biomedical Imaging
- Development of Biomechanical Devices
- Device and Technique Development for Biomedical Imaging
- Digital Geometry Algorithms for Computational Vision and Visualization
- Experimental Biomechanics
- Gait & Posture Mechanics
- Multiscale Analysis in Biomechanics
- Neuromuscular Biomechanics
- Numerical Methods for Living Tissues
- Numerical Simulation
- Software Development on Computational Vision and Biomechanics
- Grid and High Performance Computing for Computational Vision and Biomechanics
- Image-based Geometric Modeling and Mesh Generation
- Image Processing and Analysis
- Image Processing and Visualization in Biofluids
- Image Understanding
- Material Models
- Mechanobiology
- Medical Image Analysis
- Molecular Mechanics
- Multi-Modal Image Systems
- Multiscale Biosensors in Biomedical Imaging
- Multiscale Devices and Biomems for Biomedical Imaging
- Musculoskeletal Biomechanics
- Sport Biomechanics
- Virtual Reality in Biomechanics
- Vision Systems

More information about this series at http://www.springer.com/series/8910

Durai Pandian · Xavier Fernando
Zubair Baig · Fuqian Shi
Editors

Proceedings of the International Conference on ISMAC in Computational Vision and Bio-Engineering 2018 (ISMAC-CVB)

Volume 1

 Springer

Editors
Durai Pandian
SCAD Institute of Technology
Palladam, India

Xavier Fernando
Department of Electrical
and Computer Engineering
Ryerson University
Toronto, ON, Canada

Zubair Baig
School of Computer and Security
Science
Edith Cowan University
Joondalup, WA, Australia

Fuqian Shi
Wenzhou Medical University
Wenzhou, China

ISSN 2212-9391 ISSN 2212-9413 (electronic)
Lecture Notes in Computational Vision and Biomechanics
ISBN 978-3-030-00664-8 ISBN 978-3-030-00665-5 (eBook)
https://doi.org/10.1007/978-3-030-00665-5

Library of Congress Control Number: 2018954619

This Springer imprint is published by the registered company Springer Nature Switzerland AG
The registered company address is: Gewerbestrasse 11, 6330 Cham, Switzerland

Contents

Digital Image Watermarking Using Sine Transform Technique

S. N. Prajwalasimha, S. Sonashree, C. J. Ashoka and P. Ashwini

Abstract In this chapter, digital image watermarking using sine transformation has been introduced in the frequency domain. In the proposed technique, the secret image is first compressed using a sine transformation and then embedded into the host image. The resultant images obtained from the proposed algorithm are subjected to various security attacks and the results are compared with other existing algorithms. The results obtained are better compared to the existing techniques.

1 Introduction

Authentication and information security play a vital role in multimedia communication. The digital watermarking technique provides both authentication and security to host data as well as a watermark, respectively. Many algorithms are designed to provide security to the information by encrypting the secret data. In order to provide authentication to the host, an information should be embedded into it [1–3]. Visible and invisible watermarking techniques are the fundamental classifications. While embedding the watermark into the host, the size of the watermark is a major aspect. Large the size of the watermark, more the vulnerability to the noise [4]. Irreversibility is the major problem among the conventional techniques, which leads to the large difference between the original and recovered watermark. This is unacceptable to the high authentication sectors like military and forensics. Reversible watermarking techniques provide the solution to the above-stated problem and are classified based on compression, differential expansion, and histogram shifting. In the compression-based algorithms, the secret image or watermark is first subjected to compression and then embedded into the host. Celik et al. proposed a lossless least significant bit (LSB)-based embedding algorithm [5]. In this technique, the compressed watermark is embedded into the LSB of the host. The correlation between the host and watermarked data considerably differs to some extent due to embedded information

S. N. Prajwalasimha (✉) · S. Sonashree · C. J. Ashoka · P. Ashwini
Department of Electronics & Communication, ATME College of Engineering, Mysuru, India
e-mail: prajwalasimha.sn1@gmail.com

© Springer Nature Switzerland AG 2019
D. Pandian et al. (eds.), *Proceedings of the International Conference on ISMAC in Computational Vision and Bio-Engineering 2018 (ISMAC-CVB)*, Lecture Notes in Computational Vision and Biomechanics 30,
https://doi.org/10.1007/978-3-030-00665-5_1

1

in the host. A unique difference expansion based watermarking technique has been described by Tian [6]. In this approach, the embedded watermark in the host is very sensitive to noise. Histogram shifting method was proposed by Ni Z. et al., where the embedding process was done based on the peak values in the histograms of host and watermark [7]. Due to the histogram shifting, the pixel intensity variations can be observed in the watermarked data which intern affects the correlation between the host and watermarked data.

The proposed technique is a reversible, lossless and compression-based approach by which desired outcomes are noticed compared to popular existing techniques. Along with this, it supports secrete key authentication at the initial stage of the algorithm and provides the first stage security.

2 Proposed Scheme

The proposed algorithm consists of two phases: Embedding the watermarking into the host image and watermark extraction from the host image. Figure 1 shows the flow diagram of the proposed watermarking scheme.

2.1 Embedding the Watermark

Step 1 A Grayscale 256×256 image is taken as host image for the embedding process.

Step 2 By getting the desired secrete key, the watermark of size 128×128 is subjected for sine transformation.

$$W|(x, y) = \text{DST}[W(x, y)] \quad 1 < x, y < 128 \tag{1}$$

where

$W(x, y)$ is the input watermark.
$W|(x, y)$ is the transformed watermark.

Step 3 A complete watermark of size 256×256 is composed by concatenating the transformed watermark back to back and side by side.

$$W||(x, y) = \begin{bmatrix} W|(x, y) & W|(x, y) \\ W|(x, y) & W|(x, y) \end{bmatrix} \tag{2}$$

where

$W|(x, y)$ is the transformed watermark of size 128×128.

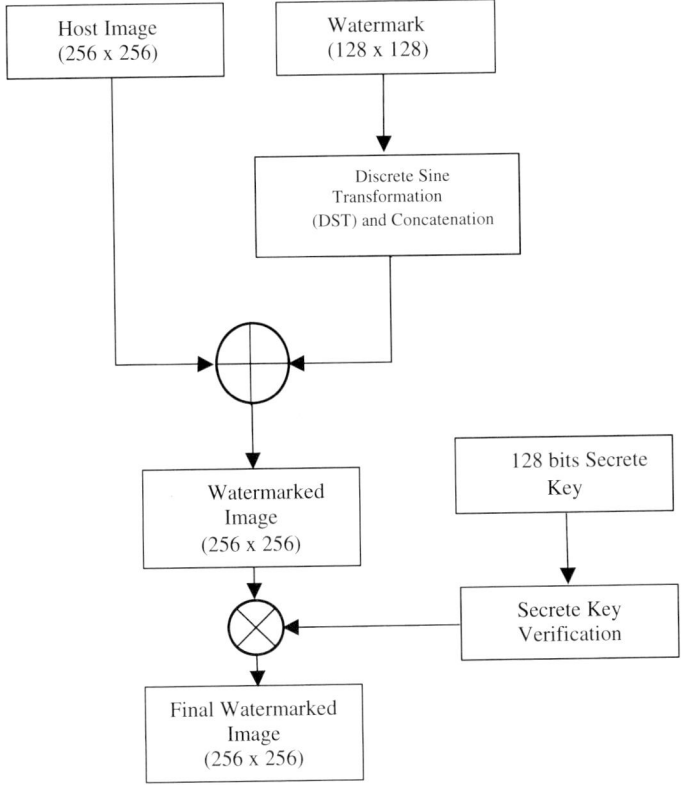

Fig. 1 Flow diagram of the proposed watermarking scheme

$W||(x, y)$ is the concatenated watermark of size 256×256.

Step 4   The arithmetic addition is performed between concatenated watermark and the host to get the final watermark.

$$W, (x, y) = H(x, y) + W||(x, y) \quad 1 < x, y < 256 \qquad (3)$$

where

$W, (x, y)$ is the Watermarked image.
$H(x, y)$ is the host image of size.
$W||(x, y)$ is the transformed watermark.

2.2 De-watermarking

Step 1   Watermarked image is subjected to filtering in the first stage to reduce the noise contents.

Step 2   The filtered image is subjected to arithmetic subtraction with the host image to get the watermark in the transformed phase.

$$W'||(x, y) = W, (x, y) - H(x, y) \quad 1 < x, y < 256 \qquad (4)$$

where

$W, (x, y)$ is the watermarked image.
$H(x, y)$ is the host image of size.
$W'||(x, y)$ is the transformed watermark.

Step 3   The obtained watermark of size 256×256 is separated by de-concatenating the transformed watermark.

$$W'||(x, y) = \begin{bmatrix} W'|(x, y) & W'|(x, y) \\ W'|(x, y) & W'|(x, y) \end{bmatrix} \qquad (5)$$

where

$W'|(x, y)$ is the transformed watermark of size 128×128.
$W'||(x, y)$ is the concatenated watermark of size 256×256.

Step 4   By getting the desired secret key, the obtained watermark of size 128×128 in the transformed phase is subjected for inverse sine transformation.

$$W'|(x, y) = \text{IDST}[W||(x, y)] \quad 1 < x, y < 128 \qquad (6)$$

where

$W||(x, y)$ is the transformed watermark.
$W'|(x, y)$ is inverse sine-transformed watermark.

Step 5   Comparison is made between the actual secret key and the input key and based on the desired combination, the final watermark is made visible.

3 Experimental Results

MATLAB2013a software is used for the analysis of the proposed algorithm. Three standard images are considered for the analysis with a secret image. The performance analysis involves Peak Signal-to-Noise Ratio (PSNR), Mean Square Error (MSE),

Table 1 The comparison of PSNR between the host and watermarked images

Images	PSNR (dB)
Lena	49.29 (DWT) [8]
	29.589 (DCT) [8]
	37.27 (HFT) [8]
	41.28 (IMD-WC-T) [8]
	40.6926 (Genetic algorithm) [9]
	51.6317 (Proposed algorithm)
Cameraman	40.2608 (Genetic algorithm) [9]
	51.6033 (Proposed algorithm)
Pirate	52.3612 (Genetic algorithm) [9]
	53.3748 (Proposed algorithm)

Table 2 The comparison of mean square error (MSE) between host and watermarked images

Images	MSE
Lena	1 (DWT) [8]
	26 (DCT) [8]
	12 (HFT) [8]
	5 (IMD-WC-T) [8]
	0.4466 (Proposed algorithm)
Cameraman	0.4495 (Proposed algorithm)
Pirate	0.2990 (Proposed algorithm)

Table 3 The comparison of correlation between the host and watermarked images

Images	Correlation
Lena	1 (DWT) [8]
	0.984 (DCT) [8]
	0.994 (HFT) [8]
	0.999 (IMD-WC-T) [8]
	1 (Proposed algorithm)
Cameraman	1 (Proposed algorithm)
Pirate	1 (Proposed algorithm)

Correlation (NC), and Watermark to Document Ratio (WDR) between host image and watermarked image. In all the above tests, the proposed system has given much better results compared to existing algorithms as tabulated in (Tables 1, 2, 3 and 4; Figs. 2, 3, 4 and 5).

Table 4 The comparison of watermark to document ratio between host and watermarked (WDR) images

Images	WDR (dB)
Lena	−69.8801 (Proposed algorithm)
Cameraman	−69.8516 (Proposed algorithm)
Pirate	−71.6231 (Proposed algorithm)

Fig. 2 Watermark

Fig. 3 Host image

Fig. 4 Watermarked image

Fig. 5 Recovered watermark

4 Conclusion

In the proposed algorithm, the watermark is subjected for sine transformation and then embedded into the host image. The watermarked image is then subjected to various security tests. By this method, the embedded watermark is very less affected by robust attacks. It is observed that very high PSNR is achieved for the watermarked image compared to most popular existing techniques. Very less mean square error is observed and it very close to the ideal value. It is also noticed that the correlation between the host and watermarked images is equal to unity, which is observed only in the DWT technique. The statistical analysis indicates that the proposed algorithm is a better alternative to the existing techniques. And also the algorithm utilizes very less computational time for execution. Further, a hybrid technique can be developed by implanting both sine and cosine transformation to get better results.

References

1. Ahmaderaghi B, Kurugollu F, Del Rincon JM, Nekrasov D, Bouridane A (2018) Blind image watermark detection algorithm based on discrete Shearlet transform using statistical decision theory. IEEE Trans Comput Imaging 4(1):46–59
2. Brandão AS, Jorge DC (2016) Artificial neural networks applied to image steganography. IEEE Latin Am Trans 14(3):1361–1365
3. Prajwalasimha SN (2018) Pseudo-Hadamard Transformation-Based Image Encryption Scheme. Integrated Intelligent Computing, Communication and Security, information & communication technology, Studies in Computational Intelligence 771, Springer, pp 575–583
4. Prajwalasimha SN, Pavithra AC (2018) Digital Image Watermarking based on Successive Division. In: Proceedings of IEEE international conference on communication and electronics systems, pp 31–35
5. Celik MU, Sharma G, Tekalp AM, Saber E (2005) Lossless generalized-LSB data embedding. IEEE Trans Image Process 14(2):253–266
6. Tian J (2003) Reversible data embedding using a difference expansion. IEEE Trans Circuits Syst Video Technol 13(8):890–896
7. Ni Z, Shi YQ, Ansari N, Su W (2006) Reversible data hiding. IEEE Trans Circuits Syst Video Technol 16(3):354–362
8. Salama AS, Mokhtar MA (2016) Combined technique for improving digital image watermarking. In: Proceedings of IEEE international conference on computer and communications, pp 557–562
9. Khanna AK, Roy NR, Verma B (2017) Digital image watermarking and its optimization using genetic algorithm. In: Proceedings of IEEE international conference on computing, communication and automation, pp 1140–1144

Logarithmic Transform based Digital Watermarking Scheme

S. N. Prajwalasimha, A. N. Sowmyashree, B. Suraksha
and H. P. Shashikumar

Abstract In this chapter, a new frequency domain transformation based digital watermarking technique has been introduced. The proposed technique adopts logarithmic transformation in order to embed the watermark image into the host image. The algorithm is subjected for various security attacks and compared with other frequently available digital watermarking schemes. The results obtained from the proposed algorithm are more satisfied compared to the existing systems.

1 Introduction

Data confidentiality and authentication are important aspects in the secured communication systems. Confidentiality can be more effectively achieved by various cryptosystems. In order to achieve both confidentiality and authentication, various steganosystems are available. As encryption, digital watermarking plays a vital role to achieve high degree of confidentiality and authentication [1–3].

Digital watermarking techniques are classified into two categories: spatial domain and frequency domain based on algorithms used for watermarking process [4, 5]. In spatial domain watermarking systems, the pixel values of the watermark image are directly altered and embedded into the host image. Least significant bit (LSB) technique is one among them. Whereas in frequency domain, the pixel values of the watermark image are first subjected for transformation and then altered before embedding into the host image. Based on applications, digital watermarking techniques are classified into two categories: robust and fragile techniques. Robust techniques are used for copyright and fingerprint protection and fragile techniques are used for data authentication and tamper detections [6]. Digital watermarking techniques are further classified into two classes: Perceptible and Imperceptible techniques based on visibility of the watermark on the host images. If the watermark is

S. N. Prajwalasimha (✉) · A. N. Sowmyashree · B. Suraksha · H. P. Shashikumar
Department of ECE, ATME College of Engineering, Mysuru, Karnataka, India
e-mail: prajwalasimha.sn1@gmail.com

© Springer Nature Switzerland AG 2019
D. Pandian et al. (eds.), *Proceedings of the International Conference on ISMAC in Computational Vision and Bio-Engineering 2018 (ISMAC-CVB)*, Lecture Notes in Computational Vision and Biomechanics 30,
https://doi.org/10.1007/978-3-030-00665-5_2

visible on the host image, the technique used is termed as perceptible. Otherwise, it is imperceptible watermark [4, 7].

The proposed technique is a frequency domain, robust and imperceptible watermarking technique, which adopts logarithmic transformation to alter the pixel values of the watermark image and then it is embedded into the host image.

2   Related Work

Embedding a watermark into the multimedia data can be done by various methods. In the genetic algorithm based watermarking technique, strength of the algorithm is decided by three major factors: invisibility, less computations and resistance to various attacks. The algorithm is tested against filtering, resizing and Gaussian noise attacks. The results prove that the algorithm is better against the above mentioned attacks on various conditions [4]. More popular frequency domain watermarking algorithms are based on discrete cosine and wavelet transformation technique (DCT and DWT). A combined watermarking technique (IMD-WCT) has been proposed by Ahmed S. Salama et al. In this, they compared the results of various digital watermarking schemes with combined technique introduced by them. The results show that combined method is not much effective against discrete wavelet transformation technique [6].

Sanjay Kumar et al. made a comparative study on various watermarking schemes and classified them on their respective domains. In addition, they compared the results obtained by various algorithms by subjecting them to various attacks [7].

In interleaving and block extraction based technique, the watermarked image is subjected for various attacks including median filtering and noising. The recovered watermark is interrupted to some extent but resistive to them [8].

Mashruha Raquib Mitashe et al. proposed an adoptive technique, which is the combination of Discrete Wavelet Transform (DWT), XieBeni clustering (XFCM) and Particle Swarm Optimization (PSO). The watermark embedding has been made by DWT, XFCM is used to locate the positions to embed the watermark into the host and PSO is used for preprocessing the host image. The algorithm is effective against various attacks but the peak signal-to-noise ratio (PSNR) obtained for the watermarked image is less compared to other frequency domain watermark approaches [9]. In the Hilbert transformation based watermarking technique, the watermark is subjected for transformation and then embedded into the host. Various attacks are made on the watermarked image and observed better results in the recovery process [10].

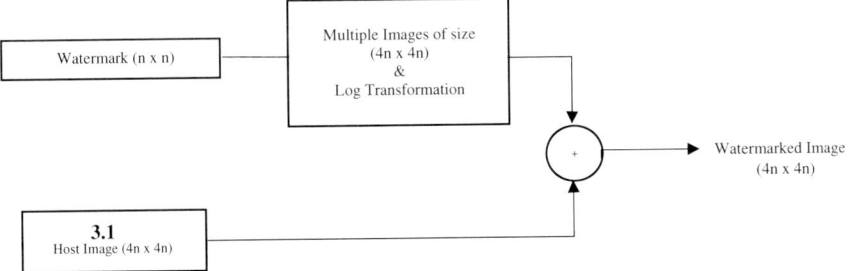

Fig. 1 Proposed model for watermark embeddeing algorithm

3   Proposed Scheme

The proposed algorithm is organized with both watermark embedding and extraction processes. A new method adopted here is logarithmic transformation. Each pixel value of the watermark is subjected for logarithmic transformation and the base is the ratio of maximum pixel value to its minimum. Figure 1 shows the proposed model of digital watermarking scheme.

3.1   Watermark Embedding Algorithm

Step 1   The input watermark image of size $n \times n$ as shown in Fig. 2 is made as multiple images of size $4n \times 4n$, which is same as the size of host image as shown in Fig. 3 ($4n \times 4n$).

Fig. 2   Watermark

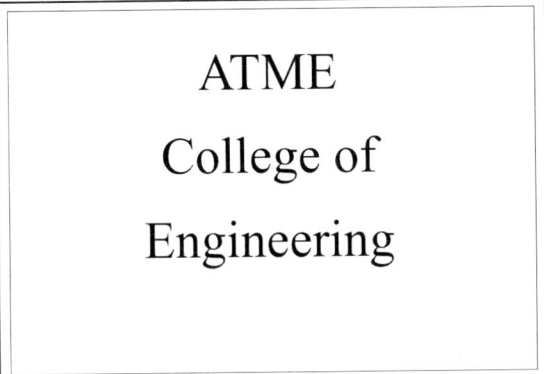

Fig. 3 Host image of Lena

Step 2   The watermark is then subjected to logarithmic transformation.

$$T(x, y) = \log_m[I(x, y)] \quad 0 < x, y < 4n \tag{1}$$

where $m = \dfrac{\max[I(x,y)]}{\min[I(x,y)]}$

$I(x, y)$ is the input watermark of size $4n \times 4n$

$T(x, y)$ is the transformed watermark of size $4n \times 4n$

Step 3   The transformed watermark and the host image are subjected to arithmetic addition.

$$W(x, y) = H(x, y) + T(x, y) \quad 0 < x, y < 4n \tag{2}$$

where $W(x, y)$ is the watermarked image of size $4n \times 4n$

$H(x, y)$ is the host image of size $4n \times 4n$

$T(x, y)$ is the transformed watermark of size $4n \times 4n$

The resultant watermarked image is of size $n \times n$ as shown in Fig. 4.

Fig. 4 Watermarked image

3.2 Watermark Extraction Algorithm

Step 1 The watermarked image is subjected for filtering in the first stage to reduce the noise contents.

Step 2 The filtered image is then subtracted with the host image to get the transformed watermark.

$$W'(x, y) = W^f(x, y) - H(x, y) \quad 0 < x, y < 4n \quad (3)$$

where $W'(x, y)$ is the same as $T(x, y)$ size $4n \times 4n$

$H(x, y)$ is the host image of size $4n \times 4n$
$W^f(x, y)$ is the transformed watermark after filtering of size $4n \times 4n$

Step 3 The transformed watermark is subjected to antilogarithmic transformation to get the original watermark.

$$I'(x, y) = m^{[W'(x,y)]} \quad 0 < x, y < 4n \quad (4)$$

where $m = \dfrac{\max_{[I(x,y)]}}{\min_{[I(x,y)]}}$

$I'(x, y)$ is the recovered watermark of size $4n \times 4n$

The resultant image is divided into eight $n \times n$ images and all together added to get the final retrieved watermark.

The recovered watermark is shown in Fig. 5.

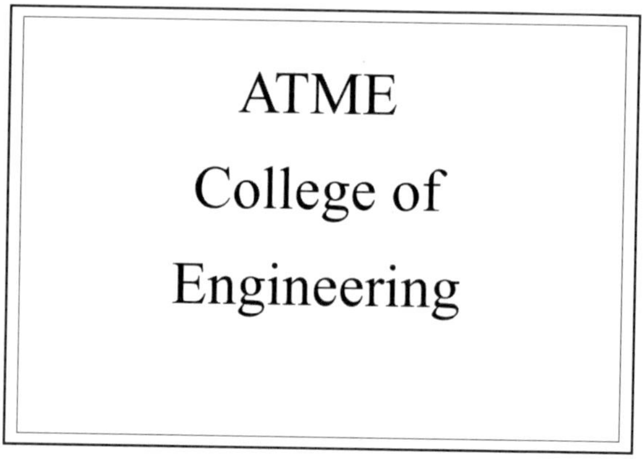

Fig. 5 Recovered watermark

4 Experimental Results

Three standard images of Lena, Cameraman, and Pirate are considered for the performance analysis [4]. The proposed algorithm is designed and implemented in MATLAB R2013a using Intel i3 processor @ 1.7 GHz, 4 GB DDR RAM and Windows 8 OS. The host images are considered with size 512×512 and the watermark of size 128×128. The performance analysis involves peak signal-to-noise ratio (PSNR), mean square error (MSE), Correlation (NC), and watermark to document ratio (WDR) between host image and watermarked image. In all the above tests, the proposed system has given much better results compared to existing algorithms as tabulated in Tables 1, 2, 3, 4 and 5.

Table 1 The comparison of PSNR between host and watermarked images

Images	PSNR (dB)
Lena	49.29 (DWT) [6]
	29.589 (DCT) [6]
	37.27 (HFT) [6]
	41.28 (IMD-WC-T) [6]
	40.6926 (Genetic algorithm) [4]
	54.0098 (Proposed algorithm)
Cameraman	40.2608 (Genetic algorithm) [4]
	54.0098 (Proposed algorithm)
Pirate	52.3612 (Genetic algorithm) [4]
	54.0098 (Proposed algorithm)

Table 2 The comparison of mean square error (MSE) between host and watermarked images

Images	MSE
Lena	1 (DWT) [6]
	26 (DCT) [6]
	12 (HFT) [6]
	5 (IMD-WC-T) [6]
	1.0372 (Proposed algorithm)
Cameraman	1.0372 (Proposed algorithm)
Pirate	1.0372 (Proposed algorithm)

Table 3 The comparison of correlation between the host and watermarked images

Images	Correlation
Lena	1 (DWT) [6]
	0.984 (DCT) [6]
	0.994 (HFT) [6]
	0.999 (IMD-WC-T) [6]
	1 (Proposed algorithm)
Cameraman	1 (Proposed algorithm)
Pirate	1 (Proposed algorithm)

Table 4 The comparison of watermark to document ratio between host and watermarked (WDR) images

Images	WDR (dB)
Lena	−68.9512 (Proposed algorithm)
Cameraman	−68.7148 (Proposed algorithm)
Pirate	−66.2388 (Proposed algorithm)

Table 5 Computational time of the proposed algorithm for standard images

Images	Computational time in seconds
Lena	0.308817
Cameraman	0.319627
Pirate	0.325353

5   Conclusion

In the proposed algorithm, the multiple images of the watermark are first transformed using logarithmic transformation and then embedded into the host image. By this method, the embedded watermark is very less affected by robust attacks. It is observed that very high PSNR is achieved for the watermarked image compared to most popular existing techniques. Very less mean square error is observed and it is almost equal to that of DWT technique. It is also noticed that the correlation between the host and watermarked images is equal to unity, which is observed in DWT technique. The statistical analysis indicates that the proposed algorithm is a better alternative to the exiting techniques. And also the algorithm utilizes very less computational time for execution. Further, the proposed technique can be combined with DWT to improve the results for the future research.

References

1. Nezhadarya E, Wang ZJ, Ward RK (2017) Robust image watermarking based on multiscale gradient direction quantization. IEEE Trans Inf Forensic Secur 6(4):1200–1213
2. Ahmaderaghi B, Kurugollu F, Del Rincon JM, Bouridane A (2018) Blind image watermark detection algorithm based on discrete shearlet transform using statistical decision theory. IEEE Trans Comput Imaging 4(1):46–59
3. Prajwalasimha SN (2018) Pseudo Hadamard transformation based image encryption scheme. In: Integrated intelligence computing, communication and security. Studies in computational intelligence, Springer, Chapter 58, Vol 771, pp 375–384
4. Khanna AK, Roy NR, Verma B (2017) Digital image watermarking and its optimization using genetic algorithm. In: Proceedings of international conference on computing, communication and automation (ICCCA2016), pp 1140–1144
5. Prajwalasimha SN, Shetter A, Swapna H (2018) Digital image watermarking using tenth root of exponential function. In: Proceedings of IEEE international conference on recent trends in electronics, information & communication technology, pp 634–637
6. Salama AS, Mokhtar MA (2016) Combined technique for improving digital image watermarking. In: Proceedings of 2nd IEEE international conference on computer and communications, pp 557–562
7. Kumar S, Dutta A (2016) Performance analysis of spatial domain digital watermarking techniques. In: Proceedings of international conference on information communication and embedded system (ICICES)
8. Bi H, Zhao C, Liu Y, Li N (2016) Digital watermarking based on interleaving extraction block compressed sensing in Contourlet domain. In: Proceedings of 9th international congress on image and signal processing, biomedical engineering and informatics (CISP-BMEI), pp 766–770
9. Mitashe MR, Habib ARB, Razzaque A, Tanima IA, Uddin J (2017) An adaptive digital image watermarking scheme with PSO, DWT and XFCM. In: Proceedings of IEEE international conference on imaging, vision & pattern recognition (ICIVPR)
10. Agarwal R, Santhanam MS, Srinivas K (2016) Digital watermarking: an approach based on Hilbert transform. In: Proceedings of international conference on computing, communication and automation (ICCCA), pp 1035–1042

A Survey of Medical Imaging, Storage and Transfer Techniques

R. R. Meenatchi Aparna and P. Shanmugavadivu

Abstract Medical data keeps growing with the growing number of scans every year. Patient experience plays a vital role in development of healthcare technologies. The speed with which the data can be accessed when the patient really wants to get diagnosed be it the same hospital or a different hospital becomes a very important requirement in future healthcare research. With growing amount of modality techniques and size of the captured images, it is very important to explore the latest technologies available to overcome bottlenecks. With (Computed Tomography) CT and Magnetic Resonance Imaging (MRI) modalities increasing the number of slices and size of the image captured per second, the diagnosis becomes accurate from the radiology perspective, but the need to optimize storage and transfer of the images without losing vital information becomes obviously evident. In addition security also plays an important role. There are various problems and risks when it comes to handling medical images because it is of key use to diagnose a disease which may be life threatening for the patient. There are evidences of radiologists waiting for the data for a considerable time to access the data for diagnosis. Hence time and quality plays a very important role in healthcare industry and it is major area of research which has to be explored. This scope of this survey is to discuss about the open issues and techniques to overcome the existing problems involved in medical imaging and transfer. This survey concludes the few optimization techniques with the medical imaging and transfer applications. Finally, limitation and future scope of improving medical imaging and transfer performance is discussed.

R. R. Meenatchi Aparna · P. Shanmugavadivu (✉)
Department of Computer Science and Applications, Gandhigram Rural Institute—Deemed
University, Gandhigram, Dindigul, Tamilnadu, India
e-mail: psvadivu67@gmail.com

R. R. Meenatchi Aparna
e-mail: meenscience@gmail.com

© Springer Nature Switzerland AG 2019
D. Pandian et al. (eds.), *Proceedings of the International Conference on ISMAC
in Computational Vision and Bio-Engineering 2018 (ISMAC-CVB)*, Lecture Notes
in Computational Vision and Biomechanics 30,
https://doi.org/10.1007/978-3-030-00665-5_3

1 Introduction

The aim of this study is to analyze the latest trends in medical imaging techniques and find how they can contribute to improve the medical image handling in terms of quality and performance. The processor speed, CPU (Central Processing Unit) utilization. Graphical Processing Unit (GPU) programming, Multiprocessing algorithms have proved to be a lot useful in other domains and has to be explored in full to improve the overall handling of images in a hospital. The following technologies are analyzed to find how they can contribute to improvement in performance and quality. Disadvantages of such techniques also open up areas of research which can become vital in future.

2 Medical Imaging Overview

In medical imaging data is highly important since it deals with patient privacy. This data is used for finding the patients disease and help him save his life. So data is important at that particular moment and at that particular stage of the disease. When this data is to be handled there are certain rules and regulations followed by the service providers, like Health Insurance Portability And Accountability Act (HIPAA).

There are standardization of the protocol for medical image exchange and storage. Most widely followed standard is Digital Imaging and Communication in Medicine (DICOM).

Yaorong et al. describe that image sharing by using CDs are a burden for the patients, and image sharing by networks increases the patient safety issues [1]. As explained in [2] it becomes inevitable to assure the privacy of the patient and integrity of the data when the data travels through different medium across geographical boundaries. This restriction is very important since we deal with medical data which becomes an integral part of patient's privacy information.

Hence it is very important to provide a high-quality image, within a specific time. With a highly accessible infrastructure for the hospital to cope up with the increase in medical data over the years and increased number of diseases which require scanning, currently the radiologists have to deal with terabytes of data unlike before. It adds to increasing number of cases [2]. Security of PHI is one of the topmost concerns of healthcare industry today. When patient data is processed, transported or archived, the security vulnerability loopholes play a major drive for the hackers to access the data.

There are certain representations where a non-DICOM image like SR reports can be stored, but the scope of this paper will be to address the problems dealt with DICOM images.

3 Healthcare Data

These are the data *volume* (size), typically in the petabyte range (1 PB = 10^{15} bytes), data *heterogeneity*, including (un)formatted, ASCII/Binary, (un)structured, and the data *velocity*, or data *derivative*, which captures the change, transfer, and discovery of raw and derived data [3–5].

4 Security of Healthcare Data

With the increasing trends of cloud computing there are some basic security and functional privacy issues come into picture especially with medical imaging workflows. The security of a medical imaging software is assured by allowing a particular user to access a particular patient data according to Health Insurance Portability and Accountability Act (HIPAA).

The major risks involved in medical images are

1. When the medical images are exposed to cloud, it is highly needed to adhere to security norms.
2. Required to anonymize the data.
3. Most of the cloud vendors according to current trend compress the data to achieve better efficiency but in a medical domain Compression is usually avoided to ensure data clarity
4. Eavesdropping the network is possible

Medical imaging requirements are highly risk related because of the following possibilities

- Complexity of Medical requirement
- Adherence to various international Standards
- Multimodality testing
- Risk involved in failure of the software (Affecting Human life)
- Loss of data cannot be entertained even to minimal extent
- Continuous evolution of the Technology with various sized and quantified images.

5 Whooping Increase in Healthcare Data

The volume, diversity, and velocity of biomedical data are exponentially increasing providing petabytes of new neuroimaging and genetics data every year [6]. Medical imaging solutions across the world generate humungous amount of data every second.

Based on the growing trend, it was estimated that over 100,000 terabytes of data will be performed in the United States during year 2014, which will generate petabytes of data [7]. When there is increase in the volume of data, medical image processing and analysis are computationally expensive [8].

6 DICOM

DICOM is the standard by which medical images are represented. The standard not only defines about the structure of the data but also about the messages used to exchange medical images. Each DICOM file has 128 bytes of header information and a set of tags t define a particular IoD (Information Object Definition). The IoDs are composed of a set of tags and values. The Tags are classified as mandatory type 1 tag or optional Type 2 tag.

There are various DIMSE (Dicom Messaging Service) service object pairs supported by every medical application which deals with message exchange mechanism of DICOM. The following are the major messages [9] used.

1. C ECHO
2. C MOVE
3. C STORE
4. C FIND

7 Latest Trends in Medical Imaging

The following technologies are mainly explored in correlation with handling medical images by researchers across the world:

1. Cloud Computing
2. Multicore
3. Big Data
4. Mobile Computing
5. GPU Programming

A. *Cloud computing*

Cloud computing addresses the problem of storing a large amount of data across different locations. Even though the same can be achieved by different techniques, the main attraction here is "pay as u use". This can help the medical imaging vendors to reduce the cost to large extent.

At the same time the inherent advantage of cloud, a form of cloud in which the required hardware can be used based on necessity, i.e., "Infrastructure As A Service" will be of great help for the medical imaging vendors in terms of scalability and redundancy.

Although there are a lot of advantages by using cloud in medical imaging, there are certain problems when it comes to sharing medical images in server. As explained in [2], HIPAA(Health Insurance Portability and Accountability Act) plays a vital role in medical informatics. Every cloud vendor who supports the medical imaging applications should assure that the security rules are followed without any problem.

The security issues with cloud are

1. User access control for a particular data
2. Data reliability
3. Patient privacy issues
4. Data loss that might occur during anonymization and conversions.

Inspite of all these issues since the FDA (Food and Drug Administration) mandates the medical image data retention to 25 years [10], cloud will become the default solution for medical image archives [11]. Offsite backup, disaster recovery, high resource availability, and mobile device security and support are some of the greatest advantages that cloud can offer for medical imaging fraternity [12].

However, privacy and security concerns have slowed adoption of cloud storage, according to Nahim Daher, an analyst at consulting firm Frost & Sullivan. Cloud storage vendors store data at multiple sites, and the provider "doesn't know where the data is sitting and doesn't have direct oversight into who is looking at it," Daher says [13].

B. Multicore

The basic difference between an multiprocessor OS and a traditional OS being the management of process execution on different CPUs [14] creates new opportunities to do a lot of image processing operations at a faster pace which will again make the scans more faster and hence exposure to rays minimal.

With multicore technology being adapted to various industries, medical imaging can use multicore processing for process intensive tasks like image processing operations, 3D generation of models, simulation algorithms, etc. Sanjay et al. explains about how a multicore can be used effectively for image processing algorithms which will really speed up the algorithms [15].

There are some operations in which multicore becomes quite complex like sequential image viewing like a 2D or 3D movie, etc. Dev explains how a multicore Digital Signal processors (DSP s) inside the medical imaging equipment makes way for treatment to move towards patient instead of patient approaching the hospitals. This paves way for greater telemedicine opportunities. An application which is multi-threaded can be mapped to multicore, with each core performing an important part of the functionality.

In CT, Mike explains that the core count can go up to 1000, since there is a growing trend towards faster and more scan [16]. There are no proved evidences of multicore being used for image storage or transfer operations, which is still open for research.

C. *Big Data*

Big Data products like Cassandra really help the way the data is stored. With high and efficient data access, data redundancy also is addressed by some platforms where data is stored simultaneously at different locations and Data can be retrieved even when one location is down.

The main requirement when choosing a particular Big Data Technology is the updating of data should be done appropriately so that data consistency is maintained. With huge quantities of complex and high-quality data [17], medical data becomes obviously a candidate for Big Data.

The various algorithms used by the Big Data analytics pave way for diagnosis of the disease at a faster pace since it provides greater access to wider data. IN current time, Big Data analytics are explored hugely in US market for medical data [18] for further developments. Medical data include heterogeneous, multi-spectral, incomplete and imprecise observations (e.g., diagnosis, demographics, treatment, prevention of disease, illness, injury, and physical and mental impairments) derived from different sources using incongruent sampling [19].

There are a lot of companies starting to explore big data for medical images like explained in [20] mainly for the two reasons

1. Predictive analysis
2. Machine Learning

With huge amount of history medical data available, it would be possible in future to find out the disease that a patient might get in future, or make the machine to give data about the severity of a particular epidemic in a particular region. Big Data also helps to store unstructured data. This made medical image industry players to move towards Big data. Another reason was that medical data may consist of non-image data which is huge like waveforms, reports, etc.

D. *Mobile computing*

With latest mobile computing technology leading to smartphone support, medical images can also be shared through mobiles so that the radiologists can access the image "on the go" and help for investigation. With the tremendous development in telemedicine and teleradiology, the extent to which images are shared has increased over the last decade.

The main problems concerned with mobiles are in security of the patient images. All security issues related to mobile networks hold good for the medical images in Mobile computing technology also and hence can lead to access control issues.

E. *GPU programming*

This technology has seen a tremendous growth in last few years and research is extensively done to improve the usage of GPUs in programming. The main idea of this technology is that the GPU (graphical processing unit) which sits idle during most of the time can be utilized for intensive operations since there are more number of cores in a GPU. It is more powerful when compared to CPU and quite a set of constructs are available to implement GPU programming like CUDA, OpenMP 4.0, etc. Anders et al. describe that the denoising is performed in 8 min in GPU on contrary to CPU architectures where the performance was 50 min [21]. In a medical imaging application whenever the graphical operations are not done, the GPU can be effectively used for workflows like handling the data especially in image transfer workflows. The applications which use Geometric transformations and high level of mathematical transformations can very well use the GPU which is very much capable of doing multiprocessing operations [22].

Jeyarajan et al. describe about the mapping of a motion estimation algorithm to GPU architectures and explain the performance gains [23]. With developing languages such as CUDA it is possible to achieve high performance especially when the operations can be made parallel.

F. *Vendor Neutral Archives*

Vendor Neutral Archive represents the storage of medical data in a non proprietary format and provides for seamlessly sharing across organizations.

The main functionality of VNA is to decouple the vendor specific PACS from the DICOM network and replace it with a neutral module which can easily convert the images stored in Archive to DICOM and process further. By this means dependency on PACS and native format is completely prevented. PACS from one vendor can be easily replaced with another vendor.

8   Existing Works on Latest Technologies

Technology	Author, Year and References	Problem addressed
Cloud	Kagadis et al. [24]	Applicability of advanced cloud computing in medical imaging
	Karthikeyan et al. [25]	Emergency healthcare sector in an umbrella with physical secured patient records with SaaS
	Dai et al. [26]	Review extant cloud-based services in bioinformatics, classify
	Ya et al. [27]	Cloud-based Hospital Information Service Center
	Liu et al. [28]	Three-layer hybrid cloud
MultiCore	Niendorf et al. [29]	Cardiovascular MRI slowdown
	Lecron et al. [30]	Vertebra Detection and Segmentation in X-Ray Images, slowdown
	Xu et al. [31]	Iterative reconstruction algorithm
	Hofmann et al. [32]	FDK algorithm
	Mittal et al. [33]	CPU-GPU Heterogeneous Computing
GPU	Howison [34]	GPU ray casting
	Massanes et al. [35]	Multi-GPU proceeding
	Olmedo et al. [36]	Image processing
	Westhoff [37]	Segmentation
	Weinlich et al. [38]	Raycasting
VNA	Kumar et al. [39]	PACS neutrality
	Cook [40]	VNA
	Gray [41]	VNA

9   Proposed Work

GPU programming is already proved to be more efficient in terms of graphical processing. Our experiments would analyze the use of GPU in creation of DICOM messages post-image processing so that the performance can be improved. The main idea being using GPU is about 60% of the time in transfer is used for message preparation. So ideally if this part is made to be executed in GPU, there would be a heavy performance improvement.

The modules used in the transfer would be message creation module, sending module on the sending end and the receiving module and demultiplexing module on the receiving end in a network. Figure 1 shows how this would happen in a network.

Our proposal would be to move the message creation module to GPU so that the sending module becomes very balanced to use the network efficiently.

Figure 2 illustrates the proposed workflow.

The proposed workflow will create the messages in GPU and transfer it using the CPU. The next step of this experiment would be to use the multicore in the CPU as well by the sending module (Fig. 3).

Fig. 1 Normal workflow

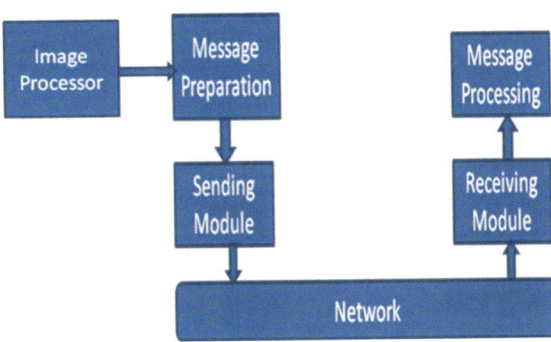

Fig. 2 Proposed workflow

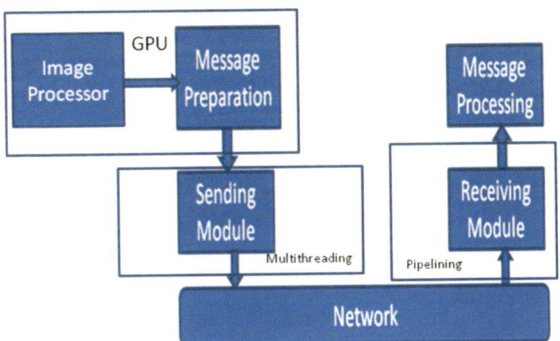

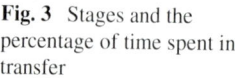

Fig. 3 Stages and the percentage of time spent in transfer

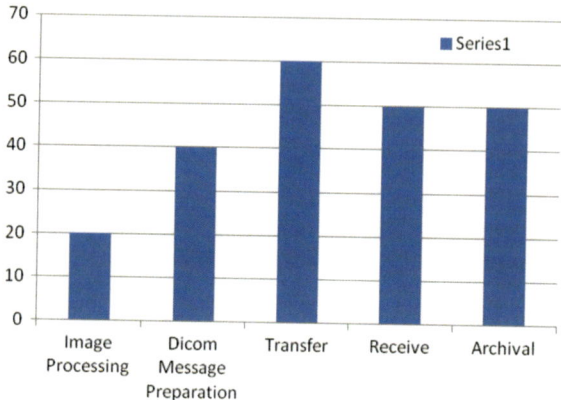

A. *Hardware Configuration*

The specification of the hardware used.

GPU: NVIDIA GeForce 920 M
Speed: 1.4 GHz
Multiprocessor count: 2
Number of cores: 385 (2×192)
Memory: 4 GB
Threads: 1024 per Block
Language: CUDA 7.0, OpenMP

B. *Tools Used*

The following are the list of tools used for this experiment:

1. OpenMP
2. C
3. C++
4. NVidia GPU
5. Pentium processor
6. JDICOM

10 Results

Since the GPU which is idle for most of the time is used for the CPU intensive operation, it leads to a better performance in the transfer workflow. The heterogeneous combination of multicore CPU and GPU implementation gives higher performance when compared to conventional transfer method. This overcomes the disadvantages of multithreading in which the program runs in the same processor in chunks by using time slicing and GPU programming, the memory swap that happens between GPU and CPU to communicate the output.

11 Conclusion

The major advantage of the above workflow is that we are using the most powerful processing units for image processing and message preparation and CPU for network send. A combination of hybrid strategy would provide a multi stage performance improvement.

With advancements in various medical imaging technologies, the area of medical imaging transfers has to be explored adapting to the new trends. Increasing the performance of medical transfers is a challenge which needs to be explored. This research will confirm performance improvement achieved by using heterogenous model for medical imaging and transfer.

References

1. Ge Y, Ahn DK, Unde B, Gage H, Carr JJ (2013) Patient-controlled sharing of medical imaging data across unaffiliated healthcare organizations. J Am Med Inf Assoc 20(1):157–163.http://www.ncbi.nlm.nih.gov/pmc/articles/PMC3555338/
2. Medical imaging analytics (2015) (Online) Available https://www.research.ibm.com/haifa/dept/imt/mia.shtml. Accessed 17 Aug 2015
3. Foster K, Spicer M, Nathan S (2011) IBM infosphere streams: assembling continuous insight in the information revolution. International Technical Support Organization, San Jose, CA
4. Howe D et al (2008) Big data: the future of biocuration. Nature 455(7209):47–50
5. Lynch C (2008) Big data: how do your data grow? Nature 455(7209):28–29
6. Dinov ID, Petrosyan P, Liu Z, Eggert P, Zamanyan A, Torri F, Macciardi F, Hobel S, Moon SW, Sung YH, Toga AW (2014) The perfect neuroimaging-genetics-computation storm: collision of petabytes of data, millions of hardware devices and thousands of software tools. Brain Imaging Behav 8(2):311–322
7. Prepare for disaster & tackle terabytes when evaluating medical image archiving (2008) Frost & Sullivan. http://www.frost.com
8. Rodger JA (2015) Discovery of medical big data analytics: improving the prediction of traumatic brain injury survival rates by data mining patient informatics processing software hybrid hadoop hive. Inf Med Unlocked 1:17–26
9. DICOM standard 2015, Message Exchange, NEMA
10. Retention and storage of images and radiological patient data. Dated February 2008. https://www.rcr.ac.uk/docs/radiology/pdf/ITguidance_Retention_storage_images.pdf
11. Liu BJ, Cao F, Zhou MZ, Mogel G, Documet L (2003) Trends in PACS image storage and archive. Comput Med Imaging Graph 27

12. Shah D, Kollaikal P Top trends in medical imaging technology (Online) Available: http://www.citiustech.com/uploads/article/pdf/top-trends-in-medical-imaging-technology-89.pdf. Accessed 22 Jan 2018
13. Healthcare in cloud: a Storage solution or security risk. http://www.advisory.com/daily-briefing/2013/04/10/health-care-in-the-cloud-a-storage-solution-or-security-risk. Date 10 Apr 2013
14. Kagadis GC, Langer SG (2012) Informatics in medical imaging. CRC Press, Boca Raton
15. Saxena S, Sharma N, Sharma S (2013) Image processing tasks using parallel computing in multi core architecture and its applications in medical imaging. Int J Adv Res Comput Commun Eng 2(4)
16. Hinds M (2009) White paper on "Power up: moving toward parallel processing in medical imaging compute systems
17. Cowan B (2015) Big data medical imaging (Online) Available http://nihi.auckland.ac.nz/sites/nihi.auckland.ac.nz/files/pdf/informatics/bigdata/Big%20Data%20Medical%20Imaging%20-%20Brett%20Cowan%206.pdf. Accessed 17 Aug 2015
18. PRNewsWire (2015) US medical imaging industry leaps firmly into the big data realm (Online) Available http://www.prnewswire.com/news-releases/us-medical-imaging-industry-leaps-firmly-into-the-big-data-realm-300105491.html. Accessed 17 Aug 2015
19. Dinov ID (2016) Volume and value of big healthcare data. J Med Stat Inform 4:3. https://doi.org/10.7243/2053-7662-4-3
20. Ridley EL (2015) http://www.auntminnie.com. Israeli start-up eyes big-data tools for imaging analysis (Online) Available https://mail.google.com/mail/u/0/#inbox/14f3c7be7d177547?projector=1. Accessed 17 Aug 2015
21. Eklund A, Andersson M, Knutsson H (2011) True 4D image denoising on the GPU. Int J Biomed Imaging 2011
22. Shams R, Sadeghi P, Kennedy RA, Hartley RI (2010) A survey of medical image registration on multicore and the GPU. IEEE Sign Process Mag 27
23. Thiyagalingam J, Goodman D, Schnabel JA, Trefethen A, Grau V (2011) On the usage of GPUs for efficient motion estimation in medical image sequences. Int J Biomed Imag 2011
24. Kagadis GC, Kloukinas C, Moore K, Philbin J, Papadimitroulas P, Alexakos C, Nagy PG, Visvikis D, Hendee WR (2013) Cloud computing in medical imaging. Med Phys 40(7):070901. https://doi.org/10.1118/1.4811272
25. Karthikeyan N, Sukanesh R (2012) Cloud based emergency health care information service in India. J Med Syst 36(6):4031–4036. https://doi.org/10.1007/s10916-012-9875-6
26. Dai L, Gao X, Guo Y, Xiao J, Zhang Z (2012) Bioinformatics clouds for big data manipulation. Biol Direct 28(7):43. https://doi.org/10.1186/1745-6150-7-43 discussion 43
27. Yao Q, Han X, Ma XK, Xue YF, Chen YJ, Li JS (2014) Cloud-based hospital information system as a service for grassroots healthcare institutions. J Med Syst 38(9):104. https://doi.org/10.1007/s10916-014-0104-3
28. Liu L, Chen W, Nie M, Zhang F, Wang Y, He A, Wang X, Yan G (2016) iMAGE cloud: medical image processing as a service for regional healthcare in a hybrid cloud environment. Environ Health Prev Med 21(6):563–571
29. Niendorf T, Sodickson DK (2006) Parallel imaging in cardiovascular MRI: methods and applications. NMR Biomed 19(3):325–341
30. Lecron F, Mahmoudi SA, Benjelloun M, Mahmoudi S, Manneback P (2011) Heterogeneous computing for vertebra detection and segmentation in X-ray images. Int J Biomed Imaging 2011, Article ID 640208
31. Xu M, Thulasiraman P (2011) Mapping iterative medical imaging algorithm on cell accelerator. Int J Biomed Imaging 2011, Article ID 843924

32. Hofmann J, Treibig J, Hager G, Wellein G (2013) Performance engineering for a medical imaging application on the intel Xeon Phi accelerator (online) https://arxiv.org/pdf/1401.3615.pdf. Accessed 17 Dec 2013
33. Mittal S, Vetter JS (2015) A survey of CPU-GPU heterogeneous computing techniques. ACM Comput Surv (CSUR), 47(4), Article No. 69
34. Howison M (2010) Comparing GPU implementations of bilateral and anisotropic diffusion filters for 3D biomedical datasets. In: SIAM conferences of imaging science
35. Massanes F, Cadennes M, Brankov JG (2011) Compute-unified device architecture implementation of a block-matching algorithm for multiple graphical processing unit cards. J Electron Imaging 20(3):1–10
36. Olmedo E, Calleja J, Benitez A, Medina MA (2012) Point to point processing of digital images using parallel computing. IJCSI Int J Comput Sci Issues 9(3):1–10
37. Westhoff AM (2014) Hybrid parallelization of a seeded region growing segmentation of brain images for a GPU cluster. In: Proceedings of the international conferences on architecture of computing systems
38. Weinlich A, Keck B, Scherl H, Kowarschik M, Hornegger J (2008) Comparison of highspeed ray casting on GPU using CUDA and OpenGL. In: Proceedings of the international workshop on new frontiers in high-performance & hardware-aware computing, pp 25–30
39. Tapesh Kumar Agarwal, Sanjeev (2012) Vendor neutral archive in PACS. Indian J Radiol Imaging 22(4):242–245
40. Cook R Is VNA the future of image delivery? (online) http://www.healthcareitnews.com/news/should-you-use-vna-whats-vna
41. Gray M The bridge from PACS to VNA scale out (online) https://www.emc.com/collateral/hardware/white-papers/h10699-bridge-from-pacs-to-vna-wp.pdf

Majority Voting Algorithm for Diagnosing of Imbalanced Malaria Disease

T. Sajana and M. R. Narasingarao

Abstract Vector borne diseases like malaria fever is one of the most elevating issues in medical domain. Accurate identification of a patient from the given set of samples and classification becomes one of the challenging task when dealing with imbalanced datasets. Many conventional machine learning and data mining algorithms are shows poor performance to classify skewed distributed data because they are trained very well with the majority class samples only. Proposing an ensemble method called majority voting defined with a set of machine learning algorithms namely decision tree—C4.5, Naive Bayesian and K-Nearest Neighbor (KNN) classifiers. Classification of samples can be done based on the majority voting of classifiers. Experiment results stating that voting ensemble method shows classification accuracy of 95.2% on imbalanced malaria disease data whereas dealing with balanced malaria disease data voting ensembler shows 92.1% of accuracy. Consequently voting shows 100% classification report on precision, Recall and F1-Score on imbalanced malaria disease data sets whereas on balanced malaria disease data voting shows 96% of Precision, Recall and F1-Score metrics.

1 Introduction

Mosquito bite diseases like Malaria is one of the frightful diseases facing by many people throughout the world wide. Due to the climate changes and food habits still the disease has good influence on remote areas [1, 2]. Regarding health reports of various organizations the effect of vector borne diseases still become a challenging issue in medical domain [3] and Identification of an effected patient within time is very important among the set of patients data [3, 4]. Even in other applications like

T. Sajana (✉) · M. R. Narasingarao
Department of Computer Science & Engineering, K L E F, Vaddeswaram, Guntur, India
e-mail: sajana.cse@kluniversity.in

M. R. Narasingarao
e-mail: Ramanarasingarao@kluniversity.in

© Springer Nature Switzerland AG 2019
D. Pandian et al. (eds.), *Proceedings of the International Conference on ISMAC in Computational Vision and Bio-Engineering 2018 (ISMAC-CVB)*, Lecture Notes in Computational Vision and Biomechanics 30,
https://doi.org/10.1007/978-3-030-00665-5_4

31

identification of oil spills in satellite images, spam emails detections are also facing the problem of classification of minority class samples [5, 6].

Malaria—caused by the bite of female mosquito called "Anopheles" which affects the red blood cells directly that causes the damage of liver functionality. There are mainly four types of malaria parasites namely *Plasmodium falciparum*, *Plasmodium vivax*, Plasmodium Malaria and *Plasmodium ovale* and additionally to these parasites *Plasmodium knowlesi* also to be considered for the identification of malaria disease [7, 8, 17].

Many authors are proposed various data mining and machine learning algorithms for the classification of malaria disease. But, all these traditional algorithms are well performed when there is balanced class distribution present in between the samples. But, in the area of imbalanced class problem all these algorithms are biasing towards majority class samples only which becomes a crucial task in identification of effected patient called minority class sample. Hence, consider ensemble methods which are designed with a set of classifiers and trained on given samples very effectively that inference the classifiers for the exact classification of effected patients [9, 10, 19–23].

The rest of the paper stated that history on imbalanced malaria disease and balanced malaria disease datasets which was described in Sect. 2. Section 3 explores the methodology for classification of both malaria disease data using voting ensemble method and experimental outcomes are discussed in Sect. 4.

2 Literature Review

Identification of an effected patient within time and diagnosing plays a vital role. Many authors are suggested, proposed conventional machine learning on equal distribution of class samples of malaria disease data as stated below:

- Pandit and Anand [11] proposed Artificial Neural Networks for the identification of effected erythrocytes.
- Bbosa et al. [12] suggested rule based classification for identification of effected malaria patient.
- Wu and Wong [13] proposed a neural network model effective diagnosing of malaria disease.
- Tsai et al. [14] suggested images segmentation method for detection of malaria parasites.
- Rahmanti et al. [15] investigated on parasite classification using image processing and KNN techniques.
- Charpe and Bairagi [16] identified the parasites stage using image processing and classification techniques.
- Somasekar and Reddy [17] derived a method for identification of effected erythrocytes based on Adaptive median filter, edge enhancement and Fuzzy C-Means clustering techniques.

But he traditional machine learning algorithms are biased towards the majority class samples only. Hence, due to the nature of imbalanced distribution, classification of minority class samples becomes a burning issue suggested by various authors [9, 10, 18–23].

Researchers are proposed many methods to handle the class imbalance problem im many applications including malaria disease. Consider the history of class imbalance problem in different applications:

- Guo Haixiang et al. presented a plethora of methods for handling imbalanced data sets in the era of many application domains [24].
- Salma Jamal developed a Predictive model of anti-malarial molecules inhibiting apicoplast formation for an imbalanced malaria disease dataset using Cost sensitive Naive Bayesian, Random Forest and J48 algorithms and found that Random Forest algorithm produces better accuracy [25].
- Bruno B. Andrade et al. suggested that severe stage of malaria disease causes to reducing of inflammatory cytokines which is a high level imbalance class problem in medical domain [26].
- Rashmi Dubey proposed an ensemble frame work for classification of Alzheimer's disease [27].
- Wing W. Y. Ng et al. Suggested Diversified Sensitivity-Based Under sampling method for handling different imbalanced datasets [28].
- Yazan F. et al. Experimentally investigated 2080 cardiac patients data using various classifiers for better disease diagnosing [29].
- Jia Pengfei et al. Developed a new sampling method called Distinct—Borderline algorithm for balancing the class distribution of imbalanced data sets [30].
- N. Poolsawad et al. Conducted a comparative study between set of classifiers like Multilayer Perceptron (MLP), RBFN, SVM, Decision Tree and Random forest algorithms on LIFELAB dataset and found that RF algorithm produces better accuracy [22].
- V. Garcia et al. İnestigated various methods for imbalanced data sets handling [31].
- Jaree Thongkam et al. proposed C-Support vector machine method for prediction of imbalanced breast cancer data set [32].
- Xing-Ming Zhao et al. Proposed an ensemble classifiers for protein classification. Conducted a two ways of comparative study between balanced data and unbalanced data using individual classifiers & ensemble classifiers and concluded that ensemble classifiers are shows best classification performance [33].
- Victoria López et al. proposed an ensemble framework on different datasets by conducting a comparison study between C4.5, SVM and instance based learning methods [34].
- Li Ma et al. investigated that CURE-SMOTE (Clustering Using Representatives—Synthetic Minority Oversampling Technique) is best sampling method over the Border line SMOTE1, safe level SMOTE, C-SMOTE, k-means SMOTE techniques and then suggested Random Forest, a hybrid algorithm for feature selection and parameter optimization [35].

3 Proposed Methodology

In the era of class imbalance problem many different methods are defined for classification of samples especially samples belongs to the minority class samples. But, the difficulty with these methods like data sampling is they can over fit the data or loss some valuable sample also which tends to misclassify the affected patient when compared with un affected patient. Hence, Ensemble methods are best methods for classification of imbalanced datasets because set of classifiers are grouped together to increase the classifiers accuracy performance and also they can reduce data over fitting which is one of the drawback of data handling techniques for imbalanced data. Another advantage is error rate will be automatically reduced or simply null. Consider the proposed methodology design as shown in Fig. 1 which describes voting ensembler a combination of Decision tree—C4.5, Naive Bayesian and KNN classifiers as stated below.

Consider the Decision tree - C4.5 algorithm:
Input: data set D.
Output: A Decision tree.
Method:

- Construct a node N if instances in D are belongs to same class, C, then Return N as a leaf node labelled with the class C;
- If attribute _ list is empty then
- Return N as a leaf node labelled with the majority class in D;
                                                              // majority voting
- Apply attribute_selection_method (D, attribute _ list) to find the "best" splitting.
- Let node 'N' hold the tuples of partition D.
- Calculate information gain to classify a tuple in D is given by

$$\mathrm{Info}(D) = -i^* \log 2(p_i)$$

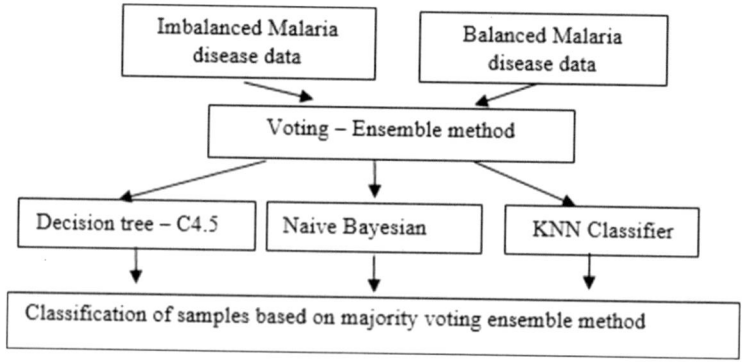

Fig. 1 Methodolgy for classification of samples using voting ensembler

- Select node N with maximum information gain.
- Find the expected information required to classify a tuple D based on partitioning of an attribute set A { $a_1, a_2, a_3, a_4, ..., a_v$ } is measured by

$$Info_A(D) = \sum_{j=1}^{v} \frac{|D_j|}{|D|} \times Info(D_j)$$

- Branch the sub trees for j
- let Dj be the set of data tuples in D satisfying outcome j;
- Amount of information attained by A is defined by

$$Gain(A) = Info(D) - Info_A(D)$$

- Apply normalization method called split information to overcome the bias problem with information gain by an attribute A

$$SplitInfo_A(D) = -\sum_{j=1}^{v} \frac{|D_j|}{|D|} \times \log_2\left(\frac{|D_j|}{|D|}\right)$$

- Select the attribute with highest gain ratio,

$$GainRatio(A) = \frac{Gain(A)}{SplitInfo(A)}$$

- if Dj is NULL
- Then branch a node with majority class in D.
- Else grow the sub tree by creating a node which is selected by decision tree.
- Apply Pessimistic pruning method for tree pruning to find the error rate using training data set.

Consider the Naive Bayesian classification algorithm:
Input: dataset D.
Output: Classification of sample.
Method:

- D: A dataset with training tuples.
- $X = (x_1, x_2, x_3, ..., x_m)$ is an n-dimensional vector each vector tuple of D defined over 'n' attributes of $A_1, A_2, A_3, ... A_n$.
- $C_1, C_2 ... Cm$: set of classes.
- Find the maximum posterior probability of each sample belongs to class 'C_i' if and only if, $P(C_i|X) > P(C_j|X)$ for $1 \leq j \leq m, j! = i$.
- By Bayesian theorem calculate maximized posterior hypothesis,
- $P(C_i|X) = [P(X|C_i). P(C_i)] / P(X)$ where $P(X)$ is constant.
- Predict the class of X based on class C_i if and only if $P(X|C_i) P(C_i) > P(X|C_j) P(C_j)$ for $1 \leq j \leq m, j \neq i$.

Consider the KNN Classifier algorithm as follows:
Input: dataset D.
Output: Classification of given data.
Method:

- Find the distance between points x and x_i i.e., $d(x, x_i)$ where $i = 1, 2, \ldots n$.
- Select initial 'm' distances from the set of increasing order of distances.
- Obtain k points w.r.t the $m = k$ distances.
- For $k \geq 0$, k_i defines the no of points belongs to i^{th} class.
- For each $k_i > k_j \ \forall \ i \neq j$ assign 'x' in class'i'.

Dataset : Gathered 165 tuples data (real dataset) defined as follows:

Attributes	Actual range
Age	1–20 years
Haemoglobin	M13.0–18.0/F11.0–16.5 g%
RBC	3.80–5.80 millions/cumm
Hct	35.0–50.0%
Mcv	80–97 fl
Mch	26.5–33.5 pg/cells
Mchc	31.5–35.0%
Platelets	1.50–3.90 lakhs/cumm
WBC	3500–10,000 cells/cumm
Granuls	43.0–76.0%
Lymphocytes	17.0–48.0%
Monocytes	4.0–10.0%
Malaria	Yes/No

Consider the samples distribution of collected Malaria Dataset as described in Table 1.

Table 1 Samples distribution and imbalanced % of malaria disease dataset

No. of tuples	Parameters	Minority/effected samples	Majority/uneffected Class samples	Class (effected/uneffected)	% Class (effected, uneffected)
165	13	5	160	(Positive, Negative)	(3.1, 96.9)

4 Results and Discussions

Proposing a well-defined algorithm of ensemble method namely voting a combination of set of classifiers Decision tree—C4.5, Naive Bayesian and KNN which are trained with common set of samples to the frame work as shown in Fig. 1 and then performance of these algorithm measured with various metrics like accuracy, sensitivity, specificity and F1-score which are derived from confusion matrix as defined Table 2.

Presenting a combination of machine learning algorithms like Decision Tree—C4.5, Naive Bayesian and KNN Classifiers as an ensemble method of Voting which classifies the samples based on majority voting. Voting ensemble algorithm increases the classification accuracy performance of imbalanced malaria disease data i.e., 95.2% when compared with balanced malaria disease data as 92.1% which is as shown in Fig. 2. But unlike other ensemble methods voting of C4.5, Naive Bayesian, KNN Classifiers shows good bias variance at each k-fold cross validation of both imbalanced and balanced malaria disease datasets which is shown in Fig. 3 and also shows 100% performance on skewed malaria disease data w.r.t Precision, Recall and F1-Score as shown in Figs. 4, 5 and 6.

Table 2 Confusion Matrix

		Predicted	
		Effected/minority class	Uneffected/majority class
Actual	Effected class	TP	FN
	Uneffected class	FP	TN

Fig. 2 Classification accuracy % of both imbalanced and balanced malaria disease data using voting ensemble algorithm

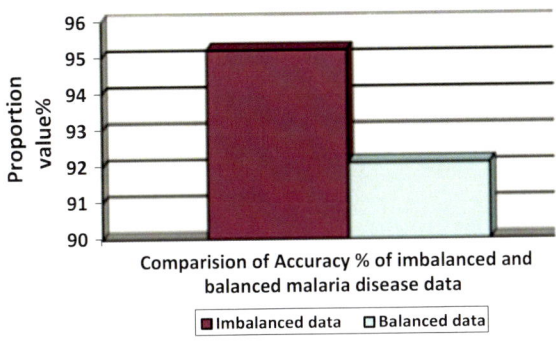

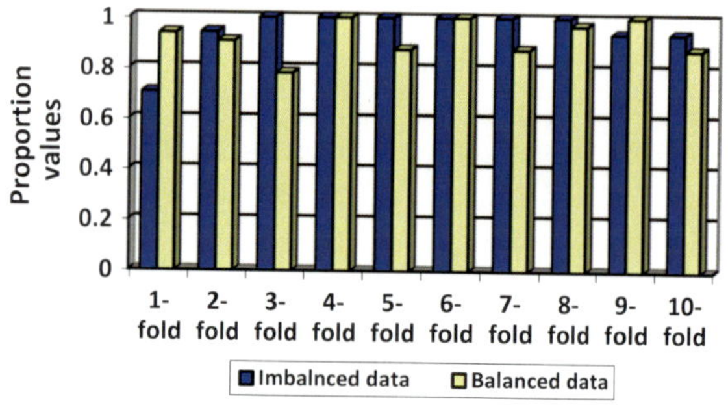

Fig. 3 Cross validation
results of voting ensemble
algorithm on both
imbalanced and balanced
malaria disease datasets

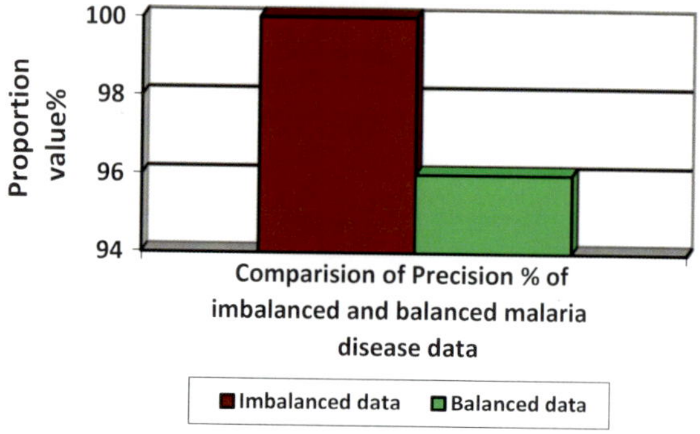

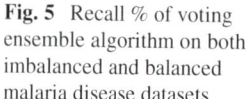

Fig. 4 Precision % of voting
ensemble algorithm on both
imbalanced and balanced
malaria disease datasets

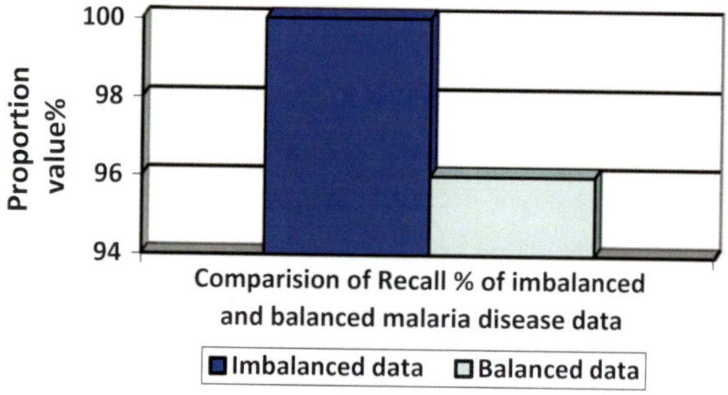

Fig. 5 Recall % of voting
ensemble algorithm on both
imbalanced and balanced
malaria disease datasets

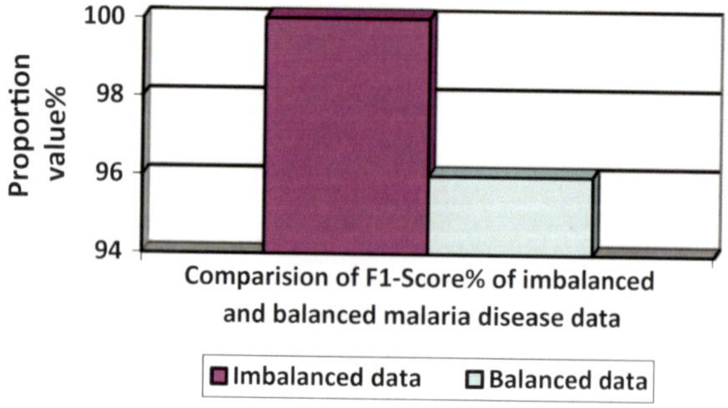

Fig. 6 F1-Score % of voting
ensemble algorithm on both
imbalanced and balanced
malaria disease datasets

5   Conclusion

Vector borne diseases namely Malaria has high impact on world population today.

Even though vaccinations are providing still neonatals and children are effected by this disease. Hence proper diagnosing within stipulated time is very important other wise sometimes it causes to death also. So, accurate identification of patient and diagnosing within time plays very difficult task in medical domain. We are proposing an ensemble method called voting of C4.5, Naive Bayesian and KNN classifiers to classify the minority class samples especially for a better prediction system.

References

1. Bui TQ, Pham HM (2016) Web based GIS for spatial pattern detection: application to malaria incidence in Vietnam. Bui Pham Springer Plus 5(1014):1–14
2. MacLeod DA, Jones A, Di Giuseppe F, Caminade C, Morse AP (2015) Demonstration of successful malaria forecasts for Botswana using an operational seasonal climate model. Environ Res Lett 10:044005, 1–11 (IOP Publishing)
3. Rahman MZ, Roytman L, Kadik A, Rosy DA (2015) Environmental data analysis and remote sensing for early detection of dengue and malaria. In: Proceedings of SPIE, vol 9112, pp 1–9
4. WHO Malaria Report (2016) http://www.who.int/mediacentre/factsheets/fs387/en/
5. Pengfei J, Chunkai Z, Zhenyu H (2014) A new sampling approach for classification of imbalanced data sets with high density. In: IEEE—BigComp, pp 217–222
6. Ditzler G, Polikar R (2012) Incremental learning of concept drift from streaming imbalanced data. IEEE Trans Knowl Data Eng, pp 1–30
7. Nugroho HA, Akbar SA, Murhandarwati EEH (2015) Feature extraction and classification for detection malaria parasites in thin blood smear. In: IEEE 2nd international conference on information technology, computer and electrical engineering (ICITACEE), pp 197–201
8. Das DK, Maiti AK, Chakraborty C (2015) Automated system for characterization and classification of malaria-infected stages using light microscopic images of thin blood smears. J Microsc 257(3):238–252
9. Ruiz D, Brun C, Connor SJ, Omumbo JA, Lyon B, Thomson MC (2014) Testing a multi-malaria-model ensemble against 30 years of data in the Kenyan highlands. Malaria J 13:206, 1–14
10. Smith T, Ross A, Maire N, Chitnis N, Studer A, Hardy D, Brooks A, Penny M, Tanner M (2012) Ensemble modeling of the likely public health impact of pre-erythrocytic malaria vaccine. PLOS Med 9(1):1–20
11. Pandit P, Anand A (2016, August) Artificial neural networks for detection of malaria in RBCs. ArXiv: 1608.06627)
12. Bbosa F, Wesonga R, Jehopio P (2016) Clinical malaria diagnosis: rule based Classification statistical prototype. Springer Plus 5:939
13. Wu C, Wong PJY (2016) Multi-dimensional discrete Halanay inequalities and the global stability of the disease free equilibrium of a discrete delayed malaria model. Adv Differ Equ 2016:113
14. Tsai M-H, Tsai M-H, Yu S-S, Chan Y-K, Jen C-C (2015) Blood smear image based malaria parasite and infected-erythrocyte detection and segmentation. Transactional Processing Systems. J Med Syst 39:118. https://doi.org/10.1007/s10916-015-0280-9
15. Rahmanti FZ, Ningrum NK, Imania NK, Purnomo MH (2015, November) *Plasmodium vivax* classification from digitalization microscopic thick blood film using combination of second order statistical feature extraction and K-Nearest Neighbour (K-NN) classifier method. In: IEEE

4th international conference on instrumentation, communications, information technology, and biomedical engineering (ICICI-BME), Bandung, pp 2–3

16. Charpe KC, Bairagi V (2015) Automated malaria parasite and there stage detection in microscopic blood images. In: IEEE sponsored 9th international conference on intelligent systems and control (ISCO)

17. Somasekar J, Reddy BE (2015) Segmentation of erythrocytes infected with malaria parasites for the diagnosis using microscopy imaging. Comput Electr Eng, pp 336–351 (Elsevier)

18. Cameron E, Battle KE, Bhatt S, Weiss DJ, Bisanzio D, Mappin B, Dalrymple U, Hay SI, Smith DL, Griffin JT, Wenger EA, Eckhoff PA, Smith TA, Penny MA, Gething PW (2015) Defining the relationship between infection prevalence and clinical incidence of *Plasmodium falciparum* malaria. Nat Commun 6:8170, 1–10

19. Krawczyk B (2016) Learning from imbalanced data: open challenges and future directions. Prog Artif Intell, pp 1–12

20. Deng X, Zhong W, Ren J, Zeng D, Zhang H (2016) An imbalanced data classification method based on automatic clustering under-sampling. IEEE Trans, pp 1–8

21. Ali A, Shamsuddin SM, Ralescu AL (2013) Classification with class imbalance problem: a review. Int J Adv Soft Comput Appl 5(3):1–30

22. Poolsawad N, Kambhampati C, Cleland JGF (2014) Balancing class for performance of classification with a clinical dataset. In: Proceedings of the World Congress on engineering, vol 1, pp 1–6

23. Rahman MM, Davis DN (2013) Addressing the class imbalance problem in medical datasets. Int J Mach Learn Comput 3(2):224–228

24. Haixiang G, Yijing L, Shang J, Mingyun G, Yuanyue H, Bing G (2016) Learning from class-imbalanced data: review of methods and applications. Expert Syst Appl, pp 1–49

25. Jamal S, Periwal V, Scaria V (2013) Predictive modeling of anti-malarial molecules inhibiting apicoplast formation. BMC Bioinform 14:55, 1–8

26. Andrade BB, Reis-Filho A, Souza-Neto SM, Clarencio J, Carmargo LMA, Barral A, Barral-Netto M (2010) Severe *Plasmodium vivax* malaria exhibits marked inflammatory imbalance. Malaria J 9:13, 1–8

27. Dubey R, Zhou J, Wanga Y, Thompson PM, Ye J (2014) Analysis of sampling techniques for imbalanced data: An n = 648 ADNI study. Elsevier Neuro Image 87:220–241

28. Ng WWY, Hu J, Yeung DS, Yin S, Roli F (2015) Diversified sensitivity-based under sampling for imbalance classification problems. IEEE Trans Cybern, pp 1–11

29. Roumani YF, May JH, Strum DP, Vargas LG (2013) Classifying highly imbalanced ICU data. Health care Manag Sci 16:119–128

30. Pengfei J, Chunkai Z, Zhenyu H (2014) A new sampling approach for classification of imbalanced data sets with high density. In: IEEE transaction, pp 217–222

31. Garcia V, Sanchez JS, Mollineda RA (2012) On the effectiveness of preprocessing methods when dealing with different levels of class imbalance. Knowl Based Syst 25:13–21 (Elsevier)

32. Thongkam J, Xu G, Zhang Y, Huang F (2009) Toward breast cancer survivability prediction model through improving training space. Expert Syst Appl 36:12200–12209 (Elsevier)

33. Zhao X-M, Li X, Chen L, Aihara K (2007) Protein classification with imbalanced data. Wiley InterSci 70:125–1132

34. López V, Fernandez A, Garcia S, Palade V, Herrera F (2013) An insight into classification with imbalanced data: empirical results and current trends on using data intrinsic characteristics. Inf Sci 250:113–141 (Elsevier)

35. Ma L, Fan S (2017) CURE-SMOTE algorithm and hybrid algorithm for feature selection and parameter optimization based on random forests. BMC Bioinform 18:169

Automatic Detection of Malaria Parasites Using Unsupervised Techniques

Itishree Mohanty, P. A. Pattanaik and Tripti Swarnkar

Abstract The focus of this paper is towards comparing the computational paradigms of two unsupervised data reduction techniques, namely Auto encoder and Self-organizing Maps. The domain of inquiry in this paper is for automatic malaria identification from blood smear images, which has a great relevance in healthcare informatics and requires a good treatment for the patients. Extensive experiments are performed using the microscopically thick blood smear image datasets. Our results reveal that the deep-learning-based Auto encoder technique is better than the Self-organizing Maps technique in terms of accuracy of 87.5%. The Auto encoder technique is computationally efficient, which may further facilitate its malaria identification in the clinical routine.

1 Introduction

Malaria remains a life-threatening serious infectious disease and considered neglected, with roughly 200 million cases and 429,000 deaths per year according to the World Malaria Report 2016 [1]. The most severe and lethal form is caused by the genus Plasmodium that is transmitted and affects the red blood cells (RBCs) of the blood samples through the bites of infected female Anopheles mosquitoes. Various qualitative clinical assessments are being done, which involves manual counting of infected RBCs, microscopic diagnosis, and drug effectiveness in order to bring a control on the serious disease. The risk burden of being infected with malaria and developing the disease is still accounted due to lack of clinical and research exper-

I. Mohanty (✉) · P. A. Pattanaik · T. Swarnkar
Department of Computer Science & Engineering, S 'O' A (Deemed to be University),
Bhubaneswar 751030, India
e-mail: itishreemohantym@gmail.com

P. A. Pattanaik
e-mail: priyadarshiniadyashapattanaik@soa.ac.in

T. Swarnkar
e-mail: triptiswarnakar@soa.ac.in

© Springer Nature Switzerland AG 2019
D. Pandian et al. (eds.), *Proceedings of the International Conference on ISMAC in Computational Vision and Bio-Engineering 2018 (ISMAC-CVB)*, Lecture Notes in Computational Vision and Biomechanics 30,
https://doi.org/10.1007/978-3-030-00665-5_5

tizes, widespread of fake and substandard medicines, erroneous manual enumeration of malaria diagnosis, time-consuming visual diagnosis, handling of huge unlabelled data, lack of advanced machine learning tools and many more [2, 3].

Interestingly, many automated computer vision techniques have been developed for medical diagnosis that is useful for doctors to detect malaria parasite in microscopically thick blood smear images high magnification multi-views. These advanced techniques can help to improve diagnostic processing with less cost and reduce the clinical effort of pathologists. Considering, the above issues and approaches, we have proposed a quantitative comparative study of two unsupervised techniques for the diagnosis of malaria parasite cell samples using their microscopic blood smear high magnification images [4]. In this work, two types of the unsupervised Auto encoder (AE) and Self-organizing Maps (SOM) techniques are applied for segmentation of blood cells. The purpose of our paper is to handle the most challenging phase of malaria detection by segmenting the complex and varying shapes along with overlapping cells.

Figure 1 states the contribution of this paper where we present the latest systematic update SOM technique in image analysis and deep learning AE for malaria identification. The major contributions of our work are summarized as follows:

- We have developed a framework to compare between two unsupervised algorithms in terms of their computational performance, yielding to learn more robust and abstract semantic features for malaria parasite identification.
- Our experimental results show that our comparative learning framework can learn better deep representation features for diagnosis and improve the malaria prediction accuracy.
- The superior performance of deep-learning-based AE technique is one of the fast pre-processing algorithms, which has the ability to improve the image quality and helps in better feature representation using deterministic approach.

The remainder of this paper is summarized as follows: Sect. 2 describes the prior arts in the field of malaria parasite identification analysis; Sect. 3 describes the comparison study of two methodologies; Sect. 4 presents the details of the dataset used in the experimental setup and results in analysis, and Sect. 5 includes the conclusion of the comparative study of this work.

2 Prior Art

To our best knowledge, there are prior methods for diagnosing malaria from microscopic blood smear images [3], generally, follow processes including cell segmentation along with the extraction of useful features using computer vision classification techniques to classify infected and noninfected microscopic blood smear images. Das et al. [4] briefly describes the computational microscopic imaging methods on segmentation and feature extraction for malaria parasite identification. Jan et al. [5] reviews the complete overview of malaria diagnosis process of manual assessment

INPUT IMAGE

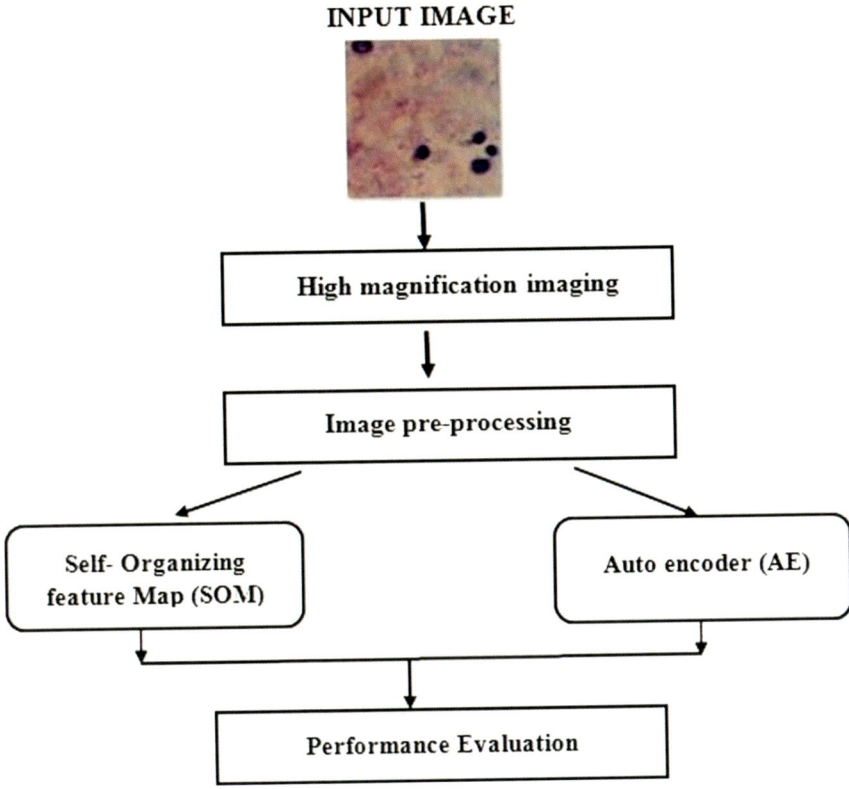

Fig. 1 Overview of our comparative paradigms of two unsupervised data reduction techniques, namely auto encoder and self-organizing maps to detect malaria parasite from microscopic blood smear images

and automated assessment in blood smear images. Devi et al. [6] states the different computational techniques used to detect the erythrocyte features for malaria detection in blood smear images. Poostchi et al. [7] describes the image analysis techniques used for imaging, image pre-processing, cell segmentation, and classification for detecting infected malaria slides for both thin and thick blood smear images. The paper discusses the latest developments in deep learning and smartphone technology towards malaria diagnosis. Shen et al. [8] proposes an image coding scheme by using the stacked autoencoder to achieve lossless compression for malaria infection diagnosis. The proposed deep learning autoencoder model provides a higher compression than other lossless methods like JPEG-LS, JPEG 2000 and CALIC. Bustamam et al. [9] state a clustering self-organizing map method (SOM) for papillomavirus detection which causes cervical cancer disease. Cervical cancer disease is one of the most dangerous diseases and SOM technique forms a clustering process for its identification with two of HPV in the first cluster and 16 others in the second cluster.

Input Layer   **Output Layer**

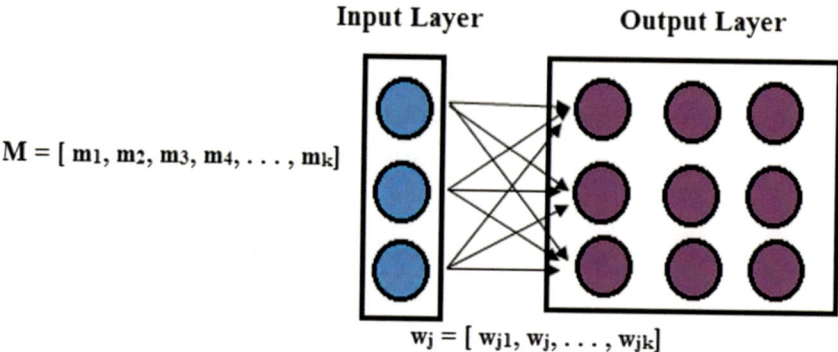

$M = [\, m_1, m_2, m_3, m_4, \ldots, m_k]$

$w_j = [\, w_{j1}, w_j, \ldots, w_{jk}]$

Fig. 2 Self-organizing feature map (SOM) structure

This paper describes the two unsupervised techniques been used in detail below.

3 Methods

The original microscopic blood smear malaria image dataset contains a significant amount of redundant information. However, to gain good computational quality of classification accuracy, image segmentation, and denoising have to be done in order to extract useful information of infected and noninfected blood components. Each color image is converted into a grayscale image through thresholding method and denoising is followed to improve the image quality. These are the data pre-processing steps to be carried followed by the two neural network techniques.

3.1 Self-organizing Map (SOM)

The Kohosen's self-organizing map is an unsupervised self-organizing artificial neural network composed of the input layer and output layers creating a static grid cell of fixed size. SOM technique forms a feedforward network structure with a computational layer in rows and columns from a higher dimensional input space to lower dimensional space [10]. The input layer carries input patterns to each individual nodes in the output layer through the weight matrix structure [11].

In Fig. 2, $M = [m_1, m_2, m_3, m_4, \ldots, m_k]$ is input vector and the weight vector that connect i-th node of the input layer and j-th node of output layer w_{ij} ($j = 1, 2, 3, \ldots, S$) is followed by

$$w_j = \left[w_{j1}, w_j, \ldots, w_{jk} \right] \tag{1}$$

The weights on the input neurons are randomly initialized and the weights are updated by two basic methods. The first method (Best matching unit takes all) includes the neuron whose weights are close to the input vector components gets modified in such a way that the weights get updated as possible to the input vector. The second method (Best matching unit takes most) includes the neuron whose weight is same as the input value gets the most priority. This way its weights and neighboring neurons weight are updated for a given number of epochs or satisfy the given stop criteria.

The update methods of the weight vector $w_q(t)$ at time t is,

$$w_q(t+1) = w_q(t) + Þ(t) \, (M - w_q(t)) \tag{2}$$

where $Þ(t)$ learning coefficient value at time t.

3.2 Autoencoder

Autoencoder is a typical fully connected encoder-decoder architecture [8, 12] where it reconstructs its input vector 'M' at the output 'S' using deterministic approach. Let us consider a set of data samples $M = [m_1, m_2, m_3, m_4, ..., m_k]$ in the input vector where $m_i \in R^d$, the training objective of an autoencoder is to minimize the reconstruction error.

$$1 = \sum_k ||mk - mk^{\cdot\cdot}||^2 \tag{3}$$

where m_k and $m_k^{\cdot\cdot}$ are the input vectors and the hidden layer includes both the encoding and decoding processes of

$$\begin{cases} h_k = f(W_{mk} + b) \\ m_k^{\cdot\cdot} = f(W' h_k + b') \end{cases} , \tag{4}$$

where $h_k \in R^n$ as the compact representation and W, W', b, b' as the weights of encoding layer, decoding layer, and the bias terms respectively. The sigmoidal function in this paper can be represented using $f(.)$ activation function,

$$f(o) = \frac{1}{1 + \exp(-o)} \tag{5}$$

4   Experimental Setup

To demonstrate the effectiveness and importance of the comparative model, we conducted experiments using the publicly available 1182 thick blood smear malaria images dataset consisting of 750×750 pixels obtained from Android smartphone to a Brunel SP150 microscope from Makerere University. The images were randomly splitted into training and testing images collected from different patients and the full dataset is provided in this link: http://air.ug/index.html.

4.1   Evaluation Criteria

To evaluate the comparative techniques, quantitative evaluation criteria includes calculation of accuracy, sensitivity, and specificity using various proportions of true results. Accuracy is the degree of veracity proposition test of true outcomes from either true positive (TP) or true negative (TN), in the samples. The performance measure calculation is illustrated

$$\text{Accuracy} = (TP + TN)/(TP + TN + FP + FN) \tag{6}$$

$$\text{Sensitivity} = (TP)/(TP + FN) \tag{7}$$

$$\text{Specificity} = (TN)/(TN + FP) \tag{8}$$

where TP = True Positive, TN = True Negative, FP = False Positive, FN = False Negative.

4.2   Results and Discussion

In this comparative paradigm, we presented the experiments with a patient level by randomly dividing the whole dataset into 70% training and rest 30% in testing. As per Fig. 1, this approach of using blood smear image cytological criteria for diagnosis involves (i) SOM technique used to segment the cell components present in blood smear images. (ii) Quantification and deep extraction of cell features are done using deep learning technique named Autoencoder. Various morphometric features from each of the segmented blood smear image includes color histogram, color autocorrelogram, area granulometry, gradient descent, relative shape measurements, hu moments, scale invariance, number of colors of diagnosis the malaria parasite infection in erythrocyte, nuclear density, nucleocytoplasmic, Euler number are, fractal dimension and color channel histogram, phase of image, skewness, kurtosis, standard deviation, Sobel histogram, flat texture, co-occurrence matrix and run-length matrix.

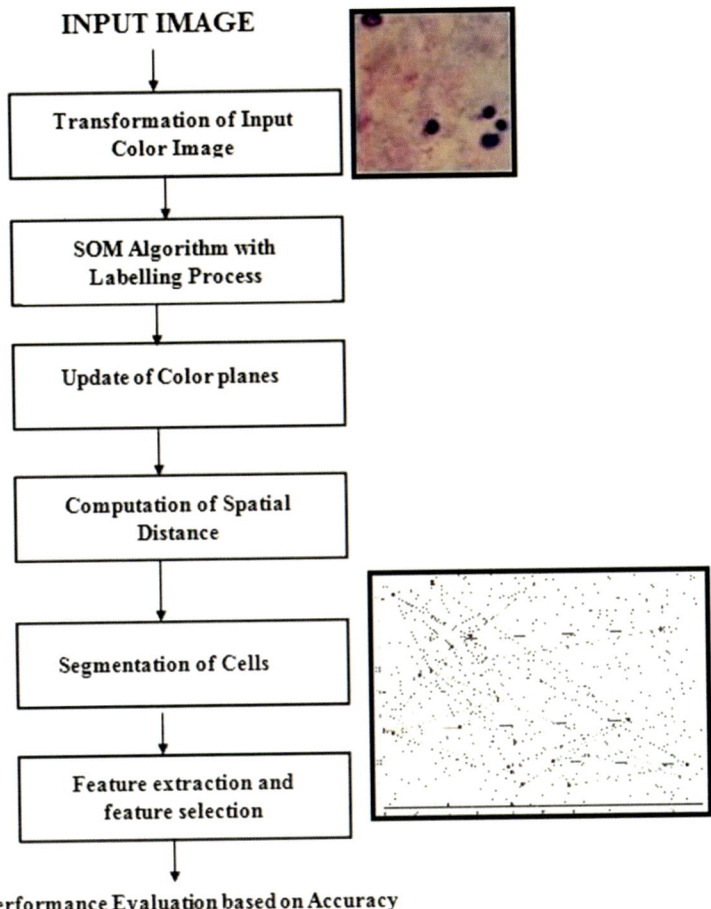

Fig. 3 Conceptual illustration for identifying malaria in the blood smear image using self-organizing feature map (SOM) technique

In SOM technique, each identified blood smear blood cell is segregated from other neighboring blood cells based on minimum area. To optimize the technique performance, we have extracted the malaria features from already computed SOM technique. For blood component classification, we have used texture, color, and shape as the important features for differentiating infected and noninfected blood smear images. The healthy blood smear images contain no nucleus in RBCs whereas unhealthy blood smear images contain the central pale area with some abnormalities in cells as shown in Fig. 3. Using the SOM technique, the spatial distance between blood components is calculated and performance based on accuracy is measured. This technique delivers an accuracy of 0.79 with a sensitivity of 0.80 and specificity of 0.78 in detecting malaria-infected blood smear images (Fig. 4).

INPUT IMAGE **OUTPUT IMAGE**

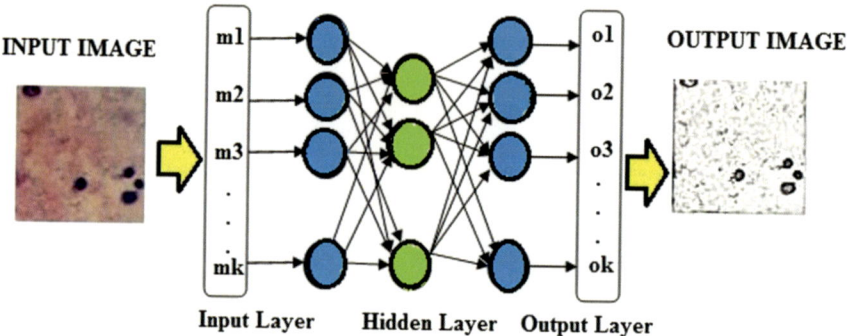

Input Layer   Hidden Layer   Output Layer

Fig. 4 Autoencoder (AE) structure

For simulation purpose, we have used MATLAB 2016B: an open source library. In this section, we have performed on a workstation with Intel Core i5 CPU and 16 GB of memory. The input to the Auto encoder technique is a mat file of 1182 images of pixel intensities, which further gets converted to vector input layers after layers. The autoencoder model is an unsupervised pre-trained model with a learning rate of 0.05 and the number of epochs set to be 1000 [8]. In the first layer or input layer, AE model generates 12,500 input nodes and in the output section of the decoder layer, 30 useful features were noted. The models are all trained from scratch and are running with an epoch of 1000. Through 1000 epochs, 30 useful features were considered for representing infected and noninfected blood smear images. Considering the input nodes and output nodes, performance assessment based on accuracy is done and explained further. This AE technique delivers an accuracy of 0.875 with a sensitivity of 0.84 and specificity of 0.80 in detecting malaria-infected blood smear images.

5   Conclusion

In this work, we have presented an efficient comparative approach based on malaria identification that focuses on the benefits of using two unsupervised neural network based models. The comparative models learn robust and hierarchical representations from unlabelled data and their experimental results show the effectiveness in identifying malaria parasite in blood smear images. Table 1 presents classification accuracy performance of both the techniques along with the details of training and test data in each experiment.

The blood smear based cell sample classification achieves the accuracy of 79% using SOM technique which is less than the classification accuracy of 87.5% using AE technique. This shows the comparative framework ability to identify the infected cells as well as noninfected cells in blood smear images.

Table 1 Accurancies of the two different neural network techniques to identify malaria in blood smear images

Techniques	Accuracy	Sensitivity	Specificity
SOM technique	0.79	0.80	0.78
Auto encoder	0.875	0.84	0.80

References

1. WHO (2016) Malaria microscopy quality assurance manual-version 2. World Health Organization
2. WHO (2016) World malaria report 2016. World Health Organization
3. Tek FB, Dempster AG, Kale I (2009) Computer vision for microscopy diagnosis of malaria. Malaria J 8(1):153
4. Das D, Mukherjee R, Chakraborty C (2015) Computational microscopic imaging for malaria parasite detection: a systematic review. J Microsc 260(1):1–19
5. Jan Z, Khan A, Sajjad M, Muhammad K, Rho S, Mehmood I (2017) A review on automated diagnosis of malaria parasite in microscopic blood smears images. Multimedia Tools Appl 77:1–26
6. Devi SS, Sheikh SA, Laskar RH (2016) Erythrocyte features for malaria parasite detection in microscopic images of thin blood smear: a review. Int J Interact Multimed Artif Intell 4(2):34–39
7. Poostchi M, Silamut K, Maude R, Jaeger S, Thoma G (2018) Image analysis and machine learning for detecting malaria. Transl Res 194
8. Shen H, Pan WD, Dong Y, Alim M (2016) Lossless compression of curated erythrocyte images using deep autoencoders for malaria infection diagnosis. In: Picture Coding Symposium (PCS), pp 1–5
9. Bustamam A, Aldila D, Fatimah, Arimbi MD (2017) Clustering self-organizing maps (SOM) method for human papillomavirus (HPV) DNA as the main cause of cervical cancer disease. In: AIP conference proceedings, vol 1862, no 1, pp 30–155
10. Corral JA, Guerrero M, Zufiria PJ (1994) Image compression via optimal vector quantization: a comparison between SOM, LBG and k-means algorithms. In: 1994 IEEE international conference on neural network. IEEE World Congress on computational intelligence, vol 6, pp 4113–4118
11. Marghescu D, Rajanen MJ (2005) Assessing the USE of the SOM technique in data mining. In: Databases and applications, pp 181–186
12. Razzak MI, Naz S, Zaib A (2018) Deep learning for medical image processing: overview, challenges and the future. In: Classification in BioApps, pp 323–350

Study on Different Region-Based Object Detection Models Applied to Live Video Stream and Images Using Deep Learning

Jyothi Shetty and Pawan S. Jogi

Abstract There is a plenty of very interesting problems in the field of computer vision, from the very basic image classification problem to 3d pose estimation problem. One among the many interesting problems is object detection, which is the computer capability to accurately identify the multiple objects present in the scene (image or video) with the bounding boxes around them and the appropriate labels indicating their class along with the confidence score indicating the degree of closeness with the class. In this work, we have discussed in detail different types of region-based object detection models applied on both live video stream and images.

1 Introduction

Object detection has been one of the hot topics in the area of computer vision. Computer vision aims at enabling vision (ability to see) capability to the machines. Different tasks under computer vision includes (1) Image classification which basically involves classifying the given input image accurately to one of the many classes with the confidence score indicating the degree of closeness with the class. (2) Localization is similar to classification but in addition, it involves identifying the exact location of the objects present in the image with the bounding box. (3) Object detection is one of the major tasks under computer vision, which involves localizing distinct objects present in the image. (4) Image segmentation is the next stage of object detection, which performs a pixel-by-pixel segmentation of multiple objects present in the query image. Figure 1 shows different tasks under computer vision. The emergence of deep learning has completely changed the traditional methods of performing computer vision tasks. In 2012 Alex Krizhevsky, Geoffrey Hinton et al. presented a paper "ImageNet classification with deep convolutional neural network", a convolutional

J. Shetty · P. S. Jogi (✉)
NMAMIT, Nitte, Karkala, India
e-mail: pawansnj03@gmail.com

J. Shetty
e-mail: jyothi_shetty@nitte.edu.in

© Springer Nature Switzerland AG 2019
D. Pandian et al. (eds.), *Proceedings of the International Conference on ISMAC in Computational Vision and Bio-Engineering 2018 (ISMAC-CVB)*, Lecture Notes in Computational Vision and Biomechanics 30,
https://doi.org/10.1007/978-3-030-00665-5_6

Fig. 1 Different computer
vision tasks

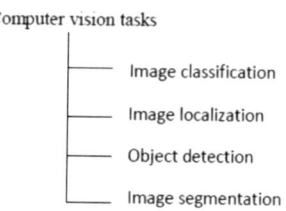

neural network (CNN or ConvNet) based deep learning model which reached the state of the art in the task of performing object detection and won the annual ImageNet challenge (International event on performing computer vision tasks), dropping the error rate from 26 to 15%. This was the first ever successful deep learning model.

Neural networks played a remarkable role in the area of computer vision. A model named ConvNet is performing extremely well in the field of computer vision. ConvNets are the biologically inspired models based on the theory proposed by the researchers D. H. Hubel and T. N. Wissel, motivated by the detailed study of the working of human brain cells called neurons (the exact working of a human brain is still an unsolved problem). They demonstrated how the human brain would interpret images or videos by the simulation of multiple layers of billions of clusters of interconnected neurons, which are responsible for getting a more abstract view of the objects, this turned the researchers to develop the similar pattern recognition models for computer vision tasks. Object detection finds its application in ADAS (Advanced Driver Assistance System), advanced robotics (Humanoids), helping visually challenged people, surveillance system, pedestrian detection, face recognition, military system, healthcare, security system, visual search, pose recognition, gaming, space research, and many more.

The paper is structured as follows. In Sect. 2, we have discussed the evolution of region-based object detection techniques, Sect. 3 is about a detailed study on different object detection models applied on live videos and images, Sect. 4 draws the conclusion.

2 Region-Based Approaches for Object Detection Models

The evolution of region-based approaches changed the direction of performing computer vision tasks. Ross Girshreik (researcher at Facebook AI) and his team at UC Berkeley invented one of the most effective models for performing computer vision tasks. The objective is very simple, for an image the model must be able to identify all the objects present in it by putting the bounding boxes around them along with the appropriate labels. There are two types [1] (1) Single-stage object detection model and (2) Two-stage object detection model. Figure 2 shows different region based convolutional neural network models.

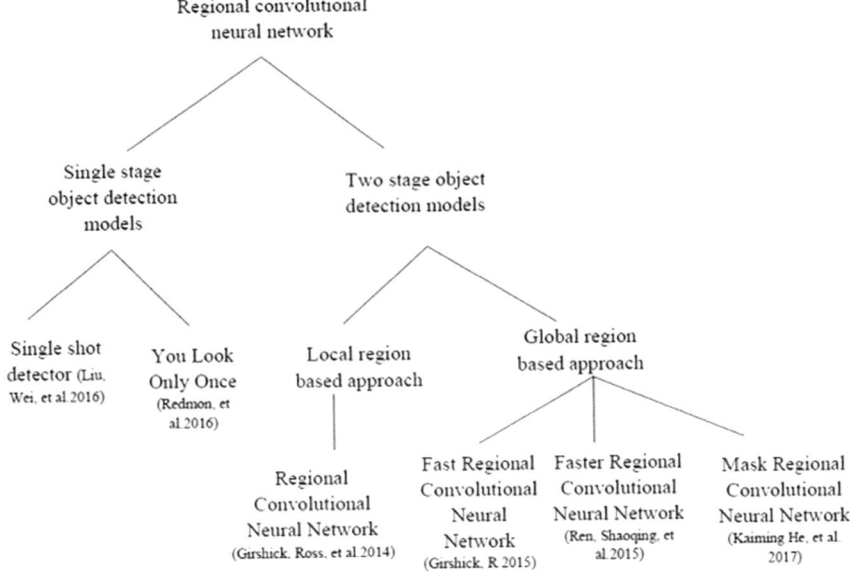

Fig. 2　Different types of region-based convolutional neural network

2.1　*Single-Stage Object Detection Models*

Single-stage object detection models are the feed-forward neural network, which can perform object detection. These kind of models are not using any kind of object proposal generating methods (techniques for identifying the different parts of the image which probably contain the objects) and resampling stages, instead, all these stages are combined in a single stage so that the computation cost can be reduced which will increase the performance. The whole purpose of single-stage object detection model is speed, however, they lack accuracy. Some of the examples of single-stage object detection models are Single-Shot Detector (SSD) and You Look Only Once (YOLO).

Single-Shot Detector Model (SSD) Single-Shot Detector (SSD) [2] is a single pipelined deep neural network model for performing object detection. This model is having the capacity to combine the predictions generated from the different feature maps with a variable resolution to handle objects of distinct size. SSD gets rid of separate steps for performing object proposal generation and resampling, instead, it binds all these in a single framework. Elimination of the object proposal generation resulted in the decrease in computation cost and increase in performance to the greater extent. This model can be directly integrated with other systems that require object detection module. It is faster but less accurate when compared with the two-stage object detection models. SSD performed well on datasets like MSCOCO, PASCAL VOC.

You Look Only Once Model (YOLO) You Look Only Once (YOLO) [3] is a single deep neural network, which is capable of predicting the bounding boxes for the objects and finding out the probability of the class to which it belongs from the whole image, since the entire process involves only one network, end-to-end optimization is possible. YOLO is very fast when compared with any other object detection models, the base YOLO model is capable of processing 45 frames per second, whereas Fast YOLO processes 150 frames per second. YOLO is capable of processing real-time streaming videos with less than 25 ms of latency. This model produced competitive accuracy as two-stage object detection models. YOLO detection mechanism involves three stages, (1) Resizing the query image to 448 × 448 pixels. (2) Running single convolution neural network. (3) Classifying the objects by using model's confidence. YOLO performed well on VOC dataset. However, YOLO failed to reach the accuracy level of two-stage object detection models.

2.2   Two-Stage Object Detection Models

The first stage of the two-stage detection model is using object proposal generation methods (techniques for identifying different regions of the image which probably contain the objects) followed by decision refinement. Some well-known methods like regional convolutional neural network used selective search to generate object proposals, and ConvNets to extract and feed features to the classifier. The second phase involves refining the boundary boxes, identifying the fake detections and rescoring the boxes based on the other objects detected on the image. This is a very complex process and is very difficult to optimize since each of the individual components must be trained independently. Two-stage object detection models are having very high accuracy but very difficult and slow to train when compared with the single-stage object detection models.

Regional Convolutional Neural Network Model (R-CNN) Regional convolution neural network (R-CNN) [4] is a local region-based convolutional neural network. This model is based on identifying the different regions that could possibly contain object followed by the running classifier in order to identify the different objects. This method combines two key features (1) Using high-capacity ConvNets to identify the objects and (2) Supervised learning for the objects in a case when there is a shortage of the labeled training data to get efficient results. This system (1) Takes the query image of size 227 × 227 pixels (2) extracting approximately 2000 bottom-up region proposals by using selective search (3) finding the features for each of the proposal (bypassing the local regions of the image) using large CNN's, and then (4) classification is performed by using linear support vector machine (LSVM). This was a very complex task and the performance of the R-CNN was very poor because it should perform convolution for every region. R-CNN produced good results on ILSVRC 2013 dataset. The main drawback of R-CNN was its low speed. Some of the disadvantages of R-CNN model are listed below. (1) Training involves multiple stages (2) Training is not cost efficient. (3) Object detection is slow.

Fast Regional Convolutional Neural Network Model (Fast R-CNN) Fast regional convolution neural network (Fast R-CNN) [5] is the next version of the original R-CNN. Fast R-CNN consumes less computation time and improves the detection accuracy, due to the following reasons (1) Feature extraction is performed on the image before generating the object proposals thus it will be enough running one CNN over the entire image (passing the global or full image for feature extraction) instead of running 2000 ConvNets over 2000 proposed regions. (2) Using softmax instead of SVM for classification. Then for every object, a pooling layer called Region of Interest (RoI) will extract a fixed-length feature vector from the feature map. Thus, obtained feature vectors are fed onto a series of fully connected layers. This made the model not only fast, but also end-to-end trainable. In Fast R-CNN the object proposal generation algorithms like selective search (SS) played a very crucial role (even though it was less efficient). Advantages of the proposed model are (1) Higher detection accuracy. (2) Training is a single stage. Fast R-CNN reached the state of the art on VOC07, 2010, and 2012.

Faster Regional Convolutional Neural Network Model (Faster R-CNN) Faster regional convolution neural network (Faster R-CNN) [6] is the part 3 of R-CNN series. The whole idea of this model is to replace the selective search algorithm (one of the major drawbacks of Fast R-CNN is its low speed) and to make the model end-to-end trainable. The core idea behind Faster R-CNN model was that region proposals depended on the features of the image that were already calculated with the forward pass of the ConvNets. They have used the same ConvNets results for region proposal generation instead of running selective search algorithm. Region proposal network (RPNs) were introduced. RPN is a fully convolution neural network on top of the features of the ConvNets, which is capable of generating the bounding boxes for the object along with the scores indicating the degree of closeness with the class to which it belongs. The proposed system was cost efficient and produced very high accuracy on PASCAL VOC dataset.

Mask Regional Convolutional Regional Neural Network (Mask R-CNN)
Mask regional convolutional neural network (Mask R-CNN) [7] aimed at performing pixel level segmentation. Mask R-CNN is built on top of the Faster R-CNN with the additional step called mask. It takes a pixel and predicts whether it s a part of the object or not. Mask is a fully convolutional neural network built on top of the ConvNet based feature map. Workflow of mask R-CNN is very simple it will take CNN feature map as the input and outputs the matrix with 1's in all location in case the pixel belongs to the same object or 0 otherwise. Mask R-CNN performed well on MSCOCO dataset.

3 Experiments and Results

Creating the deep learning models with the ability to localize and identify the multiple objects present in the image with the highest accuracy is considered as one of the challenging problems to succeed. Google released object detection framework which is built on top of tensorflow (deep learning library released by Google Brain Team). This helped researchers and developers to build their own custom object detection models. It allows us to use pretrained models for performing computer vision tasks. These models are having mAPs (mean average precision), which tell us how sensitive the model is for objects of interest and how well the model handles the fake detections. As the mAP value increases the models will tend to have more accuracy, higher the accuracy the cost of execution increases. These models are trained on MSCOCO dataset. MSCOCO (Microsoft Common Object in Context) is one of the most widely used datasets for performing computer vision tasks. This dataset contains approximately 330,000 images along with associated labels, which belong to 90 of the most commonly found objects.

3.1 Object

In this experiment, we have detection on images performed the object detection using SSD Mobilenet V1 COCO (single-stage object detection model) and Faster R-CNN Resnet101 COCO (two-stage object detection model) and applied it on both images and live video stream. We found that SSD was able to output the result in less time with low accuracy having only one bounding box around the objects, whereas Faster R-CNN model generated results with more accuracy by putting multiple bounding boxes on objects but it took more time than SSD. We have shown results in Fig. 3.

3.2 Object Detection on Live Video Stream

We have performed object detection on a live video stream by using TensorFlow and OpenCV. OpenCV is a library for processing the images as well as videos. It supports C++, C, Python, and Java. OpenCV is used for performing all kind of video and image analysis like face recognition and detection, license plate reading, advanced robot-vision, and many more. We have performed this experiment by using SSD and Faster R-CNN models results are shown in Fig. 4, SSD model was unable to identify all the objects present in the image, whereas Faster R-CNN model accurately identifies all the objects present in the image.

Faster R-CNN   **SSD**

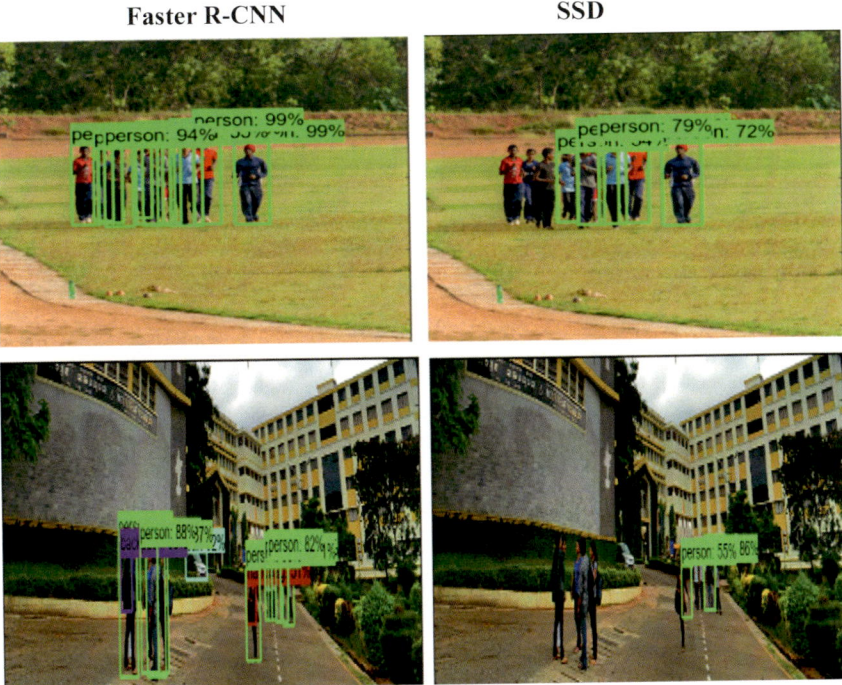

Fig. 3 Comparison of SSD and faster R-CNN models

Faster R-CNN   **SSD**

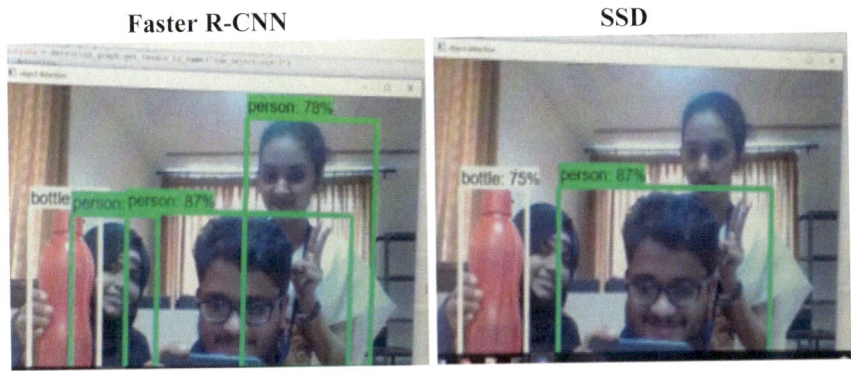

Fig. 4 Comparison of SSD and faster R-CNN model in the live video

3.3 Images Versus Time Comparison

In this experiment, we have taken a total of 500 images (in 100 intervals), all images are <1 mb and performed object detection using SSD and Faster R-CNN models. As expected SSD was able to produce the results quickly with less accuracy whereas,

Fig. 5 Number of images
versus time graph

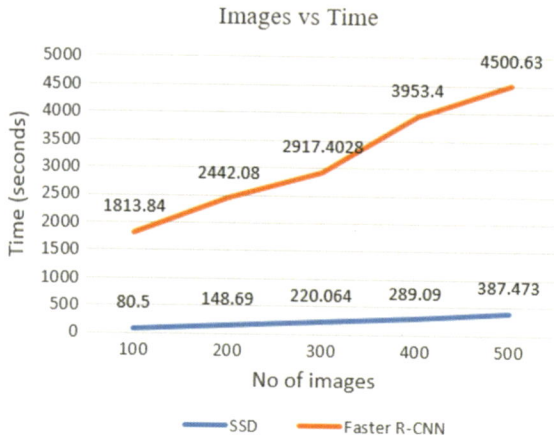

Faster R-CNN was able to produce results which are highly accurate but it took very long time when compared with SSD. We plotted images versus time graph. Results are shown in Fig. 5. We also found that as the image size increases the models took more time to output the result.

3.4 Accuracy Comparison Between SSD and Faster R-CNN

Accuracy is a very important factor for any machine learning models. Here we have investigated the accuracy level of different convolutional neural network based object detection models. We have two types of object detection models one focusing on accuracy (Two-stage object detection models. Ex: Faster R-CNN) and another one on speed (Single-stage object detection models. Ex: SSD). So, it is very important to choose the model which best suits our application. We have evaluated the accuracy based on the number of accurately identified objects among the total number of objects present in the image. We have examined that, as the accuracy level of the model increases the speed will decrease drastically [8]. So it is very clear that, designing a model which is good at both accuracy and speed is very hard. We have shown our results on live videos in Fig. 6, followed by on images in Fig. 7. We have performed all the experiments on Ubuntu 16.04 LTS core i3 processor with 8 GB RAM and 1 TB hard disk having a processor speed of 2.00 GHz and tensorflow 1.4.

4   Conclusion

In this work, we have discussed in detail about the different types of region-based object detection models. There exists a trade-off between speed and accuracy among

Fig. 6 Accuracy
comparison on videos

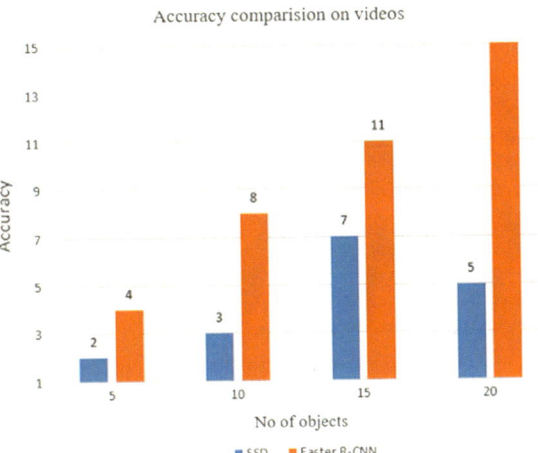

Fig. 7 Accuracy
comparison on images

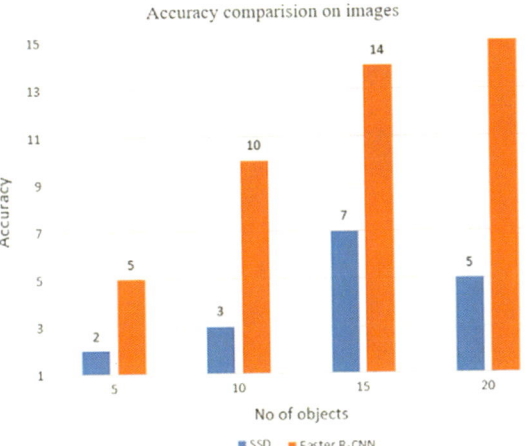

different types of deep learning based object detection models. This made us to investigate about the speed and accuracy of various object detection algorithms on both images as well as on live videos which will help researchers and developers to choose the detection architecture depending upon their requirement.

Acknowledgements We are thankful to NMAM Institute of Technology and Engineering Nitte, for allowing us to take the images required for performing the experiment.

References

1. Huang J, Wang D, Wang X (2017) Novel single stage detectors for object detection
2. Liu W, Anguelov D, Erhan D, Szegedy C, Reed S, Fu CY, Berg AC (2016) Ssd: single shot multibox detector. In: European conference on computer vision. Springer, Cham, pp 21–37
3. Redmon J, Divvala S, Girshick R, Farhadi A (2016) You only look once: unified, real-time object detection. In: Proceedings of the IEEE conference on computer vision and pattern recognition, pp 779–788
4. Girshick R, Donahue J, Darrell T, Malik J (2014) Rich feature hierarchies for accurate object detection and semantic segmentation. In: Proceedings of the IEEE conference on computer vision and pattern recognition, pp 580–587
5. Girshick R (2015) Fast r-cnn. In: Proceedings of the IEEE international conference on computer vision, pp 1440–1448
6. Ren S, He K, Girshick R, Sun J. (2015) Faster R-CNN: towards real-time object detection with region proposal networks. In: Advances in neural information processing systems, pp 91–99
7. He K, Gkioxari G, Dollár P, Girshick R (2017) Mask r-cnn. arXiv preprint arXiv:1703.06870
8. Huang J et al (2017) Speed/accuracy trade-offs for modern convolutional object detectors. IEEE CVPR

Digital Video Copy Detection Using Steganography Frame Based Fusion Techniques

P. Karthika and P. Vidhyasaraswathi

Abstract An effective and exact technique for copying video location in a huge dataset is the utilization of video pictures. We have exactly picked the shading format description, a minimal and powerful casing-based description to make pictures which are additionally determined by vector quantization (VQ). We recommend a new nonmetric length measure to discover the likeness among the question and a dataset video picture and tentatively demonstrate its better execution over other length measures for exact copy identification. The effective look cannot be executed for high-dimensional information utilizing a nonmetric distance measure with accessible ordering systems. Consequently, we create a novel search algorithm in the view of precompiled distances and new dataset reduce systems yielding reduced recovery times. We perform different things with colossal dataset recordings. For singular questions with a normal span of 60 s (around half of the normal dataset video duration), the copy videos are recovered in 0.032 s, on Intel Xeon with CPU 2.33 GHz, with a high exactness of 98.5%.

1 Introduction

The productions of the digital image are an essential and test errand in shared correspondence and different strategies are utilized to produce advanced pictures. There are three objectives of system or data security, for example, secrecy, uprightness, and accessibility (CIA). Privacy implies that data is secure and not accessible to the unapproved individual. Honesty alludes to the precision of data and accessibility implies that data is in accessing time to the approved individual. Picture protection is not adequate for dependable data like content, sound, video, and computerized pictures.

P. Karthika (✉) · P. Vidhyasaraswathi
Department of Computer Applications, Kalasalingam Academy of Research and Education, Krishnan Koil, Tamil Nadu, India
e-mail: karthikasivamr@gmail.com

P. Vidhyasaraswathi
e-mail: vidhyasaraswathi.p@gmail.com

© Springer Nature Switzerland AG 2019
D. Pandian et al. (eds.), *Proceedings of the International Conference on ISMAC in Computational Vision and Bio-Engineering 2018 (ISMAC-CVB)*, Lecture Notes in Computational Vision and Biomechanics 30,
https://doi.org/10.1007/978-3-030-00665-5_7

There are numerous methods to secure pictures including watermarking, advanced watermarking, reversible watermarking, cryptography, steganography, and so on. In this paper, video duplicate identification utilizing steganography is described. We proposed a half and half production of data to approach that is a combination of steganography and frame-based fusion techniques. A short presentation of every procedure has been discussed in the following sections.

A. **Steganography**

Invisible communication has been possible through the steganography. In steganography, the first image is hidden within the cowl image to masquerade the trespasser or hacker and therefore the resulted image is termed stego image. To discover the hidden pictures, a corporation should actively monitor network traffic that is time and processor intensive. However, those that are at home with the network's traditional traffic patterns will merely rummage around for changes like the accumulated movement of enormous pictures across the network, which can warrant additional, careful investigation. It is also informed to have and actively enforce a security policy that clearly outlines acceptable usage, what information will and cannot be sent across the network and the way it should be protected. Also, limit unauthorized programs, the utilization of unauthorized encoding and steganography within the geographical point and think about limiting the scale of mailboxes.

B. **Video copy detection**

It is the method of work lawlessly traced videos by analyzing and examination them to original content today the unlawful issues making on audio, video and text into numerous forms through the net. The electronic communication net is everything is visible and accessible to each user. Therefore, image security of knowledge could be a necessary and vital task varied techniques victimization the copy video detection for steganography is that the art of concealment the very fact communication is going down, by concealment data in different data many various carrier file formats may be used, however digital pictures area unit the foremost in style thanks to their frequency on the net. This paper introduces two new ways, wherever one is fusion and another one is steganography area unit combined to write in code the info still on hide the info in another medium through image process.

C. **Fusion Techniques**

The image combinations of the extraordinary information from every particular picture are melded structure. A consequential excellence image is healthier than any knowledge of the pictures. Image fusion technique will be roughly categorized into two.

1. Abstraction area Fusion Technique
2. Spatial and Temporal Domain Fusion Technique

   Abstraction area Fusion Technique: A pair of remodel domain fusion and abstraction domain techniques, we have a tendency to specifically manage the image pixels values square measure controlled to accomplish made-up outcomes. In return house techniques, the image is first moved into the return space.

Spatial and Temporal Domain Fusion Technique: We have a tendency to reduce image pixels and manage them to get the specified yield. In remodel Domain technique and Fourier Transformation, square measure is applied to the image to urge the specified yield. A number of the primitive ways in image fusion square measure averaging technique is to pick the most and minimum. It is a derived composition generated from a sequence of pictures and may be obtained by understanding geometric relations between the photographs.

D. Context Analysis for Retrieval

Relevant data has been effectively talked about from various perspectives, going from the spatial, transient, shape setting, design, and relevant setting. Labels and areas are two normally utilized setting data for picture recovery. Social connections have pulled in the considerations which are utilized to contemplate the client-to-client, and client-to-photograph relations. The client setting and interpersonal organization setting were additionally used to comment on pictures. In some current investigations, the combination of substance and setting is a typical method to enhance the execution. The semantic setting prompted from the discourse transcript encompassing a key casing was joined with the visual catchphrases to enhance the execution of close copy picture recovery. A setting chart was developed at the record point for video seek reranking, in which setting alludes to the properties depicting who, where, when, what, and so forth, of web archives. Labels, clarification, migrations, and visual information have been used to enhance the recovery data for Flickr. In labels, area and substance investigation (shading, surface, and neighborhood focus) were utilized to recover the pictures of topographical-related milestones. The vast majority of the specified works are basically in light of the picture sharing site Flickr. Be that as it may, there has been small research investigating the setting data for video sharing sites, for example, YouTube. It stays vague whether relevant assets are likewise successful for web recordings. Specifically, the combination of substance and setting data for close copy web video end has not been genuinely tended to.

2 Literature Survey

Video duplicate and comparability discovery have been effectively considered for its potential to seeking [1], subject following [2], and copyright security [1]. Different methodologies utilizing distinctive highlights and coordinating calculations have been proposed. Among existing methodologies, numerous accentuate the quick distinguishing proof of copy recordings with smaller and solid worldwide highlights. These highlights are by and large alluded to as marks or fingerprints which outline the worldwide measurement of small level highlights. Run of the mill highlights incorporate shading, movement and ordinal mark [3, 4] and model-based mark [1, 5]. These worldwide component-based methodologies are appropriate for recognizing relatively indistinguishable recordings, and can identify minor altering in the spatial and fleeting space [1, 3–5, 9]. Our examination of an assorted arrangement of well-known web recordings demonstrates that there are around 20% correct copy recordings among all close copy web recordings [6]. It is normal for web clients to transfer correct copy recordings with insignificant change. This features the need of an approach for quick location of copy recordings. Be that as it may, world-wide highlights end up plainly inadequate when managing video duplicates covered in layers of altering beauty care products. For these more troublesome gatherings, low-level highlights at the fragment or attempt level are useful to encourage nearby coordinating [4, 6–8, 9]. Commonly, the granularity of the fragment level coordinating, the adjustments in the worldly request, and the addition/erasure of edges all add to the similitude score of recordings. Contrasted with signature-based techniques, portion-level methodologies are slower, however, fit for recovering estimated duplicates that have experienced a generous level of altering. At a larger amount of multifaceted nature, more troublesome copies with modification in foundation, shading, and lighting, need considerably more perplexing and solid highlights at district stage. Highlights, for example, shading, surface, and shape can be separated at the key casing level, which thus could be additionally portioned into various area units. Notwithstanding, the issue of division unwavering quality and the granularity choice brings into question the adequacy of these methodologies. As of late, nearby intrigue focuses (key focuses) are appeared to be helpful for close copy and duplicate recognition [1, 3–6, 41]. Nearby focuses are striking neighborhood areas recognized over picture scales, which find neighborhood locales that are open-minded to geometric and photometric varieties [7]. Striking areas in each key edge can be separated with neighborhood point indicators and their descriptors are for the most part invariant to nearby changes. Keypoint based nearby element identification approach stays away from the deficiency of worldwide and portion level highlights, and gives a precise estimation notwithstanding for pictures experienced extreme changes. Albeit neighborhood focuses have generally been recognized as solid and vigorous highlights, productively coordinating extensive measure of nearby focuses remains a troublesome issue. Late arrangements incorporate utilizing ordering structure and quick close copy following with heuristics [1]. On a very basic level, the errand of close copy discovery includes the estimation of repetition and curiosity, which has been investigated in content data recovery [2, 7]. The curiosity recognition approaches for records and sentences chiefly center around vector space replica and factual dialect models to quantify the level of oddity communicated in words. Question pertinence and data oddity have been consolidated to re-rank the reports/pages by utilizing graph [2] and dialect models [3]. To the most excellent of our insight, there is little research on close copy video recognition and re-positioning for substantial scale web video investigates [1].

3 Problem Analysis

1. Local element based CBCD calculations have demonstrated better recognition rate, and however, extraction of nearby highlights alongside their coordinating procedure have noteworthy time prerequisites.
2. Video preprocessing done by CBCD frameworks incorporates expulsion of dark outskirt, picture-in-picture, and camcording impacts. Because of such preprocessing, the worldwide highlights can adequately manage intense changes.
3. Applied idea of sack with DCT-sign based component, which is typically utilized with nearby highlights. Accordingly to applying the ideas of nearby highlights with worldwide ones can productively build strength of worldwide highlights against different changes.
4. As worldwide highlights are not ready to adapt to geometric changes, these worldwide highlights can be proficiently joined with neighborhood highlights to reinforce them against both photometric and geometric changes.
5. Motion highlights are utilized to recognize recordings and however are not vigorous to content additions or different impediments which hinder the movement from being caught. Changes including revolution will alter the course of movement vectors and provide poor outcomes.
6. In expansion to vigor and separating capacities, the extricated highlight vector ought to be sufficiently smaller to perform a quick coordinating task, as reduced mark requires least storage room and performs likeness estimation in less calculation time.

4 Implementation

- The objective of undertaking to segment each reference video and inquiry video into short video portions
- A small video fragment is a gathering of comparable sequential edges
- It is not necessary that the fragments be a similar length and each section can choose one of its frames as a representative frame.

Steps for Novel Search Algorithm

For the most extreme movement relocations of h7, the proposed NS calculation uses an inside one-sided seek design with nine inspection focuses on a 5×5 window in the initial step rather than a 9×9 window in the 3SS. The focal point of the pursuit window is then moved to the point with least piece contortion measure. The pursuit window size of the following two stages relied upon the area of the base BDM (Block Distortion Measure) focuses. On the off chance that the base BDM point is found at the focal point of the inquiry window, the search will go to the last advance (Step 4) with 3×3 look window. Something else, the inquiry window measure is kept up in 5×5 for stage 2 or stage 3. In the last advance, the pursuit window is

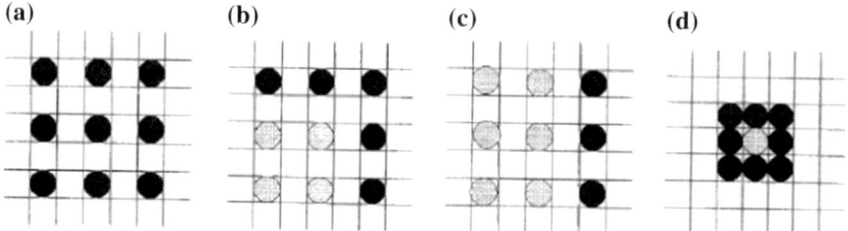

Fig. 1 **a** Nine-checking points pattern, **b** block distortion measure, **c** searching pattern strategy, **d** the Least BDM Point among these nine searching points

decreased to 3 × 3 and the inquiry stops at this little hunt window. The NS algorithm is summarized as follows

1. **Step 1**: A least BDM point is established from a nine-checking focuses design on a 5 × 5 window situated at the focal point of the 15 × 15 seeking region as appeared in Fig. 1a. In the event that the base BDM point is established at the focal point of the search window, go away to Step 4; generally go away to Step 2.

2. **Step 2**: The investigate pixel size is maintained in 5 × 5 to investigate and the outline will depend on the location of the previous least BDM point.

   (a) If the least BDM (block distortion measure) point is positioned at the area of the previously investigated pixel for five supplementary inspection points as shown in Fig. 2b are used.
   b) If the least BDM point is positioned at the center of horizontal or vertical axis of the earlier investigated pixel for three supplementary inspection points as shown in Fig. 2c are used. If the least BDM point is found at the middle of the investigate window, go away to Step 4; otherwise go away to Step 3.

3. **Step 3**: The searching model approach is equal as Step 2, but to finish it will go away to Step 4

4. **Step 4**: To investigate the pixel is reduced to 3 × 3 as shown in Fig. 2d and the direction of minimum BDM point among these nine searching points considered to the overall motion vector.

We can locate from the algorithm transitional strides of the NS might be left out and after that bounced to the last advance with a three by three pixel if whenever the base BDM point is situated at the focal point of the inquiry window. In light of this four-advance inquiry design, we can plaster the entire 12 × 12 uprooting window level just little pursuit windows, 4 × 4 and 3 × 3, are utilized. There are covered checking focuses on the 4 × 4 look skylight in the stage 2 and stage 3, along these lines the aggregate number of read-through focuses is fluctuated from (8 + 9) = 17 to (9 + 6 + 6 + 8) = 29. The most pessimistic scenario is the computational prerequisite of the NS is 29 square equivalents, it will occur for the evaluation of expansive development. Two cases of NS appeared in the various inquiry ways.

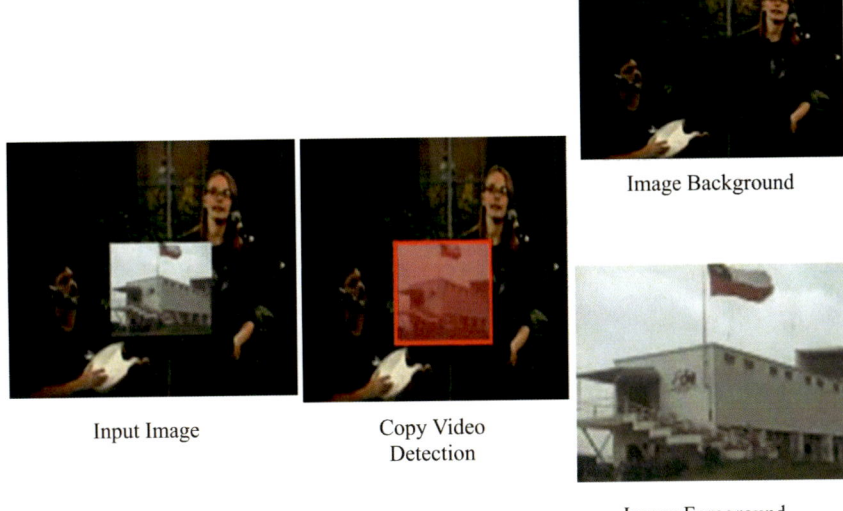

Input Image   Copy Video   Image Background
                     Detection

Image Foreground

Fig. 2 Preprocessing task: A query video with PIP (Picture-in-Picture), detected and new query videos

For the high pursuit way, an aggregate of 25 inspection focuses is utilized with the assessed movement vector equivalent to $(3, -8)$. For the most pessimistic scenario container of the lesser seek way, the quantity of checking focuses necessary is 29 and the assessed movement vector is equivalent to $(-8, 8)$. We can locate that the computational multifaceted nature of NS (Novel Search) is just two square equal more than the Three Step Search while 6 piece coordinates not exactly the N3SS (New Three-Step Search) in the worst case.

5   Result Analysis

The objective of this preprocessing task is to divide every reference video and query video into short video segments. A small videotape segment is a collection of similar repeated frames and is not necessary that the segments be the matching length. Optionally, every segment can be choosing a single of its frames as a delegate frame. We separate a video into segments instead of attempt since dividing a video few segments and various corpora enclose duplicate shorter construction for division necessary in a single attempt. We just select keyframes for quality extraction and we will use an entire segment instead of just delegate frames. In our execution, we first separate the video into segments of fixed time-span, and then we join two repeated segments when all their frames are almost matching between them.

6 Conclusion

We propose the TASC (Transformation-Aware Soft Cascading) to sort out different multimodal locators in a falling and change mindful way, which is relied upon to accomplish high discovery precision while limiting the handling time. One productive usage is likewise created by using three ordinarily utilized multimodal highlights to develop four unique chains. A detection-on-copy-unit is TASC component is presented in which settles on the choice of duplicate location relying upon the closeness between their majority comparable CUs instead of the video-level comparability. To do as such, we likewise suggest a copy unit look calculation to discover a couple of copy units-based restriction calculation to locate the exact areas of duplicate sections with the attested CUs as the inside. To deal with the issue that the duplicates and non-duplicates are perhaps straightly indivisible in the component. We present an adaptable delicate choice limit technique in the TASC and after that suggest a bi-edge learning calculation for tough choice and use SVM classifier to delicate edge in view of the SIFT key point coordinating for delicate choice. Because of its incredible handling execution, the TASC is fit for fulfilling different prerequisites in down to earth duplicate discovery submission. The TASC-based framework will utilize this innovation in the video web index to distinguish the semantically and outwardly indistinguishable copies from video query items. For the future work, we plan to additionally advance the execution and adaptability of the TASC applications.

References

1. Liu H, Hong L, Xue X (2013) A segmentation and graph-based video sequence matching method for video copy detection. IEEE Trans Knowl Data Eng 25(8):1706–1718
2. Jiang M, Tian Y, Huang T (2012) Video copy detection using a soft cascade of multimodal features. In: Proceedings of the IEEE international conference on multimedia and expo (ICME'12), pp 374–379
3. Haitsma J, Kalke T (2012) A highly robust audio fingerprinting system. In: Proceedings of the international symposium on music information retrieval, pp 107–115
4. Tasdemir K, Cetin AE (2014) Content-based video copy detection based on motion vectors estimated using a lower frame rate. In: Proceedings of signal, image and video processing. Springer, Berlin, pp 1049–1057
5. Lei Y, Luo W, Wang Y, Huang J (2012) Video sequence matching based on the invariance of color correlation. IEEE Trans Circuits Syst Video Technol 22(9):1332–1343
6. Esmaeili MM, Fatourechi M, Ward RK (2011) A robust and fast video copy detection system using content-based fingerprinting. IEEE Trans Inf Forensics Secur 6(1):213–226
7. Barrios JM, Bustos B (2011) Competitive content-based video copy detection using global descriptors. Multimed Tools Appl https://doi.org/10.1007/s11042-011-0915-x (Springer Science+Business Media)
8. Song J, Yang Y, Huang Z, Shen HT, Hong R (2013) Multiple feature hashing for large scale near-duplicate video retrieval. IEEE Trans Multimedia 15(8):1997–2008

Color Image Encryption: A New Public Key Cryptosystem Based on Polynomial Equation

P. K. Kavitha and P. Vidhya Saraswathi

Abstract Image security is the need of today's era. The security task is being convoluted as there are a variety of data to secure as different levels of data security is needed based on the data type. Encryption is the maximum expedient approach to guarantee the safety of images over public networks. We have proposed a new public key cryptosystem based on polynomial equation. This approach has been desired because of their resistance to the cryptanalysis attacks. Parameters such as correlation coefficient, entropy, histogram analysis are used for the efficiency of our proposed method. The theoretic and simulation result provides a high quantity of protection. Thus, it is suitable for practical use in secure communication.

1 Introduction

Due to the enormous growth of network era, numerous facts can be transmitted and posted fluently and quick. To store the resources, hundreds of digital images of some particular application fields additionally had been transmitted through the public network. There can be plenty of secret and sensitive records contained in the images [1]. So, we have to encrypt those digital images earlier before transmission through the open network. The protection of digital images has been a focal point and one of the important significant branches in the information protection area. Because of its unique features such as lively and intuitionist, image has been one of the favored forms expressed through network. An image has high redundancy and strong correlation between pixels. So, tough security is needed for storage and transmission of images through networks. Most of this information is collected and saved on electronic computers and transferred all over the network [1]. If eavesdroppers

P. K. Kavitha (✉) · P. Vidhya Saraswathi
Department of Computer Applications, Kalasalingam Academy of Research and Education,
Krishnankoil 626126, Tamil Nadu, India
e-mail: pkkavitha78@gmail.com

P. Vidhya Saraswathi
e-mail: vidhyasaraswathi.p@gmail.com

© Springer Nature Switzerland AG 2019
D. Pandian et al. (eds.), *Proceedings of the International Conference on ISMAC in Computational Vision and Bio-Engineering 2018 (ISMAC-CVB)*, Lecture Notes in Computational Vision and Biomechanics 30,
https://doi.org/10.1007/978-3-030-00665-5_8

found the secret images regarding enemy positions then security could lead to dec-lination of clash. Protecting images is an ethical and authorized requirement. A lot of images are usually used in one kind of outstanding procedure. Image encryption is a vital function within the field of data hiding. Image encryption method prear-ranged an image by applying some mathematical formulas, so no one gets the original image [2]. In this paper, color image encryption is done using polynomial equation. The cryptography process requires some algorithms for encrypting the data. Two categories of cryptography are symmetric and asymmetric [2]. In symmetric key cryptography, both the sender and the recipient use the similar secret key and both of them know the key. In sender side, the data is encrypted using the secret key and the recipient decrypts the information using the same secret key. Key depends on the nature of the key. Examples of symmetric key methods are data encryption standard (DES), advanced encryption standard (AES), international data encryption algorithm (IDEA). In asymmetric key cryptography, a couple of keys used are public and pri-vate keys. It has six substances. There are plain texts, encryption algorithm, public key, private key, cipher text, and decryption algorithm. Examples of asymmetric key methods are Diffie–Hellman Key Exchange [3], RSA Encryption Algorithm, Elliptic Curve Cryptography (ECC), and ElGamal Cryptosystem [2]. In this work, we follow an innovative public key cryptosystem corresponding to polynomial equation for more security of color images.

The rest of this paper is structured as follows. Section 2 describes the related work. Section 3, in brief, explains the concept of polynomials that is the core of this report. Section 4 describes the results and discussion of the proposed scheme. Section 5 produces experimental results and evaluation parameters to validate the proposed color image encryption algorithm. Section 6 concludes the paper.

2 Related Work

Seyedzade et al. [4] presented a precise set of rules for image encryption primarily based on SHA-512. The crucial idea of the set of regulations is to use one half of image data for encryption of the alternative half of the image equally. Individual characteristics of the set of rules are high protection, immoderate sensitivity and excessive pace that may be applied for encryption of grayscale and color images. The goal is to increase the image entropy. Hashim et al. [5] selected a public key cryptosystem known as the ElGamal Cryptosystem. The ElGamal cryptosystem over a primitive root of a huge excessive is utilized in messages encryption inside the unfastened privacy guard software application, present day-to-day variations of pretty good privacy, and other cryptosystems. Here, change of the cryptosystem is via making use of it over gray and color images. That is probably by using remodeling an image into its corresponding matrix using MATLAB program. The encryption and decryption algorithms is applied over it. This new modification should make the cryptosystem greater immune toward a few destiny assaults considering that breaking this cryptosystem relies upon on fixing the discrete logarithm hassle and it is actually now not feasible with massive pinnacle numbers. Khachatrian et al. [6] added a public key encryption and virtual signature system based on permutation polynomials is evolved. It is proven that for the new device to obtain a comparable safety with conventional public key structures based totally on both Discrete logarithm or Integer factorization issues, significantly a lot of processing period is wanted to ensue in a large acceleration of public key operations. Sokouti et al. [7] carried out a new GGH encryption algorithm, where GGH is a new public key cryptosystem that is corresponding to the closest vector problem. It encrypts records in the matrix form and makes it appropriate for image encryption. Tentative research converting one pixel in GGH encryption does not have enough effect at the whole encryption output, so a new method is used which uploads padding to the original image in advance than making use of GGH encryption method. Then the forward and backward snail tour XORing into the original image is applied. Based on the very last consequences which correspond to the proposed approach, converting one pixel affects the entire image and additionally influences the cipher image pretty right this is encrypted with the resource of GGH set of rules. Dwivedi et al. [8] have been planned to offer at ease cryptosystem to transmit images. A new curved scrambling method expanded the complexity of key generation method, consequently difficult to decode via the hacker. Also diffusion technique adjusts the corresponding pixel values and hence fend off repeated permutation. All these strategies assist to layout a greater complexity and comfy cryptosystem to transmit the data over the insecure community. Wagh et al. [9] offered a unique image encryption set of rules based on permutation polynomials over integer rings which makes a strive to conquer the restrictions of current strategies. Here, the authentic image is scrambled by way of applying permutation polynomial to its rows and columns. The estimated changed price in a wide variety of pixels and Unified Averaging Changing Intensity measures are near to theoretical values and are comparable to present encryption strategies.

3 Proposed Methodology

The proposed method is a new public key cryptosystem based on polynomial perspective. Original image with public key image produces an encrypted image. In the recipient side, to get the authentic image receiver's own private key image is used. The experimental effects proved that this new cryptosystem is further competent when compared with common existing techniques used for image encryption.

3.1 Polynomial

A polynomial of degree n inside variable x is a function defined by means of

$$f(x) = k_0 + k_1 x_1 + k_2 x_2 + \cdots + k_{n-1} x_{n-1} + k_n x_n$$

Here, n is the degree of the polynomial k_n = the leading coefficient, k_0 is the constant term, k is real numbers and also the coefficients of the polynomial.

3.2 Encryption Phase

In this section, we discuss the algorithm design and implementation of the proposed image encryption and decryption method using polynomial. Row and column of an image is depicted as r, c. The row and column value is mentioned as $r = c = 256$. The decryption scheme is in direct contrary to an encryption procedure. The proposed approach is based on public key cryptosystem and for this reason, encryption and decryption require a pair of keys.

Algorithm for Encryption

1. Input: (original image, public key image, private key image)
2. Compute: Mean value of private key image
3. Unimodular matrix (private key image, mean value)
4. If pixel value of private key image > mean value
5. Set the unimodular matrix value as 1
6. Else
7. Set the unimodular matrix value as 0
8. End
9. Use the polynomial equation to choose the pixel in the unimodular matrix
10. For r = 1: (size of the original image, 1)
11. For c = 1: (size of the original image, 2)
12. Obtain (first pixel, second pixel) from original image
13. Choose the same index as (first pixel, second pixel) from public key image
14. Perform XOR between original image and public key image
15. Goto next pixel based on polynomial graph
16. Repeat steps 10–14 until all of the plain image pixels are processed
17. End
18. End
19. Output: encrypted image.

4 Result Analysis

The planned polynomial-based approach has been developed in MATLAB R2013a software. Color image is taken as an input image. It has been resized into 256×256 pixels. Pick another image as a public key image and resized into 256×256 pixels. Then, the private key image could be read and resized into 256×256 pixels. Then these three images changed into grayscale images. Computed the mean value

Fig. 1 Color images **a** Original image. **b** Public key image. **c** Private key image. **d** Polynomial in unimodular matrix. **e** Encrypted image

of a private key image. Based on this mean value, generate the unimodular matrix (i.e., 0s and 1s). Suppose MV is the mean value of the private key image. UM is the unimodular matrix. $[X, Y]$ is the pixel value of unimodular matrix.

$$\text{If } PR > \text{MV then} [X, Y] = 1, \text{ else } [X, Y] = 0.$$

That is the value set as 1, if the mean value is greater than private key image value and the value set as 0, if it is less than the pixel value of the private key image. Then the unimodular matrix is created. Now, the new polynomial-based approach can be almost carried out to an original image. Figure 1 demonstrate the step by step encryption of color image using our proposed approach.

In the encryption phase, the public key image is used. From the public key image, take an index value from the polynomial graph in a unimodular matrix image. The same index is taken from the original image and calculate the value of encrypted image by using the original image pixel value and the public key pixel value. Then an encrypted image is produced by XORing the two values from the original image and public key image.

Private key is used to decrypt an encrypted image. On behalf of decryption, the recipient receives an encrypted image and use the contrary approach of the private key

(a)   **(b)**   **(c)**

Fig. 2 Decryption process **a** Encrypted image. **b** Private key image. **c** Decrypted image

image to an encrypted image to get the original image. Here, XORing two values are obtained from the encrypted image and the private key image depends on the polynomial graph in unimodular matrix. The decrypted image is as shown in Fig. 2c.

5 Performance Measurement

The proposed polynomial-based encryption approach is tested on MATLAB R2013a software. Proposed method should defy along with recognized attacks like cipher-text attack, known-plain text, brute-force attack, and differential attack. Safety measures analysis plays on the proposed approach simultaneously with the statistical analysis and differential analysis as follows.

5.1 Histogram Analysis

Histogram graph depicts the wide variety of pixels in an image at dissimilar intensity values observed in an image. The histogram will show 256 numbers displaying the distribution of pixels among the one's of the grayscale values. Proposed encryption approach's competence is checked with the aid of the distribution of grayscales in an encrypted image.

It needs to be uniform and notably distinctive from histograms of an Original Image. From Fig. 3a–d are the histograms of the authentic input image and an encrypted image. Histogram of an encrypted image is identical and widely contradictory from an original image. It doesn't give any evidence to utilize any statistical attack on the proposed polynomial-based image encryption approach.

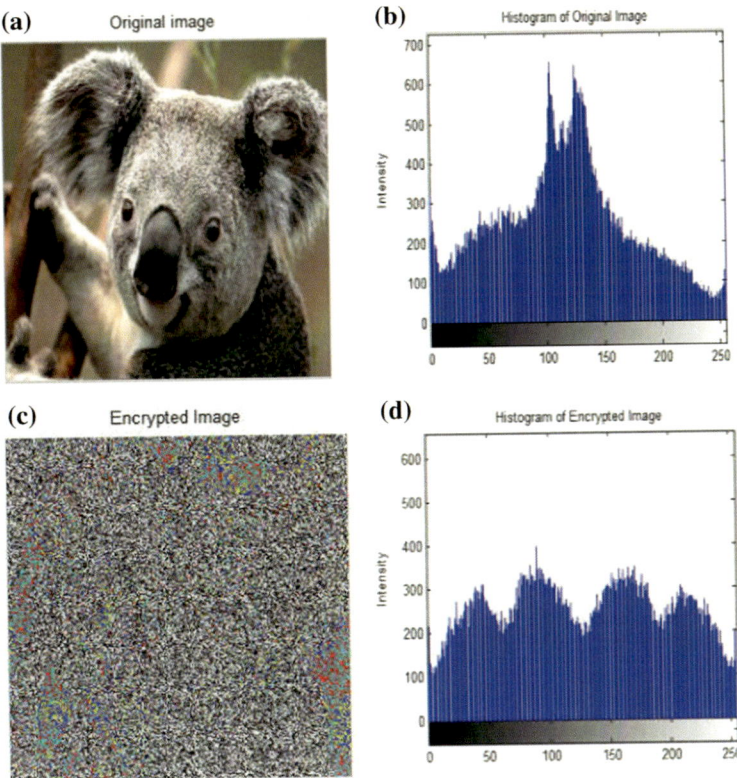

Fig. 3 **a** Input image (Koala). **b** Histogram of input image. **c** Encrypted image of Koala. **d** Histogram of encrypted image

Table 1 Mean values of original and encrypted image for RGB channels

Test image: Koala	Original	Encrypted
Red mean	125.40	127.50
Green mean	114.74	127.40
Blue mean	101.97	127.35

5.2 RGB

From the outcome, the original image and encrypted image histograms and their equal histograms for Red, Green, and Blue are as shown in Fig. 4. The encrypted image is again and again spread and hugely unlike from the histograms of an input Image. Mean values of original and encrypted image for RGB channels displayed in Table 1.

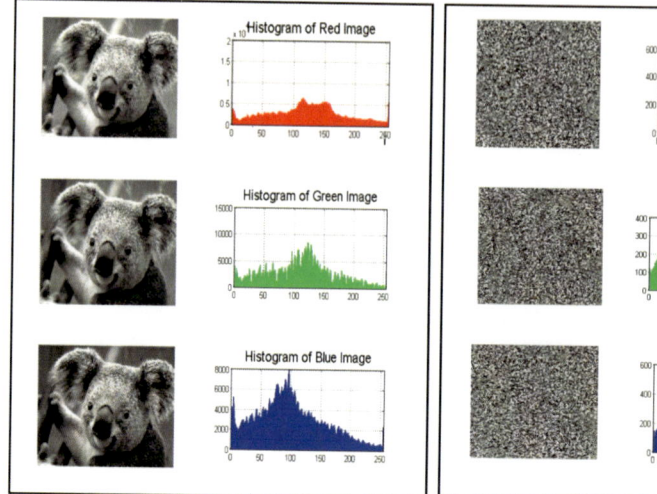

Fig. 4 RGB histograms for original and encrypted Koala image

5.3 Correlation Analysis

Correlation is a quantity of the link between two variables or pixels in an image. Adjacent pixel correlation means, if two pixels from an image taken are the nearest pixels in an image, then say that these two pixels are strongly related. If the two pixels are distant, then they are not as much of correlated. Correlation among the adjacent pixels value can be very high. The competent approach can diminish the relationship between the pixels. First, prefer 2000 pixel pairs from an input image. Figure 4 suggests the connection part of adjacent pixels in the input image and cipher image (koala image of size 256×256) for the personalized encrypted image. Conversely, the pair of adjacent pixels in an input image is extremely interrelated. Correlation coefficient of all pair can be calculated using the formula:

$$Cor = \frac{k\left(\sum (m, n) - \sum m \sum n\right)}{\sqrt{\left[k\left(\sum (m)^2 - \left(\sum m\right)^2\right)\right]\left[k\left(\sum (n)^2 - \left(\sum n\right)^2\right)\right]}}$$

Here,

Cor	Correlation Coefficient
k	The number of pairs of data
$\sum mn$	Range of the products of paired data
$\sum m$	Sum of the m statistics
$\sum n$	Sum of n data
$\sum m^2$	Sum of Squared m

Table 2 Correlation coefficient and entropy value

Test image: Koala	Input image	Encrypted image
Horizontal	0.9460	−0.0071
Vertical	0.9568	−0.0058
Diagonal	0.9240	0.0041
Entropy	7.8292	7.9539

$\sum n^2$ Sum of Squared n

Correlation coefficients are 0.9460 and −0.0071, respectively, for input image and encrypted image. These two values are distant from each other. Related outcome for vertical and diagonal directions was attained like in Table 2.

5.4 *Entropy Value*

It shows the quality of an image. Entropy is expressed in terms of bits. Entropy value of an input image (Koala) and encrypted image is like in Table 2. Use the following formula for entropy value, we have:

$$E = \sum_{i=0}^{k-1} g(y) \log_2(g(y))$$

Here,
$E =$ Entropy Value, $k =$ Input image gray value from 0 to 0.255 and $g(y) =$ Probability of the amount of symbol y

6   Conclusion

A new polynomial-based public key cryptosystem is proposed in this paper and implemented in MATLAB R2013a programming software. The conclusion can be made from the results and discussions are done. This proposed polynomial-based color image encryption algorithm gives more security compared to the existing methods. This paper offers a solution to fulfill the increasing requirements for secure image transmission through networks. Polynomial-based cryptosystem supports encryption of various image sizes. Experimental analysis shows that our new polynomial-based image encryption technique is efficient.

References

1. Gonzalez W (2008) Digital image processing, 2nd edn. Prentice Hall, Englewood Cliffs
2. Stallings W (2014) Cryptography and network security, principle and practice, 6th edn
3. Tambe Chaitali A, Gadilkar Rupali S, Pawar Vimal T (2017) A survey on new image encryption algorithm based on Diffie-Hellman and singular value decomposition. Int J Innov Res Comput Commun Eng
4. Seyedzade SM, Atani RE, Mirzakuchaki S (2010) A novel image encryption algorithm based on hash function. In: 6th Iranian conference on machine vision and image processing, 27–28 Oct 2010, pp 1–6
5. Hashim HR, Neamaa IA (2014) Image encryption and decryption in a modification of ElGamal cryptosystem in MATLAB. Int J Sci: Basic Appl Res 14(2):141–147
6. Khachatrian G, Kyureghyan M (2014) A new public key encryption system based on permutation polynomials. In: IEEE international conference on cloud engineering
7. Sokouti M, Zakerolhosseini A, Sokouti B (2016) Medical Image encryption: an application for improved padding based GGH encryption algorithm. Open Med Inf J
8. Dwivedi N, Gupta RK, Agarwal S (2017) Image encryption using curved scrambling and diffusion. Int J Eng Technol
9. Wagh NB, Kolhekar M (2017) A novel approach utilizing permutation polynomials over integer rings as a cryptological application for effective encryption of digital images., IJCSE 5(1). E-ISSN 2347–2693

Thermal Imaging of Abdomen in Evaluation of Obesity: A Comparison with Body Composition Analyzer—A Preliminary Study

S. Sangamithirai, U. Snekhalatha, R. Sanjeena and Lipika Sai Usha Alla

Abstract Changes in body composition parameters lead to obesity progression which has an impact on body metabolism. The aim and objectives of the study was: (i) to estimate body composition parameters such as fat-free mass, skeletal muscle mass, lean body mass, subcutaneous fat mass using body composition analyzer; (ii) and to measure mean surface temperature of various regions of the body such as abdomen, chest, forearm front and back, shank front and back using thermal imaging method. Adult volunteers of age 20–29 years were participated in this preliminary study. The body composition parameter such as fat mass, fat free mass, muscle mass, and skeletal muscle mass was measured using body composition analyzer. The thermal imaging of the forearm, abdomen, and shank region was obtained and the skin surface temperature was measured in both normal and obese. The FFM negatively correlated with mean surface temperature at the abdomen region and found to be highly significant with $p < 0.05$. In this study, BCA parameters correlated significantly with the skin surface temperature measured at forearm and abdomen region in obese subjects.

1 Introduction

Obesity refers to abnormal fat accumulation in the body, which leads to impairment to the health [1]. According to the World Health Organization (WHO) criteria, the

S. Sangamithirai · U. Snekhalatha (✉) · R. Sanjeena · L. S. U. Alla
Biomedical Engineering Department, SRMIST, Kattankulathur, Tamil Nadu, India
e-mail: sneha_samuma@yahoo.co.in

S. Sangamithirai
e-mail: sangamithirai.s.kumar@gmail.com

R. Sanjeena
e-mail: sanjeenar.2297@gmail.com

L. S. U. Alla
e-mail: sophiadumbledore@yahoo.com

© Springer Nature Switzerland AG 2019
D. Pandian et al. (eds.), *Proceedings of the International Conference on ISMAC in Computational Vision and Bio-Engineering 2018 (ISMAC-CVB)*, Lecture Notes in Computational Vision and Biomechanics 30,
https://doi.org/10.1007/978-3-030-00665-5_9

79

simple tool to evaluate the obesity is through calculating the body mass index (BMI). The BMI >30 kg/m^2 is considered as obese as per WHO criteria. The major risk factors of obesity include diabetes, cardiovascular disorder (CVD) and cancer, sleep apnea, and depression. The prevalence of obesity worldwide is 13% in the year 2016 [2]. By the year 2030, the increase in obesity level is expected to be 47%, 39%, and 35% in the US, Mexico, and England, respectively [3]. The prevalence of obesity in India is 9.3% in the male population and 12.6% in the female population [4].

The different diagnostic methods for the obesity assessments are as follows: BMI, waist circumference, skinfold thickness, waist to hip ratio, bioelectric impedance method, dual energy X-ray absorptiometry, computerized tomography (CT) and magnetic resonance imaging (MRI). Body composition parameters estimate the fat mass, fat free mass, muscle mass and segmental muscle mass from the subjects using body composition analyzer. Several Researchers have performed the body composition analysis in obese subjects using bioimpedance analysis method and estimated the fat mass, fat free mass, BMI, BMR, waist circumference, hip circumference, waist–hip ratio, etc. [5–8].

Thermal imaging is a noninvasive, noncontact imaging modality, which works on the principle of our body emits radiation which is sensed by infrared cameras. The thermal imaging method is used in a variety of applications in the medical field such as in assessment of breast cancer [9], diabetic mellitus [10], rheumatoid arthritis [11], cardiovascular disorders [12], temporomandibular joint disorders [13], vascular disorders [14], ocular [15] and skin diseases [16], etc. Infrared thermography is used as a research tool for assessment of obesity in various regions of the body.

The body surface temperature gets reduced in obese people due to loss of metabolic heat, because of core body temperature stimulates the vasoconstriction which is high in obesity [17]. Savastano et al. conducted the study on obesity to study the relationship between the adiposity and hand and abdomen temperature in both normal and obese adults [17]. Chudeka et al. evaluated the changes in body composition parameters and body surface temperature measured using thermal imaging in females after abdominal liposuction [18]. Chudeka et al. investigated the level of visceral and subcutaneous adipose tissue which affects the core body surface temperature and performed the thermal mapping of skin surface temperature in obese women [19]. Jalil et al. compared the body temperature of the abdomen, neck, and hand using thermal imaging in obese and normal subjects during oral glucose tolerance test [20].

The aim and objectives of the study was: (i) to estimate body composition parameters such as fat free mass, skeletal muscle mass, lean body mass, subcutaneous fat mass using body composition analyzer; (ii) and to measure mean surface temperature of various regions of body such as abdomen, chest, forearm front and back, shank front and back using thermal imaging method.

2 Methodology

2.1 Participants

Volunteer adults (age: 20–29 years) participated in this study. It was explained that the study investigated the relation between body composition parameters and thermal images in adults. None of the participants were undergoing treatment related to obesity or were using medication that affected heat balance in the body. The consent of all participants were obtained.

2.2 Experimental Design

Each volunteer visited the Medical Instrumentation Laboratory in the Department of Biomedical Engineering, SRMIST, Kattankulathur, India. First, height and weight were measured and body composition parameters such as fat mass, fat free mass, muscle mass, skeletal muscle mass were analyzed. Second, the participants were told to place their forearm on a table of height of about 2 feet. A thermal camera was set at a focal length of 600 mm from the table. Regional body temperatures were assessed in the forearm, shank and abdomen areas. The room was temperature controlled (21 °C) and usual comfortable clothing was worn by the subjects.

2.3 Regional Body Temperature Measurements

The thermal camera used a 640 * 512 array of sensors sensitive to changes in temperature. For further off line analysis, the thermal images acquired were stored on a computer. Participants were advised to stay still during imaging, to keep motion artifacts to a minimum.

During forearm imaging, subjects were seated on a chair with their arm placed on a black background to minimize drafts. Infrared images were acquired in each region for 10 min duration. During abdominal imaging, a curtain was drawn to lessen drafts while the participants lay in a supine position on the table. Their clothing was adjusted in such a way as to allow complete abdominal exposure from the level of the xiphoid process to 2.24 cm above the symphysis pubis. A 5 min acclimatization period was set with the patient at rest, after which, sequential thermographic images were acquired for 5 min with his or her abdomen exposed. The shank area was imaged after asking the participants to sit on a table 2 feet above the floor. Lens was focused on the exposed region and images were acquired for 5 min.

Statistical Package for the Social Sciences (SPSS) software was used to process the infrared images collected. Temperature values were extracted from each region of interest (ROI) for the sequential images. The subcutaneous fat deposit in the abdomen was selected depending on the most consistent region among the participants. A rectangular ROI (100 pixels total) 1.2 cm subadjacent to the umbilicus was analyzed.

2.4 Body Composition Measurements

Measurement of body composition in humans acts as a response to conditions such as obesity, where body fat and BMD allow for clinical diagnosis. The body composition analyzer is a single frequency device using single-point load weighing system in the scale platform for different segments of the body such as right arm, left arm, right leg, and left leg. The measurements for fat, total body water (TBW), body mass index (BMI), fat free mass (FFM), basal metabolic rate (BMR), visceral fat rate (VFR), extracellular water (ECW), muscle mass, bone mass, and fat mass were recorded.

3 Results and Discussion

The demographic variables and BCA parameters such as muscle mass and fat mass at hand and leg for normal, overweight, and obese subjects were given in Table 1. The basic demographic variables such as BMI, FFM, and BMR are increased by 50%, 12.5%, and 10.2%, respectively, in obese subjects compared to normal subjects. The muscle mass at right arm shows the highest significant difference of 17.76% between normal and obese subjects compared to other regions such as left arm, left leg and right leg. Similarly, the fat mass at right arm depicts the greater difference 62.3% in obese compared to normal subjects in other regions.

The mean skin surface temperature measured at the forearm, abdomen, and shank region for normal, overweight, and obese were listed in Table 2. The decreased temperature of about -2.78% exhibits in the obese subject compared to normal in the forearm region. Similarly, the lesser temperature was observed in obese subjects compared to normal in abdomen and shank region. Among the three regions studied, abdomen shows the highest temperature difference between the normal and obese subjects.

The correlation between the BCA parameters and thermal imaging parameters in obese subjects were depicted in Table 3. The FFM negatively correlated with mean surface temperature at the abdomen region and found to be highly significant with $p < 0.05$. Similarly the other BCA parameters such as BMR, Muscle mass, bone mass, and fat mass significantly produced a negative correlation with the mean surface temperature. In the case of forearm region, FFM, BMR, muscle mass, bone mass, and fat mass correlated with average temperature at the significance level $p < 0.01$. But the BCA parameters had not correlated significantly with the mean surface temperature at the shank region.

Table 1 Body composition parameters for normal, overweight, and obese subjects

BCA	Baseline variables			% Difference
		Normal	Obese	
Parameters	AGE	24 ± 6.9	23.62 ± 6.45	−1.608
	BMI	20.78 ± 2.39	41.5 ± 4.2	50
	FFM	36.05 ± 5.71	41.22 ± 7.8	12.5
	BMR	1126.25 ± 1.77	1255 ± 207.8	10.2
Muscle mass	LA	1.53 ± 0.3	1.83 ± 0.42	16.4
	RA	1.583 ± 0.4	1.925 ± 0.43	17.76
	LL	6.608 ± 0.9	7.55 ± 1.925	12.47
	RL	6.75 ± 0.98	7.725 ± 1.6	14.45
Fat mass	LA	0.59 ± 0.28	1.46 ± 0.4	60
	RA	0.533 ± 0.27	1.412 ± 0.4	62.3
	LL	3.15 ± 0.96	5.837 ± 1.3	46
	RL	3.2 ± 1.0027	5.787 ± 1.22	44.7

Table 2 Mean skin surface temperature measured at the forearm, abdomen, and shank region for normal, overweight and obese

Region of interest	Normal ($N = 11$)	Obese ($N = 10$)	% Difference
Forearm	33.21 ± 1.6	32.31 ± 1.81	−2.78
Abdomen	33.17 ± 1.03	32.86 ± 1.16	0.94
Shank	32.65 ± 1.371	32.61 ± 1.396	0.12

Table 3 Correlation between the BCA and temperature parameters at forearm, abdomen, and shank region

BCA parameters	Forearm	Abdomen	Shank
FFM	−0.104	−0.058	−0.078
BMI	−0.037	−0.13	−0.028
FFM	−0.337*	−0.578**	−0.208
BMR	−0.3805*	−0.592**	−0.205
Muscle mass	−0.3045*	−0.599**	−0.164
Bone mass	−0.416*	−0.614**	−0.231
Fat mass	−0.381*	−0.223	−0.115

*$p<0.05$
**$p<0.01$

(a)   **(b)**

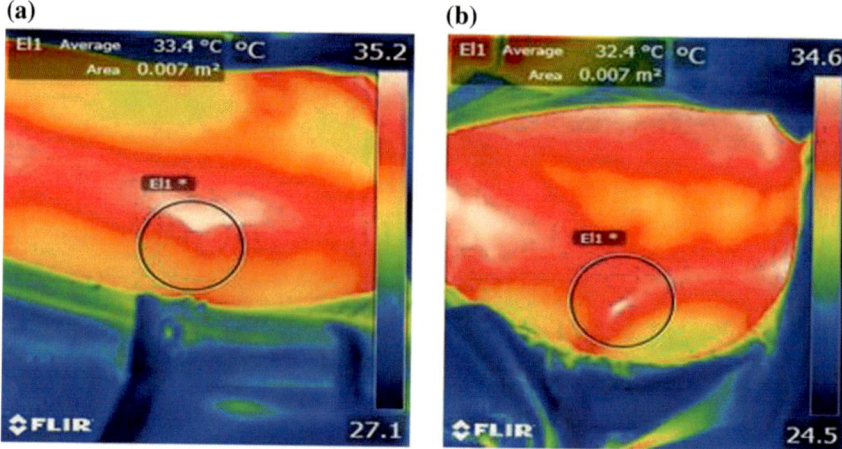

Fig. 1 **a** and **b** represent the forearm of normal and obese, respectively

(a)   **(b)**

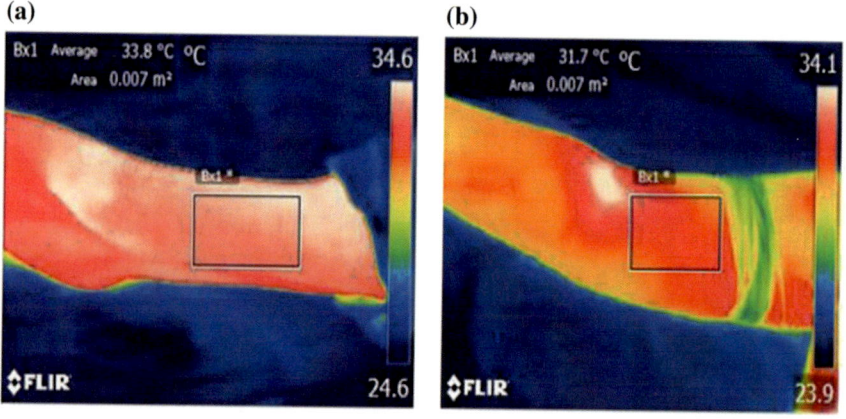

Fig. 2 **a** and **b** indicate abdomen of normal and obese, respectively

Figure 1a represents temperature measured at forearm by keeping circular ROI as fixed. The average temperature obtained at forearm region for the normal subject was 31.7 °C, whereas the measured temperature for the obese subject at forearm region depicts 31.4 °C as mentioned in Fig. 1b. Figure 2a indicates the temperature measured at abdomen region for normal subject in which the average temperature displays 32.6 °C. Figure 2b illustrates the average temperature as 31.7 °C in abdomen region for obese subject. Figure 3a depicts the average temperature as 31.3 °C measured at shank region for normal subject, whereas for obese subject the average temperature measured at shank region was 30 °C.

Savastono et al. obtained the mean surface temperature measured at abdomen region for normal and obese subjects as 33 ± 0.3 °C and 31.8 ± 0.3 °C, respectively [18]. They found the negative correlation ($\gamma = -0.45$, $p < 0.005$) between the fat

(a) **(b)**

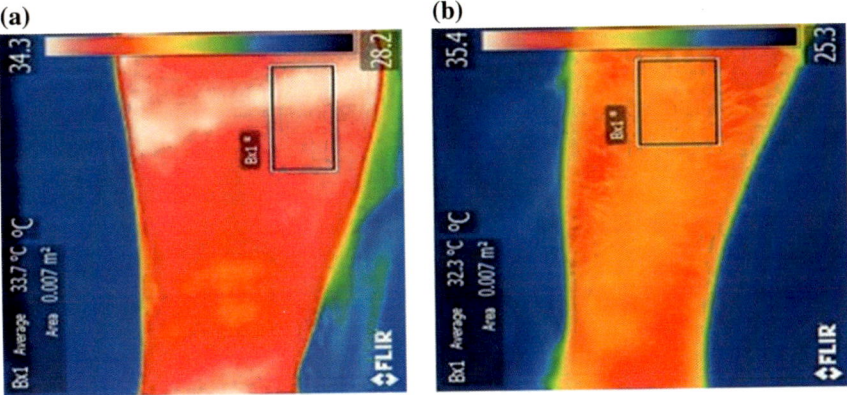

Fig. 3 **a** and **b** depict the temperature measured in the shank region of normal and obese, respectively

mass and abdominal temperature. Also, they found the significant difference between the two groups such as normal and obese in the analysis of BCA parameters such as fat mass and fat free mass. Chudecka et al. correlated the BCA parameters such as visceral fat mass, subcutaneous fat mass, and skeletal muscle mass with the abdomen temperature measured by infrared imaging [21]. They obtained the significant negative correlation between the visceral fat mass and abdominal temperature as $r = -0.7, p = 0.001$, and between the subcutaneous fat mass and abdominal temperature as $r = -0.62, p = 0.006$. In our study, significant negative correlation achieved between the fat free mass and abdominal temperature ($r = -0.57, p < 0.01$), and between muscle mass and abdominal temperature ($r = 0.59, p < 0.01$). This negative correlation is due to as BCA parameters increases in obese subjects, the temperature measures at abdomen region decreases. The reason behind was heat dissipation occurs at adipose tissues in nearby fat deposition regions in the abdomen.

4 Conclusion

In this study, BCA parameters correlated significantly with the skin surface temperature measured at forearm and abdomen region in obese subjects. Among the three regions studied, abdomen region shows greater temperature difference between the normal and obese subjects. The muscle mass and fat mass measured at forearm region exhibit higher value in obese subjects compared to normal. The BMI, BMR, FFM increased in obese subjects compared to normal and produces a significant difference. Hence in young age population, obesity exists due to sedentary lifestyle and leads to increased risk of causing other complications such as diabetes, CVD, etc. Hence, dietary control and maintaining the metabolic rate prevents obesity in young adults.

Acknowledgements The authors would like to express their sincere gratitude to Head, Biomedical Engineering Department SRMIST for providing the infrastructure and equipment facility.

References

1. Ofei F (2005) Obesity-a preventable disease. Ghana Med J 39:98–101
2. http://www.who.int/mediacentre/factsheets/fs311/en/
3. https://www.oecd.org/els/health-systems/Obesity-Update-2017.pdf
4. Nagendra K, Nandini C, Belur M (2017) A community based study on prevalence of obesity among urban population of Shivamogga, Karnataka, India. Int J Community Med Public Health 4:96–99
5. Shishkova A, Petrova P, Tonev A, Bahlova P, Softov O, Kalchev E (2007) Analysis of body composition in overweight and obese women using bioimpedance (BIA) system. J IMAB 13:8–12
6. Ayvaz DNC, Kilinc FN, Pac FA, Cakal E (2011) Anthropometric measurements and body composition analysis of obese adolescents with and without metabolic syndrome. Turk J Med Sci 41:267–274
7. Namwongprom S, Rerkasem K, Wongthanee A, Pruenglampoo S, Mangklabruks A (2014) Relationship between body composition parameters and metabolic syndrome in young Thai adults. J Clin Res Pediatr Endocrinol 6:227–232
8. Hunma S, Ramuth H, Miles-Chan JL, Schutz Y, Montani JP, Joonas N, Dulloo AG (2016) Body compostion-derived BMI cut-offs for overweight and obesity in Indians and Creoles of Mauritius: comparison with Caucasians. Int J Obes 40:1906–1914
9. Han F, Shi G, Liang C, Wang L, Li K (2015) A Simple and efficient method for breast cancer diagnosis based on infrared thermal imaging. Cell Biochem Biophys 71:491–498
10. Sivanandam S, Anburajan M, Venkatraman B, Menaka M, Sharath D (2012) Medical thermography: a diagnostic approach for type 2 diabetes based on non-contact infrared thermal imaging. Endocrine 42:343–351
11. Snekhalatha U, Rajalakshmi T, Gopikrishnan M, Gupta N (2017) Computer-based automated analysis of X-ray and thermal imaging of knee region in evaluation of rheumatoid arthritis. Proc Inst Mech Eng, Part H: J Eng Med 231:1178–1187
12. Jayanthi T, Anburajan M, Menaka M, Venkatraman B (2014) Potential of thermal imaging as a tool for prediction of cardiovascular disease. J Med Phys 39:98–105
13. Clemente M, Coimbra D, Silva A, Aquiar Branco C, Pinho JC (2015) Application of infrared thermal imaging in a violinist with Temperomandibular disorder. Med Probl Perform Art 30:251–254
14. Bagavathiappan S, Saravanan T, Philip J, Jayakumar T, Raj B, Karunanithi R, Panicker TMR, Paul Korath M, Jagadeesan K (2009) Infrared thermal imaging for detection of peripheral vascular disorders. J Med Phys 34:43–47
15. Kawali AA (2013) Thermography in ocular inflammation. Indian J Radiol Imaging 23:281–283
16. Gurjarpadhye AA, Parekh MB, Dubnika A, Rajadas J, Inayathullah M (2015) Infrared imaging tools for diagnostic applications in dermatology. SM J Clin Med Imaging 1:1–5
17. Jongh D, Serne RT, Ijzerman RG, De vries RG, Stehouwer C (2004) Impaired microvascular function in obesity: implications for obesity-associated micro-angiopathy hypertension and insulin resistance. Circulation 109:2529–2535
18. Savastano DM, Gorbach AM, Eden HS, Brady SM, Reynold JC, Yanovski JA (2009) Adiposity and human regional body temperature. Am J Clin Nutr 90:1124–1131
19. Chudecka M, Dmytrzak A, Lubkowska A (2016) Changes in selected Morphological parameters and body composition as well as Mean body surface temperature assessed by thermal imaging in women after abdominal liposuction. Central Euro J Short Sci Med 14:21–26

20. Jalila B, Hartwigb V, Moronia D, Salvettia O, Benassia A, Jalile Z, Guiduccib L, Pistoiac L, Tegrimic TM, Galvanc AQ, Iervasib G, Abbateb AL (2017) Near Infrared and thermal imaging of normal and obese women during oral glucose tolerance test (OGTT). In: 14th International workshop on advanced infrared technology and applications, Quebec city, Canada, pp 144–148
21. Chudecka M, Lubkowska A, Podhorodecka AK (2014) Body surface temperature distribution in relation to body composition in obese women. J Therm Biol 43:1–6

A Review on Protein Structure Classification

N. Sajithra, D. Ramyachitra and P. Manikandan

Abstract A massive amount of sequence data is gradually produced by the genome projects that have to be annotated in terms of structure, molecular, and biological functions. In structural genomics, the aim is to resolve several protein structures in an efficient way and to exploit the solved protein structures for assigning the biological function to theoretically solved protein structures. In earlier stages, the protein structures are classified manually in a successful manner and now it suffers from updating problem because of the high throughput of recently solved protein structures. To overcome this issue, several data mining techniques have been examined for the structural classification of the protein world. This review article presents an overview of the existing classification techniques, databases, tools, and performance metrics used for evaluating the performance of protein structure classification algorithms.

1 Introduction

It is particularly a fact for the proteins, whose functions are honestly related to their 3D structures [1, 2]. Protein structure classification is used to identifying the hierarchal clusters of proteins with related structures and it plays a fundamental crisis in computational biology [3]. In molecular biology, the high-throughput technologies have generated a huge sum of data about the function, structure, and evolution of biological macromolecules at the genome level. To identify the interactions of these molecules in the cell, it is essential to classify the molecules into significant categories that are associated with the existing biological data [4]. It is essential to classify the

N. Sajithra (✉) · D. Ramyachitra · P. Manikandan
Department of Computer Science, Bharathiar University, Coimbatore, India
e-mail: sajithramidhun@gmail.com

D. Ramyachitra
e-mail: jaichitra1@yahoo.co.in

P. Manikandan
e-mail: manimkn89@gmail.com

© Springer Nature Switzerland AG 2019
D. Pandian et al. (eds.), *Proceedings of the International Conference on ISMAC in Computational Vision and Bio-Engineering 2018 (ISMAC-CVB)*, Lecture Notes in Computational Vision and Biomechanics 30,
https://doi.org/10.1007/978-3-030-00665-5_10

protein structures for comparing the protein structures and also it can be used to easily identify and group similar folds and families of protein structures. One of the major advantages of classifying the protein structures is to introducing some wisdom of order to the increasing volume of available protein structural data. The obstacle occurs in classifying the protein structures is the reality that the structures are frequently made up of distinct globular proteins [5].

In recent days, the computational techniques in structural genomics are well found that proteins often have similar folds, even if their protein sequences are apparently dissimilar [6]. A prototype computational platform named Homologous Structure Finder (HSF) has the capability of incorporating a variety of comparison methods and parameterizations. HSF platform has been validated by more complex classification tasks and it successfully performed an automatic comparison and classification of the set of 3D protein structures [7]. Most of the common methods are used for comparing and classifying the virus proteins based on sequence information [8]. Nowadays, it is easier to identify the similar proteins that share a similar function based on their structure pretty than on their sequences [9]. In this article, the new emerging techniques and tools to classify the protein structures are studied. The increasing protein structure database helps both biological and computational scientists to predict structural similarity, functions, and progress the diagnosis and treatment of diseases. The remaining sections of the paper are organized as follows. Section 2 deals with the challenges in the structural classification of proteins. The algorithms which are used to classify the protein structures are discussed in Sect. 3. Section 4 provides the tools involves in classifying the protein structures. Section 5 discusses the protein structure databases. The performance measures are discussed in Sect. 6 and conclusion is provided in Sect. 7.

2 Challenges in Protein Structure Classification

The major challenge in structural biology is to classify well-known protein structures to replicate their evolutionary, structural, and functional relatedness. The relationships between the protein structures are personified as hierarchical classification schemes in Structural Classification of Proteins (SCOP) [10]. But, there are some challenges in classifying the protein structures in a hierarchical manner [11]. On the other hand, the lower levels have a propensity to be dense and overlapping to contain the ubiquitous functional similarity and evolutionary relationships. In order to organize the complex universe of known protein structures, human expertise has been playing a crucial role. For a major number of proteins, there are differences in classification among any two release versions of SCOP and a huge number of proteins remain listed under the Not a True Class level. Hence, an enthusiastic significance is needed in the automation of protein structure classification [12].

3 Techniques for Protein Structure Classification

There are several "-omics" projects such as genomics, functional genomics [13], and structural genomics [14] that are producing a huge amount of information and it has been stored into databases. One of the key successes in human ability is to analyze and organize the biological information. In other words, the human ability is to classify the proteins based on the sequence, structure, and functions to perform classification. Some of the techniques which are used to perform the protein structure classification are shown in Fig. 1.

The automated classification of protein structure is performed by using different algorithms from different categories of supervised algorithms on a set of 11,336 pairs of protein domains with 35% sequence identity [15].

To analyze the [16] protein structures Gauss metrics method is used and this protein will be formed as clusters. The advancement of Gauss metric is scaled Gauss metric that is used to maintain the database and it will provide the automatic classification. This method has high speed and it provides better performance.

In the Protein Data Bank [17], the graph-based classification method provides the primary overview of the entire protein complex and allows nonredundant sets to be derived at diverse levels in detail. This reveals that among one-half and two-thirds of recognized protein structures are multimeric, depending on the stage of accepted redundancy. The protein structures in provisions of the topological arrangement of their subunits are analyzed and find that they form a tiny number of arrangements compared with all theoretically possible ones.

The 3D shape-based approach [18] is tested with three classification methods to examine the performance of this approach.

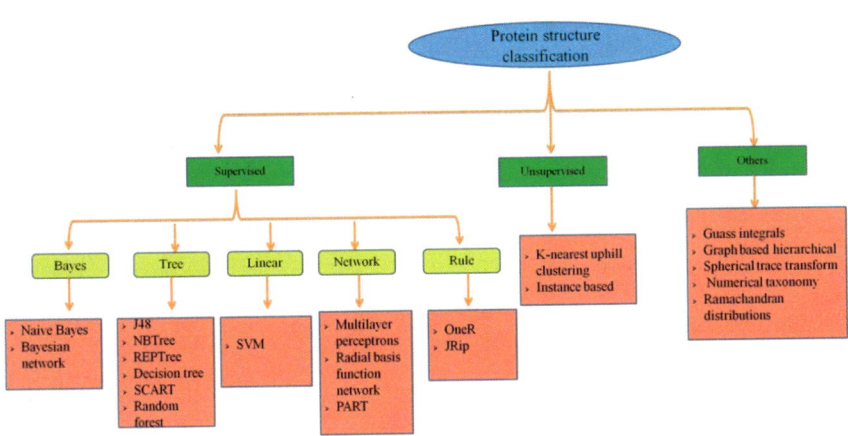

Fig. 1 Classifications of protein structure techniques

A novel density-based K-nearest uphill clustering technique can efficiently reduce noisy pairwise structure similarities of proteins and identifies the density peaks as cluster centers [19].

The illustration of the crisis [20] in an attribute-based method allows the purpose of numerous well-established machine learning algorithms.

The usability of the random forest algorithm [21] in classifying larger domains is verified by applying it to domains consisting of four, five and six Secondary Structure Elements.

A numerical taxonomy is introduced for automatic classification of protein structures and identification of evolutionary relationships [22]. The topology of protein space is explored using structural similarity. Penetrating for clusters of structural neighbors of proteins, where the members constantly share numerous functional attributes directs to an optimal partitioning of protein structure space. This clustering matches up well to the analog or homolog boundaries drained by biologists, with applications in the creation of functional hypotheses in nonhypothesis-driven structural genomics efforts.

A nonparametric technique is developed for the combined estimation of numerous bivariate density functions for a group of populations of protein backbone angles [23].

4   Tools

Protein structure classification algorithms and tools are mainly used for identifying the structural similarities between the proteins. In recent years, the algorithms which are used for automatically identifying the structural relationships have been a sizzling area in computational biology [24]. Numerous techniques are used for handling the shifts in protein secondary structure orientations and the wide indels among distant homologs such as COMPARER [25], DALI [26] and SSAP [27]. Several resources are available for classifying the protein and its families such as HAMAP [28], TIGRAMs [29], PANTHER [30], SFLD [31], PFAM [32], SMART [33], InterPro [34], PRINTS [35], CATH [36], NCBI-CDD [37], ECOD [38], and SCOP [39].

5   Databases

Several databases has been used for analyzing the problem of protein structural classification that includes Protein Data Bank (PDB) [40], Universal Protein Resource [41], Database of Secondary Structure of Protein (DSSP) [42], Structural Classification of Proteins (SCOP) [10], SCOP2 (a successor of SCOP) [43] and Class, Architecture, Topology, Homology (CATH) [36]. From the above databases, SCOP and CATH have become precious resources in protein structural classification research.

6 Performance Metrics

In general, the performance measures are chosen based on the algorithms used in that research work. Some of the common measures that can be used in protein structure classification algorithms are Accuracy, True Positive rate (TP), False Positive rate (FP), Mathews Correlation Coefficient (MCC), F-Measure and it can be calculated by using Eqs. 2, 3, 4, 5 and 6, respectively. The overall accuracy is the percentage of correctly predicted proteins and it is calculated by using Eq. 1.

$$\text{Overall Accuracy} = \frac{\text{Number of correctly predicted proteins}}{\text{Total number of proteins in the database}} \tag{1}$$

$$\text{Accuracy} = \frac{(TP + TN)}{(TP + FP + FN + TN)} \tag{2}$$

$$\text{True Positive/Recall} = \frac{TP}{TP + FN} \tag{3}$$

$$\text{False Positive} = \frac{FP}{FP + TN} \tag{4}$$

$$\text{Matthews Correlation Coefficient} = \frac{TP \times TN - FP \times FN}{\sqrt{(TP + FP)(TP + FN)(TN + FP)(TN + FN)}} \tag{5}$$

$$\text{F-Measure} = 2 * \frac{\text{Precision} * \text{Recall}}{\text{Precision} + \text{Recall}} \tag{6}$$

$$\text{Precision} = \frac{TP}{TP + FP} \tag{7}$$

where,

True Positive (TP)	included class objects with corrected
False Positive (FP)	incorrectly included class objects with uncorrected
True Negative (TN)	excluded class objects with corrected
False Negative (FN)	excluded class objects with uncorrected

ROC analysis is used to compare the rate of True Positive as a function of the rate of False Positive and it is scored with the area under the corresponding curve. The ROC value of 1 indicates that all TPs are identified first and this match to the ideal measure [9].

7 Conclusion

Performing an automated classification of protein structures remains a challenging problem in computational biology and it is necessary to develop an accurate method for classifying the protein structures from the structural data. Some of the techniques which are used to perform the protein structure classification are discussed in this review article. And also this article provides a brief note on tools and databases, which can be used to perform protein structure classification. Finally, this article discussed some common performance metrics for evaluating the performance of protein structure classification algorithms. Protein structure classification is still an attractive research area with a bunch of interesting possibilities for additional improvements in terms of efficiency and accuracy.

Acknowledgements The authors like to thank the Department of Science and Technology (DST), New Delhi (DST/INSPIRE Fellowship/2015/IF150093) for the financial support under INSPIRE Fellowship for this research work.

References

1. Richardson J (1981) The anatomy and taxonomy of protein structure. Adv Protein Chem 34:167
2. Branden C, Tooze J (1991) Introduction to protein structures. Garland Publishing, New York
3. Kolodny R et al (2013) On the universe of protein folds. Annu Rev Biophys 42:559–582
4. Ouzounis CA et al (2003) Classification schemes for protein structure and function. Nat Rev Genet 4(7):508–519
5. Hadley C, Jones DT (1999) A systematic comparison of protein structure classifications: SCOP, CATH and FSSP. Structure 7(9):1099–1112
6. Pastore A, Lesk AM (1990) Comparison of the structures of globins and phycocyanins: evidence for evolutionary relationship. Proteins 8(2):133–155
7. Ravantti J et al (2013) Automatic comparison and classification of protein structures. J Struct Biol 183(1):47–56
8. Palmenberg et al (2009) Sequencing and analyses of all known human rhinovirus genomes reveal structure and evolution. Science 324:55–59
9. Le Q et al (2009) Structural alphabets for protein structure classification: a comparison study. J Mol Biol 387(2):431–450
10. Murzin AG et al (1995) Scop: a structural classification of proteins database for the investigation of sequences and structures. J Mol Biol 247:536–540
11. Govindarajan S et al (1999) Estimating the total number of protein folds. Proteins: Struct Funct Bioinform 35:408–414
12. Andreeva et al (2008) Data growth and its impact on the SCOP database: new developments. Nucleic Acids Res 36:D419–D425
13. Burley S et al (1999) Structural genomics: beyond the human genome project. Nat Genet 23:151–157
14. Hieter P, Boguski M (1997) Functional genomics: it's all how you read it. Science 278:601–602
15. Jain P et al (2009) Supervised machine learning algorithms for protein structure classification. Comput Biol Chem 33(3):216–223
16. Røgen P, Fain B (2003) Automatic classification of protein structure by using Gauss integrals. Proc Natl Acad Sci U S A. 100(1):119–124
17. Levy ED et al (2006) 3D complex: a structural classification of protein complexes. PLoS Comput Biol 2(11):e155
18. Daras P et al (2006) Three-dimensional shape-structure comparison method for protein classification. IEEE/ACM Trans Comput Biol Bioinform 3(3):193–207
19. Cui X, Gao X (2017) K-nearest uphill clustering in the protein structure space. Neurocomputing 220:52–59
20. Leon F et al (2009) Performance analysis of algorithms for protein structure classification. In: 2009 IEEE 20th international workshop on database and expert systems application. https://doi.org/10.1109/dexa.2009.17. ISBN: 978-0-7695-3763-4

21. Jain P, Hirst JD (2010) Automatic structure classification of small proteins using random forest. BMC Bioinform 11:364
22. Dietmann S, Holm L (2001) Identification of homology in protein structure classification. Nat Struct Biol 8(11):953–957
23. Najibi SM et al (2017) Protein structure classification and loop modeling using multiple Ramachandran distributions. Comput Struct Biotechnol J 8(15):243–254
24. Swindells MB et al (1998) Contemporary approaches to protein structure classification. BioEssays 20(11):884–891
25. Sali A, Blundell TL (1990) Definition of general topological equivalence in protein structures. A procedure involving comparison of properties and relationships through simulated annealing and dynamic programming. J Mol Biol 212:403–428. https://doi.org/10.1016/0022-2836(90)90134-8
26. Holm L, Sander C (1993) Protein structure comparison by alignment of distance matrices. J Mol Biol 233:123–138. https://doi.org/10.1006/jmbi.1993.1489
27. Taylor WR, Orengo CA (1989) Protein structure alignment. J Mol Biol 208:1–22
28. Pedruzzi I et al (2013) HAMAP in 2013, new developments in the protein family classification and annotation system. Nucleic Acids Res 41:D584–D589
29. Haft DH, Selengut JD, White O (2003) The TIGRFAMs database of protein families. Nucleic Acids Res 31:371–373
30. Mi H, Muruganujan A, Thomas PD (2013) PANTHER in 2013: modeling the evolution of gene function, and other gene attributes, in the context of phylogenetic trees. Nucleic Acids Res 41:D377–D386
31. Akiva E et al (2013) The structure–function linkage database. Nucleic Acids Res 42:D521–D530
32. Finn RD et al (2014) Pfam: the protein families database. Nucleic Acids Res 42:D222–D230
33. Letunic I, Doerks T, Bork P (2015) SMART: recent updates, new developments and status in 2015. Nucleic Acids Res 43:D257–D260
34. Hunter S et al (2012) InterPro in 2011: new developments in the family and domain prediction database. Nucleic Acids Res 40:D306–D312
35. Attwood TK et al (2012) The PRINTS database: a fine-grained protein sequence annotation and analysis resource—its status in 2012. Database 2012:bas019
36. Sillitoe I et al (2015) CATH: comprehensive structural and functional annotations for genome sequences. Nucleic Acids Res 43:D376–D381
37. Marchler-Bauer A et al (2013) CDD: conserved domains and protein three-dimensional structure. Nucleic Acids Res 41:D348–D352
38. Cheng H et al (2014) ECOD: an evolutionary classification of protein domains. PLoS Comput Biol 10:e1003926
39. Andreeva A et al (2007) Data growth and its impact on the SCOP database: new developments. Nucleic Acids Res 36:D419–D425
40. Bernstein FC et al (1977) The protein data bank. Eur J Biochem 80:319–324
41. Consortium, U (2008) The universal protein resource (UniProt). Nucleic Acids Res 36:D190–D195
42. Kabsch W, Sander C (1983) Dictionary of protein secondary structure: pattern recognition of hydrogen-bonded and geometrical features. Biopolymers 22:2577–2637
43. Andreeva A et al (2014) SCOP2 prototype: a new approach to protein structure mining. Nucleic Acids Res 42:310–314

Emotion Analysis Through EEG and Peripheral Physiological Signals Using KNN Classifier

Shourya Shukla and Rahul Kumar Chaurasiya

Abstract Emotions are the characteristics of human beings which are triggered by the mood, temperament or motivation of an individual. Emotions are nothing but the response to the stimuli that are experienced by the brain. Any changes in one's emotional state results in changes in electrical signals generated by the brain. The emotions can be explicit or implicit, i.e. either emotion may be expressed or remain unexpressed by the individual. As these emotions are experienced by the individual as the result of the brain stimulus, we can observe Electroencephalogram (EEG) signal to classify the emotions. Some of the physiological signals may also be taken into account as any change in emotional state result in some physiological changes. For the analysis, we have used the standard DEAP dataset for emotion analysis. In the dataset, the 32 test subjects are shown with 40 different 1-minute music videos and the EEG and other physiological signals are recorded. On the basis of the Self-Assessment Manikins (SAM), we classify the emotion state in the valence arousal plane. The K-Nearest Neighbour classifier is used to classify the multi-class emotions as higher/lower levels of the valence arousal plane. The comparison of KNN with other classifiers depicts that KNN has produced best average accuracy of 87.1%.

1 Introduction

Electroencephalogram is the electrical signals produced by the brain. EEG signal are used to analyse the emotional state of the subject under study. Emotions are triggered in the brain by stimuli such as video signals or audio signals etc., [1]. Depending upon the mood, temperament and motivation of the subject, different emotions are experienced. The brain works in different frequency bands, namely delta, theta, alpha, beta and gamma [2, 3]. Beta (12–30 Hz) band is emitted when someone is conscious

S. Shukla (✉) · R. K. Chaurasiya (✉)
National Institute of Technology, Raipur, India
e-mail: shourya.shukla2000@gmail.com

R. K. Chaurasiya
e-mail: rkchaurasiya@nitrr.ac.in

© Springer Nature Switzerland AG 2019
D. Pandian et al. (eds.), *Proceedings of the International Conference on ISMAC in Computational Vision and Bio-Engineering 2018 (ISMAC-CVB)*, Lecture Notes in Computational Vision and Biomechanics 30,
https://doi.org/10.1007/978-3-030-00665-5_11

97

and alert, and the subject is thinking or concentrating. Alpha band (8–12 Hz) is most active when the subject is in the state of physical and mental relaxation but in aware and conscious. Theta (4–7 Hz) band is associated with daydreaming or sleepy state. It is also called the creative state. The delta band (0.1–4 Hz) is the lowest frequency state related to deep sleep. At the upper highest frequencies, that is the gamma band (>30 Hz) the subject id in deep meditation, mainly found significant in Buddhist monks.

The human–human interaction is easier as human are more aware of the sentiments of the human they are interacting, but in the case of the Brain–Computer interaction (BCI), it is more complicated as the computer is unaware of the emotions of the human subject. Here, we present a method which uses a signal-based approach by extracting information from the EEG and peripheral physiological signals. These peripheral physiological signals are useful in predicting the emotional state of the subject as they complement EEG signals in the emotional analysis. These peripheral signals are significant in different frequency bands and hence, the Discrete Wavelet Transform (DWT) of the signals is taken into consideration. The DWT of a signal splits the signal into the higher frequency detail (D) and the lower frequency approximation (A) coefficients [2].

The features are extracted from the wave decomposed signals, and the KNN classifier is used to classify the emotions as low valence low arousal, low valence high arousal, high valence low arousal and high valence high arousal. The valence indicates the pleasing level of the brain. A high valence level indicates happy or elated emotion, whereas a low valence indicates sad or stressed behaviour. The arousal indicates activeness of the brain. A higher arousal indicates alert or excited response, whereas the lower level of arousal indicates uninterested or bored response of an individual.

2 Related Works

The EEG signals were first recorded by the English scientist Richard Caton [4] in the year 1875. The study of the EEG signals was first explored by Hans Berger in 1920 [4]. The first study of emotion and physiological signals goes back to the year 1941 by Hadley, J. M. in which the author described the relationship between the EEG and peripheral physiological signals while performing multiplications of varying difficulties [5]. Plutchik, R first associated with high sound intensities on the performance feelings and physiology of the subject [6]. Moon et al. proposed the method for video preferences based on the extracted using EEG-based responses quadratic–discriminant–analysis-based model using BP features [7]. There are many proposed ways to classify the emotion of the subject using EEG. Ekman and Friesen, 1987 were the first to propose the six emotions existing that can be classified using facial signals [8]. The valence arousal plane is used to separate different emotional state as proposed by Sander Koelstra et al. (2011). A Database for Emotion Analysis using Physiological Signals (DEAP dataset) uses the EEG and other physiological signals to classify the emotions in the valence arousal plane using F1 score and naïve Bayes classifier is performed [9].

The peripheral physiological signals play an important role in determining the emotion as they complement the EEGs and provide the information about the subject's reaction to a stimuli. Torres-Valencia et al. suggested a multimodal emotion recognition using the DEAP dataset by Hidden Markov Model (HMM) using the Galvanic Skin Resistance (GSR) and Heart Rate (HR) [10]. Ramasamy et al. defined the heart–brain interaction through the EEG and ECG involving the emotion through biofeedback system [11]. Li et al. related the EEG with peripheral physiological signals such as Electrooculogram (EOG), Electrocardiogram (ECG), Electromyogram (EMG), skin temperature variation and electrodermal activity in a brain–computer interface system to measure the attention level in ubiquitous environment [12].

3 Materials and Methods

3.1 Dataset for Experimental Analysis

In the last few years, the BCI is one of the most studied topics in the field of machine learning. The DEAP dataset is being used for research purpose by many scholars. We are also using the DEAP dataset which is available for the research work. An End-User License Agreement (EULA) is acquired to access the data. The dataset consists of the records of 32 patients (subjects). Out of 32 subjects, 22 subjects' facial recording is provided. Each subject's file consists of data and label file. The data file has the 32-channel EEG recordings along with 8 peripheral physiological signals. The physiological signals which are used to understand the emotional state of the subject are Galvanic Skin Resistance (GSR), Respiration Amplitude (RA), Skin Temperature (ST), Electrocardiogram (ECG), Blood Volume Pressure (BVP) and Electromyogram (EMG) of the zygomaticus and trapezoidal muscles. The eyeball movement is captured by the Electrooculogram (EOG) signal. The signals are sampled at the rate of 512 Hz and the peripheral physiological are further down-sampled to 256 Hz. To record the EEG and other signals, visual stimuli are used. 40 videos were selected using the web-based survey and the stimulating 1 minute of each video is shown to the subject. The data is hence of the dimension $40*40*8064$, for 40 min, 40 signals for each minute and 8064 samples of each signal. 22 subjects' facial video was also recorded using SONY DCR-HC27E camcorder.

The EEG signals were obtained using a 10–20 electrode system. The EEG has 32-channel electrode having a 10–20 system with odd number of electrode placed on the left hemisphere and electrode with the number placed on the right hemisphere of the brain. The dataset used the Bio Semi Active Two System for recording the EEG signals.

The labelling of the data is done by the rating given by the individual. The process is named Self-Assessment Manikins (SAM). In the labelling process, after completion of every 1-min video, the subjects are asked to rate the video between 1 and 9, where 1 being the lowest and 9 being the highest rating. The subject is asked to move the cursor horizontally and click on the rating bar to give the scores on the arousal and valence parameters. To define a binary class system, midpoint threshold were taken on the rating scale of 1–9 for arousal and valence. Other SAMs like dominance, liking and familiarities were also assessed. The dominance rating represents the feeling of being empowered, whereas for liking scale, thumbs up or thumb down symbols were used. Familiarity rating suggests that how well the subject knows or remembers the video stimulus. The labelling of data is done as high valence high arousal, high valence low arousal, low valence high arousal and low valence low arousal. Thus, multi-class classification is performed on basis of these four classes.

3.1.1   Preprocessing, Feature Extraction and Classifier Methodology

The DEAP dataset contains records of 32 test subjects, and the large dataset makes it difficult to work on the integrated data. We segmented the data for each minute recording, i.e. time signals were obtained for each minute video. Therefore, each subject's psychophysiological signals are divided into 40 segments. Then for each minute, we find the discrete wavelet transform coefficient. These coefficients are time signals with ascending order of frequencies. The length of each frequency band is given in the length coefficient matrix. The DWT gives the details and approximations as response to high-pass and low-pass filter, respectively. The five-level DWT is obtained for the EEG signals of the frequency range 0–30 Hz. The five-level DWT is also used for the peripheral physiological signals as the EMG signals works on the higher frequency range of 4–40 Hz.

3.1.2   Feature Extraction Using DWT

EEG being a non-stationary signal, Fourier transform is not a suitable transform for it. Hence, to find the feature matrix, five-level DWT is used. The DWT decomposes the wavelets in the time–frequency domain. The DWT decomposes the signal according to the increasing frequency bands. The frequency decomposition is obtained by selecting odd and even samples of the signal and then, these samples are passed through low-pass and high-pass filter, respectively, further, the filtered signals are downsampled by a factor of 2. The resulting high-pass downsampled signals is the detail signals and low-pass downsampled signals are the approximate signals. The multilevel DWT is performed by applying recursively DWT to the $(n-1)$th approximate samples [13]. For each minute video, the five-evel DWT is applied on the psychophysiological signals, hence each signal is decomposed into five detail levels D1–D5 and an approximate A5, features are then extracted and classified by KNN classifier. As different signals works in different frequency ranges (EEG in the 0–30 Hz, whereas trapezoidal muscles in 4–40 Hz), all the DWT coefficients are taken into account.

Feature extraction is performed on the decomposed wavelets to study the EEG and other physiological signals in different frequency ranges. The EEG features extracted for classification of the emotional state of the subject are mentioned below:

(1) First feature consists of the logarithmic values of power spectral densities of EEG samples of mean values of each band.
(2) Second feature consists of the difference in power spectral densities of corresponding right and left hemispheric electrode of the EEG signal.

The GSR measures the resistance of the skin. It is related to the amount of perspiration the subject has while watch the video stimulus. The degree of perspiration provides information about the nervousness and anxiety of the subject. The resistance decreases with increase in perspiration. Lang et al. suggested that the arousal is correlated with the mean value of GSR [14]. The following features are extracted from the GSR signal:

(1) The average values of each band D1–D5 and A5 are extracted.
(2) The mean values of derivative of GSR is extracted for each band.
(3) The mean values of derivative for negative values only. Hence, average decrease rate during perspiration is evaluated.
(4) Ratio of number of negative sample to the total number of samples in each band.
(5) Number of local minima of each band.
(6) Zero-crossing rate of the GSR.

The BVP is measured using plythesmograph. It is the measure of the pressure by which heart exerts pressure into the arteries. The plythesmogarph is attached to thumb of the subject. The BVP can be used to find the heart rate and heart rate variability. The heart rate and its variability are correlated to the emotional state as the faster heart rate indicates subject under stress. The higher blood pressure indicates the sense of fear or surprise. The following are the feature extracted from the blood volume pressure of the subject undergoing visual stimuli:

(1) Average value of BVP.
(2) Standard deviation of BVP.
(3) Average value and standard deviation of the heart rate (heart rate is identified with the help of local maxima, i.e. the heart beat and the heart beats per minute is the heart rate).
(4) Heart rate variability (Iwona Cygankiewicz et al., suggested that HRV reflects beat to beat changes in R-R interval. Heart rate changes may occur as a response to the physical and mental stress [15]. And hence, related to the emotional state of the subject under study).
(5) Inter-beat interval.
(6) Energy of the blood volume level in each frequency band.
(7) Energy ratio of blood volume level in consecutive frequency bands.
(8) Power spectral densities of blood volume level.

The next physiological signal is the RA. The RA is recorded as the speed of respiration depends on the emotional state, a slow RA represents relaxation and fast or irregular respiration implies a feeling of fear or anger. The RA has high correlation with the arousal. The feature extracted from RA is as follows:

(1) Inter-band difference in energy of respiration signal.
(2) Average respiration signal.
(3) Mean of derivatives of each frequency band.
(4) Standard deviation of RA. (This feature shows variation in respiration amplitude which in turn shows the change in mood of the patient).
(5) Range or greatest breath time taken by the subject.
(6) Spectral centroid of the respiration frequency bands.
(7) Breathing rate.
(8) Power spectral density of respiration amplitude.

ST is also used to extract the emotion and hence the following features are extracted:

(1) Average value of each band of temperature.
(2) Average of its derivatives are extracted to find the mean rate of change of ST
(3) Power spectral densities

Muscle signals were obtained in the DEAP dataset to extract information about the facial expression and shoulder movement. The facial muscle plays an important role in expressing one's emotions. Hence, EMG of zygomaticus muscle was recorded. The shoulder muscle movement is implies laughter (happy) emotion or anger as implicit tags. The features extracted are:

(1) Energy of the muscle signals.
(2) Mean of the EMG.
(3) Variance of the zygomaticus and trapezoidal muscles.

The eyeball movement and tracking is involved in predicting one's emotions. The blinking rate is decreased to a large extent when the person experiences high arousal. The EOG signal can be related to anxiety of the person also. The features presented by the DEAP dataset are:

(1) Energy of the EOG signals in each band.
(2) Mean of the EOG signal in each frequency band.
(3) Variance in eyeball movement.
(4) Blinking rate of the eye. (The blinking rate is determined by the detectable peaks of the EOG).

3.2 Classification Using KNN Classifier

The KNN is a supervised learning algorithm, hence, it works on the given dataset directly. For any new instance, predictions are made on the search of the K most proximate instances. Here in the case of KNN, these instances are known as "Neighbours". To measure the parameter of the neighbour, we are using the Euclidean distance. The K-nearest neighbour's Euclidean distance is calculated with respect to the instance data X. The majority voting is performed and the instance X is allotted the class of majority of K-nearest neighbour's class [16, 17].

4 Experiments and Results

In this paper, we have used a supervised machine learning approach to classify EEG and peripheral physiological signals to differentiate between different emotional states. The EEG and peripheral physiological signals were wavelet decomposed to five-level DWT using Haar wavelets. Thirty-four features are extracted using the detail and approximate subbands of the decomposed wavelets. Vectors are formed by concatenating these features. This feature matrix is classified using KNN classifier with fivefold cross-validation method. This method is proposed to find the accuracy of the emotional analysis using psychophysiological signals.

Different pilot experiments were done and $K = 10$ was chosen based on the performance in terms of accuracy.

The accuracy is computed using the confusion matrix. The multi-class confusion matrix is $4 * 4$ matrix with true positive are placed at $i = j$ cell of A_{ij} confusion matrix. The sensitivity and specificity is calculated using confusion matrix.

$$\text{SENSITIVITY} = \frac{\text{TRUE POSITIVE}}{\text{TRUE POSITIVE} + \sum_N \text{FALSE NEGATIVE}} \tag{1}$$

$$\text{SPECIFICITY} = \frac{\text{TRUE NEGATIVE}}{\text{TRUE NEGATIVE} + \sum_N \text{FALSE POSITIVE}} \tag{2}$$

The multi-class accuracy is obtained from the sensitivity and specificity of the classified results:

$$\text{ACCURACY} = \frac{\text{SENSITIVITY} + \text{SPECIFICITY}}{2} \tag{3}$$

The four classes are classified with the KNN classifier with 87.1% accuracy.

In Fig. 1, the confusion matrix of KNN classifier with data points is inserted as true class versus predicted class is shown. The diagonal elements of the $4 * 4$ matrix depicts that the data points are accurately predicted, whereas the other elements of the matrix depicts the missclassification of the data points.

In Fig. 2, the confusion matrix is shown in terms of percentage of data points accurately classified or misclassified. In Fig. 3, the region of convergence is shown for the KNN classifier.

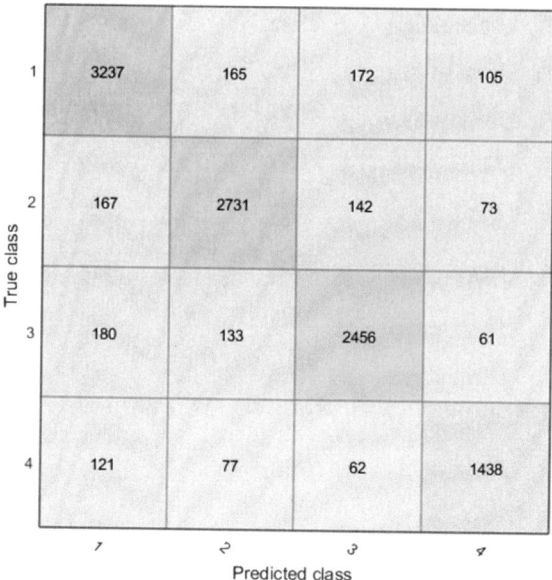

Fig. 1 Confusion matrix in terms of data points of KNN classifier

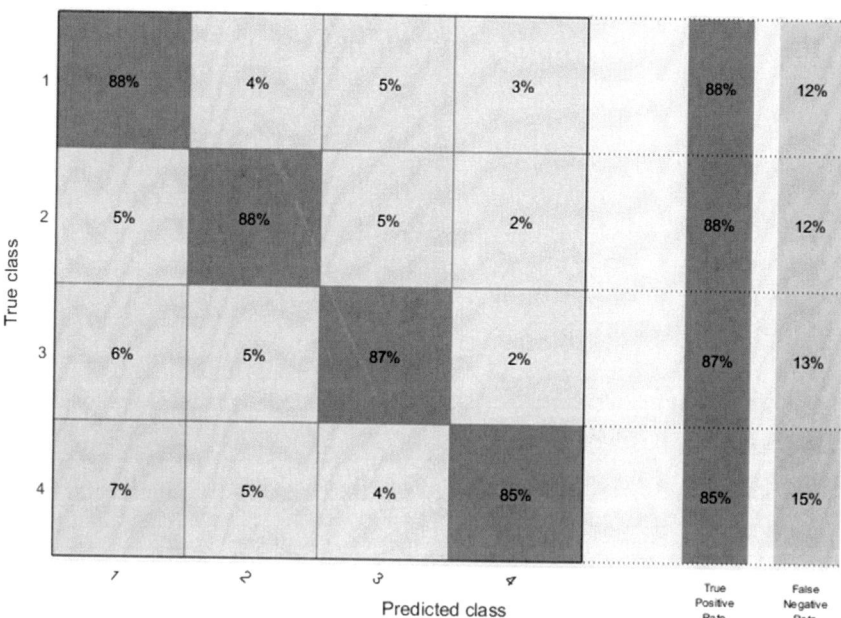

Fig. 2 Confusion matrix in terms of percentage accuracy of KNN classifier

Fig. 3 Region of convergence curve for KNN classifier

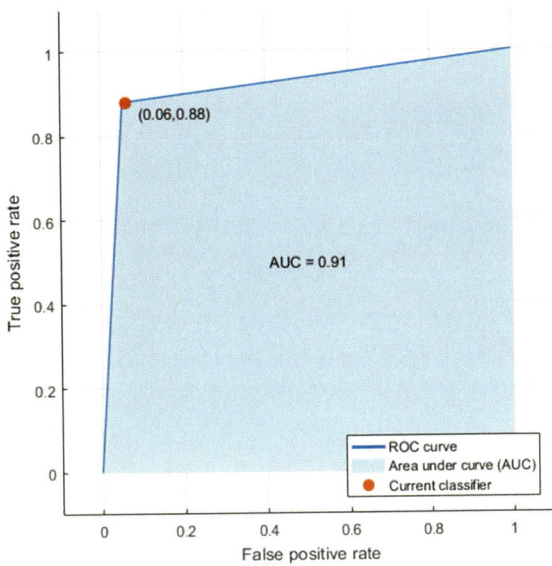

5 Conclusion

In this paper, we have proposed a DWT-based approach to the emotion analysis of the humans using video stimuli. The proposed approach trains the psychophysiological signals using the five-level DWT and then, extracting features such as mean, standard deviation, power spectral density, average decrease rate during decay time, heart rate and heart rate variability and eye blinking rate to classify the emotion expressed explicitly or even implicit emotions are detected. The KNN classifies the emotions as into four classes that are high valence high arousal, high valence low arousal, low valence high arousal and low valence low arousal. A good classification accuracy of 87.1% is obtained using KNN classifier suggest that this algorithm is suitable for training data to classify the emotions using visual stimuli.

Use of weighted KNN with feature reduction using principal component analysis will also be tried to reduce the system complexity. In future, the focus will be to use deep learning, Random forest etc., to further improve the performance.

References

1. Dasdemir Y, Yildirim E, Yildirim S (2017) Analysis of functional brain connections for positive–negative emotions using phase locking value. Cogn Neurodyn 11(6):487–500
2. Jahankhani P, Kodogiannis V, Revett K (2006) EEG signal classification using wavelet feature extraction and neural networks. In: International symposium on modern computing, 2006. JVA'06. IEEE John Vincent Atanasoff 2006. IEEE, New York, pp 120–124

3. Başar E, Başar-Eroglu C, Karakaş S, Schürmann M (2001) Gamma, alpha, delta, and theta oscillations govern cognitive processes. Int J Psychophysiol 39(2):241–248

4. Haas LF (2003) Hans Berger (1873–1941), Richard Caton (1842–1926), and electroencephalography. J Neurol Neurosurg Psychiatry 74(1):9

5. Hadley JM (1941) Some relationships between electrical signs of central and peripheral activity: II. During 'mental work'. J Exp Psychol 28(1):53

6. Plutchik R (1959) The effects of high intensity intermittent sound on performance, feeling and physiology. Psychol Bull 56(2):133

7. Moon J, Kim Y, Lee H, Bae C, Yoon WC (2013) Extraction of user preference for video stimuli using EEG-based user responses. ETRI J 35(6):1105–1114

8. Ekman P, Friesen WV, O'sullivan M, Chan A, Diacoyanni-Tarlatzis I, Heider K, Scherer K (1987) Universals and cultural differences in the judgments of facial expressions of emotion. J Pers Soc Psychol 53(4):712

9. Koelstra S, Muhl C, Soleymani M, Lee JS, Yazdani A, Ebrahimi T, Pun T, Nijholt A, Patras I (2012) Deap: a database for emotion analysis; using physiological signals. IEEE Trans Affect Comput 3(1):18–31

10. Torres-Valencia CA, García HF, Holguín GA, Álvarez MA, Orozco Á (2015) Dynamic hand gesture recognition using generalized time warping and deep belief networks. In: International symposium on visual computing. Springer, Cham, pp 682–691

11. Ramasamy M, Varadan VK (2017) Study of heart-brain interactions through EEG, ECG, and emotions. In: Nanosensors, biosensors, info-tech sensors and 3D systems 2017, vol 10167. International Society for Optics and Photonics, p 101670I

12. Li Y, Li X, Ratcliffe M, Liu L, Qi Y, Liu Q (2011) A real-time EEG-based BCI system for attention recognition in ubiquitous environment. In: Proceedings of 2011 international workshop on ubiquitous affective awareness and intelligent interaction. ACM, New York, pp 33–40

13. Hoa LT, Anh ND (2007) Orthogonal-based wavelet analysis of wind turbulence and correlation between turbulence and forces. Vietnam J Mech 29(2):73–82

14. Lang PJ, Greenwald MK, Bradley MM, Hamm AO (1993) Looking at pictures: affective, facial, visceral, and behavioral reactions. Psychophysiology 30(3):261–273

15. Cygankiewicz I, Wranicz JK, Bolinska H, Zaslonka J, Zareba W (2004) Relationship between heart rate turbulence and heart rate, heart rate variability, and number of ventricular premature beats in coronary patients. J Cardiovasc Electrophysiol 15(7):731–737

16. Zhang ML, Zhou ZH (2007) ML-KNN: a lazy learning approach to multi-label learning. Pattern Recogn 40(7):2038–2048

17. Parthasarathy G, Chatterji BN (1990) A class of new KNN methods for low sample problems. IEEE Trans Syst Man Cybern 20(3):715–718

Enhancement of Optical Coherence Tomography Images: An Iterative Approach Using Various Filters

M. Saya Nandini Devi and S. Santhi

Abstract Speckle is an important noise in the optical coherence tomography (OCT) images that play a vital role in the degradation of the visual quality of images and makes it difficult to assess the quality of the images. In order to improve the quality, filtering is essential which removes the noises and reproduce the OCT images. This proposed work addresses various filters like Mean, Median, Adaptive Median, Gaussian, and Wiener in order to enhance the OCT image. This paper gives estimate of various filtering techniques on the basis of the performance indices calculated from the experimental results. The results of this research work suggest the filter suitable for OCT images depending on the Cross-Correlation, Mean Square Error-(MSE) and Peak Signal-to-Noise Ratio-(PSNR) which are very much used as performance indices in medical image applications and its analysis. Initially, the filter is tested with standard medical image and performance measures are calculated. Next, noise-corrupted OCT image is applied as an input to the filter for the purpose of denoising and finally, a comparative analysis was performed.

1 Introduction

Speckle is a typical noise in engineering that corrupts the details of the image. It is the initial factor that restricts contrast resolution of images, thereby restraining detectability of small, low-contrast objects and makes the images usually complex for interpreting by nonspecialist. This noise affects the visual impact of the optical coherence images and leads to a problem in diagnosing the images. Speckle noise is most prevalent among tomography images and is acquired by the pixels during the scanning or imaging phase of the image acquisition Phenomena. The incident and

M. Saya Nandini Devi · S. Santhi (✉)
Department of Electronics and Instrumentation Engineering, Annamalai University,
Chidambaram, India
e-mail: santhi.sathyamurthy@gmail.com

M. Saya Nandini Devi
e-mail: nandini46@gmail.com

© Springer Nature Switzerland AG 2019
D. Pandian et al. (eds.), *Proceedings of the International Conference on ISMAC in Computational Vision and Bio-Engineering 2018 (ISMAC-CVB)*, Lecture Notes in Computational Vision and Biomechanics 30,
https://doi.org/10.1007/978-3-030-00665-5_12

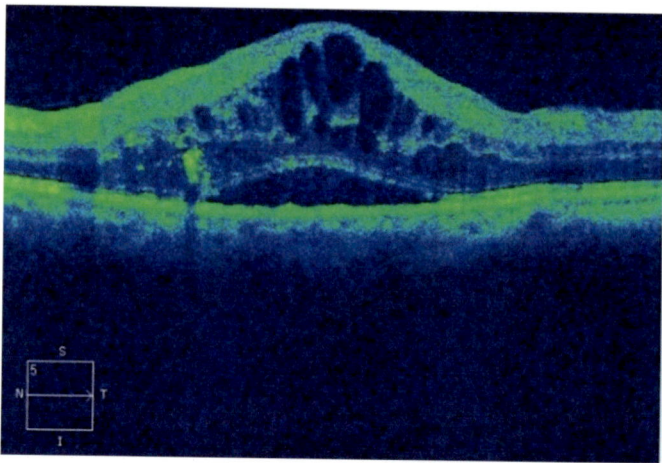

Fig. 1 Sample OCT image

reflected beams of the imaging devices cause a backscattering effect on the pixels of the image and appear as tiny speckles of noise particles which severely degrade the quality of the image and even causing loss of critical data if not detected and treated early. It is a high speed, high-resolution three-dimensional clinical diagnostic images in guiding the various procedures in the medical field. The coherence imaging modalities suffer from the speckle noise.

Speckle noise is similar to salt and pepper noise except that it is multiplicative in nature, unlike its counterpart which is additive in nature. This multiplicative noise causes increased grainy appearance due to its multiplicative nature and should be filtered early during the preprocessing phases. By assuming the linear polarization and speckle pattern which is sophisticated, speckle perhaps modeled as multiplicative noise that follows unit mean Rayleigh distribution. Figure 1 depicts the randomly selected medical image and the sample of OCT image. Analysis algorithms, 2-D display, 3-D volume rendering, and image processing effective applications are restricted by this. This requires reduction without affecting features of OCT images. To carry out the parallel valuation of speckle reduction algorithms on the basis of quality estimate is the key objective of this paper.

2 Related Work

A survey has been made to find the filter performance in OCT images. Literature [1] depicts the function of the 3D median filter in the filtering process of the noise in an OCT image. The author addresses the median filter as the widely used filter in image processing as it has the power to eliminate the noise in the edges and reduces the blurring effects particularly for the images of large dimensions. In this article, the median filter is applied for reducing the speckle noise and further, it was processed for the histogram intensity equalization. Literature [2] describes the use of noise-adaptive wavelet thresholding filter in the noise removal process in OCT images. The work has concluded with the performance evaluation of SNR and preserving structural by the quantity β, compared to conventional wavelet domain thresholding and Gaussian filtering. The finding from the article [3] describes the various denoising filters for decreasing speckle noise, in the dermatology B scans with edge preserving potential. The results are obtained by the author using the Block matching and enhanced sigma filters which are considered as a 2D filter in the denoising. The research work [4] depicts the attributes of the speckle reduction in the area of interpretation and diagnosis of spectral optical coherence retinal image. In the work, anisotropic diffusion filtering has been executed by measuring the parameters of structural similarity index for preserving the edges in that image. Literature [5] describes the de-speckling of Multiframe Optical Coherence Tomography based on anisotropic diffusion for similarity comparison between frames.

Article [6] depicts the work of spatial technique for speckle reduction in OCT. On the basis of finding the most likely intensity values at certain locations, the algorithm has been presented. The algorithm presented is based on finding the most likely intensity value for each pixel at a specific location in the image. The research paper [7] presented method basing on the scattering model for the calculation of attenuation coefficient in OCT profiles. The phantom experiments used in this have the ability for estimating accurately attenuation coefficients in case of uniforms and even layered. The literature [8] depicts the function of Wavelet denoising filter in the filtering process of multiframe OCT data. The wavelet coefficients detail is scaled with the weights, averaged, and transformed back. Literature [9] describes the Non-Local Mean [NLM] filter with double Gaussian anisotropic kernels removes speckle noise within OCT image. In this research, filters are compared with the various filters namely the NLM, Median, Bilateral, and the Wiener filters. Literature [10] investigated speckle noise as well as motion artifact and described their relation. Article [11] describes Gray Level CO-occurrence Matrix (GLCM) feature extraction and Back-Propagation Neural Network (BPNN) classification is used for reduction of speckle noise in an OCT image. The result shows better performance in the noise identification. Literature [12] detailed the Multi-Photon Tomography (MPT) and OCT systems to acquire the deviant skin lesions and the 3D images for the normal. The purpose of the paper is visualizing the 3D morphology of various layers of the skin and to detect characteristic features with the help of MPT and OCT. The research paper [13] presents an automatic segmentation of retinal layers in OCT images in case of spectral domain with the help of sparsity-based denoising, graph theory, S-GTDP, and the support vector machines. Literature [14] describes repetitious advancement on the basis of bilateral filtering in order to decrease the speckle in case of multiframe OCT data. Conventional bilateral filter applied for enhancing the OCT data and after that, the bias caused by the noise is decreased from every frames that are filtered. In [15], Multiscale Sparsity Based Tomographic Denoising (MSBTD) algorithms remove the speckle noise in an SDOCT image. This technique illustrates efficiently the simultaneous reduction of noise and interpolation of missing data.

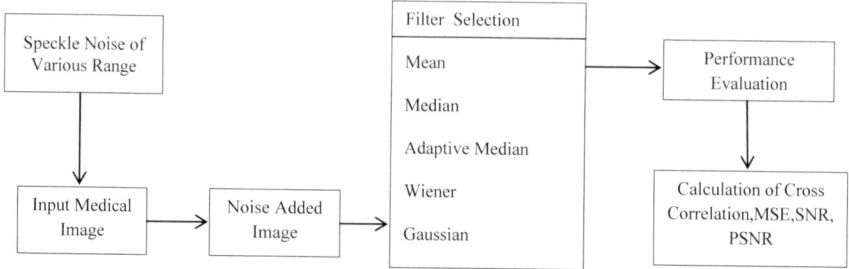

Fig. 2 Proposed model flow diagram

3 Proposed Work

The proposed work investigates the performance of five different filters employed for decreasing speckle noise of OCT image. Speckle noise and its interference were calculated based on the general model and it is given as

$$Dn(nf) = c(nc) \cdot p(np) + s(ns) \qquad (1)$$

where Dn is the observed data with noise, p is multiplicative, and s is additive component of speckle noise in n data that affects noiseless constant data set. To handle data in the speckle noises, the components are generally decoupled into logarithmic space. The noisy image in the process contains the pixel values which are much greater than the constant input image which already used as a reference. To estimate the noise parameters, the patches and the pixels are calculated so that a correlation coefficient value is obtained. The filers which are used in the proposed model reduce the noises based on the estimation process. In general, the Gaussian noise is close to multiplicative component of speckle. Hence, the distribution of the speckle can be related as

$$D(nf) = N(nf|\gamma) = \frac{a}{\sqrt{n\pi\alpha_i}} e^{-n(f-1)\frac{n}{2\alpha^2}} \qquad (2)$$

where nf is deviation factor in the noise present in the entire image (Fig. 2).

3.1 Wiener Filter

Wiener estimates the local mean and the variance of a sliding window of size $n \times m$ pixels and each pixel is located on the ith row and the kth column of the image and it generates a new pixel consisting of estimated pixel values and it is calculated as

$$f(i, k) = \mu + \frac{\sigma^2 - \beta^2}{\sigma^2}(f(i, k) - \mu) \tag{3}$$

where μ is the mean, σ^2 is the variance and β^2 is the local variance of the sliding window of $n \times m$ pixels and each pixel is located in the ith row and the kth column of the image and it generates the pixel value as

$$\beta^2 = \frac{1}{mn}\left\{\frac{f^2 \times M}{mn} - \frac{f \times M}{mn}\right\}^2 \tag{4}$$

where M indicates the matrix of ones with the same dimension as the sliding window indicating the convolution operation.

3.2 Median Filter

This is the most used nonlinear filter for removing the speckle noise and other noises and it preserves the other details in the image. The filter works by replacing the pixel of an image by median value belonging to the pixel with nearest neighbors. Mathematical model of this filter is given as

$$f_{i,k} = \text{med}\left(n\left(G_{i,k}\right)\right) \tag{5}$$

where $n\left(G_{i,k}\right)$ gives the neighborhood function with M centered at pixel location i, k.

3.3 Adaptive Median Filter

The Adaptive Median Filter is an advanced method of filtering process compared to standard median filter. The filter performs the spatial process to determine affected pixels in the image by comparing the pixel values with its neighbor pixels. The noise pixel was identified by observing the majority of pixels and other different pixels. These noises are replaced by median pixel values of neighborhood values. The process of adaptive median filter is described as follows

Level A: A1 = Zmed – Zmin

A2 = Zmed – Zmax

if A1 > 0 *AND A2* < 0, *go to level B*

else increase the window size

fwindowsize < *Smax*, *repeatlevelA*

else output Zxy

Level B: B1 = Zxy – Zmin

B2 = Zxy – Zmax

if B1 > 0 *AND B2* < 0, *output Zxy*

else output Zmed

3.4 Gaussian Filter

The Gaussian filter function is used in the research area as for defining the probability distribution of the noise and the possible values give the space which varying the values from negative to positive. A two-dimensional Gaussian function is generally used for the filtering process and the filter works based on point spread function. Using convolution, the 2D Gaussian function performs the distribution function. Distribution verge close to zero around three standard deviations away from the mean. 99% belonging to the distribution comes close to the three standard deviations. It implies that the general restriction of the kernel size is containing values not beyond three standard deviations close to the mean. Kernel coefficient sampled from 2D Gaussian function is given as

$$G(x, y) = \frac{1}{2\pi\sigma^2} e^{-\frac{x^2 - y^2}{2\sigma^2}} \tag{6}$$

An effective yet simple method of filtering out noise components from an image is the averaging or means filtering which is simple in construction and consumes least computational time and complexity. It usually utilizes a kernel or mask-based approach of an image $m \times n$ which is moved over the entire image in a horizontal or vertical scanning manner. The filter works on a replacement basis where the noisy pixel is replaced by the mean of the surrounding pixels resulting in the filtered image. However, the PSNR could be visualized as to be of a very high value as the replaced pixel represents only the averaged values of the nearby pixels.

3.5 Mean Filter

The Average or Mean Filter is an easy sliding window spatial filter which substitutes centered value inside the window by "average" of every pixel values inside window.

$$\text{Meanfilter}(x_1 \ldots x_n) = \frac{1}{N} \sum_{t=1}^{N} x_i \tag{7}$$

3.6 Performance Evaluation

The common metrics are the SNR and the PSNR of the filtered image. The PSNR of an image is given by

$$\text{PSNR} = 10 \times \log_{10} \frac{255^2}{\text{MSE}} \tag{8}$$

where "MSE" is the mean square error and is given by

$$\text{MSE} = (\text{sum}(\text{sum}(\text{err.} * \text{err})))/(M * N) \tag{9}$$

The implementation of the proposed method is as follows:

- Initialize the filter and generate variable correlation coefficient values based on speckle noises added in the original medical image.
- Correlation value for speckle of n values = corr2(Original Image, Filtered Image)
- MSE = (sum(sum(err. * err)))/(M * N)
- SNR = 10 * log 10((1/M * N) * sum(sum(A * B))/(MSE))
- PSNR = 10 * log 10(255 * 255/MSE)
- Compare the values of MSE, SNR, PSNR
- Repeat the previous steps for OCT image and compare the performance of filters.

4 Results and Discussion

The proposed speckle noise reduction process is tested on a 1.7 GHz processor with 2 GB RAM running Windows 7 and coded by Matlab 15. An OCT image is taken into account for the identification of suitable filter for the removal of speckle noise. The process starts from developing Matlab code for the various filters such as Mean filter, Median, Adaptive Median, Wiener, and Gaussian filter and analyzing their performance for a test image. The test image considered is MRI axial neck image and speckle noise with amplitude ranging from 10 to 100% was added to it. The noise added images were passed as input to the filters to identify a better filter for the removal of speckle noise. The performance measures such as correlation coefficient, mean square error, SNR, and PSNR were calculated for the test image to validate the filter algorithms. It has been observed that the Gaussian filter performs greatly for MRI image. Next, the filter algorithms were tested with OCT image that naturally has speckle noise embedded in it the process of its acquisition. The performance measures were again calculated to identify a suitable filter for the removal of speckle noise. It is observed that for an OCT image wiener filter performs well in removing speckle noise and is proposed for our future work of developing detection and classification of macular diseases on using OCT images (Figs. 3, 4, 5 and 6; Tables 1 and 2).

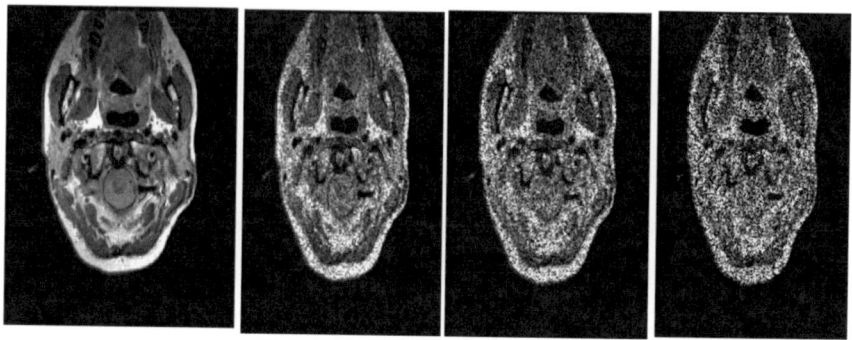

Fig. 3 Results obtained with special noise added for various value to medical image (Axial neck MRI): (1) Original image, (2) 10% speckle noise added, (3) 60% speckle noise added, (4) 100% speckle noise added

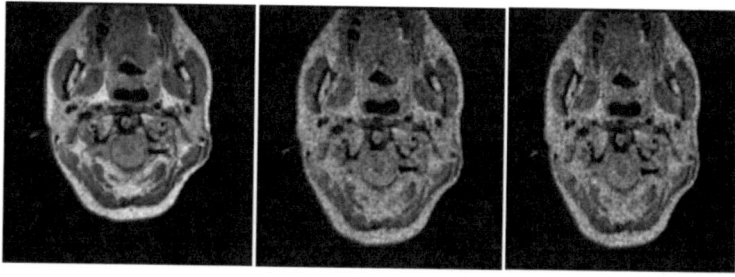

Fig. 4 Results acquired with Gaussian filter when applied for the medical image (Axial neck MRI): (1) 10% filtered output, (2) 60% filtered output, (3) 100% filtered output

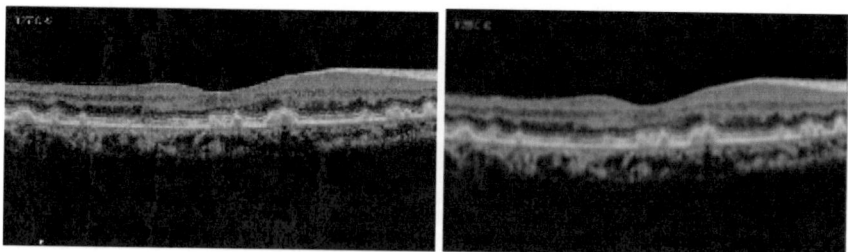

Fig. 5 Results acquired with Gaussian filter when applied for the OCT image: (1) Speckle noise image; (2) Filtered output image

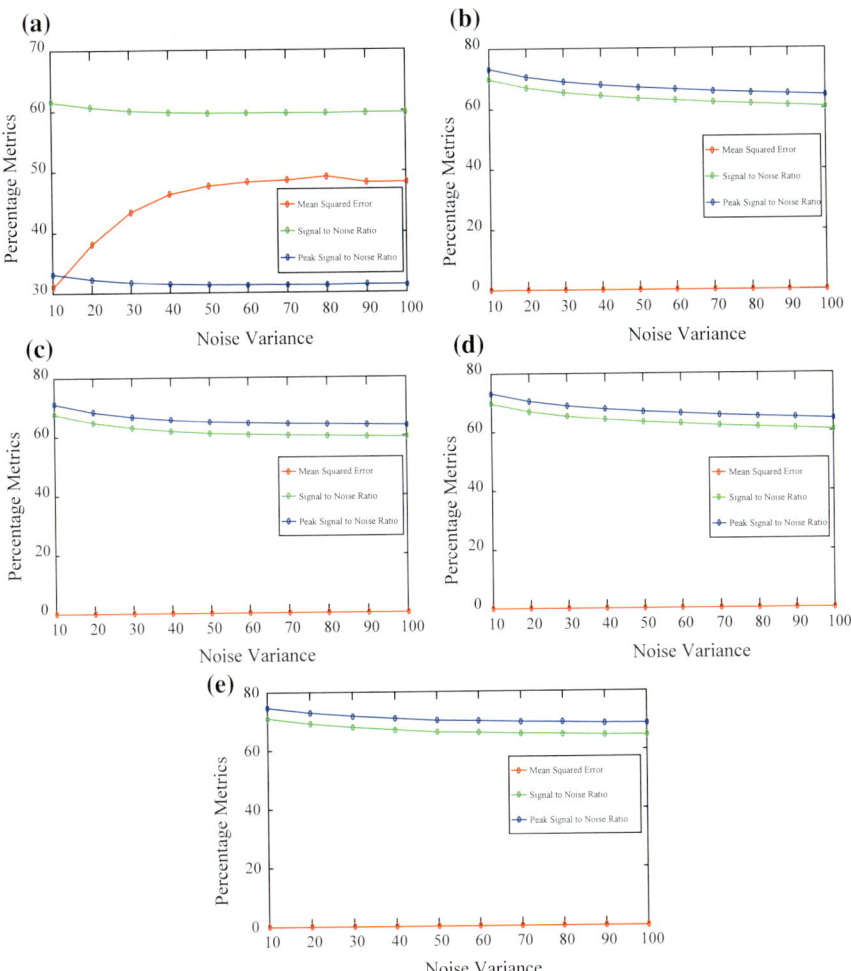

Fig. 6 Performance metrics for filters and its comparison for medical image (Axial Neck MRI): **a** Mean filter **b** Median filter **c** Adaptive median filter **d** Wiener filter **e** Gaussian filter

Table 1 Comparison of cross-correlation values for the input medical image

Filters	10%	20%	30%	40%	50%	60%	70%	80%	90%	100%
Mean	0.97	0.95	0.93	0.92	0.90	0.89	0.87	0.86	0.84	0.83
Median	0.94	0.90	0.86	0.83	0.82	0.80	0.80	0.79	0.79	0.78
Adaptive median	0.98	0.97	0.96	0.95	0.94	0.93	0.93	0.92	0.91	0.91
Wiener	0.96	0.92	0.89	0.86	0.84	0.82	0.81	0.80	0.79	0.78
Gaussian	0.98	0.97	0.97	0.96	0.95	0.95	0.95	0.94	0.94	0.93

Table 2 Performance measurement values for the input OCT image

Filters	MSE	SNR	PSNR
Mean	0.0018	70.4935	75.5302
Median	0.0024	69.3073	74.3760
Adaptive median	5.6705e−04	75.6791	80.5946
Wiener	4.0974e−04	77.0844	82.0058
Gaussian	0.0025	69.0584	74.1606

5 Conclusion

This research paper examined various filter performance characteristics for process-ing OCT images in order to remove speckle noise that is inherently present in these type of images. Algorithms for various filters such as mean, median, adaptive median, Wiener, and Gaussian were developed through Matlab coding. The performance of the various filters was first validated for the normal medical image and then the filter algorithms were applied to optical coherence tomography image. The process started with the speckle noise addition at various levels to the normal medical image and then the performance measures were calculated. Then, the various filters were applied with OCT images which contained speckle noises and performance measures such as SNR, PSNR, and Mean square error were calculated. It is observed that Gaussian serves better for medical image and Wiener filter is superior for OCT image.

Acknowledgements We would like to thank Dr. M. Ravishankar, director of the Nethralayam and Senior Consultant of the Rajan Eye Care Hospital, Chennai for providing the Optical Coherence Tomography Images.

References

1. Gocławski J, Sekulska-Nalewajko J (2016) A new idea of fast three-dimensional median filtering for despeckling of optical coherence tomography images. Image Process Commun 20(3):25–34
2. Zaki F, Wang Y, Su H, Yuan X, Liu X (2017) Noise adaptive wavelet thresholding for speckle noise removal in optical coherence tomography. Biomed Opt Express 8(5):2720
3. Gómez-Valverde JJ, Ortuño JE, Guerra P, Hermann B (2015) Evaluation of speckle reduction with denoising filtering in optical coherence tomography for dermatology. IEEE, vol 2, No 1, pp 494–497
4. Padmasini N, Abbiraime KS, Yacin SM (2014) Speckle noise reduction in spectral domain optical coherence tomography retinal images using anisotropic diffusion filtering. IEEE, vol 1, No 3, pp 244–248
5. Bian L, Suo J, Chen F, Dai Q (2015) Multiframe denoising of high-speed optical coherence tomography data using interframe and intraframe priors. J Biomed 20(3)
6. Avanaki MR, Cernat R, Tadrous PJ, Tatla T, Podoleanu AG, Hojjatoleslami S (2013) Spatial compounding algorithm for speckle reduction of dynamic focus OCT images. IEEE Photonics Technol Lett 25(15):1439–1442

7. Vermeer K, Mo J, Weda J, Lemij H, de Boer J (2014) Depth-resolved model-based reconstruction of attenuation coefficients in optical coherence tomography. Biomed Opt Express 5(1):322–337
8. Mayer MA, Borsdorf A, Wagner M, Hornegger J, Mardin CY, Tornow RP (2012) Wavelet denoising of multiframe optical coherence tomography data. Biomed Opt Express 3(3):572–589
9. Aum J, Kim J, Jeong J (2015) Effective speckle noise suppression in optical coherence tomography images using nonlocal means denoising filter with double Gaussian anisotropic kernels. Appl Opt 54(13):D43–D50
10. Rajabi H, Zirak A (2016) Speckle noise reduction and motion artifact correction based on modified statistical parameters estimation in OCT images. Biomed Phys Eng Express 2
11. Mittal A (2013) Automatic noise identification using GLCM properties. Int J Adv Res Comput Sci Softw Eng 3:943–947
12. Alex A, Weingast J et al (2013) Three dimensional multiphoton/optical coherence tomography for diagnostic applications in dermatology. J Biophotonics 6(4):352–362
13. Srinivasan PP et al (2014) Automatic segmentation of up to ten layer boundaries in SD-OCT images of the mouse retina with and without missing layers due to pathology. J Biomed Opt Express 5(2):348–365
14. Sudeep PV, Niwas SI, Palanisamy P, Rajan J, Xiaojun Y, Wang X, Luo Y, Liu L (2016) Enhancement and bias removal of optical coherence tomography images: an iterative approach with adaptive bilateral filtering. Comput Biol Med 71:97–107
15. Fang L, Li S, McNabb RP, Nie Q, Kuo AN, Toth CA, Izatt JA, Farsiu S (2013) Fast acquisition and reconstruction of optical coherence tomography images via sparse representation. IEEE Trans Med Imaging 32(11):2034

Hybrid Method for Copy-Move Forgery Detection in Digital Images

I. J. Sreelakshmy and Binsu C. Kovoor

Abstract Digital image authenticity is significant in many social areas. Image forgery detection becomes a challenging task. Copy-move forgery is one of the tampering techniques which is frequently used, part of the image is copied and pasted to other parts of the same image. This paper proposes a new method for copy-move forgery detection. Proposed method integrates both block-based and keypoint-based forgery detection. Host image is first divided into blocks and keypoints are extracted from each image block. Blocks are compared based on the keypoints in them. Number of similar keypoints identified from a pair of blocks exceeds a preset threshold, then those block pair is matched. Matched blocks are considered as the forged region and Output is displayed after neighbour pixel merging and morphology operations. The accuracy of the method is calculated and analysed with different images.

1 Introduction

In this digital world, digital images play a crucial role. Majority of the information is shared in digital image form. Every second, millions of images are uploaded and shared through social media. Digital images are now considered as a valid evidence even in the courts. Though we are very much depending on the images, the credibility of these images is still a question. Because of the widespread availability of photo editing tools, a person without any technical skill can modify the images. Manual analysis is infeasible on this huge volume of images. So forensic tools are needed to be more advanced to check the integrity of the digital images.

Forgery detection mechanisms are mainly of two types, active and passive. Active methods like Watermarking, Digital signature need to embed information at the time of creation, then only it can be verified in the future. Passive method, on the other

I. J. Sreelakshmy · B. C. Kovoor (✉)
Cochin University of Science and Technology, Kochi, India
e-mail: binsu.kovoor@gmail.com

I. J. Sreelakshmy
e-mail: 93sreelakshmy@gmail.com

© Springer Nature Switzerland AG 2019
D. Pandian et al. (eds.), *Proceedings of the International Conference on ISMAC in Computational Vision and Bio-Engineering 2018 (ISMAC-CVB)*, Lecture Notes in Computational Vision and Biomechanics 30,
https://doi.org/10.1007/978-3-030-00665-5_13

119

hand, can verify digital images without any prior knowledge about the image. So it does not need advanced equipment, as well as the quality of the image, does not get degraded because of the data embedding as in Active method. These advantages make the recent researches tend to be more on to Passive forgery detection area.

One of the most common passive forgery detections is copy-move image forgery detection. In Copy-move forgery, some part of the image is copied and pasted to other parts of the same image. Copy-move forgery can easily hide any objects in the scene or it can create multiple clones of a particular object in the image, thus the meaning of the image gets altered entirely after forgery. Existing copy-move forgery detection techniques can be categorized into two, keypoint-based and block-based methods. Here, we are discussing a hybrid method of the above two. Host image is first divided into blocks as in block-based method and like keypoint-based method, keypoints are used for block matching. Matched blocks are considered as the forged region and exact regions are extracted after neighbour pixel analysis and morphology operations. The accuracy of the method is calculated and analysed with different variations of the image also.

The remaining contents of the paper are presented in the following manner. Next section deals with the previous work related to copy-move image forgery detection. Section 3 explains the proposed method, Sect. 4 contains Result and Analysis, and in the end, we have conclusion and references.

2 Related Work

Existing Copy-move forgery detections can be either block-based or keypoint-based method. In block-based methods, the host image is divided into segments and for each segment, block features are computed. Pair of blocks whose feature vector similarity exceeds the threshold value is identified as forged regions. In keypoint-based methods, keypoints are extracted from the image and feature descriptor is computed for every keypoint. These feature descriptors are compared to find duplicated regions.

Initially, most of the works are done using block-based methods. All of them describe the same workflow, differ only in the method chosen to compute the block feature vector. Fridrich et al. [1] proposed extraction of Discrete Cosine Transform(DCT) coefficients from each block for feature vector and checking the segment pairs with same DCT coefficients detected as suspected regions. Principal Component Analysis (PCA) is used by Popescu and Farid [2], in this paper, blocks with similar PCA components identified as forgery regions. Mahdian and Saic [3] uses Blur moment invariants with PCA reduction. After the overlapped segmentation of the image, for each block, a feature vector is constructed with 24 blur moment invariants. Principal component transformation is used to reduce the feature vector dimension. The similarity of the blocks is then identified by comparing the feature vectors. K-d tree is used for an efficient searching neighbour blocks. If the neighbourhood of two similar blocks is also similar then those pairs are labelled.

Luo et al. [4] proposed intensity components as a feature vector. The image is divided into small overlapping blocks and characteristic features (7 intensity components) of each block is extracted. Comparing the feature vector of each block with one another and blocks which have the same shift vector and the difference in their characteristic coefficients is less than the threshold value are marked as similar blocks. Another variation of DCT method is discussed in [5]. DCT applied to each block and feature reduction by circular block representation and matching reduced 4 vector feature set for forgery detection. DWT can also use to compute the block feature vector [6]. DWT is applied to the image and select only the low-frequency portions from DWT result. Low-frequency regions are segmented and apply DCT on each segment. DCT coefficient features extracted from each block and lexicographically sorted and find the correlation coefficient to match the features. Bayram et al. proposed Fourier Transform coefficients as block feature vector [7]. But all of these block-based methods are not robust to any type of transformations and compression. Also, the complexity gets large as the image size increases.

In order to overcome the drawbacks of block-based methods keypoint-based methods are introduced. Because of the transformation invariant property of the keypoints, it can detect the duplicated regions even after scaling and rotation. Huang et al. proposed Scalar Invariant Feature Transform (SIFT) algorithm for keypoint extraction [8]. SIFT keypoints are extracted from the image. Forgery is detected by matching the keypoints. Match value exceeds the given threshold then those points are marked as forged ones. Bo et al. [9] followed the same procedure but instead of SIFT, they use Speeded-Up Robust Features (SURF) algorithm to retrieve the keypoints. An improved keypoint method is introduced by Ardizzone et al. [10], keypoint extraction as well as segmentation are done for the detection. After the keypoint extraction, image is segmented to triangles based on these keypoints. Inner angles, colour and mean vertex descriptors are compared to find similar triangles and thereby detecting forgery. Finally, RANSAC filter is applied to remove false matches. Keypoint-based methods cannot detect the forgery if the region is too small or forged regions are homogeneous. We cannot find enough keypoints in those cases [11].

3 Proposed Method

To detect copy-move forgery in digital images, we introduce an improved hybrid procedure of block-based and keypoint-based method. So we can make use of the advantages of both the methods. First, the image is segmented into overlapping blocks as in block-based method, then keypoints are extracted using SURF algorithm from each block. And we are comparing these keypoint descriptors to detect forgery region as in keypoint-based method. Finally, apply some morphology operations and adding similar neighbour to get a more accurate region. The basic workflow is depicted in Fig. 1.

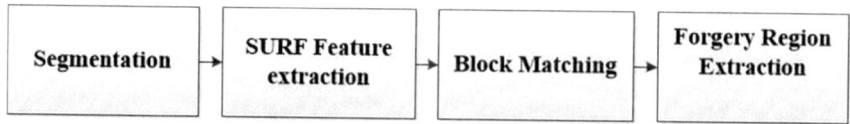

Fig. 1 Work flow of the proposed model

3.1 Segmentation

Simple Linear Iterative Clustering (SLIC) algorithm is used to segment the image into overlapping blocks. Even though SLIC is a well-known k-mean clustering-based algorithm for image segmentation, its segmentation is so much depending on the superpixel size value inputted by the user. So the user must be careful while selecting the superpixel size value. To get a perfect segmentation, for different images its suitable superpixel size value also differs. Our method computes the superpixel value adaptively, thereby reducing the extra burden on the user. First examine the texture of the image, if it is a smooth image, pixel size will be set as a large value, or set a small value if the image is a detailed one. Existing relation between texture and frequency of the image help us to understand whether the host image is smooth or detailed one. An image with the majority of the frequency energy account of low-frequency energy considered as a smooth image, otherwise detailed image. DWT using Haar wavelets is used to calculate the percentage of low-frequency energy in an image and thereby computing the corresponding superpixel size. With this superpixel size value, SLIC segmentation can be carried out [12].

Haar wavelet transform is applied in 4 levels. Low-frequency energy is calculated from the approximate coefficient results from the fourth stage of DWT. Diagonal, vertical, and horizontal coefficients of all levels are used to find out the high-frequency energy.

$$E_{LF} = \sum |CA_4| \tag{1}$$

$$E_{HF} = \sum_i \sum |CD_i| + \sum |CH_i| + \sum |CV_i|, \tag{2}$$

where CA_4—Approximation coefficients at the fourth level of DWT;
CD_i, CH_i and CV_i—Detailed coefficients at the ith level of DWT, $i = 1, 2, ..., 4$.

Percentage of low-frequency energy (P_{LF}) can be computed easily from this E_{LF} and E_{HF} values and based on the obtained P_{LF} value we can set superpixel size accordingly, using the following equations:

$$P_{LF} = \frac{E_{LF}}{E_{LF} + E_{HF}} \cdot 100\% \tag{3}$$

$$S = \begin{cases} \sqrt{0.02 \times M \times N} & P_{LF} > 50\% \\ \sqrt{0.01 \times M \times N} & P_{LF} \leq 50\% \end{cases} \tag{4}$$

3.2 SURF Keypoint Extraction

After segmenting the images using the computed superpixel value, the second phase is the extraction of keypoints from each of these image segment. SIFT and SURF are the commonly used algorithms for keypoint extraction. The accuracy of SLIC is slightly greater than SURF. But SURF can perform as good as SIFT in small regions as well as it is faster than SIFT. Here we have to extract the keypoints not from the entire image but from the individual segments only. Because of the small image regions, it is better to use SURF. So SURF algorithm is used here for keypoint extraction. SURF is carried out mainly in a two steps, Keypoints detection and calculate descriptor for each identified keypoints. These descriptors are used in the next stage of the matching process.

3.3 Block Matching

To detect duplicated regions, we have to find out the similar blocks. Similarity measure of all possible pair of blocks is computed and the block pair whose similarity measure exceeds the preset threshold T_B is considered as suspected regions. Similarity measure of a block pair is the number of matching keypoints in them. Keypoints are matched based on its descriptors. If the number of keypoints matched among two blocks greater than T_B, and the distance between them is greater than threshold T_D then we have marked those keypoints as suspected points. Neighbour segments of a uniform portion of the image naturally have some similar keypoints and because of this, they are wrongly detected as a suspected region. Distance threshold T_D will remove this type of false positives.

3.4 Forgery Region Extraction

At the end of block matching, we got all the duplicated points. We have to extract the region of forgery from these marked points. The image is segmented with the same SLIC algorithm with a larger superpixel size, in order to make very small segments. Now the segments containing marked feature points are considered as suspected regions of forgery. There may be chances that, to miss out some of the forged pixels because of the segmentation procedure. So check the neighbour pixels of these suspected regions. Merge the neighbour pixels with similar characteristics. Similar pixels are identified here by checking the average of RGB values of each pixel and for any pair of pixels, if the difference in average is not more than preset threshold T_{NP}, we treat them as similar characteristic pixel pair. So the neighbour pixels are compared in this way and the similar characteristics neighbours are also merged into the suspected region. Resulting forgery regions are displayed after some morphologic operations.

4 Result and Analysis

A series of experiments were conducted to evaluate the effectiveness and robustness of the proposed image forgery detection scheme. In the following experiments, the image dataset given in [13] is used and the proposed method was tested, implemented in MATLAB R2013a. Figure 2 shows the test image and output obtained using the proposed method.

The accuracy of the detection result is evaluated at the pixel level based on its ground truth image. Main characteristics used for accuracy analysis are precision and recall. The probability that the detected regions are relevant is termed as Precision and probability that the relevant regions are detected is Recall. According to the randomness of test images, F-score is used to test accuracy. F-score is calculated using the following equation:

$$F1score = 2 \cdot \frac{precison \cdot recall}{precision + recall} \tag{5}$$

(a)

(b)   (c)

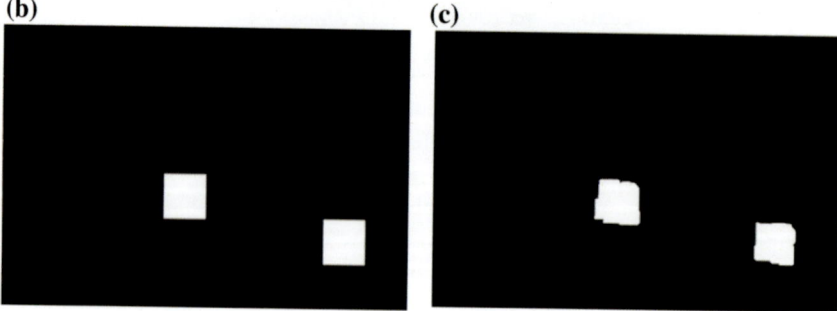

Fig. 2 **a** Input image **b** Ground truth of corresponding image **c** Forgery detection output

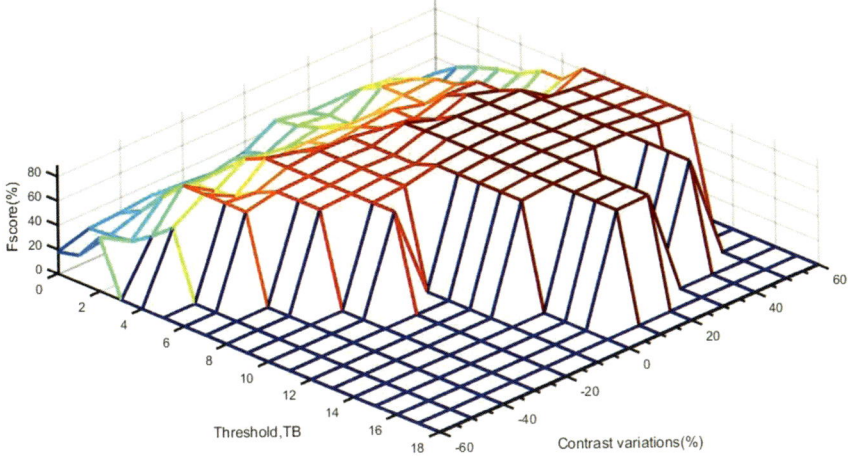

Fig. 3 F-score value against contrast variations and block matching threshold

4.1 Effect of Contrast Variations

A total of 30 images were tested by varying the contrast. The F-score of the test corresponding to the variation of contrast between -60 and $+60\%$ and block matching threshold (T_B) values between 0 and 18 is given in Fig. 3. The F-score is found to be minimum for all values of contrast where the threshold is less than or equal to 0 and greater than or equal to 18. F-score is positive over contrast between 0 and 40% when the threshold is 10–16. But in low-contrast images, we get the F-score value only in small values of threshold (between 2 and 6). This indicates that threshold and contrast significantly influence the performance of the model. On high-contrast versions of the image, we have got more accurate results, if we set threshold to be a large value. Similarly, threshold should be set minimal to process low-contrast images. In real time, if the model is used for outdoor and the detection is to be made for only certain hours of the day, the threshold can be approximately set corresponding to the contrast level of the scenery to be detected.

The proposed model uses keypoints to identify the duplicated regions. Analysis of regions will be better if there are more number of keypoints. The number of keypoints obtained in each contrast level is observed in Fig. 4. As the contrast increases, the number of keypoints extracted by the SURF algorithm also increases. So high-contrast images are more suitable for this method whereas low-contrast image results with less accuracy because they did not get enough keypoints for the evaluation. To make low-contrast images to work well in this method, we have to do a preprocessing stage, which increases the percentage of contrast level. Then, we can detect the forgeries more accurately in low-contrast images also.

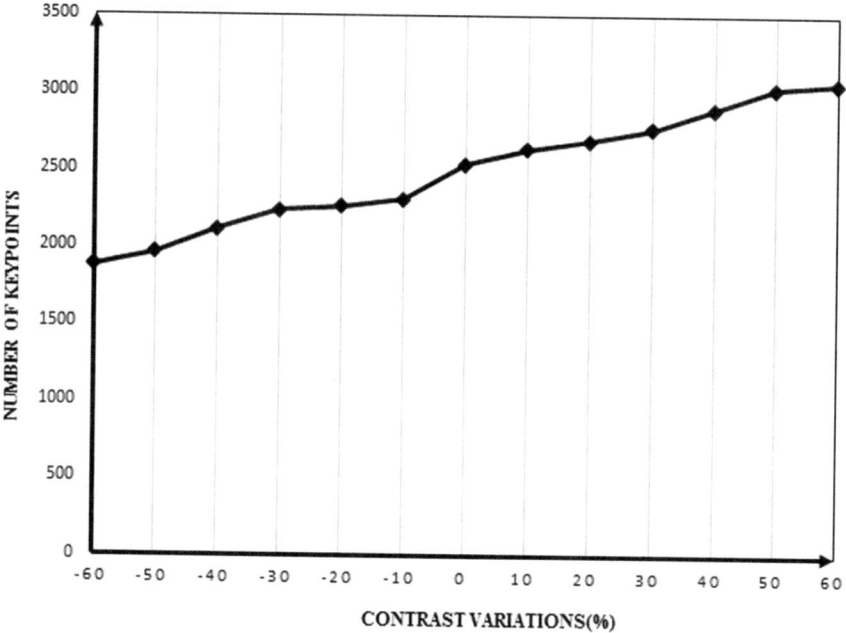

Fig. 4 Relation between contrast variation and number of keypoints

5 Conclusion

In this digital era, digital images are easily tampered and the detection of forgeries
are becoming more difficult. An improved method to detect copy-move forgery is
discussed in this paper. Host image is divided into overlapping blocks and keypoints
are extracted from each block. To detect the duplicated regions within the image, we
have to find similar blocks. All pair of blocks is compared based on the keypoints
in them, and the matched blocks are identified as suspected regions. To get a more
accurate result similar neighbour pixel merging and morphologic operations are also
performed on the suspected regions. Model is analysed for different types of images
and discovered various ranges of the threshold to detect the forgery in real time.

References

1. Fridrich AJ, Soukal BD, Lukas AJ (2003) Detection of copy-move forgery in digital images.
   In: Proceedings of digital forensic research workshop
2. Popescu AC, Farid H (2004) Exposing digital forgeries by detecting duplicated image regions.
   Department Computer Science, Dartmouth College, Technology Report TR2004-515
3. Mahdian B, Saic S (2007) Detection of near-duplicated image regions. Comput Recogn Syst
   2:187–195

4. Luo W, Huang J, Qiu G (2006) Robust detection of region-duplication forgery in digital image. In: 18th international conference on pattern recognition, vol 4. IEEE, New York
5. Cao Y (2012) A robust detection algorithm for copy-move forgery in digital images. Forensic Sci Int 214(1):33–43
6. Hayat K, Qazi T (2017) Forgery detection in digital images via discrete wavelet and discrete cosine transforms. Comput Electr Eng
7. Bayram S, Sencar HT, Memon N (2009) An efficient and robust method for detecting copy-move forgery. In: IEEE international conference on acoustics, speech and signal processing
8. Huang H, Guo W, Zhang Y (2008) Detection of copy-move forgery in digital images using SIFT algorithm. In: Pacific-Asia workshop on computational intelligence and industrial application, PACIIA'08, vol 2. IEEE, New York
9. Bo X, Junwen W, Guangjie L, Yuewei D (2009) Image copy-move forgery detection based on SURF. In: Proceedings of IEEE international conference on multimedia information network security (MINES), pp 889–892
10. Ardizzone E, Bruno A, Mazzola G (2015) Copymove forgery detection by matching triangles of keypoints. IEEE Trans Inf Forensics Secur 10(10):2084–2094
11. Yu L, Han Q, Niu X (2016) Feature point-based copy-move forgery detection: covering the non-textured areas. Multimedia Tools Appl 75(2):1159–1176
12. Pun C-M, Yuan X-C, Bi X-L (2015) Image forgery detection using adaptive over-segmentation and feature point matching. IEEE Trans Inf Forensics Secur 10(8):1705–1716
13. Christlein V, Riess C, Jordan J, Riess C, Angelopoulou E (2012) An evaluation of popular copy-move forgery detection approaches. IEEE Trans Inf Forensics Secur 7:1841–1854

Gait Recognition Using Normal Distance Map and Sparse Multilinear Laplacian Discriminant Analysis

Risil Chhatrala, Shailaja Patil and Dattatray V. Jadhav

Abstract In visual surveillance applications, gait is the preferred candidate for recognition of the identity of the subject under consideration. Gait is a behavioral biometric that has a large amount of redundancy, complex pattern distribution and very large variability, when multiple covariate exist. This demands robust representation and computationally efficient statistical processing approaches for improved performance. In this paper, a robust representation approach called Normal Distance Map and multilinear statistical discriminant analysis called Sparse Multilinear Discriminant Analysis is applied for improving robustness against covariate variation and increase recognition accuracy. Normal Distance Map captures geometry and shape of silhouettes so as to make representation robust and Sparse Multilinear Discriminant Analysis obtains projection matrices to preserve discrimination.

1 Introduction

Automated identification and recognition of the individual person in a surveillance environment with natural setting, is so hard problem that not a single gait recognition system has been reported to be working in challenging real world conditions. A large portion of the literature is dedicated to important aspects of gait recognition, so as to make it a realizable solution. The aspects like segmentation, pre-processing, gait representation schemes and pattern recognition algorithms are widely studied to understand diverse aspect requirement for gait processing.

The main contribution is as follows.

1. A new robust gait representation scheme that make use of boundary curvature information over a complete gait cycle called as Normal Distance Map is proposed.

R. Chhatrala (✉) · S. Patil
Rajarshi Sahu College of Engineering,
Savitribai Phule Pune University, Pune, Maharashtra, India
e-mail: therisil@gmail.com

D. V. Jadhav
Directorate of Technical Education, Mumbai, Maharashtra, India

© Springer Nature Switzerland AG 2019
D. Pandian et al. (eds.), *Proceedings of the International Conference on ISMAC in Computational Vision and Bio-Engineering 2018 (ISMAC-CVB)*, Lecture Notes in Computational Vision and Biomechanics 30,
https://doi.org/10.1007/978-3-030-00665-5_14

2. The representation preserves the natural tensorial structure along with spatio-temporal and structural information of gait.
3. The Extracted features provide a new feature space that addresses covariates and is found to be robust for gait recognition.
4. Sparse Multilinear Laplacian Discriminant Analysis for tensor objects is used to improve discrimination capability and increase recognition rate.

This paper is structured as follows. After reviewing the literature work in Sect. 2, the gait representation scheme based on boundary curvature information over a complete gait cycle called as Normal Distance Map is presented in Sect. 3. Section 4 presents feature extraction and pattern recognition using Sparse Multilinear Laplacian Discriminant Analysis followed by experimentation in Sect. 5. Sections 6 and 7 gives Discussion and Conclusion respectively.

2   Review of Literature

Comprehensive review of the published techniques and strategies for gait as a biometric can be found in the work of Makihara et al. [1], Sivarathinabala et al. [2], Zhang et al. [3], Boulgouris et al. [4] and Wang et al. [5]. The widely researched areas of the gait recognition system is gait representation, feature dimensionality reduction and classification.

The pioneer approaches for gait descriptor are Gait Energy Image (GEI) [6], Shifted Energy Image (SEI) [7], Gait Entropy Image (GEnI) [8], Gait flow image [9], Frequency-Domain Features [10], Depth Gradient Histogram Energy Image (DGHEI) [11] and Histogram of boundary normal vector (HoNV) using local Gauss Maps [12].

3   Gait Representation

In most recent work; it is observed that, the spatio-temporal variation exhibited by gait is mainly represented by shapes and kinematic variation averaged over gait period. The work of Tang et al. [13] and El-Alfy et al. [12, 14] inspired us to use local curvatures of a silhouette contour obtained from the geometry of the silhouettes and distance transform, together to capture boundary and area information. This feature descriptor is called Normal Distance Map (NDM). In all further discussion, Normal Distance Map (NDM) as suggested by El-Alfy et al. [14] is employed as gait feature representation. Following subsection gives a brief review of work from El-Alfy et al. [12, 14].

3.1   Histogram of Boundary Normal Vector

The histograms of boundary normal vectors were introduced by El-Alfy et al. [12]. The method focuses on curvature of contours extracted from the geometry of the silhouettes. Local Gauss maps are used to link Unit vectors normal to a surface to its curvature [15, 16]. The key advantage of doing this is to come to conclusion that, "If the normal vectors magnitude is fixed and scaled to unity, then contours extracted from 'parallel' geometric curvatures need to have same histograms." [12]. This causes the descriptor to be robust to covariate that affect gait itself.

3.2   Gauss Maps

A Gauss map g is a mapping function that maps each point p from a surface $M \in R^3$ to the unit sphere S^2, with the unit vector n_p normal to M at p.

$$g : M \rightarrow S^2 \qquad (1)$$
$$p \mapsto n_p$$

The normal vectors to flat surface M has no variation between them and is always parallel to each other. On the other hand, variation is seen for an "overly" curved surface. Hence, the mapping g is used to model the curvature of the surface. The Gaussian curvature k is defined as:

$$k = \lim_{\Omega \to 0} \frac{\text{area of } g\,(\Omega)}{\text{area of } \Omega} \qquad (2)$$

The total curvature of Ω is defined by:

$$\text{total curvature of } \Omega = \int_{\Omega} k dA \qquad (3)$$

where dA is a surface element on M.

Since, Gaussian curvature of a surface $M \in R^3$ is invariant under local isometries, globally computed surface cannot always be used to discriminate. Hence, Gauss maps are defined locally, in the form of patches or cells and for each small surface or cell, local curvatures are computed.

Fig. 1 Third order tensorial
binary silhouette video
sequence (GSV)

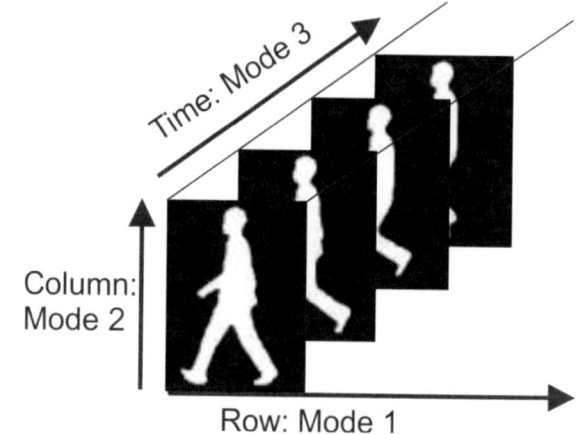

Time: Mode 3

Column:
Mode 2

Row: Mode 1

3.3 Normal Distance Map as Tensor

In order to compute NDM, the approach of El-Alfy et al. [14] is used. In this paper, the final feature descriptor is represented in the form of matrix for each frame. The gait cycle consists of multiple frames exhibiting inherent variation in pixel distribution due to kinematic motion, it is represented as third order binary gait silhouette volume (GSV) as shown in Fig. 1. Once, NDM for each frame from GSV is computed separately, it is repeated for all frames in GSV and as a result third order tensor representation based on NDM is obtained. It is then further processed by using statistical tensor based dimensionality reduction technique called as SMLDA [17]. Following section gives a brief overview of SMLDA.

4 Sparse Multilinear Laplacian Discriminant Analysis

It is a multilinear discriminant subspace learning derived from previous work [17]. The weighted Laplacian scatter difference along with sparsity constraint derives, sparse matrices $\mathbf{W}_k \in R^{m_k \times D_k}$, $m_k < D_k$, $k = 1, 2, 3$ that map $\mathcal{X}$ in to reduced tensor space.

$$\mathcal{Y} = \mathcal{X} \times_1 \mathbf{W}_1^T \times_2 \mathbf{W}_2^T \times_3 \mathbf{W}_3^T \in R^{m_1 \times m_2 \times m_3} \tag{4}$$

The key idea for SMLDA is to provide discrimination; when samples are from different classes and minimize variation, when samples are from the same class. It has sparsity constraints in the form of L_1 and L_2 norms penalty. The objective function is given by

$$J\left(\mathbf{W}_k\right)\big|_{k=1}^{N} = \min tr\left(\mathbf{W}_k^T\left(\mathbf{LS}_w^{(k)} - \lambda_k * \mathbf{LS}_B^{(k)}\right)\mathbf{W}_k\right) + \alpha_k\left\|W_k\right\|^2$$

$$+ \sum_j \beta_{kj}\left|w_{kj}\right| \text{ Subject to } W_k^T W_k = I_k \qquad (5)$$

where $|\cdot|$ and $\|\bullet\|$ denote L_1 and L_2 norm respectively, λ_k is weight parameter.

The projection matrices are computed from the gallery NDM sequence by training. Locality constrained group sparse representation (LGSR) classifier [18] is used to classify the projected low dimension features. Recognition accuracies are computed with Rank 1 ("R1") correct classification rate.

5 Experimental Evaluation

The proposed approach is validated using USF benchmark [19] and OU-ISIR data sets [20]. The recognition rate in terms of absolute as well as relative correct classification rates (CCR) are reported.

5.1 Experimentation Settings

The gait video sequence is preprocessed by background subtraction, binarization, spatial normalization, gait period detection and temporal alignment processes. As suggested in El-Alfy et al. [14], NDM is computed cell wise for each frame and the same procedure is repeated for all frames in one gait cycle. The processing for the gallery and probe sequence is repeated to obtain processed NDM tensor. Once the tensorial NDM sequence is available for both gallery and probe sets, similarity matrix based on tensor coding length is computed. The approach of minimum reconstruction error is used in the locality constrained group sparse representation (LGSR) is used for classification of probe binary gait tensor in to one from a gallery.

5.2 Experiments on OU-ISIR Dataset

OU-ISIR is the worlds largest and widely preferred available gait database with 1872 females and 2135 males totaling more than 4007 subjects with maximum diversity of covariate.

Experiment on OU-ISIR dataset A As suggested by El-Alfy et al. [12] all guidelines for experimental protocol are followed. The entire database is divided in to five sets as A-55 to A-85 and A-All. Table 1 shows the performance of the proposed technique.

Table 1 Recognition rate for OU-ISIR dataset

Dataset	No. of Subjects	GEI	HONV	NDM [14]	NDM SMLDA (TCL)
A-55	3706	84.7	91.6	94.1	95
A-65	3770	86.6	92.1	95	96.2
A-75	3751	86.9	93.3	95.7	97
A-85	3249	85.9	93.0	95.9	96.6
A-All	3141	94.2	97.5	98.1	97.2
Average		87.6	93.5	95.8	96.4

5.3 Experiment on USF Database

The USF dataset by Sarkar et al. [19] has a total of 1870 gait sequences captured in an outdoor environment for 122 subjects. Entire dataset is divided into two categories, one for the gallery and another category as twelve probe sets (A–L) having diverse variations: like shoe, surface, view points, carrying condition and time. Table 2 shows the performance of proposed technique.

Table 2 Rank-1 correct classification rate (%) for USF dataset

Probe set	Probe size	Baseline [19]	CGI Fusion [21]	Gabor-PDF + LGSR [18]	Gabor + RSM + HDF [22]	NDM + SMLDA (TCL)
A	122	73	91	95	100	100
B	54	78	93	93	95	97
C	54	48	78	89	94	91
D	121	32	51	62	73	78
E	60	22	53	62	73	75
F	121	17	35	39	55	60
G	60	17	38	38	64	50
H	120	61	84	94	97	87
I	60	57	78	91	99	93
J	120	36	64	78	94	93
K	33	3	3	21	42	45
L	33	3	9	21	42	45
Average		43	56	65	77	76

6 Discussion

Comparing the results in Tables 1 and 2, the following observations can be drawn:

1. When single-template-based or matrix based gait representation approach is compared with the third-order tensor representation, it outperforms and the average correct recognition rate is much improved. The simplest justification is the preservation of inherent structural information.
2. The covariate leads to partial feature corruption problems. The proposed representation preserves inherent correlation by retaining tensor based structural information. Simply by avoiding vectorization, robustness are improved by restricting corruption of features.
3. Tensor coding length (TCL) approach uses natural correlation of pixels from their spatial locations to reduce contamination thereon.

6.1 Timing Analysis

The timing analysis of the discriminative feature extraction scheme and the iterative projection-method-based optimization procedure of SMLDA is done by measuring the amount of time required for execution. The code of our method is run on a PC with an Intel Core i7 3.5 GHz processor and 16GB RAM. For USF dataset, the training time for the tensor coding length approach is reported as 300 s and query time as 0.42 s.

7 Conclusion

In this paper, a robust representation approach called Normal Distance Map and multilinear statistical discriminant analysis called Sparse Multilinear Discriminant Analysis is applied for improving robustness against covariate variation and increase recognition accuracy. NDM captures geometry and shape of silhouettes so as to make representation robust and SMLDA obtains projection matrices that preserve discrimination.

References

1. Makihara Y, Matovski DS, Nixon MS, Carter JN, Yagi Y (2015) Gait recognition: databases, representations, and applications. Wiley Online Library
2. Sivarathinabala M, Abirami S, Baskaran R (2017) A study on security and surveillance system using gait recognition. In: Intelligent techniques in signal processing for multimedia security. Springer, Berlin, pp 227–252

3. Zhang Z, Hu M, Wang Y (2011) A survey of advances in biometric gait recognition. In: Chinese conference on biometric recognition. Springer, Berlin, pp 150–158

4. Boulgouris NV, Hatzinakos D, Plataniotis KN (2005) Gait recognition: a challenging signal processing technology for biometric identification. IEEE Signal Process Mag 22(6):78–90

5. Wang J, She M, Nahavandi S, Kouzani A (2010) A review of vision-based gait recognition methods for human identification. Digit Image Comput: Tech Appl pp 320–327

6. Han J, Bhanu B (2006) Individual recognition using gait energy image. IEEE Trans Pattern Anal Mach Intell 28:316–322

7. Huang X, Boulgouris NV (2012) Gait recognition with shifted energy image and structural feature extraction. IEEE Trans Image Process 21:2256–2268

8. Bashir K, Xiang T, Gong S (2009) Gait recognition using gait entropy image. In: In 3rd international conference on crime detection and protection, London, UK

9. Lam THW, Cheung K, Liu JN (2011) Gait flow image: a silhouette-based gait representation for human identification. Pattern Recogn 44:973–987

10. Makihara Y, Sagawa R, Mukaigawa Y, Echigo T, Yagi Y (2006) Gait recognition using a view transformation model in the frequency domain. In: European conference on computer vision. Springer, Berlin, pp 151–163

11. Hofmann M, Bachmann S, Rigoll G (2012) 2.5 d gait biometrics using the depth gradient histogram energy image. In: 2012 IEEE fifth international conference on biometrics: theory, applications and systems (BTAS). IEEE, New York, pp 399–403

12. El-Alfy H, Mitsugami I, Yagi Y (2014) A new gait-based identification method using local gauss maps. In: Asian conference on computer vision. Springer, Berlin, pp 3–18

13. Tang S, Wang X, Lv X, Han TX, Keller J, He Z, Skubic M, Lao S (2012) Histogram of oriented normal vectors for object recognition with a depth sensor. In: Asian conference on computer vision. Springer, Berlin, pp 525–538

14. El-Alfy H, Mitsugami I, Yagi Y (2017) Gait recognition based on normal distance maps. IEEE Trans Cybern

15. Gauss KF (1902) General investigations of curved surfaces of 1827 and 1825

16. Hazewinkel M (2001) Encyclopaedia of mathematics, vol 13. Springer, Berlin

17. Chhatrala R, Patil S, Lahudkar S, Jadhav DV (2017) Sparse multilinear Laplacian discriminant analysis for gait recognition. Pattern Anal Appl pp 1–14

18. Xu D, Huang Y, Zeng Z, Xu X (2012) Human gait recognition using patch distribution feature and locality-constrained group sparse representation. IEEE Trans Image Process 21(1):316–326

19. Sarkar S, Phillips P, Liu Z, Vega IR, Grother P, Bowyer K (2005) The humanid gait challenge problem: data sets, performance, and analysis. IEEE Trans Pattern Anal Mach Intell 27:166–177

20. Iwama H, Okumura M, Makihara Y, Yagi Y (2012) The ou-isir gait database comprising the large population dataset and performance evaluation of gait recognition. IEEE Trans Inf Forensics Secur 7(5):1511–1521

21. Wang C, Zhang J, Pu J, Yuan X, Wang L (2010) Chrono-gait image: a novel temporal template for gait recognition. In: European conference on computer vision. Springer, Berlin, pp 257–270

22. Guan Y, Li CT, Roli F (2015) On reducing the effect of covariate factors in gait recognition: a classifier ensemble method. IEEE Trans Pattern Anal Mach Intell 37(7):1521–1528

Comparative Study and Analysis of Pulse Rate Measurement by Vowel Speech and EVM

Ria Paul, Rahul Shandilya and R. K. Sharma

Abstract The paper presents two noncontact pulse rate measurement techniques from vowel speech signals and Eulerian Video Magnification. The proposed methods use signals those are neither audible nor visible to naked eyes. The signals are recorded and their characteristic plots and spectrum analysis by Short-Time Fourier Transform reveal some peaks from which pulse rate can be calculated. The methods are then compared with the conventional methods where the accuracy differs by only 3.9% for vowel speech and by 0.4% for Eulerian Video Magnification. The Bland–Altman plot for the techniques shows that both are acceptable as they lie between ±1.96 Standard Deviation. The data collected from the methods are processed in MATLAB and also implemented on FPGA using serial communication by RS232.

1 Introduction

Heart rate is the number of times of contraction and expansion of heart per minute. Heart rate depends on many factors. A normal resting heart beats at 60–100 Beats Per Minute (BPM). Heart rate depends on the body's need to absorb oxygen and therefore varies during physical activity, various emotional stage, fitness, medications, temperature, etc. Heart rate is traditionally measured by counting arterial pulsation by fingers. The heart activity can be detected by electrocardiogram using 12 lead electrodes. The oxygen saturation of blood and pulse rate is measured from com-

R. Paul (✉) · R. Shandilya · R. K. Sharma
School of VLSI Design and Embedded Systems, National Institute of Technology Kurukshetra, Kurukshetra, Haryana, India
e-mail: ria.paul1993@gmail.com

R. Shandilya
e-mail: rss.nitk@gmail.com

R. K. Sharma
e-mail: mail2drrks@gmail.com

© Springer Nature Switzerland AG 2019
D. Pandian et al. (eds.), *Proceedings of the International Conference on ISMAC in Computational Vision and Bio-Engineering 2018 (ISMAC-CVB)*, Lecture Notes in Computational Vision and Biomechanics 30,
https://doi.org/10.1007/978-3-030-00665-5_15

mercial oximeters. These conventional technologies are significant in monitoring of cardiac patients.

The conventional heart rate monitoring methods are time-consuming and costly. The patient needs to visit the clinic for checkup. In times of emergency when the condition of the patient is very critical and there is no clinic nearby, the conventional methods become inconvenient. Also, for a cardiac patient who cannot afford regular heart checkup, proposed the methods which become very convenient as it requires only a microphone or a camera. This paper discusses about design and comparative analysis of two noninvasive methods by which one can easily detect pulse rate and heart condition without visiting a clinic. The first method is vowel speech [1, 2] signal processing recorded by using standard microphones. The second method is Eulerian Video Magnification (EVM) [3, 4] by video and image processing of red-plane intensity of video streaming from finger tip by passing LED light through it.

The proposed methods are then implemented on FPGA by sending collected data via RS232 from MATLAB and storing them on BRAM. The stored data are then analyzed and pulse rate is detected. Both the methods require the same architecture. The only difference is that for the first method data is collected by recording of vowel such as "e" by a standard microphone and the second method requires a camera and a LED. All the components are easily available, affordable, and convenient. The accuracy of the second method that is Eulerian Video Magnification is very high and can be easily used in times of emergency.

2 Architecture

2.1 Pulse Rate Measurement Using Vowel Speech Signals

Human speech signals contain biological and semantic information. The speech outputs are results of activities of sound energy source like larynx and supralaryngeal vocal tract. According to the "source-filter theory of speech production", speech output is the result of two-stage process (i) generation of sound energy with some spectrum and (ii) filtration or modification of source sound by vocal tract. The larynx has muscles which contain blood vessels and as part of the cardiovascular system, these vessels have connection to the heart. Heartbeat is also related with length and volume of vocal cord. Thus, heartbeat can easily be detected from human speech. In our experiment, we record the vowel "e" and plot the spectrum for heartbeat analysis. Recording rooms require high sound quality and less noise. Walls of porous materials like sponges, wood, and melamine create less noise than concrete walls. The pulse rate is detected from the spectrum by the given steps.

Steps to measure pulse rate from fingertip are shown in Fig. 1 as:

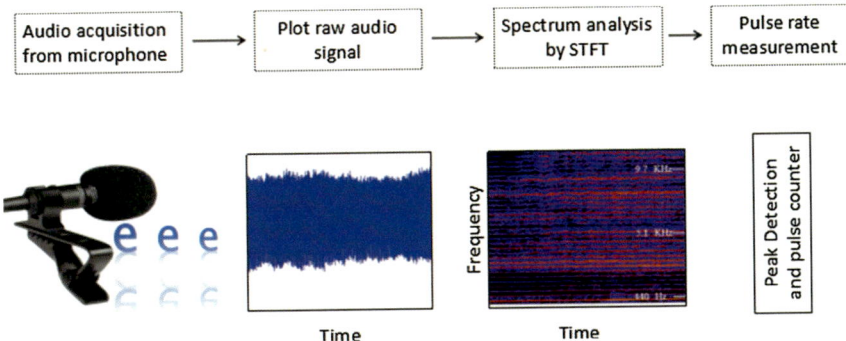

Fig. 1 Block diagram of pulse rate measurement from vowel speech recording

Step 1—The speech of any vowel is recorded using a microphone. The microphone of a mobile phone can also be used. The vowel recorded in this experiment is "e".

Step 2—The plot of the audio signal is done with respect to time but this does not give any information about frequency. The graph is again analyzed for obtaining frequency components.

Step 3—The speech signal is nonstationary and changes over time. If we apply DFT or FFT over long window, no information is obtained about frequency. Therefore, Fourier transform is applied over short period of time because over short duration or short window, speech signal can be considered stationary. The size of the window used should be less than the R-R interval of heartbeat. The windows must be tapering like Hann, Hamming and also overlapping. The technique used is called Short-Time Fourier Transform (STFT) [5, 6]. The STFT also filters unnecessary noises from the speech signal. The total number of peaks over total gives pulse rate per second. The pulse rate per second is multiplied by 60 to get pulse rate or heart Beats Per Minute (BPM). Equation (1) gives formula for STFT where $x[n]$ is the signal and $w[n]$ represents the window used.

$$X(n, \omega_0) = \sum_{m=-\infty}^{\infty} \left(x[m]e^{-j\omega_0 m}\right) w[n-m] \tag{1}$$

2.2 Pulse Rate Measurement Using Eulerian Video Magnification

By using camera, the proposed approach was implemented. This camera is used to absorb the light from blood samples. That oxygenated blood will move to the next step. Based on the intensity of the light, it can estimate the volume of blood. Based on the following steps, pulse rate can be absorbed.

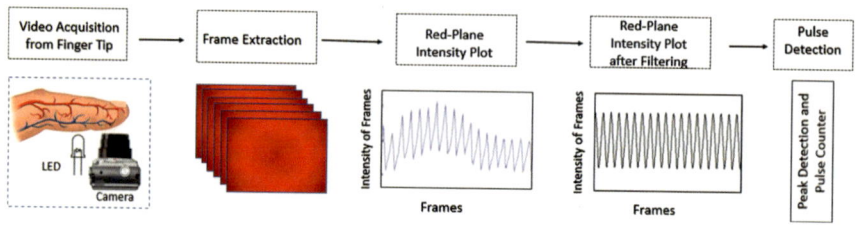

Fig. 2 Block diagram of pulse rate measurement from EVM

Fig. 3 Block diagram for FPGA implementation of the techniques

Steps to measure pulse rate from fingertip are shown in Fig. 2 as:

Step 1—By using the camera, it captures the fingerprint and it will produce the video that will be frames per second and it is very minute. This process is called Eulerian Video Magnification (EVM) [7] and is applied to fingertip of our hands or earlobe to measure pulse rate. These regions are chosen because they do not have any bones to block the light and measurement can be done accurately.

Step 2—Frame extraction process will be done through each frame of the video.

Step 3—The high-pass filters remove the low-frequency variations including non-linearity and low-pass filters remove the high-frequency noises. The plot is further made linear by using detrend( ) function in MATLAB. Now, the peaks are detected from the filtered data. The total number of peaks over total time of video streaming gives pulse rate per second. The pulse rate per second is multiplied by 60 to get pulse rate or heart Beats Per Minute (bpm).

2.3   Design for Implementation of the Methods on FPGA

The audio or video signal is stored in MATLAB which sends the data to FPGA via RS232 by serial communication [8]. The BRAMs on the FPGA are programmed to store the data. The FPGA then processes the data from BRAM and after detecting the peaks from the given data, the pulse rate per minute is measured.

Steps to measure pulse rate by FPGA implementation are shown in Fig. 3 as:

Step 1—The audio or video signal required for the heartbeat analysis are obtained from microphone or camera and are sent for MATLAB analysis in PC. The raw data

Fig. 4 Flow chart for pulse
rate measurement from peak
detection in Verilog

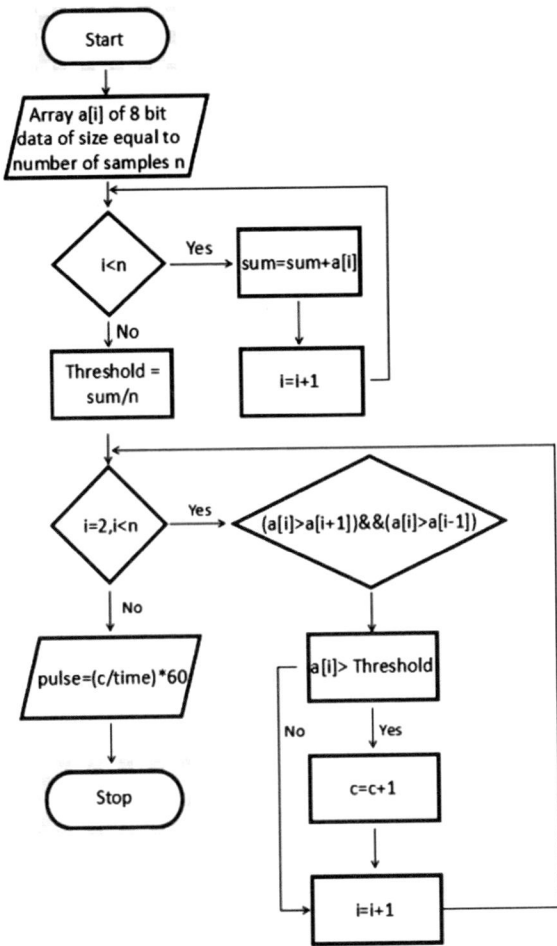

is analyzed by MATLAB where noises and false peaks are removed. The filtered
data are now sent to FPGA using serial communication by RS232.

Step 2—The BRAM on FPGA are programmed to store the received data. The
stored data are then processed by FPGA to find out the peaks in the data. The flowchart
for peak detection and pulse rate measurement using Verilog on FPGA is shown in
Fig. 4.

Step 3—The pulse rate is shown on LCD display of FPGA and also sent back to
PC for display.

3   Results and Discussion

The techniques are processed in MATLAB and then implemented on FPGA Spartan 3E kit. The microphone used is Beyer dynamic 716413 MMX 102iE with sensitivity of 104 dB.

3.1   *Pulse Rate Measurement by Vowel Speech Recording*

The raw data signal is plotted first and then the spectrum analysis is done using Short-Time Fourier Transform which is further analyzed for peak detection and pulse rate measurement. The recording of vowel "e" is done for 10 s and the plot of the signal versus time is shown in Fig. 5. The plot does not give any direct information about pulse rate. Therefore we go for spectrum analysis of the audio signal.

The spectrum analysis done by Short-Time Fourier Transform in Fig. 6 shows the peaks and it can be further analyzed to show the peaks distinctively as samples. The total number of peaks over total time gives the pulse rate per second. Equation (2) gives pulse per minute.

$$\text{Pulse Rate} = \frac{\text{Number of peaks detected}}{\text{Duration of vowel speech}} \times 60 \tag{2}$$

3.2   *Pulse Rate Measurement by Eulerian Video Magnification*

300 frames of video stream are acquired. The red-plane intensity of each frame of the video is plotted in Fig. 7, and the graph clearly shows the peaks of intensity variation that represents heartbeat with time. The plot is then filtered to remove false peaks and it is also made linear.

Fig. 5 Vowel speech signal versus time

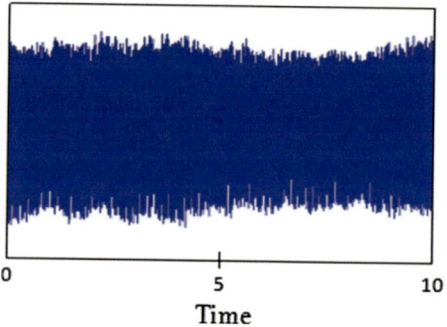

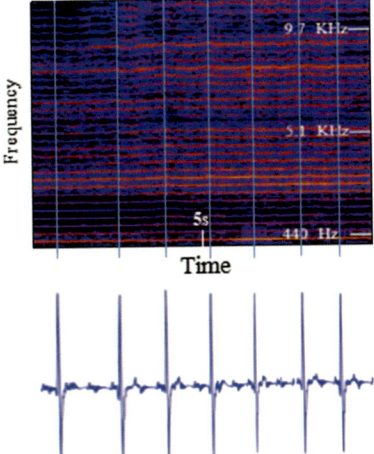

Fig. 6 STFT of vowel speech signal for spectrum analysis

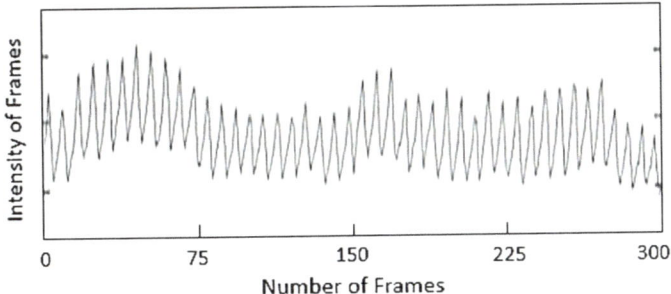

Fig. 7 Red-plane intensity graph of video stream captured from fingertip

The plot in Fig. 8 shows the intensity graph after passing through the low-pass and high-pass filters and removing nonlinearity.

The pulse per minute is given by Eq. (3). The pulse rate using commercial oximeter, vowel speech technique and Eulerian Video Magnification for 10 subjects are compared in Table 1. The average accuracy of the proposed vowel speech technique is 96.1% and average accuracy of proposed Eulerian Video Magnification technique is 99.6%. Therefore, Eulerian Video Magnification is more accurate and can be easily used as it has 99.6% accuracy. The vowel speech method is also acceptable as it has more than 95% accuracy.

$$\text{Pulse Rate} = \frac{\text{Number of peaks detected}}{\text{Duration of video stream}} \times 60 \qquad (3)$$

Figure 9 shows Bland–Altman plot [9, 10] of ten subjects for the above two techniques. Bland–Altman plot is used to compare the agreement between two methods.

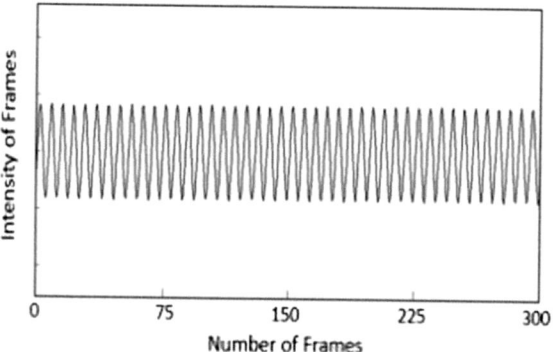

Fig. 8 Red-plane intensity graph after filtering

Table 1 Comparison of pulse rate recorded by proposed techniques

Subject	Age	Weight	Pulse rate (/min) Using commercial oximeter (A)	Pulse rate (/min) Using vowel speech (B)	Error (%) in vowel speech ((A-B)/A) × 100	Pulse rate (/min) Using EVM (C)	Error (%) in EVM ((A-C)/A) × 100
1	23	50	77	80	3.9	78	1.3
2	24	48	89	87	2.25	89	0
3	35	75	69	71	2.9	69	0
4	21	80	87	90	3.45	86	1.15
5	45	85	83	80	3.6	83	0
6	25	70	64	61	4.69	63	1.56
7	22	45	91	87	4.4	92	1.1
8	23	58	72	76	5.55	73	1.39
9	25	55	80	76	5	80	0
10	24	60	92	89	3.26	92	0

The plot shows the difference in parameters versus average of the parameters. The plot shows pulse rate of each of the ten subjects by vowel speech and Eulerian Video Magnification. It is observed that all the values fall within ±1.96 Standard Deviation which means the methods have a good agreement and thus both the methods can be used for pulse rate measurement.

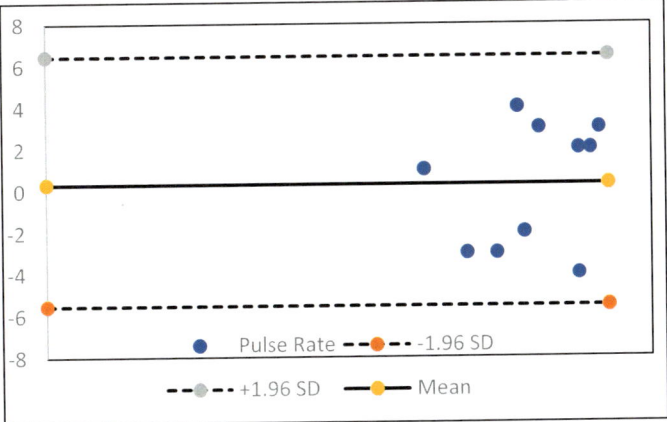

Fig. 9 Bland–Altman plot for pulse rate measurement by vowel speech and EVM

4 Conclusion

The noncontact pulse rate measurement techniques are especially suited in cases of emergency for heart patients. They also help to keep a track of the heart's condition for cardiac patients at no cost. The vowel speech method is less reliable compared to traditional oximeters but can be easily used as they have 96.1% accuracy. The Eulerian Video Magnification technique has 99.6% accuracy and is easily reliable and authentic. The techniques discussed in the paper only need a smartphone as it has both LEDs and camera which is needed for EVM technique. The vowel speech technique needs an earphone with speakers which are very commonly available. Thus, the techniques require zero cost. The vowel speech technique can also be used to analyze different emotional stages [11]. However, these methods are not applicable for patients with artificial heart and pacemakers. The successful implementation of the techniques on FPGA assures that in future a device can be designed for pulse rate measurement and heartbeat analysis using vowel speech recording and Eulerian Video Magnification.

References

1. James AP (2015) Heart rate monitoring using human speech spectral features. Hum-Centric Comput Inf Sci 5(1):33
2. Mesleh A et al (2012) Heart rate extraction from vowel speech signals. J Comput Sci Technol 27(6):1243–1251
3. Sukaphat S et al (2016) Heart rate measurement on Android platform. In: 2016 13th international conference on electrical engineering/electronics, computer, telecommunications and information technology (ECTI-CON). IEEE, New York, pp 1–5

4. He X, Goubran RA, Liu XP (2016) Wrist pulse measurement and analysis using Eulerian video magnification. In: 2016 IEEE EMBS international conference on biomedical and health informatics (BHI). IEEE, New York, pp 1–4
5. Aubel C, Stotz D, Bölcskei H (2018) A theory of super-resolution from short-time Fourier transform measurements. J Fourier Anal Appl 24(1):45–107
6. Zhang G (2018) Time-phase amplitude spectra based on a modified short-time Fourier transform. Geophys Prospect 66(1):34–46
7. He X, Goubran RA, Liu XP (2014) Using Eulerian video magnification framework to measure pulse transit time. In: 2014 IEEE international symposium on medical measurements and applications (MeMeA). IEEE, New York, pp 1–4
8. Pokharkar S (2015) FPGA based design and implementation of ECG feature extraction. Int J Adv Found Res Sci Eng (IJAFRSE) 1(12)
9. Altman DG, Bland JM (2017) Assessing agreement between methods of measurement. Clin Chem 63(10):1653–1654
10. Matsuura N et al (2017) Bland–Altman analysis for method comparisons. Adv Mod Med 354
11. Ryskaliyev A, Askaruly S, James AP (2016) Speech signal analysis for the estimation of heart rates under different emotional states. In: 2016 International conference on advances in computing, communications and informatics (ICACCI). IEEE, New York

Real-Time Input Text Recognition System for the Aid of Visually Impaired

B. K. RajithKumar, H. S. Mohana, Divya A. Jamakhandi, K. V. Akshatha, Disha B. Hegde and Amisha Singh

Abstract It is estimated that 285 million people globally are visually impaired. A majority of these people live in developing countries and are among the elderly population. Reading is essential in daily life for everyone. Visually impaired persons can read only by use of special scripts specially designed for them such as Braille language. Further, only trained people can read and understand. Since every product does not provide the product information on product cover in Braille, the present work proposes an assistive text reading framework to help visually impaired persons to read texts from various products/objects in their daily lives. The first step in implementation captures the image of the required by extracting frames from real-time video input from the camera. This is followed by preprocessing steps which includes conversion to grey scale and filtering. The text regions are further extracted using MSER followed by canny edge detection. The text regions from the captured image are then extracted and recognized by using Optical Character Recognition software (OCR). The OCR engine Tesseract is used here. This extracts the text of various fonts and then sizes can be recognized individually and then combined to form a word. Further, producing audio output by using Text to Speech module. The result obtained is very much comparable with other existing methods with better time efficiency. The real-time input is taken and passed through the algorithm which applies filters and removes noise then later image is passed through MSER, OCR, Canny edge detection to get the final audio output.

B. K. RajithKumar (✉) · D. A. Jamakhandi · K. V. Akshatha · D. B. Hegde · A. Singh
Department of Electronics and Communication Engineering, R V College of Engineering, Bengaluru, India
e-mail: Rajith.bkr@rvce.edu.in

K. V. Akshatha
e-mail: Akshatahakv.ec16@rvce.edu.in

H. S. Mohana
Department of Electronics and Instrumentation Engineering, Malnad College of Engineering, Hassan, India
e-mail: Hsm@mcehassan.ac.in

© Springer Nature Switzerland AG 2019
D. Pandian et al. (eds.), *Proceedings of the International Conference on ISMAC in Computational Vision and Bio-Engineering 2018 (ISMAC-CVB)*, Lecture Notes in Computational Vision and Biomechanics 30,
https://doi.org/10.1007/978-3-030-00665-5_16

147

1 Introduction

Although there are many, but not all things/products which have the text printed in a manner which an average blind person can recognize [1]. While a lot of visually impaired seek assistance of their fellow pedestrians in acquiring data from public information boards, the need of the hour is the availability of cost-effective, robust and accurate module which can identify the text from an image captured and give the output in a way that can be utilized comfortably by a visually impaired person, which makes them confident and independent enough so as they make their own way in this fast-growing society. We have thus developed a module which can take images from a real-time video camera, identify text, and give the output as an audio file so as to aid the visually impaired person with the information on public information boards.

The problem in achieving the task is to automatically localize and detect the text from real-time captured images with different backgrounds [2]. The existing Optical Character Recognition (OCR) technique can identify text from a clean picture of a sheet, the challenge in the current situation is to recognize text from the images with extreme noise, text with various fonts and sizes. This has been tackled by adopting various image filtering techniques. These techniques involve Canny edge detection, maximally stable extremal regions, and OCR. These techniques help us to eliminate noise in every stage. The images from the camera are captured in the form of frames and are processed real time to extract text and convert the obtained text to speech signals.

2 Related Works

Most text detection algorithms utilize a similar flow of operations to extract the regions of interest, followed by the classification of the text. These operations or methods include the identification of Maximally Stable Extremal Regions (MSER), Canny Edge Detection, and some form of OCR [3–5]. These have been determined to increase the efficiency of test recognition as compared to other methods when using novel OCR techniques [4]. There are three existing methods of text recognition: (1) texture-based method [6]; (2) Connected Component (CC) based method [7, 8]; and (3) hybrid method [9]. In texture based method the text region is considered to have a different distinguishable texture from the background. A classifier can be trained to identify the existence of text on the image. If images are more complicated, texture segmentation scheme is not sufficient. Connected component based method carries out connected component analysis and then the grouping of character candidates into text. Noncharacters are erased by Conditional Random Fields (CRFs).

First, MSER approaches are applied to a scene image to detect a large region of noncharacter which employs MSER as a basic Connected Component. The implementation of MSER may be carried out by first performing a luminance threshholding on the image in question. The extracted regions give rise to various grouping hypotheses, which can be narrowed down by examining the properties of different regions using region filtering. The paper [3] has made use of region filtering to identify the properties of different regions using pixel values. These properties can be used to separate the image into sub-images. The relationship of regions with respect to colour similarity, shape similarity can also be considered [5]. To enhance the performance, MSER must be combined with canny edge detection, which outlines the text characters and can be optimized by varying threshholding values. A Canny edge is the weak edge of local maxima of the gradient of the natural image. The noisy regions of the image can be filtered out by performing reliable geometric checks on the connected components [4]. OCR is one of the most successful methods of character recognition [3]. The entire OCR is mainly classified into two categories: traditional OCR and object recognition based. For traditional OCR-based methods, various binarization methods have been proposed to get the binary image which is directly fed into the off-the-shelf OCR engine. On the other hand, object recognition based methods assume that scene character recognition is quite similar to object recognition with a high degree of intra-class variation for scene character recognition; these methods directly extract features from original images and use various classifiers to recognize the character. OCR engines output in the form of text. In conclusion, the overall benefit of combining the methods of MSER, Canny Edge detection and OCR are evident in the cleaner output obtained. In [5, 10], their novel OCR technique obtained a precision of 89.3% and was 77.47% successful in f measure. Overall studies show that the combined usage of these methods leads to a cleaner image that is ideal for OCR and conversion into text. In this proposed work, similar flow was used, by implementing MSER techniques followed by canny edge detection, as these papers have shown the benefit of pairing these methods together. Hence, an OCR engine for character recognition was used.

3 Algorithm

Step 1: Real-time video recording is done. Frames are captured from this video to process and extract text.

Step 2: The captured 3D image is then converted into a 2D grey scale image.

Step 3: Bilateral filter is applied to remove the background noise from the image.

Step 4: MSER is applied to get the sharp edges like text regions change in pixel intensities.

Step 5: Canny edge detection is done to detect all the sharp edge boundaries mainly to extract the characters from the image.

Step 6: Dilation and intensification are done to fill in the above-drawn edges.

Step 7: This image is given as an input to the OCR engine which gives an array of characters present in the picture.

Step 8: This string is then passed to the TTS engine that converts text to speech and this audio is then fed into the earphones to help the disabled.

4 Methodology and Implementation

The block diagram of the proposed module is shown in Fig. 1.

4.1 Image Conversion

Processing a 2D array is simpler than processing a 3D array, the captured frames have three channels red, green and blue. So to convert this to grey scale (2D array) standard NTSC conversion, this calculates the effective luminance of each pixel. Instead, the Ibgr2color function of open CV can also be used shown in Figs. 2 and 3.

$$I_{gray} = 0.2989 * I_{rgb}(,, 1) + 0.5870 * I_{rgb}(,, 2) + 0.1140 * I_{rgb}(,, 3) \tag{1}$$

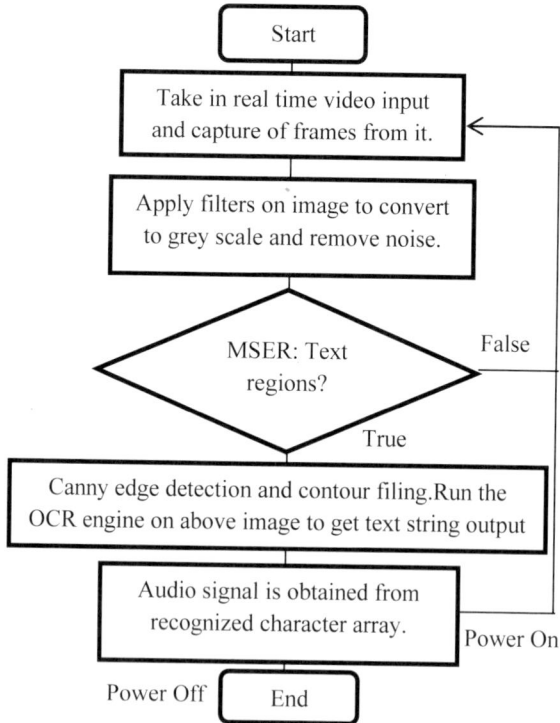

Fig. 1 Block diagram of the proposed system

Fig. 2 Colour (3D) to grey scale conversion (2D)

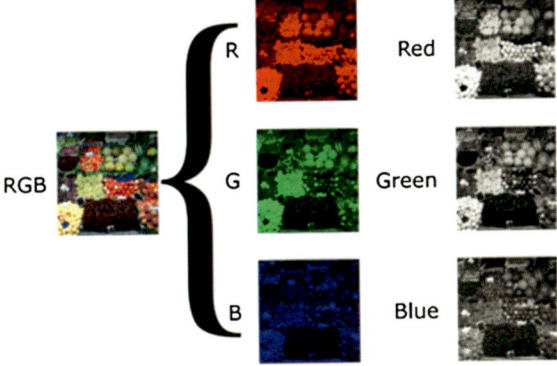

Fig. 3 MSER applied image

4.2 Bilateral Filter

A bilateral filter is an image smoothing and noise removal tool which replaces the intensity of each pixel with a weighted average of intensity values from nearby pixels. Gaussian distribution is the principles behind this weighted average. These weighted averages depend on both the Euclidean distance of pixels and on the radiometric differences, therefore preserving sharp edges. The following filters are equations are given below in (2) and (3).

The bilateral filter can be defined as

$$I^{\text{filtered}}(x) = \frac{1}{W_p} \sum_{x_i \in \Omega} I(x_i) f_r(\|I(x_i) - I(x)\|) g_s(\|x_i - x\|), \tag{2}$$

where the normalized term is

$$W_p = \sum_{x_i \in \Omega} f_r(\|I(x_i) - I(x)\|) g_s(\|x_i - x\|) \tag{3}$$

4.3 MSER

On these images, MSER is applied to detect the regions of text and the images without text regions are discarded to reduce further processing on such images.

MSER is a method for blob detection in the images; it is a stable connected component of some grey level sets of the image. MSER depends on the threshold of the image, if we give them some threshold value the pixels below that threshold value are 'white' and all those above or equal are black. MSER detect the objects and all the objects can be filled with different colours in this process some of the regions include the extra background pixels.

Mathematically, let $Q1,\ldots, Qi - 1, Qi,\ldots$ be a series of nested extremal regions $(Qi, Qi + 1)$. Extremal region $Qi*$ is maximally stable if and only if $q(i) = |Qi +``\ Qi - ``|/|Qi|$ has a local minimum at $i*$. (Here $| |$ denotes cardinality). The equation checks for regions that remain stable over a certain number of thresholds. If a region $Qi+``$ is not significantly larger than a region $Qi-``$, region Qi is taken as a maximally stable region. It proceeds by first sorting the pixels by intensity. After sorting, pixels are marked in the image, and the list of growing and merging connected components and their areas is maintained using the union-find algorithm in practice these steps are very fast.

Those are removed in the Canny edge detection process.

The procedure of implementation of MSER includes the following four steps:

i.   Simple luminance threshholding: First of all sweep threshold of intensity from black to white performing a simple luminance threshholding of the image.
ii. Then the connected components are extracted ('Extremal Regions').
iii. A threshold is found when an extremal region is maximally stable.
iv. Finally, the regions descriptors as features of MSER are received.

4.4 Canny Edge Detection

The selected images go through canny edge detection is applied which marks the sharp boundaries of the region thus drawing an outline for the text. Canny edge detection is a tool used to extract sharp contours that is intense pixel densities, or regions with a high gradient from the background objects that do not have sharp variations. This marks these sharp variations in pixel thus reducing the amount of data to be processed on the image.

Canny edge processing has mainly five steps:

i.   Noise is removed by applying a Gaussian filter to the image.
ii. Pixel intensity and the gradients of all regions in the image is found.
iii. Spurious response to edge detection is eliminated by non-maximum separation.
iv. Potential edges are determined by double thresh holding.

v. All weak edges and edges not connected to strong edges are suppressed to get the final edge marking; this is called edge marking using hysteresis.

Image noise easily affects edge detection; therefore it is necessary to filter out the noise to prevent the wrong detection caused due to noise. This filter slightly removes noise before edge detection. The filter kernel of size $(2k+1) * (2k+1)$ is as follows.

$$H_{ij} = \frac{1}{2\pi\sigma^2} \exp\left(-\frac{(i-(k+1))^2 + (j-(k+1))^2}{2\sigma^2}\right); \quad 1 \le i, \ j \le (2k+1) \quad (4)$$

The intersection of the Canny edge and the MSER is most likely to be a region containing text. These edges and detected text region are dilated to complete open boundaries. These completed boundaries are then filled in this step intensifies the text and thus it can be sent as input to the OCR engine.

4.5 Optical Character Recognition

OCR is an engine which has a huge library of all possible characters and using images that have discrete text characters involved in them it gives a string or character array as output. The intensified images are then passed through the OCR engine which gives the text from the image. OCR requires pre-processing which is done by the above filters like greyscale filter, bilateral filter, MSER, and canny edge. There are two basic types of core OCR algorithm, which may produce a ranked list of candidate characters. Matrix matching involves comparing an image to a stored character (glyph) on a pixel-by-pixel basis. This relies on the input glyph being correctly isolated from the rest of the image, and on the stored glyph being in a similar font and on the same scale. This technique could be better implemented with a predefined set of fonts than on a complete never style of characters. Feature extraction is also a type of OCR implementation which decomposes glyphs into "features" like lines, closed loops, line direction, and line intersections. The recognition process is made computationally efficient by reducing the dimensionality by extracting features. The extracted features are compared with a vector-type representation which may also reduce one or more glyph prototypes.

4.6 Text to Speech Engine

The text output from the OCR engine is then passed to text to speech conversion engine 'TTS' which converts this to the speech signal and this is given as output to the speaker. TTS is a python library which we use to convert text to speech. Any text input is converted into speech with default language as English, further using more libraries other language output can also be developed.

5   Results and Discussions

The following algorithm works with an efficiency of 90% (after 15 trials). Some bright strains of light during video capture can disrupt the functioning by erroneously detecting some text characters. The below figures are the output of the algorithm after each stage of image processing and also the final text output of module that gets converted to speech.

Figure 4 shows the frame captured from the real-time video input. The value of each pixel is a single sample representing only an amount of light, that is, it carries only intensity information for processing; this is shown in Fig. 5. Figure 6 shows the application of bilateral filter which sharpens the text and removes noise. Then MSER is applied which is depicted in Fig. 7. This method used for blob detection in images. It extracts a comprehensive number of corresponding image elements contributes to the wide-baseline matching. Canny edge detection is performed on the MSER output to get the output as shown in Fig. 8. Further dilation and intensification filters are applied to get the input to the OCR engine as shown in Figs. 9 and 10. Figure 11 shows the string output of the OCR engine which is fed to the Text to Speech Engine.

Fig. 4 Frame from the captured real-time video

Fig. 5 Greyscaled image

Fig. 6 Bilateral filter application

Fig. 7 MSER application

Fig. 8 Canny edge detection

Fig. 9 Dilation

Fig. 10 Intensified image
(input to OCR)

5.1 *Efficiency Based on Varying Parameters*

Distance: Efficiency was observed to decrease by a small amount with increasing distance. It is highly dependent on the resolution of the camera or imaging device as more blur is known to reduce the accuracy of MSER technique. The efficiency reduced from 95% at 5 m to 90% at 20 m to 80% at 40 m distance.

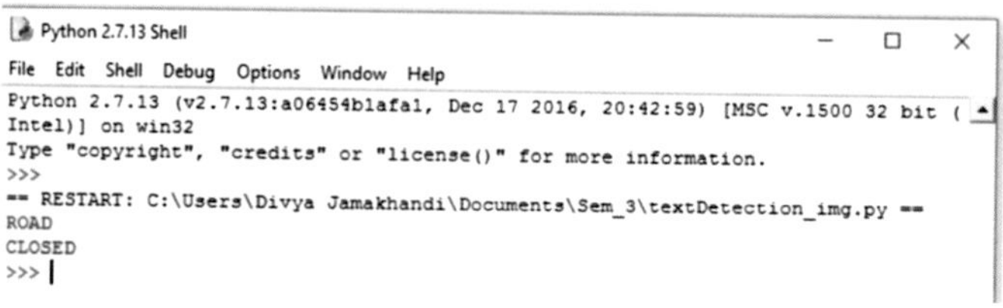

Fig. 11 Output of OCR (input to TTS)

5.2 Brightness Levels

The ambient light in the captured image varies the output due to improper edge detection. Concentrated bright spots in the image were found to obscure the edges of characters in uneven light settings. This affected similarly shaped characters more significantly than others, i.e., 'c' and 'o', 'f' and 't', 'P' and 'R', etc., due to partial obscurement of the characters. Efficiency was found to be maximum in natural light with a decrease in settings with an artificial source pointed directly or indirectly on the sign. Low light setting also brought a decrease in efficiency, but this was also dependent on the camera used.

6 Conclusion

The proposed module uses a camera, embedded on spectacles which serve as visual assistance for visually impaired. With the proposed module, visually impaired would be able to commute easily from one place to another, independently without seeking much assistance. Despite a small reduction in accuracy with increasing distance, it shall not affect the effectiveness of the module in real-world situations. The efficiency was seen to be maximum in natural light, which is the normal use-case for such a module. The future expansion would be implementing a GPS module in the system so that if the user feeds in the destination, using voice recognition it would detect the place and give commands to the destined places. With various kinds of inputs being given, considerable accuracy was achieved during the testing phase. The sequence of elimination of non-text frames and noise elimination prove to be a good choice in real-time text recognition system. The text to speech conversion library that was used gives appreciable speech output from the text output of numerous levels of filters of text recognition system. Overall, the approach of providing an aid to visually impaired through a technological product which can read out the essential text on the public information boards would help millions of visually impaired to be on par with the rest of the world with the help of this low-cost, reliable, and user-friendly product which can recognize text from public information boards.

References

1. Strotthe T et al (1997) Mobility of blind and elderly people interacting with computers. National Institute for the Blind, report on the MOBIC project. http://www.tiresias.org/reports/mobicf.htm
2. Real Time Text Detection and Recognition on Hand Held Objects to Assist Blind People. In: 2016 international conference on automatic control and dynamic optimization techniques (ICACDOT), International Institute of Information Technology (I^2IT), Pune
3. Venkateswarlu K, Velaga SM. Text detection on scene images using MSER
4. Islam MR, Mondal C, Azam MK, Syed A, Islam MJ Text detection and recognition using enhanced MSER detection and a novel OCR technique
5. Gómez L, Karatzas D. MSER-based real-time text detection and tracking
6. Kim KI, Jung K, Kim JH (2003) Texture-based approach for text detection in images using support vector machines and continuously adaptive mean shift algorithm. IEEE Trans Pattern Anal Mach Intelligence 25(12):1631–1639
7. Koo HI, Kim DH (2013) Scene text detection via connected component clustering and nontext filtering. IEEE Trans Image Process 22(6):2296–2305
8. Srivastav A, Kumar J (2008) Text detection in scene images using stroke width and nearest-neighbor constraints. In: TENCON IEEE region 10 conference, pp 1–5
9. Zhou G, Liu Y, Tian Z, Su Y (2011) A new hybrid method to detect text in natural scene. In: 18th IEEE international conference on image processing (ICIP), pp 2605–2608
10. Gómez L, Karatzas D (2014) MSER-based real-time text detection and tracking. In: 22nd international conference on pattern recognition (ICPR), pp 3110–3115

Compromising Cloud Security and Privacy by DoS, DDoS, and Botnet and Their Countermeasures

Martin K. Parmar and Mrugendrasinh L. Rahevar

Abstract Today, every firm either academic or private or government sectors are using the cloud as a platform to store and communicate information over the internet. Lots of data are being exchanged using various applications software and services. Services are integrated and reused the information over WWW. To effectively deal with all information, people are moving on cloud computing platform because of improving the cost of hardware and software, only pay what you use and setting up infrastructure easily and available for 24 × 7. Even though lots of advantages of sharing data on a cloud platform, their obvious question arise is that are those data secure and maintain its privacy by the service provider? Data privacy and security is an essential thing in today's world. And to provide security and privacy over cloud computing is often challenging part for many organizations. In this paper, we represent some security challenges for cloud platform as services. We state with two security attack using DDoS and Botnet, and also show some countermeasure.

1 Introduction

Today, use of cloud network providers is rapidly increasing in almost all the sectors of business. There is huge demand for information storage and after that, it can be shared easily anywhere anytime. Services providers of cloud are providing services as SAAS, PAAS, and IAAS [1]. Increasing usages of cloud computing also produce some issues such as security and privacy issues, availability, reliability, and so on. One of the major concerns here is the compromising cloud security and privacy by exploiting malicious attacks. In this paper, we mainly focus effect of DoS, DDoS, and Botnet attack (Fig. 1).

M. K. Parmar (✉) · M. L. Rahevar
U & P U. Patel Department of Computer Engineering, CSPIT, CHARUSAT,
Changa, Gujarat, India
e-mail: martinparmar.ce@charusat.ac.in

M. L. Rahevar
e-mail: mrugendrarahevar.ce@charusat.ac.in

© Springer Nature Switzerland AG 2019
D. Pandian et al. (eds.), *Proceedings of the International Conference on ISMAC in Computational Vision and Bio-Engineering 2018 (ISMAC-CVB)*, Lecture Notes in Computational Vision and Biomechanics 30,
https://doi.org/10.1007/978-3-030-00665-5_17

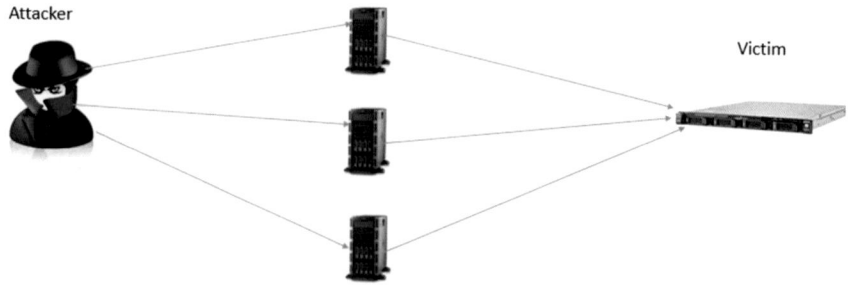

Fig. 1 DDos attack

The major effect of DDoS attacks are given below [2]:

- Unavailability of target servers.
- Increase traffic which slow down network performance.
- Significant increase in large number of emails, queries, etc.
- Make an entire network (wired or wireless) unavailable by frequent disconnection.

1.1 Advantages of Cloud Computing

- **Flexible costs**: Compare to traditional computing where a company needs to purchase expensive hardware and software whether it is required or not in future, by setting up cloud architecture, one can demand hardware and software as per requirements and also easily upgrade the system. So by avoiding large expenditure on hardware and software, a cloud can improve cost by matching your current requirements.
- **On-demand**: Looking toward current requirements as well as future requirements, your cloud environment grows. Adding and removing cloud service is on-demand [3].
- **Focus on your business**: It is difficult to monitoring infrastructure 24/7 which is time-consuming and expensive. A cloud can manage solution like monitoring your infrastructure and keeping your data secure. In addition, a cloud can provide creative and practical solutions to keep your IT infrastructure working effectively.
- **Availability of Data**: Information on a cloud is available every time anywhere. A user can access and operate data which is available in cloud storage safely using a mobile phone or computer.

1.2 Disadvantages of Cloud Computing

Technical outages

Cloud services entirely depend on Internet connectivity. You can get services efficiently if your Internet connectivity is proper. You would not be able to access any of your applications, resources or information from the cloud, if your Internet connection is offline.

Cloud Security and Privacy

By using cloud computing environment, we give our valuable information to cloud service providers. All the confidential information of the company is now stored in cloud storage and also can be accessed from anywhere using the internet. Thus, trusting on service providers are the biggest question. To minimize the risk of trusting on third-party cloud service providers, we should choose appropriate and the most reliable cloud service provides.

Interoperability

Even if cloud vendor gives surety about reliable and flexible cloud environment, it has been found so difficult to migrate or change service providers. As the infrastructure and storage structure is different than other service providers.

Security Issues in Cloud Computing

Data security is a big concern. There are lots of data stored on a cloud platform. Moreover, these data are shared or exchanged among different consumers over the heterogeneous environment. That is why the security, as well as privacy of user's data, should be maintained in a proper way. Even though there are enough service level agreements (SLAs) between provider and consumer but still there are certain issues such as authentication, authorization, confidentiality, loss of information, security breaches, and other vulnerability in the system [4]. These issues may be caused by internal or external intruders who willingly or not but causes serious damages and result in the loss of a piece of information.

There are mainly two issues exist with cloud security and privacy.

1. No any control over data
2. Dependency on the cloud service provider.

Issues due to no control over data are given below:

1. There is lack of transparency between user and service provider. Because users may not be aware of their data which are going to use by providers and how, when, why, and where they will use [5].
2. Sensitive data such as sharing private information over social media applications like videos or photos. Most providers often do mining or research of user data. In this case, also users may not aware of sharing of their private information among customer of providers.
3. There are certain applications in mobile devices which are driven by cloud providers and without noticing of users, their private information may be used or shared. So, the users are always in a risk about the transparency of their data since they think all information is stored locally between mobile and servers.
4. Sometimes, the connection between the cloud provider and consumer are compromised by intruders by eavesdropping sensitive information. There are also

DNS spoofing and denial of service attack which often make end resources available for users.

5. Entire control is shifted to cloud providers so trust management between cloud providers and customers are very hard to believe. If a customer wants to make some verification, then appropriate login should be provided.

6. There are so many data scattered around and one important aspect is that delectation of those data. For example, picture or image type of data, there should be some protection mechanism or guarantee that once data will be deleted and then it will not exist any other place.

7. Cloud computing should be provided through secure and reliable communication to end user. There should not be any other third-party involve between provider and service consumers.

Issues with the dependency on the Cloud Service provider are given below:

1. The availability is the biggest issue. If the services which are provided by cloud computing providers are down, then it would be difficult for customers for accessing data. Shutting down the services create so many problems and make data inaccessible.

2. There is not any contract between users and some service provides, therefore, users are unaware of some misuse of information [5].

3. Cloud services are provided by so many services providers which target different customers sometimes, so it is difficult to change providers.

2   Attacks on Cloud Computing

2.1   DoS Attack

Denial of service (DoS) is a kind of attack in which the end resources will become inaccessible and unavailable to end users. Unlike, other attacks in which your secure data is being stolen or eavesdropping and intercepting any information, DoS is only intended to make your end resources such as server or website or any other resources, unavailable [6].

DoS Attack Prevention:

There are some steps to follow for prevention of DoS attack:

- Monitor your HTTP traffic and find unexpected behavior by installing a firewall.
- Keep an eye on social networking or any other discussion forum which gives hints about incoming DoS attack.
- Use third-party service which helps to minimize DoS attacks.
- Make your firewall policy strong to block unwanted traffic.

2.2 DDoS Attack

Distributed denial of service (DDoS) attack tries to make an end resources or a network resource unattainable to users. A DDoS attack uses multiple connected devices which are often called botnets. Using botnets, they interrupt the end resources and make them down [7].

There are basically two types of DDoS attack

1. Application layer DDoS attack
   Main aim of the attacks is sending n number of requests to target machine which makes the end machine overloaded. Furthermore, it causes high processing power of CPU and memory usage that conclusively suspend or fail down the application. Application layer attacks like HTTP floods, slow attacks, and zero-day assaults which targets vulnerabilities in operating system, websites and communication protocols.
2. Network layer DDoS attack
   Using botnets, these attacks are typically launched. The main goal behind network layer attack is to expend the end resource's upstream bandwidth and result in network saturation.
   Network layer attacks like UDP floods, SYN floods, DNS amplification and IP fragmentation are typically used to overwhelm server by sending number of packages.

DDoS attacks can be performed by making target infrastructures and services such as DNS servers unavailable to users. It makes false DNS request by botnet devices.

DDoS Attack Prevention:

On-Premise Appliances

The on-premise approach to DDoS protection uses hardware appliances deployed inside a network, placed in front of protected servers. Such appliances usually have advanced traffic filtering capabilities. Typical mitigation appliances can be effectively used to filter out malicious incoming traffic. This makes them a viable option for stopping application layer attacks.

However, several factors make it unfeasible to rely on appliances:

Scalability remains an issue. The ability of the hardware to handle large amounts of DDoS traffic is capped by a network's uplink, which is rarely more than 10 Gbps (burst).

On-premise appliances need to be manually deployed to stop an attack. This impacts time to response and mitigation, often causing organizations to suffer downtime before a security perimeter can be established.

Finally, the cost to purchase, install and maintain hardware is relatively high—especially when compared to a less costly and more effective cloud-based option. This makes mitigation appliances an impractical purchase unless an organization is obligated to use on-premise solutions (e.g., by industry-specific regulations).

Off-premise, Cloud-based Solutions

Off-premise solutions are either ISP-provided or cloud-based. ISPs typically offer only network layer protection, while cloud-based solutions provide additional filtering capabilities required to stop application layer attacks. Both offer virtually limitless scalability, as they are deployed outside of a network and are not constrained by the previously identified uplink limitations.

Generally, off-premise mitigation solutions are managed services. They do not require any of the investment in security personnel or upkeep required by DIY solutions and on-premise hardware. They are also significantly more cost-effective than on-premise solutions while providing better protection against both network and application layer threats.

Off-premise solutions are deployed either as an on-demand or always-on service, with most market-leading vendors offering both options.

On-demand option

Enabled by BGP rerouting, the on-demand option stops network layer attacks—including those directly targeting the origin server and other components of core network infrastructure. These include SYN or UDP floods, which are volumetric attacks designed to clog network pipes with fake data packets.

Always-on option

The always-on option is enabled through DNS redirection. It stops application layer assaults attempting to establish TCP connections with an application in an effort to exhaust server resources. These include HTTP floods, DNS floods and various low and slow attacks (e.g., Slowloris).

2.3 Comparison of DoS Versus DDoS [8]

Following are the comparison about effects of DoS and DDoS attack.

DoS	DDoS
✓ Use dedicated computer or Internet connection to attack	✓ Use multiple machines and connections to attack
✓ Target system is overloaded by sending many request by one machine	✓ Target systems are overloaded by sending many requests by different machines
✓ Compare to DDoS, it is less difficult as attack is launched from one machine	✓ It is difficult to identify legitimate traffic than attacker's traffic as attacks are launched from different machines
✓ There is no any malware there	✓ Using BotNets, thousands of machined can be infected

Finally, DoS attack can be stopped using right security policy but DDoS attack can become a headache to identify legitimate traffic.

2.4 Botnet DDoS Attacks

It is also known as "Zombie Army". A botnet attack is launched from the group of connected machines with the internet. It mainly infects machines by injecting malicious malware in distributed or heterogeneous environment. Once all the machines become infected, it is called bot and then, the entire network can be controlled by remote location [9].

A botnet attack can be performed by different purpose such as information stealing (financial, user credentials for login, and relive any online identity). The botnet attack is mostly used with DDoS attack in which it tries to infect as many machines as possible and break down the entire network or target machines without knowledge of user [10].

3 Proposed Model for Mitigating DDoS Attack

The DDoS mitigation basically protects a destination target from a distributed denial of service (DDoS) attack.

A DDoS mitigation process can be broadly defined by following four stages:

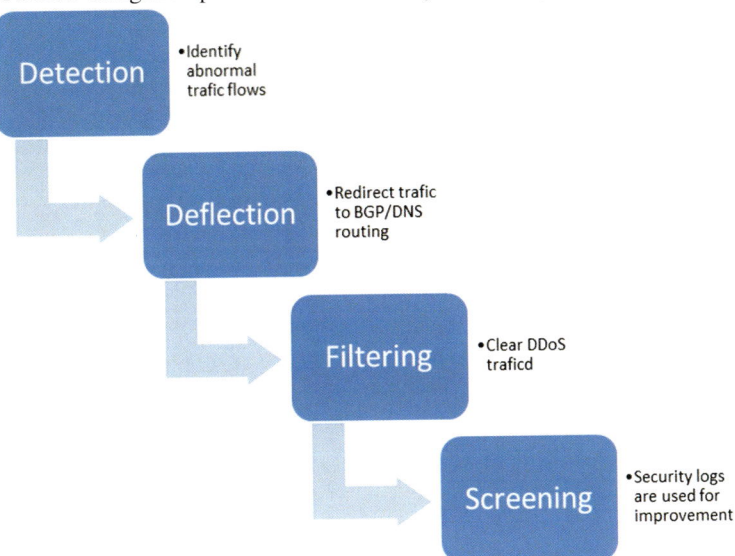

1. **Detection**: In this stage, the main focus is on recognizing different stream of traffic diversity which triggers the action for DDoD attack. It is up to an individual who may have the skill to identify an attack impressively. It would be great if the attack is detected instantly.
2. **Deflection**: After detection, the traffic is filtered and discarded or it is diverted from target machine.
3. **Filtering**: By observing and analyzing, one can recognize particular patterns which will make differentiate among legitimate traffic, DDoS-Bot and malicious request. The main thing is that to provide complete transference by identifying, detecting and block attack without any knowledge of users.
4. **Screening**: It is a very crucial point to review all possible security logs and collect detail information for any malicious action. It is good practice which can improve to identify further attack in future.

3.1 Network Layer Mitigation Techniques

Various service providers have different methods of protecting from network layer DDoS attacks, some of which are not preferable than others:

- **Null routing**: In normal routing, each package directs to destination IP but in null routing, data goes nowhere. It means null routing contains IP address which does not exist. This way, it creates lots of junk data in network.
- **Sinkholing**: Sinkhole redirects traffic from real destination machine to another machine which is known as sinkhole. It is used for good purpose and also bad. Sinkhole measures data flow in the network so it can be used as a research tool. But, sometimes, attacker can redirect traffic from legitimate user.
- **Scrubbing:** An improvement on an arbitrary sinkhole, scrubbing routes all ingress traffic through a security service. Malicious network packets are identified based on their header content, size, type, a point of origin, etc. The challenge is to perform scrubbing at an inline rate without causing lag or otherwise impacting legitimate users [11].

3.2 Application Layer Mitigation Techniques

Being much stealthier than their network layer counterparts, application layer DDoS attacks typically mimic legitimate user traffic to evade security measures. To stop them, your solution should have the ability to profile incoming HTTP/S traffic, distinguishing between DDoS bots and legitimate visitors.

During mitigation service trials, testing its application layer defenses is essential. Effective filtering uses cross-inspection of HTTP/S header content and behavioral patterns, in addition to IP and autonomous system number (ASN) information. Many security services also use different types of challenges, such as testing each request for its ability to parse JavaScript and hold cookies [12].

It is equally important to verify that the service does not overuse CAPTCHAs, "delay pages" and other such filtering methods that only serve to annoy legitimate visitors.

3.3 Mitigating Botnet DDoS Attacks

3.3.1 Network Layer Attacks

In the event of a network layer attack, we can prepare a model which provides dynamic resource overprovisioning—offering nearly limitless, on-call scalability. In addition to, using reverse proxy helps mask original service IPs which is the first line of protection against direct-to-IP attacks [13].

In an event where target IPs are already known, and the masking effect is insufficient,

1. Enforce routing policies using BGP announcements which ensure that all incoming traffic travels through legitimate server first, where it undergoes deep packet inspection and filtration process [14].
2. After first step, only clean traffic is permitted to reach the source via a secure, two-way GRE tunnel.
3. During the mitigation process, the security system continually monitors the IP addresses and sortie patterns.
4. Lastly, suspicious data is transmitted to DDoS threat database and this will benefit to all of service consumers. Because of that, it generates quick response for any malicious activity and also prepare to identify for any future botnet attack [15].

3.3.2 Application Layer Attacks

Analyzing and observing traffic data from legitimate, behavior of network, and number of request to the server can be useful for mitigating application layer DDoS attack.

3.3.3 DNS-Targeted Attacks

Webmasters can set their authoritative domain name server, while DNS zone file management remains independent of the network.

- All inbound DNS queries first reach to legitimate server, where malicious requests are automatically filtered out. Only legitimate ones are allowed to pass through, enabling smooth traffic flow at all times.

4 Conclusion

Cloud computing has tremendous advantages of data storing and make data available every time. Service providers are providing effective solution of on-demand services at reduced cost. Thus, every enterprise, private sectors, government sectors, and other firms use cloud platform for their business growth. As a result, there is obvious question might be raised about certain issues like information security and privacy. Moreover, attackers always try to do malicious activities such as make end system unavailable by overloading server or infect the thousands of machines by injecting malwares. In this paper, we have demonstrated some issues of unavailability of resources using DoS, DDoS, and Botnet attacks. Dos attack can be launched from one machine and by applying proper security policy, it can be prevented. But, attacking from multiple sources particularly in cloud environment where every resource is geographically distributed and in addition, identifying legitimate resources from bots, are very difficult and time-consuming task.

In this paper, we presented a proposed model for mitigating DDoS attack and DDoS using Botnet attack. By using detection, deflection, filtering and screening, we can lower down the change of DDoS attack and also caught fake requests from bots. But still, it is difficult to deal with number of fake requests using bots and identifying right controller who command the bots. So in the future, we can use such an effective technique to handle the number of bots in cloud environment.

References

1. Deshmukha RV, Devadkar KK Understanding DDoS attack & its effect in cloud environment. In: 4th international conference on advances in computing, communication and control (ICAC3'15)
2. Rai A, Challa RK (2016) Survey on recent DDoS mitigation techniques and comparative analysis. In: Second international conference on computational intelligence & communication technology. IEEE, New York
3. Wang B, Zheng Y, Lou W, Hou YT (2015) DDoS attack protection in the era of cloud computing and software-defined networking. www.elsevier.com/locate/comnet (Elsevier)
4. Somani G, Gaur MS, Sanghi D, Conti M, Buyya R (2017) DDoS attacks in cloud computing: issues, taxonomy, and future directions. www.elsevier.com/locate/comcom (Elsevier)
5. Zissis D, Lekkas D (2012) Addressing cloud computing security issues. Future Gen Comput Syst 28(3):583–592. https://doi.org/10.1016/j.future.2010.12.006
6. Balobaid A, Alawad W, Aljasim H (2016) A study on the impacts of DoS and DDoS attacks on cloud and mitigation techniques. In: 2016 international conference on computing, analytics and security trends (CAST) College of Engineering Pune, India. Dec 19–21, 2016
7. Alosaimi W, Zak M, Al-Begain K (2015) Denial of service attacks mitigation in the cloud. In: 9th international conference on next generation mobile applications, services and technologies
8. Incapsula. https://www.incapsula.com/ddos/denial-of-service.html
9. Yan Q, Yu FR (2015) Distributed denial of service attacks in software-defined networking with cloud computing. In: Security and privacy in emerging networks. IEEE, New York
10. Liu X, Yang X, Lu Y (2008) To filter or to authorize: network-layer DoS defense against multimillion-node botnets. In: ACM SIGCOMM computer communication review. ACM, New York

11. Bhardwaj A, Subrahmanyam GVB, Avasthi V, Sastry H, Goundar S (2016) DDoS attacks, new DDoS taxonomy and mitigation solutions—a survey. In: International conference on signal processing, communication, power and embedded system (SCOPES). IEEE, New York
12. Khor SH, Nakao A (2011) DaaS: DDoS mitigation-as-a-service. In: IPSJ international symposium on applications and the internet. IEEE, New York
13. Bawany NZ, Shamsi JA, Salah K (2017) DDoS attack detection and mitigation using SDN: methods, practices, and solutions. In: Review article—computer engineering and computer science. Springer, Berlin
14. Wong F, Tan CX (2014) A survey of trends in massive DDoS attacks and cloud-based mitigations. Int J Netw Secur Appl (IJNSA) 6(3):57–71
15. Farahmandian S et al (2013) A survey on methods to defend against DDoS attack in cloud computing. Proc Recent Adv Knowl Eng Syst Sci 6:185–190

A New Automated Medicine Prescription System for Plant Diseases

S. Sachin, K. Sudarshana, R. Roopalakshmi, Suraksha, C. N. Nayana
and D. S. Deeksha

Abstract In the current situation, agriculture is facing a wide number of problems
to address the increasing global population. Also, the plant diseases affect the production and quality of crops. Specifically, plant disease severity identification is the
most important problem in the agricultural field which can avoid the excess use of
pesticides and minimize the yield loss. In the existing systems, no methodology exists
to identify the disease severity and to prescribe the required quantity of medicines
to be sprayed. In order to solve this problem, an automated medicine prescription
system is proposed in this paper, which takes the images from the uncontrolled environment, enhances, and preprocesses the images received for the identification of
disease. Precisely, in the proposed framework, k-means and SVM algorithms are
used for clustering and disease identification tasks, respectively. Experimental setup
and snapshots of results demonstrate the performance of the proposed system, by
means of indicating the severity of the identified disease.

1 Introduction

India is an agricultural country, wherein a large portion of the population relies upon
farming. Indian economy mainly stands on agriculture, since over 58% of income
comes through agricultural segment [1]. In the current situation, agriculture faces
wide number of problems due to the increasing global population and the plant
diseases affecting the production and quality of the crops. However, agriculture is
influenced by various other climatic factors such as drought, inordinate rainfall, and

S. Sachin · K. Sudarshana (✉) · R. Roopalakshmi · Suraksha · C. N. Nayana · D. S. Deeksha
Alvas Institute of Engineering and Technology, Moodbidri, Mangaluru 574225, India
e-mail: kerenalli@gmail.com

S. Sachin
e-mail: shettysachin1996@gmail.com

R. Roopalakshmi
e-mail: drroopalakshmir@gmail.com

D. Pandian et al. (eds.), *Proceedings of the International Conference on ISMAC
in Computational Vision and Bio-Engineering 2018 (ISMAC-CVB)*, Lecture Notes
in Computational Vision and Biomechanics 30,
https://doi.org/10.1007/978-3-030-00665-5_18

sudden changes in temperature, increase in pollution factors, and release of industrial as well as chemical wastage into agricultural lands [2].

Out of many factors which affect the production, the plant diseases are the more significant. Hence, detecting the plant disease plays a more significant role. Plant diseases are the main reason for the reduction in the agricultural product quality [3]. Disease is a sick situation of the healthy state of the plant by which major biological process of the plant life. Current methods for detecting the plant diseases require the naked eye observations by the experts, which is inefficient, costlier, and time-consuming.

Farmers in many countries do not have adequate facilities or even ideas that they can contact to experts, which makes the consulting experts even cost expensive. In the current context, there a need for monitoring the large forms automatically. Machine vision-based detection of the diseases symptoms on the plant leaves makes the disease severity detection easier and cost-effective [4]. In this way, it is possible to automatically detect plant diseases using plant leafs images for various image processing operations.

The paper is further organized as follows: Sect. 2 discusses about existing work, and in Sect. 3, a description on an integrated novel technique for automated medicine prescription system is proposed and implemented. Section 4 concludes the current work followed by future enhancements.

2 Related Work

Most of the initial symptoms of plant diseases are microscopic and hence diagnosis is limited by the human vision. Recently, in 2015, Pujari et al. [5] presented a study on the image processing techniques. In this work, symptoms of various fungal diseases are affected on different crops. The above method is complex and suffers from various in terms of high variability in outdoor conditions and general symptoms.

Khirade and Patil [6] used image processing techniques for diseased part of the plant leaf segmentation, where the features of infected leaf are extracted and classified. However, the accuracy was less. In 2012, Revathi and Hemalatha proposed a HPCCDD using leaf spot images. They are captured by farmers through smartphones [7]. The Sobel and Canny filters were used to identify the edges. The HPCCDD algorithm is used for image analysis and disease classification. It used only single leaf images to obtain the disease severity.

In 2013, Husin et al. [8] described a technique in which chilly plant diseases visual symptoms are identified using the analysis of colored images. The input image was enhanced to preserve information of the affected pixels before extraction. This method takes more time to detect the disease and hence failed to render the expected performance.

Sanjay and Nitin [9] proposal was based on k-means clustering to solve low-level image segmentation. SGDM method was used for extracting statistical texture features. İsolating the picture from the unwanted scenes was difficult. In 2009, Smith and Camargo [10], set out an image processing method to identify the visual symptoms of plant diseases from colored images analysis and it fails to extract the diseased region parameters and image classification.

In 2012, Di Cui et al. [11], proposed method for multispectral images. It is used to identify the plant leaf rust, along with disease severity. The parameters used were color distribution and ratio of infected area (RIA) or rust color index (RCI). A reference database can be built for automated rust severity detection.

Similarly, Tian and Zhang [12] used hyperspectral imaging for comparative examination on downy mildew disease of cucumber plant.

Without bothering the illness level, the pesticides are sprayed unconditionally. In 2017, Mude et al. [13] described a technique for the pesticide spraying only of plant got affected; otherwise not. Here, picture handling procedures were applied to discover the illness. It also suffers from less severity identification.

Automatic spraying of pesticides shall reduce the health issues caused by the overexposure to farmers. In 2012, Husin et al. [14] proposed a method to detect cotton leaf diseases efficiently using image processing techniques. A tool implemented in MATLAB is used to detect types of diseases on leaf. A robot is used to spray the pesticides once the disease is detected.

To outline, the existing literature fails to focus on disease severity, accuracy of results, and time complexity. In order to overcome from the above issues, we propose a disease detection model.

3 Proposed Framework

Figure 1 represents the proposed system, which consists of six phases, namely image acquisition, image enhancement, preprocessing, segmentation, feature extraction, classification, and the seventh step is medicine spraying. In the proposed framework, image is acquired through the camera which is associated with the robot. The robot movement is controlled by the Android application through Bluetooth. On the other hand, it also controls the camera movement to capture the images. The view that is created by the camera is shown on the system to catch the proper parts of the plant.

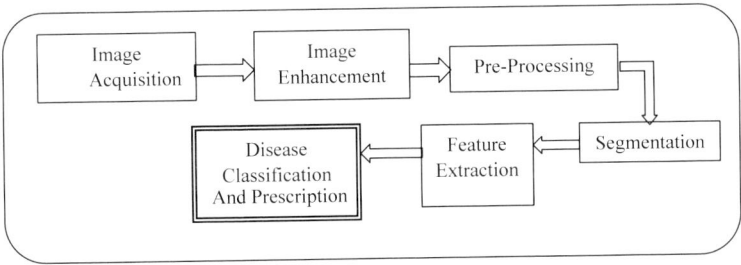

Fig. 1 Block diagram of proposed automated medicine prescription system

In the preprocessing step, the input RGB image is changed over into grayscale image. The contrast of the image is improved using adaptive histogram equalization. After the enhancement, noise is removed from the input image. The resultant images are utilized for further processing. K-means clustering algorithm is used to form different clusters. In this method the images are divided into three clusters in which cluster containing affected region is considered for further processing. Gray-level Co-occurrence Matrix is utilized as a part of the proposed model for the feature extraction, which extracts 13 features including mean, standard deviation, entropy, RMS, variance, smoothness, kurtosis, skewness, IDM, contrast, correlation, energy, and homogeneity, which are used for the classification process. SVM algorithm is used in this method to classify the disease. It maps the extracted features with the database stored to identify the disease type. Once the disease type is identified, its severity can be identified by using the following formula,

$$\text{Disease severity} = \text{Affected pixels}/\text{Total pixels} \qquad (1)$$

Once the disease severity is identified, then it is compared with the database information to identify the pesticide and quantity of pesticides to be sprayed over the affected region of the plant. The same information shall be forwarded to the robot, which will find the required pesticide and spray it over the affected region.

4 Experimental Setup

Figure 2 depicts the experimental setup of different components of the system and their respective connections. Arduino Uno-based control system handles the activities of the robot and pesticide spraying. Arduino is connected to servo motor to control the camera.

Arduino receives the instruction from the Android application through Bluetooth. L293D motor driver is connected to the Arduino to handle the robot movement. Relay is connected to various pesticides which handles the spraying system.

5 Results and Discussion

Figure 3 represents the snapshot of preprocessing of diseased leaf images. The captured leaf image is loaded and its contrast is enhanced. In visual perception, contrast is determined by the difference in the color and brightness of the object with other objects. After this step image segmentation takes place, in which the leaf image is divided into multiple clusters and the cluster that is needed for the processing is selected. Figure 4 shows the snapshot of clustering of the leaf images. Clustering is used to separate the group of objects. It also helps to separate the healthy region of the image from the diseased one. One resultant image is selected to identify the type

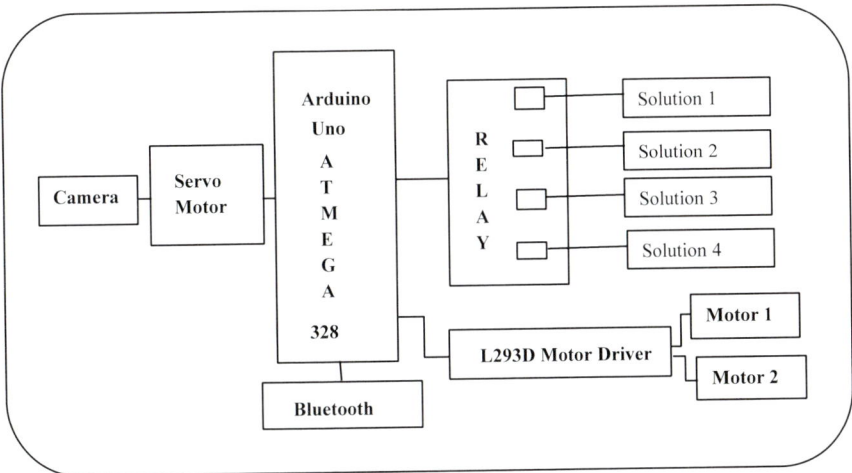

Fig. 2 Experimental setup

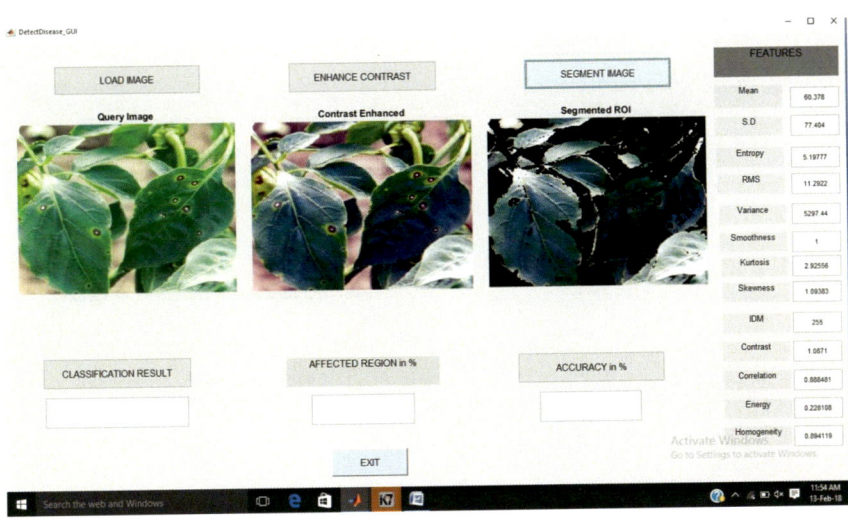

Fig. 3 Snapshot showing preprocessing of diseased leaf

of disease. In Fig. 4, the original image which is divided into three clusters and the cluster that contains the diseased part is selected for the further processing. Figure 5 shows the snapshot of disease severity details of the diseased leaf image. Precisely, features that are necessary for the identification of the disease are extracted from the segmented image. These features are shown in Fig. 5.

Once the features are extracted, disease is identified in Fig. 5 and it identifies the disease which is popularly known as Cercospora leaf spot and its severity is

Fig. 4 Snapshot showing clustering of the diseased leaf image

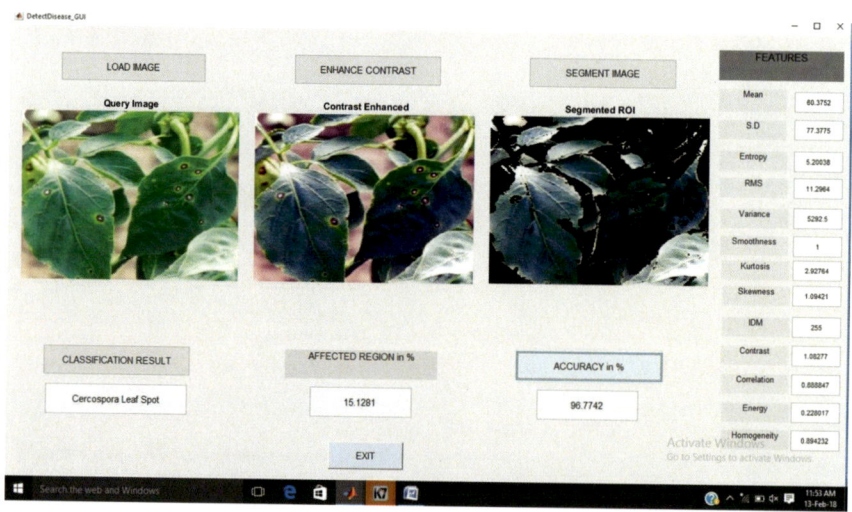

Fig. 5 Snapshot showing disease severity details of the diseased leaf image

15.1281%. Accuracy of the identified leaf disease is 96.77%, which illustrates the good performance of the proposed framework.

6 Conclusion

The proposed framework is useful for farmers and horticulturists by providing useful information related to plant leaf diseases and their identification techniques. The proposed system, can also control, schedule, and monitor all the agricultural recourses, which offers a great aid to reduce the time and cost to the farmer. This system helps the farmers, by reducing their efforts and also increases the production of crops. This method can be further enhanced, by incorporating the features such as multiple plant multi-disease detection system and developing a system for long-distance disease detection.

References

1. Agriculture in India: Information about Indian agriculture its importance. http://www.ibef.org/industry
2. Factors that affect the distribution of agriculture. https://www.s-cool.co.uk/a-level/geography/agriculture/revise-it/factors-that-affect-the-distribution-of-agriculture
3. The impact of plant disease on food security. www.mdpi.com/journal/agriculture/special_issues/plant_disease
4. Singh V, Misra AK (2017) Detection of plant leaf diseases using image segmentation and soft computing techniques. Elsevier Inf Process Agric 4(1):41–49
5. Pujari JD, Yakkundimath R, Byadgi AS (2015) Image processing based detection of fungal diseases in plant. In: Elsevier international conference and communication technologies (ICICT), pp 1802–1808
6. Khirade SD, Patil AB (2015) Plant disease detection using image processing. In: Proceedings of international conference on computing communication control and automation, pp 768–771
7. Revathi P, Hemalatha M (2012) Classification of cotton leaf spot diseases using image processing edge detection techniques. In: Proceedings of IEEE international conference on emerging trends in science, engineering and technology, pp 169–173
8. Husin Z, Md Shakaff AYB, Aziz AHBA (2013) Feasibility study on plant chilli disease detection using image processing techniques. In: Proceedings of international conference intelligent systems, modelling and simulation. IEEE, New York, pp 291–296
9. Sanjay BD, Nitin PK (2013) Agricultural plant leaf disease detection using image processing. Int J Adv Res Electr Electron Instrum Eng 2(1):599–602
10. Camargo A, Smith JS (2009) An image-processing based algorithm to automatically identify plant disease visual symptoms. Elsevier Biosyst Eng 17(1):9–21
11. Cui D, Zhang Q, Li M, Hartman GL, Zhao Y (2010) Image processing methods for quantitatively detecting soybean rust from multispectral images. Elsevier Biosyst Eng 22(4):186–193
12. Tian Y, Zhang L (2012) Study on the methods of detecting cucumber downy mildew using hyperspectral imaging technology. Elsevier Int Conf Phys Biomed Eng 11(5):743–750
13. Mude S, Naik D, Patil A (2017) Leaf disease detection using image processing for pesticide spraying. Int J Adv Eng Res Dev 4(4):1–5
14. Husin ZB, Md Shakaff AYB, Aziz AHBA, Farook SM (2012) Plant chilli disease detection using the RGB color model. In: Proceedings of research notes in information science (RNIS), pp 291–296

A Comparative Assessment of Segmentations on Skin Lesion Through Various Entropy and Six Sigma Thresholds

Srinivasan Sankaran, Jason R. Hagerty, Muthukumaran Malarvel, Gopalakrishnan Sethumadhavan and William V. Stoecker

Abstract We present four entropy-based methods for colour segmentation within a lesion in a dermoscopy image for classification of the image as melanoma or benign. Four entropy segmentation methods are based on Tsallis, Havrda and Charvat, Renyi and Kapur entropy measures. Segmentation through Six Sigma threshold as preprocessor is also evaluated by this assessment approach. The proposed methods are inspired by two clinical observations about melanoma. First, colours within a lesion provide the most useful measures for melanoma detection; second, the disorder in colour variety and arrangement provides the best assessment of melanoma colours. These observations lead to the hypothesis that colour disorder is best measured by entropy. The five different models for colour splitting studied with SSIM measures taken from each region in the colour-split image for segmentation assessment. Based on the score helps to understand segmentation region assessment effectively.

1 Introduction

Melanoma causes most skin cancer deaths, with an estimated 9730 people expected to die of melanoma in the United States in the year 2017 [1]. Melanoma is fully curable if diagnosed early. The number of benign lesions biopsied exceeds melanomas by over a factor of 10, yet many cases of melanoma are incorrectly diagnosed by dermatologists [2–4].

The dermoscopy imaging method has been reported to be an important tool in the early detection of melanoma [5–8]. Studies have shown that dermoscopy increases the diagnostic accuracy over clinical visual inspection in the hands of experienced

S. Sankaran · M. Malarvel · G. Sethumadhavan (✉)
SASTRA Deemed University, Tirumalaisamudram, India
e-mail: sgk@mca.sastra.edu

J. R. Hagerty · W. V. Stoecker
S&A Technologies, Rolla, MO, USA

S. Sankaran
HCL Technologies Limited, Chennai, India

© Springer Nature Switzerland AG 2019
D. Pandian et al. (eds.), *Proceedings of the International Conference on ISMAC in Computational Vision and Bio-Engineering 2018 (ISMAC-CVB)*, Lecture Notes in Computational Vision and Biomechanics 30,
https://doi.org/10.1007/978-3-030-00665-5_19

179

physicians [9–11]. Even higher diagnostic accuracy has been reported recently for automated digital image analysis [3, 4].

Colours are believed to provide the most critical features in melanoma diagnosis, especially when combined with a location in the lesion to capture colour chaos [12–14]. Analytic colour descriptors usually employ the red, green and blue (RGB) colour space. Of all analytic descriptors for melanoma region detection, these colours were evidenced as the most significant descriptors. Ferris et al. evidenced that the top three features through statistical derivatives for melanoma were all colour descriptors from colour histograms and colour asymmetry [4]. Rubegni et al. found that the top critical feature for melanoma detection was red asymmetry [15], capturing the "colour island" chaos [16]. Other colour features for melanoma detection are a variation of hues [17], analytical colour analysis of variegation [18], RGB colour channel statistical parameters [19], spherical and La*b* colour coordinate features [20], and the number of colours of concern present within the skin lesion [21]. Colour quantization was performed using the median split colour algorithm [22]. Crisp and fuzzy colour histogram techniques were used to identify melanoma colours [23, 24]. The improved colour contrast was used to better separate colours [25]. A colour palette mimicking that used by dermatologists was developed [26].

Based on Otsu and various entropy functions, image segmentation at each possible threshold in the grey image is investigated through randomness measure for its evaluation [27]. The idea of multilevel segmentation was a key item conceived through this paper. Based on the analysis, results show an outperforming behavior of generalized entropy measure of Havrda function in comparison with any other existing entropy functions. Sankaran et al. [28] developed segmentation using Six Sigma for malarial parasitaemia image and derived infectious cells based on Kapur entropy measure.

In this paper, four entropy-based colour splitting and one statistical method to capture colour variety and arrangement are presented. The rest of the paper is organized as follows. Section 2 explains the entropy-based colour segmentation algorithms used in this research. Section 3 discusses segmentation generated using Six Sigma as a threshold. Section 4 explains the assessment methodology. Section 5 elaborates about tool and methodology applied. Section 6 describes the results and discussion.

2 Entropy-Based Threshold Segmentations

A total of four different entropy-based colour segmentation algorithms are used in this study. Entropy is a measure of uncertainty associated with information of a source. Secretion of melanin in the affected skin region is being measured using the amount of colour variation present in that region; and this colour distortion is also highly uncertain due to the inherent property of the disease. In general, distribution of uncertainty data can be assessed through probability distribution using the normalized histogram on various colour channels. Based on the colour distortion of colour channels in melanoma images, various entropy methods guide to derive appropriate threshold to segment the lesion region in the image. This uncertainty in colour spread information on various channels is sourced for Tsallis entropy [29], which is defined as

$$S_\alpha = \frac{1}{\alpha - 1}\left(1 - \sum_{i=1}^{n} p_i^\alpha\right) \tag{1}$$

where P_i is the normalized histogram defined as $P_i = n_i/n$ and n be the total number of pixels and n_i be the number of pixels with grey value.

Let $f(x,y)$ be the image of spatial resolution $M \times N$ and bit level resolution k with the grey scale $[0, L-1]$ where $L = 2k$. Let, $t = 0$ to $L - 1$ be the probability distribution with the pixel values ranging from 0 to t as object pixels (O) and the pixels values ranging from $t+1$ to $L - 1$ as background pixels (B). Then,

$$H(O,t) = -\sum_{i=0}^{t} P_i \ln(P_i) \tag{2}$$

$$H(B,t) = -\sum_{i=t+1}^{L-1} P_i \ln(P_i) \tag{3}$$

Image histogram carries important information about the content of an image and can be used for discriminating the abnormal tissue from the local unhealthy background. Havrda and Charvat have defined entropy [30] of a discrete finite probability distribution as follows:

$$H = \frac{1}{1-\alpha}\left(\sum_{i=0}^{L-1} P_i^\alpha - 1\right) \tag{4}$$

where α is a positive real parameter used to derive suitable threshold using this nonlinear equation. Thus, a one-parameter generalization of Shannon's entropy is Havrda and Charvat entropy. Also, it can be shown that Shannon's entropy is the limiting case of Renyi's entropy, when $\alpha \to 1$.

$$H(O,t) = \sum_{i=0}^{t} \frac{1}{1-\alpha}\left(\sum_{i=0}^{t} P_i^\alpha - 1\right) \tag{5}$$

$$H(B,t) = \sum_{i=t+1}^{L-1} \frac{1}{1-\alpha}\left(\sum_{i=t+1}^{L-1} P_i^\alpha - 1\right) \tag{6}$$

If *foreground* and *background* pixels have the same grey level histogram, then the parameter α is critical in finding Havrda and Charvat and Renyi entropy thresholds.

The typical melanoma image has colour variation in both the lesion and background skin. The Kapur threshold [31] divides the image histogram into two probability distributions, one for lesion the other for the background. The Kapur entropy threshold maximizes the sum of entropies of background and lesion.

$$H(O, t) = -\sum_{i=0}^{t} \frac{p_i}{P_t} \ln \frac{p_i}{P_t} \tag{7}$$

$$H(B, t) = -\sum_{i=t+1}^{L-1} \frac{p_i}{1 - P_t} \ln \frac{p_i}{1 - P_t}; \quad \text{where } P_t = \sum_{i=0}^{t} p_t \tag{8}$$

The Rényi [32] entropy is important in statistics as it provides an index of diversity, which can measure the greater number of colours found in melanomas [14, 26]. Renyi entropy sums the probability of pixel grey levels for foreground and background. The probabilities of the pixels at each grey level are combined, as in Eq. (9) using mathematical operations. The final threshold is the maximum priori combination value.

$$R = \frac{1}{1 - \alpha} \ln \left(\sum_{i=0}^{L-1} P_i^\alpha \right) \tag{9}$$

where $\alpha (\neq 1)$ is a positive real parameter which proportionately determines the suitable threshold value.

$$H(O, t) = \sum_{i=0}^{t} \frac{1}{1 - \alpha} \ln \left(\sum_{i=0}^{t} P_i^\alpha \right) \tag{10}$$

$$H(B, t) = \sum_{i=t+1}^{L-1} \frac{1}{1 - \alpha} \ln \left(\sum_{i=t+1}^{L-1} P_i^\alpha \right) \tag{11}$$

$E(t)$ can be derived as the sum of $H(O,t)$ and $H(B,t)$. And the corresponding maximum value of $E(t)$ received as l_E^* where $E = T_S$, HC, K and R are the selected Tsallis, Havrda and Charvat, Kapur and Renyi entropy-based threshold, respectively. Equations 12 and 13 states the procedure in getting these suitable thresholds.

$$E(t)(Ts(t) = HC(t) = K(t) = R(t) = H(O, t) + H(B, t) \tag{12}$$

$$l_E^* = \max_{0 \le t \le L-1} E(t) \tag{13}$$

3 Segmentation Through Six Sigma Threshold

Dr. W. A. Shewhart proposed that variability does exist on all repetitive process through Control Chart. This methodology is applied to lesion region where variability does exist. For this purpose, to generate samples, the image is split into a rectangular window of sizes 4×4, 4×5, 5×4 and 5×5 into four different patterns. Based on variable $\bar{X}$, Control Chart equations on each window splitting; Upper Control Limit $(UCL = \bar{\bar{x}} + A_2 \bar{R})$, Centre line $(CL = \bar{\bar{x}})$ and Lower Control Limit $(LCL = \bar{\bar{x}} - A_2 \bar{R})$ where $\bar{R}$ is average range of samples; $\bar{x}$ is population mean and A_2 [33] is constant, applied [28, 34] to derive threshold which is used on the image for segmentation.

4   Assessing Segmented Regions Through SSIM Index

Mean Structural SIMilarity (SSIM) measures the image quality by computing the similarity between images. SSIM [35] is designed to improve the traditional methods such as peak signal-to-noise ratio (PSNR) and mean squared error (MSE). The resultant SSIM index is a decimal value. SSIM is calculated between the two windows, F and G, of common window size $n \times n$. The common window size is set as 3×3 in this study. A higher value of SSIM indicates that the resultant image has more similarity to the original image. Also, the formulation in Eq. (14) given below satisfies a number of properties

(a)   Guarantees **symmetry**, meaning that $SSIM(F, G) = SSIM(G, F)$; thus also ensure post swapping on the original and coded images
(b)   Ensures **boundedness**, in the sense that $SSIM(F, G) < 1$.
(c)   Unique **maximum**, meaning that $SSIM(F, G) = 1$ if and only if $F = G$.

$$SSIM(F, G) = \frac{(2\mu_F \mu_G + C_1)(2\sigma_{FG} + C_2)}{(\mu_F^2 + \mu_G^2 + C_1)(\sigma_F^2 + \sigma_G^2 + C_2)} \tag{14}$$

where μ_F is the average of F; μ_G is the average of G; α_F^2 and α_G^2 are the variance of F and G respectively; σ_{FG} is the covariance; C_1 and C_2 are stabilizing parameters. Generally, $C_1 = 0.05$ and $C_2 = 0.05$. Recall that the covariance between F and G is defined as

$$\sigma_{FG} = \frac{1}{L-1} \sum_{i=0}^{L-1} (F_i - \mu_F)(G_i - \mu_G) \tag{15}$$

Their mean SSIM is defined as

$$SSIM(F, G) = \frac{1}{W} \sum_{j=1}^{W} (F_j G_j) \tag{16}$$

where W is the number of local windows of size 3×3 in the image.

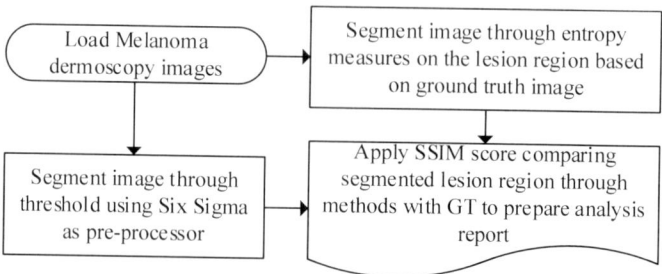

Fig. 1 Methodology for generating segmentation and assessment through SSIM

5 Tools and Methodology

For malignant melanoma, appropriate imaging of skin lesions is visioned as a priority role for diagnostic aid. In this proposed method uses digitized dermoscopy skin lesion images in RGB colour space with a similar size of 1024×768 pixels. The image set composed from 1636 digital dermoscopy images from by Dr. William V. Stoecker, Stoecker & Associates, USA. Images obtained using non-polarized contact dermoscopy techniques. These images may contain hair which was preprocessed using SharpRazor, a state-of-the-art tool. The methodology for generating results is explained in Fig. 1.

6 Results and Discussion

Segmented region through entropy and Six Sigma was assessed through SSIM, it is based on their score results with respect to image plane listed in Table 1. It was evidenced that for all the entropy measure their SSIM score falls for a maximum number

Table 1 Based on 1636 melanoma images samples and their related SSIM score on various segmented methods aligned on similar value ranges

SSIM score	Kapur	Harda and Charvat	Renyi	Tsallis	Six Sigma
Within >=0 and <=0.1	56	44	5	17	3
Within >0.1 and <=0.2	56	72	21	78	9
Within >0.2 and <=0.3	88	95	67	143	18
Within >0.3 and <=0.4	190	181	160	197	97
Within >0.4 and <=0.5	308	321	300	342	245
Within >0.5 and <=0.6	*371*	*359*	*384*	*354*	360
Within >0.6 and <=0.7	265	273	332	247	**396**
Within >0.7 and <=0.8	202	199	246	181	305
Within >0.8 and <=0.9	90	83	110	70	174
Within >0.9 and <=1	10	9	11	7	29

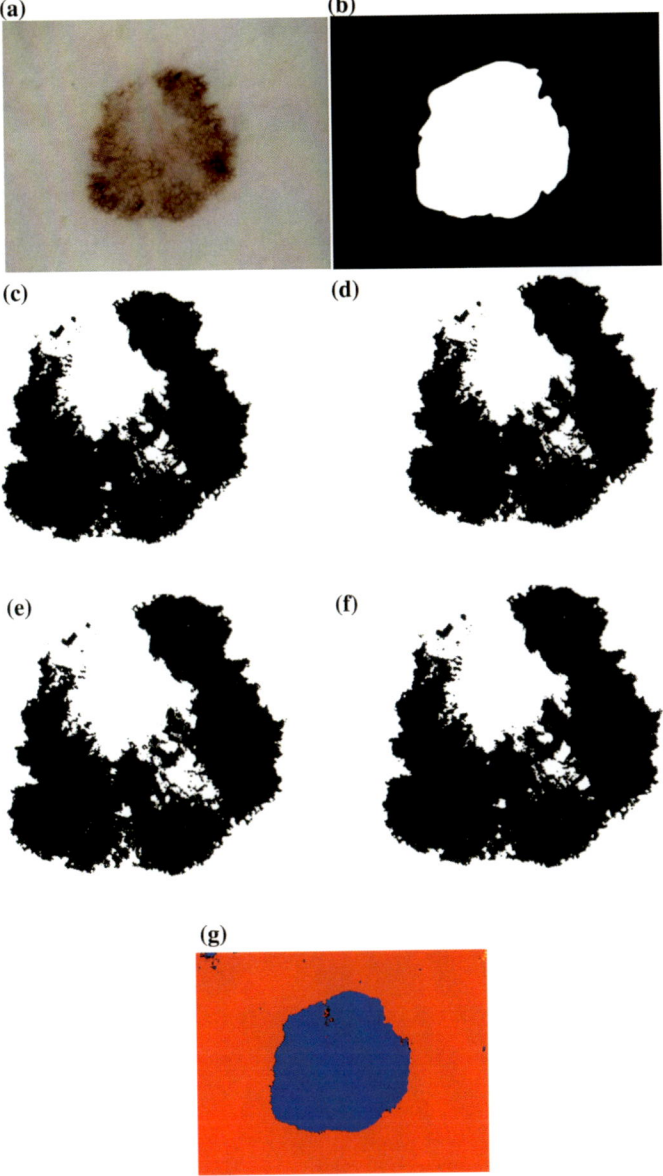

Fig. 2 Segmented image results on image post applying threshold through different methods. **a** Malignant melanoma dermoscopy image; **b** Manual marked lesion region (GT) image; segmented image through **c** Havrda and Charvat entropy; **d** Kapur entropy; **e** Renyi entropy **f** Tsallis entropy and **g** Segmented based on Six Sigma based threshold as preprocessor

of images within $\{> 0.5, \leq 0.6\}$. In case if Six Sigma score falls maximum within $\{> 0.6, \leq 0.7\}$ which shows that Six Sigma tries to give better optimized segmented region in comparison with Ground Truth (GT) image than the entropy measures. Figure 2 demonstrates sample malignant melanoma image, its expert identified GT and corresponding segmented results through proposed methods. Future work was planned to assess the applicability of various thresholds for detecting the internal borders with metrics for quantifying the spreading of disease. Results of various samples and the executable codes are made available at https://drive.google.com/open?id=1Z0a9frSZB2YImYHld3qvWi38QBHlNKNt.

Acknowledgements This publication was made possible by SBIR Grants R43 CA153927-01 and CA101639-02A2 of the National Institutes of Health (NIH). Its contents are solely the responsibility of the authors and do not necessarily represent the official views of the NIH. Further, the authors would like to acknowledge the support rendered by the management of SASTRA Deemed to be University, Tirumalaisamudram, India.

This work is a by-product of generic image processing tool named "Bhadraloka" being developed (Dot Net platform) in SASTRA for the project titled "Development of techniques for processing radiographic images for automated detection of defects" with the funding assistance from Board of Research in Nuclear Science (BRNS), Department of Atomic Energy, Government of India (No. 2013/36/40-BRNS/2305).

Also, SS would like to thanks HCL Technologies Limited in supporting all through this research work.

References

1. Siegel RL, Miller KD, Jemal A (2017) Cancer statistics. CA Cancer J Clin 64(1):9–29
2. Weinstein DA, Konda S, Coldiron BM (2017) Use of skin biopsies among dermatologists. Dermatol Surg 43(11):1348–13657
3. Esteva A, Kuprel B, Novoa RA et al (2017) Dermatologist-level classification of skin cancer with deep neural networks. Nature 542(7639):115–118
4. Ferris LK, Harkes JA, Gilbert B et al (2015) Computer-aided classification of melanocytic lesions using dermoscopic images. J Am Acad Dermatol 73(5):769–76
5. Pehamberger H, Binder M, Steiner A et al (1993) In vivo epiluminescence microscopy: improvement of early diagnosis of melanoma. J Invest Dermatol 100:356S–362S
6. Soyer HP, Argenziano G, Chimenti S, Ruocco V (2001) Dermoscopy of pigmented skin lesions. Eur J Dermatol 11(3):270–277
7. Soyer HP, Argenziano G, Talamini R, Chimenti S (2001) Is dermoscopy useful for the diagnosis of melanoma? Arch Dermatol 137(10):1361–1363
8. Stolz W, Braun-Falco O, Bilek P, Landthaler M, Burgdorf WHC, Cognetta AB (eds) (2002) Color atlas of dermatoscopy. Wiley-Blackwell, Hoboken
9. Braun RP, Rabinovitz HS, Oliviero M, Kopf AW, Saurat JH (2002) Pattern analysis: a two-step procedure for the dermoscopic diagnosis of melanoma. Clin Dermatol 20(3):236–239
10. Boldrick JC, Layton CJ, Nguyen J, Swetter SM (2007) Evaluation of digital dermoscopy in a pigmented lesion clinic: clinician versus computer assessment of malignancy risk. J Am Acad Dermatol 56(3):417–421
11. Perrinaud A, Gaide O, French LE, Saurat JH, Marghoob AA, Braun RP (2007) Can automated dermoscopy image analysis instruments provide added benefit for the dermatologist? A study comparing the results of three systems. Br J Dermatol 157(5):926–933

12. Mishra NK, Celebi ME (2016) An overview of melanoma detection in dermoscopy images using image processing and machine learning. arXiv preprint arXiv:1601.07843
13. Rosendahl et al (2012) Dermatoscopy in routine practice: 'Chaos and clues'. Aust Fam Physician 41(7):482
14. Friedman RJ et al (1985) Early detection of malignant melanoma: the role of physician examination and self-examination of the skin. CA Cancer J Clin 35(3):130–151
15. Rubegni P et al (2015) Computer-assisted melanoma diagnosis: a new integrated system. Melanoma Res 25(6):537–542
16. Andreassi L et al (1999) Digital dermoscopy analysis for the differentiation of atypical nevi and early melanoma: a new quantitative semiology. Arch Dermatol 135(12):1459–1465
17. Landau M et al (1999) Computerized system to enhance the clinical diagnosis of pigmented cutaneous malignancies. Int J Dermatol 38(6):443–446
18. Umbaugh SE et al (1989) Automatic color segmentation of images with application to detection of variegated coloring in skin tumors. Eng Med Biol Mag IEEE 8(4):43–50
19. Aitken JF et al (1996) Reliability of computer image analysis of pigmented skin lesions of Australian adolescents. Cancer 78(2):252–257
20. Ercal F et al (1994) Neural network diagnosis of malignant melanoma from color images. IEEE Trans Biomed Eng 41(9):837–845
21. Ganster H et al (2001) Automated melanoma recognition. IEEE Trans Med Imaging 20(3):233–239
22. Kaushik RHC et al. (2013) The median split algorithm for detection of critical melanoma color features. In: International conference on computer vision theory and applications (VISAPP), pp 492–495
23. Stanley RJ et al (2007) A relative color approach to color discrimination for malignant melanoma detection in dermoscopy images. Skin Res Technol 13(1):62–72
24. Almubarak HA et al (2017) Fuzzy color clustering for melanoma diagnosis in dermoscopy images. Information 8(3):89
25. Sabbaghi Mahmouei SA et al. (2015) An improved colour detection method in skin lesions using colour enhancement. In: Australian biomedical engineering conference (ABEC 2015)
26. Madooei A et al (2013) A colour palette for automatic detection of blue-white veil. In: Color and imaging conference, vol 2013, no 1, pp 200–205
27. Tiwari R, Sharma B (2016) A comparative study of Otsu and entropy based segmentation approaches for lesion extraction. In: Conference: 2016 international conference on inventive computation technologies (ICICT)
28. Sankaran S, Malarvel M, Sethumadhavan G, Sahal D (2017) Quantitation of malarial parasitemia in giemsa stained thin blood smears using six sigma threshold as preprocessor. Optik Int J Light Electr Opt 145:225–239. ISSN 0030-4026, http://dx.doi.org/10.1016/j.ijleo.2017.07.047
29. Comparison of Shannon, Renyi and Tsallis Entropy used in Decision Trees, Tomasz Maszczyk and Wlodzislaw Duch
30. Ja Havrda, František Charvát (1967) Quantification method of classification processes: concept of structural a-entropy. Kybernetika 03(1):30–35
31. Kapur JN, Sahoo PK, Wong AKC (1985) A new method for Gray-level picture thresholding using the entropy of the histogram. Comp Vis Graphics Image Process 29:273–285. https://doi.org/10.1016/0734-189X(85)90125-2
32. Rényi A (1961) On measures of entropy and information. In: Proceedings of the fourth Berkeley symposium on mathematical statistics and probability, Volume 1: contributions to the theory of statistics, pp 547–561. University of California Press, Berkeley, California. http://projecteuclid.org/euclid.bsmsp/1200512181
33. http://web.mit.edu/2.810/www/files/readings/ControlChartConstantsAndFormulae.pdf
34. Sankaran S, Sethumadhavan G (2013) Quantifications of asymmetries on the spectral bands of MALIGNANT melanoma using six sigma threshold as preprocessor. In: Third international conference on computational intelligence and information technology (CIIT 2013), Mumbai, pp 80–86. https://doi.org/10.1049/cp.2013.2575

35. Wang Z, Bovik AC, Sheikh HR, Simoncelli EP (2004) Image quality assessment: from error visibility to structural similarity. IEEE Trans Image Process 13(4):600–612. https://doi.org/10.1109/tip.2003.819861

Impact of Speckle Filtering on the Decomposition and Classification of Fully Polarimetric RADARSAT-2 Data

Sivasubramanyam Medasani and G. Umamaheswara Reddy

Abstract Decomposition and classification are vital processing stages in polarimetric synthetic aperture radar (PolSAR) information processing. Speckle noise affects SAR data since backscattered signals from various targets are coherently integrated. Current study investigated the impact of speckle suppression on the target decomposition and classification of RADARSAT-2 fully polarimetric data. Speckle filters should suppress the speckle noise along with the retention of spatial and polarimetric information. The performance of improved Lee–Sigma, intensity-driven adaptive neighborhood (IDAN), refined Lee, and boxcar filters were assessed utilizing the spaceborne dataset, that is, fully polarimetric RADARSAT-2 C-band SAR data for the Mumbai region, India. The effect of speckle suppression on target decomposition was analyzed in this study. Different speckle noise suppression techniques were applied to RADARSAT-2 dataset, followed by Yamaguchi three-component and VanZyl decompositions. The obtained findings revealed that the improved Lee–Sigma filter demonstrated better volume scatterings in forest areas and double bounce in urban areas than the other techniques considered in the analysis. Additionally, the efficacy of the different speckle suppression techniques listed above was assessed. The effectiveness of the speckle filtering algorithm was evaluated by applying the Wishart supervised classification to the filtered and unfiltered data. IDAN, boxcar, refined Lee, and improved Lee–Sigma filters were assessed to find the classification accuracy improvement. A considerable amount of improvement was observed in the classification accuracy for mangrove and forest classes. Minimal enhancement was detected for settlement, bare soil, and water classes.

S. Medasani (✉) · G. Umamaheswara Reddy
Department of Electronics and Communication Engineering, Sri
Venkateswara University College of Engineering, Sri Venkateswara University,
Tirupati, Andhra Pradesh, India
e-mail: medasani7@gmail.com

G. Umamaheswara Reddy
e-mail: umaskit@gmail.com

© Springer Nature Switzerland AG 2019
D. Pandian et al. (eds.), *Proceedings of the International Conference on ISMAC in Computational Vision and Bio-Engineering 2018 (ISMAC-CVB)*, Lecture Notes in Computational Vision and Biomechanics 30,
https://doi.org/10.1007/978-3-030-00665-5_20

189

1 Introduction

Nowadays SAR imaging, a coherent technique for microwave remote sensing, is used for producing large-scale two-dimensional surface reflectivity images of the Earth; these images are of high resolution. Solar illumination does not influence the performance of microwave remote sensing sensors because of the active operation mode; thus, images can be taken during the day or night. Furthermore, the effects of smoke, fog, clouds, and rain can be avoided by working in the microwave frequencies. Emerging polarimetric SAR (PolSAR) [1, 2] images increase the applicability of SAR remote sensing and provide an additional dimension to the SAR information. PolSAR images have various applications including obtaining information of soil moisture; determining surface roughness, mapping snow and ice; and classifying land cover. These obtained factors are used in various areas including forestry, urbanism, agriculture, industry, and geology.

Currently, using polarimetry for radar remote sensing has achieved substantial attention. The use of detected microwave scattering by volume and surface structures to extract the physical information is essential for radar remote sensing applications. The coherency matrix $[T]$ of dimension 3×3 is the most crucial matrix measured by polarimetric SAR systems. This matrix illustrates the local differences observed in the scattering matrix. Moreover, an operator of the lowest order is appropriate for extracting polarimetric parameters of distributed scatterers when either additive or multiplicative noise or both are present. In radar remote sensing, a multiplicative statistical description is needed by several targets of concern because of the combined random vector scattering effects and coherent speckle noise from volume and surface. To achieve the classification or inversion of scatter data of such targets, a scattering mechanism that is dominant or average should be generated. The aforementioned averaging procedure engenders the notion of distributed targets, which exhibit a unique structure, unlike pure single or stationary targets.

The most crucial application of PolSAR is the terrain and land-use classification. For the supervised and unsupervised classification of terrains, several algorithms are developed. Moreover, for the supervised classification of terrains, training sets were selected for each class on the basis of scattering contrast differences or ground truth maps observed in PolSAR images. The PolSAR response for all images was ingrained in three complex and three real parameters—total nine parameters. Because of the high dimensions of PolSAR data, the selection of training sets may become difficult in case the ground truth maps are unavailable. In contrast, the classification of the unsupervised terrain involves the automatic classification of images by finding clusters on the basis of a specific criterion. Classification methods based on information obtained from the scattering mechanisms include the support vector machine, eigenvalue-based or model-based decompositions, neural networks, and maximum likelihood algorithm with the complex Wishart distribution of the coherency or covariance matrix. This study examined the effect that speckle suppression has on the target decomposition and classification of fully polarimetric RADARSAT-2 data.

Table 1 Specifications of RADARSAT-2

Frequency band	Mode	Incidence angle (°)	Date of acquisition	Polarization	Slant range resolution (m)
C	FQ22 (Ascending right)	41.7	February, 16 2011	Quad Pole	8.0

FQ fine quad

Section 2 provides RADARSAT-2 specifications. Section 3 discusses the procedure of the proposed work; Sect. 4 explains the results. At the end of this paper, Sect. 5 provides the overall conclusions.

2 Dataset and Study Area

The spaceborne dataset considered in the current study constitutes fully polarimetric RADARSAT-2 C-band SAR data obtained for Mumbai. Specifications of the RADARSAT-2 sensor are given in Table 1.

3 Methodology

3.1 Assessment of Speckle Suppression Techniques by Using Fully Polarimetric RADARSAT- 2 Data

Speckle noise affects SAR information since backscattered signals from various targets are coherently integrated. For visual and quantitative evaluation of speckle filters, refined Lee, boxcar, intensity-driven adaptive neighborhood (IDAN), and improved Lee–Sigma filters were applied to the fully polarimetric RADARSAT-2 C-band SAR data (Fig. 1). The aforementioned filters were quantitatively compared using various parameters [3–5] such as the edge preservation index (EPI), mean square error (MSE), speckle suppression index (SSI), peak signal-to-noise ratio (PSNR), bias, standard deviation to mean ratio (SD/M), and mean and standard deviations of the ratio image.

3.2 Impact of Speckle Suppression on the Target Decomposition

Target decomposition techniques [1, 6] are classified on the basis of the following factors: (i) the coherent decomposition of the scattering matrix (Krogager and

Fig. 1 Assessment of various speckle suppression techniques

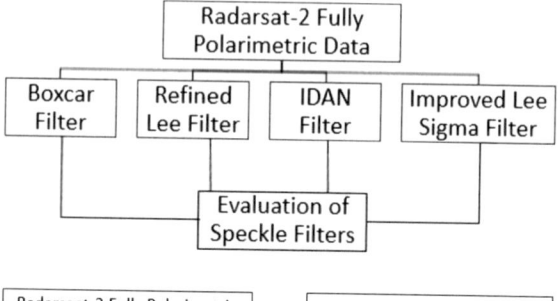

Fig. 2 Impact of speckle suppression on the target decomposition

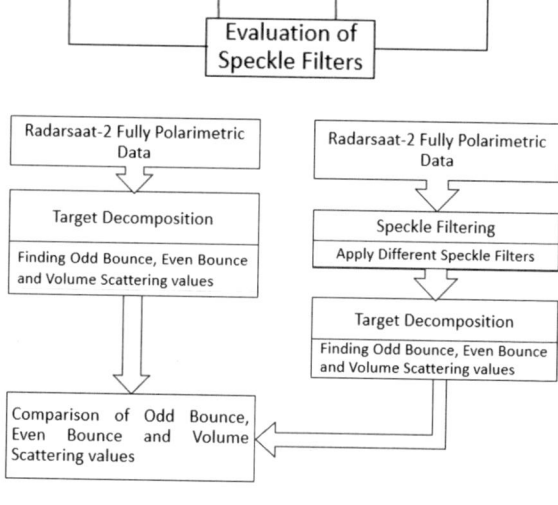

Cameron), (ii) the dichotomy of the Kennaugh matrix (Huynen and Barens), (iii) the model-based decomposition of the coherency or covariance matrix (Freeman and Durden, and Dong), and (iv) the analysis of eigenvalue or eigenvector of the coherency or covariance matrix (Cloude, VanZyl, Cloude, and Pottier). In this study, the Yamaguchi three-component [7, 8] and VanZyl decompositions [9] were applied for analyzing the speckle filtering impact on decomposition. In [10] various speckle suppression techniques are evaluated using hybrid polarimetric data (RISAT-1) and also the impact of speckle suppression on the target decomposition of hybrid polarimetric data is presented. Figure 2 presents the block diagram of the method employed to determine the influence of speckle suppression on the target decomposition of fully polarimetric RADARSAT-2 data.

3.3 Impact of Speckle Suppression on Classification

The most crucial application of PolSAR is the land-use and terrain classification. Several algorithms have been generated for the unsupervised and supervised terrain classification. The PolSAR classifications, both unsupervised and supervised, are generally influenced by averaging or filtering processes such as the multi-look technique that is employed to reduce speckle noise. The Wishart classification [11–13] is a renowned technique utilized to obtain the supervised classification of PolSAR

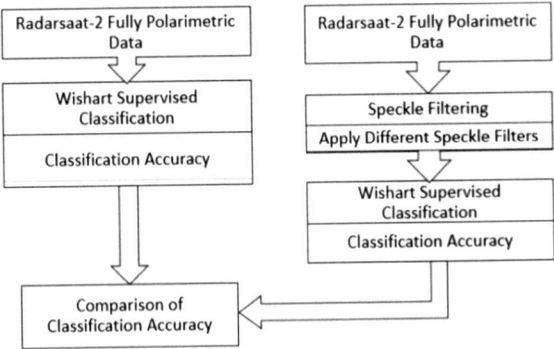

Fig. 3 Impact of speckle suppression on classification

image data. The size of the sliding window size [14] for boxcar and refined Lee filters is a crucial parameter for speckle suppression. Speckle filters are evaluated using hybrid polarimetric data (RISAT-1) and also the impact of speckle filtering on the classification of hybrid polarimetric data is presented in [10]. The proposed flowchart for the comparison of the Wishart supervised classification is displayed in Fig. 3.

4 Results and Discussion

4.1 Assessment of Speckle Suppression Techniques by Using Fully Polarimetric RADARSAT- 2 Data

For the visual and quantitative evaluation of the speckle filters, various aforementioned filters were applied to fully polarimetric data for Mumbai. Figure 4 displays the results of RADARSAT-2 C-band data.

After the visual examination, the filtered images were quantitatively compared. Various parameters were used for the quantitative comparison including MSE, PSNR, EPI, SD/M, bias, SSI, and the standard deviation and mean of the ratio image. The highest values of PSNR and EPI represent optimal performance of the filters. Moreover, the lowest values of MSE, SD/M, bias, and SSI indicate optimal performance of the filters. The standard deviation and mean of the ratio image should be as small as possible and approximately equal to one, respectively.

The visual interpretation indicates that the boxcar filter degrades the spatial resolution because it does not include the internal mechanism for preserving the spatial resolution or details. The spatial resolution preservation for the refined Lee filter exhibited improvement; however, a considerable smoothing effect was still remarked in the filtered data. IDAN filter uses adaptive neighborhoods as a spatial support, and

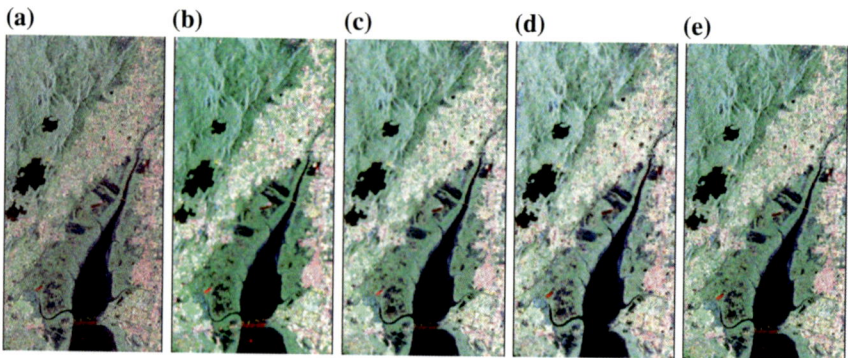

Fig. 4 Assessment of various speckle suppression techniques by using RADARSAT-2 data for the Mumbai region. **a** Original Pauli RGB and **b** boxcar, **c** refined Lee, **d** IDAN, and **e** improved Lee–Sigma filters

Table 2 Assessment of speckle suppression techniques (7 × 7 window) by using RADARSAT-2 data for the Mumbai region

Filter/Parameter	MSE	PSNR	EPI	SD/M	Bias	SSI
Boxcar	208.6176	24.9373	0.1078	0.4149	0.0570	0.8115
Refined Lee	203.6277	25.0424	0.1989	**0.4139**	0.0575	**0.8095**
IDAN	202.4071	25.0685	**0.2190**	0.4154	0.0592	0.8124
Improved Lee–Sigma	**198.8244**	**25.1461**	0.2072	0.4192	**0.0544**	0.8199

Bold text indicates better value or performance

these neighborhoods were derived from the intensity information. Due to the use of adaptive neighborhoods, the speckle noise was substantially reduced. However, the contour and fine details were preserved, thus avoiding the blurring effect. The improved Lee–Sigma filter, where the estimate of the minimum mean square error (MMSE) was used as a priori mean, exhibited an excellent preservation of fine details and strong targets.

The quantitative comparison of the various aforementioned filters employed in this study by using RADARSAT-2 data for the Mumbai region is given in Tables 2 and 3. Table 2 illustrates that the improved Lee–Sigma filter exhibited an improved performance in terms of MSE with the lowest value of 198.8244, PSNR with the highest value of 25.1461, and bias with the lowest value of 0.0544. The IDAN filter had superior performance in terms of EPI and had the highest value of 0.2190 than the other filters considered in this study. SD/M (0.4139) and SSI (0.8095) for the refined Lee filter were the lowest, respectively, compared with the other filters employed. For the ratio image, Table 3 displays that the boxcar filter exhibited the mean value of 0.9546 (~1), and the standard deviation (0.5963) for the improved Lee–Sigma filter was the lowest.

Table 3 Assessment of speckle suppression techniques (7 × 7 window) by using RADARSAT-2 data for the Mumbai region and the ratio image

Filter/Parameter	Mean	Standard Deviation
Boxcar	**0.9546**	0.9133
Refined Lee	0.7994	0.6813
IDAN	0.6891	0.7966
Improved Lee–Sigma	0.8031	**0.5963**

Bold text indicates superior value or performance

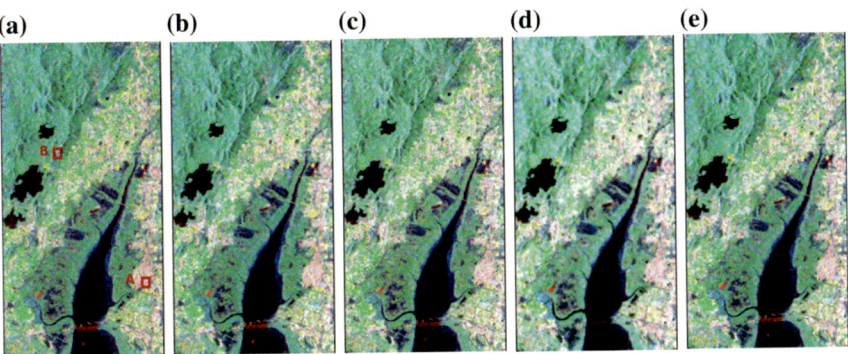

(a)   **(b)**   **(c)**   **(d)**   **(e)**

Fig. 5 Yamaguchi three-component decomposition images before and after employing the various speckle suppression techniques on RADARSAT-2 data for the Mumbai region. **a** Original data and **b** boxcar, **c** refined Lee, **d** IDAN, and **e** improved Lee–Sigma filters

4.2 Impact of Speckle Suppression on the Target Decomposition

Current study examined the influence of despeckling on the target decomposition. The speckle filters, including refined Lee, boxcar, improved Lee–Sigma, and IDAN filters, were applied to the RADARSAT-2 dataset. The Yamaguchi three-component and VanZyl decompositions were then conducted on the dataset. Figures 5 and 6, respectively, display the Yamaguchi three-component and VanZyl decomposition images before and after applying the various speckle suppression techniques on RADARSAT-2 data for the Mumbai region. In Figs. 7 and 8, the percentage of volume, even bounce, and odd bounce scatterings are displayed, respectively, by conducting the Yamaguchi three-component and VanZyl decompositions for RADARSAT-2 C-band data in the (A) urban and (B) forest areas, which are shown in Figs. 5a and 6a before and after employing the various speckle suppression techniques.

The variation in volume, single bounce, double-bounce scatterings of data for the urban area is displayed in Figs. 7a and 8a before and after employing the various speckle suppression techniques to conduct the Yamaguchi three-component and VanZyl decompositions, respectively, for RADARSAT-2 data. Double-bounce

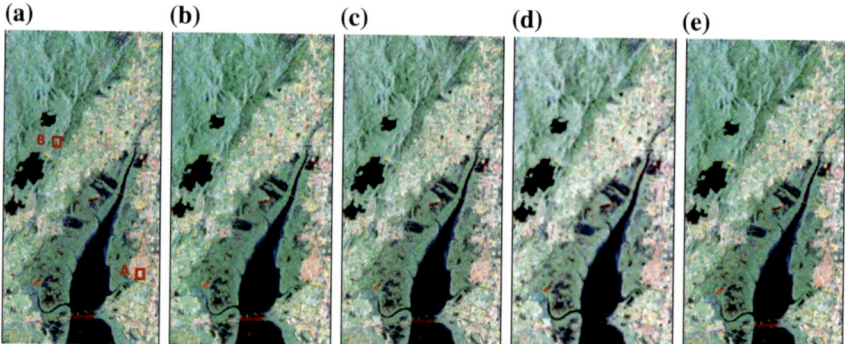

Fig. 6 VanZyl decomposition images before and after employing the various speckle suppression techniques on RADARSAT-2 data for Mumbai region. **a** Original data and **b** boxcar, **c** refined Lee, **d** IDAN, and **e** improved Lee–Sigma filters

Fig. 7 Variation in the volume, double bounce, and single bounce scatterings by conducting the Yamaguchi three-component decomposition before and after applying the various speckle filters on RADARSAT-2 data for the Mumbai region in the **a** urban and **b** forest areas

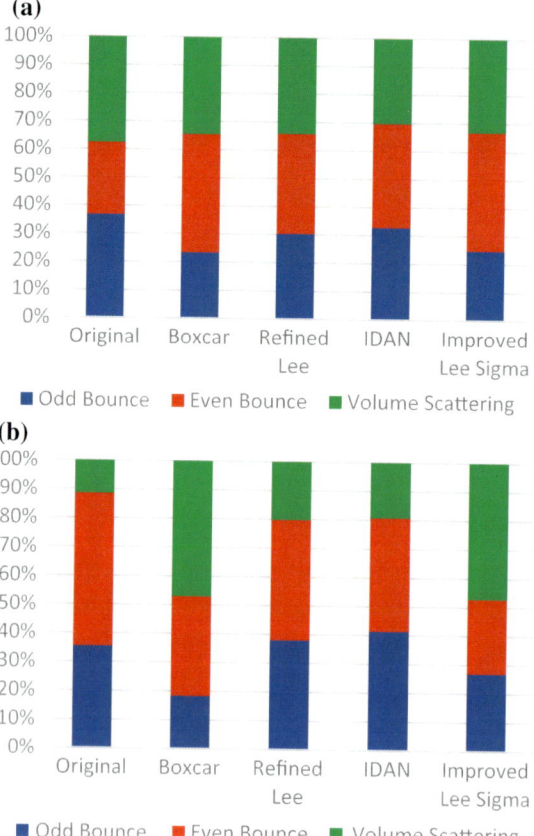

Fig. 8 Variation in volume, single bounce, and double-bounce scatterings by conducting VanZyl decomposition before and after employing the various speckle suppression techniques on RADARSAT-2 data for the Mumbai region in the **a** urban and **b** forest areas

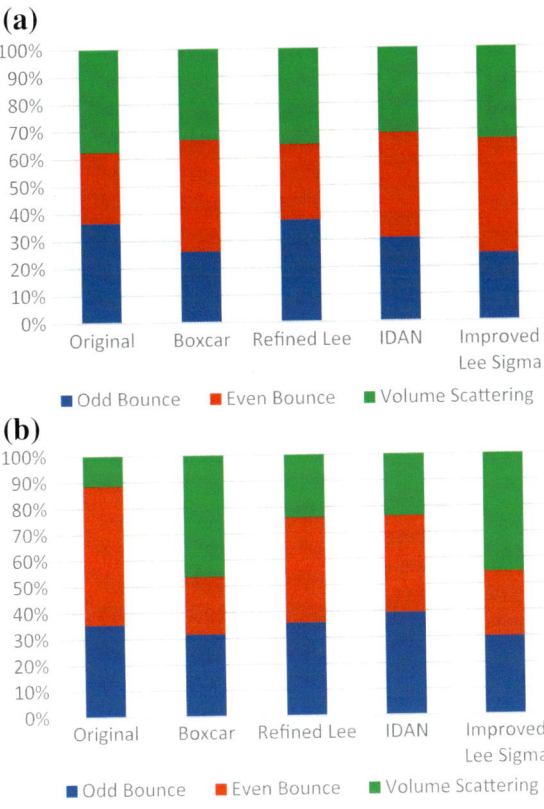

scattering for all filters considered in this study for RADARSAT-2 dataset was considerably developed. However, for the boxcar filter, the edges exhibited more blurring than the other filters considered in this study because of indiscriminate averaging. The edge preservation was better in IDAN and improved Lee–Sigma filters than that in the refined Lee filter. The variation in double-bounce scattering in the refined Lee filter was less than that in the IDAN filter. Double-bounce scattering for the improved Lee–Sigma filter improved marginally compared with the IDAN filter. Figures 7b and 8b show the variation in volume, double bounce, and single bounce scatterings in the forest area before and after the application of different speckle suppression techniques for RADARSAT-2 data. All filters considered in the study exhibited improved volume scattering. However, refined Lee, IDAN, and improved Lee–Sigma filters exhibited better edge preservation than that of the boxcar filter.

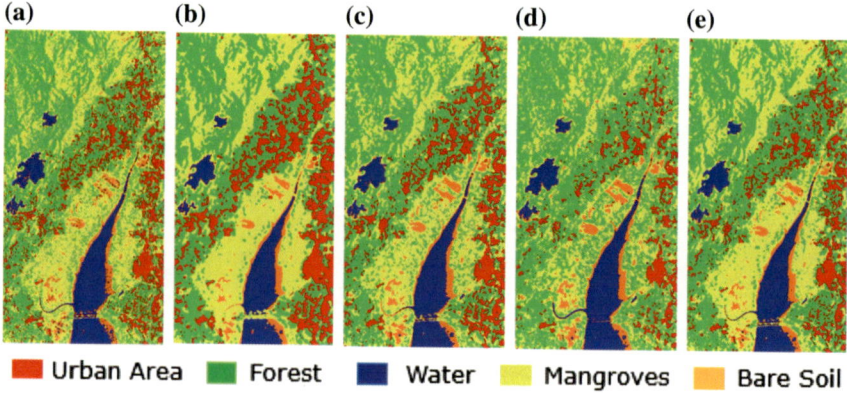

Fig. 9 Classification images of RADARSAT-2 data for the Mumbai region. **a** Original data and **b** boxcar, **c** refined Lee, **d** IDAN, and **e** improved Lee–Sigma filters

4.3 Impact of Speckle Suppression on Classification

An improvement in the accuracy of the classification due to the speckle suppression of fully polarimetric data is the essential target of current study. In this study, the impact of speckle suppression on classification accuracy was examined utilizing the fully polarimetric data. Various speckle suppression techniques—IDAN, boxcar, refined Lee, and improved Lee–Sigma were employed on fully polarimetric RADARSAT-2 data, followed by a Wishart supervised classification.

Figure 9 displays the Wishart supervised classification images of RADARSAT-2 C-band data for the Mumbai region before and after employing different speckle suppression techniques. For refined Lee, boxcar, and improved Lee–Sigma filters, the window size was 7×7. Table 4 indicates the variation in the classification accuracy before and after employing the various speckle suppression techniques for the quantitative comparison of different classes and the overall accuracy. Speckle filters considered in this study exhibited improved classification accuracy of the classes of the unfiltered data. The extent of improvement was different for all classes. The classification accuracy of several classes depends on the speckle suppression technique employed. RADARSAT-2 C-band data for the Mumbai region exhibited higher improvement in the classification accuracy for the forest and mangrove classes than the water, bare soil, and settlement classes.

For refined Lee and boxcar filters, the sliding window size is a essential parameter for speckle suppression. The result obtained by changing the window size was assessed for the refined Lee and boxcar filters by utilizing RADARSAT-2 C-band data. The classification accuracy of the speckle suppressed data was evaluated by changing the size of the window from 3×3 to 11×11. The effects of boxcar and refined Lee filters for the moving window sizes of several classes considered in the study are given in Tables 5 and 6, respectively, for RADARSAT-2 data.

Table 4 Assessment of the classification accuracies of various speckle suppression techniques (7 × 7 window) by using RADARSAT-2 data for the Mumbai region

Class	No filter	Boxcar	Refined Lee	IDAN	Improved Lee–Sigma
Settlement	98.21	100	100	100	100
Forest	63.33	69.41	69.41	67.92	70.45
Water	96.44	100	100	100	100
Mangrove	88.83	99.19	89.82	93.56	94.83
Bare soil	97.55	98.57	100	99.80	99.18
Overall accuracy	88.43	93.79	92.45	92.37	93.03

Table 5 Assessment of classification accuracies of the boxcar filter with various window sizes by using RADARSAT-2 data for the Mumbai region

Class	No Filter	3 × 3	5 × 5	7 × 7	9 × 9	11 × 11
Settlement	98.21	100	100	100	100	100
Forest	63.33	62.87	64.98	69.41	74.68	79.54
Water	96.44	99.42	99.81	100	100	100
Mangrove	88.83	91.76	95.15	99.19	99.03	99.35
Bare soil	97.55	97.34	97.75	98.57	98.57	97.95
Overall accuracy	88.43	90.11	91.73	93.79	94.74	95.61

Table 6 Assessment of classification accuracies of the refined Lee filter with various window sizes by using RADARSAT-2 data for the Mumbai region

Class	No filter	3 × 3	5 × 5	7 × 7	9 × 9	11 × 11
Settlement	98.21	98.74	99.08	100	100	100
Forest	63.33	61.81	64.14	69.41	73.93	76.35
Water	96.44	98.45	99.25	100	100	100
Mangrove	88.83	90.11	91.18	92.82	95.64	98.71
Bare soil	97.55	96.23	98.77	100	100	100
Overall accuracy	88.43	89.15	90.30	92.45	93.93	95.12

Tables 5 and 6 indicate that the refined Lee filter requires a window of a larger size for obtaining an accuracy level identical to that of the boxcar filter. However, the edge preservation of the refined Lee filter was better than the boxcar filter. The percentage improvement of the relative classification accuracies for RADARSAT-2 C-band data by using the refined Lee and boxcar filters on the unfiltered data are given in Table 7 for the window size of 7 × 7. The assessment indicates a considerable amount of enhancement in the accuracy of the classification for the mangrove and forest classes. A minimal enhancement was seen for the water, bare soil, and settlement classes.

Table 7 Percentage of relative improvement in the classification accuracy for the original data by using the boxcar and refined Lee speckle filters

Filter/Class	Settlement	Forest	Water	Mangrove	Bare Soil
Boxcar	2	10	4	12	1
Refined Lee	2	10	4	5	3

5 Conclusions

Fully polarimetric RADARSAT-2 data was used to evaluate IDAN, boxcar, improved Lee–Sigma, and refined Lee filters. The boxcar filter degrades the spatial resolution because it does not comprise the internal mechanism for preserving spatial resolution or details. The refined Lee filter improved the spatial resolution preservation; however, the filtered data was considerably smooth. The spatial strategy was improved using the IDAN filter; this filter satisfactorily preserved the spatial resolution and details, although some data were evidently smoothed. The computational efficiency of improved Lee–Sigma and refined Lee filters should be similar because both algorithms are based on local statistics. The improved Lee–Sigma filter, where the estimate of MMSE is used as the priori mean, exhibited an excellent preservation of fine details and strong targets.

In the current study, the impact of speckle suppression on the target decomposition was assessed using fully polarimetric RADARSAT-2 data. A significant amount of enhancement was observed in volume and double-bounce scatterings in forest and urban areas for all the speckle suppression techniques deliberated in this study. The speckle reduction impact on the classification accuracy was assessed using fully polarimetric data. The speckle suppression techniques considered in this study exhibited enhancement in the classification accuracy of classes of the unfiltered data. The extent of improvement was different for all classes. The classification accuracy of various classes depends on the filtering technique employed. RADARSAT-2 C-band data for the Mumbai region exhibited higher enhancement in the classification accuracy for the mangrove and forest classes than that for the water, bare soil, and settlement classes. For refined Lee and boxcar filters, the moving window size is an essential parameter for speckle suppression. RADARSAT-2 C-band data was used to evaluate the result of the changing sliding window size for the refined Lee and boxcar filters. The classification accuracy of data with suppressed speckle was evaluated by varying the size of the window from 3×3 to 11×11. The refined Lee filter requires a window with a larger size for achieving classification accuracy identical to that of the boxcar filter. However, the edge preservation performance of the refined Lee filter is better than that of the boxcar filter.

Acknowledgements The authors are grateful to Space Application Centre, ISRO, India for giving the opportunity to carry out research work and providing the data under TREES. The authors are thankful to Dr. Anup Kumar Das, SAC, ISRO for providing the guidance to conduct the research. The authors are thankful to Dr. C. V. Rao, NRSC, ISRO for his constant support and encouragement. The authors are grateful to the Centre of Excellence and Department of Electronics and Communication

Engineering at Sri Venkateswara University College of Engineering for providing the resources. Furthermore, the authors are thankful to Mr. P. Anil Kumar, Mr. C. Raju, Mr. N. Chintaiah, and research scholars for the valued discussions and encouragement. The authors would like to thank the European Space Agency for providing the open-source software and the experimental data of the PolSARpro project.

References

1. Lee JS, Pottier E (2009) Polarimetric radar imaging: from basics to applications. CRC Press, Cleveland
2. Cloude SR (2009) Polarisation applications in remote sensing. Oxford University Press, Oxford
3. Foucher S, López-Martínez C (2014) Analysis, evaluation, and comparison of polarimetric SAR speckle filtering techniques. IEEE Trans Image Process 23(4):1751–1764
4. Argenti F, Lapini A, Alparone L, Bianchi T (2013) A tutorial on speckle reduction in synthetic aperture radar images. IEEE Geosci Remote Sens Mag 1:6–35
5. Di Martino G, Poderico M, Poggi G, Riccio D, Verdoliva L (2014) Benchmarking framework for SAR despeckling. IEEE Trans Geosci Remote Sens 52(3):1596
6. Cloude SR, Pottier E (1996) A review of target decomposition theorems in radar polarimetry. IEEE Trans Geosci Remote Sens 34(2):498–518
7. Freeman A, Durden S (1998) A three-component scattering model for polarimetric SAR data. IEEE Trans Geosci Remote Sens 36(3):963–973
8. Yamaguchi Y, Moriyama T, Ishido M, Yamada H (2005) Fourcomponent scattering model for polarimetric SAR image decomposition. IEEE Trans Geosci Remote Sens 43(8):1699–1706
9. Van Zyl JJ (1992) Application of Cloude's target decomposition theorem to polarimetric imaging radar data. In: Proceedings SPIE conference on radar polarimetry, San Diego, CA, vol 1748, pp 184–212
10. Medasani S, Umamaheswara Reddy G (2018) Speckle filtering and its influence on the decomposition and classification of hybrid polarimetric data of RISAT-1. Remote Sens Appl: Environ Soc 10:1–6
11. Lee JS, Grunes MR, Kwok R (1994) Classification of multi-look polarimetric SAR imagery based on complex Wishart distribution. Int J Remote Sens 15(11):2299–2311
12. Lee JS, Grunes MR, Ainsworth TL, Li-Jen D, Schuler DL, Cloude SR (1999) Unsupervised classification using polarimetric decomposition and the complex Wishart classifier. IEEE Trans Geosci Remote Sens 37(5):2249–2258
13. Ferro-Famil L, Pottier E, Lee JS (2001) Unsupervised classification of multifrequency and fully polarimetric SAR images based on the H/A/Alpha-Wishart classifier. IEEE Trans Geosci Remote Sens 39(11):2332–2342
14. Shitole S, De S, Rao YS, Mohan BK, Das A (2015) Selection of suitable window size for speckle reduction and deblurring using SOFM in polarimetric SAR images. J Indian Soc Remote Sens 43(4):739–750

Estimation of Precipitation from the Doppler Weather Radar Images

P. Anil Kumar and B. Anuradha

Abstract Estimating the rainfall from Radar observation plays an important role in the hydrological research. The Radar Rainfall plays a fundamental role in weather modeling and forecasting applications. Doppler Weather Radar (DWR) is used estimating the rainfall within 120 km from the Radar station. Rainfall intensity data obtained from the Surface Rainfall Intensity (SRI) product of DWR has been validated with the rain gauges located at Automatic Weather Station (AWS) data. Image processing methods such as edge detection and color identification are used to extract the rainfall from the SRI product. Time series rainfall over a particular location is compared with the AWS data using statistical parameters like correlation coefficient and Squared Pearson coefficient. The experimental results convey that the proposed method yields the high amount of accuracy. Graphical User Interface is developed to extract the point rainfall and time series rainfall over different locations within the range of Radar.

1 Introduction

Rain Gauges are used to measure the rainfall at different points in the field, the spatial variation of rainfall cannot be measured due to lack of sufficient rain gauges. In view of the practical problem of installing rain gauges at different stations, Radar has found to be an alternative. Radar has been used in predicting the weather over the last four to five decades. Radar can provide information such as rain, reflectivity, wind speed, and direction in both time and space varying domains. The underlying principle of radar is the Rayleigh Scattering in which scattering of electromagnetic radiation occurs by particles smaller than the wavelength of the radiation [1]. The meteorological radars are characterized by their operating electromagnetic frequencies such as X, C, or S bands. The weather Radars look deeper into the system to provide information such as Intensity, Rain Rate, Vertical Extent, and Drop Size Distribution. The Primary

P. Anil Kumar (✉) · B. Anuradha
Department of ECE, SVU College of Engineering, SV University, Tirupati, India
e-mail: anilkumar417@gmail.com

© Springer Nature Switzerland AG 2019
D. Pandian et al. (eds.), *Proceedings of the International Conference on ISMAC in Computational Vision and Bio-Engineering 2018 (ISMAC-CVB)*, Lecture Notes in Computational Vision and Biomechanics 30,
https://doi.org/10.1007/978-3-030-00665-5_21

203

products of the DWR are Reflectivity (Z), Radial Velocity (V), and Spectrum Width (W). The Secondary products derived from Primary products such as Surface Rainfall Intensity (SRI) and Precipitation Accumulation (PAC) products are among them.

Radar Rainfall data derived from S-band DWR was used as input to hydrological modeling. Hydrologic Engineering Center-Hydrological modeling System (HEC-HMS) simulation results are calibrated and validated with the observed data [2]. Raindrop scatters pulses back to the receiver by the process of reflection, refraction, and scattering. The reflected signal strength and the time delay between the pulses define the distance of the target. Rainfall intensity is estimated from the reflectivity with the help of Marshall–Palmer Equation as shown in Eq. 1. The value of the constants decides the type of geographic area and precipitation such as convective rain, thunderstorm, orographic rain, stratiform and snow. The parameters for the Marshall–Palmer are fine-tuned for different geographical conditions and different climatic conditions [3, 4]. The values of a and b are 267 and 1.345 for the secondary product SRI of S-band DWR Chennai. The Reflectivity information was extracted from the DWR MAX (Z) product using image processing techniques. The time series Reflectivity extracted over a particular location was plotted and cloud vertical profile was obtained [5].

$$Z = N(D)D^6 = a\, R^b \tag{1}$$

where

Z is the measured radar reflectivity in mm^6/m^3
D is the drop diameter in mm
$N(D)$ is the number of drops of given diameter per cubic meter
R is the rainfall rate at the ground level in mm/h
a, b are constants which depend on latitude, longitude, season, and types of rainfall.

2   Data

The DWR SRI images from the Indian Meteorological Department (IMD) located at Chennai were downloaded continuously for every 10 min since November 2015 and a database is maintained (http://www.imd.gov.in). The AWS Data for the Chennai and surrounding locations were collected.

3   Methodology

The DWR SRI image which contains the convective activity within the range of DWR is shown in Fig. 1. The resolution of DWR SRI is 0.4 km/pixel. It covers a range

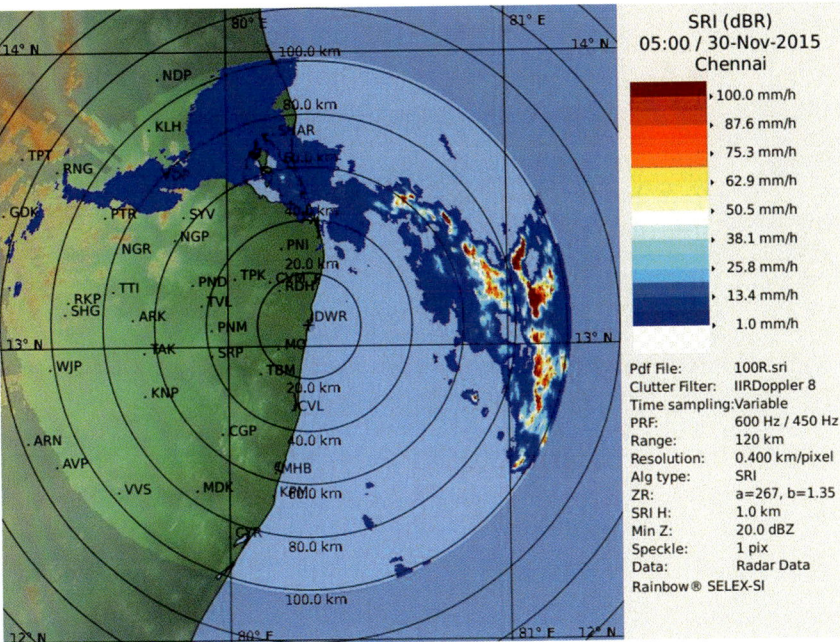

Fig. 1 Convective SRI image

of 120 km from the radar center. The minimum detectable reflectivity for rainfall calculation is 20 dBZ for the SRI product. The SRI product displays the value of rainfall intensity over each pixel of the cylindrical volume in its plane view. The color bar provided on the right side of the image gives information about the intensity of precipitation over a certain area of the image. The blue color corresponds to 1 mm/h and red color indicates the rainfall intensity of 100 mm/h. Similarly, each color in the color bar signifies the different intensity level of precipitation. The flowchart for rainfall extraction from SRI image is shown in Fig. 2.

An SRI image with zero convective activity at all locations is considered as a reference image (I_{ref}). An image which contains convective activity or the image which is of interest is considered as an image of interest (I_{int}). The images which are downloaded from the IMD website contains salt and pepper noise. Median filter which has the characteristics of scale-invariant nonlinear smoother is used to eliminate the noise by preserving the edges. Image preprocessing is done on the image of interest to make it suitable for further processing [6]. The convective image is subtracted from the reference image to extract only the convective portion of the image [5]. The gray level image is then converted to binary image using soft thresholding algorithm using the mean value of the image as the threshold value. The convective portion of the binary image will be white and nonconvective portion will be black. Morphological operations such as image opening and image closing depend on the ordering of the pixel values but not on the intensity of the pixel values. The size and

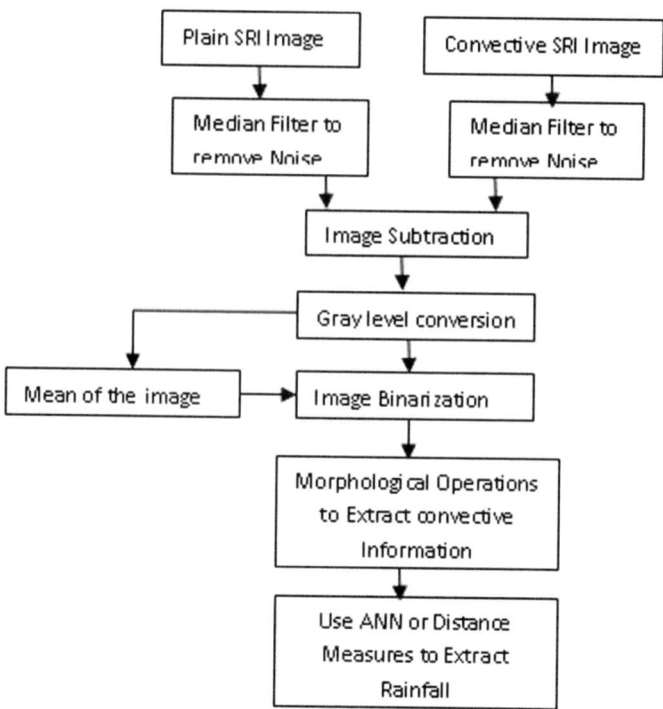

Fig. 2 Flow chart for the extraction of rainfall from the DWR SRI image

shape of the structuring element are sensitive to shapes in the input image. Image opening eliminates the small islands and sharp peaks. Image closing merges the narrow breaks and eliminates the holes. The morphological operated binary image is multiplied with noise eliminated original image which presents only convective activity of the image as shown in Fig. 3.

The extracted convective portion is compared against the color bar to extract the rainfall intensity over a particular location of the image [7]. The rainfall is extracted from the SRI image using various distance measures such as Euclidean, Standard Euclidean, City block, Minkowski, Chebychev, Mahalanobis, cosine, and correlation. This method presents an accuracy of 94%. Artificial Neural Networks are trained with color bar values to extract the rainfall intensity. Backpropagation algorithm is used to adjust the hidden layer and output layer weights. The ANN is trained with different backpropagation algorithms such as Levenberg–Marquardt, Gradient descent, Gradient Descent with Momentum, Conjugate gradient, Variable Learning Rate with Momentum, and resilient back propagation [8]. The Levenberg–Marquardt algorithm given the best output with minimum error and an accuracy of 98%.

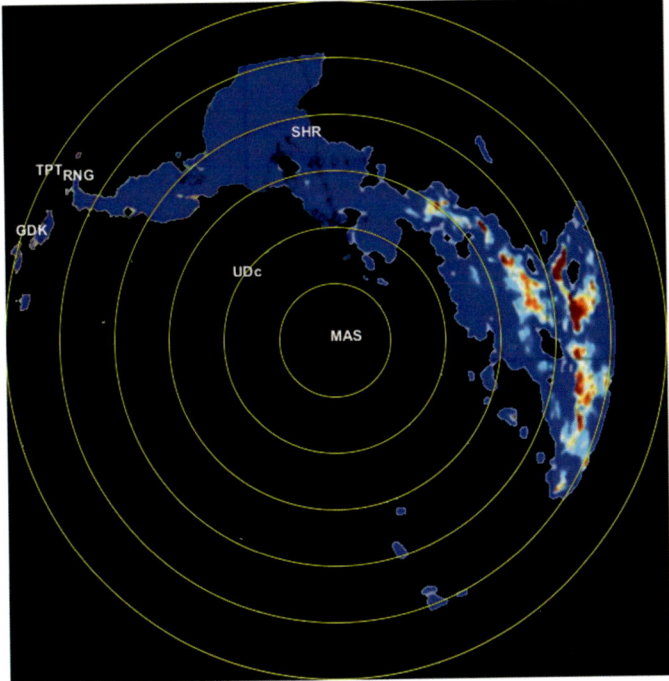

Fig. 3 Convective portion extracted from SRI image

4   Results and Discussions

The DWR SRI images were compared with the reference image to obtain the convective portion of the image by eliminating the static background is shown in Fig. 3. The green circles with increments of a 20 km radius are drawn with Chennai as the center. The static background image has no convective activity implies zero rainfall rate. The low-level clouds precipitate to a little extent. The deep convective clouds precipitate thunderstorm rain to a larger extent.

A Graphical User Interface (GUI) is Developed for Precipitation estimation from the DWR SRI product in MATLAB. It has a provision for rainfall estimation of a single image and a time series estimation of rainfall. GUI has the option to estimate the rainfall over a location (Latitude–Longitude) or click-based on the image. GUI fetches the location information from Google based on the lat-long entered in it. The Sample GUI for single image and time series extraction of rainfall is shown in Figs. 4 and 5. The images will be updated on the website for every 10 min. The images downloaded from the official website may not be continuous, and this is due to lack of updating the information. The lack of image or nonconvective activity over a location is treated as a NaN (Not a Number) for representation. The correlation coefficient is calculated without considering these NaN values.

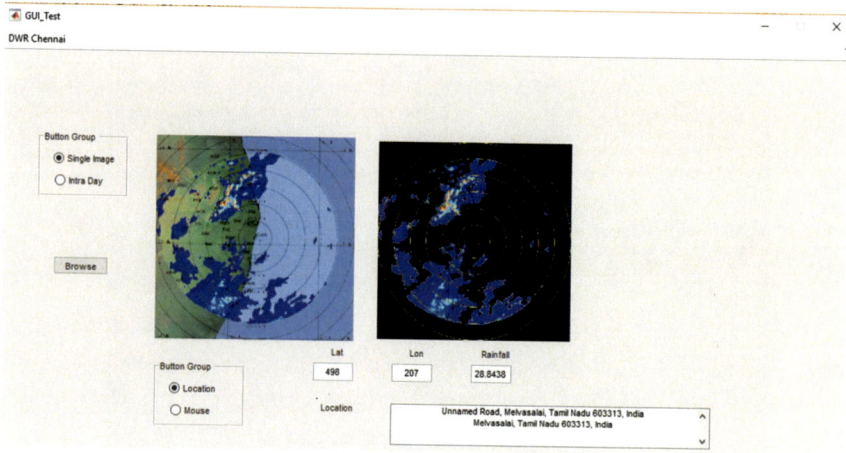

Fig. 4 GUI for a single image

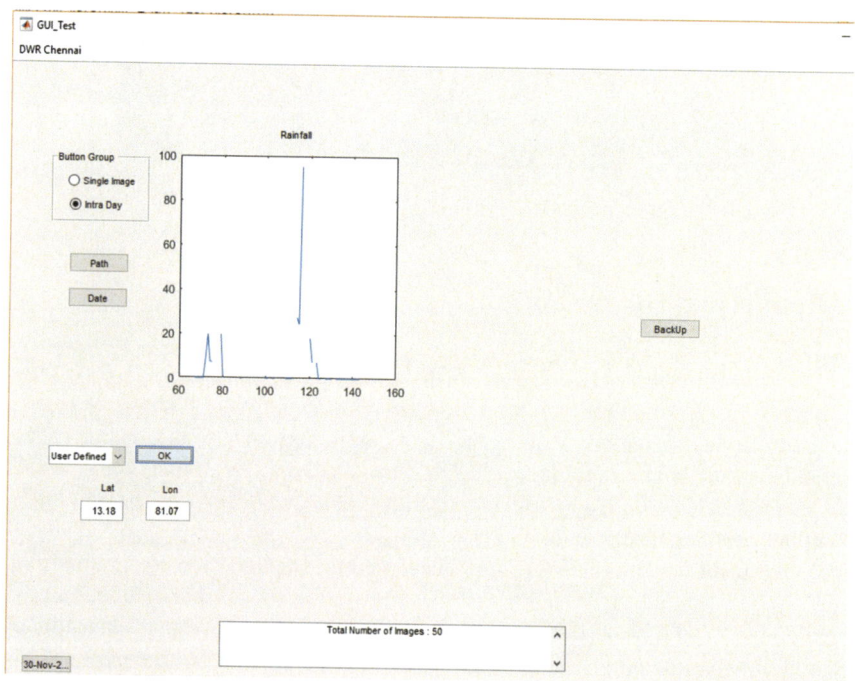

Fig. 5 GUI for a time series intraday

The rainfall from different AWS stations surrounding Chennai is collected and compared with the DWR estimated precipitation [2]. The metrics used in this paper are correlation coefficient. Correlation Coefficient determines the relationship

Table 1 Correlation coefficient between AWS and DWR

Date	Station	Correlation coefficient
3rd June 2016	SHAR	0.76515
	Chennai	0.69587
27th May 2016	SHAR	0.78128
	Chennai	0.73183
29th Nov 2015	SHAR	0.78424
	Chennai	0.79156
30th Nov 2015	SHAR	0.77818
	Chennai	0.76183

between the local average rainfall and rainfall of AWS and DWR derived rainfall is given in Eq. 2. It can be interpreted as the proportion of variance in y to the variance in x. The range of coefficient values is between -1 and 1. If the value of correlation coefficient is negative, it means as x increases y decreases in the same proportion. If the value of the correlation coefficient is positive, then x and y are positively correlated. If the value is zero, then there is no relation between input and output variables [9].

$$r = \frac{\sum (x - \bar{x})(y - \bar{y})}{\sqrt{\sum (x - \bar{x})^2 \sum (y - \bar{y})^2}} \tag{2}$$

The correlation coefficient between AWS and DWR extracted rainfall from SRI product are tabulated in Table 1 which shows the high degree of similarity. The correlation value is less due to the nonavailability of the DWR images at sometimes because of the problem in reception of the data due to glitches in the system and artifact effect arises due to temperature variation while receiving the images. The DWR station in Chennai has to update the images in the database for every 10 min, so a total of 144 images should be available per day. The station will not update the images in the web database. The nonavailability of images present wrong interpretation of the results. The nonavailability of the images will be represented as Not A Number (NAN), so there will be gaps in the time series image as shown in Fig. 5.

5 Conclusion

In this paper, precipitation is estimated from the DWR SRI images using image processing techniques. A GUI was developed for this application is user-friendly, which extracts the point rainfall or time series rainfall over a particular location. The extracted rainfall from the DWR was compared with the AWS data and a reasonable correlation coefficient was achieved. The method presented here is robust to noise.

Acknowledgements We are thankful to Mr. Kannan from IMD Chennai, Mr. Yesubabu from NARL Gadanki, and Mr. M. S. Arunachalam from IIT Chennai, who helped us in understanding the DWR images and deriving the algorithm.

References

1. Doviak RJ, Zrnik DS (1992) Doppler radar and weather observations, 2nd edn
2. Josephine VS, Mudgal BV, Thampi SB (2014) Applicability of Doppler weather radar based rainfall data for runoff estimation in Indian watersheds—A case study of Chennai basin. Sadhana 39(4):989–997
3. Suresh R, Ravichandran PK, Gupta JP, Thampi SB, Kalyanasundaram S, Rajesh Rao P (2005) On optimum rain rate estimation from a pulsed Doppler weather radar at Chennai. Mausam 56:433–446
4. Marshall JS, Palmer W (1948) The distribution of raindrops with size. Adv Geophys 2:1–56
5. Anil Kumar P, Anuradha B, Arunachalam MS (2017, July) Extraction of time series convective cloud profile from Doppler weather radar MAX (Z) product using a novel image processing technique. Int J Adv Eng Res Dev 4(7)
6. Gonzalez RC, Woods RE (1992) Digital image processing. Addison-Wesley, Reading, MA, p 716
7. Anil Kumar P, Anuradha B (2017) Estimating reflectivity of DWR images by analysing different colour spaces through distance measures. Adv Comput Sci Technol 10(8):2191–2200
8. Taravat A, Del Frate F, Cornaro C, Vergari S (2014) Neural networks and support vector machine algorithms for automatic cloud classification of whole-sky ground-based images. IEEE Geosci Remote Sens Lett 12(3):666–670
9. Kaur A, Kaur L, Gupta S (2012) Image recognition using coefficient of correlation and structural similarity index in uncontrolled environment. Int J Comput Appl 59(5)

Text-Independent Handwriting Classification Using Line and Texture-Based Features

T. Shreekanth, M. B. Punith Kumar and Akshay Krishnan

Abstract This paper addresses the problem of making a machine recognize the writer by means of the handwriting. It delineates the preprocessing methods used to enhance handwritings, so as to ease the process of feature extraction. It discusses six statistical texture-based features that characterize a handwriting. Once these features are extracted, a nearest neighbor approach is used to classify a sample handwriting into one of those in the database. The methods are verified on a self-compiled database, and a performance evaluation is also performed. This method can be used to identify an unknown handwriting and is in specific demand in the forensic domain. Unlike other biometric identification methods, handwriting-based identification is the least intrusive.

1 Introduction

The problem of classifying handwritings is one that has had a long history in research [1]. Earlier research focused on extracting data from handwritings [2–4]. In recent times, the problem of identifying the author has become just as significant as that of identifying data in the writing. In applications like forensics, it is often necessary to identify the writer based on the handwriting obtained from crime sites. Other applications include signature and document verification [5]. With applications that call for accuracy, it is very necessary that the solutions to such problems are reliable.

Handwriting recognition is characterized into two methods: text-dependent and text-independent. Text-dependent handwriting recognition involves the detection of

T. Shreekanth · A. Krishnan (✉)
Department of ECE, SJCE, Mysore, India
e-mail: akshay.krishnan.30@gmail.com

T. Shreekanth
e-mail: shreekanth_t@sjce.ac.in

M. B. Punith Kumar
Department of ECE, PESCE, Mandya, India
e-mail: punithpes@gmail.com

© Springer Nature Switzerland AG 2019
D. Pandian et al. (eds.), *Proceedings of the International Conference on ISMAC in Computational Vision and Bio-Engineering 2018 (ISMAC-CVB)*, Lecture Notes in Computational Vision and Biomechanics 30,
https://doi.org/10.1007/978-3-030-00665-5_22

the characters used in the handwriting database before the writer can be recognized. This also requires the segmentation of characters as a preprocessing step. Once the characters have been segmented, they are classified. The differences between an individual's alphabet and the general alphabet are used to identify the handwriting.

In text-independent handwriting, the recognition is done without comparing the characters in the sample with those in the database. For this, texture-based statistical features are extracted from the dataset. These features are then used by a classifier to identify the sample image. The classifier may use nearest-neighbor approach or may be based on neural networks.

In this paper, we first present certain methods to enhance the quality of an image for handwriting recognition. This is followed by certain features that are extracted from the image in order to be fed to the classifier. The paper then discusses the nearest neighbors approach used for classification. This is followed by a discussion on the results obtained and future work.

2 Literature Survey

The preprocessing techniques used in handwriting recognition are similar to those of optical character recognition [6]. The difference is that in handwriting recognition the difference in handwriting (inter-class distance) is enhanced whereas in character recognition systems, the same is minimized. The preprocessing techniques used are as follows:

Converting to grayscale

Our implementation uses texture-based statistical features from images. Since the texture is independent of color, the classification and feature extraction is done on binary images. To convert a color image to a binary image, it is first converted to grayscale and is then thresholded.

Otsu's thresholding

Thresholding is used to convert the image to a binary form in which the foreground information is represented by black pixels and background information is represented by white pixels. Features can be easily extracted from binary images, once all the background information is isolated. The threshold that is selected plays a vital role is the thresholding process. In our implementation, Otsu's method [7] is used to select the threshold.

Connected component Labeling

Connected components labeling [8] is used to extract all the connected pixels of an image. Within a connected component, all pixels share the same intensity. The connected component labeling algorithm assigns a label (starting from 0) to each component. This step is necessary because it is much easier to apply feature extraction on selected connected components (letters or words) than on the image itself.

The algorithm works by scanning an image, pixel-by-pixel (both vertically and horizontally) in order to identify connected pixel regions V. (In our implementation, $V = \{255\}$).

Connectivity in this algorithm can be defined in terms of 8-neighbors or 4-neighbors. Our implementation uses 8-neighboured connected components. This is chosen owing to the sophisticated curves involved in handwritings.

Based on this information, the labeling of p occurs as follows:

- If all eight neighbors are 0, a new label is assigned to p.
- If one of the neighbors has an intensity value of 1, assign its label to the pixel.
- If more than one neighbors have an intensity value of 1, any value is assigned to p, and a note of this position is made.

Contour extraction using Moore neighborhood

The Moore neighborhood based tracing algorithm [9] that is used to extract contours is summarized as follows:

Input: A connected component X.
Output: A list Y consisting of the boundary pixels of each connected component.
Let Moore's neighborhood of pixel a in X be M(a).
The pixel being used, c is in M(p)

- Y is initialized to an empty set.
- The image is scanned until a black pixel in X is found.
- This pixel is added to Y.
- This point is set as the current point, say, s.
- Select the pixel from which s was entered.
- Let c be the next clockwise pixel.
- While c is not equal to s, loop:

> If c is zero:
>> Add c to Y
>> Let p be c
>> Move the current pixel c to the pixel from which p was entered.
> Else:
>> Move on to the next clockwise pixel.

Feature extraction: Research in handwriting recognition has brought to the forefront many features, most of them being statistical. Table 1 shows a summary of the features that have been used so far. Our solution uses statistical texture-based features that can uniquely identify a handwriting based on the research by [10].

Classification: The features corresponding to a sample handwriting can be classified as belonging to one from the database using the nearest neighbor approach or neural networks. Our implementation uses a nearest neighbor approach based on Euclidean distances.

Table 1 Literature survey of developments in handwriting recognition

Sl. No.	Paper name	Authors	Features used	Classifier
1	Writer identification using edge-based directional features [11]	Marius Bulacu Lambert Schomaker Louis Vuurpijl	Edge direction, edge hinge direction, run lengths, autocorrelation	Nearest neighbor, using Euclidean distances
2	Writer identification using text line based features, IEEE 2001 [12]	U. V. Marti, R. Messerli, H. Bunke	Relative heights of three main writing zones, slant angle of writing	K-Nearest neighbor, feed forward neural network
3	Biometric personal identification based on handwriting [13]	Yong Zhu, Tieniu Tan, Yunhong Wang	Gabor filters	Weighted euclidean distance
4	Individuality of handwriting [14]	Sargur N. Srihari, Sung-Hyuk Cha, Hina Arora, Sangjik Lee	Measures of pen pressure, measures of writing movement, measures of stroke formation, slant and proportion, paragraph and word level features	Nearest neighbor rule
5	Text-independent writer identification and verification using textural and allographic features [10]	Marius Bulacu, Lambert Schomaker	Direction and hinge probability distribution function, run length, autocorrelation, allograph level features	Nearest neighbor
6	Preprocessing and feature extraction for a handwriting recognition system [6]	T. Caesar, J. M. Gloger, E. Mandler	Peak area, upper line area, medium area, below baseline area, baseline area	N/A

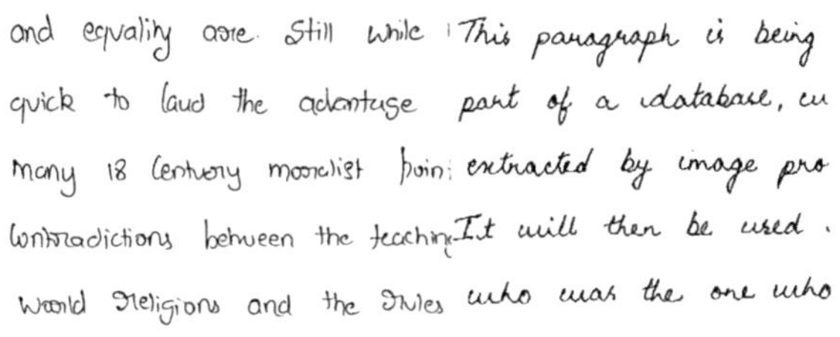

Fig. 1 Two handwritings from the database: before and after thresholding

3 Methodology

The process of text-independent handwriting recognition used in this paper can be divided into three subprocesses:

1. preprocessing of handwriting images.
2. Feature extraction.
3. Classification.

Preprocessing of handwriting images: Preprocessing is carried out as a series of spatial domain techniques listed as under:

Conversion to gray scale: Color images are converted to gray scale since color information is irrelevant to our approach and processing of gray scale images are much easier.

Thresholding: Features are easily extracted by separating the background information from the foreground. This is done by associating a 1 with the text and a 0 with the background, using Otsu's thresholding algorithm [7]. Our implementation later converts the text to a pixel value of 255 and the background to 0. This eases the processing and helps distinctly view the text. Figure 1 shows two handwritings from the database, one with thresholding and the other without it.

Connected components labeling: Since processing of all the text in an image of a paragraph is cumbersome, it is preferred to segment different portions of the paragraph and analyze them one after another. Depending on the handwriting, connected component labeling may segment words, or individual letters. A label is associated to each connected component at the end of this stage.

Contour extraction: In order to extract features like slant and roundness of the handwriting, contours need to be processed. This method is inspired by [9], where the features extracted are probability distribution functions from contours.

Feature extraction: The performance of a classification depends on the number and relevance of features extracted. The features extracted much be non-redundant. That is, each feature must convey information that is not conveyed by another. The features taken together must also provide enough information to distinguish one handwriting class from another. Taking into consideration various aspects of a handwriting, the following texture-based features are extracted.

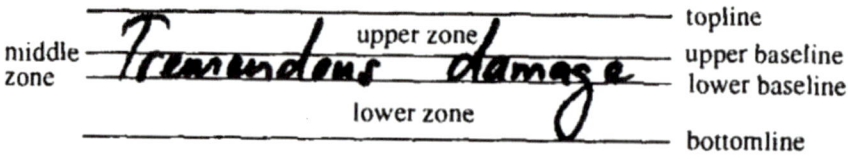

Fig. 2 Writing zones and bounding lines [12]

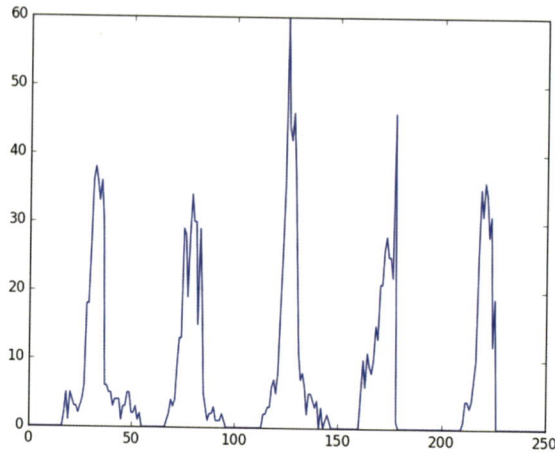

Fig. 3 The line profile/horizontal histogram of a handwriting with 5 lines of text, used for extracting the line-based features

1. Three text line based features

The features are mainly based on visible characteristics of the writing, such as the height of the three main writing zones, the width of the characters, the slant of the writing, and the text's legibility. These features can also be observed and interpreted by humans.

The features are based on the height of the three main writing zones, which are determined by the topline, the upper baseline, the lower baseline, and the bottomline (see Fig. 2). To determine these lines, a horizontal projection p of the text line image is computed. In the horizontal projection (as in Fig. 3), there is a spike between the upper baseline and the lower baseline, and relatively small values (distinguished by a threshold 't'. t is relative and varies from image to image. Therefore, it is set as a fraction of the peak value between the upper and lower baselines) between the lower baseline and bottomline, and the topline and upper baseline. Between the bottomline of one line and topline of the next, the value is zero.

$$x = |\text{topline} - \text{upper baseline}| \tag{1}$$

$$y = |\text{upper baseline} - \text{lower baseline}| \tag{2}$$

$$z = |\text{lower baseline} - \text{bottomline}| \tag{3}$$

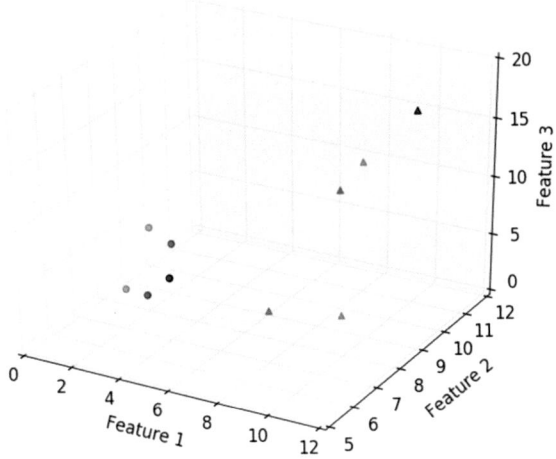

Fig. 4 Line-based features corresponding to two different handwritings

The values of x, y, and z however are scale variant. To achieve scale invariance, their relative values are considered as features. These are

$$f1 = x/y \qquad (4)$$
$$f2 = y/z \qquad (5)$$
$$f3 = x/z \qquad (6)$$

The three features $f1$, $f2$, and $f3$ extracted for 8 different samples belonging to two handwritings (4 sample per handwriting), produced results as shown in Fig. 4.

2. **Contour-based probability distribution functions as features:**

Further, in this section, we describe the extraction methods for the texture-level features [10]. In these features, the handwriting is merely seen as a texture described by some probability distributions.

1. **Writing slant**: The slant of any handwriting is perhaps the most obvious feature that sets apart different writers. In order to capture the slant of a handwriting in a feature, we resort to contours.

In order to extract this feature, we follow these steps:

- For each contour in the image:
- For each point p1 along the contour:
- Select a point p2 that is a minimum number (k) away from p1 along the contour.
- K is decided based on the thickness of the handwriting and in our implementation, it is set to 4.
- Calculate the angle made by the line joining p1 and p2 with the horizontal.

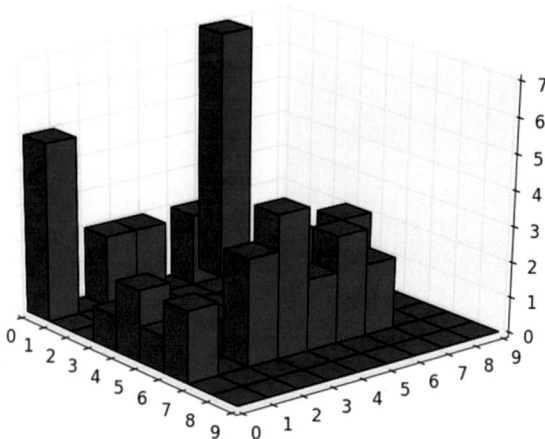

Fig. 5 Representation of the two-dimensional probability distribution function for one handwriting

- For computational ease, this angle is divided into groups of 20.

A number of contours corresponding to a particular angle are stored in a histogram which is normalized to give a probability distribution function, which is used as the fourth feature $f4$.

1. **Handwriting curvature**: This feature is used to identify the curvature of the handwriting. The following steps are to be followed for extracting this feature:

   - For each contour in the image:
   - For each point p1 along the contour:
   - Select a point p2 that is a minimum number (k) away from p1 along the contour.
   - Calculate the angle made by the line joining p1 and p2 with the horizontal, let it be A.
   - Select a point p3 that is a minimum number (k) away from p1 along the contour in the other direction.
   - Calculate the angle made by the line joining p1 and p2 with the horizontal, let it be B.
   - For computational ease, A and B are divided into groups of 20.
   - Increment the value for the particular group in the joint probability distribution function $P(A, B)$.

This joint probability distribution (Fig. 5) is used as the fifth feature $f5$.

1. **Directional Co-Occurrence of handwriting:**

This feature is used to identify the roundness of the handwriting. The following steps are used to extract this feature:

- For each contour in the image.

Table 2 Performance evaluation of the handwriting recognition algorithm

Number of handwritings in database	100
Number of samples tested	30
Number of correct recognitions	26
Number of incorrect recognitions	4
Accuracy of the method	86.67%

- For each point p1 along the contour.
- Select a point p2 that is horizontally opposite to along the contour in the other direction. This is done by identifying all other points along the contour with same row-index and removing the ones that are redundant.
- Find the contour direction for both p1 and p2, let them be A and B.
- For computational ease, A and B are divided into groups of 20.
- Increment the value for the particular group in the joint probability distribution function $P(A, B)$.
- This joint probability distribution constitutes the fifth feature $f5$.

Classification: The final stage of handwriting recognition is the classification of any handwriting sample into a class that is defined by the handwritings in the database. Our implementation uses a classifier that minimizes the variance between the sample and the classes in the database, in terms of the six features identified in the previous step.

4 Results

The database used for handwriting recognition in our implementation comprised 100 handwritings of which 6 were self-compiled and the others were from the IAM handwriting database [15]. Once the algorithm was implemented, it was tested using different handwriting samples of the same authors. The results are summarized in Table 2.

5 Conclusion and Future Work

In order to increase the accuracy of the handwriting recognition system, more features can be incorporated into the system. Psychology has suggested characteristics of a person's handwriting based on the pressure applied on the paper [14]. These manifest in the form of the thickness of the handwriting. This paper considers only those features based on the overall texture of the handwriting. If the thickness were to be evaluated, the information obtained can be used to further improve the performance of the classifier.

The paper focuses on off-line handwriting recognition in which the handwriting can be recognized only if a written paragraph is fed as a sample. An online recognition system would be able to detect the writer, during the process of writing itself. This can be used in tablets and other such touchscreen devices. Future work shall focus on extending our methods to on-line handwriting recognition.

References

1. Raj A, Chaudhary A (2016) A review of personal identification using handwriting. Int J Eng Sci Comput
2. Feature extraction and identification of handwritten characters, Computer Technology Application Key Laboratory of Yunnan Province, Kunming University of Science & Technology, Kunming, China
3. Nath RK, Rastogi M (2012) Improving various off-line techniques used for handwritten character recognition: a review. IJCA 49(18)
4. Saady YE, Rachidi A, El Yassa M, Mammass D (2011) Amazigh handwritten character recognition based on horizontal and vertical centerline of character. IJAST 33(17):33–50
5. Nguyen V, Blumenstein M (2011) An application of the 2D gaussian filter for enhancing feature extraction in off-line signature verification. In: 2011 international conference on document analysis and recognition (ICDAR), pp 339–343
6. Preprocessing and Feature Extraction for a Handwriting Recognition System. 0-8186-4960-7/93 Q 1993 IEEE
7. Kurita T, Otsu N, Abdelmelek N (1992) Maximum likelihood thresholding based on population mixture models. Pattern Recognit 25(10):1231–1240
8. He L, Chao Y, Suzuki K, Wu K (2009) Fast connected-component labeling. Pattern Recognit (Elsevier)
9. Seo J, Chae S, Shim J, Kim D, Cheong C, Han T-D. Fast contour-tracing algorithm based on a pixel-following method for image sensors
10. Bulacu M, Schomaker L (2007) Text-independent writer identification and verification using textural and allographic features. IEEE Trans Pattern Anal Mach Intell 29(4)
11. Bulacu M, Schomaker L, Vuurpijl L (2003) Writer identification using edge-based directional features. In: Proceedings of 7th international conference on document analysis and recognition (ICDAR 2003), IEEE Press
12. Marti U-V, Messerli R, Bunke H (2001) Writer identification using text line based features. Institut fur Informatik und angewandte Mathematik, 0-7695-1263-1/01 0, IEEE
13. Zhu Y, Tan T, Wang Y. Biometric personal identification based on handwriting (unpublished)
14. Srihari SN, Cha S-H, Arora H, Lee S (2002) Individuality of handwriting. J Forensic Sci 47(4)
15. Marti U, Bunk H (2002) The IAM-database: an english sentence database for off-line handwriting recognition. Int J Doc Anal Recognit 4:39–46

A Unified Preprocessing Technique for Enhancement of Degraded Document Images

N. Shobha Rani, A. Sajan Jain and H. R. Kiran

Abstract The field of Document Image Processing has encountered sensational development and progressively across the board relevance lately. Luckily, propels in PC innovation have kept pace with the fast development in the volume of picture information in different applications. One such utilization of Document picture preparing is OCR (Optical Character Recognition). Pre-preparing is one of the pre-imperative stages in the handling of record pictures which changes the archive to a frame reasonable for ensuing stages. In this paper, various preprocessing techniques are proposed for the enhancement of degraded document images. The algorithms implemented are adept at handling variety of noises that include foxing effect, illumination correction, show through effect, stain marks, and pen and other scratch marks removal. The techniques devised works based on noise degradation models generated from the attributes of noisy pixels which are commonly found in degraded or ancient document images. Further, these noise models are employed for the detection of noisy regions in the image to undergo the enhancement process. The enhancement procedures employed include the local normalization, convolution using central measures like mean and standard deviation, and Sauvola's adaptive binarization technique. The outcomes of the preprocessing procedure is very promising and are adaptable to various degraded document scenarios.

N. Shobha Rani (✉) · A. Sajan Jain · H. R. Kiran
Department of Computer Science, Amrita School of Arts and Sciences,
Amrita Vishwa Vidyapeetham, Mysuru, India
e-mail: n_shobharani@asas.mysore.amrita.edu

A. Sajan Jain
e-mail: jain.sajan2@gmail.com

H. R. Kiran
e-mail: kiranhr1993@gmail.com

© Springer Nature Switzerland AG 2019
D. Pandian et al. (eds.), *Proceedings of the International Conference on ISMAC in Computational Vision and Bio-Engineering 2018 (ISMAC-CVB)*, Lecture Notes in Computational Vision and Biomechanics 30,
https://doi.org/10.1007/978-3-030-00665-5_23

1 Introduction

Records that can be handled by OCR incorporate antiquated, matured archives, for example, chronicled books, machine printed archives, for example, reminders, letter specialized reports, and books, hand composed archives, for example, individual letters, addresses on postal mail, notes in the edges of records, online written by hand archives, and so on.

Among all the document types, ancient documents of machine printed type or handwritten types are highly affected with noise due to aging [1] and hence in need of intensive preprocessing. The degradations incur due to noise are categorized as pen marks, stain marks, scratch marks, scan motion blur, bleed-throughs, nonuniform illumination, show through effects and foxing effect, etc. [2, 3].

Preprocessing of document image varies based on the document context leading to the design of algorithms suiting the document contexts. Quite a good number of works are reported in this area including text strokes crossing removal [3], graphical component detection and removal [4], adaptive document binarization [5], preprocessing for handwritten documents [6], page layout analysis [7], neural network based document preprocessing [8], stain removal [9], enhancement of degraded document quality [10], and many other methods. The details of some of the important works are as summarized subsequently.

Chang et al. [11] had proposed an approach for picture denoising and pressure which utilizes a versatile, information-driven limit by means of wavelet soft thresholding. The limit was inferred in a Bayesian structure, and the earlier utilized on the wavelet coefficients was the summed up Gaussian conveyance. The experiments depicts that BayesShrink, was regularly within 5% of the Mean Squared Deviation of the best delicate thresholding benchmark with the picture expected. Hsia et al. [12] had propped a strategy to denoise the interfering or obvious watermarks from foundation data for written by hand archives utilizing versatile histogram adjustment to stifle undesirable meddling strokes and acquired an unmistakable picture utilizing versatile directional lifting-based discrete wavelet changes. The trial comes about portray that Haar wavelet was not the ideal technique in light of the fact that the first strokes were broken, however, the meddling strokes were expelled. Ntogas et al. [13] had proposed a binarization method for pre-filtered historical manuscripts images consisting of five discrete steps from data acquisition, conversion of RAW files to JPEG, cropping and converting to gray scale and then denoising using various filters in spatial domains, followed by thresholding and morphology. Experimentations were performed on Byzantine historical manuscripts. Kitadai et al. [14] had proposed a similarity evaluation technique for character designs with missing shape parcels utilizing nonlinear normalization and adjusted the formats for every trial of the recovery. The result of the strategy indicated changes of the recovery precision up to 72%. Shirai et al. [15] had concocted a strategy for anisotropic morphological enlargement by means of certain smoothing to restore the corrupted character states of binarized pictures. They connected a smoothing technique not to the double picture but rather to the separation changed picture, and after that reconverted it by binarization. Additionally, depicted a strategy for reestablishing the flood of strokes, utilizing verifiable smoothing for the morphological task in order to evacuate loud examples out of sight and interface the fractionated character strokes. Lu et al. [16] had contributed a binarization procedure for sectioning content from seriously lit up archive pictures by using smoothing polynomial surface for the shading estimation and pay, which delivered a generally consistently lit up record picture that could be binarized by some worldwide thresholding strategies. Experimentation demonstrated that the system was tolerant to the varieties in content size and report differentiate could binarize severely lit up archive picture. Kavallieratou and Antonopoulou [17] had built up a recursive calculation for improvement of recorded archives by utilizing mixes of entangled picture handling strategies by considering spatial qualities of the report pictures. Normal issues of verifiable reports like foundation difference stains and straightforwardness were overwhelmed by their technique in correlation with Bernsen's, Niblack's, and Otsu's strategies for binarization and record improvement. Wolf [18] had proposed another technique for blind document bleed-through removal attributed of three phases: making of a twofold MRF display with a solitary perception field, at that point plan of an iterative streamlining calculation in light of the base cut/most extreme stream in an entire rebuilding process. The technique is assessed on checked record pictures of the eighteenth century and 83.23% review and 74.85% accuracy rates were asserted. Shi and Govindaraju [19] had proposed a standardization calculation appropriate for recorded report pictures utilizing a versatile direct capacity to estimate the uneven foundation because of the uneven surface of the archive paper, matured shading and light wellspring of the cameras for picture lifting. The calculation adaptively caught the foundation with a "best fit" straight capacity and standardized regarding the estimation. Garain et al. [20] had contrived a system that isolates the closer view and foundation in low-quality record pictures experiencing different sorts of debasements including examining clamor, maturing impacts, uneven foundation, or frontal area, and so forth. The calculation is likewise versatile to handle those issues of uneven enlightenment and nearby changes or non-consistency in foundation and frontal area hues. The approach was basically intended for preparing of shading archives and assessment results demonstrate that the strategy could remove lines and words with correctness's of around 84 and 93%, individually. Most promising works are reported in the literature suiting the requirements of a wide variety of documents of varying layouts and contexts. The algorithms devised are mostly adaptable with gray scale/monochrome image based operations. It is noticed that commonly found problems of degradation types addressed include illumination correction, bleed-through removal, aging effects removal, and other binarization-related issues. Preprocessing techniques devised are based on adaptive thresholding, morphological operations based and normalization techniques. In the perspective of improvising the preprocessing outcomes, a novel technique works based on the knowledge of noise degradation models of various degradation types persisting in the noisy document images. The working of proposed techniques is discussed subsequently.

2 Proposed Technique

The techniques for preprocessing of degraded document types include the creation of noise degradation models, detection of foxing effects (noisy regions) in the images, nonuniform illumination correction, show through effect removal, pen, scratch, and other stain mark removals.

2.1 Creation of Noise Degradation Models

Image enhancement is a nontrivial process which is highly subjective and bound by the context of image contents. Image restoration is the process of remodeling the actual geometrical properties of an image based on the prior knowledge of noisy phenomenon of a degraded image. Creation of noise degradation models based on the noisy patterns in an image is one the image restoration techniques [21]. In the proposed method, Gaussian and Rayleigh noise degradation models along with image averaging and harmonic mean image restoration filters are utilized for the creation of noise degradation models. Figure 1 depicts some of the noise patterns administered for the creation of noise degradation models.

Figures 2, 3, and 4 represent the sample documents considered for experimentation.

The gray levels of the noisy patterns of the image are exploited for estimation of its statistical features. Let v represent the Gaussian random variable and also a gray level, μ indicates the mean of average value of v and σ is the standard deviation, and then the Gaussian noise degradation estimate is given by (1).

$$p(v) = -\frac{1}{\sqrt{2\pi}\sigma} e^{(v-\mu)^2}/2\sigma^2 \qquad (1)$$

Fig. 1 Instances of noise patterns

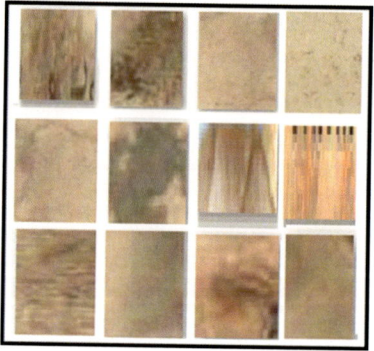

Fig. 2 Degraded document image—type 1

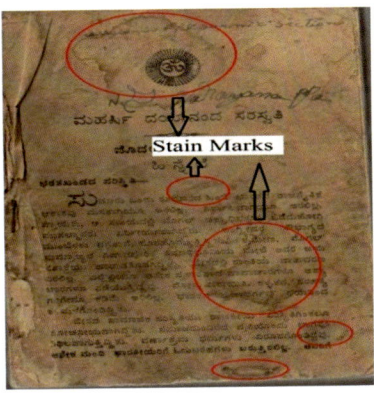

Fig. 3 Degraded document image—type 2

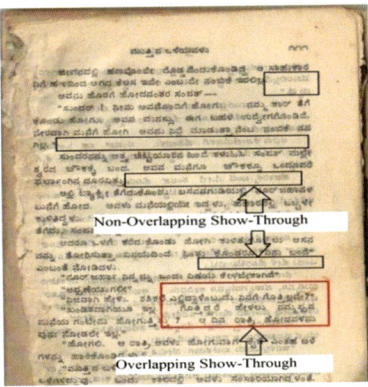

Similarly, the Rayleigh noise degradation estimate with respect to a random variable v is given by (2).

$$p(v) = \begin{cases} \frac{2}{b}(v - a)e^{-(v-a)^2/b}; & v \geq a \\ 0; & v < a \end{cases} \tag{2}$$

Further, the harmonic mean filter a image averaging is employed as one of the prominent features for creation of feature vector at the later level. If $g(x, y)$ is the pixel location for which the average gray level with respect to a region of K pixels, then average gray level at location $g(x, y)$ is given by (3).

$$\overline{g}(x, y) = \frac{1}{K} \sum_{i=1}^{K} g_i(x, y) \tag{3}$$

where $\overline{g}(x, y)$ is the average gray level computed over a neighborhood of K pixels with respect to a pixel $g(x, y)$ at location i.

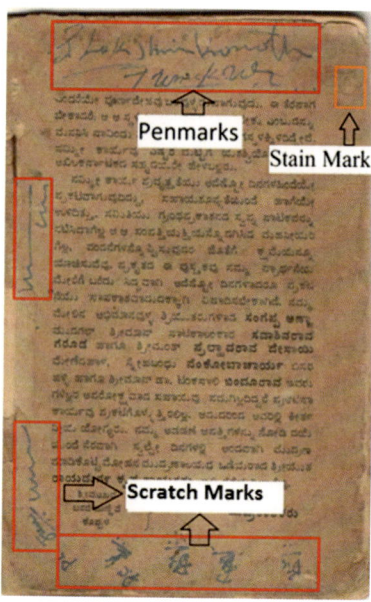

Fig. 4 Degraded document image—type 3

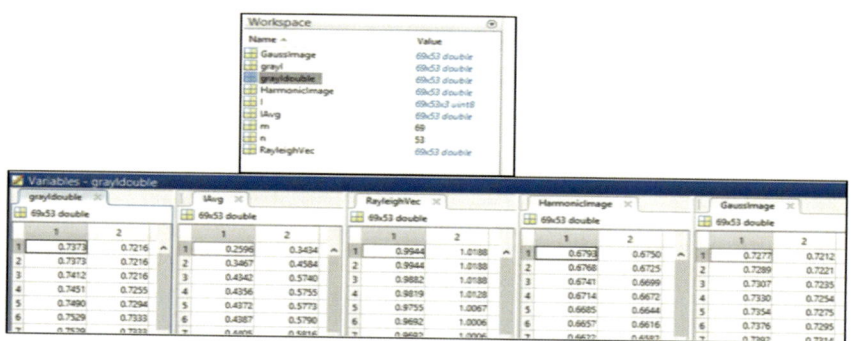

Fig. 5 Features extracted from noisy patterns

The harmonic mean filtering operation is given by (4)

$$\overline{g}(x, y) = \frac{mn}{\sum\limits_{(s,t)\in S_{xy}} \frac{1}{g(s,t)}}$$ (4)

where mn represents the size of rectangular neighborhood S_{xy}, $g(s, t)$ is the pixel at location (s, t) and $\overline{g}(x, y)$ is the harmonic mean at location $g(s, t)$.

Figure 5 presents the features extracted from the noisy patterns of the images.

2.2 Detection of Foxing Effect

The features extracted corresponding to noise degradation models and mean filters are further subject to detection of noisy patterns through convolution technique [22]. The features extracted are convolved over the image for the detection of foxing effects in the image. If the features computed during convolution matches with feature vectors of degradation models, then the regions will be considered as noisy.

Figure 6 represents the detection of foxing effects for an instance of an image through convolution and Fig. 7 shows the other types of noisy patterns detected using convolution.

Convolution is carried out with rectangular neighborhood of dimensions 15×15. Once the noisy regions are detected, the removal of detected noise is accomplished by comparing the mean and standard deviation range of noisy pixels.

Let μ and σ represent the mean and standard deviation, then the probability that an arbitrary pixel $p(x, y)$ to get assigned with the value between 0 and 1 is given by (5).

Fig. 6 The noisy region detected is encircled

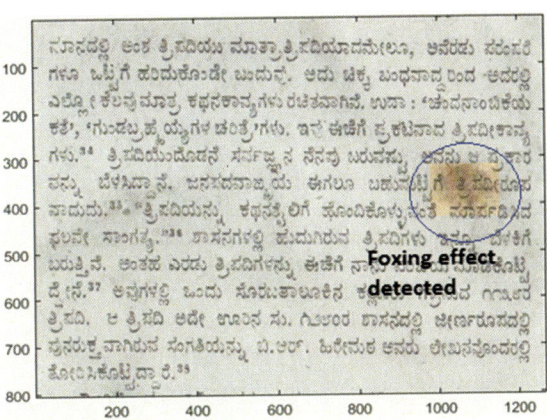

Fig. 7 Samples of detected noise patterns

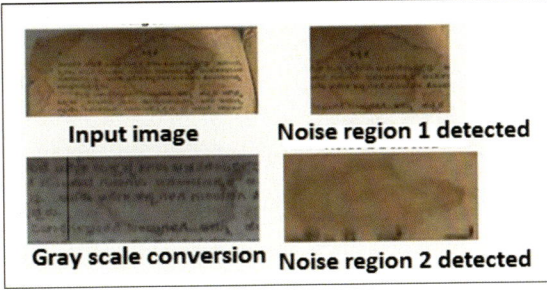

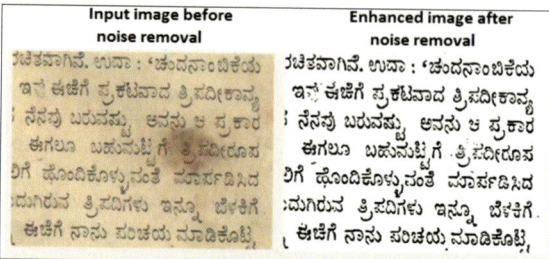

Fig. 8 Removal of foxing effect

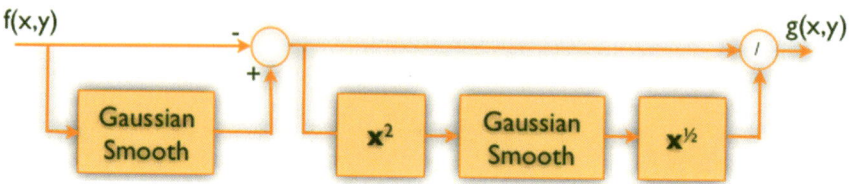

Fig. 9 Procedure of local spatial smoothing

$$p(x, y) = \begin{cases} 1 \\ 0 \end{cases} \tag{5}$$

where 1 indicates the background and 0 indicates the text pixels. It is expected $p(x, y) = 1$ subject to $0.3 < \mu[p(x, y)] < 1$ otherwise $p(x, y) = 0$.

Figure 8 depicts the outcome of foxing noise removal.

2.3 Nonuniform Illumination Correction

The occurrences of illumination inconsistencies within the document are one of the factors that results in noise coverage of textual regions. In the proposed method, correction of illumination is achieved using local spatial smoothing technique [23]. Local spatial smoothing is one of the most commonly used preprocessing techniques for medical image enhancement. Local spatial smoothing produces the effect of low pass filtering by attenuation of high-frequency components in an image. As in the current scenario, the illumination nonuniformities are existent due to the overriding of low-frequency regions (background) by high-frequency elements, therefore it is attempted to perform illumination correction using local spatial smoothing technique. Gaussian filtering [24] is the core functionality of spatial smoothing which is as presented in Fig. 9.

If $f(x, y)$ represents the input image, $m_f(x, y)$ and $\sigma_f(x, y)$ are the local mean and local standard deviation estimated on $f(x, y)$, then the output $g(x, y)$ is given by (6).

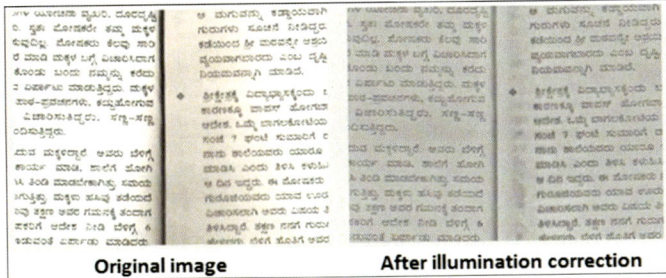

Fig. 10 Results of illumination correction

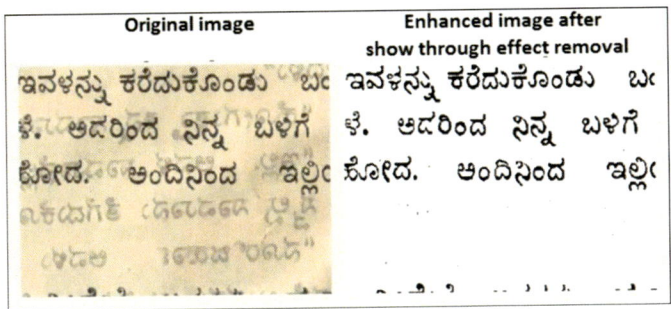

Fig. 11 Removal of show through effect

$$g(x, y) = \frac{f(x, y) - m_f(x, y)}{\sigma_f(x, y)} \qquad (6)$$

Figure 10 depicts the outcome of illumination correction.

2.4 *Removal of Show Through Effect, Pen Marks and Others*

Elimination of show through effect, pen/scratch marks and stains due to aging in document images are also performed using local normalization through local spatial smoothing technique. Figures 11, 12, and 13 show the results of local normalization.

3 Experimental Analysis

It is well known, the evaluation of preprocessing algorithmic outcomes is subjective to the satisfaction of user, hence the performances of proposed techniques are eval-

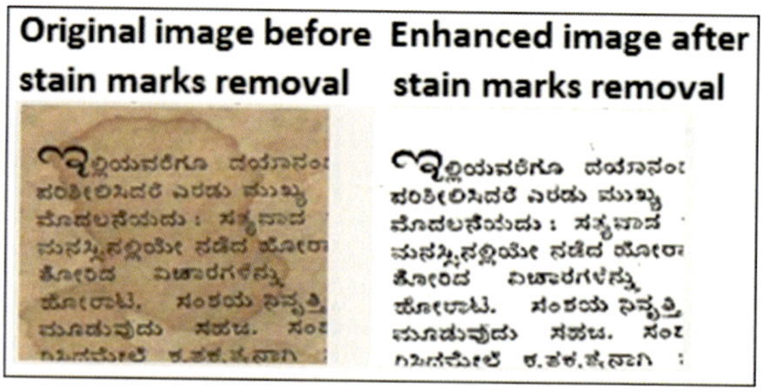

Fig. 12 Removal of stain marks

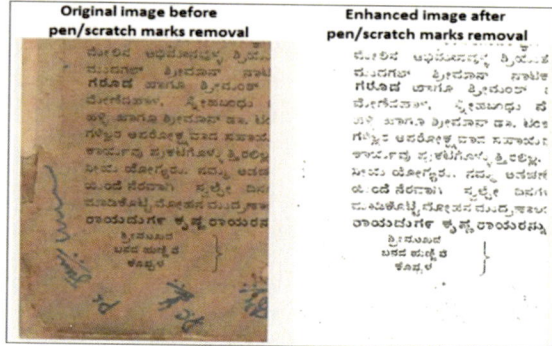

Fig. 13 Removal of pen/scratch marks

Table 1 Performance of preprocessing techniques

Type of noise pattern	No. of documents experimented	No. of documents with rating "Good"	No. of documents with rating "Moderate"	No. of documents with rating "Poor"
Foxing effect	72	33	28	11
Illumination correction	85	40	32	13
Show through effect	75	30	35	10
Pen/scratch marks	90	56	26	8
Stain removal	150	112	35	3

uated through subject studies. Around 50 scorers are used to rate the performance of outcomes obtained. All the datasets employed for experimentation are collected from Kannada library, where the context of works is related to poetry and literature of eighteenth and nineteenth centuries. A collection of more than 500 documents are scan converted to document images with a resolution of 300 dpi. The performance of outcomes is rated qualitatively into good, moderate and poor classes. The performance metrics of ratings provided by scorers are tabulated in Table 1 (Figs. 14 and 15).

Fig. 14 Accuracy of foxing effect removal

Fig. 15 Accuracy of illumination correction

Fig. 16 Accuracy of show through effect

3.1 Conclusion

A total of 472 documents with various noise patterns were experimented and the results showed a good amount of accuracy in preprocessing all different kinds of noises. Stain mark removal showed the highest accuracy with 75% of the testing documents being rated as good. Followed by, pen/scratch mark removal, illumination correction, foxing effect removal and show through effect removal with 62, 47, 46, and 40% of the documents being rated good, respectively (Figs. 16 and 17).

4 Conclusion

The preprocessing of historical documents is different when compared with other type of images like simple text documents of machine printed or typewritten images, etc. In this research, the focus is on developing some generic preprocessing algorithms

Fig. 17 Accuracy of
pen/scratch marks

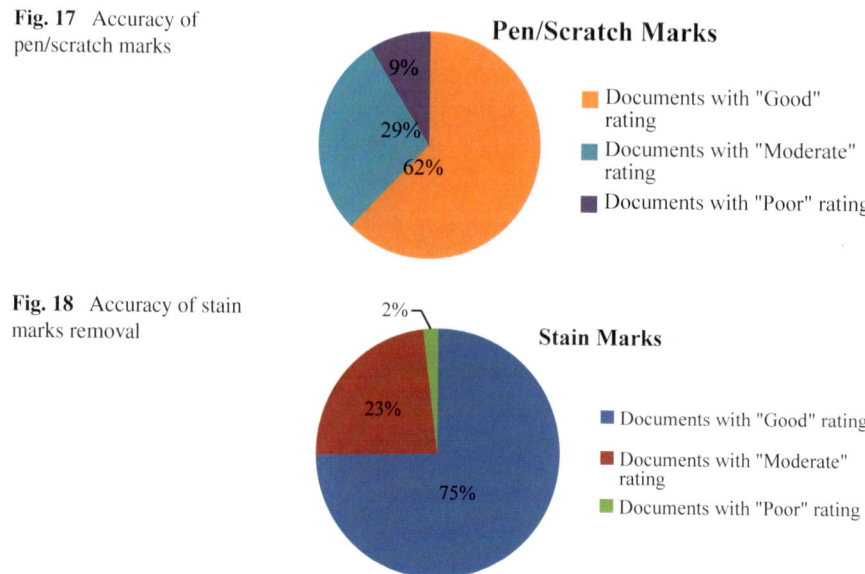

Fig. 18 Accuracy of stain
marks removal

specific for ancient document images. Specifically, the problems include the elimination of noises, identification and removal of stains, scratch marks, pen marks or irrelevant portions of text which are not in understandable format neither by humans nor by machines. The proposed technique employs noise models like Gaussian, Lucy Richardson, Rayleigh, etc., and extracts features from the noise imaged followed by the training of the degraded images and detection using convolution and correlation. The experimentation was carried out on the South Indian language historical document images. The results were shown in the form of screenshots which showed clear visuals of noise being eliminated from the degraded Kannada historical documents (Fig. 18).

References

1. Gupta MR, Jacobson NP, Garcia EK (2007) OCR binarization and image pre-processing for searching historical documents. Pattern Recogn 40(2):389–397
2. Farooq F, Govindaraju V, Perrone M (2005) Pre-processing methods for handwritten Arabic documents. In: Proceedings of the eighth international conference on document analysis and recognition, 2005, pp 267–271. IEEE
3. Rani NS, Vasudev T (2018) An efficient technique for detection and removal of lines with text stroke crossings in document images. In: Proceedings of international conference on cognition and recognition, pp 83–97. Springer, Singapore
4. Rani DANS, Vineeth P, Ajith D (2016) Detection and removal of graphical components in pre-printed documents. Int J Appl Eng Res 11(7):4849–4856
5. Gatos B, Pratikakis I, Perantonis SJ (2006) Adaptive degraded document image binarization. Pattern Recogn 39(3):317–327

6. Farooq F, Govindaraju V, Perrone M (2005) Pre-processing methods for handwritten Arabic documents. In: Proceedings eighth international conference on document analysis and recognition, 2005, pp 267–271. IEEE
7. O'Gorman L (1993) The document spectrum for page layout analysis. IEEE Trans Pattern Anal Mach Intell 15(11):1162–1173
8. Rehman A, Saba T (2014) Neural networks for document image preprocessing: state of the art. Artif Intell Rev 42(2):253–273
9. Gatos B, Ntirogiannis K, Pratikakis I (2009) ICDAR 2009 document image binarization contest (DIBCO 2009). In: 10th international conference on document analysis and recognition, 2009. ICDAR'09, pp 1375–1382. IEEE
10. Kavallieratou E, Stamatatos E (2006) Improving the quality of degraded document images. In: Second international conference on document image analysis for libraries, 2006. DIAL'06, 10-pp. IEEE
11. Chang SG, Yu B, Vetterli M (2000) Adaptive wavelet thresholding for image denoising and compression. IEEE Trans Image Process 9(9):1532–1546
12. Hsia CH, Hoang HG, Tu HY (2015) Document image enhancement using adaptive directional lifting-based wavelet transform. In: 2015 IEEE international conference on consumer electronics-Taiwan (ICCE-TW), pp 432–433. IEEE
13. Ntogas N, Veintzas D (2008) A binarization algorithm for historical manuscripts. In: WSEAS Proceedings of the international conference on mathematics and computers in science and engineering, no. 12. World Scientific and Engineering Academy and Society
14. Kitadai A, Nakagawa M, Baba H, Watanabe A (2012) Similarity evaluation and shape feature extraction for character pattern retrieval to support reading historical documents. In: 2012 10th IAPR international workshop on document analysis systems (DAS), pp 359–363. IEEE
15. Shirai K, Endo Y, Kitadai A, Inoue S, Kurushima N, Baba H et al (2013) Character shape restoration of binarized historical documents by smoothing via geodesic morphology. In: 2013 12th international conference on document analysis and recognition (ICDAR), pp 1285–1289. IEEE
16. Lu SJ, Tan CL (2007) Binarization of badly illuminated document images through shading estimation and compensation. In: Ninth international conference on document analysis and recognition, 2007. ICDAR 2007, vol 1, pp 312–316. IEEE
17. Kavallieratou E, Antonopoulou H (2005) Cleaning and enhancing historical document images. In: International conference on advanced concepts for intelligent vision systems. Springer, Berlin, pp 681–688
18. Wolf C (2010) Document ink bleed-through removal with two hidden markov random fields and a single observation field. IEEE Trans Pattern Anal Mach Intell 32(3):431–447
19. Shi Z, Govindaraju V (2004) Historical document image enhancement using background light intensity normalization. In: Proceedings of the 17th international conference on pattern recognition. ICPR 2004, vol 1, pp 473–476. IEEE
20. Garain U, Paquet T, Heutte L (2006) On foreground—background separation in low quality document images. IJDAR 8(1):47
21. Kanungo T, Haralick RM, Phillips I (1993) Global and local document degradation models. In: Proceedings of the second international conference on document analysis and recognition, 1993. IEEE, pp 730–734
22. Lee J-S (1980) Digital image enhancement and noise filtering by use of local statistics. IEEE Trans Pattern Anal Mach Intell 2:165–168
23. Ord JK, Getis A (1995) Local spatial autocorrelation statistics: distributional issues and an application. Geogr Anal 27(4):286–306
24. Young IT, Van Vliet LJ (1995) Recursive implementation of the Gaussian filter. Sig Process 44(2):139–151

An Efficient Classifier for P300 in Brain–Computer Interface Based on Scalar Products

Monica Fira and Liviu Goras

Abstract In this paper, a simple but efficient method for detection of P300 waveform in a Brain–Computer Interface (BCI) is presented. The proposed method is based on computing scalar products between the waveforms to be classified and a P300 pattern. Depending on the degree of concentration of the subject and the number of trails, rates of recognition between 85 and 100% have been obtained.

1 Introduction

The Brain-Computer Interface (BCI) is a communication system where the messages or commands that a person sends to the outside world do not go through the normal pathways of the brain and peripheral muscle. In fact, in a BCI system built to function based on EEG signals, messages are encoded in the EEG activity. The purpose of a BCI system is to provide its user with an alternative way of acting on the world by extracting the encoded activity from the EEG signal.

A BCI system is built from the following blocks or component parts: (a) a set of sensors that record the neuronal activity; (b) a signal processing block that extracts features; and (c) a translation algorithm that creates commands to operate an external device. Another aspect that needs to be considered is that the loop needs to be filled with feedback from the external device to the BCI user. Therefore, a BCI system is a closed loop running in real time. In the case of BCI communications systems, the external device serves as a means of communicating the user with the external environment.

A large number of BCI systems have been reported in the literature over theyears. Their operating principles use different neuronal characteristics, so they differ as a way of using neuronal information collected from sensors. Thus, noninvasive EEG

M. Fira (✉)
Institute of Computer Science, Romanian Academy, Iasi, Romania
e-mail: mfira@etti.tuiasi.ro

L. Goras
"Gheorghe Asachi" Technical University of Iasi, Iasi, Romania

© Springer Nature Switzerland AG 2019
D. Pandian et al. (eds.), *Proceedings of the International Conference on ISMAC in Computational Vision and Bio-Engineering 2018 (ISMAC-CVB)*, Lecture Notes in Computational Vision and Biomechanics 30,
https://doi.org/10.1007/978-3-030-00665-5_24

235

signals are use slow cortical potentials [1], motor potentials [2], event-related synchronizations and desynchronizations [3, 4], steady-state evoked potentials [5], and P300 potentials [6].

Depending on the way of communication between the human brain and the system, there are two classes of BCI systems, namely: dependent and independent. Dependent BCIs use the activity in the brain's normal output pathways to generate the brain activity (e.g., EEG) required for the system to function. An independent BCI does not depend on the brain's pathways (i.e., peripheral nerves or muscles); activity in such brain output is unnecessary to create the brain activity (e.g., EEG) required to execute a certain command [6, 7]. The independent BCIs are of greater theoretical interest than dependent BCIs [8–11]. This increased interest in independent BCI systems is due to the fact that they provide the brain with completely new output pathways comparative with dependent BCIs. Besides, for people with the most severe neuromuscular disabilities, who may lack all normal output channels (including extraocular muscle control), independent BCIs are likely to be more useful [12, 13].

Until now, P300-based BCI systems have been extensively studied and it has been concluded that from a psychological point of view, P300 evoked potentials that help maintain active memory when external stimuli are upgraded [7]. It has been shown that the P300 potential amplitude is directly proportional to the attention given to fulfilling a particular task [14, 15], and is associated to memory performance. The amplitude of the P300 pulse can also be seen as a reflection of the extent to which the information received from the outside is brain-processed in the form of representations of stimuli that generated the impulse. Amplitude variations are a measure of the degree to which information is processed. In short, there is a clear correlation between the impulse amplitude and the quality of the information processing. The pulse duration is the one that measures the classification speed, [16]. Tests have shown that the duration of the impulse varies from one subject to another and is a measure of the rapidity with which subjects allocate and maintain attention resources. There are pathologies in which P300 pulse duration is elevated, [15], and this is still an argument that it can measure cognitive abilities.

BCI based on the P300 can be done not only by illuminating letters or figures but also pictures or icons. The physiological mechanism of generating the P300 evoked potential is the same as in the case of the spelling paradigm, a visual stimulus. Thus, Hoffmann et al. [17] present a paradigm with the illumination of six pictures and on their web page, they also provide the EEG signals collected on 8 human subjects, with and without disabilities. They test two classification methods, namely, BLDA and FLDA as classifiers for P300 waveform detection.

In this paper, we present a simple and efficient method of classifying P300 evoked potentials by illuminating some pictures. To test the algorithm, we used the database available on the Internet at [18].

2   Materials and Methods

2.1   Experimental Schedule, Subjects, and Database

Experimental setup and schedule. Human subjects were shown six images on a laptop monitor. These images were: a television set, a telephone, a lamp, a door, a window, and a radio. The images were illuminated randomly, with only one image at a time. Image has been presented to the subject for 100 ms, and no more images have been lit in the next 300 ms, i.e., the inter-stimulus interval was 400 ms. The EEG signal was collected using the international standard 10–20 at 2048 Hz sampling rate and 32 electrodes placed [17].

For each subject, there were four sessions, two sessions a day and the time between the first session and the last session was less than two weeks. For each session, there were six rounds, one round for each image.

The subjects silently counted how often a picture appeared a prescribed image. After each run, the subjects were asked the result of counting. This has been done to monitor the performance of subjects [17].

Subjects. Human subjects were both normal and disabled, i.e., subjects 1–5 were people with handicaps and subjects 6–9 were normal. The disabled subjects were all wheelchair-bound but had varying communication and limb muscle control abilities [17].

Database. Four seconds after the alert tone, a random sequence of blinkers was started and the EEG registered. The flash sequence was randomized in block mode, which means that after six flashes, each image lit once, and after 12 flashes, each picture was blown twice, etc. The number of blocks was randomly chosen between 20 and 25. On average, 22.5 blocks of six intermittent were displayed in a single step, i.e., a cycle consisted of 22.5 target studies (P300) and $22.5 * 5 = 112.5$ nontarget (non-P300). The duration of a cycle was about one minute, and the duration of a session (including the installation of electrodes and short breaks between rounds) was about 30 min. A session covered, on average, 810 trials, and all data for one subject consisted, on average, of 3240 trials [17].

2.2   The Method

The first step consists of data preparation stage, namely the transformation of the EEG signals into sets of signals to be further processed. The second stage consists in choosing the EEG channels on which the classification will be made. This is followed by the step of *building a pattern* of the P300 waveform based on the EEG signals on the channels that were selected in the previous step. The next step is to classify the signals based on P300 pattern by calculating the *scalar product* between the pattern and each new EEG signal that is to be classified, and then establishing that the P300 has the signal corresponding to the illumination of the picture having the highest scaled product. In other words, at this stage, our classification problem appears as a 1 out of 6 ranking: which one of the 6 pictures has the highest probability to be the desired picture. For this, we will calculate the scalar product between the P300 pattern and the EEG signal corresponding to each photo illumination, and the decision was based on the maximum of the 6 scalar product values. In the following, we will detail each stage separately.

Preprocessing stage: For preprocessing EEG data, we chose to keep the same way of filtering, segmenting, windsorizing, and normalizing as Hoffmann used in [17]. Thus for data *filtering*, a 6th order forward–backward Butterworth bandpass filter was used. Cut-off frequencies were set to 1.0–12.0 Hz. Then, the EEG signals were downsampled from 2048 to 32 samples. For *trial extraction*, the following methodology was chosen: from the EEG signals trials of duration 1000 ms were extracted. The starting point of the trial was the stimulus onset, i.e., at the beginning of the intensification of an image, and end point was 1000 ms after the stimulus onset. For *scaling data* a normalization which takes into account the mean and the standard deviation, namely data normalization has been chosen. Several channel configurations have been tested namely:

- 4 electrodes:
- 6 electrodes:
- 8 electrodes:
- 16 electrodes:
- all electrodes (32 electrodes):
- (see Fig. 1)

The pattern construction stage: The samples from the selected channels were concatenated into vectors called EEG classification vectors. The vectors thus created are divided according to the session in two sets, namely, a training set and a test set. For the training set, we also used the information about the illumination of the picture; i.e., we extracted from this set only those EEG vectors that come from illuminating a picture to be watched (that is, a picture that at least theoretically must contain P300). Once these vectors are extracted (i.e., vectors containing P300), we will build a P300 pattern by mediating all of these vectors. The pattern size is given by the number of channels selected, i.e., $32 * Nc$, where Nc denotes the number of channels.

Classification stage: The classification stage consists in calculating the scalar product between the P300 pattern built at the previous stage and all the EEG vectors in the test set. We note that several trials were run (between 20 and 25, but in this paper we only used the first 20) and that a trial consists of lighting all 6 pictures. This means that in fact each of the 6 pictures has been illuminated 20 times. Thus, for each picture, in testing stage, we have more values of the scaled products (obtained from scalar product by P300 pattern and EEG signal corresponding to illumination picture), then we add all the values corresponding to each picture and thus obtain a single number, which is actually the sum of all the scalar products of the picture with

Fig. 1 The configuration
with all electrodes

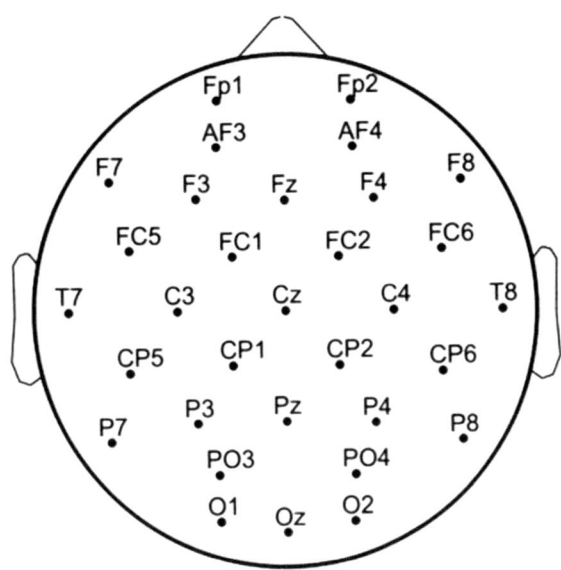

the P300 pattern. Then to classify a picture of the six possible ones, we considered
the highest value as belonging to the desired picture.

3 Experimental Results

Figure 2 shows the classification results for all subjects and for all channel configurations tested. It is found that for most subjects, maximum results are obtained when
the signals from all channels are used. Subjects 6 and 9 are exceptions, but subject 6
reported that he accidentally focused on the wrong stimulus during one run in session
1, which explains the weaker results. Also, lower performance for subject 9 may be
fatigue since somewhat lower performance is restricted to session 4.

There are differences between disabled subjects (1–4) and normal subjects (6–9).
Thus, for normal subjects, very good classification rates are obtained even with a
smaller number of channels and the classification rate reaches 100%. For subjects
with disabilities, the results were much lower than those obtained for normal subjects.

In Table 1, the classification results according to the number of trials are presented.
This result was obtained for all channels. In general, increasing the number of trials
improves the classification rate. This aspect is also observed if we look at the average
values written in the last row of the table.

In Table 2, the average classification for all subjects versus channels configurations
for a number of 20 trials is presented. There is a slight increase in the average value
with the increase in the number of channels, but it is found that for more than 8

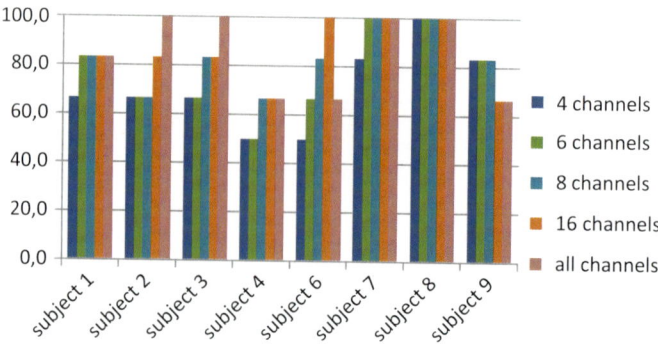

Fig. 2 Classification results versus subjects for several configurations channels for 20 trials

Table 1 Classification results versus subjects for several configurations channels for some trials
No. repetitions

	5 trials	10 trials	15 trials	20 trials	
Subject 1	66,7	83,3	100,0	83,3	All channels
Subject 2	50,0	100,0	100,0	100,0	All channels
Subject 3	66,7	50,0	83,3	100,0	All channels
Subject 4	33,3	33,3	66,7	66,7	All channels
Subject 6	83,3	66,7	66,7	66,7	All channels
Subject 7	83,3	100,0	100,0	100,0	All channels
Subject 8	100,0	83,3	100,0	100,0	All channels
Subject 9	33,3	66,7	66,7	66,7	All channels
Mean of all subjects	64,6	72,9	85,4	85,4	

Table 2 Average for all subjects versus channels configurations for 20 trials

Average for all subject versus channels configurations	
70,8	4 channels
77,1	6 channels
83,3	8 channels
85,4	16 channels
85,4	all channels

channels the growth is small and it is not justified to increase the number of channels to increase calculation time or the necessary resources.

4 Conclusions

This paper has presented a new and efficient P300 detection method of BCI system. With the proposed method good results for a large number of test trials are obtained. It has been found that the increased number of channels helps to improve the results. The accuracy classification obtained is 85% using EEG signals on all channels and 20 testing trials.

Acknowledgements We want to thank all human subjects who have voluntarily participated in experiment and Ulrich Hoffmann and his team for permission to use the EEG data available on the Internet.

References

1. Birbaumer N, Ghanayim N, Hinterberger T, Iversen I, Kotchoubey B, Kübler A, Perelmouter J, Taub E, Flor H (1999) A spelling device for the paralyzed. Nature 398:297–298
2. Mason SG, Birch GE (2000) A brain-controlled switch for asynchronous control applications. IEEE Trans Biomed Eng 47:1297–1307
3. Pfurtscheller G, Flotzinger D, Kalcher J (1993) Brain-computer interface: a new communication device for handicapped persons. J Microcomput Appl 16:293–299
4. Wolpaw JR, McFarland DJ, Neat GW, Forneris CA (1991) An EEG-based brain-computer interface for cursor control. Electroencephalogr Clin Neurophysiol 78:252–259
5. Jones KS, Middendorf M, McMillan GR, Calhoun G, Warm J (2003) Comparing mouse and steady-state visual evoked response-based control. Interact Comput 15:603–621
6. Farwell LA, Donchin E (1988) Talking off the top of your head: a mental prosthesis utilizing event-related potentials. Electroencephalogr Clin Neurophysiol 70:510–523
7. Donchin E, Spencer K, Wijesinghe R (2000) The mental prosthesis: assessing the speed of a P300-based brain-computer interface. IEEE Trans Rehab Eng 8(2)
8. Sutton S, Braren M, Zubin J, John ER (1965) Evoked correlates of stimulus uncertainty. Science 150:1187–1188
9. Donchin E (1980) Presidential address. Surprise!...Surprise? Psychophysiology 18:493–513
10. Fabiani M, Gratton G, Karis D, Donchin E (1987) Definition, identification and reliability of the P300 component of the event-related brain potential. In: Ackles PK, Jennings JR, Coles MGH (eds) Advances in psychophysiology, vol 2. JAI Press, New York, pp 1–78
11. Polich J (1999) P300 in clinical applications. In: Niedermeyer E, Lopes da Silva FH (eds) Electroencephalography: basic principles, clinical applications and related fields, 4th edn. Williams and Wilkins, Baltimore, pp 1073–1091
12. Rosenfeld JP (1990) Applied psychophysiology and biofeedback of event-related potentials (brain waves): historical perspective, review, future directions. Biofeedback Self Regul 15:99–119
13. Coles MGH, Rugg MD (1995) Event-related potentials: an introduction. In: Rugg MD, Coles MGH (eds) Electrophysiology of the mind: event-related brain potentials and cognition. Oxford University Press, New York
14. Kramer AF, Strayer DL (1988) Assessing the development of automatic processing: an application of dual-track and event-related brain potential methodologies. Biol Psychol 26:231–267
15. O'Donnell BF, Friedman S, Swearer JM, Drachman DA (1992) Active and passive P3 latency and psychometric performance: influence of age and individual differences. Int J Psychophysiol 12:185–187

16. Polich J (1986) Attention, probability, and task demands as determinants of P300 latency from auditory stimuli. Electroencephalogr Clin Neurophysiol 63:251–259
17. Hoffmann U, Vesin JM, Ebrahimi T, Diserens K (2008) An efficient P300-based brain–computer interface for disabled subjects. J Neurosci Methods 167(1):115–125 (15 Jan 2008)
18. http://infoscience.epfl.ch/record/101093

Detection of Weed Using Visual Attention Model and SVM Classifier

Manda Aparna and D. Radha

Abstract Agriculture is one of the provenances of human ailment in this heavenly body. It plays an extrusive role in the economy. Flourishing crops are a constituent of agriculture. Weeds are the additional plants to the crop. Removal of weeds is a challenging job for the farmers as it is a periodic, time–consuming, and cost-intensive process. Different ways to remove those weeds are by hand labor, spraying pesticides and herbicides, and machines but with their own disadvantages. The software solution can overcome these drawbacks to an extent. The main concern in software is in the identification of weeds among the crops in the field. The proposed system helps in detection of weeds in the agriculture field using computer vision methods. The method works with a dataset of crops and weeds. The plants are identified as salient regions in visual attention model and the identified plants are classified as crops or weeds using support vector machine classifier.

1 Introduction

Weed infestation is a universal complication in agriculture that negatively influences crop yielding. Despite, the method of herbicides constantly in a farmland has concluded in serious environmental contamination. The utmost trivial measures are definiteness spraying with choosy pesticides to preserve the assurance of drinking water and to diminish environmental collision in agriculture. Researchers have become increasingly vigilant of the crucial lead precision agriculture (PA) will play in the future [1]. Due to weeds random property, discrete approaches and methods for automatic weed recognition by machine perception have been proposed. Machine perception for discriminatory weeding or choosy herbicide sprinkling completely

M. Aparna · D. Radha (✉)
Department of Computer Science & Engineering, Amrita School of Engineering, Amrita Vishwa Vidyapeetham, Bengaluru, India
e-mail: d_radha@blr.amrita.edu

M. Aparna
e-mail: mandaaparna76@gmail.com

© Springer Nature Switzerland AG 2019
D. Pandian et al. (eds.), *Proceedings of the International Conference on ISMAC in Computational Vision and Bio-Engineering 2018 (ISMAC-CVB)*, Lecture Notes in Computational Vision and Biomechanics 30,
https://doi.org/10.1007/978-3-030-00665-5_25

243

confides on the capability of the ideology to evaluate weed images. Despite the image, segmentation may be favored as the presently primary step for detecting the objects using machine vision. Its unsteadiness arising from instability of the outdoor operating environment is a central interference to influence precisely the instability and rapidly abstract region-of-interest features from the immense measure of in-field image evidence is a stimulating complication.

The human's visual entity has a visual attention appliance that assists individuals promptly which prefers the vastly significant enlightenment from a scene. Featuring visual attention and simulating eye saccades are very profitable measures in computer vision. Recently, advancing reckoning visual attention representations are used to replicate the human visual appliance has been inducing enhanced interest in the domain of computer vision. Based on this, the foremost goal of this analysis is to detect plant as salient regions using visual attention mechanisms and computer vision theory in the outdoor operating environment and differentiate them as weed and crop.

For differentiating them as weed and crop here, we are using support vector machine as a classifier. Support vector machine (SVM) is a superintended machine learning innovation, which can be pre-owned for both classification and regression confrontations. Mainly it is issued for the classification process.

2 Related Work

In earlier days, the most common technique of removing weed in the crop was by manual monitoring in which a man was separately appointed for cross-checking of weed and crop and removing it by hand in the case if it was a weed. Nowadays, disparate image processing approaches have been adopted for the removal of weed in the agricultural farmland. Some previous works are listed as below. In earlier days, weeds were identified by simple image processing methods. The techniques involved were texture-based weed detection algorithm, probabilistic neural networks [PNN] and many more techniques were involved which was complex and time-consuming. In this algorithm, greenness identification and morphological operations were involved. This morphological operation considers many mathematical operations. This classification algorithm contains noise [2].

Weed detection is done using image processing and clustering analysis. It involves edge, texture properties, and clustering analysis. The results are not convincing in which the edges are not well defined. It is computationally expensive. Non-globular clusters [3] are not suitable in this case.

Other methods used for weed detection are Morphological functions, Otsu thresholding and Artificial Neural Networks (ANN). Here, morphological operations are used only for segmentation. In Otsu thresholding method, the performance degrades, when the global distribution of target image and background varies widely. The network structure of ANN is complex comparatively [4].

Image processing methods like canny edge detection, threshold algorithm, and morphological operations are also used for weed detection. This was a time-consuming process and selection of threshold is crucial because the wrong selection may result in over or under-segmentation [5]. In automated weed assortment including local pattern-relied texture descriptors, they used texture-based classification method and the classifier support vector machine (SVM). In this, the result is not convincing, as there is no preprocessing and because of it, noise is generated. A massive number of support vectors is desired from the training firm to execute regulation responsibility [6].

In weed detection system, the consequence of automatic systematization of broad and narrow weed utilizing characteristic vector were extricated by a composition of Gabor filter and FFT, and the classification employing the support vector machine (SVM). The weed classification routine comprises three processes like pre-transforming, feature extrication, and executing. Different parameters of SVM have to be analyzed for getting better results and a large number of images in the dataset have to be used for training [7]. The other way is vision-planted classifier in which Bayes and support vector machines, image segmentation, and decision-making are involved in detection of weed. The similarity in the texture and spectral signatures of the cereal crop and the weed leads to difficulty in weed detection [8].

The proposed system uses the visual attention model, which makes the salient region to be highlighted, and the classification is based on the salient region. The background and foreground are separated and classification is performed using support vector machine. This system works well in classifying broadleaf plants as crops and narrow leaf plants as weeds.

3   Proposed System

3.1   Visual Attention Model (VAM)

In modern lifetime, analyzers have determined that adopting human visual attributes to disclose objective could gain affirmative conclusions. The judicious attention property of human vision benefits inhabitants to pursue the salient object instantly and accurately from desolated locality without getting affected by complicated environment. Ultimately, the current saliency model structure usually needs an aggregation of the elementary visual features to propagate saliency maps, such as intensity, orientation and color information, and so on. This map provides a precise aspect of the visual environment, affirming salient locations in the visual field. The preference of visual features for saliency detection relies upon the evident application. Whereas assorted visual features contribute a discrete improvement to saliency detection, an assured feature may be capable in one case but fragile in another. Providing an input image, the first processing step consists of dissolving this input into an assortment of the apparent features: namely intensity, color, and orientation. The feature maps are created for various features as shown in the Fig. 1.

Fig. 1 Saliency mapping in
visual attention model

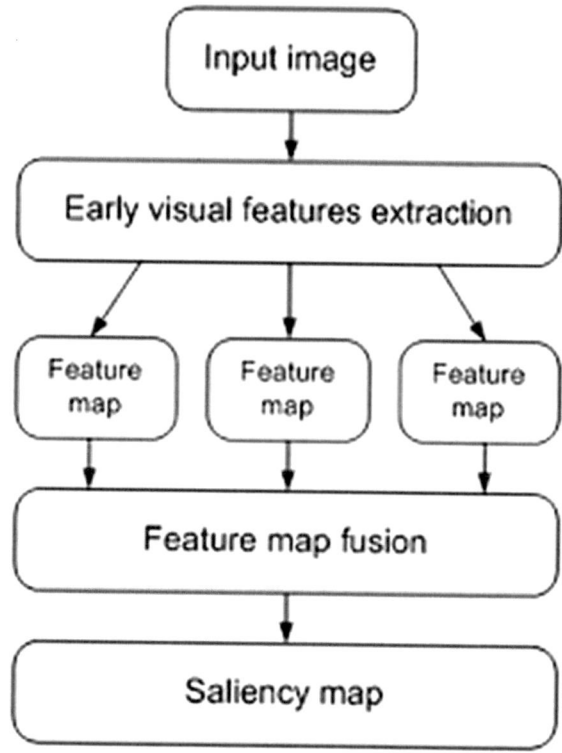

3.2 *Support Vector Machine (SVM)*

The SVM approach is a current literature setup for coordination employment. Support
vector machine (SVM) is an approach for constructing an optimal binary (2-class)
classifier. Figure 2a displays a dual-class issue by multifold available hyperplanes
splitting the pair data sets that are not naturally superlative. In Fig. 2b, an ideal
splitting hyperplane (OSH) is exhibited which produces the utmost limit (dashed
border) intervening the pair of data sets. SVM discovers this OSH by inflating the
limit among the classes. SVM initially converts prescribed information into a superior
dimensional zone by path of a kernel function and then builds a linear OSH among the
pair of classes in the commuted slot. Those evident vectors adjacent to the established
line in the transfigured field are labeled as the support vectors (SV). SVM is a relative
employment of the approach of "structural risk minimization" addressing to gain
subsided possibility of principle inaccuracy. To elevate SVMs, a kernel objective is
preferred [9].

In this analysis, the radial basis function (RBF) kernel [10] is preferred for
selection of SVM parameter. RBF kernel is ultimately utilized to overcome SVM

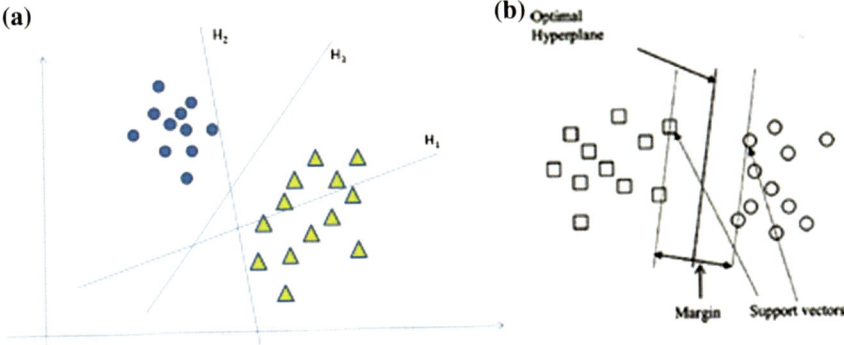

Fig. 2 a Abundant feasible distributing hyperplanes isolating the couple of groups. **b** Ideal dividing hyperplanes and the ultimate border. The circles and squares exhibit elements of two groups

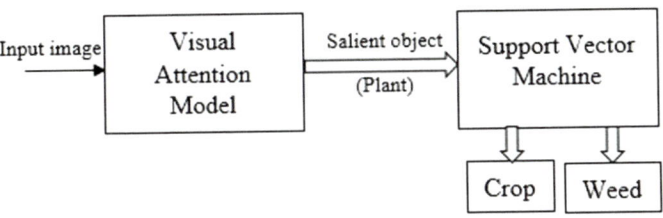

Fig. 3 Suggested system

classification issues. The generalized representation of the suggested system consists of the computational part as displayed in Fig. 3.

In the prospective system, the input image is given to the visual attention model, which results in extracting the saliency map of the image. This saliency map undergoes binarization that helps in locating the shape of the region of interest as a plant. Finally, the support vector machine classifies the plant into weeds or crops.

4 Implementation

The implementation consists of three modules namely visual attention model, binarization, and support vector machine. The plant image is given as input to the visual attention method. In visual attention, several processes are involved to abstract the region of interest needed. The primary process involved is feature extrication [11]. Feature extrication is a type of scale degradation that smoothly imitates the captivating element of an image as a solid attribute point. This access is appropriate when an image immensity is enormous and an abbreviated feature embodiment is recommended to instantly finish tasks such as image matching and image retrieval. The next process is saliency mapping. In computer vision, a saliency map is an image that displays each pixel's individual characteristic. The intention of a saliency map [12] is to elucidate and/or compressing the delegation of an image into the commodity that is more purposeful and effortlessly to estimate. The objective of saliency map is to embody the eminence or saliency in whole region of visual field by a scalar measure. It is used to lead the choosing of the attended region depending on the geographical allocation. The aggregation of feature maps supports bottom-up input to the saliency map.

Binarization is similar to segmentation for the identification of salient object from salient region. The main goal of binarization is separation of background and foreground of the plant image [13]. It is processed by using threshold. By using this threshold, the salient region of the plant image is obtained [14].

Finally, the classification of the plant object as crop or weed is identified by SVM classifier from the dataset. SVM classifier is trained with the training dataset to acquire needed splitting hyperplanes.

The dataset considered for the analysis are manually collected images with 31 broadleaf images and 47 narrow leaf images [15]. Broadleaf images are taken as crops and narrow leaf images are considered as weed. 55 images are randomly selected for training module and 31 broadleaf images, and 47 narrow leaf are used for the testing module. The results obtained from the training dataset are used to test the images and classify it as crop or weed. The implementation is shown in Fig. 4.

5 Results and Analysis

The plant image is given as input to visual attention model in order to obtain saliency map, binaralized saliency map, and salient region. Saliency map of input plant image exhibits each pixel's exclusive characteristic. Saliency mapping is a set of contours extricated from the image. Each of the pixels in the region is identical relative to some qualities such as color, intensity, and orientation. The objective of saliency map is to elucidate an input image into an entity for easier analysis. Binaralized saliency mapping is computed in which the foreground and background are shown in different colors. It is used to detect boundaries of an image. Saliency maps are composed at various scales. These maps are combined pixel-wise to achieve the eventual saliency map. The salient region of plant image is resolved as the local contrast of an image region relative to its surrounding at different scales. Finally, the salient region is detected. Figure 5 shows the computation process of the visual attention model.

The SVM classifier is used for identification of weeds and crops. The training and testing images are in a predefined folder. Here, 55 images are used for training and 78 images are used for testing. The Gabor kernel and Fourier transform (FT) carry out the filtering of these images and input real-time image. Filtering is a method for altering or embellishing an image. Gabor kernel filter is a linear filter applied for texture analysis which fundamentally determines any precise frequency content in the image in any particular directions in a bounded region around the analysis region. The Fourier transform (FT) is a crucial image processing tool, which is used

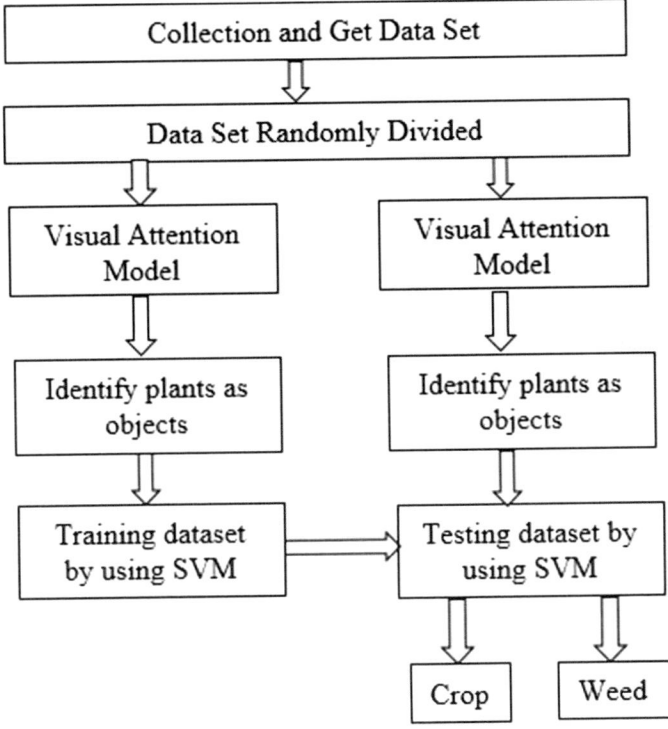

Fig. 4 SVM testing and training modeling framework

Table 1 Results of crop input images

Total number of crop images	Number of crop images detected as crops correctly	Number of crop images detected as weeds (Incorrect)	Percentage of detecting crops correctly
31	20	11	64.5

to dissolve an image into its sine and cosine parts. In the Fourier, concern image, each point shows specific frequency contained in the spatial domain image. In this paper, assumption of weed is by 0 and crop is by 1. All these values of images are stored in an array. SVM classifier performs the identification of weed and crop from input image by comparing the features of the training images. Figure 6 is a sample crop image given as input for visual attention model and the SVM classifier for crop or weed identification. Figure 7 is a screenshot of the result obtained for the input image shown in Fig. 6.

A sample weed image shown in Fig. 8 as input for visual attention model and the SVM classifier for crop or weed identification. Figure 9 is a screenshot of the result obtained for the weed input.

The results of crops images as input is shown in Table 1.

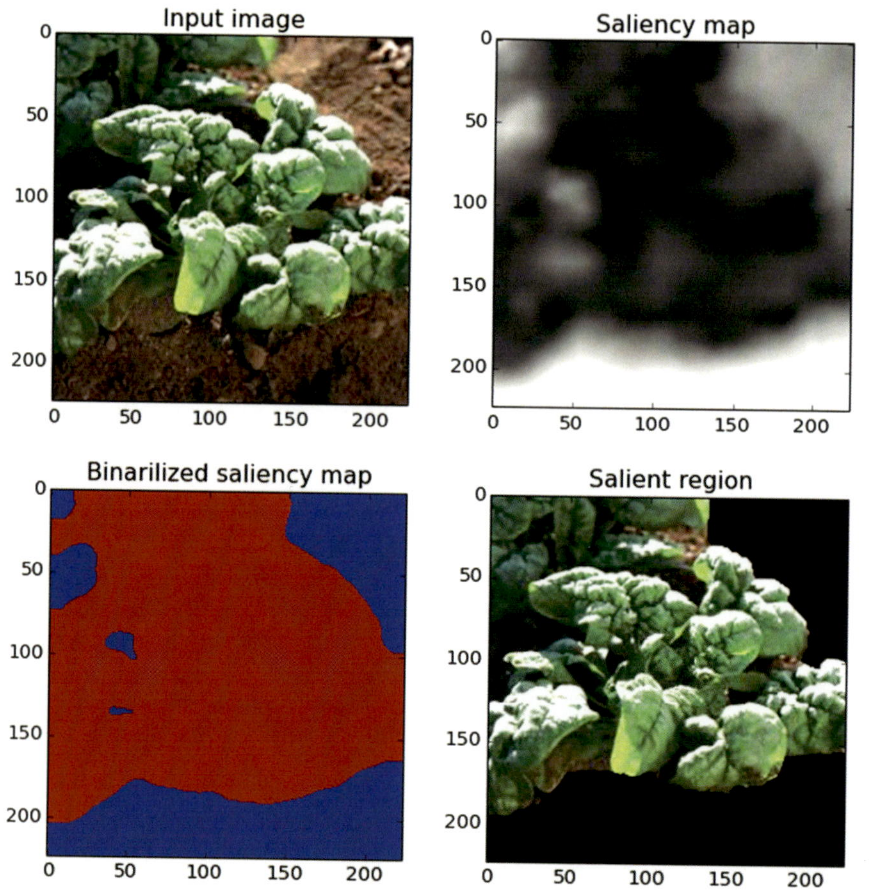

Input image

Saliency map

Binarilized saliency map

Salient region

Fig. 5 Computation process

Fig. 6 Crop image

The results of weeds images as input is shown in Table 2.

```
                                                                    Python 3.6.4 Shell
File  Edit  Shell  Debug  Options  Window  Help
Python 3.6.4 (v3.6.4:d48eceb, Dec 19 2017, 06:04:45) [MSC v.1900 32 bit (Intel)] on win32
Type "copyright", "credits" or "license()" for more information.
>>>
==================== RESTART: F:\prjt\testFrida - Crop.py ====================
[2018-03-19 14:18:31,193      INFO] Shape of the test data after transform is : (3, 8)
[2018-03-19 14:18:31,198      INFO] Confusion matrix:
 [[0 1]
 [0 2]]
(1, 8)
[2018-03-19 14:18:31,607      INFO] Shape of the real life test data after transform is : (1, 8)
[1]
crop is detected
>>>
```

Fig. 7 Detection of crop

Fig. 8 Weed image

```
                                                                    Python 3.6.4 Shell
File  Edit  Shell  Debug  Options  Window  Help
Python 3.6.4 (v3.6.4:d48eceb, Dec 19 2017, 06:04:45) [MSC v.1900 32 bit (Intel)] on win32
Type "copyright", "credits" or "license()" for more information.
>>>
==================== RESTART: F:\prjt\testFrida - Crop.py ====================
[2018-03-19 14:15:35,349      INFO] Shape of the test data after transform is : (3, 8)
[2018-03-19 14:15:35,354      INFO] Confusion matrix:
 [[0 0]
 [2 1]]
(1, 8)
[2018-03-19 14:15:35,758      INFO] Shape of the real life test data after transform is : (1, 8)
[0]
weed is detected
>>>
```

Fig. 9 Detection of weed

6   Conclusion

Removal of weeds is a challenging job for the farmers as it is a periodic, time—consuming, and cost-intensive process. The competence of spotting and analyzing crops and weeds could spark the evolution of sovereign vision-instructed agricultural

Table 2 Results of weed input images

Total number of weed images	Number of weed images detected as weeds correctly	Number of weed images detected as crops (Incorrect)	Percentage of detecting weeds correctly
47	40	7	85.1

contraptions for site-peculiar herbicide employment. In this paper, we have suggested a categorization type model which uses visual attention model and support vector machine(SVM) to organize the plants in the images as crop or weed in order to scale down the exaggerate treatment of herbicides in agricultural systems. Analysis of the result exposes that the SVM acquires good percentage of accuracy in identifying the weeds. This type of classification can be used for the agricultural robot to distinguish the crops and weeds which in turn can help in removal of weeds.

References

1. Wu L (2017) Detection of the salient region of in-field rapeseed plant images based-on visual attention model. In: 2017 2nd Asia-Pacific conference on intelligent robot systems (ACIRS). IEEE, New York, pp 33–36
2. Mohanapreethi J, Sujaritha M (2017, May) Texture-based weed identification system for precision farming. Int J Sci Eng Res 8(5). ISSN 229-5518
3. Sukumar P, Ravi S (2016) Weed detection using image processing by clustering analysis. Int J Emerg Technol Eng Res (IJETER) 4(5):14–18
4. Ayswarya R, Balaji B, Balaji R, Ramya R, Arun S (2017, March) Weed detection in agriculture using image processing. Int J Adv Res Electr, Electron Instrum Eng 6(3)
5. Deepa S, Hemalatha R (2015) Weed detection using image processing. Int J Res Comput Appl Robot 3(7):29–31
6. Ahmed F, Kabir MH, Bhuyan S, Bari H, Hossain E (2014) Automated weed classification with local pattern-based texture descriptors. Int Arab J Inf Technol 11(1):87–94
7. Ishak AJ, Mustafa MM, Tahir NM, Hussain A (2008) Weed detection system using support vector machine. In: International symposium on information theory and its applications, 2008, ISITA 2008. IEEE, New York, pp 1–4
8. Tellaeche A, BurgosArtizzu XP, Pajares G, Ribeiro A (2007) A vision-based classifier in precision agriculture combining Bayes and support vector machines. In: IEEE international symposium on intelligent signal processing, 2007, WISP 2007. IEEE, New York, pp 1–6
9. Auria L, Moro RA (2008) Support vector machines (SVM) as a technique for solvency analysis
10. Ahmed F, Bari ASMH, Hossain E, Al-Mamun HA, Kwan P (2011) Performance analysis of support vector machine and Bayesian classifier for crop and weed classification from digital images. World Appl Sci J 12(4):432–440
11. Venkataraman D, Mangayarkarasi N (2016) Computer vision based feature extraction of leaves for identification of medicinal values of plants. In: 2016 IEEE international conference on computational intelligence and computing research
12. Sriram B, Sai Hemanth Reddy K, Santhosh Kumar S, Sikha OK (2017) An improved levelset method using saliency map as initial seed. In: International conference on signal processing and communication (ICSPC'17)—28th & 29th July 2017
13. Malemath VS, Hugar SM (2016, May-June) A new approach for weed detection in agriculture using image processing techniques. Int J Adv Sci Tech Res 3(6). ISSN 2249-9954

14. Cheng M-M, Warrell J, Lin W-Y, Zheng S, Vineet V, Crook N (2013) Efficient salient region detection with soft image abstraction. In: 2013 IEEE international conference on computer vision (ICCV). IEEE, New York, pp 1529–1536
15. Anjali Rani KA, Supriya P, Sarath TV (2017) Computer vision based segregation of carrot and curry leaf plants with weed identification in carrot field. In: Proceedings of the IEEE 2017 international conference on computing methodologies and communication (ICCMC)

Design and Development of Scalable IoT Framework for Healthcare Application

Siddhant Mukherjee, Kalyani Bhole and Dayaram Sonawane

Abstract With increasingly fast-paced life and alarmingly high rate of chronic ailments in general population, there is a need for quickening the current process of healthcare monitoring, especially in emergency situations. The recent developments in communication technology, especially in the field of Internet of Things (IoT) have enhanced the accessibility of such systems. This can be achieved by transmitting/uploading the data of various health parameters acquired by different physiological sensors with wireless sensor networks onto the cloud platform. This data later can be accessed by the concerned medical authorities when required for diagnosis. In this work, we focus on the development of the scalable IoT framework for monitoring the physiological parameters of the patient. The customized MATLAB-based GUI is designed to perform real-time analysis of sensor's data which is used for continuous monitoring of vital parameters of the body. We have developed a wearable band which can be worn as a wristband by the patient. The band consists of temperature, pulse, and ECG sensors those are used to transmit the vital parameters of the patient integrated with the ultra-low-power battery-operated Texas Instruments MSP430 microcontroller with CC110L sub-1 GHz RF wireless transceiver. The concept is successfully demonstrated by transmitting three physiological parameters wirelessly over 100 m distance as well as over the cloud platform.

1 Introduction

Wireless sensor networks are growing as a platform which enhances the relationship between the devices used by several communities for generations and devices which constantly make the use of Internet and maintain the organization of data related to

S. Mukherjee · K. Bhole · D. Sonawane (✉)
College of Engineering Pune, Shivajinagar, Pune, India
e-mail: dns.instru@coep.ac.in

S. Mukherjee
e-mail: mukherjeesg16.instru@coep.ac.in

K. Bhole
e-mail: kab.instru@coep.ac.in

© Springer Nature Switzerland AG 2019
D. Pandian et al. (eds.), *Proceedings of the International Conference on ISMAC in Computational Vision and Bio-Engineering 2018 (ISMAC-CVB)*, Lecture Notes in Computational Vision and Biomechanics 30,
https://doi.org/10.1007/978-3-030-00665-5_26

255

showcasing the condition of the environment through means of the physical parameters collected by a single station. Applications vary from the use of such networks to highlight problems related to a particular societal cause ranging from manufacturing and health care to retail business and home automation. Wireless sensor networks or WSNs are budding technology with an immense amount of potential to change the way people are living their lives. WSNs have been used to enable better data collection in scientific studies, create more effective strategic military defenses and monitor factory machinery [1]. Wireless sensors can be classified on the basis of readiness for field deployment, scalability, and cost. Readiness for field deployment measures maturity for field deployment in terms of economic and engineering efficiency. The requirement for the sensor to be scalable to distributed environmental monitoring tasks requires that the sensors are small and inexpensive and architecture needs to be equally capable enough to accommodate or scale-up to many distributed systems. Sensors are deployed in large numbers and a prerequisite for the deployment of such sensors is that cost will drop and also energy requirement for maintaining the healthy state of a scalable architecture incorporating such sensors should be low. Sensor networks represent a significant advantage over traditional invasive methods of monitoring and also represent a more economical method for conducting long-term studies than traditional methods [2].

Sensor networks used for medical applications mostly range their use either as an implantable or as a wearable. Implantables can be inserted into a human body, whereas wearables are devices that are used to be worn on the surface of the skin or to the nearest portion of the body. Body area networks can collect information about an individual's health, fitness, and energy expenditure [3, 4]. Scalability is a very important and crucial issue in the design of routing protocols for WSNs. Scalability involves sensitivity and range of sensors, communication bandwidth of the radio and power usage. The software issues include reliability of data transfer, development of a methodology to maintain the management of a large volume of data and scalable algorithms for viewing and analyzing the data in real time. However, there needs to be a proper balance between the increased number of nodes and the rise in complexity to manage such increase in nodes. Internet of things or IoT is a new Internet paradigm based on the fact that there will be many more things than human connected to the Internet.

Jiao et al. (2017) adopted CC2420 transceiver chip which used 2.4 GHz radio frequency band for wireless transmission of sensor data to the remote site [5]. We have used CC110l transceiver chip which supports 915–920 MHz radio frequency bands in this project for communicating sensor values between two transceivers. However, CC110l consumes less receiver current (i.e., 14 mA) compared to CC2420 (i.e., 12 mA). Gupta et al. (2010) designed a MATLAB GUI for acquisition and online analysis of ECG signal along with an embedded system [6]. In this project, Matlab GUI was used for acquisition and analysis of temperature, pulse, and ECG sensor at the same time as it helps in real-time analysis of raw data coming from these sensors, i.e., one can understand if the patient is suffering from cardiac arrhythmia by analyzing his/her QRS complex with respect to time interval in milliseconds at which the signal is recorded. Moreover, one can switch quickly among various buttons to visualize sensor data separately or all at the same time which indeed offers an instantaneous view of physical parameters of the body.

This paper introduces the development of such framework for three sensors which can be scaled up to eight sensors depending upon the application. The system will make use of local RF band for transmission and reception of sensor data and the data will be sent to a web-based dashboard using Wi-Fi protocol. This paper also mentions the use of MQTT as an application layer to send the data and its advantages over other protocols. This paper also presents a real-time user interface developed in Matlab to easily monitor sensors irrespective of their number and represents the data graphically as well as numerically and simultaneously store data onto a file which keeps a record of such data. The paper includes the following sections: Sect. 2 provides detailed information about system architecture, Sect. 3 highlights results obtained after implementation of the system architecture and incorporating sensors with it and discussions upon it, and Sect. 4 offers conclusions and future scope of this project along with references.

2   System Architecture

The system consists of blocks that represent the whole architecture which is as follows:

- Sensor Block
- Transmitter Block
- Receiver (Gateway) Block
- Data Acquisition Block.

The Transmitter Block communicates with Receiver (Gateway) Block via local RF protocol and Receiver (Gateway) Block transfers such data to the web-based dashboard over Wi-Fi which can be viewed on a personal computer or on any device that supports Wi-Fi connectivity. This architecture enables any sensor connected to the transmitter block to communicate with the central station which supports Wi-Fi connectivity independently. Figure 1 represents the system architecture, the figure also shows a connection flow among different components and how they communicate with each other.

2.1   Sensor Block

The Sensor Block consists of sensors which are used for monitoring the health status of an individual. Sensors used for this purpose are temperature sensor, pulse sensor, and ECG electrodes. LM35 is used to sense and measure the temperature of the body. LM35 is a temperature sensor IC which measures temperature in the range of 0–100 °C which also covers body temperature range of 4–40 °C and is calibrated

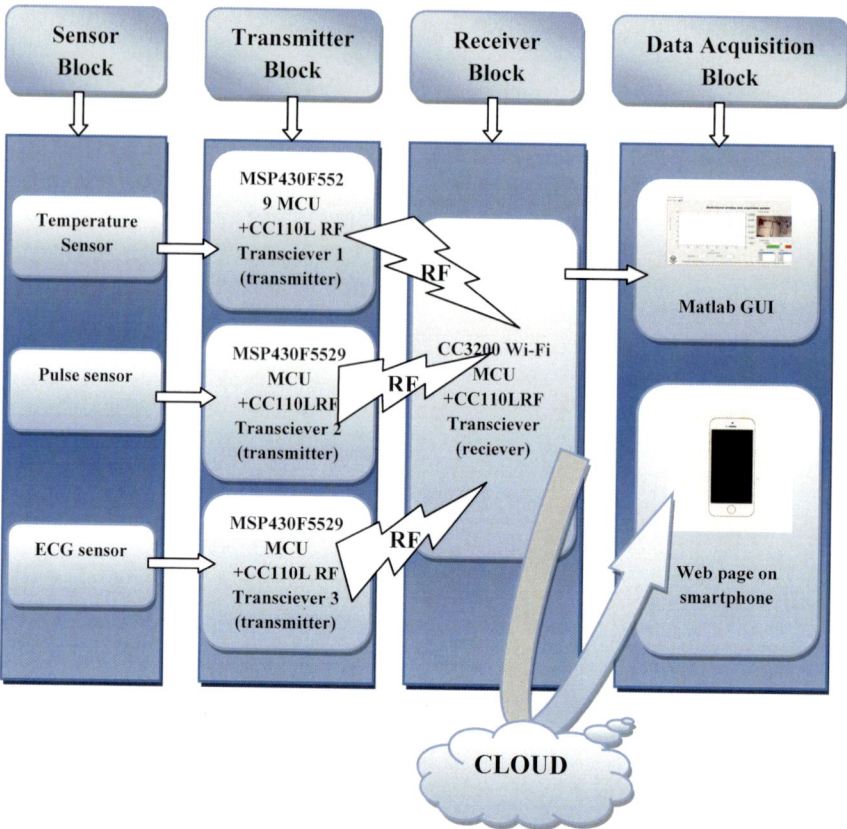

Fig. 1 Block diagram of a proposed system architecture

and connected to an analog channel of MSP430F5529 microcontroller unit as shown in Fig. 2. Pulse sensor is used to calculate the BPM or Beats Per Minute which is connected to the analog channel of separate MSP430F5529 microcontroller unit as shown in Fig. 2. Pulse sensor unit consists of a green led and a low-cost ambient light photosensor (photodiode) and placed at the tip of the finger or near the veins on an arm. Pulses are calculated via reflective method, i.e., light from green led falls on the surface and gets reflected as soon as the pulse is detected (blood cells are pumped into the blood vessels and forms a pulse). BPM is calculated by calculating the difference between the peak of two consecutive pulses and multiplying by 60 s. ECG electrodes were used as the third sensor to measure ECG by placing three Ag–AgCl electrodes at the right arm, left arm and right leg, respectively. These electrodes were connected to the AD8232 analog front-end ECG module which consists of analog high-pass filter and low-pass filter to remove the noise from ECG signal which are presented as a result of 50 Hz baseline interferences. The AD8232 module was connected to the analog channel of the third MSP430F5529 microcontroller unit as shown in Fig. 2.

(a) **(b)**

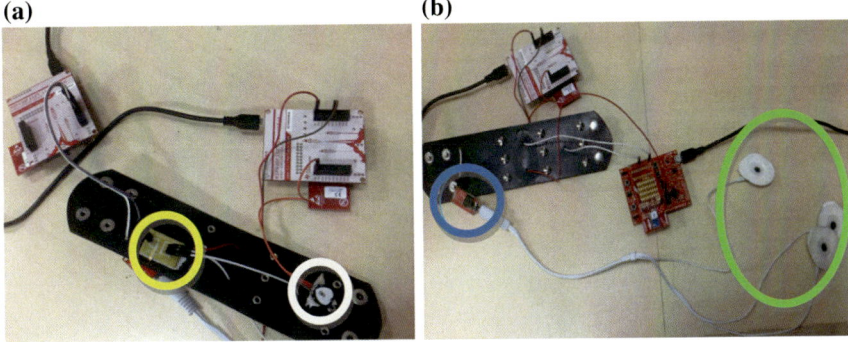

Fig. 2 Health band showing **a** temperature sensor encircled in yellow and pulse sensor encircled in white and **b** AD8232 ECG module encircled in blue and ECG electrodes encircled in yellow

2.1.1 MSP430 and Its Applications in Health Care

Devices used for monitoring personal health requires the availability of some of the key features like ultra-low-power consumption, battery life which decides form factor, reduced size, and so microcontrollers need to be chosen accordingly. The main reason behind using MSP430 for healthcare applications is that it is designed for an ultra-low-power application because it utilizes a flexible clocking system and a variety of operating modes designed to reduce power consumption and hence, extending battery life which can be said by the fact that the current drawn during active mode is in the range of 100–200 μA/MHz and less than 1 μA during a standby mode and can also operate down to 1.8 V, further improving a superior power specification [7]. One can achieve a 10-year life with 1 Ah battery using MSP430 as it provides 10 μA average power budget for the entire system which can be useful for developing both portable as well as implantable personal healthcare devices [8].

2.2 Transmitter Block

The Transmitter Block consists of Texas Instruments MSP430F5529 MCU or microcontroller unit to which Anaren's CC110L Air Transceiver Booster Pack is attached. CC110L Air Transceiver Booster Pack is connected to MSP430F5529 MCU via SPI interface and is programmed as a transmitter. This transmitter sends the relevant sensor data in the form of RF packets and uses sub-1 GHz RF, i.e., 915–920 MHz. Sub-1 GHz has several benefits over 2.4 GHz radios of being low power and give coverage up to 1–4 km and can pass through walls much better than 2.4 GHz [9]. In this project, three such transmitters to which three sensors were connected are shown in Fig. 2a, b. However, a number of sensors connected to the system can be increased or scaled up to eight sensors depending upon the application.

2.3 Receiver (Gateway) Block

The Receiver (Gateway) Block consists of Texas Instruments CC3200 Wi-Fi Micro-controller Unit which features Wi-Fi CERTIFIED[TM] chip and uses IEEE 802.11 bgn which means it supports 900 MHz–2.4 GHz single band frequency band [10]. CC3200 Wi-Fi MCU also comes with Enhanced IoT Networking Security and has been optimized for low power. In this project, CC110L Air Transceiver Booster Pack is connected to CC3200 Wi-Fi MCU via SPI interface and programmed as a receiver. This receiver accepts the RF packet sent by the transmitter depending upon the pre-requisite that transmitter and receiver should have the same channel ID and the same initial address. The received sensor packet from the receiver goes to the separate 12-bit ADC channels of CC3200 Wi-Fi MCU chip and such data can be acquired by gaining access to the serial port of CC3200 Wi-Fi MCU. As sensor data can be accessed from the ADC channels, we can increase the number of sensors connected to Transmitter Block depending upon ADC channels present in CC3200 Wi-Fi MCU which makes this IoT architecture scalable and more efficient to be used.

2.4 Data Acquisition Block

The Data Acquisition Block comprises a real-time based user interface which is utilized to visualize the collected data by gaining access to the serial port of CC3200 Wi-Fi MCU. MATLAB GUI was designed which can connect to the serial port of the CC3200 Wi-Fi MCU and allows visualization of data graphically as well as numerically and simultaneously store data on the folder onto which MATLAB GUI was made and also consist of different channels which can show the individual status of sensors and their data. Further, CC3200 Wi-Fi MCU was programmed to connect to a Wi-Fi network by adding credentials of that Wi-Fi network and send data to the web-based dashboard using MQTT protocol as an application layer. MQTT or Message Queue Telemetry Transport is a publish/subscribe protocol that runs on the top of the transport layer. This protocol is better for IoT as it provides lower network bandwidth and increases the lifetime of battery-run devices [11]. MQTT protocol requires the name of the topic under which sensor data will be sent to the MQTT broker and data can be accessed on a web-based dashboard which can be seen from anywhere on an internet webpage via PC or smartphone having excellent Internet connectivity.

3 Results and Discussions

Results are obtained after implementing the architecture as shown in Fig. 1 and sensor data from health band were visualized on MATLAB GUI as well as on web-based dashboard shown in Figs. 3, 4, 5 and 6.

Results obtained from sensors via data acquisition using MATLAB GUI shows the freedom of visualizing important medical information from data by clicking on any one of the three channels and analyzing all such information in real time. We have also learned that instead of relying on patient monitors and physical presence of a physician for the analysis of physical parameters of a patient, a physician can continuously monitor the patient's health from his smartphone having internet connectivity and give his/her expert advice to the patient from anywhere.

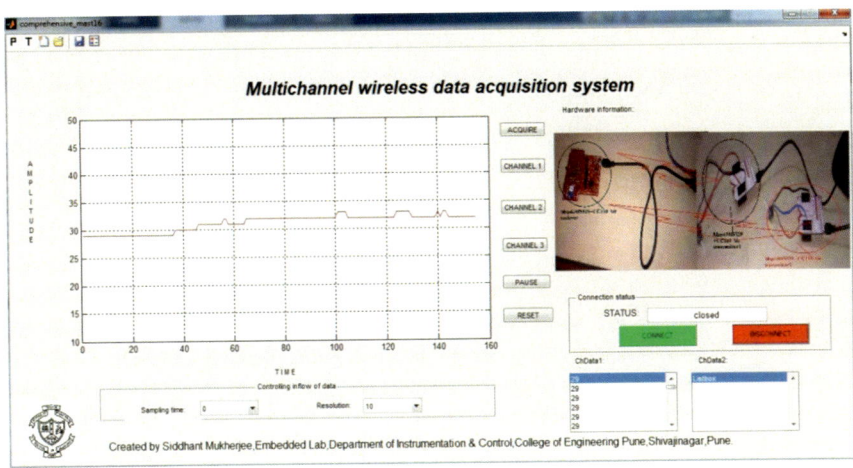

Fig. 3 Temperature sensor showing the temperature of the body in MATLAB GUI

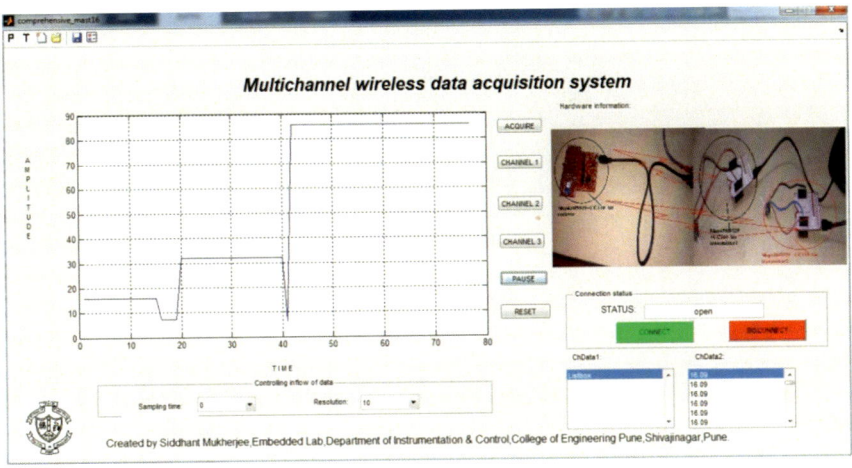

Fig. 4 Pulse sensor showing BPM of the person in MATLAB GUI

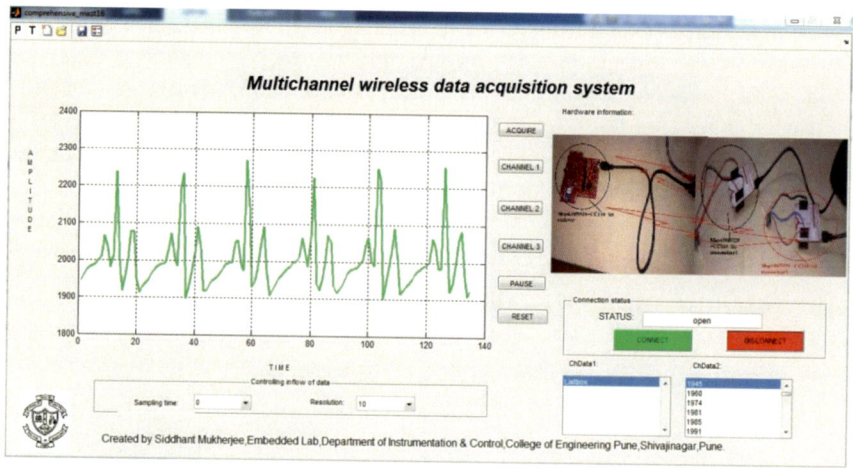

Fig. 5 ECG electrodes showing ECG of the person in MATLAB GUI

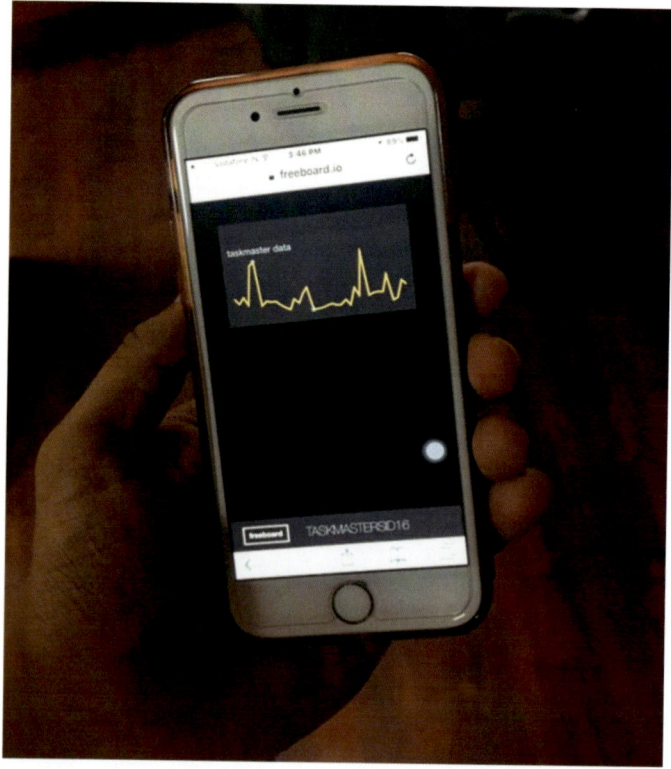

Fig. 6 Sensor data displayed on a webpage on an IOS-based smartphone

4 Conclusions and Future Scope

Data from body temperature, pulse, and ECG sensors were collected and visualized using MATLAB GUI and webpage and the IoT framework for wireless sensor network designed for healthcare application (health band) was successfully tested for its scalability, i.e., ability to accommodate utilization of around eight sensors. In the future, we have planned to work on power optimization to extend battery life and increase the durability of the wristband. It is also planned to incorporate two-finger ECG onto a wristband as well as extraction of the features derived from the received physiological parameters of a patient. We look forward to visualizing the possibility of coming up with the commercialized model of a wristband having three sensors which will be communicating the vital parameters of a patient over local wireless networks as well as over the cloud platform.

References

1. Culler DE, Mulder H (2004) Smart sensors to network the world. Sci Am 290:52–59
2. Mainwaring A, Polastre J, Szewczyk R, Culler D, Anderson J (2002) Wireless sensor networks for habitat monitoring. In: WSNA '02 Proceedings of the 1st ACM international workshop on wireless sensor networks and applications, pp 88–97
3. Peiris V (2013) Highly integrated wireless sensing for body area network applications. SPIE Newsroom
4. O'Donovan T, O'Donoghue J, Sreenan C, Sammon D, O'Reilly P, O'Connor KA (2009) A context-aware wireless Body Area Network (BAN) (PDF). In: Pervasive computing technologies for healthcare
5. Jiao C, Cheng G, Tang M, Chen S (2016) Portable monitoring instrument of the physiological parameter based on MSP430 microcontroller. In: 2016 IEEE advanced information management, communicates, electronic and automatic control conference (IMCEC), pp 1800–1803
6. Gupta R, Bera JN, Mitra M (2010) Development of an embedded system and MATLAB-based GUI for online acquisition and analysis of ECG signal. Elsevier J Int Measur Confederation 43:1119–1126 (2010)
7. Dishongh TJ, McGrath M (2010) WSN technologies: microcontrollers, Wireless sensor networks for healthcare applications. Artech House, Boston, pp 16–34
8. Sridhara SR. Ultra-low power microcontrollers for portable, wearable, and implantable medical electronics. Texas Instruments, Inc. https://pdfs.semanticscholar.org/6da8/7a8faa359a61a68e856c347b8f52ac143a89.pdf
9. Freitas E, Azevedo A (2016) Wireless biomedical sensors networks: the technology. In: Proceedings of the 2nd world congress on electrical engineering and computer systems and science (EECSS'16), Budapest, Hungary, pp 134-1–134-8
10. Official IEEE 802.11 working group project timelines. 2016-03-23. Accessed from the original on 2016-04-07
11. Karagiannis V, Chatzimisios P, Vazquez-Gallego F, Alonso-Zarate J (2015) A survey on application layer protocols for the internet of thing. In: Transaction on IoT and cloud computing 2015, pp 1–10

Template-Based Video Search Engine

Sheena Gupta and R. K. Kulkarni

Abstract The exponential increase in video-based information has made it challenging for users to search specific video from a huge database. In this paper, template-based video search engine is proposed to improve the retrieval efficiency and accuracy of search engines. To begin with, the system splits the video sequence into eight key frames and then the fused image is created. The visual features like color and texture are extracted from the fused image and stored as complete feature set in a database. Now, the query clip is selected from the query database and then the template image is selected from the fused query image. The template query image features are compared with stored feature database using various similarity measures. The relevant retrieval experiments show that template-based video search engine using wavelet-based feature extraction gives better result in terms of average precision and recall using Euclidean distance as a similarity measure.

1 Introduction

The search for video content over the web still seems to be extremely difficult even after the success of different web search engines [1]. Generally, search engines index the metadata of videos and search them by text. However, in most cases, text-based search output is not relevant in case as videos are a spatiotemporal entity [2]. Based on the research, fused-image-based video retrieval system is an effective and efficient retrieval technique with minimum retrieval time. The fused image is a single image representing the characteristics of all key frames, and hence reduces the computational complexity of the system. However, it shows good results in terms of average

S. Gupta · R. K. Kulkarni (✉)
Department of Electronics & Communication Engineering, VESIT, Chembur, Mumbai, Maharashtra, India
e-mail: ramesh.kulkarni@ves.ac.in

S. Gupta
e-mail: sheena.gupta@ves.ac.in

© Springer Nature Switzerland AG 2019
D. Pandian et al. (eds.), *Proceedings of the International Conference on ISMAC in Computational Vision and Bio-Engineering 2018 (ISMAC-CVB)*, Lecture Notes in Computational Vision and Biomechanics 30,
https://doi.org/10.1007/978-3-030-00665-5_27

precision and recall when the key frames of query are highly correlated but when they are not correlated then the results are not so satisfactory [3].

This paper presents an effective template-based video search engine which rectifies the above limitation. Fused image generation and template image selection are the two important representation schemes of query video clip used in the proposed system. So, by selecting the template image from fused query image, the retrieval search is more focused, and hence it increases the efficiency of fused-image-based video search engine. The proposed method gives higher precision and recall values for complex query video clips and the retrieval efficiency of the proposed system depends upon the template image selected.

The paper is structured as follows: The proposed system block diagram is given in Sect. 2. The framework of template-based video search engine is presented in Sect. 3. The simulation results and performance evaluation is given in Sect. 4. Finally, the conclusion and future scope are illustrated in Sect. 5.

2 Template-Based Video Search Engine

2.1 Registration Phase

In registration phase, the sample video is selected from stored video database. Then, the key frames are extracted followed by fused image creation from the sample video. The features like color moments and texture are extracted from that fused image and stored as a complete feature vector of the respective videos. This procedure is repeated for all videos stored in a video database and then the complete feature vector database is built.

2.2 Query Execution Phase

On the query side, the user gives the query video clip as input and eight key frames are extracted and fused to form a query image. From that fused query image, template image is selected and then the visual features like color moments and texture are extracted and stored as a complete feature set of that respective query video. Now, this query feature set is compared with complete feature set stored in a database using different similarity measures. Finally, the top five videos are retrieved from the videos collection. The selection of template image from fused image which contains multiple key frames in a single image that plays an important role in retrieval efficiency of system. The block diagram of the proposed template-based video search engine is shown in Fig. 1.

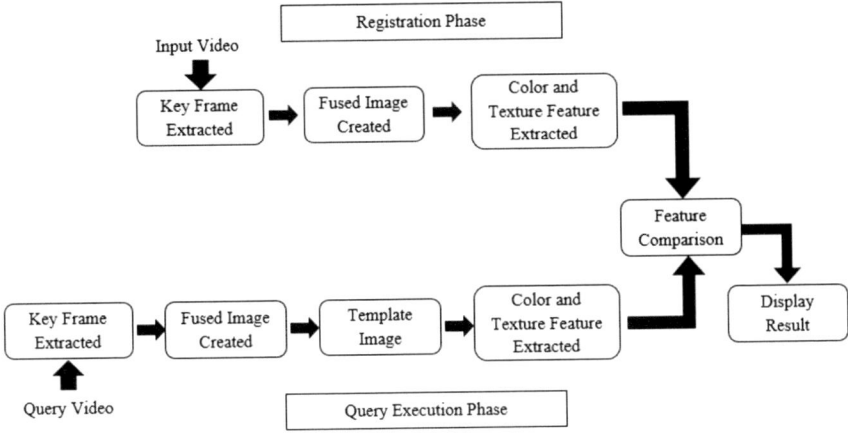

Fig. 1 Proposed framework of template-based video search engine

3   Framework of Proposed Methodology

Feature-based approach for template matching is well suited when both reference and template images had more correspondence with respect to features and control points [4, 5]. The process is explained in two parts.

3.1   *Feature Extraction Using Wavelets and Color Moments*

Wavelets provide multiresolution capability, good energy compaction, and adaptability to human visual characteristics and are now being adopted for various applications like industrial supervision of gearwheel, speech recognition, computer graphics, and multifractal analysis [6, 7]. Haar transforms are used to extract features as they give high energy compaction in transformed domain. The combination of energy and standard deviation is stored as a feature vector of that respective template image.

The Haar wavelet's mother wavelet function $\varphi(t)$ can be described as [8] follows:

$$\varphi(t) = \begin{cases} 1, & 0 \le t \le \frac{1}{2} \\ -1, & \frac{1}{2} \le t \le 1 \\ 0, & \text{otherwise} \end{cases} \tag{1}$$

And, its scaling function $\varphi(t)$ can be described as

$$\varphi(t) = \begin{cases} 1, & 0 \le t \le 1 \\ 0, & \text{otherwise} \end{cases} \tag{2}$$

Color Feature. Color moment is one of the popular techniques widely used by researchers. It basically describes the color distribution of an image through its moment [9]. The three central moments mean, standard deviation, and skewness are considered for representing the feature vector [10].

3.2 Feature Extraction Using Gabor Filter and Color Moments

Gabor filters are a group of wavelets, with each wavelet capturing energy at a specific frequency and a specific direction. The scale (frequency) and orientation tunable property of Gabor filter make it especially useful for texture analysis [11]. The filters of a Gabor filter bank are designed to detect different frequencies and orientations. From each filtered image, Gabor features can be calculated and used to retrieve images. For a given image $I(x, y)$, the discrete Gabor wavelet transform is given by a convolution [11] as

$$W_{mn} = \sum_{x1} \sum_{y1} I(x1, y1) gmn * (x - x1, y - y1) \tag{3}$$

where $*$ indicates complex conjugate and m, n specify the scale and orientations of wavelet, respectively. A feature vector f (texture representation) is created using mn as the feature components. M scales and N orientations are used and the feature vector is given in Eq. 4

$$f = \left[\sigma_{00}, \sigma_{01}, \sigma_{02} \cdots \sigma_{(M-1)(N-1)} \right] \tag{4}$$

4 Simulation Results and Performance Evaluation

4.1 Experimental Results

Template-based video search engine has been implemented using MATLAB R2017a and the performance of the proposed system is analyzed using evaluation metrics including precision, recall, retrieval efficiency, and retrieval time. The experiments are performed on a dataset of 40 videos. The collected videos contain the categories of mountain, street, ocean, and plane.

Table 1 Performance metrics for different proposed methods using Euclidean distance

Euclidean distance

Proposed methods	Query category	Avg. precision (%)	Avg. recall (%)	Avg. retrieval efficiency (%)	Avg. retrieval time (s)
Method P₁	Q1	68	34	100	49.03
	Q2	44	22	100	51.83
	Q3	66	33	90	56.96
	Q4	44	22	100	49.26
Method P₂	Q1	52	26	70	52.31
	Q2	58	34	100	48.50
	Q3	46	23	90	48.26
	Q4	64	32	90	46.50
Method P₃	Q1	40	20	80	55.18
	Q2	64	32	100	64.86
	Q3	44	22	90	61.99
	Q4	56	28	80	58.84

Method P₁ is fused-image-based video search engine
Method P₂ is template-based video search engine using wavelet-based feature extraction
Method P₃ is template-based video search engine using Gabor filter feature extraction

Fig. 2 Average precision for different query categories using Euclidean distance

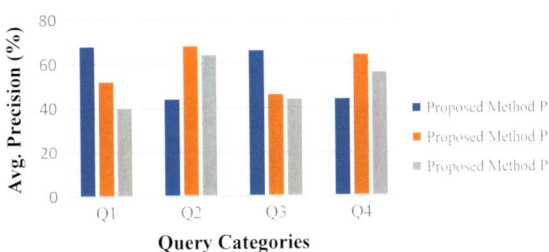

Figure 6a, b shows one query video clip and fused image of the query video and the template image with top five video retrieval results are shown in Figs. 7 and 8. In Tables 1, 2, and 3, the average precision, recall, retrieval efficiency, and retrieval time is calculated using three similarity measures. Figures 2, 3, and 4 show the graphical representation of average precision for all query categories (Fig. 5).

From Tables 1, 2, and 3, it is clear that the percentage of average precision and recall is higher in Proposed Method P_2 (template-based video search engine using wavelet-based feature extraction) for different query categories with Euclidean distance as a similarity measure. It is also clear that the Proposed Method P_2 gives highest precision for street (Q2) and plane (Q4) categories where the key frames are not correlated with each other, and hence the results are better in comparison to Proposed Method P_1 (fused-image-based video search engine). By selecting template image from a fused image which contains multiple key frames in a single image, the search is more focused about that particular object. The retrieval time of Proposed Method P_2 is also lesser because wavelet transform gives higher energy compaction which results in reduction in feature vector size and also reduces the retrieval time.

Table 2 Performance metrics for different proposed methods using correlation distance

Correlation distance					
Proposed methods	Query category	Avg. precision (%)	Avg. recall (%)	Avg. retrieval efficiency (%)	Avg. retrieval time (s)
Method P_1	Q1	36	18	70	51.61
	Q2	50	25	100	51.34
	Q3	58	29	90	53.64
	Q4	50	25	100	57.91
Method P_2	Q1	58	29	60	48.75
	Q2	48	24	90	52.42
	Q3	28	14	60	44.91
	Q4	52	26	70	54.52
Method P_3	Q1	42	21	60	53.20
	Q2	46	23	100	60.77
	Q3	42	21	90	53.64
	Q4	50	25	90	60.51

Fig. 3 Average precision for different query categories using correlation distance

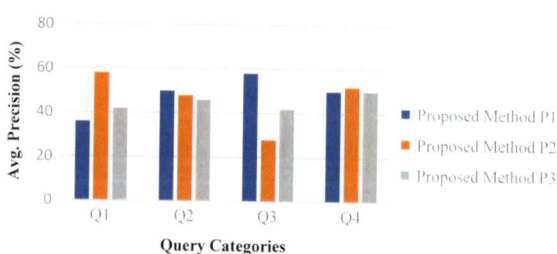

Table 3 Performance metrics for different proposed methods using Minkowski distance

Minkowski distance					
Proposed methods	Query category	Avg. precision (%)	Avg. recall (%)	Avg. retrieval efficiency (%)	Avg. retrieval time (s)
Method P₁	*Q1*	*66*	*33*	*100*	*51.61*
	Q2	52	26	100	51.34
	Q3	*56*	*28*	*80*	*53.64*
	Q4	44	22	100	57.91
Method P₂	Q1	52	26	70	49.87
	Q2	*64*	*32*	*100*	*52.69*
	Q3	38	19	70	49.05
	Q4	*60*	*30*	*100*	*54.02*
Method P₃	Q1	42	21	80	58.83
	Q2	64	32	100	61.29
	Q3	42	21	90	69.66
	Q4	52	26	70	56.17

Fig. 4 Average precision for different query categories using Minkowski distance

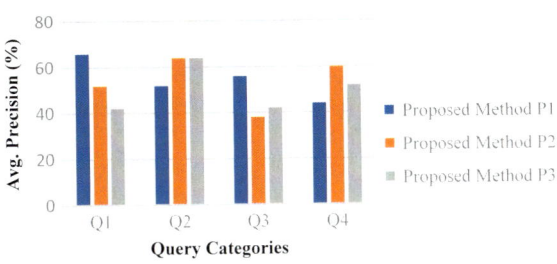

It is observed from Fig. 5 that Proposed Method P_3 (template-based video search engine using Gabor-filter-based feature extraction) gives higher average retrieval efficiency in comparison to Proposed Method P_1 but the retrieval time is more because Gabor wavelet transform requires large space for storage, and hence increases the computational time.

4.2 Performance Evaluation

The proposed method is analyzed using three similarity measures, viz., Euclidean, correlation, and Minkowski distance.

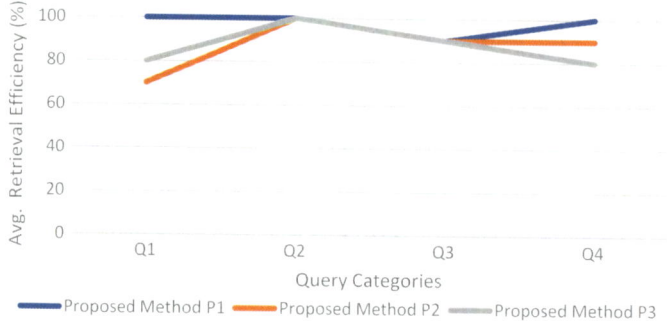

Fig. 5 Average retrieval efficiency for different query categories using Euclidean distance

Fig. 6 **a** Proposed framework for template-based video search engine. **b** Key frames extraction for the query input video clip

Fig. 7 Shows template image for feature extraction

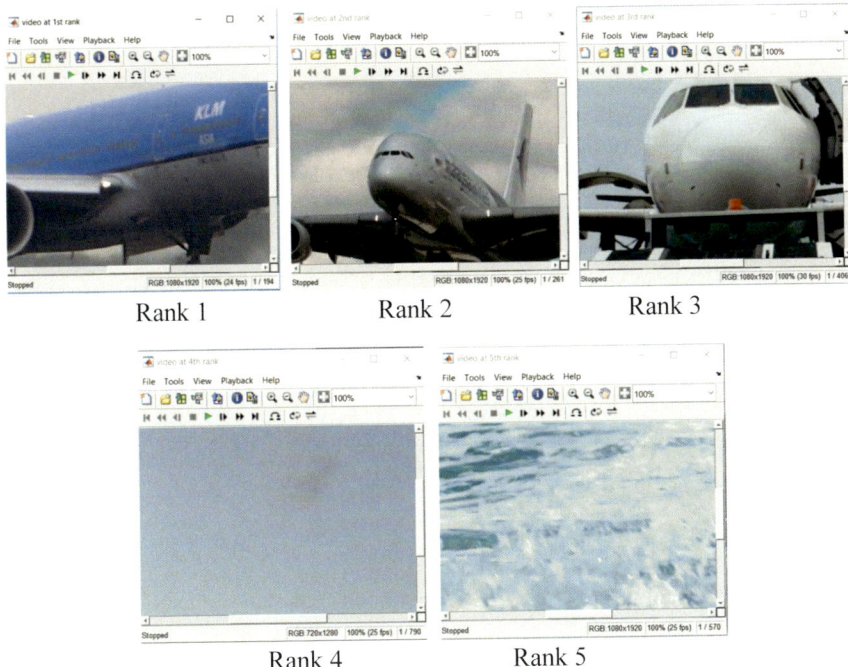

Rank 1   Rank 2   Rank 3

Rank 4   Rank 5

Fig. 8 Shows the top five ranked videos of the query video shown in Fig. 6a

5   Conclusion and Future Scope

Different methods of template-based video search engine have been tested and the average precision, recall, retrieval efficiency, and retrieval time are analyzed. Best precision is given by template-based video search engine using wavelet-based feature extraction (Proposed Method P_2) with Euclidean distance as similarity measures along with less retrieval time. So, by choosing template image from fused query image, the retrieval search is more focused and accurate. The video search engine using template matching is also compared with fused-image-based video search engine and result shows there is improvement in precision values for complex query video clips.

In future, the accuracy of the proposed system can be increased further by increasing number of key frames and by extracting more visual features.

References

1. Shanmugam TN, Rajendran P (2009) An enhanced content-based video retrieval system based on query clip. Int J Res Rev Appl Sci 1(3). ISSN:2076-734X, EISSN:2076-7366
2. Hu W, Xie N, Li L, Zeng X, Maybank S (2011) A survey on visual content-based video indexing and retrieval. IEEE Trans Syst Man Cybern Part C Appl Rev 41(6):797–819
3. Gupta S, Kulkarni RK (2017) Fused image based video search engine. Int J Video Image Process Netw Secur (IJVIPNS-IJENS) 17(5) 174305-2929 (October)
4. Mahalakshmi T, Muthaiah R, Swaminathan P (2012) Review Article: An overview of template matching technique in image processing. Res J Appl Sci Eng Technol 4:5469–5473. ISSN: 2040-7467
5. Nazil P, Kumar D, Bhardwaj I (2013) An overview on template matching methodologies and its applications. Int J Res Comput Commun Technol 2(10):988–995
6. Thepade SD, Yadav N (2015) Novel efficient content based video retrieval method using cosine-haar hybrid wavelet transform with energy compaction. In: International conference on computing communication control and automation. IEEE. 978-1-4799-6892
7. Deepa T, Girisha H (2014) Image compression using Hybrid wavelet Transform and their Performance Comparison. Int J Mod Eng Res (IJMER) 4(6):6–12. ISSN: 2249–6645
8. Kekre HB, Thepade SD, Gupta S (2013) Content based video retrieval in transformed domain using fractional coefficients. Int J Image Process (IJIP) 7(3):238–274
9. Padmakala S, Anandha Mala GS, Shalini M (2011) An effective content based video retrieval utilizing texture, color and optimal key frame features. In: International conference on image information processing (ICIIP 2011). IEEE. 978-1-61284-861
10. Geetha P, Narayanan V (2011) An effective video search re-ranking for content based video retrieval. IEEE. 978-1-4673-0131
11. Ansari A, Mohammed MH (2015) Content based video retrieval system-methods, techniques, trends and challenges. Int J Comput Appl 112(7) (Feb)

Gray-Level Feature Based Approach for Correspondence Matching and Elimination of False Matches

R. Akshaya and Hema P. Menon

Abstract Matching of interest points (feature points) is a basic and very essential step for many image processing applications. Depending on the accuracy of the matches, the quality of the final application is decided. There are various methods proposed to tackle the problem of correspondence matching. In this paper, a method that makes use of the textural features, mainly gray-level features with respect to a pixel's neighborhood has been discussed for point matching with an emphasis on its use for image registration. Feature points are obtained from the images under consideration using SURF and are matched using gray-level features. Then, the false matches are removed using a graph-based approach.

1 Introduction

With a lot of images that are captured nowadays, to analyze them or to make a meaningful conclusion from it, these images should be aligned first. This process of aligning the images with one as a reference image is called Registration [1]. During Registration, all the images are mapped to a common coordinate system. The images can be different due to various reasons like different viewpoints, captured at different depths, taken at different timings, or due to the changes in the capturing modes [2]. Generally, Registration is divided into two types called Rigid and Non-rigid registration. Rigid registration is used in the case when the is no change in the objects size and shape and Non-Rigid registration is used in case if there is a change in the size and shape of the object. Registration is used in various applications like Medical images, fusion, Super-resolution, shape analysis, biometrics, 3D reconstruction, etc., [3–7].

R. Akshaya · H. P. Menon (✉)
Department of Computer Science and Engineering, Amrita School of Engineering,
Amrita Vishwa Vidyapeetham, Coimbatore, India
e-mail: p_hema@cb.amrita.edu

R. Akshaya
e-mail: cb.en.p2cvi16001@cb.students.amrita.edu

© Springer Nature Switzerland AG 2019
D. Pandian et al. (eds.), *Proceedings of the International Conference on ISMAC in Computational Vision and Bio-Engineering 2018 (ISMAC-CVB)*, Lecture Notes in Computational Vision and Biomechanics 30,
https://doi.org/10.1007/978-3-030-00665-5_28

275

In simple words, Registration involves identifying the points that are the same from both the images, define a transformation function to align the target image with respect to the source image. The image that is used as a reference image is called the source image, it is with respect to this image, all other images are aligned. The other image that is aligned with respect to the source image is called the target image. The first step in Registration is very important, that is finding the points that are the same or the corresponding points in the images. This is the basic step that decides the success of Registration. Correspondences in the image are found using feature descriptors. This paper mainly discusses how to match the points and remove the points that are falsely matched [8–10].

2   Related Work

Finding out corresponding points are almost similar to a search problem, if a point is there in the first image, a match for it is searched in the second. There are different methods that have been proposed to find the match, for example, pixelwise search, which is computationally huge, feature-based methods where features are extracted and match these to find the corresponding points or graph methods to do the same as the structure of the object remains the same in both the images. The matching can also be done using the neighborhood of the pixels, like the template matching or the using textural features of the neighborhood. This could possibly not be done on every point as first it is computationally complex and second the pixels in the neighborhood will mostly remain the same approximately. So, feature points that are unique to the image are extracted. SURF is one such method. Speeded-up robust features is a local feature descriptor. It is a much faster version of the SIFT extractor. The DOG is used in case of SIFT and LOG with a box filter is used in case of SURF. The reason why SURF is faster is that convolution of a box filter is as simple as calculating the integral images and can be done simultaneously at all scales [11]. The textural features from a neighborhood can be extracted using the Gray-Level Co-occurrence Matrix. The GLCM is calculating features using the statistical methods. It essentially uses the spatial relation between the pairs of pixels at different orientations to define a matrix. For example, if the pixel pair that is next to each other are considered, then the angle between them is $0°$, and likewise for different angles. From this GLCM, various features like contrast, energy, homogeneity, etc., are extracted [12].

There are a lot of papers that discuss how to match the points in the given image. Patricio et al. proposed a method where matching between the pair of images, which was done for stereo vision matching. The matching between the images was done based on the correlation computed between the point and the neighborhood. The pixels that were not similar to the neighborhood were avoided for computation, a method of the adaptive neighborhood is defined, which is based on the similarity of the pixel with its neighborhood. The size and shape of the neighbourhood are decided based on the content it has. The algorithm is compared with algorithms like the SAD, HIR, SMW, has reduced error rate [13]. Silva et al. have proposed a method to find out the corresponding interest points based on the Morse complex. It uses a topological operator called the Local Morse Context (LMC) to mainly explore the structural information. It mainly reduces the incorrect matches by giving a confidence number. Here, the interest points are defined by a Morse cell adding more importance to the neighborhood pixels using subcomplex of a complex, where standard graph solution is used. Eigen decomposition is used to solve the problem of inexact graph matching to handle noises and inconsistencies [14].

3 Implementation

3.1 Proposed Method

In the proposed method, the corresponding matches in both images were found using the points extracted using descriptors. The general flow diagram of the method is shown in Fig. 1.

Fig. 1 Above are the general steps in the proposed algorithm

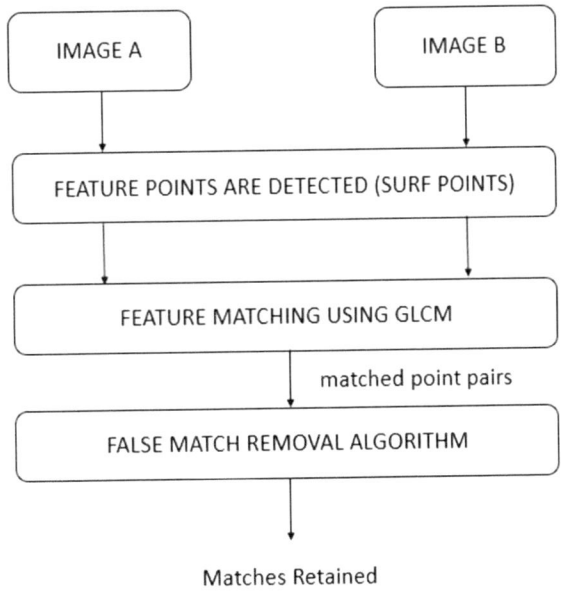

3.2 Dataset Details

The datasets contain a set of 6 ppm images of 640 × 800 resolution. These images are taken at different Viewpoints. And the other data is also a set of 6 ppm images of size 1000 × 7000 [15].

3.3 Methodology

The descriptor that is used to extract the point here is SURF. SURF feature descriptor is applied to both the images and a set of unique points is got. Say we have p_1 points from image A and p_2 points from image B. p_1 and p_2 need not be essentially the same. So, we select the image with the lesser number of points from images A and B. Since we are finding a match for all the points that are extracted, using the image with more number of points will lead to mapping a point to more of number of points from which arriving at the correct match might be difficult.

For all the points that are extracted using the Surf features, GLCM is calculated on a neighborhood of 11 * 11 and an angle of 0°. Once the GLCM is obtained, features like the correlation, variance, energy, contrast, mean, etc., are calculated for it. The gray features of every point in the image with a lesser number of features points is compared with the features of every other point on the other image. The comparison is done using the Euclidean distance between the features. The point is then matched with the point on the other image which has got the minimum difference when compared to all other points. This procedure is then done for all the points present, the point that is matched with the previous point will be still available for the points that come after that due to the fact that the previous match could be a false one. So, now we have p matches, [min $(p1, p2)$].

All the matches need not be essentially the true matches. So, we apply an algorithm proposed by Zakharov et al. [16] for the removal of false matches. A matrix of distance is calculated, G. And singular value decomposition is done on this distance matrix. The eigenvalue matrix is then replaced with an identity matrix to get a matrix P. A matrix of correspondence, K is also created where matrix has the value equal to one when the corresponding indices of the points match, for example when point with the index 1 in the first set of points match with the point having index 2 in the second set the (1, 2) position of the matrix is equal to one. Then a dot product is applied on the matrices P and K. If the value crosses a particular threshold then the points at the corresponding indices are said to be true matches. The threshold set in this experiment is 0.5, so after performing a dot product if the value crosses 0.5 then its considered as corresponding points. The above algorithm is applied to every 5% of the points to get the results. Thus, the points corresponding points and the false matches are identified.

4 Experimental Results

The experiments were performed on a set of images from the dataset. The output images are shown below. Figure 2 shows the images that show first 50 matched points using GLCM features without the removal of false matches. Figure 3 shows the output of the first 50 points matched after false match removal. Figure 4 shows the final output of the image 1 and 2 in the dataset [15]. Figure 5 is the final output of the images 1 and 2 in the dataset (Fig. 6).

Table 1 shows the number of points matched using the GLCM algorithm and points retained after false match removal in the images of dataset [15].

Fig. 2 Matched points (50) got after matching from GLCM

Fig. 3 Output of matching first 50 points of images

Table 1 Showing the number of pairs extracted and the number remaining after

Sample image pairs considered	Number of matched points	Number of matched retained after false match removal
Pair 1	1463	240
Pair 2	1463	158
Pair 3	1463	10
Pair 4	1463	7

Fig. 4 Output for corresponding points matching in images

Fig. 5 Output for corresponding points matching of images

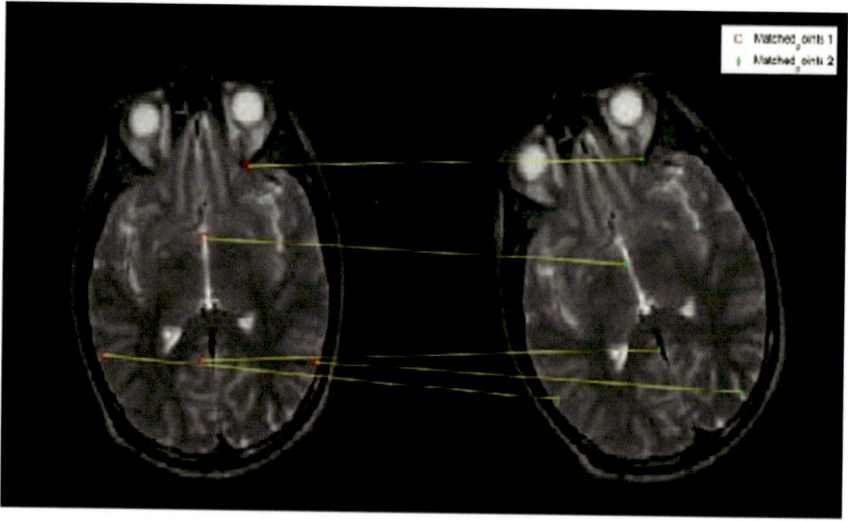

Fig. 6 Output for corresponding points matching of images (BRAIN)

3	3
1	3

2	1
1	6

3	1
1	4

Fig. 7 Confusion matrix for the brain, painting and bikes dataset, respectively

4.1 Analysis

Table 1 shows the points that were retained by the algorithm as true matches. To evaluate the performance of the algorithm, further analysis has to be carried out. There might be matches that are true and removed or vice versa, so it is necessary to analyze the results using True positives, True negatives, False positive and False negatives. Analyzing the above results visually is difficult as it is very clustery. Hence, a subset of points is used to make the analysis easy. Figure 7 given is the confusion matrix of different datasets.

4.2 Inference

Experiments have been performed with the image 1 in the dataset as the reference image and the matched points remain a constant because the minimum number of points from the two images are considered and mostly image 1 had the minimum number of points. The observation made from the results are, the false matches that were present in the image after matching were removed on the application of the false match removal algorithm. And the number of points that were obtained after the false match removal reduced as the viewing angle deviated a lot with respect to the reference image. This can be seen in the above-tabulated results too. The viewing angle increases from image 2 to 5. From the confusion matrix, it can be seen that most of the false matches are removed. The above algorithm works better in case of textural images, whereas in the case of medical images brain MRI the algorithm also removes most of the true matches. Thus, the points obtained can be given to any registration algorithm to register the images.

5 Conclusion

This paper presents a method to match the interest points from the images and remove the false matches from them using the textural features from the images for a particular neighborhood. The features extracted help us to define the area around the pixel of interest and matches the pixel with the same sort of features in the other image. These matched feature points are given to an algorithm that removes the false matches present in the image. As a future work, the false matches that are still existing will be eliminated to make the end result much better.

References

1. Zitova B, Flusser J (2003) Image registration methods: a survey. Image Vis Comput 21(11):977–1000
2. Brown LG (1992) A survey of image registration techniques. ACM Comput Surv (CSUR). 24(4):325–376
3. Ong EP, Xu Y, Wong DW, Liu J (2015) Retina verification using a combined points and edges approach. In: 2015 IEEE international conference on image processing (ICIP), 27 Sept 2015, pp 2720–2724. IEEE
4. Menon HP, Narayanankutty KA (2015) Comparative performance of different perceptual contrast fusion techniques using MLS. Int J Biomed Eng Technol 18(1):52–71
5. Shwetha R, Rajathilagam B (2015) Super resolution of mammograms for breast cancer detection. Int J Appl Eng Res 10(1):21453–21465
6. Huang X, Zhang J, Fan L, Wu Q, Yuan C (2017) A systematic approach for cross-source point cloud registration by preserving macro and micro structures. IEEE Trans Image Process 26:3261–3276
7. Arathi T, Parameswaran L (2014) Image reconstruction from 2D stack of MRI/CT to 3D using shapelets. Int J Eng Technol (IJET). 6(1):2595–2603
8. Jain V, Li X (2004) Point matching methods: survey and comparison. Project report for CMPT 8888
9. Menon HP, Nitheesh AS (2017) Structural matching of control points using VDLA approach for MLS based registration of brain MRI/CT images and image graph construction using minimum radial distance. In: The international symposium on intelligent systems technologies and applications. Springer, Cham, pp 356–369
10. Menon HP (2017) An analysis on the influence that the position and number of control points have on MLS registration of medical images. In: International symposium on signal processing and intelligent recognition systems. Springer, Cham, pp 47–56
11. Bay H, Tuytelaars T, Van Gool L (2006) Surf: speeded up robust features. In: European conference on computer vision. Springer, Berlin, pp 404–417
12. Mohanaiah P, Sathyanarayana P, GuruKumar L (2013) Image texture feature extraction using GLCM approach. Int J Sci Res Publ 3(5):1
13. Patricio MP, Cabestaing F, Colot O, Bonnet P (2004) A similarity-based adaptive neighborhood method for correlation-based stereo matching. In: 2004 international conference on image processing, 2004. ICIP'04, vol 2. IEEE, pp 1341–1344
14. da Silva RD, Schwartz WR, Pedrini H, Pulido J, Hamann B (2015) A topology-based approach to computing neighborhood-of-interest points using the Morse complex. J Vis Commun Image Represent 30:299–311
15. Image Details. http://www.robots.ox.ac.uk/~vgg/data/data-aff.html
16. Zakharov AA, Tuzhilkin AY, Zhiznyakov AL (2015) Finding correspondences between images using descriptors and graphs. Procedia Eng 1(129):391–396

A New Approach for Image Compression Using Efficient Coding Technique and BPN for Medical Images

M. Rajasekhar Reddy, M. Akkshya Deepika, D. Anusha, J. Iswariya and K. S. Ravichandran

Abstract Medical images produce a digital form of human body pictures. Most of the medical images contain large volumes of image data that is not used for further analysis. There exists a need to compress these images for storage issues and to produce a high-quality image. This paper discusses an image compression using Back Propagation Neural network (BPN) and an efficient coding technique for MRI images. Image compression using BPN produces an image without degrading its quality and it requires less encoding time. An efficient coding technique—Arithmetic coding is used to produce an image with better compression ratio and redundancy is much reduced. Back-Propagation Neural Network with arithmetic coding gives the better results.

1 Introduction

Image compression is a compression technique which reduces storage space of an image without degrading the quality of the on diagnosis of diseases, hence it is important to maintain the image quality [1]. An efficient compression technique is required for the medical image to produce a high-quality image [2]. The MRI images are very large in size so it is necessary to reduce the cost of storage to increase the transmission speed [3].

The Back-Propagation Neural Network (BPN) is the most widely used neural network for image compression [4]. An image is partitioned into blocks without overlapping and fed into neural network [5]. The blocks are used to train the neural network and gives the better reconstructed image [6].

Arithmetic coding is a form of entropy encoding used in all type of data compression [7]. Arithmetic coding encodes the frequently seen symbols with fewer bits [8]. Flexibility is an advantage of arithmetic coding [9].

M. Rajasekhar Reddy (✉) · M. Akkshya Deepika · D. Anusha · J. Iswariya · K. S. Ravichandran
School of Computing, SASTRA Deemed University, Thanjavur, India
e-mail: rajasekharmanyam04@gmail.com

© Springer Nature Switzerland AG 2019
D. Pandian et al. (eds.), *Proceedings of the International Conference on ISMAC in Computational Vision and Bio-Engineering 2018 (ISMAC-CVB)*, Lecture Notes in Computational Vision and Biomechanics 30,
https://doi.org/10.1007/978-3-030-00665-5_29

283

In this paper, an image is compressed using BPN and arithmetic coding. An image is compressed using BPN and further compression is done by using arithmetic coding [10]. The final image gives the better Peak Signal-to-Noise Ratio (PSNR).

2   Existing Methods

2.1   Back-Propagation Neural Network

There are three layers in neural network. The size of input and output layers are same [11]. The hidden layer is fully interconnected with input and output layer [12]. Image encoding is done for each pixel of the block which is transmitted to the receiving side where reconstruction is done [13]. Compression is achieved at the hidden layer and decompression is achieved in output layer [14]. The two-activation function used in neural network is "tensing" for hidden layer and "purelin" for output layer (Fig. 1).

The neural network is initialized and trained by using the learning rate and goal with a set of desired values [5]. The default performance measures are mean square error. There will be a decrease in performance of the neural network when number of weights increases [15]. Neural network-based image compression reduces the search space and encoding further improve computation time [16].

2.2   Arithmetic Coding

Arithmetic coding is used in all types of data compression algorithms. It has some advantages over Huffman coding where it gives better compression ratio and redundancy is much reduced [17]. Unlike Huffman coding, discrete number of bits for each

Fig. 1 Back-Propagation
Neural Network

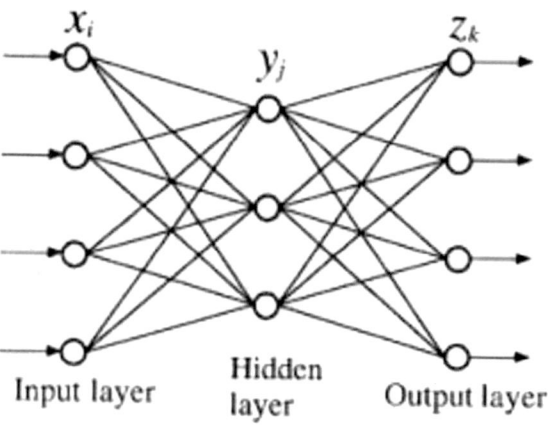

x_i y_j z_k

Input layer   Hidden layer   Output layer

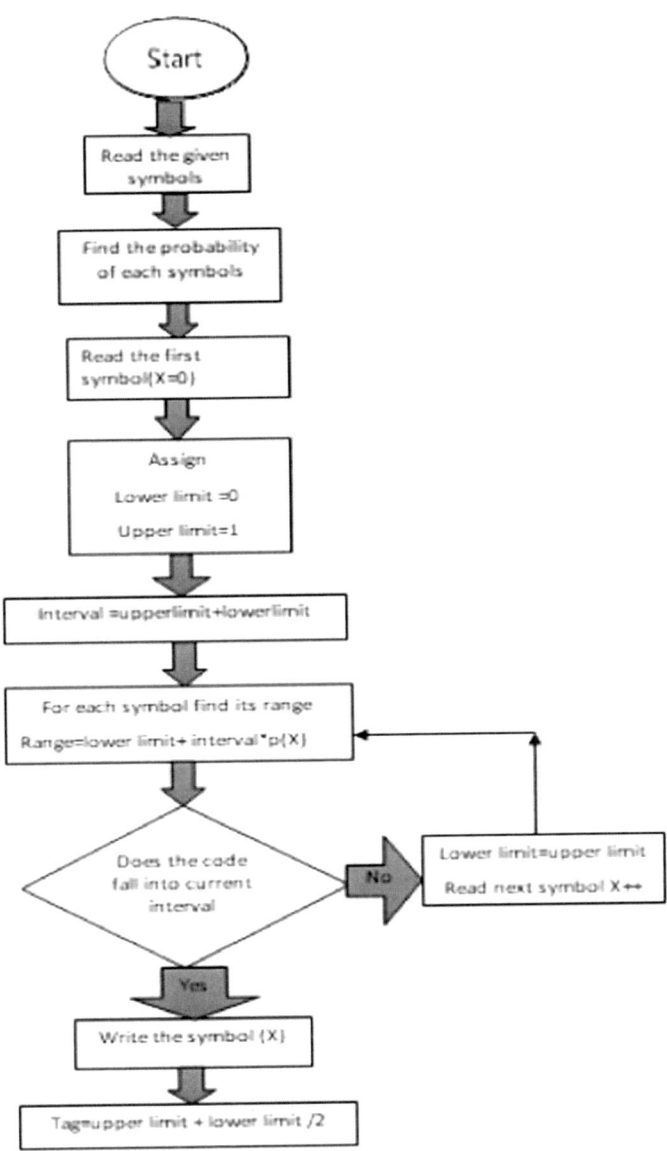

Fig. 2 Arithmetic coding

code is not used in arithmetic coding [8]. Each symbol is assigned with an interval, and the probability of each symbol is calculated [18]. Starting with lower limit 0 and upper limit 1, each interval is divided into several subintervals [7] (Fig. 2).

BEGIN
Lower limit = 0; upper limit = 1; range = 1;
While (symbol! = terminator)
{
Get (symbol);
Lower limit = lower limit + range * Range_lowerlimit(symbol);
High = lower limit + range * Range_upperlimit (symbol);
Range = upper limit − lower limit;
}
Output code word
Tag = lower limit + upperlimit/2;
END. [9]

3 Proposed Method

Image compressing is done using BPN and arithmetic which gives better results.

3.1 Procedure

1. Read an input medical image.
2. Resize it to Standard Image size and normalize the image.
3. The image is partitioned into nonoverlapping block and it is fed into neural network as input.
4. Initialize the weights and targets.
5. Set the training parameters.
6. Train the neural network.
7. Apply arithmetic coding to the resultant image.
8. Save the arithmetic codes.
9. Convert these codes using inverse arithmetic coding.
10. Obtain the reconstructed image.
11. Compute the compression ratio.
12. Calculate the PSNR value.
13. Note the execution time (Fig. 3).

4 Performance Measure

The parameters used to measure performance are Mean Square Error (MSE) and Peak Signal-to-Noise Ratio (PSNR).

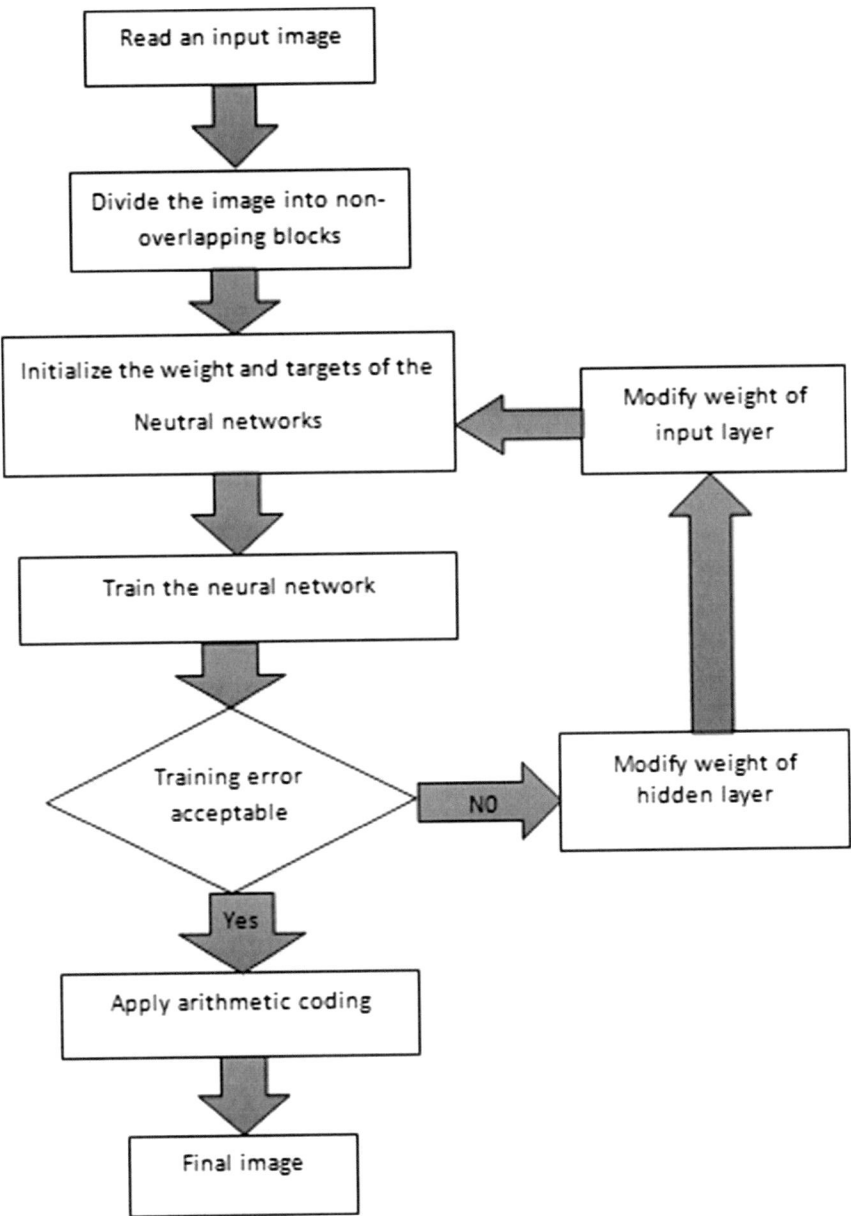

Fig. 3 Back-Propagations Neural Network and arithmetic coding

$$MSE = 1/MN \sum_{y1=1}^{m} \sum_{x1=1}^{n} \left[I(x1, y1) - I'(x2, y2) \right]^2,$$

where $M \times N$ is an image.

PSNR is defined as

$$PSNR = 20 \, \log_{10}\big[255/\text{sqrt(MSE)}\big]$$

Quality of a reconstructed image is commonly measured by PSNR value [19].

5 Results

See Table 1, Figs. 4 and 5.

Table 1 PSNR values and execution time for existing and proposed techniques

Input image	PSNR using BPN (dB)	PSNR using BPN and arithmetic (dB)	ET using BPN (s)	ET using BPN and arithmetic (s)
Image (1)	29.906	52.41	108.633	121.68
Image (2)	30.41	53.70	79.246	118.98
Image (3)	30.42	52.74	105.216	124.189
Image (4)	30.44	54.11	79.894	129.17
Image (5)	30.41	53.65	127.434	112.00
Image (6)	31.05	54.34	117.893	130.88
Image (7)	31.46	54.99	113.16	97.40
Image (8)	31.63	54.88	76.396	130.65

Fig. 4 Input image (5)

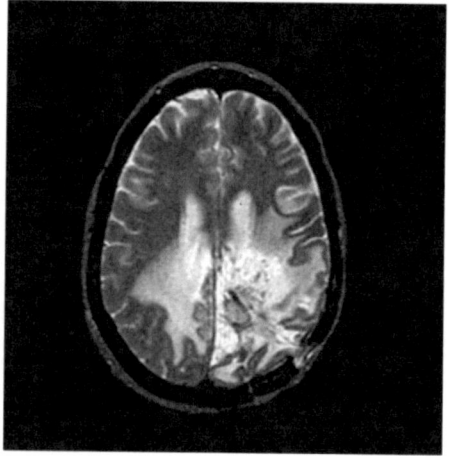

Original Image

Fig. 5 Reconstructed image (5)

Final Image

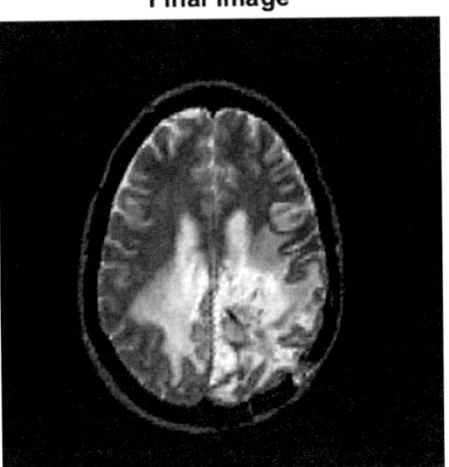

6 Conclusions

A new approach for image compression using BPN and arithmetic coding for medical images produces a better quality image. The PSNR value of the proposed method is increased as compared with an existing one. Thus, the hybrid image compressing using BPN and arithmetic coding gives better results than image compressing using BPN alone.

References

1. Dougherty G (2009) Digital image processing for medical applications, 1st edn. Cambridge University Press, Cambridge, pp 9–10
2. Shapiro JM (1993) Embedded image coding using zero trees of wavelet coefficients. IEEE Trans Sig Process 41(12):3445–3462
3. Maha Lakshmi GV, Rama Mohana Rao S (2012) A novel algorithm for image compression based on fractal. Image Compression EJSR 85(4):486–499. ISSN: 1450–216X, Eurojournals
4. Savkovic-Stevanovic J (1994) Neural networks for process analysis and optimization: modeling and application. Comput Chem Eng 18(11–12):1149–1155 (14 Ref.)
5. Soliman HS, Omari M (2006) A neural networks approach to image data compression. Appl Soft Comput 6(3):258–271
6. Maha Lakshmi GV, Rama Mohana Rao S (2013) A novel algorithm for image compression based on fractal and neural networks. Int J Eng Innov Technol (IJEIT) 3(4):8–15
7. Kavitha V, Easwarakumar KS (2008) Enhancing privacy in arithmetic coding. ICGST-AIML J 8(I):23–28
8. Masmoudi A, Masmoudi A (2013) A new arithmetic coding model for a block-based lossless image compression based on exploiting inter-block correlation. Sig Image Video Process 1–7
9. Witten IH, Neal R, Gleary JG (1987) Arithmetic coding for data compression. Commun ACM 30(4):520–540

10. Vilovic I (2006) An experience in image compression using neural networks. In: Proceedings of the 48th international symposium ELMAR-2006 focused on multimedia signal processing and communications, June 2006. IEEE Press, Zadar, Croatia, pp 95–98

11. Meyer-Bäse A, Jancke K, Wismüller A, Foo S, Martinetz T (2005) Medical image compression using topology-preserving neural networks. Eng Appl Artif Intell 18(4):383–392

12. Carrato S (1992) Neural networks for image compression. In: Neural networks: advancement and application, 2 edn. Galenbe Pub, North-Holland, Amsterdam, pp 177–198

13. Srikala P, Umar S (2012) Neural network based image compression with lifting scheme and RLC. Int J Res Eng Technol 1(1):13–19

14. Osowski S, Waszczuk R, Bojarczak P (2006) Image compression using feed forward neural networks—hierarchical approach. In: Natural to artificial neural computation, vol 3497 of Lecture notes in computer science. Springer, Berlin, Germany, pp 1009–1015

15. Ashraf R, Akbar M (2007) Adaptive architecture neural nets for medical image compression. In: Proceedings of the IEEE international conference on engineering of intelligent systems (ICEIS '06), Islamabad, Pakistan, pp 1–4

16. Northan B, Dony RD (2006) Image compression with a multiresolution neural network. Can J Electr Comput Eng 31(1):49–58

17. ISO/IEC 14496-2 FPDAM 1 Information Technology. Generic Coding of Audio-Visual Objects—Part 2: Visual, July 1999

18. Rubin F (1979) Arithmetic stream coding using fixed precision registers. IEEE Trans Inf Theory 25:672–675 (Data compression conference)

19. Miaou S-G, Ke F-S, Chen S-C (2009) A lossless compression method for medical image sequences using JPEG-LS and interframe coding. IEEE Trans Inf Technol Biomed 13(5):818–821

Person Identification Using Iris Recognition: CVPR_IRIS Database

Usha R. Kamble and L. M. Waghmare

Abstract In recent days iris is as one of the useful traits for authentication using biometric recognition. Almost all publicly available iris image databases contain data correspondent to heavy imaging constraints and suitable to evaluate by various algorithms (CASIA Iris Image Database [1]; Proença et al. in IEEE Trans Pattern Anal Mach Intell 32(8):1529–1535, 2010 [2]). This paper is first to prepare our own IRIS IMAGE DATABASE and make the availability of it to the new researchers. We have thus proposed our own database named as CVPR_IRIS DATABASE with not heavy imaging constraints. This set up contains Iris Image Capture IR Sensitive CCD camera by which eye images are captured in the visible lighting conditions with IRLED source at a distance. Clear images from this database are separated from noisy images by quality assessment technique. Thus proposed CVPR_IRIS DATABASE containing total 485 eye images of 49 individuals is made available for researchers concerned with the implementation of algorithms based on research in iris recognition. Second using proposed database we have proved good accuracy using 2D Discrete Wavelet Transform.

1 Introduction

Reliable personal identification using iris as a biometric is recently an active topic in the research area of pattern recognition. It is because of increase in the demand of enhanced security of human being in daily life. For last two decades many of the researchers have worked on improving accuracy, speed, storage cost, and robustness [3–15]. Most of these algorithms are categorized into four groups [14] such as phase-based methods, zero crossings representations, texture analysis, and low-intensity variations. Bertillon put forth the main idea of person identification by using iris

U. R. Kamble (✉) · L. M. Waghmare
Shri Guru Gobind Singhji Institute of Engineering and Technology, Nanded, India
e-mail: urkamble@sggs.ac.in

L. M. Waghmare
e-mail: lmwaghmare@yahoo.com

© Springer Nature Switzerland AG 2019
D. Pandian et al. (eds.), *Proceedings of the International Conference on ISMAC in Computational Vision and Bio-Engineering 2018 (ISMAC-CVB)*, Lecture Notes in Computational Vision and Biomechanics 30,
https://doi.org/10.1007/978-3-030-00665-5_30

Fig. 1 Front view of human eye

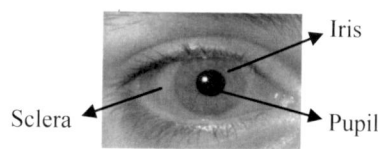

recognition. After few years John Daugman [3, 6] published his rigorous work in this area. The popularity of iris as biometrics grows because of several advantages of iris over other biometrics such as palm finger prints, hand, and face, etc. Use of iris for person identification in public sectors, government and nongovernment offices, state control lines proves that iris is most accurate from other biometrics. Currently, deployed systems rely mainly on good quality images, captured in a stop-and-stare interface, at close distances and using near infrared (700–900 nm) wavelengths [2]. Our purpose is to explore new imaging process and acquire clear images. These images are to be used for the research in iris recognition. We have large number of experiments conducted on this database and reported in the literature.

Front view of the iris structure of human eye is shown in Fig. 1 consisting of iris placed in between pupil and sclera. Iris as a biometric outperforms over other biometrics because of few reasons such as it remains stable throughout one's lifetime, its uniqueness, flexibility, non-invasive, and reliable nature. The remainder of this paper is organized as follows:

Section 2 briefly describes about related work and architecture of iris recognition. Section 3 gives description of proposed work: Preparation of THE CVPR_IRIS DATABASE and database availability. Section 4 proposes algorithm using DWT reports our experiments and discusses the results. Section 5 presents the conclusions.

2 Related Work

Daugman [3, 6] the first researcher with his lot of efforts has given to the world an algorithm based on iris codes from features using 2D Gabor Wavelets. The author has used Integro differential operators to detect the center and diameter of iris. A simple Boolean Exclusive-OR operator is used as Hamming Distance for matching. For perfect match the Hamming Distance should be equal to zero. The algorithm proves accuracy of more than 99.9%. Time required for iris identification shown is found to be less than one second. The algorithm of Boles and Boashash [4] in his research work set of one-dimensional signal is encoded by using a zero crossing transformation at different resolution levels. To determine the overall dissimilarity between two iris codes, the average of the dissimilarity at each resolution level is obtained. The author reported 100% verification and identification accuracy with the experiments conducted on 11 iris images. But the source of the testing iris images was not indicated.

In paper [7] by Li Ma, Tieniu Tan, Yunhong Wang describes a new scheme for iris recognition. Here the quality of each image in the input sequence is assessed and a clear iris image is selected from a sequence if iris images. To capture local iris bank of spatial filters, whose kernels are suitable for iris recognition is then used. A comparative study of existing methods is done on an iris image database. Database includes 2255 sequences from 213 subjects and algorithm proposed highly better performance.

2.1 Iris Recognition System

The system has acquisition and preprocessing in the first module. Second module is the image verification module which has pattern matching and identification module. The second module compares input iris mages with known irises in the database. And decides if it is in the database or not. Preprocessing of iris includes correct segmentation of iris. This segmented iris is unwrapped by using certain methods. Some of the authors reported their algorithms without normalization also.

3 Description of Proposed Work: Preparation of CVPR_IRIS DATABASE

Our own iris image acquisition setup and imaging framework is as shown in Fig. 2. It consists of camera with IR LED source and CPU. The specifications of setup of the imaging framework are given in Table 1. As illustrated in Fig. 2 this framework was installed in both natural and artificial lighting sources.

We placed several marks on the floor each after 1 m up and to 10 m from acquisition device and asked for volunteers for the image acquisition processes. When planning CVPR_IRIS database we had three basic concerns to acquire images. First, all subjects are fixed at some distance. Second varying very small imaging distances, and third having normal lighting environments. All subjects are Indian, 90% subjects are of age groups of about 20 years old and 10% are from 35 to 45 years old. The

Fig. 2 Imaging framework

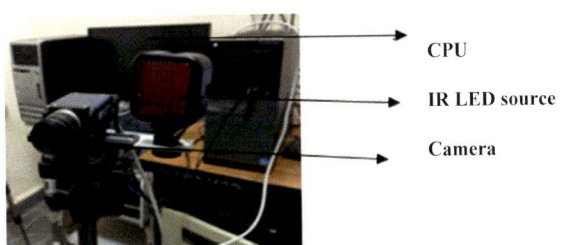

CPU

IR LED source

Camera

Table 1 Specifications of set up of imaging framework

Image capture in visible light with IR source Intel®	
Image type	BMP file format
Image dimension (number of pixels)	1024×768
Distance between camera and object	20 cm approximate, with variation of ± 3 cm
Distance between camera and infrared source	8 cm
Compression format	Bmp
Camera device	Onboard memory—8 MB Iris Image Capture IR Sensitive CCD camera, CCD Monochrome Camera
CPU	Intel Core, I7, Windows 10 Home, 8 GB, DDR4, 1 TB
Focal length	60 mm

Table 2 Proposed CVPR_IRIS DATABASE: 5 images of person 2 from left eye only

Example	Total Images	Wavelength	Varying Distances	Acquisition Device
2L1.BMP	2L2.BMP	2L3.BMP	2L4.BMP	2L5.BMP

database is constructed with 485 images from 49 individuals. For 48 individuals, we have captured five images of right and five images from left eyes. For one individual, we have captured only five images of left eye and images are numbered as 1L1.bmp to 1L5.bmp. Similarly, right and left eye images of ith person are numbered as iR1.bmp to iR5.bmp and iL1.bmp to iL5.bmp respectively.

Database availability: This will help others to obtain the same data sets and replicate your experiments. We suggest the following address reference for database availability: CVPR LAB, Department of Electronics and Telecommunication Engineering, SGGS Institute of Engineering & Technology, Nanded (MS), India.

Year = "2016–18",
Mail id: urkamble@sggs.ac.in

Table 2 shows five images of person 2 from left eye only naming L1 for person 1 left eye image. Likewise in proposed database R1 is for person 1 right image.

4 Experiments Using Proposed CVPR_IRIS DATABASE

While capturing iris images unfortunately we cannot have one image of good quality but we have captured sequence of images from single eye. The problem is to select more clear, good quality images. In the process of capturing images for our database we have some of the images that are defocused images, blurred images, motion blurred images, and partially occluded images because of upper eyelids. We have separated out all clear images. This data we are keeping for study and analysis. Though we have different database publicly available the related work for quality assessment of iris images is not enough and still it is limited.

Hence to assess and select image of good quality we have used fourier descriptors that is effective scheme by analyzing the frequency distribution of the iris image. Daugman [3, 6] in his work, iris images are assessed by measuring the total high-frequency power in the 2D Fourier spectrum. But he did not mention a detailed description of his method. Alexandre [2] and group detected 14 different noise factors that are classified into local or global category. Hence to assess and select image of good quality we have used fourier descriptors [12] that is effective scheme by analyzing the frequency distribution of the iris image.

4.1 Image Quality Assessment

All clear and properly focused iris images should have uniform frequency distribution. Fourier descriptor is defined by DR Eq. (1)

$$DR = \left[(fl+fm+fh); \frac{fm}{fl+fh} \right] \tag{1}$$

$$f_i \; 003D = \iint\limits_{\Omega} |F(u,v)| du dv \quad i = 1, 2, 3 \tag{2}$$

$$\Omega = \left\{ (u,v) \middle| f_{1i} < \sqrt{u+v^2} <= f_{2i} \right\} \tag{3}$$

$F(u, v)$ is the 2D Fourier spectrum of an iris region. fl, fm, fh are the power of low, middle and high frequency components. f_{1i} and f_{2i} are the radial frequency pairs and bound the range of the corresponding frequency components. We have selected 64×64 from right iris regions. Ideal values of d are less than $d = [6.8; 0.425]$. For larger DR values image must be discarded considering unfocussed image.

4.2 Iris Preprocessing

It consists of segmentation of iris and normalization of iris. The main objective of the segmentation step is to differentiate iris pattern from the rest of the eye image. Correctly finding the inner and outer boundaries of iris pattern is important in all iris recognition systems. We have used Daugman's Integro Differential operator for segmentation of iris.

In normalization the iris region is transformed into fixed size. We used Daugman's rubber sheet model for the normalization step.

4.3 Feature Extraction Using DWT

CVPR_IRIS database consists of image size 1024×768. For one-dimensional discrete signals 1D Discrete Wavelet Transform is mainly preferred. For images of any type two-dimensional Discrete Wavelet Transform is needed to use. 2D wavelet transform obtained by separable products of scaling functions $\emptyset$ and wavelet functions Ψ by (4)

$$C_{j+1}[k, l] = \sum_{m,n} h[m - 2k]h[n - 2]C_j[\mathrm{m, n}] \qquad (4)$$

The detail coefficient images are obtained from three wavelets given by Eqs. (5), (6) and (7).

$$\text{Vertical wavelet: } \Psi^1(t1, t2) = \emptyset(t1)\,\Psi(t2) \qquad (5)$$

$$\text{Horizontal wavelet: } \Psi^2(t1, t2) = \emptyset(t2)\,\Psi(t1) \qquad (6)$$

$$\text{Diagonal wavelet: } \Psi^3(t1, t2) = \Psi(t1)\Psi(t1) \qquad (7)$$

2D DWT is used to reduce FAR and FRR. Hence improves system efficiency. Three level decomposition of normalized image of 512×64 gives the feature vector creation of size is 1×8. Energy is computed for approximation coefficients and three detail coefficients (Horizontal, vertical and diagonal) by Eq. (8)

$$\text{Energy} = \sum_{m=0}^{M-1} \sum_{n=0}^{N-1} |X(m, n)|, \qquad (8)$$

where $X(m, n)$ is a discrete function whose energy is to be computed.

Table 3 FAR and FRR results and comparison

Algorithms	% FAR	% FRR	% Average accuracy
Daugman [6]	0.010	0.090	99.90
Ma et al. [7]	0.020	1.980	98
Sanchez-Reillo et al. [9]	0.030	2.080	97.89
Tisse et al. [8]	1.840	8.790	89.37
Proposed CVPR_IRIS Database (DB1)	3.021	7.800	91.44

4.4 Matching

Fuzzy hamming distance is metric given by Eq. (9) is used to compute similarity or match the value. For perfect match zero distance is shown and for any increase in distance mismatch is shown.

$$\text{Fuzzy Hamming Distance} = \frac{1}{N} \sum_{i=1}^{1} |Xi - Yi| \qquad (9)$$

5 Conclusions

In this paper, imaging framework and acquisition set up is prepared to acquire good quality images, hence we have announced the availability of the CVPR_IRIS database. Also we found better accuracy for proposed database as compared with Tisse [8] as shown in Table 3.

Acknowledgements The author acknowledges the technical and nontechnical support given by CVPR Lab authorized faculty and TEQIP.

References

1. CASIA Iris Image Database. http://www.cbsr.ia.ac.cn/IrisDatabase.html
2. Proença H, Filipe S, Santos R, Oliveira J, Alexandre LA (2010) IEEE Trans Pattern Anal Mach Intell 32(8):1529–1535
3. Daugman J (1993) High confidence visual recognition of persons by a test of statistical independence. IEEE Trans Pattern Anal Mach Intell 15(11):1048–1161
4. Boles W, Boashash B (1993) A human identification technique using images of the iris and wavelet transform. IEEE Trans Signal Process 46(4):1185–1188
5. Wildes RP (1997) Iris recognition: an emerging biometric technology. Proc IEEE 85(9):1348–1363

6. Daugman J (2001) Statistical richness of visual information: update on recognizing persons by iris patterns. Int J Comput Vis 45(1):25–38
7. Ma L, Wrag Y, Tam T (2002) Iris recognition using circular symmetric filters. In: 16th international conference on pattern recognition, vol 2, pp 414–417
8. Tisse C-L, Torres L, Robert M (2002) Person identification based on iris patterns. In: Proceedings of the 15th international conference on vision interface, pp 215–229
9. Sanchez-Reillo R, Sanchez-Avila C, De Martin-Roche D (2002) Iris recognition for biometric identification using dyadic wavelet transform zero crossing. In: Proceeding of IEEE 35th Carnahan international conference on security technology, pp 272–277
10. Daugman J (2004) How iris recognition works. IEEE Trans Circ Syst Video Technol 14:21–30
11. Sanchez-Avila C, Sanchez-Reillo R (2002) Iris-based biometric recognition using dyadic wavelet transform. IEEE Aerosp Electron Syst Mag 17:3–6
12. Ma L, Tan T, Wang Y, Zhang D (2003) Personal recognition based on iris texture analysis. IEEE Trans Pattern Anal Mach Intell 25(12):1519–1533
13. Poursaberi A, Araabi BN (2007) Iris recognition for partially occluded images, methodology and sensitivity analysis. EURASIP J Adv Signal Process (Springer open J) 2007, Article ID 36751:1–12
14. Sun Z, Wang Y, Tan T, Cui J (2005) Improving iris recognition accuracy via cascaded classifier. IEEE Trans Syst Man Cybern Part C Appl Rev 35(3):435–440
15. Thornton J, Savvides M (2007) Kumar V (2007) A Bayesian approach to deformed pattern matching of iris images. IEEE Trans Pattern Anal Mach Intell 29(4):596–606

Fusion-Based Segmentation Technique for Improving the Diagnosis of MRI Brain Tumor in CAD Applications

Bharathi Deepa, Manimegalai Govindan Sumithra, Venkatesan Chandran and Varadan Gnanaprakash

Abstract Diagnosing the brain tumor from Magnetic Resonance Imaging (MRI) in Computer-Aided Diagnosis (CAD) applications is one of the challenging task in medical image processing. Traditionally many segmentation methods are used to address this issue. This paper introduces a segmentation method along with image fusion. Here a Discrete Wavelet Transform (DWT) method is chosen, for image fusion followed by segmentation using Support Vector Machine (SVM) for detecting the abnormality region. The types of MRI images considered here include T1-weighted (T1-w), T2-weighted (T2-w) and FLAIR images. The various fusion combinations are T1-w and T2-w, T1-w and FLAIR, T2-w and FLAIR. Experimental results suggest that on an average, fusion-based segmented result is superior to non-fusion-based segmented result.

1 Introduction

Magnetic Resonance Imaging (MRI) is widely used for brain tumor detection. Computer-Aided Diagnosis (CAD) succors the neuro-radiologists by giving a second opinion to improve the exactitude in diagnosing MRI brain pathology. In [1, 2], different segmentation methods with preprocessing and feature extraction have been discussed and it is also useful for the radiologist to make a decision while detecting the abnormality. The classification of abnormal regions in the given input MRI brain

B. Deepa (✉)
Department of ECE, Jayaram College of Engineering and Technology, Trichy, India
e-mail: cool.deeps.143@gmail.com

M. G. Sumithra · V. Chandran · V. Gnanaprakash
Department of ECE, Bannari Amman Institute of Technology, Sathyamangalam, Erode, India
e-mail: mgsumithra@rediffmail.com

V. Chandran
e-mail: chandran@bitsathy.ac.in

V. Gnanaprakash
e-mail: gnanaprakash@bitsathy.ac.in

© Springer Nature Switzerland AG 2019
D. Pandian et al. (eds.), *Proceedings of the International Conference on ISMAC in Computational Vision and Bio-Engineering 2018 (ISMAC-CVB)*, Lecture Notes in Computational Vision and Biomechanics 30,
https://doi.org/10.1007/978-3-030-00665-5_31

image with the help of GLCM technique was explained in [3]. In [4] segmentation is carried out using spatial FCM technique for the detection of pathology. But here, fusion is used to prevent the loss of pixels without any preprocessing method in order to segment the abnormality region in the MRI brain images.

2 Related Work

Sachdeva et al. [5] segmented the tumor boundaries by identifying the weak and false edges on the brain image. Here, the hybrid methodologies such as Genetic Algorithm with Support Vector Machine (GA-SVM), and Genetic Algorithm with Artificial Neural Network (GA-ANN) were implemented for classifying the tumor region. But, this work failed to reduce the misclassification rate of tumor classification. Ng et al. [6] segmented the brain tumor by using Fuzzy C Means algorithm. Here, the membership function is defined for making different clusters. The drawback of this method is that the cluster boundaries are not well defined. Thresholding technique is most commonly used in medical image segmentation. This methodology attempts to find the threshold value for the classification of pixels into different categories like tumor and non-tumor regions [7]. Barman et al. [8] represented level set segmentation using Partial Differential Equation (PDE). K-Means is one of the simplest unsupervised algorithms that has been used for image segmentation. The given input data is clustered into different groups and the centroid (calculated by means of Euclidean distance) of one cluster is placed far away from the other cluster as possible. It fails to segment the noise corrupted images [9].

3 Proposed Work

The proposed novel technique is described in Fig. 1. It includes two steps for diagnosis of MRI brain tumor. The first step is image fusion by Discrete Wavelet Transform (DWT) and the second step is image segmentation by Support Vector Machine (SVM). The methodology used in each step has been discussed in the following sections.

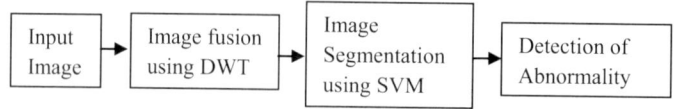

Fig. 1 Proposed methodology

In discrete wavelet transform, the given input images are decomposed into approximate (low frequency) and detail (high frequency) components [10]. Again the detail components are decomposed as horizontal, vertical, and diagonal components. The decomposition is done until the image contains the single pixel. Haar wavelet is used to decompose the image for obtaining the wavelet coefficients and maximum frequency selection rule is used for selecting the maximum absolute values for fusing the images [11]. Finally, inverse wavelet transform is used to get the fused image. Support Vector Machine (SVM) is a supervised segmentation technique used for medical images. It divides the given input fused image into resultant regions either in a rectilinear or non-rectilinear manner which is defined as a hyperplane. It is described by the training points (Supporting vector) and is given by

$$f(a) = za + q = 0 \tag{1}$$

where z is the weight vector and q is bias or threshold. If the training data is rectilinearly separable, then it is represented by $[a_i, b_i]$, $i = 1, 2, \ldots N$, where N is the number of training samples and bi belongs to $(-1, 1)$. If $b_i = 1$, it represents class 1 and if $b_i = -1$, it represents class 2 [12]. These two can be combined by the following equation:

$$b_i(z \cdot a_i + q) - 1 \geq 0 \tag{2}$$

The kernel function $f(a_i, b_i)$ is demarcated by Exponential Basis Function kernel (EBF kernel), which defines the solution to map the low dimensional space vector sets to a high dimensional vector sets in case of non-rectilinear patterns [13]. The EBF kernel is given by

$$f(a_i, b_i) = \exp\left(-\frac{\|a_i - b_i\|}{2\sigma^2}\right) \tag{3}$$

4   Result Analysis

Brain images used in this paper are taken from BRATS image database. The different types of MRI brain images like T1-weighted (T1-w), T2-weighed (T2-w) and FLAIR of same patients were analyzed for the accurate detection of abnormality. After making a literature survey, here few segmentation methods such as K-Means (KM), Thresholding (ThH), Level Set (LS), Fuzzy C Means (FCM), and Support Vector Machine (SVM) are preferred, because these methods are used mostly for medical images rather than other kinds of image. The enactment of these segmentation methods considered here is evaluated with and without fusion technique.

4.1 Sensitivity, Specificity and Accuracy

Sensitivity is defined as positive probability function also termed as true positive fraction. Specificity is known as negative probability function also termed as true negative fraction

$$\text{Sensitivity} = \frac{\text{Tpo}}{\text{Tpo} + \text{Fne}} * 100\% \tag{4}$$

$$\text{Specificity} = \frac{\text{Tne}}{\text{Tne} + \text{Fpo}} * 100\% \tag{5}$$

where Tpo is the True positive, Fne is the False negative, Tne is the True negative, and Fpo is false positive. Accuracy is calculated by

$$\text{Accuracy} = \frac{\text{Tpo} + \text{Tne}}{\text{Tpo} + \text{Fne} + \text{Tne} + \text{Fpo}} * 100\% \tag{6}$$

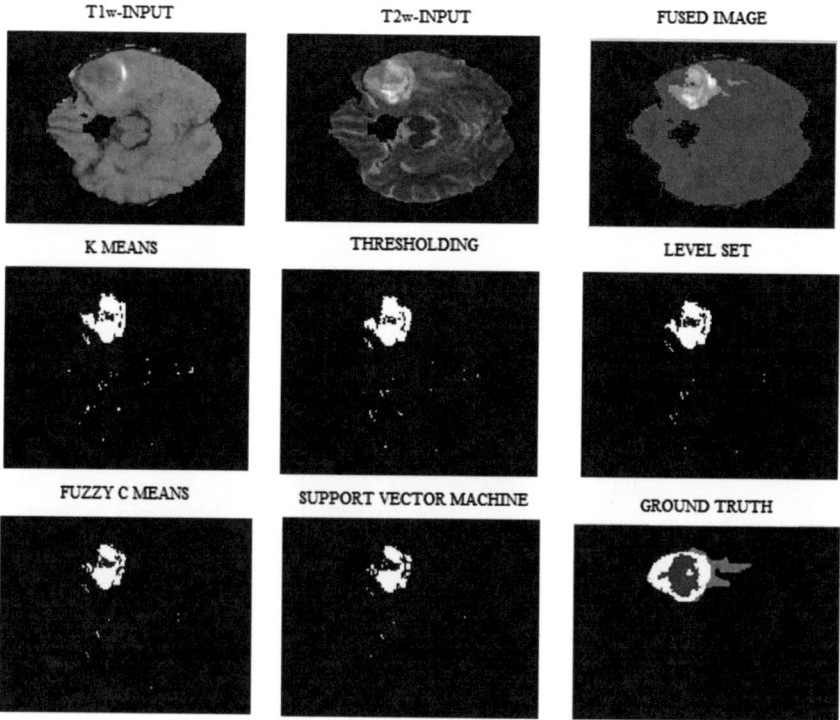

Fig. 2 Evaluation of different segmentation techniques for MRI T1-w and T2-w brain tumor image

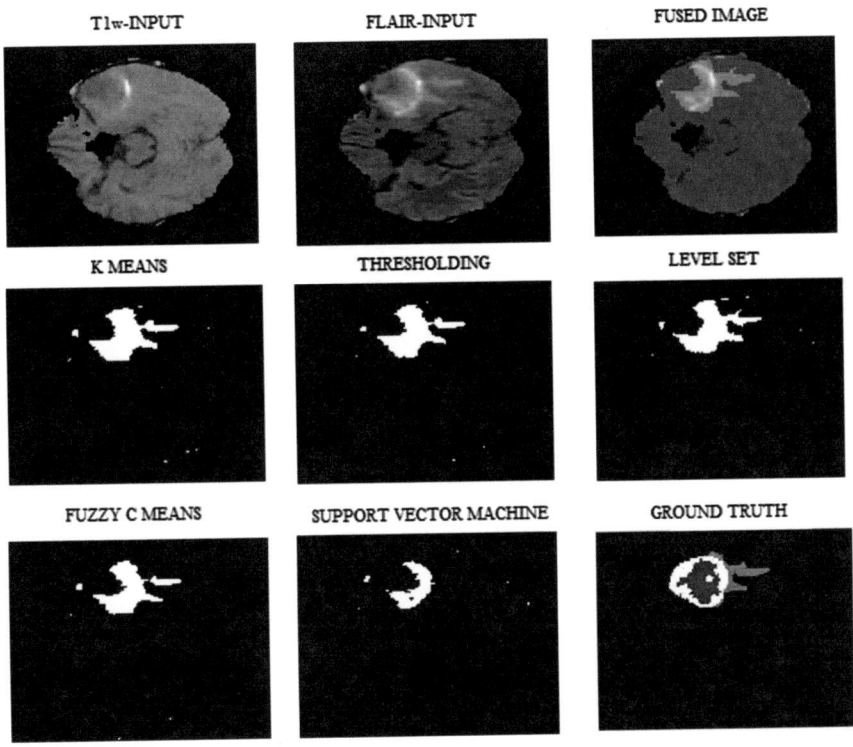

Fig. 3 Evaluation of different segmentation techniques for MRI T1-w and FLAIR brain tumor image

4.2 Positive Predictive Value (PPV) and Negative Predictive Value (NPV)

PPV and NPV for segmentation algorithm are calculated by

$$PPV = \frac{Tpo}{Tpo + Fpo} * 100\%; \quad NPV = \frac{Tne}{Tne + Fne} * 100\% \quad (7)$$

T1-w and T2-w image fusion with segmented result and its ground truth image (obtained from BRATS database) is shown in Fig. 2. The segmented result for this fusion image is better while using SVM rather than other methods. Figure 3 indicates the evaluation of various segmentation methods for the fusion of T1-w and FLAIR type MRI images. FCM and SVM results are good compared to others. T2-w and FLAIR fusion followed by segmentation is shown in Fig. 4. The result of SVM is better compared to all other abovementioned results considered here.

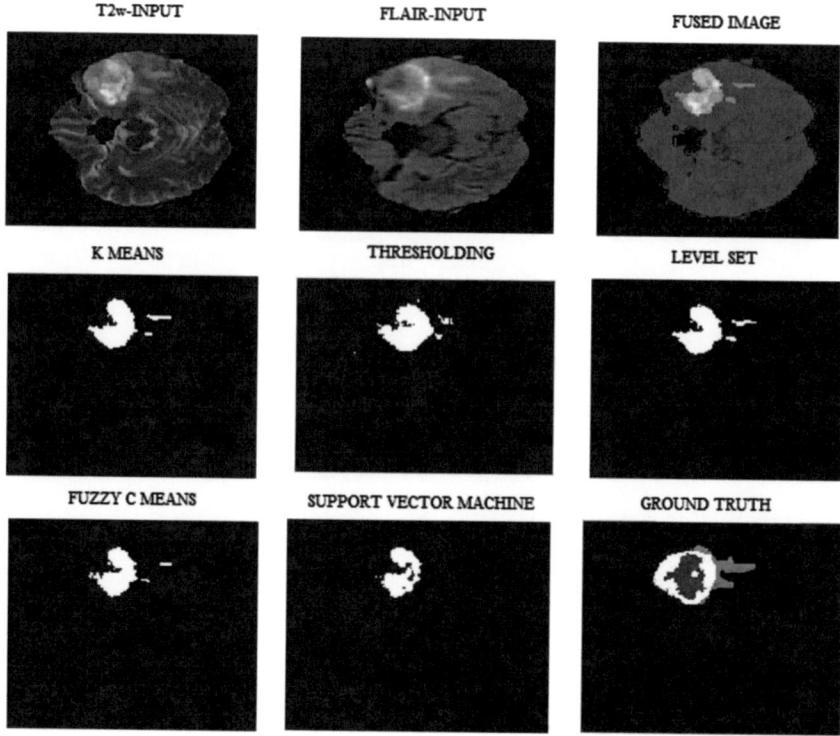

Fig. 4 Evaluation of different segmentation techniques for MRI T2-w and FLAIR brain tumor image

Figure 5 represents the performance metric for T1-w and T2-w fusion-based segmentation techniques. The obtained graphical values show that DWT fusion with SVM segmentation is giving higher result compared to others. Moreover, all the performance metric value is high in fusion-based segmentation rather than non-fusion-based segmentation. The evaluation measure for T1-w and FLAIR fused images with segmentation is shown in Fig. 6. The performance value of DWT fusion with FCM segmentation and SVM goes hand in hand. Figure 7 implicates the performance measure for fusion of T2-w and FLAIR type images followed by segmentation. It is seen that high accuracy value is for DWT fusion with SVM segmentation compared to all other techniques as discussed here. DWT with K-Means algorithm is giving inferior result.

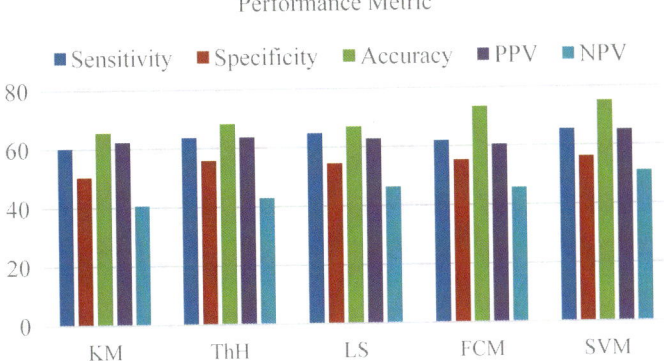

Fig. 5 Comparison of various fused segmentation techniques for MRI T1-w and T2-w brain tumor image

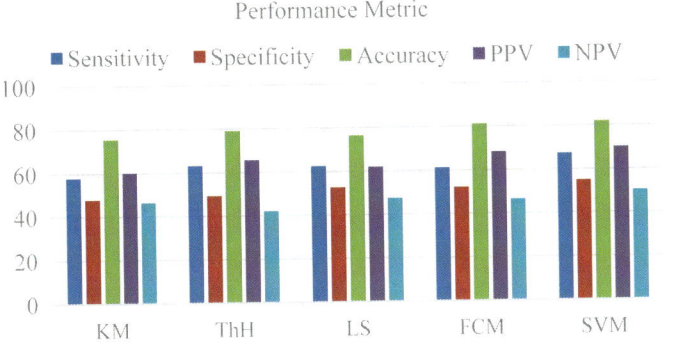

Fig. 6 Comparison of various fused segmentation techniques for MRI T1-w and FLAIR brain tumor image

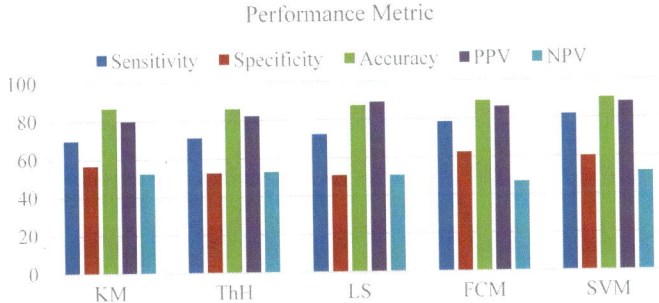

Fig. 7 Comparison of various fused segmentation techniques for MRI T2-w and FLAIR brain tumor image

Table 1 Comparitive measure of various segmentation techniques for MRI T1-w, T2-w and FLAIR brain tumor image

Input	Segmentation methods	Sensitivity	Specificity	Accuracy	PPV	NPV
T1w	KM	44.1	43.1	62.1	45.2	44.4
	ThH	44.7	42.8	63.8	44.5	45.7
	LS	43.5	44.7	63.4	46.8	46.6
	FCM	45.2	43.2	64.2	47.1	45.1
	SVM	45.9	45.7	65.5	47.8	48.2
T2w	KM	43.4	42.6	63.2	44.7	45.7
	ThH	42.5	43.9	64.8	46.2	46.4
	LS	43.8	44.7	64.6	46.9	47.5
	FCM	44.5	45.2	65.3	47.2	43.4
	SVM	45.4	46.2	66.8	48.8	46.9
FLAIR	KM	45.8	45.4	65.9	46.1	43.7
	ThH	46.1	44.4	64.6	47.5	45.4
	LS	48.7	46.5	65.7	48.2	46.5
	FCM	49.4	46.9	66.8	46.7	47.8
	SVM	50.7	48.1	67.2	49.5	45.4

Table 1 represents the performance evaluation measure like sensitivity, specificity, accuracy, Positive Predictive Value (PPV), Negative Predictive Value (NPV) for T1-w, T2-w, and FLAIR-type MRI brain tumor images by undergoing various segmentation techniques for diagnosing tumor. From the obtained tabulated values, it is seen that SVM method is giving high accuracy, sensitivity, and PPV value compared to others for all types of images.

5 Conclusion

In this paper, DWT fusion-based segmentation techniques for diagnosing brain tumor is presented. Various segmentation algorithms along with and without fusion have been examined for MRI brain images in gray scale by objective measures. From the attained results, it is seen that on an average, DWT fusion followed by segmentation methods is giving better results than non-fusion-based segmentation techniques by 20.23%. In addition, the accurate diagnosis of tumor region is achieved while using DWT fusion with SVM segmentation rather than non-fusion-based segmentation. The future work involves the introduction of novelty in wavelet transform for fusion and proposing a new hybrid segmentation technique based on intensity variation for the abnormality detection that can be finally used for classification of MRI brain tumor in CAD applications.

Acknowledgements The website link for BRATS image database is https://www.smir.ch/BRATS/ Start2013. This data set was supported for my doctoral degree purpose only. We have no conflict of interest with regard to the work presented. Ethical approval to conduct this study was obtained for my research work. Informed consent was obtained from all individual participants in the study.

References

1. Abdullah N, Chuen L, Ngah U, Ahmad K (2011) Improvement of MRI brain classification using principal component analysis. In: 2011 IEEE international conference on control system, computing and engineering (ICCSCE). IEEE, pp 557–561
2. Najafi S, Amirani M, Sedghi Z (2011) A new approach to MRI brain images classification. In: 2011 19th Iranian conference on electrical engineering (ICEE). IEEE, pp 1–5
3. Singh D, Kaur K. Classification of abnormalities in brain MRI images using GLCM, PCA and SVM. Int J Eng I:243–248 (Online). http://www.ijeat.org/attachments/File/v1i6/F0676081612.pdf2012
4. Reddy AR, Prasad E, Reddy DL (2012) Abnormality detection of brain MRI images using a new spatial FCM algorithm. Int J Eng Sci Adv Technol 2(1):1–7
5. Sachdeva J et al (2016) A package-SFERCB-"Segmentation, feature extraction, reduction and classification analysis by both SVM and ANN for brain tumors". Appl Soft Comput 47:151–167
6. Ng CR, Than JCM, Noor NM, Rijal OM (2015) Double segmentation method for brain region using FCM and graph cut for CT scan imges. In: IEEE international conference on signal and image processing applications, 978-1-4799-8996-6/15
7. Sezgin M, Sankur B (2004) Survey over image thresholding techniques and quantitative performance evaluation. J Electron Imaging 13(1):146–165
8. Barman PC, Miah MS, Singh BC, Khatun MT (2011) MRI image segmentation using level set method and implement an medical diagnosis system. Comput Sci Eng Int J 1(5)
9. Liu J, Guo L (2015) A new brain MRI image segmentation strategy based on wavelet transform and K-means clustering. IEEE 978-1-4799-8920-1-15
10. Lan T, Xiao Z, Li Y, Ding Y, Qin Z (2014) Multimodal medical image fusion using wavelet transform and human vision system. ICALIP,978-1-4799-3903-9/4. IEEE
11. Indira KP, Hemamalini R (2015) Impact of co-efficient selection rules on the performance of DWT based fusion on medical images. In: International conference on robotics, automation, control and embedded systems. ISBN 978-81-925974-3-0
12. Vijayakumar B, Chaturvedi A (2012) Automatic brain tumors segmentation of MR images using fluid vector flow and support vector machine. Res J Inf Technol 4:108–114
13. Hota HS, Shukla SP, Gulhare K (2013) Review of intelligent techniques applied for classification and preprocessing of medical image data. IJCSI Int J Comput Sci Issues 1:267–272 (Online). http://www.ijcsi.org/papers/IJCSI-10-1-3-267-272.pdf

Identification of Cyst Present in Ultrasound PCOS Using Discrete Wavelet Transform

R. Vinodhini and R. Suganya

Abstract The Polycystic Ovary Syndrome (PCOS) is an endocrine abnormality; it affects females during their reproductive cycle. It is hormone imbalance of female and it skips the menstrual cycle and makes it harder to get pregnant. The side effects of PCOS are causing blood pressure, heart disease, diabetes, obesity, etc. Thus, there is an imbalance in hormone that creates many cysts in ovary and it is called as polycystic ovary syndrome. It can be diagnosed by using ultrasound scan to identify the count, size, and severity of cyst. Preprocessing the medical image is the basic and initial step for medical image processing to remove the speckle noise, present in the ultrasound image and also helpful to create medical image applications. Compared with all other modalities of scan images, ultrasound scan is less cost-effective, but it contains more speckles due to image acquisition. Speckle is a granular noise that inherently exists in and it degrades the quality of the radar, ultrasound, and CT scan images. It also causes the difficulties for image interpretation. The aim of this research paper is to apply the modified Daubechies—discrete wavelet filters for ultrasonic scan image of PCOS for removing speckle noise for better diagnose cyst. This paper concludes the effectiveness of the proposed filter to identify cyst present in ultrasound PCOS. The following metrics—SNR, PSNR, and SSIM are used to measure the effectiveness of various categories of discrete wavelet transform.

1 Introduction

Nowadays, most of the women at the age of 18–35 are affected from hormonal imbalance called PCOS. Polycystic Ovary Syndrome is a complex problem that affects 6 out of 15 women at their reproductive age. An ovary consists of follicles

R. Vinodhini (✉) · R. Suganya
Department of Information Technology, Thiagarajar College of Engineering, Madurai,
Tamil Nadu, India
e-mail: vinodhinicse95@gmail.com

R. Suganya
e-mail: rsuganya@tce.edu

© Springer Nature Switzerland AG 2019
D. Pandian et al. (eds.), *Proceedings of the International Conference on ISMAC in Computational Vision and Bio-Engineering 2018 (ISMAC-CVB)*, Lecture Notes in Computational Vision and Biomechanics 30,
https://doi.org/10.1007/978-3-030-00665-5_32

filled with fluid structure. The gland that is situated at the base of the brain segregates Follicle Stimulating Hormone (FSH) that can secrete, and it converts the immature follicle into mature follicle and that follicle generates estrogens, and releases matured follicle and egg at the time of menstrual cycle, on the other hand, some of that follicles are changed into chocolate cyst and due to imbalanced diet, stress and depression are formed at ovary are called polycystic ovary disorder. It causes because our blood contains a high level of insulin produced by pancreases and production of male hormone (androgen) leads to infertility. It causes other abnormalities in a woman's body, such as hair fall from the scalp, growing unwanted hair in face, chest and stomach, lack of sleep, depression, obesity and also suffers from other disease such as diabetes, heart diseases and high blood pressure. It can be diagnosed by calculating Body Mass Index (BMI) of a patient with periodical review along with blood pressure, hormone level test, and ultrasonic scan to find out the size, count and severity of cyst. The purpose of this research paper is to identify these cysts at the earlier stage from the ultrasonic images by using various filtering techniques, compare them to find out which should be suitable for PCOS disorder. Filtering is the core and initial step used in image processing to enhance and remove noise from the image. It needs a higher level of decomposition filter to remove noise. The presence of speckle in a medical image would hide the necessary pathological information leading to improper result in a diagnosis.

2 Related Works

The related works narrated the importance of decomposition filters of ultrasonic images: [1]. The author proposes that anisotropic diffusion filters are used to remove the speckles of the ultrasound scan images and radar images for edge diffusion [2]. The speckled SAR and ultrasonic images are difficult to analyze due to the presence of more speckle noise, and the author uses the discrete wavelet transform filter to remove speckles of the image. In [3], the author proposes a speckle suppression technique to ultrasonic image for preprocessing [4]. The author proposes texture classification technique to differentiate the normal liver and fatty liver by using Discrete Wavelet Transform. The author [5] proposes to analyze the quality of ultrasonic image by applying various DWT filters for denoising the image by using the soft thresholding and compare the DWT filters by PSNR, SNR ratio. This paper [6] aims to compress the ultrasonic medical image based on the Region of Interest (ROI) for analyzing the patient record by using wavelet transform and their quality is measured by the metrics SNR and SSIM [7]. The author says that the quality of their ultrasonic's image is measured by using signal-to-noise ratio and they can be in the range of 8.5, 5.5, 2.2. PCOS is the endocrine disorder that affects women at the age of 18–35, and in paper [8], the author analyzes the disease by k-means algorithm and classify the number of follicles in ovary. PCOS is diagnosed by the medical record of the patient along with the ultrasound image of the abdomen of the ovary [9]. The author proposed the novel method to automate and detect the PCOS detection [10]. The excessive segregation of the insulin in the female hormone creates the segregation of androgen (male hormone). The presence of the cyst (immature follicle) in the ovary is diagnosed easily by the ultrasound image.

But all the above paper fails to address the subtle features of speckle noise contain some important information of immature follicle which causes infertility in the future. So, we planned to identify this subtle features present in the noisy ultrasound image rather than eliminating a noise.

3 Proposed Work

Automatic interpretation of ultrasound images, however, is extremely difficult because of its low signal-to-noise ratio (SNR) [11]. Image preprocessing is a technique to enhance a medical image. The ultrasonic scanning image of a patient affected from Polycystic Ovary Syndrome and uses filtering techniques to remove and analyze speckles in the image. Speckle noises tend to obscure diagnostically important features and degrade the image quality and thus increase the difficulties in diagnosis. This work is based on the major issue in ultrasound modality. This work provides knowledge about adaptive and anisotropic diffusion techniques for speckle noise removal from various types of ultrasonic images. Though much work has been done on the speckle reduction in ultrasound images in the literature, in most of the work, it is assumed that the Diffusion-based spatial filter is to suppress the noise and preserve fine details of edges in ultrasound images. But it looks to diagnose small residual features like cyst and lesion which will lead to major problems like cancer and finally lead to death. So much care should be taken to identify small pathology bearing information which is hidden by means of speckles during preprocessing. In order to focus on subtle features like immature follicles, chocolate cyst and improve edge information for ultrasound PCOs images. In this research work, we have proposed the db filter for preprocessing technique that eliminates only speckle noise and leaves subtle features which help to identify the presence of cyst or immature follicles in its earlier stages (Fig. 1).

In this proposed work, pre-process the image at the size of 222 KB and JPEG format image. From the above survey, DWT (haar, bior, db) and also with continuous wavelet filters are applied to ultrasonic image, compare the results of that image and identify which filter is suitable for filtering ultrasonic image. Comparison of filters is among metrics such as SNR, PSNR, and SSIM ratio.

4 Methodology Used

(a) Discrete Wavelet Transforms

To remove speckles from medical image, there is need of a number of decompositional layers from the survey about the filters used in ultrasonic image the above two

Fig. 1 Block diagram of the proposed work

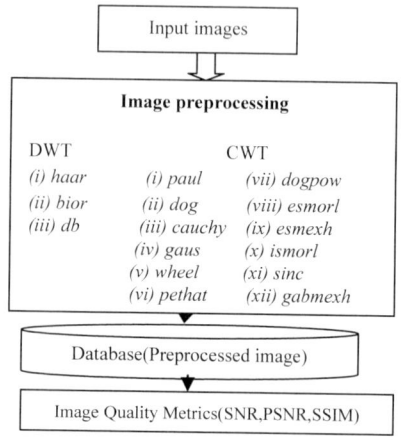

filters are mostly used: Discrete Wavelet Transform (DWT) and Continuous Wavelet Transform/anisotropic filter (CWT).

DWT:

The Discrete Wavelet Transform (DWT) has turned into an intense procedure in biomedical signal processing. The most evident distinction is that the DWT utilize scale and position esteems in light of forces of two. The values of s and τ are: $s = 2^j$, $\tau = k * 2^j$ and $(j, k) \in Z^2$ as shown in Eq. (1).

$$\psi_s, \tau(t) = \frac{1}{\sqrt{2^j}} \psi \left(x = \frac{t - k * 2j}{2^j} \right) \tag{1}$$

The essential thought behind disintegration and remaking is low-pass and high-pass separating with the utilization of down examining and up inspecting individually. The consequence of wavelet decay is progressively composed disintegrations. In this proposed work, we connected the subchannels of the DWT with the [1] delicate thresholding to denoise the picture.

(b) CWT/Anisotropic filter

The continuous wavelet transform (CWT) selects the interior eventual outcome of a pennant, $f(t)$, with deciphered and created varieties of an analyzing wavelet, $\psi(t)$. The logical state of CWT is:

$$C(a, b; f(t), \psi(t)) = \int f(t) \frac{1}{a} \psi * (t - b/a) dt \tag{2}$$

where $f(\omega)$ and $\psi(\omega)$ are the Fourier transforms of the signal and the wavelet.

We also used an anisotropic filter because it is also one of the decomposition filters and all the sub-filters of the CWT are used such as paul, dog, cauchy, gaus, wheel, pethat, dogpow, esmorl, esmexh, gaus2, gaus3, isodog, isomorl, sinc, and gabmexh.

(c) **Daubechies**

Speckles are the noise which are present at the medical image that degrades the quality of the image like ultrasonic and radar images. There are many filters to remove the noise of the image like low-pass filter, high-pass filter, adaptive filter, and Gabor filter but the ultrasonic image that contains the speckles which has the more noise and it needs some decomposition layer to remove the filters. Wavelet filter is one type of the decomposition filter. Recently, wavelet transform is used in the analysis of the image, it separates the speckles and signals into the different orientation and scales. Daubechies is the one of the sub-filter of the discrete wavelet transform. From the computed metric ratio in Table 1, db2 gives better solution compared with the other filters. It is an orthogonal wavelet filter and they are defined by the DWT. It is characterized by the maximal number of vanishing moments.

$$2^A - 1 \text{ (algebraic equation of db filter)} \tag{3}$$

where A is the number of highest vanishing moments.

It can be used by the two schemes they are using length (DA) and refer with vanishing moment (db2).

The db filter has the overlapping windows, so the result reflects the pixel intensity and it is smoother than haar filter, but it is more complex than haar. It has four wavelets and scaling coefficients. The sum of the functions of coefficient is calculated

Table 1 Comparison of various filters by metrics of ultrasound image

Filter name	Subname of filter	PSNR	SNR	SSIM
DWT	bior1.5	21.66	11.41	0.88
	db2	21.73	11.66	0.87
	haar	21.60	11.51	0.88
CWT	paul	21.59	4.71	0.82
	dog	20.59	4.79	0.91
	cauchy	21.67	1.74	0.98
	gaus	20.29	3.16	0.95
	wheel	20.70	5.79	0.93
	pethat	20.20	1.0	0.90
	dogpow	20.47	3.99	0.91
	esmorl	20.27	1.23	0.93
	esmexh	20.41	2.17	0.91
	isodog	21.52	6.15	0.98
	isomorl	20.18	11.17	0.88
	sinc	21.46	7.08	0.99
	gabmexh	20.42	1.79	0.97

by the average over the adjacent pixels increase its capabilities of image processing techniques.

4.1 Image Quality Metrics

Peak signal-to-noise ratio (PSNR):

This measure is used to compare the square error between the original and constructed image. PSNR using the term Mean Square Error in the denominator, so the PSNR rate is high means image has low error.

$$MSE = \frac{1}{mn} \sum_{i=0}^{m-1} \sum_{j=0}^{n-1} [I(i, j) - K(i, j)]^2 \tag{4}$$

The PSNR (in DB is defined as)

$$PSNR = 10 \cdot \log_{10} \left(\frac{MAX_I^2}{MSE} \right) \tag{5}$$

where MAX_1, Maximum possible pixel value of the image. MSE, Mean Square Error.

High PSNR means that the image has good quality and less error. Range of PSNR for ultrasound image is at 20.45, 20.72, 21.90.

Signal-to-noise ratio (SNR):

It is used to measure the sensitivity of the image. SNR is the ratio of average signal value μ_{sig} to the standard deviation of the signal σ_{sig}.

$$SNR = \frac{\mu_{sig}}{\sigma_{sig}} \tag{6}$$

Structural similarity index measurement (SSIM):

It is used for prediction of image quality is based on an initial uncompressed or distortion-free image as reference. The SSIM of the image is calculated on the various window sizes of the image. They measure between the two windows are x and y

$$SSIM(x, y) = \frac{(2\mu_x \mu_y + c1)(2\sigma_{xy} + c2)}{(\mu_x^2 + \mu_y^2 + c1)(\sigma_x^2 + \sigma_y^2 + c2)} \tag{7}$$

where

μ_x, μ_y the average of x and y, σ_x^2, σ_y^2 the variance of x and y
σ_{xy} the covariance of x and y and $c1 = (k_1 L)^2$, $c2 = (K_2 L)^2$

5 Result and Discussion

The experimental data of this paper are obtained from the radiology department of Meenakshi Mission Hospital in Madurai, India. We have collected nearly 20 women scan reports and conducted a small survey. Preprocessing is the root process for medical image processing and other applications by imaging. In this paper adapted to filter techniques to remove noise from an ultrasound scan image for PCOS. These can be taken of the abdomen of the patients whom are affected from PCOS periodically.

Figure 2 represents the ultrasonic scan image of the PCOS patient and using this, one can diagnose the presence of cyst (immature follicles) in the ovary and find out the severity of the disease (Fig. 3).

The Daubechies filter that it is a type of the Discrete wavelet filter is used to remove speckles with that fix the decomposition layer of filter at seven because this ultrasound image needs more number of the decomposition layer to increase the PSNR ratio, if the PSNR ratio is high the image has good quality. Comparison of the image with different decomposition layers is represented below the graph.

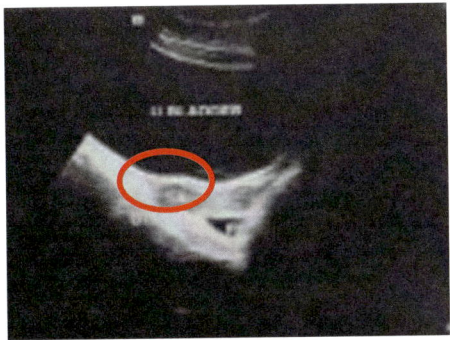

Fig. 2 PCOS ultrasonic image

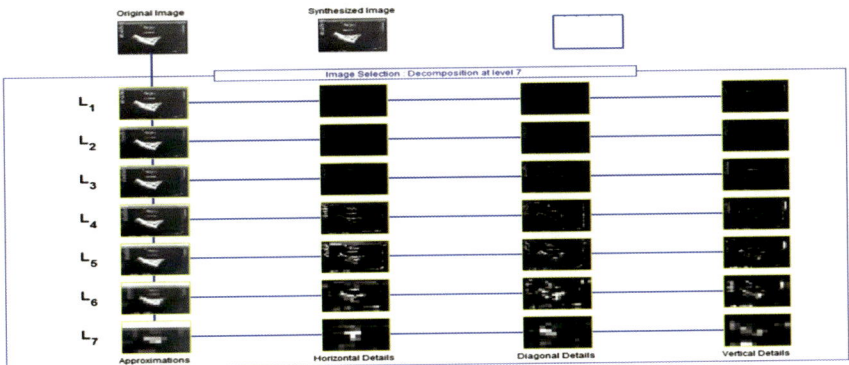

Fig. 3 db filter with decomposition layer

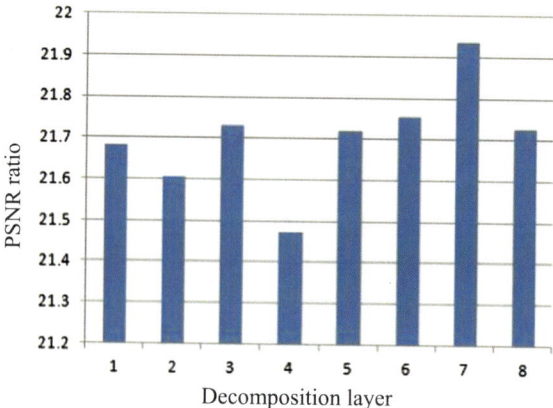

Fig. 4 PSNR value for different decomposition layer

Figure 4 represents the ultrasound image with the different level of decomposition layer, and we have considered the decomposition layer as seven for effective preprocessing, it has high value of the PSNR ratio (21.9). Approximation level, horizontal, diagonal and vertical details of the image are also determined.

Denoised images using subfilters of DWT

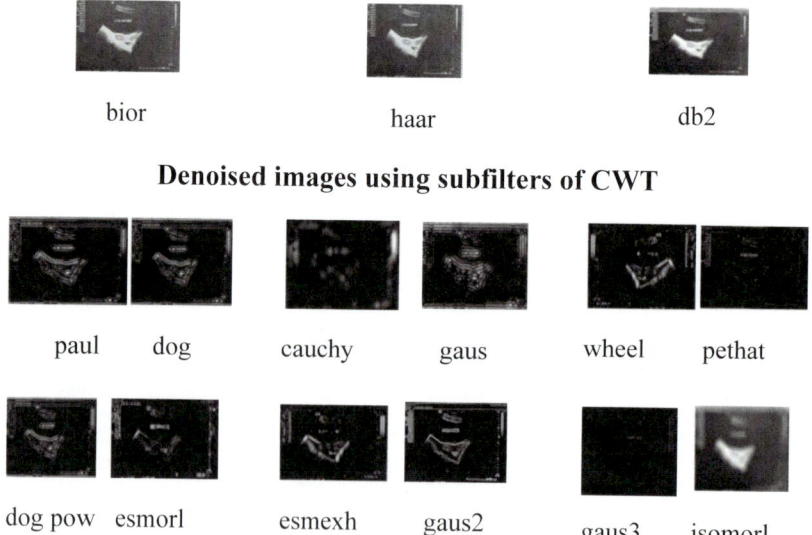

Above figures represent the denoising image of the various subfilters of wavelets. From this, we have chosen a suitable filter. It can be measured by using image quality metrics—PSNR, SNR, and SSIM. The comparison of PSNR, SNR, and SSIM for the above-filtered image is as follows.

Various filters of DWT and CWT are applied for PCOS dataset and measured by the three metrics (PSNR, SNR, SSIM): high PSNR range shows that the image has higher quality in Table 1 represents the comparison of all filters, db2 has high in ratio (21.7316). The range of SNR for ultrasound image is (2-12) and db2 also satisfies the SNR ratio as 11.6566. The range of SSIM ratios for ultrasound image is between −1 and 1, and SSIM value for db2 is 0.8673. From this, we have concluded that db2 satisfies all the metric ratio—PSNR, SNR, and SSIM. So, db2 is best for removing speckles and retains subtle features of PCOS ultrasonic image information.

6 Conclusion

The most important technique to diagnose PCOS for women at the earliest stage is ultrasound scan modality. In medical imaging, preprocessing is the initial step to remove speckle noise, and various subfilters of DWT and CWT with different decompositional layers are experimented to remove the speckles for retaining the subtle feature. Noise ratio of preprocessed image is also calculated by image quality metrics—PSNR, SNR, and SSIM. By diagnosing the preprocessed image, db2 is examined as a suitable filter for ultrasound image speckle reduction.

References

1. Yu Y, Acton ST (2002) Speckle reducing anisotropic diffusion. IEEE Trans Sig Process Soc 11(11):1260–1270
2. Choi H, Jeong J (2018) Despeckling images using a preprocessing filter and discrete wavelet transform-based noise reduction techniques. IEEE Sens Counc 99:1
3. Hao X, Gao S, Gao X (1999) A novel multiscale nonlinear thresholding method for ultrasonic speckle suppressing. IEEE Trans Med Imaging 18(9):787–794
4. Huang Y, Wang L, Li C (2008) Texture analysis of ultrasonic liver image based on wavelet transform and probabilistic neural network. In: IEEE conference
5. Thakur A, Anand RS (2005) Image quality based comparative evaluation of wavelet filters in ultrasound speckle. Digit Sig Process 15(5):455–465
6. Gupta S, Bhatia R (2007) Comparative analysis of image compression techniques: a case study on medical images. In: IEEE conference on advances in recent technologies in communication and computing
7. Omar EA, Varghese T (2011) Signal to noise ratio comparisons for ultrasound attenuation slope estimation algorithms. Nucleic Acid Res
8. Dewailly D et al (2017) Diagnosis of polycystic ovary syndrome (PCOS): revisiting the threshold values of follicle count on ultrasound and of the serum AMH level for the definition of polycystic ovaries. Oxford 26(11):3123–3129
9. Lawrence MJ, Eramian MG, Pierson RA, Neufeld E (2007) Computer assisted detection of polycystic ovary morphology in ultrasound images. In: IEEE conference computer and robot vision, 2007

10. Padmapriya B, Kesavamurthy T (2015) Diagnostic tool for PCOS classification. In: Springer conference, vol 52
11. Ramamoorthy S, Subramanian RS, Gandhi D (2014) An efficient method for speckle reduction in ultrasound liver image for e-health applications. In: International conference on distributed computing and internet technology, vol 8337, pp 311–321

Design and Development of Image Retrieval in Documents Using Journal Logo Matching

S. Balan and P. Ponmuthuramalingam

Abstract This research focuses on the detailed study of document image retrieval form World Wide Web using the existing techniques, text and image information which are extracted in the database. Retrieval based on the user's given keywords of journal papers. It fetches the information from the extracted folder and analyzing with logo matching technique using efficient SURF (Speeded-Up Robust Features) method. Precision, recall, f-measure, time, and memory are calculated to compute the accuracy of the given input. Logos are clustered using image descriptor methods using shape, edge, and color techniques. This paper proposes a new technique for logo matching in journal database depends upon the documents.

1 Introduction

Document image caries lot of text, printed, or scanned images. In the current scenario, web store tons of documents in different formats. Each format varies type and size. So the storage of information is to be text or image. Here, the process is based on journal documents (Word and Pdf) information retrieval from the web is extracted and preprocessed. In past, surveys on document images are focused on text conversion analysis, character images of text, indexing the structure of document, image abstraction via automatic, GIF images, drawings, and semantic image relations [1].

A document page is captured by using pixel-level processing method. It is separated by number of characters in the document and split into two different lines. Region values are identified by using feature values of the text and graphics of image in the given document description. For each image, binary value and threshold value are identified by using existing techniques. Some portion of text in the image is dark,

S. Balan (✉) · P. Ponmuthuramalingam
Research Department of Computer Science, Government Arts College (Autonomous),
Coimbatore, Tamil Nadu, India
e-mail: Balan.sethuramalingam@gmail.com

P. Ponmuthuramalingam
e-mail: ponmuthucbe@gmail.com

© Springer Nature Switzerland AG 2019
D. Pandian et al. (eds.), *Proceedings of the International Conference on ISMAC in Computational Vision and Bio-Engineering 2018 (ISMAC-CVB)*, Lecture Notes in Computational Vision and Biomechanics 30,
https://doi.org/10.1007/978-3-030-00665-5_33

Fig. 1 Sample IEEE logos for feature extraction

low, printed, same image with less or higher quality, extracted from we, .doc, .pdf, .HTML. So the size and clarity of image differs. Same matching image is identified by using the threshold value.

Here, Fig. 1 shows some of the sample logos of journal database is shown below. In that, IEEE images are extracted from the documents and cluster the images based on feature extraction values. For example, the logos are taken from IEEE website, Conference Template, IEEE conference advertisement, Journal Documents and Google search. This paper is categorized as follows: (1) study on Document Image retrieval (2) Survey of text and retrieval methods (3) using SURF methods (4) of performance evaluation (5) further research directions in logo matching is stated.

2 Literature Review: Survey

This survey aims to solve the various document image retrieval techniques to achieve the accuracy of the given documents. Document Image Retrieval (DIR) is used to identify the relevant documents based on the feature of the image. Documents stored in web are normally as E-books and handwritten documents. The page layout structure and text searching in documents is said to be Digital Libraries (DL). Three components of DL are as follows: indexing, querying, and similarity [2].

For example, processing step of a book is representation (paper, descriptive meta-data, pixel, structural metadata, feature and abstract repository), indexing (Database Management System, Image Database, DIR Database, and Information retrieval Engine), and Retrieval task (manual browsing, visual browsing, layout-based DIR, keyword and Full-Text). Automatic Indexing of Document Images (AIDI) describes the font indexing and document page layout structure [2].

The block diagram of DIR is query image (based on the user's query retrieves from the relevant document image), removal of noise (text image extraction such as signed documents, logo), computing feature (document image is extracted and stored in the database to represent the image attributes and values), and matching algorithm (Similarity value is compared with other images in database) [3].

Document page layout is categorized into geometric, Functional and Semantic Descriptions. Execution of the page layout is processed by Structure independent of (column, margins), logical-type dependent (address, signature), content is presentational (font, size, and style), and linguistic (meaning of block). Multiple retrieval query images are classified in the various methods: Average Query (Joint-Avg), Support Vector Machine (Joint-SVM), Average and maximum of multiple queries (AVG-MAX), and spatial re-ranking [4].

Document layout analysis based retrieval is grouped into two approaches namely top-down and bottom-up. Local content feature is based on local, global representation, and encoding. The real-time retrieval of document uses global description of image, similarity, relevance feedback, and score. The classification of document image indexing is divided into two ways: traditional indexing (Keyword spotting, OCR, New Method Signature based, Layout Structure and logo Matching) [5].

Signature matching in document is based on layout analysis, signature and hand-written text localization, binarization, noise removal, normalization, abstraction, feature value extraction, verification, and identification. Signature matching is problem of finding the similar matching from samples of signature stored in a database. It is based on value measures, detection and segmentation, structural saliency and contour grouping, and shape matching. Automatic extraction of signature retrieval based on two images namely document image signature extraction (preprocessing, finding candidate signature, extraction, and detection of feature value) and feature matching on document image retrieval on signature and document ranking [6].

Classification method used in Document Image Retrieval are KNN, ANN, DTC, MLP, CNN, and SVM. There are various different distance metrics are used as follows Canberra, Euclidean, City block, Chebychev, Cosine, Hamming and Jaccard. Indexing of documents based on signature is scanned image, segmentation, patch level classification, computing feature of signatures, touching removal and region detection [6].

Digital library consists a huge amount of documents stored in digital formats. It aims to solve the data available in the Internet is electronically categorized. The advantages of DL are physical boundary, information retrieval, preservation and conservation, resource sharing, and multiple access. DIR is based on recognition (converting document image into text) and recognition free (explicit recognition of image in relevant document) [7]. The challenges in DIR are the unavailability of OCR, degradation, poor quality, noisy images, paper quality, folding the paper, speed, cross-lingual and representation of font and encoding to find the distance matching documents in OCR uses standard distance functions (Euclidean, Manhattan, Mahalanobis, Dynamic time wrapping [8].

OCR is categorized into several parts such as character segmentation, feature extraction, and classification. Graphical document recognition mostly carries graphical information (maps and drawings) and graphical recognition (music, 2D and 3D views). OCR difficulties in graphical documents are character recognition and segmentation, separation of text, and individual text extraction. Difficulties in graphical documents area character recognition and segmentation, separation of text, and individual text extraction.

Logo similarity matching for negative shape features is identified by logo, triangular, rectangular, square, stripes, and borders. Context dependency system solves the logo matching technique consists of preprocessing the image components, brightness the values of image, improve the visual quality and adjusting the brightness values of image [9]. Some of the existing methods and dataset are used to achieve the results in the current survey as follows (Table 1).

Feature value of extraction is carried out in three ways namely color, text, and edge. Existing methods of logo spotting are based on approximate nearest algorithm, FLANN (fast library for approximate nearest neighbor), indexing, key points (query and document image) and learning. Document retrieval system query on logo matching is focused on extraction, match suing ANN, grouping in geometric filter and ranking. In document database flow is on text or non-text separation from a cluster of documents to identify the logo position [9].

Table 1 Recent survey on logo matching retrieval datasets

Authors	Method	Year	Dataset
Vaijinath et al. [9]	GLCM, ANN	2017	Organization logo
Romberg et al. [10]	Querying triangle and Edge Index	2011	FedEx logo
Dikey et al. [1]	Feature extraction	2015	Popular logo
Boonroda et al. [11]	SURF	2015	Product logo
Billa et al. [12]	Multiple descriptors	2017	Popular logo
Bharathidevi et al. [13]	SIFT descriptors	2017	Popular logo
Sawalkar et al. [14]	SIFT, RANCSAC	2014	Given images
Ramachandran et al. [7]	SURF	2015	Pair of logo

3 Proposed Evaluation of Approximate Surf Method

SURF method is used for feature detection, feature description, and feature matching. Mainly focus on integral matching and Hessian matrix. The main use is box filter techniques. It is divided into three ways namely interpoint detection, interest point description, and interest point matching.

$$\mathcal{H}(x,\sigma) = \begin{bmatrix} L_{xx}(x,\sigma) \ L_{xy}(x,\sigma) \\ L_{xy}(x,\sigma) \ L_{yy}(x,\sigma) \end{bmatrix}, \tag{1}$$

where L_{xx} and L_{yy} are approximate box filters
   To determine good approximation values:

$$\det\left(H_{\text{approx}}^{\text{SURF}}\right) = \hat{L}_{xx}\hat{L}_{yy} - \left(0.9\hat{L}_{xy}\right)^2 \tag{2}$$

To determine both values
To calculate sum of the pixels,

$$S = W - X - Y + Z \tag{3}$$

   Local Feature Detection Based on Interest Point
Algorithm is as follows:

Step 1: given input image x
Step 2: point initializing
X – Filtering the feature values
For y → {0, 1, 2, and 3} do
End for
For z: = 0 to 3 do
For m: = 1 to 3 do
X → 2° m + 1
End for
End for
Return point matching

   Figure 2 explains the general block diagram of SURF method is shown below:

3.1 Approximate SURF Method

Logo matching is based on text retrieval and image retrieval. Text retrieval is defined by document style and image retrieval is identified by using the approximate SURF method. Simple steps carried out for in journal logo database as follows:

Fig. 2 SURF method

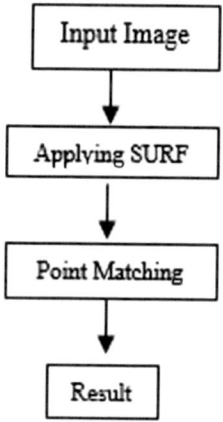

Step 1: Searching the document
Step 2: extracted and stored in separate folder
Step 3: compared and cluster the feature value within text document
Step 4: image is split into a × b windows, compute the deviation
Step 5: convert into binary image and identify the threshold value
Step 6: extract the shape, boundary, edge detection regions
Step 7: image properties are stored in separate folder
Step 8: for $n = 3$: size
For $x = 1$ to n
Find logo in extracted documents
End for
: set $m = 1$ to max $= 6$
Step 9: apply the threshold value in logo matching
Step 10: matching results are displayed
Step 11: end

4 Experimental Analysis and Discussions

To measure the effectiveness of information retrieval carried out in three steps
namely: gathering documents, testing the needs based on queries, and finding a
set of pairs for relevant or nonrelevant documents.

4.1 *Precision, Recall and F-Measure*

Precision is defined by the correct accuracy of the information system.

$$P = \text{tp}/(\text{tp} + \text{fp}) \tag{4}$$

Recall is defined by the extraction of relevant information from the system.

$$R = \text{tp}/(\text{tp} + \text{fn}) \tag{5}$$

F-measure is defined by precision and recall are opposite to each other.

$$F = 2PR/P + R \tag{6}$$

Table 2 shows that the performance evaluation of logo matching and Table 3 displays the overall measure of the user's input.

Figure 3 shows the overall measure of calculating the precision, recall, and f-measure values from ACM, Springer, IEEE, and Elsevier. Table 3 shows the performance of given input query (Fig. 4).

Here, the above measure and performance are calculated by using the given users query as automatic web page logo detection. Based on this query, it fetches the document from the web and extracted to separate folder. Using the existing techniques, text retrieval is measured and for image retrieval based on logo matching using efficient SURF method. Figure 5 shows that the probability of text retrieval for the given input. From web and matches with pair of documents.

Table 2 Performance evaluation of logo matching

Logo	Precision	Recall	F-measure	Time (ms)	Memory (Bytes)
ACM	33.58	33.15	33.37	230	235,520
Springer	37.70	38.67	38.18	141	144,384
IEEE	**41.86**	**42.88**	**42.36**	**89**	**91,136**
Elsevier	39.00	39.71	39.35	101	10,324

Table 3 Performance Measure

Text	Logo	Probability
Automatic web page logo detection	ACM	5.1
	IEEE	**7.4**
	Springer	1.8
	Elsevier	1.8

Fig. 3 Overall measure

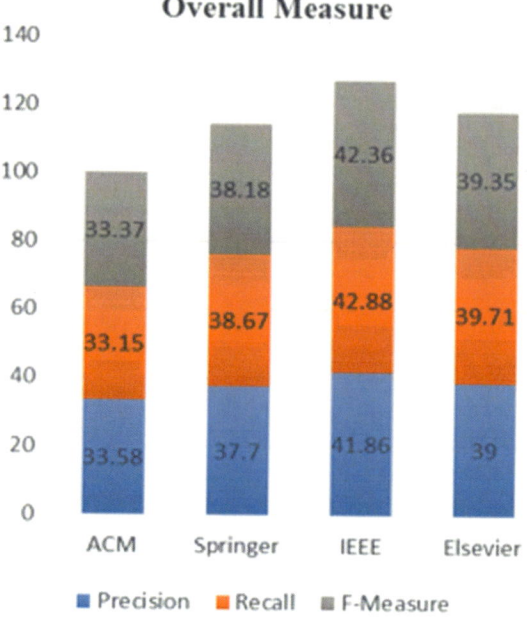

Fig. 4 Performance

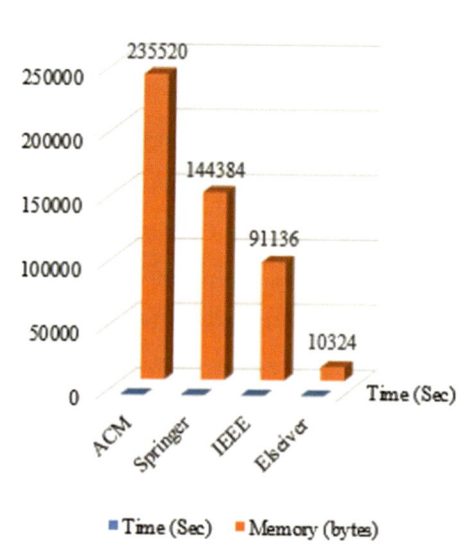

TX	RA	Prob
Automatic web page logo detection (Logo matching using semantic relevance feedback)	ACM	5.173753574536864
Automatic web page logo detection (Logo matching using semantic relevance feedback)	IEEE	7.434970306021613
Automatic web page logo detection (Logo matching using semantic relevance feedback)	Springer	1.8812752195585096
Automatic web page logo detection (Logo matching using semantic relevance feedback)	science direct	1.584615439944184
Automatic web page logo detection (Logo matching using semantic relevance feedback)	elsevier	1.8698477425411446
Automatic web page logo detection (Logo matching using semantic relevance feedback)	JCS	0.8294929229077914
Automatic web page logo detection (Logo matching using semantic relevance feedback)	IJET	1.5942492581904428
Automatic web page logo detection (Logo matching using semantic relevance feedback)	IJCA	1.1519926713228061
Automatic web page logo detection (Logo matching using semantic relevance feedback)	IJCTT	0.9072405177012266
Automatic web page logo detection (Logo matching using semantic relevance feedback)	VLDB	0.696966795424503
Automatic web page logo detection (Logo matching using semantic relevance feedback)	ICDE	1.357798896845852
Automatic web page logo detection (Logo matching using semantic relevance feedback)	IEEE ICDE	3.7114131480941177
Automatic web page logo detection (Logo matching using semantic relevance feedback)	ACM SIGMOD	0.592841265714317

Fig. 5 Text retrieval for user given query

5 Conclusion and Future Scope

This paper solves the outcome of the detailed survey in document image retrieval and extraction methods. The logo matching method is based on efficient SURF techniques. Design and analysis of techniques are processed by stepwise manner. Retrieval techniques are not only focusing the text but also finding the feature extraction cluster values of logo. The probability of given query is analyzed and tested using software prototype is found successful to achieve the accuracy of documents.

Logo image database such as ACM, Springer, Elsevier and IEEE are taken to find the precision, recall, and f-measure values are identified. Compared to other logos, the extraction from IEEE database fetches the results in sort time. The retrieval time taken as 89 ms and space allotted to store the information is 91,136 bytes. In future, the retrieval techniques of logo matching focus on the whole journal database.

References

1. Dikey KPJ (2015) Review of logo matching & recognition system based on context dependency. Int J Innov Res Comput Commun Eng 3(1):281–286
2. Keyvanpour M, Tavoli R (2013) Document image retrieval: algorithms, analysis and promising directions. Int J Softw Eng Appl 7(1):93–106
3. Marinai S, Marino E, Soda G (2007) Exploring digital libraries with document image retrieval. In: Research and advanced technology for digital libraries, pp 368–379
4. Gao H et al (2015) Focused structural document image retrieval in digital mailroom applications
5. Tavoli R (2012) Classification and evaluation of document image retrieval system. WSEAS Trans Comput. E-ISSN: 2224-2872
6. Schulz T, Sablatnig R (2015) Signature matching in document image retrieval. In: 20th computer vision winter workshop
7. Ramachandran R, Jose A (2014) Logo matching and recognition system using surf. IJRCCT 3(9):981–986
8. Roy PP, Lladós J, Pal U (2011) Multi-oriented and multi-scaled text character analysis and recognition in graphical documents and their applications to document image retrieval. Universitat Autònoma de Barcelona

9. Bhosle VV et al (2017) Multi classification technique for logo based document verification. Int J Comput Eng Appl XI(XI):1–12
10. Romberg S et al (2011) Scalable logo recognition in real-world images. In: Proceedings of the 1st ACM international conference on multimedia retrieval. ACM
11. Boonrod T, Jareanpon C, Chomphuwiset P (2015) The comparison of template matching and SURF for logo classification on product. In: Proceedings of the 3rd IIAE international conference on intelligent systems and image processing, pp 256–263
12. Billa P, Balijepalli AK, Rama Koteswara Rao I (2017) An implementation of effective logo matching and detection using multiple descriptors to enhance the resolution. Int J Comput Appl 161(5)
13. Bharathidevi B et al (2017) Logo matching for document image retrieval using SIFT descriptors. Int J Eng Res Appl 7(2):55–60
14. Sawalkar MS, Patole M (2012) Logo detection and recognition from the images as well as videos. Int J Sci Res (IJSR) 3:1233–1238

Mr. S. Balan received his Master of Computer Science from Dr. G.R.D College of Science, Coimbatore in 2009 and M.Phil. in Computer Science from Bharathiar University, Coimbatore in 2012. He is pursuing his Doctor of Philosophy in Computer Science, Government Arts College (Autonomous), Coimbatore, Bharathiar University. His research interests include Data Mining, Cloud Computing, Networks and Image Processing.

Dr. P. Ponmuthuramalingam received his Master's Degree in Computer Science from Alagappa University, Karaikudi in 1988, M.Phil. in Computer Science from Bharathidasan University, Tiruchirappalli, and Ph.D. in Computer Science from Bharathiar University, Coimbatore. He is working as Controller of Examination and Associate Professor in Computer Science, Government Arts College, and Coimbatore since 1989. His research interest includes Text mining, Semantic Web, Network Security, and Parallel Algorithms. He has contributed more than 50 papers in various National/International conference/journals.

Feature Enhancement of Multispectral Images Using Vegetation, Water, and Soil Indices Image Fusion

M. HemaLatha and S. Varadarajan

Abstract Land cover characteristics of satellite images are analyzed in this research paper. Remote sensing indices are calculated for multispectral image. In the proposed method, satellite image indices, i.e., NDVI (Normalized difference vegetation index), NDWI (Normalized difference water index), and BSI (Bare soil index), are calculated for various classes such as land, vegetation, water, and in land cover categories. All these remote sensing indices are fused to get composite bands and to enhance all features in multispectral image. This technique increases visual perception of human eye for multispectral images. Fusion plays vital role in remote sensing and medical images interpretation. In case of remote sensing, we cannot get entire information in one spectral band. So multispectral bands are combined, which leads to feature enhancement. This method depends on green (G), infrared (IR), near infrared (NIR), and short wave infrared (SWIR) bands and their fusion. Finally, error matrix is generated with reference data and classified data. The main application is to calculate vegetation, bare soil, and water indices in three land covers and to get better feature enhancement. Producer's accuracy, consumer's accuracy, commission, omission, kappa coefficient, F1score, over all accuracy, and over all kappa coefficients are calculated.

1 Introduction

Advanced ecosystems and socio economic managements are monitored by remote sensing and geographical information system. Thematic maps and general purpose maps will provide some useful information. The use of remote sensing and aerial photography has made it possible to map large areas for resource management and

M. HemaLatha (✉) · S. Varadarajan
ECE Department, S.V.U.C.E, Tirupati, A.P., India
e-mail: maddihemalatha@gmail.com

S. Varadarajan
e-mail: varadasouri@gmail.com

© Springer Nature Switzerland AG 2019
D. Pandian et al. (eds.), *Proceedings of the International Conference on ISMAC in Computational Vision and Bio-Engineering 2018 (ISMAC-CVB)*, Lecture Notes in Computational Vision and Biomechanics 30,
https://doi.org/10.1007/978-3-030-00665-5_34

exploitation. Quantitative spatial variation in handling huge data requires appropriate tools to analyze the spatial data using statistical methods or time series analysis [1].

Satellite classification is used to classify the ground data. There are many classification methods such as wavelet based, various cluster techniques, principle component analysis, and artificial neural networks. Satellite portraits are used for several applications including land cover and land use monitoring [2, 3] disaster management, soil moisture detection, snow and glacier monitoring, agriculture management, and forest fire monitoring [4, 5].

In this research paper, satellite portrait is divided into multi-classes using remote sensing indices. Various qualitative parameters are calculated. NDVI, NDWI, BSI, Producer's accuracy and user's accuracy plays very important role in this classification. Many texture-based classifications have been proposed [6, 7]. But there is problem in identifying water bodies. In the proposed technique, water bodies are identified. NDVI, NDWI, and BSI raster's are combined to get feature enhancement. Then accuracy assessment is done by error matrix. Producer's accuracy, kappa coefficient [8], omission, commission, F1score, over all accuracy, and over all kappa coefficient are calculated. Classification is done based on various remote sensing indices [9, 10].

2   Materials and Methods

Multispectral satellite image is downloaded from Bhuvan, and this image is geo-referenced in GCS WGS 1984 datum. Multispectral image is of size 1135 * 1135. This multispectral image belongs to Resourcesat-1 satellite. This image is having latitude of 6.75° and longitude of 93.5°. The characteristics of Resourcesat-1 satellite are mentioned in Table 1. The sensor used is Linear Imaging Self Scanning Sensor (LISS III). The image is visible in four bands i.e., green (G), infrared (IR), near infrared (NIR), and short wave infrared (SWIR). The spectral bands ranges for G, IR, NIR, and SWIR are (0.52–0.59) μm, (0.62–0.68) μm, (0.77–0.86) μm, and (1.55–1.7) μm, respectively. The four bands of multispectral image are fused to get false color composite (FCC) as shown in Fig. 2. This is done by composite band tool in ArcGis. The FCC image is then classified by interactive supervised classification as shown in Fig. 3. This technique does not require signature file. Maximum likelihood classification failed to classify the raster as there are huge numbers of pixels. Areas for three land covers are calculated as shown in Table 2. In this technique, some of water pixels are classified as land pixels. Due to this effect, the accuracies have been reduced shown in Table 4. Kappa coefficient and F1score are reduced as shown in Table 5 (Fig. 1 and Table 3).

In the proposed method, satellite image is given as input as shown in Fig. 1. Remote sensing indices NDVI, NDWI, and BSI are calculated by raster calculator in ArcGIS 10.3. NDVI, NDWI, and BSI maximum values are 0.778, 0.717, and 0.506, respectively. NDVI, NDWI, and BSI minimum values are 0.506, −0.613, and −0.089, respectively. NDVI, NDWI, and BSI raster's are shown Figs. 4, 5, and 6.

Table 1 Resourcesat/LISS main characteristics

Date of acquiring the image	22-3-2012
Projection	Geographic Lat/Long
Datum	WGS 84
Spatial resolution	22.5 m
Image file format	Geo-Tiff
Number of bands	4 (Band2, 3, 4, 5)
Radiometric resolution	8 bits
Temporal resolution	24 days
Swath	141 km

Table 2 Area calculation for land covers before feature enhancement

Land cover	Count	Area (km^2)
Land	125,225	63.395
Water	1191	0.594
Vegetation	2851	1.443

Table 3 Area calculation for land covers after feature enhancement

Land cover	Count	Area (km^2)
Land	107,959	54.654
Water	1385	0.701
Vegetation	3282	1.661

Table 4 Error matrix before feature enhancement

Land cover	Land	Vegetation	Water	Row total	Users accuracy (%)	Producers accuracy (%)
Land	**24**	0	24	48	50	96
Vegetation	0	**17**	1	18	94.44	85
Water	1	3	**1**	5	20	3.85
Column total	25	20	26	71		
Overall accuracy	57.53%					
Overall kappa coefficient	0.386					

Bold numbers are true value of pixels in each class which is classified

Table 5 Parameters before feature enhancement

Land cover	Commission (%)	Omission (%)	Kappa coefficient	F1score
Land	50	20	0.463	0.657
Vegetation	5.555	15	0.560	0.894
Water	80	3.846	0.580	0.064

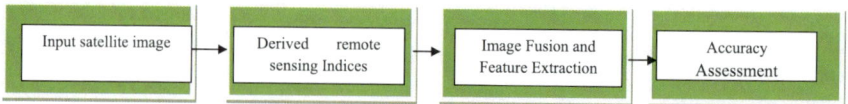

Fig. 1 Block diagram of the proposed method

Fig. 2 FCC for bands 4-3-2

Fig. 3 Classified image

Fig. 4 NDVI image

The area for three land covers is calculated shown in Table 3. All these vegetation indices raster's are combined or fused by composite bands tool in ArcGIS as shown in Fig. 7. The image features are enhanced after feature extraction. For accuracy assessment, image is classified using interactive supervised classification as shown in Fig. 8. Interactive supervised classification does not need signature file. Here, maximum likelihood classification failed to create signature file as there are more number of pixels. Ground truth data is generally obtained by field visit only, but here we can generate ground truth data by error matrix. The ground truth data is nothing but latitudes and longitudes of particular area. They are basically GPS points on earth. They are obtained by creating shape file in ArcGIS 10.3. More number of ground truth points on map is selected for better accuracy. Here, 28 points are selected in each land cover. The shape file and classified image is combined to get the combined raster. Finally, accuracy assessment is done for the three land covers, i.e., land, vegetation, and water. Producer's accuracy, user's accuracy commission, omission, kappa coefficient, F1score, overall accuracy, and overall kappa coefficients are calculated as shown in Tables 6 and 7. The proposed method has got better values compared to existing methods. Without fusion of remote sensing indices, raster's water bodies are not highlighted as shown in Fig. 3. Water has got least accuracies, kappa coefficient, F1score as shown in Tables 2 and 4. It has got more commission and omission, which is undesirable. Model calculations are shown from Eqs. 1 to 12.

Model calculations:

$$\text{Matching total} = \text{TP}_A + \text{TP}_B + \text{TP}_C \tag{1}$$

$$\text{TP} = \text{True positive, TN} = \text{True negative} \tag{2}$$

$$\text{Over all Accuracy} = \frac{\text{Matching Total}}{N} \tag{3}$$

$$\text{Matching Pixel} = \text{Diagonal elements in error matrix}$$

Fig. 5 NDWI image

Fig. 6 BSI image

$$\text{Producers accuracy} = \frac{\text{Matching Pixels}}{\text{Reference pixels}} \qquad (4)$$

$$\text{Reference Pixels} = \text{column wise}$$

$$\text{Users accuracy} = \frac{\text{Matching Pixels}}{\text{Classified Pixels}} \qquad (5)$$

$$\text{Classified pixels} = \text{row wise}$$

Fig. 7 FCC after NDVI, NDWI, and BSI fusion

Fig. 8 Classified image after fusion

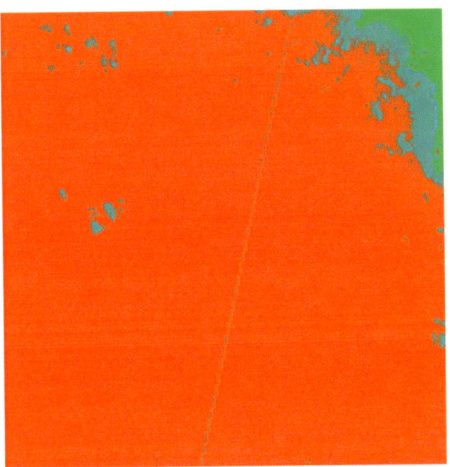

$$\text{Commission} = \frac{\text{Incorrectly classified pixels in row (TN Total)}}{\text{Total No. of pixels of the row}} \tag{6}$$

$$\text{Omission} = \frac{\text{Incorrectly classified pixels in column (TN Total)}}{\text{Total No. of pixels of the column}} \tag{7}$$

$$\text{Kappa coefficient} = N * \frac{\sum_{i=1}^{r} X_{ii} - \sum_{1=1}^{r} (X_{i+} * X_{+i})}{\left(N^2 - \sum_{i=1}^{r} X_{i+} * X_{+i}\right)} \tag{8}$$

N total number of pixels
X_{ii} sum of correct pixels
X_{i+} sum of all row total
X_{+i} sum of all column total

Table 6 Error matrix after feature enhancement

Land cover	Land	Vegetation	Water	Row total	Users accuracy (%)	Producers accuracy (%)
Land	**24**	1	9	34	70.58	96
Vegetation	0	**21**	1	22	95.45	87.5
Water	1	2	**17**	20	85	62.96
Column total	25	24	27	76		
Overall accuracy	81.58%					
Overall kappa coefficient	0.724					

Bold numbers are true value of pixels in each class which is classified

Table 7 Parameters after feature enhancement

Land cover	Commission (%)	Omission (%)	Kappa coefficient	F1score
Land	29.411	4	0.784	0.814
Vegetation	4.545	12.5	0.798	0.913
Water	15	37.037	0.797	0.723

$$F1score = \frac{2 * Precision * Recall}{Precision + Recall} \tag{9}$$

$$NDVI = \frac{NIR - IR}{NIR + IR} \tag{10}$$

$$NDWI = \frac{G - NIR}{G + NIR} \tag{11}$$

$$BSI = \frac{SWIR + IR - NIR}{SWIR + IR + NIR} \tag{12}$$

3 Conclusion

Remote sensing indices are calculated for multispectral satellite image. NDVI, NDWI, and BSI raster's are combined. By combining this raster's the features of image has been enhanced. The resultant satellite image is classified into three land covers, i.e., water, vegetation, and land. Error matrix is used to generate ground truth data. Then, quality parameters are calculated for the particular classification technique. Producer's accuracy and user's accuracy, kappa coefficient, commission, omission, F1score has been calculated. The proposed method has got better values compared to the existing technique.

Acknowledgements Authors are grateful to Bhuvan for providing satellite images to our research work.

References

1. Gonzalez RC, Woods RE (2008) Digital image processing. Prentice Hall, New Jersey
2. Lee JS, Grunes MR, Schuler DL, Pottier E, Ferro-Famil L (2006) Scattering-model based speckle filtering polar metric SAR data. IEEE Trans Geosci Remote Sens 44:176–187
3. Lillesand TM, Kiffer RW (2000) Remote sensing and image interpretation, 4th edn. Wiley, New York
4. Meddens AJ, Hicke JA, Vierling LA, Hudak AT (2013) Evaluating methods to detect bark beetle-caused tree mortality using single-date and multi-date Landsat imagery. Remote Sens Environ 132:49–58
5. Willis KS (2015) Remote sensing change detection for ecological monitoring in United States protected areas. Biol Conserv 182:233–242
6. Randon J, Hüsoy JH (1999) Filtering for texture classification: a comparative study. IEEE Trans Pattern Anal Mach Intell 21:291–310
7. Justice CO, Vermote E, Townshend JRG et al (1998) The Moderate Resolution Imaging Spectro radiometer (MODIS): land remote sensing for global change research. IEEE Trans Geosci Remote Sens 36:1228–1249
8. Hudson WD, Ramm CW (1987) Correct formulation of the kappa coefficient of agreement. Photogram Eng Remote Sens 53:421–422
9. Rokni K, Ahmad A, Selamat A, Hazini S (2014) Water feature extraction and change detection using multitemporal Landsat imagery. Remote Sens 6:4173–4189
10. Molchanov V, Chitiboi T, Linsen L (2015) Visual analysis of medical image segmentation feature space for interactive classification. In: Eurographics conference, pp 11–19

Detection of Heart Abnormalities and High-Level Cholesterol Through Iris

P. A. Reshma, K. V. Divya and T. B. Subair

Abstract Iridology is medicine technique to claim the colours, characteristics and pattern of the iris. Iridology is used to determine the existence of basic genetics, irregularities in the body, dam circulation, toxin deposition and other weakness. This paper discusses the determination of high-level cholesterol in blood and heart conditions through several stages. In this study, we examine the heart condition and cholesterol through preprocessing, segmentation, feature extraction and classification from the captured iris image. Due to the high level of cholesterol in blood, sodium ring is to be formed around the iris.

1 Introduction

The heart is an internal organ of our body. This organ provides blood supply to our body by coronary arteries. Heart disease is known as a cardiovascular disease. Now today, heart problems are increasing day by day. A number of factors that cause heart disease are smoking, high blood pressure, high level of cholesterol, diabetes, obesity or from the family history. Different symptoms and signs are occurring in heart abnormalities. One of the methods is iridology. According to iridology, certain areas of human iris represent a particular organ condition [1]. Figure 1 shows a particular organ. Iris of the eye is used to detect the condition of a body organ, genetic strength and weakness. This method believes that patterns on the iris reflect body condition [2].

P. A. Reshma (✉) · K. V. Divya
Department of Computer Science Engineering, Vidya Academy of Science and Technology, VAST, Thissur, Kerala, India
e-mail: reshmarazack01@gmail.com

K. V. Divya
e-mail: divyakv@vidyaacademy.ac.in

T. B. Subair
Technician-Classroom Technologies, Information Technology Centre, University of Sharjah, Sharjah, UAE
e-mail: sbeeravu@sharjah.ac.ae

© Springer Nature Switzerland AG 2019
D. Pandian et al. (eds.), *Proceedings of the International Conference on ISMAC in Computational Vision and Bio-Engineering 2018 (ISMAC-CVB)*, Lecture Notes in Computational Vision and Biomechanics 30,
https://doi.org/10.1007/978-3-030-00665-5_35

Fig. 1 Structure of a human eye with particular organ condition

Iris, in communication between brain and organs, has approximately 28,000 nerve network. That is, an organ irregular tries to send information about the state of the brain and this information on the pattern of the iris, as reflected in a change in the colour or characteristic [3]. Figure 1 shows some examples of signs of organ and tissue disorder.

Cholesterol is an oily substance that will not mix with the blood. This substance plant in our body cells. Our body contains two kinds of cholesterol: LDL and HDL. Both types are decisive for our body health. Blood cholesterol levels are measured by subtracting the lipoprotein profile, 9–12-h measurement of the total cholesterol level in the blood after starving with cord blood. Generally, LDL is a bad cholesterol and HDL is a good cholesterol. High cholesterol level causes heart abnormalities, diabetes, PCOS and kidney disease. In our body, the level of cholesterol increases on consuming cheese, animal foods, meats, baked goods, chocolate, deep-fried goods, etc.

The iris pattern has many distinguishing metabolism characteristics in the human body as well as there is also the ability to outplay changes [4]. Major studies with iridology between A. D. Wibawa and M. H. Purnomo diagnosis of mellitus diabetes by looking at the status of the pancreas [5], Cheng-Liang Lai and Chien-Lun Chiu's heart [6].

In general, blood cholesterol levels are measured by subtracting the lipoprotein profile. Lipoprotein profile, 9–12-h measurement of the total cholesterol level in the blood after starving with cord blood [7]. Blood cholesterol level increases light-coloured layer of sodium that begins to occur around the outer ring of iris. Increased level of cholesterol can detect with the iridology property [8]. This research will be able to contribute to the detection of high-level cholesterol and disorders of the heart through the iris image. This helps to give the information about our health disease.

2   Detection of Heart Abnormalities

The heart abnormalities detection [9] system simply illustrates that the input image is trained by preprocessing and feature extraction process. In a preprocessing step, the acquired image tests. After the feature extraction step, we assign the thresholding value and then classify whether the patient is abnormal or not.

2.1   Image Acquisition

High-resolution and specification camera are used for acquiring the iris image, for accurately getting the features of iris. We use the camera to take the image with the following specifications:

- 12.8 Megapixel
- 28–224 mm—Lens Focal Length
- Digital Zoom 4x, Optical Zoom 10x
- 1.97″–19.69″—Infinity Macro
- Back-illuminated
- Additional flashlight

The additional flashlight is used for getting the features more accurately. Figure 2 shows the difference in using an additional flashlight and not.

2.2   Preprocessing

The preprocessing stage is very important for iris recognition process. It increases the quality of the original image and determines the successful result. There are different noise removal filtering techniques available: removing the noise by linear filtering, averaging filter, median filter and an adaptive filter. In here, removing noise from the captured iris image by using a median filter.

Fig. 2 Iris image with an additional flashlight and not

(a)   **(b)**

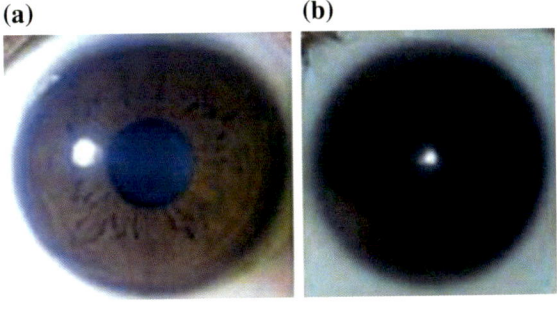

Fig. 3 Binarized iris image

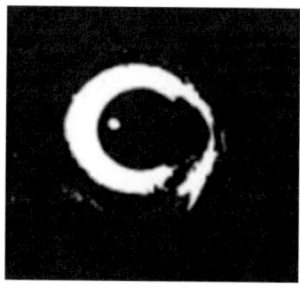

In the median filter, the output pixel value is determined by the median of the neighbourhood pixels. The median filter is better to eliminate the outliers without compressing the sharpness of the image. Here, the calculation is done on the basis of the pixel intensity values (Fig. 3).

After the noise is removed, the image is cropped. Image cropping is done by binarization of image and histogram. The binarization process converts the iris image into a binary image. The binary image is like a white and black image. The value of the white pixel and a black pixel is divided into horizontals and verticals. In cropping process, the image can be categorised as success, fail and scant. The scant image produces less perfect iris image. The full boundary of the image cannot be gained. The failed image is an incomplete one. The causes of the scant image are shadowing and improper lighting.

2.3 Feature Extraction

From the result of segmentation, the step gives the information about the disease. It is calculated by the ratio of a number of white and black pixel.

$$\text{Ratio of Black} = \text{Total Black}/\text{Total Pixels} \qquad (1)$$

$$\text{Ratio of White} = \text{Total White}/\text{Total Pixels} \qquad (2)$$

2.4 Classification

The classification is done on the basis of thresholding value. The threshold value is determined by analysis of test data. The threshold value of black is 0.50913 and white is 0.48848. First, enter the data to be classified. Then the label (normal or abnormal) is produced if the entered data value is below or above the threshold value.

3 Detection of Cholesterol

In cholesterol detection [10] process, first images are acquired from high cholesterol patient and normal patient. Then the pupil of the iris is used to determine the width of sodium ring by analysing the colour tone. The cholesterol detection process is explained below.

3.1 Image Acquisition

In cholesterol detection process, first images are acquired from high cholesterol patient and normal person. Figure 4 shows the iris image of cholesterol patient, and Fig. 5 shows the normal person.

3.2 Localization

Here localization process means finding the pupil from the iris. The iris, pupil and eye are of different colours, rather than taking advantage of the Adam edge discovery process. For finding the boundaries of the pupil by integral differential operator. The IDO was proposed by John Daugman [11]. Daugman first proposed iris recognition.

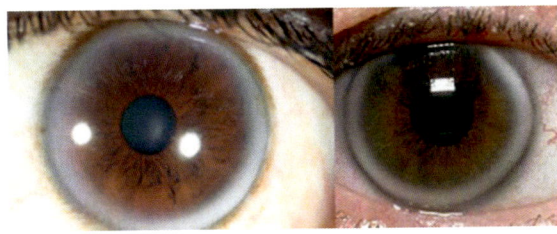

Fig. 4 Iris image of high-level cholesterol patient

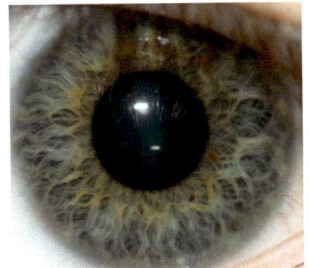

Fig. 5 Iris image of a normal person

Fig. 6 Locate the pupil boundary

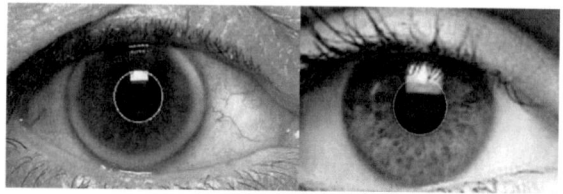

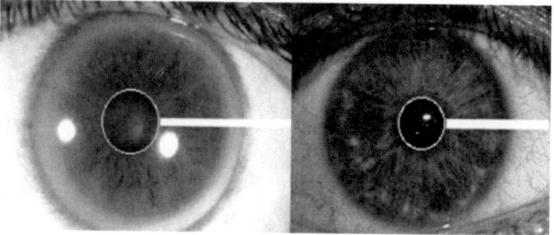

Fig. 7 Horizontal and vertical axis

3.3   Used Method

The IDO method determines the boundaries of the iris with a very high accuracy rate. The boundaries are calculated by finding the radius of pupil and radius of iris. The radius of inner and outer circles is in the range from 0.1 to 0.8 of the iris radius [12]. Equation 1 explains the IDO (integrodifferential operator).

$$\max(r, x_0, y_0) \left| G_\sigma(r) * \frac{\partial}{\partial r_{r, x_0, y_0}} \oint \frac{-I(x, y)}{2\pi r} ds \right| \tag{3}$$

where $I(x, y)$ in the (x, y) position of a hue value, x_0 and y_0 is possible to centre coordinates, r distance to the point of the possible central symbol, $G_\sigma(r)$ is a smoothing function such as a standard offset Gaussian function. By using this method, the position of the eye can be found correctly. The reflections in scanning process of the picture are ignored. Reflection of the boundaries of the eyeball is unaffected. Figure 6 simply explains it.

Sodium ring thickness is calculated by the ratio between the width of the iris and width of the sodium ring. This will be examined by taking the pupil centre as a vertical axis, and the horizontal axis starts from the right side of the eye continues to the end of the image. Figure 7 shows the pictorial representation.

Table 1 shows the sodium ring eye thickness ratio results. Picture number 4, the rate at which the database with a value of sodium ring 0.09 thinnest sodium ringed eye. The eye number five with a value of 0.23% sodium ring is the boldest look (Fig. 8).

Table 1 The ratio of the thickness

Resim Numarasi	Oran
1	0.13
2	0.19
3	0.11
4	0.09
5	0.23
6	0.16

(1)   (2)   (3)

(4)   (5)   (6)

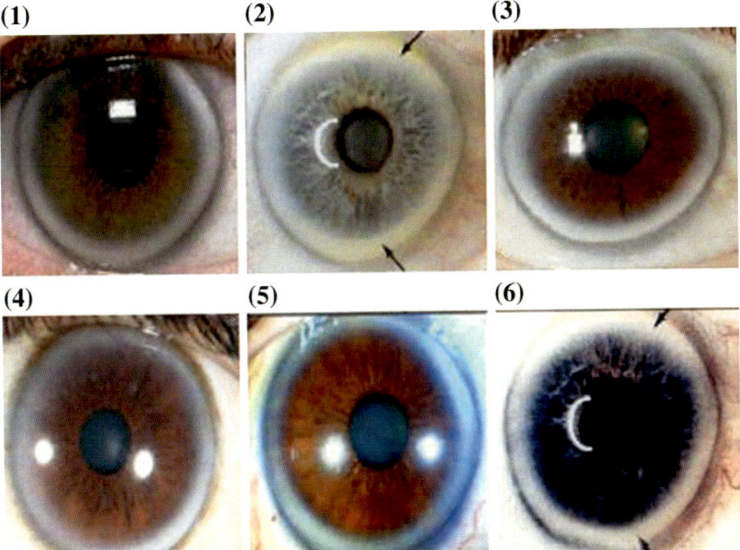

Fig. 8 Thickness of sodium ring

4   Conclusion

One of the most popular treatments is iridology. The iris image is high quality and perfectly add flashlight then the result will be accurate. The cropping process is determining the comparison of the threshold value. The ratio between the width of the iris with the width of sodium ring is computed, and evaluations have been conducted. The rate at which values are obtained as a result of evaluations for each eye is different. Therefore, iris sodium ring especially high cholesterol value is a direct relationship between can be associated. The cholesterol disease and heart abnormality are closely related, so that this research has a huge potential for future.

References

1. Hiru (2005) Iridology: Mendeteksi Penyakit Hanya dengan Mengintip Mata. Gramedia Pustaka Utama
2. Jensen B (2005) Science and practice of iridology
3. Woodward JD, Orlans NM, Higgins PT (2003) Biometrics: identity assurance in the information age. ISBN 0-07-222227-1
4. Sivasankar K, Sujaritha M, Pasupathi P, Muthukumar S (2012) FCM based iris image analysis for tissue imbalance stage identification. In: International conference on emerging trends in science
5. Wibawa D, Purnomo MH (2006) Early detection on the condition of pancreas organ as the cause of diabetes mellitus by real time iris image processing. In: IEEE Asia Pacific conference on circuits and systems
6. Lai C, Chiu C (2010) Health examination based on iris images. In: Proceedings of the ninth international conference on machine learning and cybernetics, Qingdao
7. Ramlee RA, Ranjit S (2009) Using iris recognition algorithm, detecting cholesterol presence. In: International conference on information management and engineering
8. NutritionalIridology.com (2012) Iridology charts. Retrieved 17 Oct 2013, from Nutritional Iridology. http://nutritionaliridology.com/charts.html
9. Entin Martiana K, Ridho A, Syarifa S, Afgan A (2016) Application for heart abnormalities detection through iris. IES, IEEE. 978-1-5090-1640-2/16
10. Kursat B, Kurnaz C (2016) Detection of high-level cholesterol in blood with iris analysis. IEEE. 978-1-5090-5829-7/16
11. Daugman J (2009) Iris recognition at airports and border-crossings. In: Encyclopedia of biometrics, pp 819–825
12. Daugman J (2004) How iris recognition works. IEEE Trans Circ Syst Video Technol 14(1):21–30

Wavelet-Based Convolutional Recurrent Neural Network for the Automatic Detection of Absence Seizure

Kamal Basha Niha and Wahab Aisha Banu

Abstract In this paper, the new model is proposed to automatically detect and predict absence seizure using hybrid deep learning algorithm [Convolutional Recurrent Neural Network (CRNN)] along with the Discrete Wavelet Transform (DWT) with Electroencephalography (EEG) as input. This model comprises of four steps (1) Single-channel segmentation process (2) Decomposition of segmented signal using wavelet transform (3) Extraction of relevant feature using statistical method (4) Deep learning algorithms for classification, detection, and early detection. This model enhances the feature extraction and also the overall performance by feeding the segmented data into Long Short Tern Memory (LSTM) which is one of the Recurrent Neural Network (RNN). And also the output of this network is used to calculate the extracted feature along with the classification results. The values in hidden state are used to diagnose the seizure by locating the pattern using the extracted features of time window. The proposed model achieves 100% accuracy on detection and 95% overall accuracy on early detection of normal, abnormal and absence seizure.

1 Introduction

Absence seizure is one among the types of epilepsy where its prevalence is about one percent among the population which means eight point eight per thousand population and it is higher in the rural area (one point nine percent) than urban (zero point six percent). Absence seizures occurrence lasts from short to long duration in terms of minutes. In general, its spike and wave pattern of frequency fall between

K. B. Niha (✉) · W. Aisha Banu
Computer Science and Engineering, B S Abdur Rahman Crescent Institute of Science and Technology, Chennai, India
e-mail: niha.k.cse@gmail.com

W. Aisha Banu
e-mail: aisha@crescent.education

© Springer Nature Switzerland AG 2019
D. Pandian et al. (eds.), *Proceedings of the International Conference on ISMAC in Computational Vision and Bio-Engineering 2018 (ISMAC-CVB)*, Lecture Notes in Computational Vision and Biomechanics 30,
https://doi.org/10.1007/978-3-030-00665-5_36

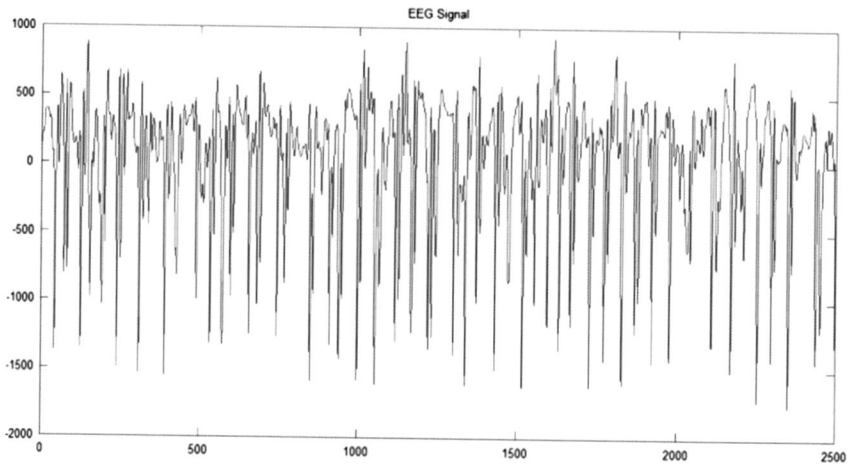

Fig. 1 Single-channel raw absence seizure EEG data

2 and 3 Hz per ms (mille second) and in terms of duration it is 5–30 s long. This patter is found in the frontal region, which may start around 4 per sec and slows down to 3–3.5 per sec then at last it decreases to 2.5 per sec. The single-channel raw seizure EEG data are shown in Fig. 1. In traditional medicine, seizures frequency has been computed manually for analysis and diagnosis are in practice. And also monitoring patients out of clinical boundary with the help of portable EEG recording systems, the same frequency seizure signals are able to obtain for in-patients are in practice. But this process is very tedious and leads to many measurement errors. Only an experienced physician can able to handle this diagnosis process. To overcome this, seizures are diagnosed by automatic detection and early detection systems [1] which has been designed using machine learning algorithms.

Among those algorithms, Conventional Neural network (CNN) and Recurrent Neural Network (RNN) play a major role. Initially, CNN was used widely in different fields for classification of input in computer vision, recognition of traffic signal, and generating image captions [2–4]. This algorithm is not only restricted to the above fields and also for signal processing domain [5]. Major advantage of this algorithm is that it outperforms the handcrafted feature extraction process [6]. In parallel, the RNN also in boom which is widely used to handle the sequence of data with varied sequence. These are most successful in speech recognition and machine translation [7–9]. When the input is in sequence with varied length, both CNN and RNN can be used. With that features are extracted from segments of data and those extracted feature form another one sequence. These sequences are taken into recurrent layer with temporal dependency can be handled in CNN using pooling mechanism. The main disadvantage in traditional CNN is to extract features they use nonlinear affine function. The advantage of this work by using CRNN is to compute the better features when compared to the traditional convolutional layer. The upcoming section briefly discusses experimental design, working method, results, and conclusion.

2 Experimental Design

The processing flow of this model is shown in Fig. 2. In that, the subjects EEG signal are recorded using clinical setup and are viewed using RMS software. Then, those signals are processed with wavelet transform and CRNN (LSTM) to obtain result.

Initially, the scalp EEG data have been recorded using clinical setup (16 channel electrodes) based on 10–20 international system. The duration of absence seizure data will last for 2 s. The sampling frequency of 512 Hz and the bandpass filter from 1.0 to 70 Hz have been used during the recording and to the power line noise notch filter have been applied as 50 Hz to remove power line noise. Duration of each recorded data is for 24 s. These recorded data are viewed using RMS software and from that single-channel (FP1-F7) data which is more reliable and sensitive [9] than other channel. Then, the data have been exported to MATLAB environment.

Fig. 2 Work flow of the CRNN system

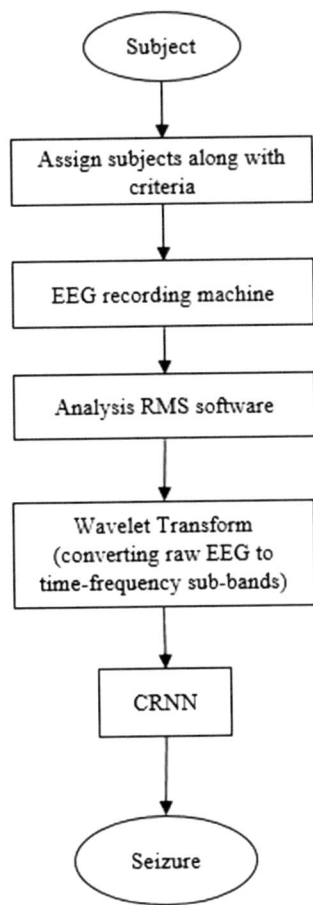

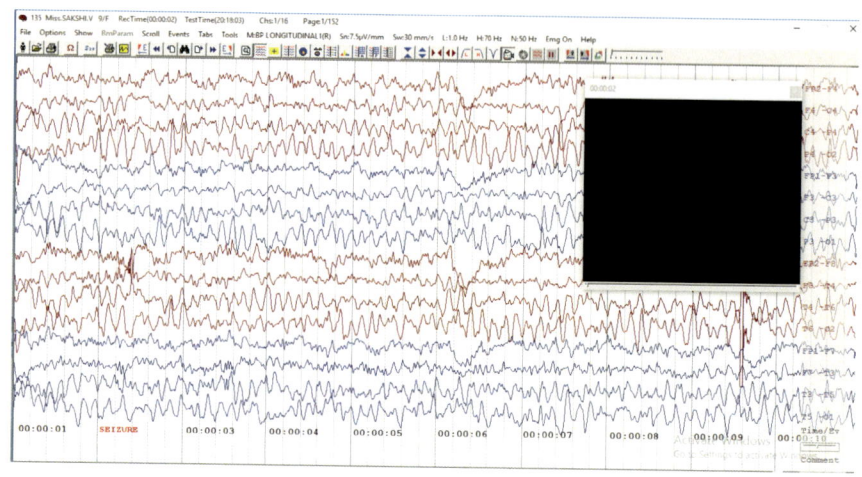

Fig. 3 RMS View of absence seizure EEG data with spike and wave

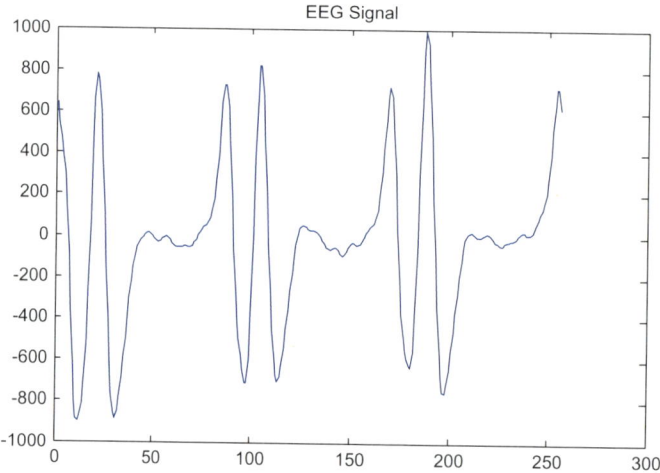

Fig. 4 Single-channel segmented absence seizure EEG data

The RMS view of absence seizure data and the segmented data have been shown in Figs. 3 and 4. The sequence of data has been segmented as 258 samples each and then these segments are undergone for time-frequency transformation using Discrete Wavelet Transform [10].

The transformed output contains five-time domain and frequency domain sub-bands. They are alpha, beta, gamma, delta, and theta. Their frequency ranges are

- Delta (0.1–3.5 Hz), Theta (4–7.5 Hz)
- Alpha (8–13 Hz), Beta (14–40 Hz)
- Gamma (greater than 40 Hz).

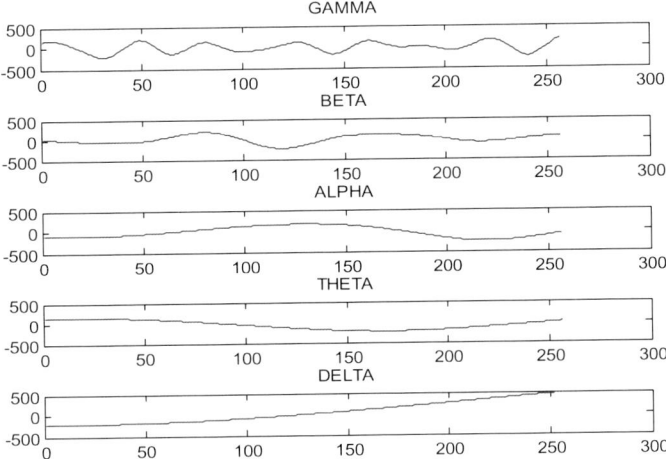

Fig. 5 Time domain absence seizure EEG data

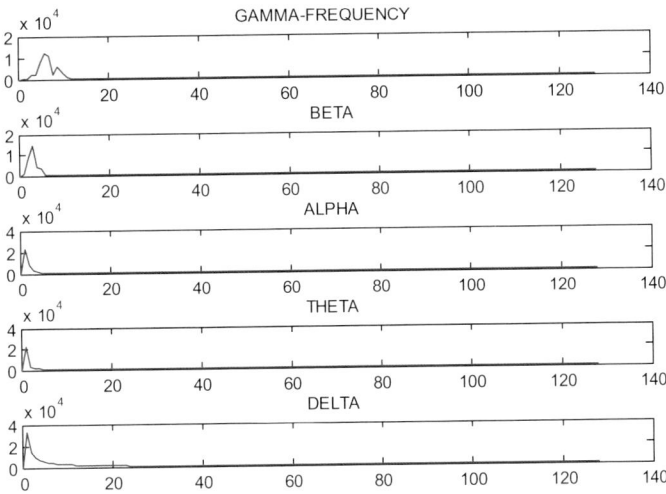

Fig. 6 Frequency domain absence seizure EEG data

These sub-bands are shown in Figs. 5 and 6. Each sub-band is processed using Convolutional Recurrent Neural Network (CRNN) for detection and early detection of seizure. Where the feature extractions are handled automatically and are classified to diagnose the seizure. Long Short-Term Memory (LSTM) has been used for early detection.

3 Method

The proposed Conversional Recurrent Neural Network in this paper extracts the EEG features by itself from the sub-bands of time domain. Initially, the absence seizure dataset comprises of varied time series of data, each sequence consists of frames and each frame has the same fixed number of features. Here, p is the sequence of data, q is the frame consist of feature vector with length r, and the features are represented as r. In this, the product of the size of the feature vector and the frame ($q * r$) is equal to the sequence of data p. The sequence of input window p comprises of $s_1, s_2, \ldots, s_n$ where the window size is equal to the product of r and s_1. Then apply the function f which extracts n features from each window which generates sequences of x' which comprises of n features with respect to each frame. After this step, the pooling mechanism has been applied to pick the maximum value among the extracted feature of each window of size $n * m_1$ from x' which comprises of $m_1, m_2, \ldots, m_n$ frames. After max pooling, the size of the frame gets transferred to windows of size $n * m_1$ to the vector of size $n * 1$. Finally, the local features of sequence p have been feed into the next layer for the classification. Where LSTM has been used as a recurrent layer in this model which generates the additional sequences they are output ($y_1, y_2, \ldots, y_{r1}$) and cell state ($c_1, c_2, \ldots, c_{r1}$) instead of hidden state to compute the features representing the windows. This model works on the 3 class classification problem (normal, ictal and absence ictal). The LSTM consists of 128 hidden layer followed by 100 hidden nodes with fully connected (f_c) layer. This layer has been applied on the continuous sequence with time distributed manner. Here the linear activation function has been used and the output is given to another 128 LSTM hidden layer. The class probability has been computed using the output of the second LSTM layer. For this, the schematics of LSTM are given below,

> Input layer_1
> LSTM_0
> Time distribution_1
> LSTM_1
> Dens_2

The characteristics of model are given below as parameters and its value, hidden unit layer_1 as 100, hidden unit layer_2 as 100, fully connected layer_2 as 100, gate activation as sigmoid, cost as cross-entropy (categorical), epoch as 200, optimizer as Adam, dropout as 40, batch size as 25. With these characteristics, the model has been trained in state full mode in which the data stored in the previous layer of internal memory propagate among samples. Where the internal memory of the LSTM has been initialized to 0 and the training has been done with Adam stochastic optimizer. The learning rate of this model is 0.001 and the model is trained using 200 epoch and validated using k-fold cross-validation.

Table 1 Confusion matrix of the LSTM

	Normal	Abnormal	Absence
Normal	12	3	0
Abnormal	0	15	0
Absence	0	0	8

4 Result

The classification accuracy has been analyzed using the performance metrics. The high accuracy classification cannot have high prediction accuracy.

To cross check that in this paper, the performance of this model has been evaluated using precision, recall, and F_1 measures. To calculate these metrics, confusion matrix (Table 1) has been created for testing data. To compute accuracy, precision, recall, and F1 score formulas are given below

$$\text{Precision} = T_p/T_p + F_p$$
$$\text{Recall} = T_p/T_p + T_n$$
$$\text{Accuracy} = T_p + T_n/T_p + F_p + T_n + F_n$$
$$F_1 = 2 * ((\text{Precision} * \text{Recall})/(\text{Precision} + \text{Recall}))$$

Calculated precision, recall, accuracy, and F_1 measures of the model are 95, 94, 95, and 94%.

5 Conclusion

In this paper, the novel RNN layer has been used for EEG absence seizure classification. Here, the RNN refers to LSTM which have been combined together with CNN to obtain a hybrid model with the accuracy of 95%. This model is very easy to train because there is no need of feature extraction and selection process. This model is less computational for fast training and better for large dataset. In future, different combinations of deep learning algorithms are used for better classification accuracy.

Acknowledgements We thank Dr. S. Velusamy, DM—Neurology, MD—Paediatrics, MBBS Neurologist, and General Physician, who has 22 years of experience for his continuous support throughout this work.

References

1. Schelter B, Winterhalder M, Maiwald T, Brandt A, Schad A, Timmer J, Schulze Bonhage A (2006) Do false predictions of seizures depend on the state of vigilance? A report from two seizure-Prediction methods and proposed remedies. Epilepsia 47(12):2058–2070

2. Krizhevsky A, Sutskever I, Hinton GE (2012) Imagenet classification with deep convolutional neural networks. In: Proceedings of advances in neural information processing systems (NIPS), Lake Tahoe, NV, pp 1097–1105

3. Ciresan D, Meier U, Schmidhuber J (2012) Multi-column deep neural networks for image classification. In: Proceedings of the IEEE conference on computer vision and pattern recognition (CVPR), Providence, RI, pp 3642–3649

4. Vinyals O, Toshev A, Bengio S, Erhan D (2015) Show and tell: a neural image caption generator. In: Proceedings of the IEEE conference on computer vision and pattern recognition (CVPR), Boston, MA, pp 3156–3164

5. Sainath TN, Weiss RJ, Senior A, Wilson KW, Vinyals O (2015) Learning the speech front-end with raw waveform CLDNNs. In: Proceedings of INTERSPEECH 2015, 16th Annual conference of the international speech communication association, Dresden, Germany, p 15

6. Amodei D, Anubhai R, Battenberg E, Case C, Casper J, Catanzaro BC, Chen J, Chrzanowski M, Coates A, Diamos G, Elsen E, Engel J, Fan L, Fougner C, Han T, Hannun AY, Jun B, LeGresley P, Lin L, Narang S, Ng AY, Ozair S, Prenger R, Raiman J, Satheesh S, Seetapun D, Sengupta S, Wang Y, Wang Z, Wang C, Xiao B, Yogatama D, Zhan J, Zhu Z (2015) Deep speech 2: end-to-end speech recognition in English and Mandarin. arXiv, preprint arXiv:1512.02595

7. Graves A, Mohamed A, Hinton GE (2013) Speech recognition with deep recurrent neural networks. In: Proceedings of the IEEE international conference on acoustics, speech and signal processing, (ICASSP), Vancouver, Canada, pp 6645–6649

8. Cho K, Van Merrienboer B, Gulcehre C, Bahdanau D, Bougares F, Schwenk H, Bengio Y (2014) Learning phrase representations using RNN encoder–decoder for statistical machine translation. In: Proceedings of the conference on empirical methods in natural language processing (EMNLP), Doha, Qatar, pp 1724–1734

9. Sutskever I, Vinyals O, Le QV (2014) Sequence to sequence learning with neural networks. In: Proceedings of advances in neural information processing systems (NIPS), Montreal, Canada, pp 3104–3112

10. Zhang B, Jiang H, Dong L (2017) Classification of EEG signal by WT-CNN model in emotion recognition system. In: International conference on Informatics and cognitive computing. IEEE

New Random Noise Denoising Method for Biomedical Image Processing Applications

G. Sasibhushana Rao, G. Vimala Kumari and B. Prabhakara Rao

Abstract Since the inception of digital image processing, noise removal in images has always been a challenge to researchers and experts of the field. Most significant of these noises are the randomly varying impulse noises developed while image is acquired. Hence, the need for methodical denoising method has led to extensive research and development of various innovative methods to remove the random valued impulse noise. For this, a method which detects and filters random valued impulse noise in medical images is employed. The method proposed in this paper uses a decision tree based impulse detector and an edge preserving filter to rebuild noise free images. This method is more efficient than the existing techniques due to its lower complexity. Different gray scale Magnetic Resonance Imaging (MRI) brain images are tested by using this algorithm and have given better Peak Signal to Noise Ratio (PSNR) than the other techniques.

1 Introduction

With the advent of digital images, productivity in science and technology has greatly increased due to better understanding capabilities, analysis, visualization and interpretation; which are not limited to a particular field but can be extended to as many

G. Sasibhushana Rao
Department of Electronics and Communication Engineering, AU College of Engineering,
Visakhapatnam 530003, India

G. Vimala Kumari (✉)
Department of Electronics and Communication Engineering, M.V.G.R College of Engineering,
Vizianagaram 535002, India
e-mail: Vimalakumari7@gmail.com

B. Prabhakara Rao
Department of Electronics and Communication Engineering, JNTUK, Kakinda 533003, Andhra
Pradesh, India

© Springer Nature Switzerland AG 2019
D. Pandian et al. (eds.), *Proceedings of the International Conference on ISMAC
in Computational Vision and Bio-Engineering 2018 (ISMAC-CVB)*, Lecture Notes
in Computational Vision and Biomechanics 30,
https://doi.org/10.1007/978-3-030-00665-5_37

as possible. However, with these advantages came the limitations of using digital images. Prominent among these include removing noise and storage problems. Noise develops in images while the image is acquired and transmitted. Presence of noise in images cripples all the advantages with using digital images as it deteriorates the quality of images making it difficult to analyse, interpret or visualize [1, 2]. This led to the development of effective denoising algorithms which target various types of noises in images. Among the discovered image noises, most arduous are the impulse noises which affect the images during their acquisition and transmission [3]. Impulse noises generally involve random occurrence or distribution of noisy pixels over the image. However, based on the values of its noisy pixels, these are divided into two categories as fixed value impulse noise and random-valued impulse noise. Fixed value impulse noise is limited to only two noisy pixel values, i.e. salt—255 and pepper—0. On the other hand, random-valued impulse noise involves noisy pixels of any value within the range [0–255] [4, 5]. Thus, there exists a need for development of an efficient denoising method for removal of random valued impulse noise in images. Mean filters perform denoising effectively, but at the cost of heavy information loss due to blurring or smoothening of the image. Median filters are preferred to preserve several important details even after denoising. Although not as much as mean filter, median filters result in information loss too [6, 7]. Hence, to avoid this switching median concept is opted. Here, the median filter is provided with an impulse detector which detects the noisy pixels prior to the filter. Thus by detecting the noisy pixels before filtering, the filter may be conditioned such that it filters only those noisy pixels that are detected thereby resulting in less information loss and better preservation of details in the image [8, 9]. Adaptive Median Filter raises the size of the processing kernel to get rid of noisy pixels. Further this filter offers best results in getting rid of noisy pixels at lower densities. Yet, the limitation of this filter is that it blurs the image at higher densities. To overcome this limitation the Adaptive Decision Based Median Filter (ADMF) [10] is proposed. A Nonlocal Means (NLM) based filtering algorithm for denoising Rician distributed MRI and utilizes the concept of self-similarity for MRI restoration [11]. An approach involving minimum absolute difference criteria for identifying the noisy pixels and replacement by mean value computed [12]. It is evident that denoising random-valued impulse noise is far more challenging than denoising salt-and-pepper noise due to its random pixel values. Of all the various methods available to denoise impulse noise, very few address the random-valued impulse noise. Thus, there exists a need for the development of an efficient denoising method for removal of random-valued impulse noise in images. Keeping in view the above requirements, in this paper a model which is of low complexity that employs a new impulse detector based on decision tree and a filter to preserve the edges to denoise random-valued impulse noise in an image is proposed.

2 Proposed Algorithm

The algorithm proposed in this paper uses a modified Decision Tree Based Denoising Method (DTBDM) to reconstruct noise-free images. DTBDM method involves two main components, viz., impulse detector and edge preserving filter. The impulse detector makes use of a 3×3 mask, to identify the noisy pixels and to activate the filter accordingly. The centre pixel is denoted by $p_{i,j}$, while its luminance is denoted by $s_{i,j}$. The neighbours are named accordingly in the range of $(i-1:i+1, j-1:j+1)$.

2.1 Impulse Detector

The impulse detector detects the corrupted pixels so as to activate the filter only for those pixels that are corrupted. Here, an impulse detector is designed based on a decision tree. The impulse detector comprises three modules, viz., segregation, perimeter and analogous. Each of these modules is independent of each other and can segregate corrupted pixels. The pixel is considered noise-free only if all three modules judge it to be uncorrupted. Several thresholds, viz., Th_1, Th_2, Th_3, Th_4, Th_5 and Th_6 are used at various places in the modules. The values of Th_1, Th_2, Th_3, Th_4, Th_5 and Th_6 are 15, 20, 45, 70, 10 and 50, respectively.

Segregation Module. Segregation logic is based on the prime assumption that the centre pixel is located on a smooth surface. Thus, to test the pixel for segregation, first, it has to be on a uniform surface. Practically, the impulse module can be easily implemented using the 3×3 mask. The mask is first split into two halves $U_{T\text{-Half}}$ and $U_{B\text{-Half}}$. Now, the nine pixels in the mask can be named as 'a, b, c, d, e, f, g, h and $s_{i,j}$', where $s_{i,j}$ denotes the centre pixel and the variables 'a to h' denote the eight neighbours of the centre pixel. $U_{T\text{-Half}}$ comprises the top four neighbours 'a to d', while, $U_{B\text{-Half}}$ comprises the four neighbours 'e to h', at the bottom of the mask. They are represented as

$$U_{T-\text{Half}} = \{a, b, c, d\} \tag{1}$$

$$U_{B-\text{Half}} = \{e, f, g, h\} \tag{2}$$

The test for uniformity can be performed by finding out the maximum possible difference in intensities of both the top and bottom halves of the mask. If either of these differences in intensity is very high, i.e. greater than a threshold, then the region is considered as a nonuniform region and the test for segregation cannot be performed. The test can be represented using the following equations

$$U_{T-\text{Halfdif}} = U_{T-\text{Halfmax}} - U_{T-\text{Halfmin}} \tag{3}$$

$$U_{B-\text{Halfdif}} = U_{B-\text{Halfmax}} - U_{B-\text{Halfmin}} \tag{4}$$

$$\text{Decision 1} = \begin{cases} \text{true,} & \text{if } (U_{T-\text{Halfdif}} \geq \text{Th\_1}) or (U_{B-\text{Halfdif}} \geq \text{Th\_1}) \\ \text{false,} & \text{otherwise} \end{cases} \quad (5)$$

If the region is found to be uniform then proceed to the test for segregation. The centre pixel is now taken into consideration and the intensity difference between the centre pixel and $U_{T\text{-Halfmax}}$, $U_{T\text{-Halfmin}}$, $U_{B\text{-Halfmax}}$ and $U_{B\text{-Halfmin,}}$ are computed, respectively. If any of these differences is higher than the threshold, then, the centre pixel is considered as an isolated pixel. Decision 2 determines whether the centre pixel is isolated or not. It is to be noted that Decision 2 is arrived at only when Decision 1 is false. Isolated pixel can be represented as

$$I_{T-\text{Half}} = \begin{cases} \text{true,} & \text{if}\left(|S_{i,j} - U_{T-\text{Halfmax}}| \geq \text{Th\_2}\right) \\ \text{or } \left(|S_{i,j} - U_{T-\text{Halfmin}}| \geq \text{Th\_2}\right) \\ \text{false,} & \text{otherwise} \end{cases} \quad (6)$$

$$I_{B-\text{Half}} = \begin{cases} \text{true,} & \text{if}\left(|S_{i,j} - U_{B-\text{Halfmax}}| \geq Th\_2\right) \\ \text{or } \left(|S_{i,j} - U_{B-\text{Halfmin}}| \geq Th\_2\right) \\ \text{false,} & \text{otherwise} \end{cases} \quad (7)$$

$$\text{Decision 2} = \begin{cases} \text{true,} & \text{if}(I_{T-\text{Half}} = \text{true}) \\ \text{or } (I_{B-\text{Half}} = \text{true}) \\ \text{false,} & \text{otherwise} \end{cases} \quad (8)$$

Perimeter Module. The perimeter module in this algorithm is one such effective mechanism that detects the edges and avoids their filtering unnecessarily. But edge pixels are always of lesser intensity differences computed along the direction of an edge. Since, the only concern is about the centre pixel, arrive at four directions $E1$ to $E4$ passing through the centre pixel using Eqs. (9–13) determine whether the centre pixel is an edge pixel or not. The directional differences are computed as follows:

$$F_{E1} = \begin{cases} \text{false,} & \text{if}\left(|a - S_{i,j}| \geq \text{Th\_3}\right) \\ \text{or } \left(|h - S_{i,j}| \geq \text{Th\_3}\right) \\ \text{or } (|a - h| \geq \text{Th\_4}) \\ \text{true, otherwise} \end{cases} \quad (9)$$

$$F_{E2} = \begin{cases} \text{false,} & \text{if}\left(|c - S_{i,j}| \geq \text{Th\_3}\right) \\ \text{or } \left(|f - S_{i,j}| \geq \text{Th\_3}\right) \\ \text{or } (|c - f| \geq \text{Th\_4}) \\ \text{true, otherwise} \end{cases} \quad (10)$$

$$F_{E3} = \begin{cases} \text{false,} & \text{if}\left(|b - S_{i,j}| \geq \text{Th\_3}\right) \\ \text{or } \left(|g - S_{i,j}| \geq \text{Th\_3}\right) \\ \text{or } \left(|b - g| \geq \text{Th\_4}\right) \\ \text{true, otherwise} \end{cases} \tag{11}$$

$$F_{E4} = \begin{cases} \text{false,} & \text{if}\left(|d - S_{i,j}| \geq \text{Th\_3}\right) \\ \text{or } \left(|e - S_{i,j}| \geq \text{Th\_3}\right) \\ \text{or } \left(|d - e| \geq \text{Th\_4}\right) \\ \text{true, otherwise} \end{cases} \tag{12}$$

$$\text{Decision 3} = \begin{cases} \text{false,} & \text{if}(F_{E1}) \text{ or } (F_{E2}) \\ \text{or } (F_{E3}) \text{ or } (F_{E4}) \text{ is true} \\ \text{true, otherwise} \end{cases} \tag{13}$$

As per decision 3, if any of the directional differences F_{E1} to F_{E4} is true, i.e. shows the existence of an edge, then, Decision 3 is made false, i.e. the centre pixel is uncorrupted, else, the centre pixel is considered as corrupted. Since a parallel logic is used, each module is expected to be independent and self-sufficient, so as to deliver the judgment onto the centre pixel.

Analogous Module. In the analogous module, to identify the noisy pixels first consider two sets of thresholds, out of which only those thresholds that are closer to the general pixel range are chosen. This improves the accuracy of the detection mechanism. To determine the thresholds, first, the nine pixels within 3×3 mask are sorted in ascending order. In ascending order, the fifth value obtained is the median represented as $K_{i,j}$, the value preceding it is the fourth value represented as $J_{i,j}$ and the value succeeding it is the sixth value represented as $L_{i,j}$. Now use these values to obtain the thresholds as

$$W_{\max} = L_{i,j} + \text{Th\_5}$$
$$W_{\min} = J_{i,j} - \text{Th\_5} \tag{14}$$

These are the first set of thresholds obtained by adding and subtracting a fixed threshold from the sixth and fourth pixels, respectively. Now, the following set of equations give the final set of thresholds

$$T_{\max} = \begin{cases} W_{\max} & \text{if}\left(W_{\max} \leq K_{i,j} + \text{Th\_6}\right) \\ K_{i,j} + \text{Th\_6}, & \text{otherwise} \end{cases} \tag{15}$$

$$T_{\min} = \begin{cases} W_{\min} & \text{if}(W_{\min} \geq K_{i,j} - \text{Th\_6}) \\ K_{i,j} - \text{Th\_6}, & \text{otherwise} \end{cases} \tag{16}$$

Here, T_{max} and T_{min} are the final thresholds that are used to test the centre pixel, while $K_{i,j}$ + Th_6 and $K_{i,j}$ − Th_6 are the second set of thresholds. Only those thresholds that are closer to the general pixel range are chosen here. Now, the final decision, whether the centre pixel is noisy or uncorrupted, is arrived at by comparing the centre pixel value with the two final thresholds T_{max} and T_{min}. If the centre pixel value lies in between the thresholds T_{max} and T_{min}, then it is considered as uncorrupted; else, it is considered as corrupted and is filtered. This decision can be expressed as

$$\text{Decision 1} = \begin{cases} \text{true,} & \text{if} \left(S_{i,j} \geq T_{max}\right) \text{ or } \left(S_{i,j} \leq T_{min}\right) \\ \text{false,} & \text{otherwise} \end{cases} \tag{17}$$

2.2 Edge Preserving Filter

The basic logic behind the working of the edge preserving filter is replacing the centre pixel with a mean value of only those pixels that are involved in the edge. To detect the presence of edge, consider eight directions within the 3×3 mask and their corresponding directional differences are D_1 to D_8. The direction with the least directional difference is more likely to have an edge. Thus, the pixels that make up that direction are used to find the mean. To avoid wrong calculation of directional difference, omit all those directions from D_1 to D_8 that include a noisy pixel during the calculation of directional differences. To determine which pixels are noisy, the W_{max} and W_{min} from the analogous module are used. In the case where all the neighbours of the centre pixel are suspected to be noisy, the weighted average of a, b and c is to be found. The estimated grey scale value of the centre pixel is given by

$$\hat{S}_{i,j} = (a + b \times 2 + c)/4 \tag{18}$$

Here, $\hat{S}_{i,j}$ denotes the estimated value of the centre pixel. While, 'a, b and c' denote the upper row neighbours of the centre pixel that are filtered previously. Generally, if none of the neighbours were found to be noisy, calculate the edge distances using the equations below

$$D_1 = |d - h| + |a - e|,$$
$$D_2 = |a - g| + |b - h|,$$
$$D_3 = |b - g| \times 2,$$
$$D_4 = |b - f| + |c - g|,$$
$$D_5 = |c - d| + |e - f|,$$
$$D_6 = |d - e| \times 2,$$
$$D_7 = |a - h| \times 2,$$
$$D_8 = |c - f| \times 2,$$

(19)

Based on which of the directional differences ends up being the $D_{\min}$, the estimated grey scale value of the centre pixel is calculated as given below

$$\hat{S}_{i,j} = \begin{cases} (a + d + e + h)/4, & \text{if } D_{\min} = D_1 \\ (a + b + g + h)/4, & \text{if } D_{\min} = D_2 \\ (b + g)/4, & \text{if } D_{\min} = D_3 \\ (b + c + f + g)/4, & \text{if } D_{\min} = D_4 \\ (c + d + e + f)/4, & \text{if } D_{\min} = D_5 \\ (d + e)/4, & \text{if } D_{\min} = D_6 \\ (a + h)/4, & \text{if } D_{\min} = D_7 \\ (c + f)/4, & \text{if } D_{\min} = D_8 \end{cases}$$

(20)

It is to be noted that whenever the median of the pixels b, d, e and g is computed, it always results in a value equal to that of the estimated value. In the case of a wrong detection of an edge or any other causes of error, the estimated value is wrongly calculated. To prevent this, the median of the pixels b, d, e and g along with estimated value is computed and use the resultant median to replace the centre pixel value. The final median value obtained, which replaces the centre pixel, is given by

$$\hat{S}_{i,j} = \text{Median}\left(\hat{S}_{i,j}, b, d, e, g\right)$$

(21)

3 Results and Discussion

Above proposed algorithm has been simulated on the MRI images of brain. The proposed algorithm has been implemented using MATLAB R2013a. Experiments are conducted on three images of size 256 × 256 namely Cavernos_Angomia, Cerebral_Hemorrhage and Glioma brain images with 8-bit pixel amplitude resolution. These images are captured from Siemens-Area MRI scanner equipment, in which image is captured on slice of thickness 1 mm × 1 mm × 1 mm by 48 multi channels with 1.5 Tesla magnetic field intensity. The advanced technology used in this equip-

Table 1 Shows comparison of restored Cavernos_Angomia brain image using PSNR for various noise densities

Method	PSNR (dB) values for Cavernos_Angomia			
	5%	10%	15%	20%
Noisy	23.47	17.12	15.56	14.16
Median	33.19	32.53	31.81	30.95
AMF	37.50	33.96	33.47	31.70
Proposed	41.89	39.05	36.95	35.32

Table 2 Shows comparison of restored Cerebral_Hemorrhage brain image using PSNR for various noise densities

Method	PSNR (dB) values for Cerebral_Hemorrhage			
	5%	10%	15%	20%
Noisy	23.33	19.37	17.35	15.07
Median	30.25	29.76	29.07	28.48
AMF	33.19	32.55	30.66	29.30
Proposed	37.45	35.32	33.76	32.27

Table 3 Shows comparison of restored Glioma brain image using PSNR for various noise densities

Method	PSNR (dB) values for Glioma			
	5%	10%	15%	20%
Noisy	21.42	19.67	17.11	15.21
Median	30.74	30.30	29.85	29.30
AMF	33.66	32.59	31.28	30.22
Proposed	37.84	36.01	34.43	33.04

ment is Magnetom Avanto-Tim technology. While simulating, random noise will be added to image to distort it. The density of noise applied on the image vary from 5 to 20%. The performance of above algorithm will be represented in terms of PSNR is given as

$$PSNR = 10 \log \frac{255^2}{MSE} \tag{22}$$

where MSE stands for mean square error. The obtained values of PSNR for a Cavernos_Angomia, Cerebral_Hemorrhage and Glioma brain images for various noise densities from 5 to 20% in the steps of 5% have been tabulated and are shown in Tables 1, 2 and 3 respectively.

From results, it can be stated that the proposed algorithm exhibits better PSNR than existing algorithms namely Median and Adaptive Median Filter (AMF). Figures 1, 2 and 3 show bar charts of different methods on various brain images in case of PSNR with respect to various noise levels. From the figures, it is observed that the proposed method has better performance than other methods in all noise levels.

Figures 4, 5 and 6 illustrate the performance of the proposed denoising algorithm for brain images Cavernos_Angomia, Cerebral_Hemorrhage and Glioma at noise density 20, 15 and 10% of random noise respectively to explore the visual quality.

Fig. 1 Evaluation of the PSNR values for various algorithms applied on Cavernos_Angomia brain image for varying noise levels from 5 to 20% of noise

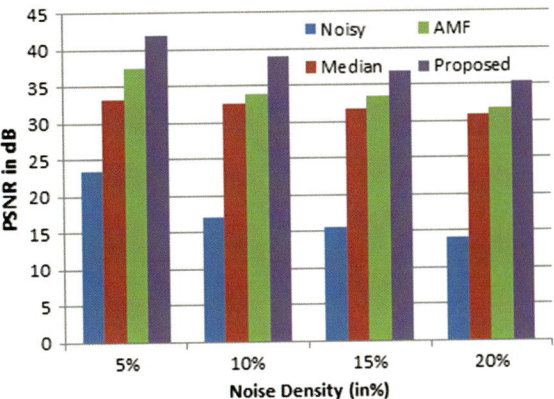

Fig. 2 Evaluation of the PSNR values for various algorithms applied on Cerebral_Hemorrhage brain image for varying noise levels from 5 to 20% of noise

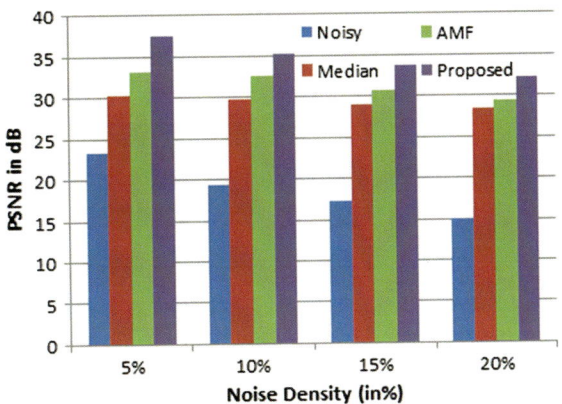

Fig. 3 Evaluation of the PSNR values for various algorithms applied on Glioma brain image for varying noise levels from 5 to 20% of noise

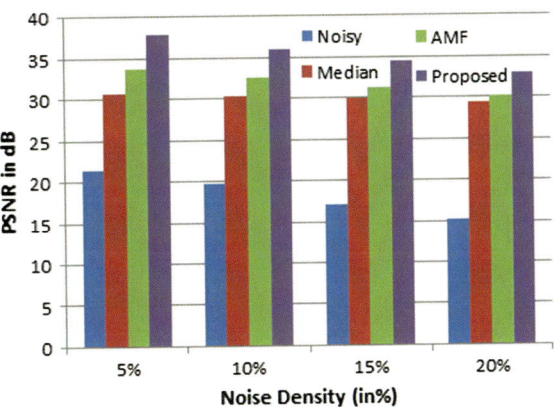

Figures 4, 5 and 6 (i) represent original brain image, Figs. 4, 5 and 6 (ii) represent the noisy image and Figs. 4, 5 and 6 (iii) represent reconstructed image of pro-

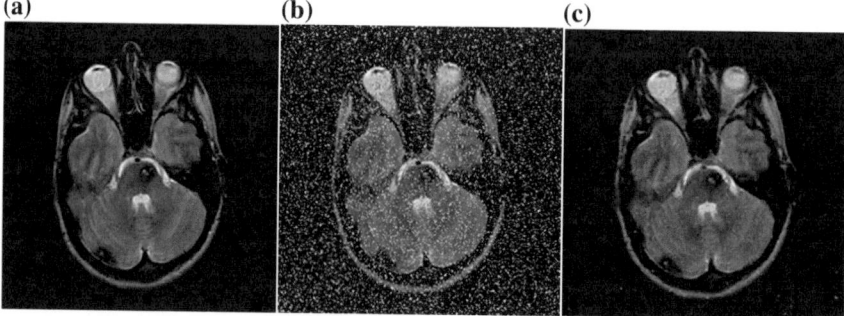

Fig. 4 Performance of the proposed algorithm for Cavernos_Angomia Brain image at a noise density of 20% Random noise **a** original image [13] **b** noisy image **c** denoised image

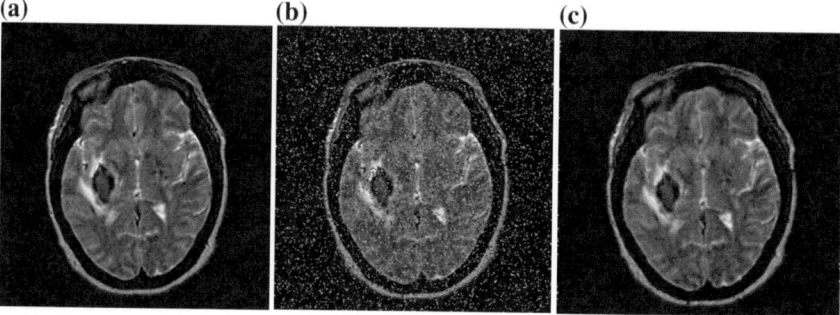

Fig. 5 Performance of the proposed algorithm for Cerebral_Hemorrhage brain image at a noise density of 15% Random noise **a** original image [13] **b** noisy image **c** denoised image

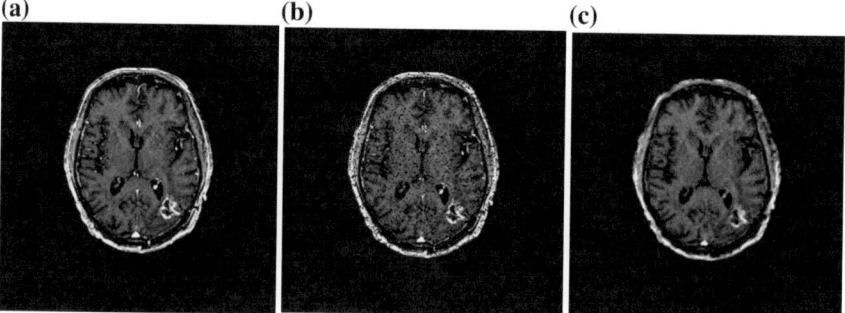

Fig. 6 Performance of the proposed algorithm for Glioma brain image at a noise density of 10% Random noise **a** original image [13] **b** noisy image **c** denoised image

posed denoising algorithm. From Figs. 4, 5 and 6 (iii), it is observed that reformed images obtained by applying the proposed algorithm preserve the edges in process of denoising random-valued impulse noise in an image.

4 Conclusions

In this paper, an decessive algorithm to denoise randomly valued impulse noise from medical images is proposed. The algorithm has been tested on various MRI brain images and produces better and accurate results. The proposed method uses an impulse detector and an edge preserving filter to denoise the corrupted pixels. The method uses simple mathematical expressions to detect and denoise impulse noise in images. Due to this simple approach the proposed method falls under the category of low complexity techniques and provides a better performance in terms of PSNR than that of the low complexity techniques, making it very helpful in telemedicine applications.

References

1. Dong Y, Xu S (2007) A new directional weighted median filter for removal of random valued noise. IEEE Sig Process Lett 14(3):193–196
2. Chen P-Y, Lien C-Y (2008) An efficient edge-preserving algorithm for removal of salt and pepper noise. IEEE Sig Process Lett 15:833–836
3. Yu H, Zhao L, Wang H (2008) An efficient procedure for removing random valued impulse noise in images. IEEE Sig Process Lett 15:922–925
4. Luo W (2006) An efficient detail-preserving approach for removing impulse noise in images. IEEE Sig Process Lett 13(7):413–416
5. Ng P-E, Ma K-K (2006) A switching median filter with boundary discriminative noise detection for extremely corrupted images. IEEE Trans Image Process 15(6):1506–1516
6. Dash A, Sathua SK (2015) High density noise removal by using cascading algorithms. In: Fifth international conference on advanced computing & communication technologies
7. Chan RH, Ho CW, Nikolova M (2005) Salt-and-Pepper noise removal by median-type noise detectors and detail-preserving regularization. IEEE Trans Image Process 14(10):1479–1485
8. Sun T, Neuvo Y (1994) Detail-preserving median based filters in image processing. Pattern Recogn Lett 15:341–347
9. Santhanam T, Chithra K (2014) A new decision based unsymmetric trimmed median filter using Eucledian distance for removal of high density salt and pepper noise from images. In: IEEE conference publications, pp 1–5
10. Suman S (2014) Image denoising using new adaptive based median filter. Int J (SIPIJ) 5(4):1–13
11. Vikrant B, Tiwari H, Srivastava A (2015) A non-local means filtering algorithm for restoration of Rician distributed MRI. In: Emerging ICT for bridging the future—proceedings of the 49th annual convention of the Computer Society of India CSI, vol 2. Springer, Cham
12. Awanish KS, Vikrant B, Verma RL, Alam MS (2014) An improved directional weighted median filter for restoration of images corrupted with high density impulse noise. In: 2014 international conference on optimization, reliability, and information technology (ICROIT). IEEE
13. https://www.frontiersin.org/articles/10.3389/fneur.2015.00033/full

Importance of LEDs Placing and Uniformity: Phototherapy Treatment Used for Neonatal Hyperbilirubinemia

J. Lokesh, K. Shashank and Savitha G. Kini

Abstract Phototherapy devices are most popularly used in medical applications to treat jaundice in infants called Neonatal jaundice. Neonatal jaundice or neonatal hyperbilirubinemia is the symptom of excessive bilirubin in blood in 60% of term babies and almost up to 70% of preterm babies. In earlier studies, it is found that the efficacy of phototherapy devices depends on spectral wavelength, body surface area, irradiance and duration of exposure. And it is important to know even importance of uniformity, peak wavelength and optimum height contribution in reducing phototherapy duration time. The paper explains the importance of uniformity and the optimum height. The uniformities for both the sets of LEDs are determined and selected for further testing. The optimum height at which the LEDs must be mounted is determined by comparing the uniformity values. The temperature is compared with the standard maximum ambient temperature up to which the infant remains healthy.

1 Introduction

Jaundice is a condition in which blood contains an excess amount of unconjugated bilirubin. Almost 60% term babies and 70% preterm babies have this condition. It is usually detected within 24–48 h of birth. Phototherapy is the widely used treatment. Earlier people would just expose the infant to sunlight for a certain period of time. But it was later found out that only light in the wavelength range of 430–490 nm was effective for the conversion of unconjugated bilirubin. In particular, 450–460 nm was found to be most effective. Most of the phototherapy devices use either FTLs or

J. Lokesh (✉) · K. Shashank · S. G. Kini
Electrical and Electronics Department, Manipal Institute of Technology, MAHE, Manipal 576104, India
e-mail: lokesh.j@manipal.edu

K. Shashank
e-mail: shanky.rao7@gmail.com

S. G. Kini
e-mail: savitha.kini@manipal.edu

© Springer Nature Switzerland AG 2019
D. Pandian et al. (eds.), *Proceedings of the International Conference on ISMAC in Computational Vision and Bio-Engineering 2018 (ISMAC-CVB)*, Lecture Notes in Computational Vision and Biomechanics 30,
https://doi.org/10.1007/978-3-030-00665-5_38

blue CFLs or a combination of both. The disadvantages of FTL devices are that the light sources give light at broader range of frequencies, thus reducing efficacy. CFLs and FTLs dissipate a large amount of heat because of ballasts. This also makes them very bulky, thus hindering mobility. Also, the lifespan of these lamps is very less, about 1000–1500 h. Apart from these, there are also health issues upon continuous exposure of infants to the mentioned light sources. The presence of mercury also raises concern because any exposure is very harmful for the infant. And the risk of exposure is very high in poorly maintained lamps.

Frequent exposure of infants to FTLs may cause Irlen Syndrome, a perception problem which affects the infant's ability to read and write in the future. In very rare cases, exposure to FTLs is also known to induce migraines. Stony Brook University researchers conducted a survey of CFLs bought from in and around their locality. The integrity of the phosphor coatings of each lamp was tested. They observed cracks in all the bulbs, through which UVA and UVC radiation were leaking out.

At Stony Brook's Advanced Energy Research and Technology Center (AERTC), the effects of these same bulbs on human skin tissue were tested. Comparisons were made with incandescent bulbs of the same intensity. It was found out that the effect of UV radiations emitted from the CFLs on human skin tissue was consistent with damage from exposure to ultraviolet radiation while there was no adverse effect of incandescent bulbs on human skin. These problems urge the need to use other sources of light for phototherapy as the effect might be magnified on an infant. But recently, LEDs are slowly replacing them because of low size and longer life. There is also no UV radiation emission and no mercury in LEDs. The major advantage of LED is that the exact wavelength required for a specific purpose can be obtained, with very small deviations in the wavelength. This paper provides the design of an LED-based phototherapy device.

2 Literature Review

Jaundice is a commonly encountered problem after birth and occurs in 60–70% of term infants and nearly all of preterm infants, including those near-term infants 35–38 weeks gestational age [1]. Nomogram for designation of risk in infants based on bilirubin concentration in blood at different ages is shown in Fig. 1.

For treatment of hyperbilirubinemia, light, in the range of approximately 400–500 nm with a peak at 460 ± 10 nm, is considered the most effective. In the current AAP guideline, intensive phototherapy is defined as the use of blue light (in the 430–490 nm band) delivered at 30 W/cm^2/nm or higher to the greatest BSA as possible. The light source must be placed at a height of 40 cm from the infant for optimal result. An irradiance level of 8–10 μW cm^{-2} nm^{-1} is considered to be normal phototherapy. Irradiance of >30 μW cm^{-2} nm^{-1} is considered to be Intensive phototherapy [1–4]. The Phototherapy enables the body to get rid of excess bilirubin through stools. It is also seen that while irradiance in the range of 30–40 μW cm^{-2}

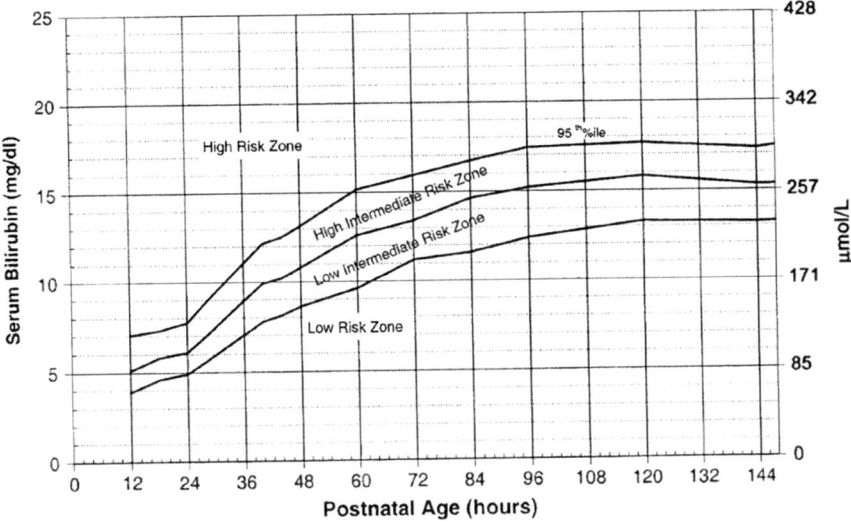

Fig. 1 Nomogram for designation of risk in infants based on bilirubin concentration in blood at different ages (*Source* AAP Guideline [1])

nm^{-1} is optimal, the rate of degradation of bilirubin is linear up to an irradiance of 55 μW cm^{-2} nm^{-1} with no saturation point [2].

The normal acceptable range of Total Serum Bilirubin (TSB) in an infant is 5.2 mg/dL [5]. Phototherapy is administered to an infant if the TSB levels are more than this value. The type of phototherapy varies between normal and intensive phototherapy. If the TSB value is <15 mg/dL, then normal phototherapy with an irradiance of 8–10 μW/cm^2/nm is administered and intensive phototherapy with an irradiance of >30 μW/cm^2/nm is administered if the TSB level is >20 mg/dL. For TSB values >25 mg/dL, blood exchange transfusion is recommended [6].

It is observed that efficiency of phototherapy is dependent on the wavelength, the comparative study between BSL (Broad spectrum light) (420–680 nm, and peak wavelength of 455 and 524 nm) and Blue LED (400–500 nm and peak of 460 nm) shows the duration of phototherapy was lower in the BSL than in the blue LED phototherapy group (15.8, 4.9 vs. 20.6, 6.0 h; $p = 0.009$) [4, 7, 8]. Blue plus green phototherapy is as effective as blue phototherapy and it attenuates irradiation-induced oxidative stress. And it is observed that the oxidative stress induced by irradiation level can be attenuated by adding green spectrum with blue and is as effective as only blue spectrum to reduce bilirubin level and tested in gunn rat model [9–13]. From the above discussion, it is seen that how the efficacy of phototherapy is dependent on wavelengths and irradiation level. In this paper, the importance of uniformity, peak wavelength and optimum height contribution in reducing phototherapy duration time and improving of efficiency of photothearpy is discussed.

3   Methodology and Validation

The prototype device is of glass of dimensions 56 cm × 49 cm × 27 cm. The inner walls and the base of the prototype are lined with light blue cardboard sheet to obtain highest possible reflectance. The base of the prototype is marked with a rectangle of 40 cm × 30 cm which is divided into 36 grid points, as shown in Fig. 2.

The LEDs are placed along the sides of the prototype and controlled with an 8 channel Optronix driver module and GUI window is used to set the input current required for the operation for good uniformity and right peak wavelength. The current, wavelength, irradiance level is obtained by testing LEDs in spectrometer (Integrating sphere). Two different set of LEDs are selected and position height is determined for better uniformity, as height changes, irradiance level and uniformity also changes. The current supplied to the LED of set 1, which are high brightness (HB) LEDs, is varied from 15 to 350 mA in steps of 10 mA and to the LED of set 2, which are low brightness (LB) LEDs, is varied from 100 to 350 mA in steps of 5 mA is shown in Tables 1 and 2, respectively, for single LED.

The selected irradiance level for test purposes is 55 μW cm^{-2} nm^{-1}. It is so selected because this is the maximum irradiance value which can be used without any change in the rate of degradation of bilirubin [1]. From Eqs. (1) and (2), we determine the required luminous flux of each LED as 7.65 Lm. From Tables 1 and 2, the current required to obtain this value of luminous flux from each HB LED is 75 and 318 mA for each LB LED.

$$\Phi_{\text{total}} = A * E_{e,\lambda} \tag{1}$$

$$\Phi = \Phi_{\text{total}}/N \tag{2}$$

where

Φ_{total} is the total radiant power;
A is the illuminated area;

Fig. 2 Depiction of grid points on prototype base

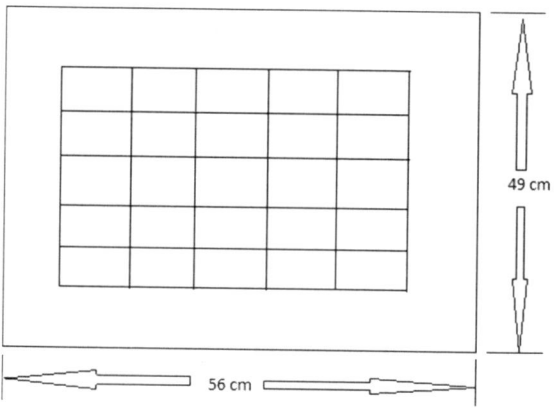

Table 1 HB LED: forward current, wavelength, and luminous flux

I_f (mA)	λ (nm)	Φ (Lm)
15	451	1.324
45	451	3.825
75	**450**	**7.663**
105	450	8.727
135	450	10.18
165	450	12.44
195	450	14.22
225	450	17.11
255	450	19.08
285	450	21.04
315	450	22.92
345	450	25.40
350	450	25.76

Table 2 LB LED: forward current, wavelength, and luminous flux

I_f (mA)	λ (nm)	Φ (Lm)
100	453	3.481
125	453	4.178
150	452	4.483
175	452	4.937
200	451	5.492
225	451	5.791
250	451	6.007
275	451	6.606
300	450	7.324
318	**450**	**7.689**
325	450	7.749
350	450	8.243

$E_{e,\lambda}$ is the irradiance level selected;

Φ is the radiant power of each LED; and

N is the number of LEDs.

Prototype: Along the length of the prototype, the LEDs are placed 20 cm from either corner. Along the width of the prototype, LEDs are placed 16 cm from either corner. They are placed at a height of 22 from the workplace. The illuminance values at the grid points on the workplace are noted. This process is done for the LED placement heights of 23 and 24 cm also and the comparison of uniformity of 2 sets of LEDs for different heights is given in Table 3 with the same number (8 Nos) of LEDs.

The uniformity is calculated from Eq. (3).

$$U = E_{\min}/E_{\text{avg}} \tag{3}$$

where

$E_{\min}$ is minimum illuminance
E_{avg} is average illuminance
U is uniformity

From Table 3, even though the LB LEDs offer higher uniformity, the current required is much higher than the current required for the HB LEDs, which is not desirable and HB LEDs are the most suitable for use because the difference between the uniformities is not large at a given height. And it is also observed that variations in numbers (Number of LEDs) gives different uniformities. The uniformity level with multiple (8, 6 and 4 nos) HB LEDs is shown in Table 4 and the illuminance distribution with 8 and 6 nos of LEDs are shown in Figs. 3 and 4, respectively. And illuminance distribution with 4 LEDs is not shown as it gives low uniformity.

The uniformity is highest at a height of 23 cm from the workplace with very little variation in the illuminance and the illuminance values are higher at the center region

Table 3 Height from the workplace, uniformity of HB LEDs, uniformity of LB LEDs

Height (cm)	U_{HBLED}	U_{LBLED}
22	0.7582	0.8588
23	0.809	0.8892
24	0.7822	0.8674

Table 4 Height from workplace, uniformities with 8, 6, and 4 LEDs

Height (cm)	U_{8LED}	U_{6LED}	U_{4LED}
22	0.7582	0.73	0.659
23	**0.809**	**0.757**	**0.664**
24	0.7822	0.721	0.631

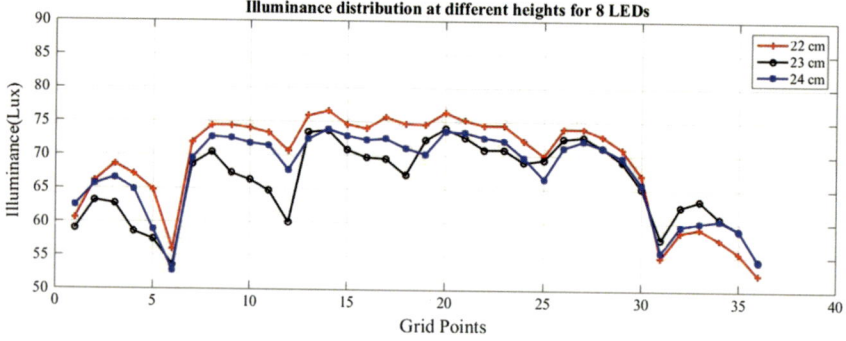

Fig. 3 Illuminance distribution at different heights for 8 HB LEDs

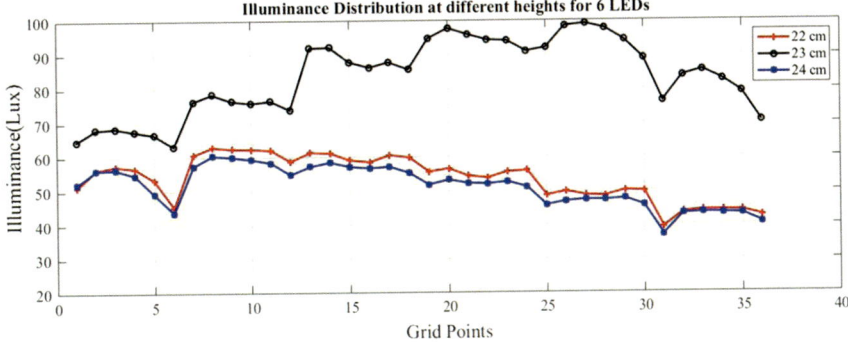

Fig. 4 Illuminance distribution at different heights for 6 HB LEDs

Table 5 Uniformity versus phototherapy: total duration (h)

	Irradiance (μW cm^{-2} nm^{-1})	Uniformity	Phototherapy: total duration (h)
Available	35	0.62	11.48
Proposed	35	0.809	11.39

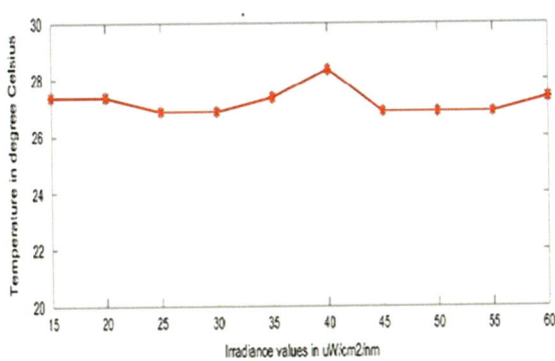

Fig. 5 Irradiance values and maximum temperature at 23 cm height

of the workplace for all the three heights (22, 23 and 24 cm), since at sufficiently greater heights, the light spread is more even at the center of the workplace and the main area of the workplace is adequately illuminated, avoiding any dark.

And case study shows uniformity will also plays an important role in reducing bilirubin level at bit less duration is shown in Table 5.

As efficacy of phototherapy is dependent on the irradiance level and irradiance level exposed is dependent on initial bilirubin, it is important to check the temperature variations for a change in irradiance level to avoid forced phototherapy and skin problems due to more irradiance. The results are shown in Fig. 5 and it is observed that there is no much changes in temperature for change in irradiance level in the proposed system as LEDs are placed at different points to maintain better uniformity. The accepted ambient temperature for an infant to not be adversely affected is 32 °C. The maximum temperature obtained at different irradiance values varies in a small interval of 26.88–28.35 °C, which is within the ambient temperature mentioned. This temperature will be even lower in an actual hospital where the room for the infants is temperature controlled with air-conditioning.

4 Conclusion

The illuminance distribution on the workplace with different heights of LED placement was found out and the most suitable position to mount LEDs was determined based on the uniformity values. Furthermore, comparisons of uniformity were done for 2 different sets of LEDs. And it is shown that importance of uniformity in reducing the treatment duration to avoid skin problems and other. The temperature at the workplace was also tested and the maximum temperature for different values of irradiance is noted, which varies in a very small interval and is less than the recommended ambient temperature. Future study is to know will peak wavelength also have an impact in reducing bilirubin level.

References

1. American Academy of Pediatrics (2004) Management of hyperbilirubinemia in the newborn infant 35 or more weeks of gestation. Pediatrics 114:297–316
2. Vreman HJ, Wong RJ, Stevenson DK (2004) Phototherapy: current methods and future directions. Semin Perinatol 28:326–333 (Elsevier Inc.)
3. Ennever JF, McDonagh AF, Speck WT (1983) Phototherapy for neonatal jaundice: optimal wavelength of light. J Pediatr 103:295–299
4. Seidman DS, Moise J, Ergaz Z et al (2003) A prospective randomized controlled study of phototherapy using blue and blue-green light-emitting devices, and conventional halogen-quartz phototherapy. J Perinatol 23:123–127
5. de Araujo MCK, Vaz FAC, Ramos JLA (1996) Progress in phototherapy. Sao Paulo Med J 114(2):1134–1140
6. Watchko JF, Maisels MJ (2003) Jaundice in low birthweight infants: pathobiology and outcome, pp F455–459
7. Pratesi S, Di Fabio S, Bresci C, Di Natale C, Bar S, Dani C (2014) Broad-spectrum light versus blue light for phototherapy in neonatal hyperbilirubinemia: a randomized controlled trial. Perinatology. ISSN 0735-1631
8. Ebbesen F, Madsen P, Støvring S, Hundborg H, Agati G (2007) Therapeutic effect of turquoise versus blue light with equal irradiance in preterm infants with jaundice. Acta Paediatr 96(6):837–841
9. Uchida Y, Morimoto Y, Uchiike T, Kamamoto T, Hayashi T, Arai I, Nishikubo T, Takahashi Y (2015) Phototherapy with blue and green mixed-light is as effective against unconjugated jaundice as blue light and reduces oxidative stress in the Gunn rat model. Elsevier Early Hum Dev 91:381–385

10. Vecchi C, Donzelli GP, Migliorini MG, Sbrana G (1983) Green light in phototherapy. Pediatr Res 17(6):461–463
11. Mohammadizadeh M, Eliadarani FK, Badiei Z (2012) Is the light-emitting diode a better light source than fluorescent tube for phototherapy of neonatal jaundice in preterm infants? Adv Biomed Res 1:51
12. Uchida Y, Morimoto Y, Haku J, Nakagawa T, Nishikubo T, Takahashi Y (2011) A comparison the therapeutic effects of blue light-emitting diodes (LED) and green fluorescence lights (green FL) in the treatment of neonatal jaundice. J Jpn Soc Premature Newborn Med 23(2):263–267
13. Itoh S, Onishi S, Isobe K, Manabe M, Yamakawa T (1987) Wavelength dependence of the geometric and structural photoisomerization of bilirubin bound to human serum albumin. Biol Neonate 51(1):10–17

Automated Glaucoma Detection Using Global Statistical Parameters of Retina Fundus Images

Prathiksha R. Puthren, Ayush Agrawal and Usha Padma

Abstract Glaucoma is an eye disorder which is prevalent in the ageing population and causes irreversible loss of vision. Hence, computer-aided solutions are of interest for screening purposes. Glaucoma is indicated by structural changes in the Optic Disc (OD), loss of nerve fibres and atrophy of the peripapillary region of optic disc in retina. In retina images, most changes appear in form of subtle variation in appearance. Hence, automated assessment of glaucoma from colour fundus images is a challenging problem. Prevalent approaches aim at detecting the primary indicator, namely, the optic cup deformation relative to the disc and use the ratio of the two diameters in the vertical direction, to classify images as normal or glaucomatous. An attempt is made to detect glaucoma by combining image processing and neural network techniques. The risk of blindness can be reduced by 50% with screening patients vulnerable to eye diseases specially glaucoma. The global statistical features of the dataset images are used to detect images as glaucoma or normal. The technique involves screening for the vital signs such as intensity values in the fundus image for detecting glaucoma in patients. The result shows the feasibility of detection of glaucoma for vulnerable patient.

1 Introduction

Glaucoma is a retinal disease and is among the leading causes which results in loss of sight. It is a degenerative optic neuropathy which results in gradual loss of retinal nerve fibre which cannot be revitalized. Thus, untreated glaucoma has a potential to cause irreparable damage to retina. This irreversible and asymptomatic nature

P. R. Puthren · A. Agrawal (✉) · U. Padma (✉)
Department of Telecommunication Engineering, RV College of Engineering, Bengaluru, India
e-mail: ayush.agrawal199508@gmail.com

U. Padma
e-mail: ushapadma2018@gmail.com

P. R. Puthren
e-mail: prathiksha.puthren@gmail.com

© Springer Nature Switzerland AG 2019
D. Pandian et al. (eds.), *Proceedings of the International Conference on ISMAC in Computational Vision and Bio-Engineering 2018 (ISMAC-CVB)*, Lecture Notes in Computational Vision and Biomechanics 30,
https://doi.org/10.1007/978-3-030-00665-5_39

Fig. 1 The chart shows the prevalence of vision impairment in India. *Source* Glaucoma Society of India, p. 55

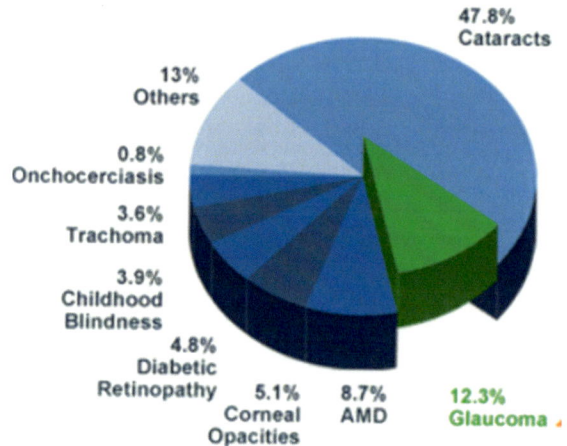

of the disease emphasizes the need for its timely detection as detecting it in early phases helps in curbing its progression through proper medication. The proliferation of glaucoma cases in recent years adds to the growing concern towards the treatment of the disease. It prevails mainly in ageing population of urban regions. It is estimated to affect 79 million people in the world by the year 2020, showing a 33% increase in the numbers within a decade. Thus, screening for glaucoma is crucial, owing to the nature, for the early detection and enabling effective treatment in early stages to prevent permanent blindness. According to the Glaucoma society of India, glaucoma is the second main cause of blindness in India which is depicted in Fig. 1. The major challenge posed by Indian population is the number. At present, 12 million people in India are affected by glaucoma which is expected to increase to 16 million by 2020. The number of patients per ophthalmologist is around 2–3 lakhs in India. Thus, apart from cost, lack of manpower in terms of skilled technicians poses major challenge in such scenarios. Cost-effective computer-based diagnostic systems can reduce this requirement to a great extent and assist medical experts in diagnosis. Automated screening systems based on retinal (fundus) image analysis can aid in reducing time and effort wasted on analysis of healthy people. Such systems can classify a given case as normal (free of glaucoma-related symptoms) or glaucomatous. Consequently, patients deemed suspect by the system need to be referred to an ophthalmologist. Thus, the categorization of retinal images as normal and glaucomatous is a problem of clinical significance in population screening.

Several techniques for glaucoma detection have been investigated by authors based on available dataset images. In [1], artificial neural network (ANN) techniques were used to detect glaucoma in images. With the help of MATLAB, the disease was classified as normal, severe and mild. Using a neural network, the parameters extracted from MATLAB are compared with standard values. The use of ANN made the detection of Glaucoma accurate as well as adaptive. Ease of operation is the main advantage in this method. In [2], authors implemented algorithm to detect glaucoma which used morphological methods to acquire two main features for identification of Glaucoma, i.e. Ratio of areas of neuro-retinal rim (NRR) in Inferior, Superior, Temporal and Nasal (ISNT) quadrants and Cup to Disc Ratio (CDR). Using this method, the achieved accuracy was 83.5%. In [3], authors reviewed several methods which are useful in identifying glaucoma. By using these techniques, there is need to develop cost-effective automatic techniques for detection of glaucoma disease precisely. In [4], algorithmic rules for robust and automatic extraction of features in colour fundus images were proposed. To locate Optic Disc, Principal Component Analysis (PCA) is used. An active shape model was designed in detection of shape of Optic Disc (OD). Based on fovea localization, fundus coordinate system was made. An approach to identify exudates was by combining edge detection and region growing was suggested. The success percentages of disc boundary detection, disc localization and fovea localization were 94, 99 and 100%, respectively. The specificity of exudates was 71% and sensitivity of the exudates detection was 100%. A screening technique for glaucoma detection by the use of super pixel categorization on optic cup segmentation and Optic Disc was developed [5]. The automated OD segmentation quality was determined by the use of a self-assessment accuracy score. For segmentation of optic cup and for obtaining histograms, location information is included to increase the performance. The results showed an overlapping error of 24.1 and 9.5% in cup and disc segmentation, respectively. In this paper, we have proposed a system for glaucoma detection by combining image processing and neural network techniques. The global statistical parameters of image such as mean, standard deviation and variance are extracted and used for classifying images as normal or glaucoma. The technique used for classifying is the back propagation neural network which achieves high accuracy for classification in several cases.

2   The Proposed Approach

In this section, the dataset, algorithm for detection of ROI and classification of retina fundus images is discussed. Figure 2 shows the generic process of detection of glaucoma.

2.1   Dataset Collection

For developing algorithm for detecting glaucoma, the first necessary step required was obtaining the applicable datasets. 14 fundus images of retina were acquired from different online databases for this purpose. Out of these, 8 images were taken from FAU database and 6 from optic-disc.org database (Fig. 3).

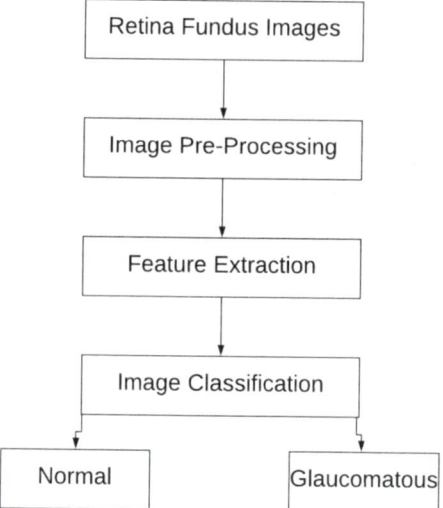

Fig. 2 Generic process of detection of glaucoma

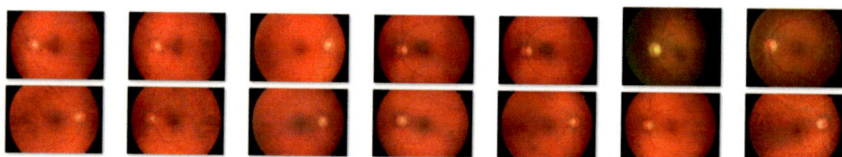

Fig. 3 Dataset images

2.2 Detection of ROI

2.2.1 Image Pre-processing

In colour retina fundus images, OD is the most intensified part having orange and pink colour. It is determined as region of interest. The boundary of this Optic Disc must be identified to separate and extract it from the image. The disc occupies around 5% pixels in coloured retina image. The cup and disc extraction was performed on full image and region of interest is located to improve accuracy and reduce cost computation. The region of interest is identified, cropped and resized from all original images as shown in Fig. 4.

2.2.2 Optic Disc and Cup Extraction from Fundus Images

For suspecting glaucoma, assessment of features is an important step, which is determined by optic cup and disc extraction. Initially, the fundus retina image was cropped

Fig. 4 Optic Disc after
extraction and resizing

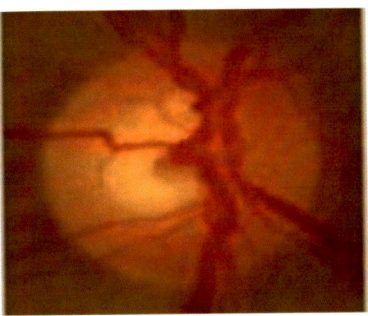

to get OD, and then resized. Then, the blood veins were removed from the image to
get a smooth image.

To accomplish this task, morphological operations known as dilation and erosion
are performed according to Eqs. (2.1) and (2.2). Dilation is used to increase the size
of objects by adding pixels to the borders of object in input image. The structuring
element 'DISC' was used for dilating the image. The dilation results in lighting
blood vessels and filling internal gaps. Increase in the size of disc also affects CDR.
Erosion of image was carried out using same sized structuring element after dilation.
Images are eroded for contrasting the boundaries of object. These operations resulted
in smooth images without blood vessels.

Dilation is defined as

$$A \oplus B = \bigcup_{b \in B} A_b \qquad (2.1)$$

Erosion is defined as

$$A \ominus B = \bigcup_{b \in B} A_{-b} \qquad (2.2)$$

where

A: Binary image and B: Structuring element.

After analysing many images, it was concluded Optic Disc had a better contrast
in V plane extract from Hue Saturation Value (HSV) images. The mean values of the
V plane images were calculated, and given as threshold to convert to binary images.
Unwanted objects which were obtained in binary image are removed by labelling
them and morphological operations are applied to remove the objects that had fewer
pixels, and hence this resulted in removing all the unwanted objects excluding Optic
Disc. Then the image boundaries were smoothened by the application of Gaussian
filter as depicted in Fig. 5.

As depicted in Fig. 5a, compared to other regions of retina fundus image, a cup
has brighter contrast. In the second step, using global threshold which chooses the
threshold to decrease the intraclass variance of the black and white pixels, the green

Fig. 5 Extraction of Optic
Disc

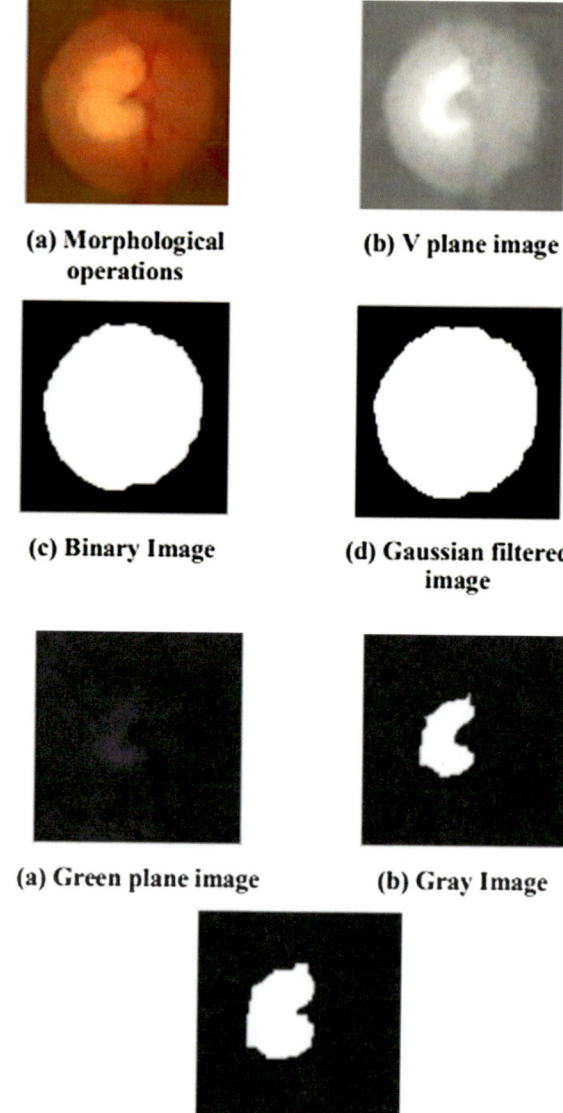

**(a) Morphological
operations**

(b) V plane image

(c) Binary Image

**(d) Gaussian filtered
image**

Fig. 6 Optic cup extraction

(a) Green plane image

(b) Gray Image

(c) Gaussian filtered image

plane is converted into grey scale image. In the third step, morphological operation
for removal of small objects is applied, same as Optic Disc but with less pixel value
as cup size is small, this is done after extracting the binary optic cup. Gaussian filter
is applied to the resultant binary image of the optic cup, to smoothen the boundaries
of optic cup, as shown in Fig. 6.

2.3 Extraction of Features

Basically, feature extraction aims at decreasing the resources needed for describing a large set of necessary data. Thus, the input image can be transformed into set of features. The features extracted from retina fundus images are based on Average Brightness (R, G, B), Standard Deviation (R, G, B) and Coefficient of Variance (R, G, B).

2.3.1 Average Brightness

The sample mean of the pixel brightness within that region is called average brightness. The average brightness (is denoted by m_a) over $\wedge$ pixels inside a region (R) is given as:

$$m_a = \frac{1}{\wedge} \sum_{(m,n) \in R} a[m,n] \qquad (2.3)$$

On the other hand, a formulae based on the brightness histogram (unnormalized), $h(a) = \wedge * p(a)$, with discrete brightness values 'a', can be used and is given by

$$m_a = \frac{1}{\wedge} \sum_a a \cdot h[a] \qquad (2.4)$$

Here, average brightness (m_a) is an estimate of mean brightness of the brightness probability distribution denoted by $h[a]$.

2.3.2 Standard Deviation

The standard deviation unbiased estimate denoted by s_a, of the brightness's within region (R) for '$\wedge$' pixels is known as sample standard deviation. It is given as:

$$s_a = \sqrt[2]{\frac{\frac{1}{\wedge - 1} \sum_{m,n \in R} (a[m,n] - m_a)^2}{1}} \qquad (2.5)$$

Using the histogram formula, the above equation can be written as

$$s_a = \sqrt{\frac{\left(\sum_a a^2 \cdot h[a] \right) - \wedge \cdot m_a^2}{\wedge - 1}} \qquad (2.6)$$

Here, standard deviation (s_a) is estimate of underlying brightness probability distribution.

2.3.3 Variance

The coefficient of variation (CV) is given by

$$CV = \frac{s_a}{m_a} \times 100\% \qquad (2.7)$$

2.4 Classification

Classification of the image is carried out based on three features, i.e. mean, standard deviation and variance of the image. Images are sent as input to the neural network which classifies based on the above features. The neural network model used is scaled conjugate gradient back propagation which is discussed in the next section.

2.4.1 Backpropagation Algorithm

Backpropagation is a supervised learning algorithm of ANNs using gradient descent. It is short for 'backward propagating of errors'. If the ANN and error function is given, this method calculates gradient of error function w.r.t neural network's weights. The calculation of the gradient happens backwards in the network. The gradient of last layer of weights is calculated first and the gradient of first layer weights is calculated last. Calculations of one layer are used to calculate gradient of the previous layer.

Backpropagation neural networks are becoming popular in the image and speech recognition applications. It is very efficient algorithm which takes advantage of specialized Graphics Processing Units (GPUs) to improve performance further.

3 Results

A multilayer perceptron backpropagation neural network is used with a hidden layer provided. Backpropagation algorithm based on 9 input neuron and 1 output neuron at the output layer was trained by a 2-layer feed forward network. 28 iterations were required to obtain the best trained network. The trained network with best validation performance had a mean square error (MSE) of 0.000000186 at epoch 28. Result obtained from the confusion matrix showed overall accuracy of 84.6% is achieved using this technique.

In Fig. 7, veins have been extracted by converting the retina fundus image to black and white image. It is mainly done so that, the optical disc which has to be cropped for further image processing does not contain any veins.

In Fig. 8, retina fundus images had been loaded. Average intensity displays the average RGB colour intensity variation. Variance displays the RGB pixels colour

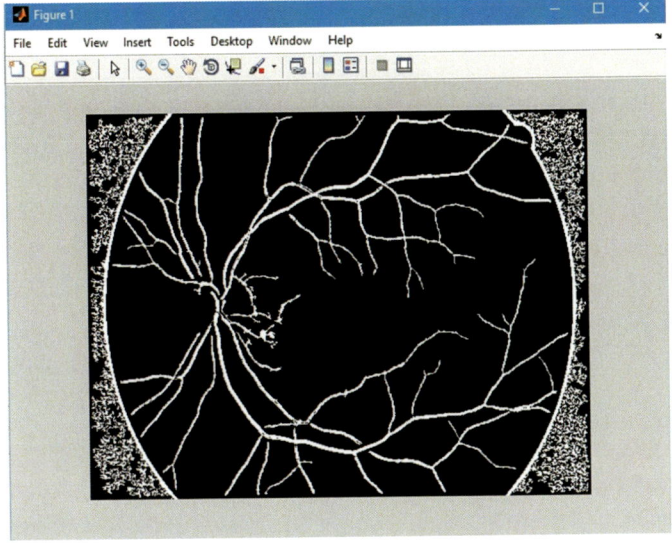

Fig. 7 Extraction of veins

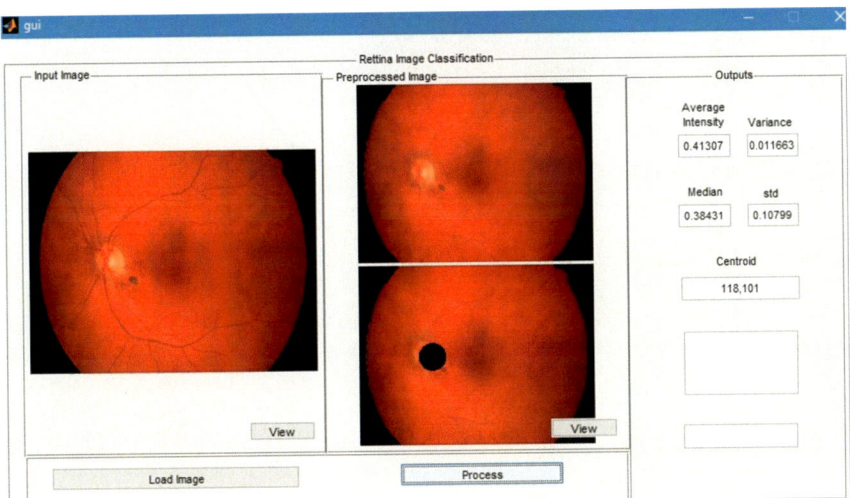

Fig. 8 Optic Disc extraction display window

variance. Median displays the median of RGB pixels intensity. 'Std' displays the standard deviation of RGB colour pixels. Centroid displays the centroid of Optic Disc.

Figure 9 contains the final GUI (graphical user interface) which displays the output whether the retinal fundus is normal or whether it is affected by Glaucoma. It displays mean values, variance values and standard deviation of RGB Colour Pixels.

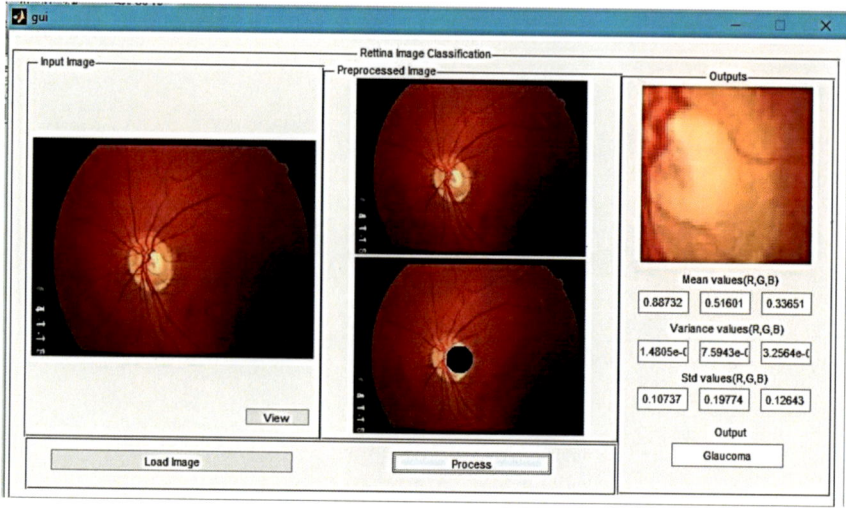

Fig. 9 Detected Glaucoma for a particular retina fundus image

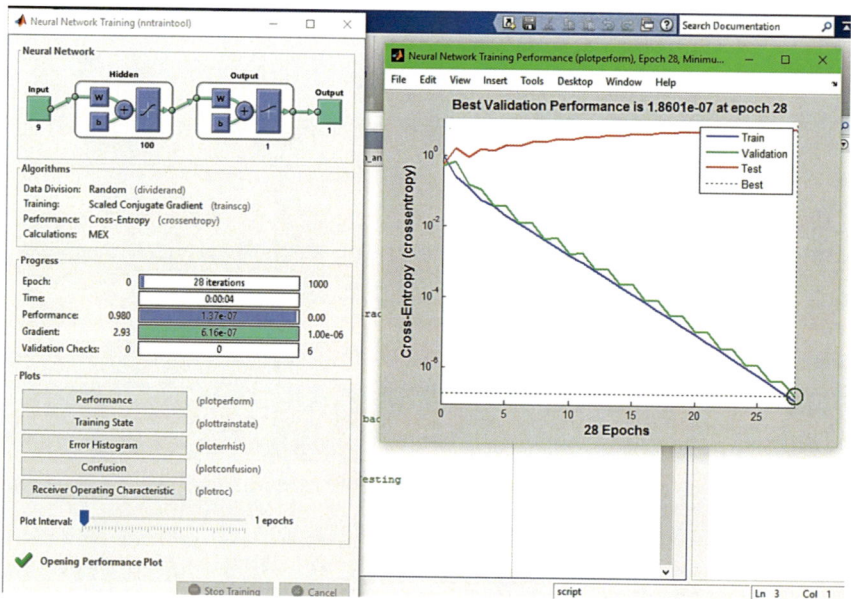

Fig. 10 Display window showing neural network training process with performance plot

The display window in Fig. 10 shows the neural network training model. The performance plot infers that the best validation performance occurs at 28th iteration.

Figure 11 displays the neural network confusion matrix.

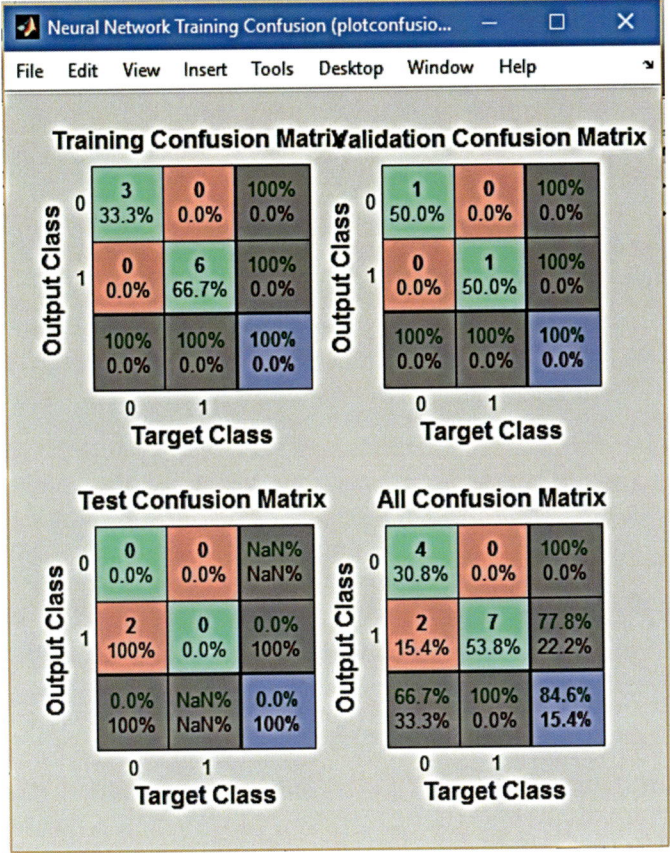

Fig. 11 Neural network training confusion matrix

4 Conclusion

In this paper, an automated glaucoma detection technique with three features, i.e. mean intensity, standard deviation and variance of the dataset images, which facilitates diagnosis of glaucoma is discussed. Morphological techniques were used to extract the features in this novel method. To observe the performance of this approach, 14 retina fundus images were used and processed. The method achieved an accuracy of around 84.6% using global statistical features of dataset images. The results are subjective, fast and consistent even though the performance is far from perfect. The results do not rely on trained glaucoma specialists or the use of specialized, costly OCT/HRT machines. The better performance of this method results in a large scale clinical evaluation and will be helpful in reporting large clinical findings in the future.

Acknowledgements This work was supported by Department of Telecommunication, R.V. College of Engineering, Bangalore, India. The authors would like to thank the Department for providing excellent facilities and timely guidance throughout the completion of the project.

References

1. Naveen Kumar B, Chauhan RP, Dahiya N (2016) Detection of Glaucoma using Image processing techniques: a review. IEEE Trans Biomed Eng 62(5)
2. Sheeba O, George J, Rajin PK, Thomas N, George S (2014) Glaucoma detection using artificial neural network. IACSIT Int J Eng Technol 6(2):158
3. Dey N, Roy AB, Das A, Choudari S (2012) Optical cup to disc ratio measurement for glaucoma diagnosis using harris corner. Eur J Sci Res 59. ISSN 1450-217X
4. Narasimhan K, Vijayarekha K, JogiNarayana KA, SivaPrasad P, Satish Kumar V (2011) Glaucoma detection from fundus image using opencv. Res J Appl Sci Eng Technol 62(5)
5. Ganesh Babu TR, Shenbagadevi S (2011) Automatic detection of glaucoma using fundus image. Eur J Sci Res. 59(1):22–32. ISSN 1450-216X
6. Zhang Z, Liu J, Cherian SN, Sun Y, Lim JH, Wong WK, Tan NM, Lu S, Li H, Wong TY (2009) Convex hull based optic cup ellipse optimization in glaucoma diagnosis. In: 31st annual international conference of the IEEE EMBS, Minneapolis, Minnesota, USA, 2–6 Sept 2009
7. Inoue N, Yanashima K, Magatani K, Kurihara T (2005) Development of a simple diagnostic method for the glaucoma using ocular fundus pictures. In: Proceedings of the 2005 IEEE engineering in medicine and biology 27th annual conference Shanghai, China, 1–4 Sept 2005
8. Li H, Chutatape O (2003) A model-based approach for automated feature extraction in fundus images. In: Proceedings of the ninth IEEE international conference on computer vision (ICCV 2003) 2-Volume Set 0-7695-1950- 4/03
9. High Resolution Fundus Image database. https://www5.cs.fau.de/research/data/fundusimages/
10. Opticdisc.org Database. http://www.opticdisc.org/library/normal-discs/page7.html

Enhanced Techniques to Secure Medical Image and Data Transit

S. Uma Maheswari and C. Vasanthanayaki

Abstract This paper displays the work related to the secured medical picture transmission in light of watermarking and encryption. Client particular watermark is installed into the LSB of unique picture. Implanting watermark in LSB does not influence the nature of picture. This watermarked picture is then encoded by utilizing a pixel repositioning calculation. Every pixel is repositioned in light of the rest of after division by number 10. This leftover portion lattice goes about as encryption key and is required at the season of unscrambling as well. Comprehensive analyses are completed on proposed approach. The outcomes demonstrate that the watermark inserted is intangible and can be effectively separated at the recipient. Likewise, the encoded picture has no visual importance with the first picture and histogram of scrambled picture is changed. Encoded picture can be decoded with no loss of data from the picture. From this decoded picture watermark can be removed which endures no misfortune in the watermark. PSNR esteems for an arrangement of medicinal pictures are fulfilling.

1 Introduction

This section talks about some fundamental ideas expected to comprehend whatever is left of the report. This part comprises of three areas: First segment portrays application space, i.e., Restorative imaging. Second segment depicts the idea of security and why we require security of restorative information. Third segment presents picture encryption and watermarking.

S. Uma Maheswari (✉)
Department of CSE, Kumaraguru College of Technology, Coimbatore, Tamil Nadu, India
e-mail: umamaheswari.s.cse@kct.ac.in

C. Vasanthanayaki
Department of ECE, Government College of Technology, Coimbatore, Tamil Nadu, India
e-mail: vasanthi@gct.ac.in

© Springer Nature Switzerland AG 2019
D. Pandian et al. (eds.), *Proceedings of the International Conference on ISMAC in Computational Vision and Bio-Engineering 2018 (ISMAC-CVB)*, Lecture Notes in Computational Vision and Biomechanics 30,
https://doi.org/10.1007/978-3-030-00665-5_40

389

1.1 Medical Imaging

Medical Imaging is the Technique and Process of making visual portrayals of insides a body, for clinical examination and restorative mediation, and also visual portrayal of the elements of a few organs or tissues. Restorative imaging looks to uncover inner structures covered up by the skin and bones and also to analyze and treat maladies.

Restorative Disease Diagnosis is finished by three ways

1. Consultation—Information which is acquired from Patients.
2. Physical Examination—Inspection, Auscultation, Measurements.
3. Medical Tests—Laboratory Analysis, Bio-flag Analysis (ECG), Image Analysis.

Picture Analysis intends to separate significant data from pictures; for the most part from Digital Images. Essentially, an Image is a rectangular exhibit of pixels. It has a distinct stature and an unmistakable width checked in pixels. The tallness of a picture is the quantity of lines and the width of a picture is the quantity of segments in a cluster. In this manner the pixel exhibit is a lattice of $n * m$ where n shows number of lines and m demonstrates number of sections.

A computerized picture is a numeric portrayal of a two-dimensional picture. Contingent upon whether the picture determination is settled, it might be of vector or raster compose. Independent from anyone else, the expression "advanced picture" normally alludes to raster pictures or bitmapped pictures. There are three sorts of computerized picture which are twofold picture, shading picture, and dim scale picture.

1. **Double Images**: Binary pictures are advanced pictures that utilizations just 1-bit to speak to every pixel which may either be 1 or 0. II.
2. **Dark-scale pictures**: Grayscale pictures utilize 8-bits to speak to every pixel of a picture. Dim-scale pictures are alluded to as monochrome pictures. They contain just dim level data and no shading data.
3. **Shading pictures**: Color picture is a computerized picture that is comprised of hued pixels every one of which holds three hues comparing to Red, Green, and Blue. Shading pictures are additionally called as RGB Images.

1.2 Picture Security

In human services industry, patients' information are moving from paper records to electronic records. Electronic patient record is a PC-based memory that can be evaluated over system. Electronic patient record contains all the human services related data of a patient and joins a few undertaking-based electronic therapeutic records particular to a patient. Electronic patient records changed the method for putting away patients' data. Electronic social insurance frameworks empower well-being suppliers' to outline and actualize particular applications that can help specialists for conclusion and treatment of patients.

Electronic patient records have made chances to human services cheats including restorative data fraud. Erroneous medicinal records may prompt restorative abuse that will make destructive results a patient. Numerous studies accept that restorative data fraud is the same as money-related wholesale fraud, however, in reality medicinal wholesale fraud has all the more crushing consequences for patients for there is an absence of plan of action for patient to revise the false passages in their therapeutic records. In actuality, false restorative data can murder a patient.

With the appearance of electronic patient records, patients' information can without much of a stretch be shared among doctors, social insurance suppliers, attendants, supporting staff, restorative research, and general human services administrations. Essentially, most medicinal services data including patients' information are not created exclusively inside a doctor/tolerant relationship, however is produced from the assorted sources, for example, non-doctor authority, nurture professionals, general well-being officers, research centers, and other auxiliary social insurance callings. The sharing and in addition dissemination of patient data empowers beneficial therapeutic research, appropriate treatment of patients, and change in medicinal services quality.

Then again, persistent records represent a test to keep up data secrecy, honesty, and accessibility—the modernized record induces that the requester has a similar equipment and programming correspondence convention and in this way empowers simple access to information. This may open ways to the unapproved parties who may deceitfully take patients' information for individual advantages, modify patients' records, and uncover patients' therapeutic history. The high openness of patients' information has made it less demanding for culprits to attack patients' data privacy, respectability, and accessibility and submit medicinal services misrepresentation.

Need of information transmission through the system has prompted the need of more noteworthy security to ensure data against various assaults. Numerous procedures are utilized to give security to data in transmission, for example, Image Encryption, Digital Watermarking, Fingerprinting, Cryptography, Digital Signature and Steganography, and so forth.

1.3 Picture Encryption or Image Encryption

It is a procedure of changing picture into either a picture which is confused or into an information of other arrangement having no reference to the first picture. The info picture to an encryption calculation is by and large alluded as a plain picture and the yield encoded picture is alluded as a figure picture. The encryption is a reversible procedure in which the encoded picture can be unscrambled to recoup the first picture. This should be possible just by the general population knowing about the calculation and the key utilized as a part of calculation to scramble the picture.

Encryption should be possible utilizing either an open key of the client or a mystery key of client, and in view of the key sort it is named

I. **Public Key encryption**: where the encryption is finished utilizing the general population key of sender and the decoding is finished utilizing the mystery key of the beneficiary.

II. **Private Key encryption**: where the encryption is finished utilizing the private key of sender and the unscrambling is finished utilizing the mix of mystery key of the sender and people in general key of recipient.

Picture encryption can be utilized to ensure pictures very still, for example, pictures put away on PCs or capacity gadgets which are being uncovered through misfortune or burglary of PCs. Picture encryption as rest shields them from being revealed and shared.

Keeping in mind the end goal to encourage mystery correspondence, picture encryption has discovered noteworthy place very touchy territories, for example, military observation, satellite data frameworks, medicinal services, private video conferencing, and so forth. A portion of the calculations utilize customary content based encryption plans to encode pictures specifically, however it is not an attainable thought because of distinction in size of content and picture, and consequently expends more measure of time.

Encryption of picture can be either a visual encryption wherein the scrambled picture will be an aftereffect of a rearranging system connected on the first picture which changes all or a noteworthy number of pixel positions in the first picture or it may be a change based plan where the picture pixels esteems will be modified in an approach to shape another picture with very surprising arrangement of pixel esteems. In the primary plan, changing the places of pixels encodes the picture just outwardly yet the properties of picture stay same for instance estimate, determination, histogram and so on, then again, in second approach changing the power esteems infers the adjustment in histogram by which the character of the picture is covered. The encryption calculations can take inputs either in square mode or in stream of information bits. Contingent upon the sort of info they can likewise be named piece figure and stream figure.

1.4 Picture Watermarking

Digital picture watermarking innovation is a rising field in software engineering, cryptography, flag preparing, and interchanges. Computerized Watermarking is proposed, by its engineers, as the answer for the need to give proprietorship assurance to the picture.

Picture watermarking is the way toward inserting information called a watermark (otherwise called Digital Signature or Tag or Label) into a picture with the end goal that watermark can be identified or separated at recipient side to make a statement about the picture. A straightforward case of advanced watermark would be an unmistakable "seal" set over a picture to distinguish the copyright of the picture. However the watermark may contain extra data including the character of the proprietor or buyer of a specific duplicate of the picture.

Picture watermarks can be isolated in two composes in view of the deceivability of the watermark in the watermarked picture

1. **Visible watermark**: A procedure where the watermark is installed in the first picture in a way that the watermark is obvious to bare eyes. The came about picture, here, is the first picture with noticeably "stamped" watermark on it.
2. **Invisible watermark**: A procedure where the watermark is installed in a way that it is perceptually undetectable to human visual framework and can be recognized just utilizing the plan utilized for implanting it. The came about picture has no visual contrasts with the first picture.

Visible watermarks are by and large utilized as a part of uses where just the possession evidence should be added to demonstrate the wellspring of picture, for example, photography magazines where picture visual quality debasement is passable, then again, the imperceptible watermarks are utilized as a part of profoundly touchy applications, for example, military applications, restorative applications, mystery administrations, and so forth., where the substance of the picture are exceptionally essential and no corruption of the picture quality is normal.

Contingent upon the level of security the watermark can be inserted in two unique spaces of the picture viz.

1. **Spatial area**: Gives the position guide of pixels in the picture.
2. **Frequency area**: Gives the power estimations of pixels in the picture.

By and large spatial area watermarking is utilized when just the proprietorship points of interest of the picture should be sent at low security level, then again recurrence space watermarking is utilized as a part of utilizations where alter location is critical at recipient and furthermore the pictures are subjected to high dangers of assaults in transmission.

The key attributes are as given beneath

i. Difficult to see: The undetectable watermark ought not be detectable to the watchers nor should the watermark debase the nature of the substance of picture.
ii. Robustness: Watermarked picture ought to be powerful to assaults for the most part changes and lossy compressions and others like scaling, editing, pivoting, and so forth.
iii. Tamper protection: Watermark ought to be impervious to altering proposed to evacuate the watermark and recuperate unique picture, any such endeavors ought to be reflected from the recouped watermark. Minimum adjustment of pixels: It must not modify countless with a specific end goal to protect an adequate visual quality.

1.5 Encryption and Watermarking

Image encryption gives high level of security, yet the authentication of innovation cannot be accomplished from the encryption as it was. Scrambled picture is secured until the point when it is in the encoded frame yet after decoding it can be assaulted either to shape new picture or to control the substance of picture. Utilizing watermarking together with encryption can resolve this issue. The idea of encryption and watermarking intends to join encryption and watermarking with each other either in a solitary stage or at two unique stages. We could first watermark the picture and after that encode it, or first scramble the picture and afterward watermark the scrambled picture. Contingent upon the strategies utilized the first picture can be unscrambled at beneficiary and watermark is removed. Both the unscrambling and extraction can be connected freely at collector or should be performed in arrangement characterized by the connected plan.

1.6 Steganography

Steganography is the specialty of concealing messages inside unsuspicious medium. The reason for steganography is to conceal the presence of a message from an outsider. Cryptography is generally utilized with steganography. Steganography works in two phases, inserting then extricating. Amid the implanting stage, a key is utilized to insert a message in a cover medium bringing about a stego-question. The stego-protest is then transmitted along open channels to its goal. At the point when the stego-question is gotten, the inserted message is removed from stegoobject utilizing the known stego-key. Steganalysis is the craft of finding and rendering such undercover message. It will probably abstain from attracting doubt to the transmission of a concealed message. Steganalysis contains two principle issues: Detection and Distortion. Disadvantage of Steganography is if the key is known then the information can be recovered effortlessly. Consequently, no security is accommodated the encoded picture. The scrambled picture can be secured by joining this steganography with visual cryptography and advanced watermarking.

2 Related Work

Number of specialists has proposed a various encryption methods having diverse highlights of each for secure picture encryption reason. This section portrays the investigation of a portion of the accessible writing identified with picture encryption and watermarking.

2.1 Techniques Used for Picture Encryption

Procedures utilized for Image Encryption AES encryption calculation is open, which brings numerous issues for its security. So to take care of the well-being issues of AES encryption calculation, Yang et al. [1] proposed an approach which enhances AES encryption calculation by methods for utilizing disorder hypothesis. Results demonstrate that the attainability and rightness of the approach are sufficient. Patel and Ragha [2] have introduced a steganography strategy that utilizations ideas of picture preparing and wavelet change systems. This procedure gives a technique to concealing mystery paired information inside a cover picture without expanding the size or dynamic scope of the picture by methods for joining cryptography and steganography. Steganography when joined with encryption gives a secured method for mystery correspondence between two gatherings. Picture is encoded by utilizing a scrambling calculation. Higher PSNR is accomplished. Jyoti and Neginal [3] presented new secure picture transmission method which makes an important mosaic picture and furthermore changes the mystery picture into a secret fragment-obvious mosaic picture of a similar size and has an indistinguishable visual appearance from the objective picture. Determination of target picture is adaptable and does not require any database creation. Unique mystery picture can be separated almost entire from the got picture. Tasneem and Bhavni [4] utilized cryptography and steganography strategies together to build the security of the information while transmitting through systems. Proposed approach utilizes content information as watermark. Before installing content into the picture it is scrambled by utilizing AES calculation with key of size of length 8. The scrambled content is installed into picture utilizing Discrete Wavelet Transform (DWT) strategy and the resultant picture is transmitted to the recipient. Watermark installed is reversible and information can be decoded at recipient. Saraf et al. [5] proposed a way to deal with consolidate of C code, Code Composer Studio and DSP processor for content encryption. Content encryption utilizes 128-piece size of key and in addition plaintext. 16 strings of size 8 are framed for encryption. AES encryption calculation in CFB mode with PKCS5 Padding technique is utilized here for picture encryption. Maitri et al. [6] have presented Byte Rotation Algorithm for record encryption and decoding with minim delay. This calculation enhances the security and decreases time for record encryption and unscrambling. Aftereffects of Byte Rotation Algorithm are practically identical with those accomplished from AES encryption calculation. To enhance record security irregular key age of 128-bit is utilized. Khandelwal and Sahu [7] proposed a story picture steganography system to stow away both picture and key in shading spread picture utilizing Discrete Wavelet Transform (DWT) and innately coordinate grouping in view of Fuzzy C-Means Clustering. Results demonstrate that there is no visual similitude between stego-picture and unique picture and the data can be recovered totally. Singh and Singh [8] introduced idea which utilizes 64-bits Blowfish Algorithm, which is intended to expand security and to enhance execution. Calculation is utilized with variable key size up to 448 bits. It utilizes Feistel organize which repeats straightforward capacity 16 times. The blowfish calculation is protected against unapproved assault and runs quicker than the well-known existing calculations. Higher number of rounds makes the approach more secure. Since Blowfish has no known security feeble focuses so far it can be considered as a phenomenal standard encryption calculation. Banu and Velayutham [9] have broken down AES Encryption calculation to give adequate levels of security to ensuring the privacy of the information. A rest booking technique is investigated to diminish the postponement of alert telecom from any sensor hub in WSNs. In particular, here two decided activity ways are utilized for the transmission of caution message alongside level-by-level counterbalance based wake-up design. On a basic occasion a caution is immediately transmitted along one of the movement ways to a middle hub, and afterward it is instantly communicated by the inside hub along another way without crash. Nurhayati and Ahmad [10] completed research to plan a use of steganography utilizing Least Significant Bit in which the message is scrambled utilizing the Advanced Encryption Standard calculation (AES). The consequence of research demonstrates the steganography is required to conceal the mystery message, so the message is not anything but difficult to know other individuals who are not qualified. Tang et al. [11] accompanied another picture encryption and steganography plot. To start with, the mystery message is encoded through the mix of another dark esteem substitution task and position stage which makes the encryption framework solid. And after that the prepared mystery message is covered up in the cover picture to satisfy steganography. Trial comes about demonstrate that the plan proposed in this paper has a high security level and better picture stego-picture quality.

Rad and Hosseini [12] proposed Grid-Based Hyper Encryption application that uses the computational assets of numerous work area PCs keeping in mind the end goal to scramble/decode substantial information documents with a standout amongst the most effective and secure encryption calculations, the propelled encryption standard. Effortlessness of utilization and elite of GBHE settles on it a perfect decision for conveying in any association that needs this usefulness.

2.2 Procedures Utilized for Medical Image Watermarking

Balamurugan and Senthil [13] consolidated three distinctive research areas in particular unique finger impression biometric, cryptosystem, and reversible watermarking. Proposed framework utilizes the unique mark biometric for validation, symmetric and also open key for cryptography process for private information and reversible watermarking for honesty. To give CIA strategy to the MDBMS, compliance SMS is sent to the patient for the helpful, that their data had achieved securely to the comparing goal. Nemade and Kelkar [14] proposed a reversible watermarking strategy for hued restorative pictures utilizing Histogram moving technique the shaded pictures is changed over to YUV picture and watermark is implanted in the Luminance (Y) segment. Histogram move strategy is likewise assessed by applying the Technique on RGB parts. Kaur and Madanlal [15] proposed a visually impaired picture watermarking calculation which in light of both the Modified Fast Haar Wavelet Transform and the Redundant Second Generation Wavelet Packet Transform. The thought behind the proposed calculation is to break down the cover picture utilizing Modified Fast Haar Wavelet Transform and Redundant Second Generation Wavelet Packet Transform as indicated by the measure of the watermark. The watermark is installed in the fine-scale groups of the Redundant Second Generation Wavelet Packet Transform of the fine-scale groups of the last Modified Fast Haar Wavelet Transform deterioration level of the host picture. Paul and Sunitha [16] proposed another technique for expanding the limit and protection of watermarking in light of round histogram tweak by taking the host framework as shading picture. By expanding the limit it can be helpful for secure transmission of tremendous databases like healing facility records through the system. Anandkumar and Mukeshgupta [17] created two calculations for implanting and extraction process. The first picture is first DWT-decayed, and after that the shading watermark is embedded. Watermark implanting is connected in various recurrence groups of the picture and PSNR and NC plots are drawn for all the recurrence groups. Bolandt and Kuannitlh [18] proposed watermarking plan in light of square DCT WT- and FFT-based calculation. The approach watermarks are intended to be imperceptible even to a cautious spectator yet contain adequate data to interestingly distinguish both the root and planned beneficiary of a picture with a low likelihood of blunder. The further improvement of powerful mistake revision codes and computerized signature methods is to be finished. Another approach of watermarking is proposed by Chang and Lu (2008) [19] which depend on installing watermark into dim level pictures as per pixel connection between them. The creator guarantees the way to deal with be lossless as far as nature of de-watermarked picture. Approach is fit for concealing expansive measure of data and can reestablish the first picture with no misfortune, and this is appeared by comes about. Likewise it has been guaranteed that the plan beats RS-installing plan and Tian's plan and consequently can essentially be connected. Shukla [20] proposed continuous copyright security calculation utilizing both obvious and undetectable watermarking plans and furthermore execution of constant picture handling procedures on Android and Embedded Platform. The idea depends on DCT and OpenCV. Pictures caught from the Smart-telephone camera are effectively watermarked utilizing this framework. The creators guarantee that DCT gives a productive strategy to executing imperceptible watermarking process. Obvious watermarking includes inserting content watermark which is strong to basic picture handling activities. Mathon et al. [21] utilized bending improvement for secure spread range watermarking for grayscale pictures. This approach of contortion minimization depends on components of transportation hypothesis and henceforth claims to accomplish solid security properties and is called as Transportation Natural Watermarking. Here, the multiresolution picture is disintegrated and joined with a multiplicative installing at dispersion level. Installing contortion and its visual effect on picture can be decreased utilizing the proposed approach and this can be found in comes about.

Peng [22] proposed a square based particular esteem decay of the picture and the discrete system strategy to install and separate watermark to and from the advanced picture. It has tackled particular esteem disintegration issue of poor heartiness. The idea depends on the neural system and solitary esteem disintegration. The creator guarantees well extraction all through the normal picture preparing and packing tasks. Kamran and Farooq [23] proposed an approach of watermarking giving great nature of picture after watermark extraction. Creator utilized component extraction and positioning to watermark pictures. The utilization of highlight positioning adds to safeguarding of data contained inside the picture. To make and implant watermark to unique picture, Particle Swarm Optimization Algorithm is utilized. The proposed plot is guaranteed to be flexible to an assortment of assaults and is affirmed by comes about demonstrated Kishor and Vankat [24] proposed a RSA-DWT-based restorative picture watermarking. Open key encryption of watermark is done before inserting watermark into the picture. This encoded watermark is then implanted into the picture utilizing DWT. Results demonstrate agreeable estimations of PSNR and zero ability to see of watermark. Dragoi and Coltuc [25] proposed a watermarking plan which depends on neighborhood expectation to accomplish contrast extension reversible watermarking. Here, a minimum square indicator is figured for every pixel on a square piece, focused on the pixel and the expectation blunder relating to that pixel is extended. This expectation mistake is then recouped at beneficiary with no extra data. Forecast setting relies upon the span of squares utilized as a part of extension. Fitting square sizes for every setting are resolved and are given in comes about. Results got by Rhombus setting are the best among the various settings. Su et al. [26] presents an approach of utilizing Scale Invariant Feature Transform and broadened pilot signals for watermarking. This approach tends to the issue of synchronization in watermark discovery at beneficiary. The watermarking signal is inserted in the invariant locales which are shaped by applying interest point extraction as scale invariant highlight change. Square frameworks are utilized to install flag. This approach accomplishes right harmony between effectiveness of flag recognition and heartiness of watermark. Installed flag are guided by pilot flag contributing towards synchronization in discovery. At discovery when a similar intrigue point is separated, the related relative parameters of matrices are acclimated to distinguish concealed pilot flag accomplishing synchronization and effective recovery of watermark and unique picture. Panyavaraporn [27] concocted an approach of utilizing wavelet change to watermark QR code pictures. The parallel watermark is implanted into a chosen sub-band which can be both of LH, HL, or HH sub-groups. The outcomes demonstrate that the plan can withstand fundamental assaults acceptably yet is not vigorous against assaults, for example, solid clamor, high-pressure geometric change and impediment and thus restricts the execution of the plan. Abdallah et al. [28] presented wavelet-based watermarking which does not require unique picture or any pilot picture to recreate watermark at decoder. The watermark is installed into the coarsest scale wavelet coefficient. Decay up to level 3 and watermark of size equivalent to sub-band is utilized. Plan utilizes quantization of wavelet coefficients in parallel way. Unlike customary wavelet-based strategies, this technique has less corruption.

Lin and Wu [29] proposed a way to deal with watermark 3D pictures. Here numerous watermarking is utilized to take care of substance insurance issue of Depth Image Based Rendering (DIBR) 3D pictures. A 3D picture contains three pictures in itself viz. left-eye picture, right-eye picture, and a middle picture. The inside picture and profundity picture created by content supplier is called as (DIBR) 3D picture. Both the, common orthogonality and request of implanting assumes imperative part water-marking 3D pictures. The upside of the approach is that it need not bother with unique information and profundity picture amid watermark discovery. Results demonstrate that the approach is secure to pressure and commotion assaults and furthermore does not get influenced by run varieties and standard separation modification up to cer-tain degree. Zhang et al. [30] did examine for watermarking on Perceptual Quality Metric in view of second order insights. Abusing the perceptual twisting with human recognition is key for strong watermarking plan.

The proposed plan of SOS performs superior to some cutting edge measurements and connects with a few databases of pictures by methods for surface concealing impact and difference affectability in Karhunen–Loeve change space. Utilization of straightforward metric guarantees fat usage and results demonstrate that the hearti-ness is made strides. Creator likewise proposes association of outsider measurements to rate the nature of watermarked picture. Tang and Hang [31] proposed utilization of Mexican Hat wavelet scale communication strategy in light of highlight focuses implanting and extraction as watermarks.

This is accomplished utilizing picture standardization which is connected to non-covered pictures independently. A succession of 16-bit signals is implanted in unique picture enhancing strength of watermarking. Reference picture is not required at indi-cator to extricate watermark. The synchronization issue is overwhelmed by utilizing visual point on picture and invariance focuses lessen the watermark look space. Guer-rini et al. [32] proposed a strategy to watermark High Dynamic Range (HDR) pic-tures having high luminance esteems. The HDR pictures are tone mapped utilizing TM calculation and after that watermarked utilizing Quantization Index Modula-tion (QIM). Watermarking framework produced for LDR pictures is utilized here in LogLuv space. The discovery of watermark relies upon the TM calculation. Weak-ness is that the plan isn't completely visually impaired and choice of squares to insert information isn't conceivable.

2.3 Strategies Utilized for Image Security

Sundari et al. [33] thought of three-method watermarking, steganography and cryp-tography to give abnormal state of security. To begin with the cover picture is be packed utilizing JPEG pressure calculation, then the message to be sent is encoded utilizing RSA encryption calculation and later the scrambled message bits and the bits of watermark are installed into cover picture. Patel and Patel [34] proposed con-solidated technique of cryptography, steganography and computerized watermarking to stow away secure picture with watermark logo inside cover picture. For this reason they utilized DCT, DWT, SVD and RSA approach. Utilizing DCT, scrambled water-mark logo (encryption performed utilizing RSA) is covered up inside secure picture bringing about Stego-picture. Gayathri and Nagarajan [35] presented approach with data stowing away in picture utilizing Zig-Zag filtering design which is more mind boggling calculation in Steganography again encoded as offers.

The offer is implanted into the host picture utilizing Least Significant Bit Insertion Technique (LSB). The plan gives more secure and significant mystery shares that are strong against various assaults. Kishor et al. [36] proposed RSA-DWT-based therapeutic picture watermarking like MRI. Watermark picture is encoded with key created RSA calculation. Scrambled patient picture is utilized as payload which is installed into Medical picture in Wavelet space. Exploratory result demonstrates that RSA-DWT shows predominant security on unsecured network. Suganya et al. [37] proposed another joint watermarking and encryption framework which ensures from the earlier and a posteriori assurance of medicinal pictures. It blends the stream figure calculation or piece figure calculation with the QIM-based watermarking system. The proposed strategy utilizes two encryption calculations in particular RC4, AES are Stream figure, Block figure calculation separately. The framework offers access to embed two messages in the spatial area and encoded space individually amid encryption process. These two messages are utilized to confirm the unwavering quality of pictures in decoding part. Li and Bai [38] built up a calculation with computerized watermarking plan which utilizes a piece of sign succession of DCT coefficients as a component of vector pictures, at same time watermark is encoded by strategic guide to upgrade classification.

3   Proposed Methodology

This part gives the portrayal about the approach we have proposed. It comprises of the means engaged with proposed approach and the stream of the approach alongside information stream in the framework. Stream outlines for the approach are likewise recorded here.

Advancement in the web, today, has improved data sharing over the more extensive region. Both the visual and printed data of a patient can undoubtedly be transmitted the whole way across the world. In any case, the simplicity of data trade has presented security issues for the data being transmitted over the web, as the data can undoubtedly be stolen and wrongfully altered at any phase without having any personality of the individual.

Pictures being shared over the web can be ensured by utilizing some safety efforts like encryption and watermarking. Work proposed here spotlights on the mean to give security to pictures being shared over the web. The approach utilizes watermarking and encryption both to give confirmation information and visual security to restorative pictures. Watermark implanting is finished utilizing LSB substitution and encryption of watermarked picture and patients' information is encoded utilizing scrambling technique. The approach is as examined beneath.

The proposed work can be isolated into four stages as tail

    i. Installing watermark in the first picture.
   ii. Encryption of Watermarked picture.
  iii. Encryption of patients' data.
  iv. Decoding of patients' picture and extraction of watermark.
   v. Decoding of patients' data.

In first stage, the watermark is implanted into unique picture to shape a watermarked picture. Both the first picture and watermark are utilized as a part of the picture arrange. Information from other arrangement with the exception of picture cannot be inserted in the first picture.

In second stage, watermarked picture in scrambled utilizing a rearranging calculation talked about later. An encryption is likewise produced in this same stage which is utilized for decoding moreover.

At that point in third stage the data of patient is encoded utilizing Byte Replacement Encryption Algorithm (BREA). All characters from the patients data graph are changed over encoded utilizing a key framed utilizing arbitrary dissemination of numbers.

At last at the recipient, got encoded watermarked picture is unscrambled utilizing the same keys utilized for encryption first and the first watermark is separated from the decoded picture. Removed watermark can be utilized to identify the legitimacy and innovation of picture. Implanting Watermark Watermarking is the way toward concealing a mystery picture or message into the first picture so as to give genuineness to the picture. Message picture can be installed in the first picture utilizing pixel substitution strategy wherein one of the bit planes of the first picture is supplanted utilizing the message picture. Pixels in the picture have 8-bit esteem speaking to their shading force. In the event that the Least Significant Bit (LSB) of these 8-bits is modified the aggregate estimation of the pixel will change just by 1. For instance, changing LSB of pixel with esteem 128 from 1 to 0 will change the incentive to 129 and outwardly a human eye can't recognize such a little distinction.

Watermark, in the proposed approach is of paired frame and is inserted by methods for supplanting the LSBs of the first picture. To start with the first LSB plane of the picture is made 0, at that point the double watermark is added to this LSB plane. LSB watermarking is one of the most straightforward strategies used to watermark pictures. It is a result of the low intricacy and high recuperation proficiency of the watermark. Watermark can simply be recuperated with no misfortune from a protected picture.

3.1 Encryption of Watermarked Picture

Watermarked picture from stage one is scrambled in this second stage. Point of encoding picture is to secure picture against stealth and control assaults. This is on account of even a slight control in the patients' therapeutic picture can cause him his life. Encryption is the way toward making an interpretation of picture into incoherent data spoke to in either picture frame or a content shape.

Approach proposed in this work utilizes disintegration of pixel esteems with a specific end goal to rearrange their positions. Encryption process, proposed here, rearranges every one of the pixels in the first picture and furthermore changes their unique pixel force esteems with a specific end goal to adjust the histogram of the last encoded picture.

Encryption calculation can be partitioned into a few phases. To reposition a pixel rest of the division of unique incentive by 10 is utilized. Not the genuine esteem but rather the remainder of division is set at the position indicated by the rest of.

Stage 1: In initial step unique pixel esteem is partitioned by 10 and both leftover portion and remainder are recorded in frameworks Q and R individually.

Stage 2: another lattice with same number of lines as that of unique picture and with sections 10 times as that of the first picture is framed, to such an extent that every one of the components take arbitrary numeric esteems from 0 to 255.

Stage 3: First component in the Q is set at position spoke to by first component of R, in new lattice. Second component from Q is put at the position spoke to by second element+10 of R.

Stage 4: Perform stage 3 for all components in first line of Q.

Stage 5: Repeat stage 4 for all sections of Q.

Stage 6: Finally change over every one of the components from scrambled lattice into their ASCII structures to accomplish a content document with encoded picture information. Components of this content document are then mapped in the scope of 0–255 by methods for utilizing their separate ASCII codes to get the last encoded picture. Here, the network with leftover portion esteems, i.e., R goes about as a key for encryption and is required to decode the scrambled picture. Figure demonstrates the picture encryption process in detail.

3.2 Encryption of Patients' Information

Patient information is contribution to the type of content. This content information is encoded utilizing Byte Replacement Algorithm. BREA calculation encodes content information examined in squares and each piece is then scrambled in parallel way. A key is produced for encryption which is additionally utilized for unscrambling moreover. BREA tests input information into strings of size 16 characters. Each such string is then reshaped into a lattice of size 4 * 4. Every one of the components are then supplanted by their equal ASCII codes. Another grid, a key framework, is produced by methods for haphazardly choosing all components remarkably from 1 to 26. Components of key network are then scaled by methods for taking mod by 2. This key network is then included with each square component by component. At that point, first component of first line is turned toward the finish of column by moving all components to left. Initial two components of second column are pivoted by moving different components towards left lastly initial three components of third line are turned toward the end. Fourth line is kept unaltered. Likewise, first component of first section is pivoted toward the end by moving different components towards up. Initial two components of second section are turned toward the end lastly initial three components from third segment are pivoted towards drawback. Fourth segment is kept unaltered.

4 Results and Discussions

Following are the depictions of the outcomes for the proposed approach. Figure 1 speaks to the real literary information entered by the patient or an agent, encoded information accomplished from the Byte Replacement Encryption Algorithm and the decoded data acquired utilizing reverse BREA.

From Fig. 2, the data entered was "Abc Def Xyz 123 789", a blended string of capitalized letters, bring down case letters, clear spaces and numbers. Encoded string accomplished is g "2bX2c z4!Dz!Be 8 9 8". It can be obviously observed that encoded does not look like with the first content specifically and is difficult to get it. It additionally does not have any standard significance.

Fig. 1 Encryption of watermarked image flow diagram

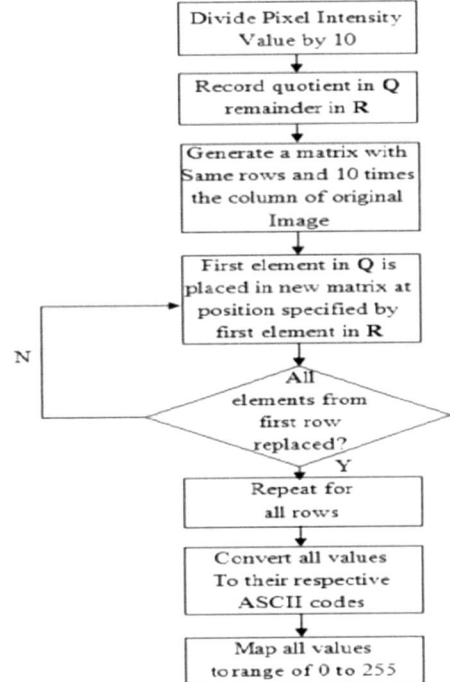

```
Command Window                                                                    ⊣ □ ◢ ✕
  Enter the Patients information
  Abc Def Xyz 123 789
  Your message
  Abc Def Xyz 123 789
  Encrypted Information

  eno_mat =

  g 2bX2o z4!Dz!Be    8  9        8

  Decrypted Information

  str_mat =

  Abc Def Xyz 123 789
```

Fig. 2 Original, encrypted and decrypted textual information

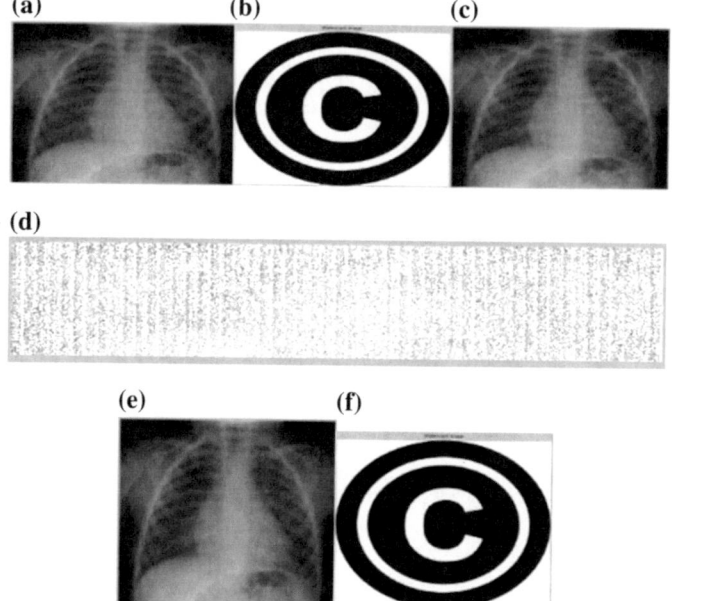

Fig. 3 a–f: Original X-Ray image, watermark image, watermarked image, encrypted image, decrypted image and extracted watermark

Unique X-Ray picture, watermark picture, watermarked picture, encoded picture, unscrambled picture and separated watermark Fig. 3 demonstrates the aftereffect of finish procedure of encryption and watermarking on picture alongside decoding and extraction comes about. It can be seen from figure (c), (d) and (e) that watermen picture can be scrambled without leaving any perceptual connection with unique picture. Likewise the expanded size presents bigger number of conceivable outcomes for false pixel repositioning decreasing the assault probability and the same scrambled picture can be decoded with no misfortune.

Again on account of the altogether changed pixel esteems histogram of unique picture does not co-relate with encoded picture PSNR Analysis Higher PSNR esteems

Table 1 PSNR analysis of images

Image	PSNR
USG image	33.5732
X-Ray image	33.5716
Brain image	34.5979
MRI image	33.6976
Kidney image	34.6006

(a)

(b)

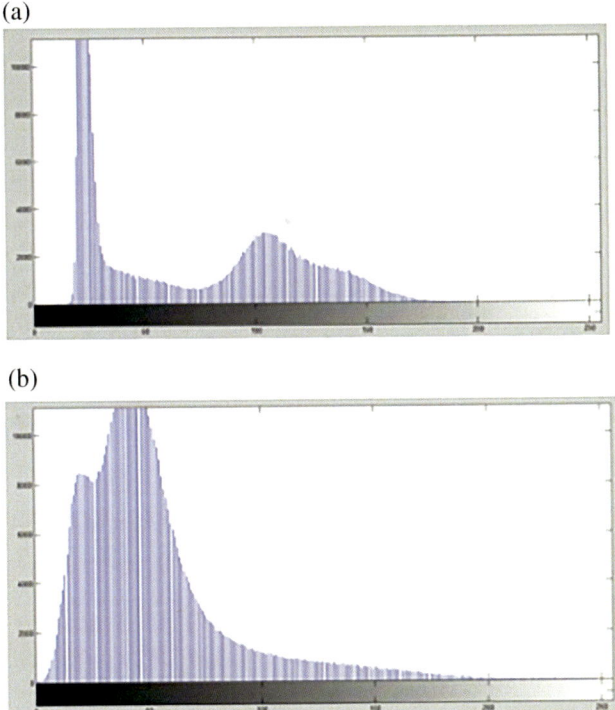

Fig. 4 **a, b** Histogram of original Sonography image and histogram of encrypted Sonography image

are wanted for better outcomes. Table 1 shows estimations of PSNR for Rib Cage Image when experienced Salt 'n' Pepper assault.

It can be plainly noted from Fig. 4 that encryption makes the picture subtle as well as changes the histogram which decreases the odds of histogram assaults, or endeavors of recovering picture from histogram.

5 Conclusion

In proposed plot watermarking of picture is done before encryption. Due to earlier watermarking the encryption is given more concentration in the approach. We have utilized LSB watermarking procedure, a simple to execute and viable, for inserting the watermark into the cover picture. Furthermore, to encrypt the picture we have a repositioning calculation which rearranges the pixels in view of the aftereffect of a division. It is then comprehended from the outcomes that the proposed approach has prevailing to watermark and scramble picture and the recovery of picture and watermark is likewise conceivable with no misfortune. An important point is that the unscrambled picture keeps up high visual quality. Additionally, the removed watermark is similar to the first one inserted and consequently can be utilized for validation and copyright assurance.

References

1. Yang Z-H, Li A-H, Yu L-L, Kang S-J, Han M-J, Ding Q (2015) An improved AES encryption algorithm based on chaos theory in wireless communication networks. In: Third international conference on robot, vision and signal processing
2. Patel K, Ragha L (2015) Binary image steganography in wavelet domain. In: International conference on industrial instrumentation and control (ICIC)
3. Jyoti RH, Neginal J (2015) A new secure image transmission technique via secret-fragment-visible mosaic image. Int J Adv Res Comput Eng Technol (IJARCET) 4(5)
4. Tasneem SF, DurgaBhvni S (2014) Secure data transmission using cryptography and steganography. Int J Emerg Trends Electr Electron 10(9)
5. Saraf KR, Jagtap VP, Mishra AK (2014) Text and image encryption decryption using advanced encryption standard. Int J Emerg Trends Technol Comput Sci (IJETTCS) 3(3)
6. Maitri PV, Sarawade RV, Deokate ST, Patil MP (2014) Secure file transmission using byte rotation algorithm in network security. In: International conference for convergence of technology
7. Khandelwal P, Sahu B (2014) Novel technique data-hiding scheme for digital image
8. Singh P, Singh K (2013) Image encryption and decryption using blowfish algorithm in MATLAB. Int J Sci Eng Res 4(7)
9. Jesima Banu A, Velayutham R (2013) Secure communication in wireless sensor networks using AES algorithm with delay efficient sleep scheduling. In: 2013 IEEE international conference on emerging trends in computing, communication and nanotechnology (ICECCN 2013)
10. Buslim N, Ahmad SS Steganography for inserting message on digital image using least significant bit and AES cryptographic algorithm
11. Tang H, Jin G, Wu C, Song P (2009) A new image encryption and steganography scheme. In: 2009 international conference on computer and communications security
12. Rad NB, Shah-Hosseini H (2008) GBHE: Grid-based cryptography with AES algorithm. In: 2008 international conference on computer and electrical engineering
13. Balamurugan G, Senthil M (2016) A fingerprint based reversible watermarking system for the security of medical information
14. Nemade H, Kelkar V (2016) Reversible watermarking for colored medical image using histogram shifting method
15. Kaur S, Lal M (2015) An invisible watermarking scheme based on Modified Fast Haar Wavelet Transform and RSGWPT. IEEE

16. Paul A, Sunitha EV (2015) Distortion less watermarking of relational databases based on circular histogram modulation. In: 2015 international conference on circuit, power and computing technologies
17. Kumar A, Gupta M (2015) Semi visible watermarking scheme based on DWT and PCA. IEEE
18. Boland FM, Kuannitlh JJ (2007) Watermarking digital images for copyright protection. Image Processing and Its Applications, 4–6 July 2007
19. Chang C-C, Lu T-C (2007) Noise features for image tampering detection and steganalysis. IEEE
20. Shukla RJ (2014) Platform independent real time copyright protection embedding and extraction algorithms on android and embedded framework. IEEE
21. Mathon B et al (2011) A parametric solution for optimal overlapped block motion compensation. IEEE
22. Peng J (2011) The research on digital watermarking algorithm based on neural networks and singular value decomposition. IEEE
23. Kamran M, Farooq M (2009) A new spatial decomposition scheme for image content-based watermarking. IEEE
24. Kishor PP, Yankat N (2014) Medical image watermarking using RSA encryption in wavelet domain. IEEE
25. Dragoi I-C, Coltuc D (2014) Local-prediction-based difference expansion reversible watermarking. IEEE Trans Image Process 23:1779–1790
26. Su P-C, Wu C-Y et al (2013) Geometrically resilient digital image watermarking by using interest point extraction and extended pilot signals. IEEE Trans Inf Forensics Secur 8:1897–1908
27. Panyavaraporn J (2013) QR code watermarking algorithm based on wavelet Symposium transform. In: 2013 13th international symposium on communications and information technologies (ISCIT), Sept 2013
28. Abdallah HA, Hadhoud MM, Abdalla Hameed (2011) Blind wavelet based image watermarking. Int J Signal Process Image Process Patter Recogn 4(1)
29. Lin Y-H, Wu J-L (2011) A digital blind watermarking for depth-image-based rendering 3D images. IEEE Trans Broadcast 57:602–611
30. Zhang F et al (2011) Spread spectrum image watermarking based on perceptual quality metric. IEEE Trans Image Process 20:3207–3218
31. Tang C-W, Hang H-M (2003) A feature-based robust digital image watermarking scheme. IEEE Trans Signal Process 51:950–959
32. Guerrini F, Okuda M, Adami N, Leonardi R (2011) High dynamic range image watermarking robust against tone-mapping operators. IEEE Trans Inf Forensics Secur 6:283–295
33. Sundari M, Revathi PB, Sumesh S (2015) Secure communication using digital watermarking with encrypted text hidden into an image
34. Patel P, Patel Y (2015) Secure and authentic DCT image steganography through DWT—SVD based Digital watermarking with RSA encryption. In: 2015 fifth international conference on communication systems and network technologies
35. Gayathri R, Nagarajan V (2015) Secure data hiding using Steganographic technique with Visual Cryptography and Watermarking Scheme. In: IEEE ICCSP 2015 conference
36. Kishor PV, Venkatraman N, Reddy SS (2014) Medical image watermarking using RSA encryption in wavelet domain. IEEE
37. Suganya G, Amudha K (2013) Medical image integrity control using join encryption and watermarking techniques. IEEE
38. Dong L, Bai H (2012) Medical image watermarking algorithm with encryption by DCT and logistic. IEEE
39. Farah T, Hermassi H, Rhoouma R, Belghith S (2012) Watermarking and encryption scheme to secure multimedia information. IEEE
40. Khan MI, Jeoti V, Malik AS, Khan MF (2011) A joint watermarking and encryption scheme for DCT based codecs. In: 2011 17th Asia-Pacific conference on communications (APCC)
41. Bouslimi D, Coatrieux G, Roux C (2011) A joint watermarking/encryption algorithm for verifying medical image integrity and authenticity in both encrypted and spatial domains

A Spectral Approach for Segmentation and Deformation Estimation in Point Cloud Using Shape Descriptors

Jajula Kalyani, Karthikeyan Vaiapury and Latha Parameswaran

Abstract In this paper, we propose a new framework for segmentation and deformation estimation in texture-less point clouds. Given a reference point cloud and a corresponding deformed point cloud, our approach first segments both the point clouds using OBB-LBS (Oriented Bounding Box-Laplace Beltrami Spectral) and estimates the semi-global dense spectral shape descriptors. These coarse descriptors identify the segments which need to be further investigated for localizing the area of deformation at a finer level.

1 Introduction and Related Work

Deformation detection and localization are quite useful in many real-world applications such as for quality check, maintenance of aircrafts, automobiles, town planning in smart cities, and surgery in medical domain. For instance, if a 3D model of an object is given and in another if the scan of the object is deformed in some locations, then detecting and estimating that change can lead to a fast quality inspection.

In this proposed approach we have used Laplace Beltrami Spectral (LBS) signatures as they are invariant to isometry. Isomerism exists in 3D models when there are similar structures on either side and it has been widely studied. In this work, we leverage a combination of OBB estimation and quantum inspired algorithms such as spectral wave kernel signatures to estimate the deformation.

J. Kalyani · K. Vaiapury
TCS Innovation Labs, Amrita School of Engineering, Amrita Vishwa Vidyapeetham,
Coimbatore, India
e-mail: cb.en.p2cvi15002@cb.students.amrita.edu

K. Vaiapury
e-mail: karthikeyan.vaiapury@tcs.com

J. Kalyani · L. Parameswaran (✉)
Department of Computer Science and Engineering, Amrita School of Engineering,
Amrita Vishwa Vidyapeetham, Coimbatore, India
e-mail: p_latha@cb.amrita.edu

© Springer Nature Switzerland AG 2019
D. Pandian et al. (eds.), *Proceedings of the International Conference on ISMAC in Computational Vision and Bio-Engineering 2018 (ISMAC-CVB)*, Lecture Notes in Computational Vision and Biomechanics 30,
https://doi.org/10.1007/978-3-030-00665-5_41

409

Various algorithms are explained for change detection in 3D and 2D space in the literature. A systematic survey of change detection algorithms of 2D space like pixel-based-, object-based-, shading model, predictive models, and hypothesis testing are explored in [1, 2] and for 3D space methods like detecting a new object in the scene and detection of change existing model are explored in [3–6]. The demerits in 2D space like illumination, temporal, and spatial constraints, shadows, atmospheric conditions can be excluded if the change detection is done in 3D space. The challenges in 3D space are due to registration errors, density difference between target and source point clouds, irregularities in the point cloud, sensitivity of the detection method to small changes and sensor noise. Various domains such as aircraft [7], building [8] and medical [9] are explored for change detection in 3D. In [10] the authors have extracted Gabor wavelet-based features for face recognition. These features can also even be extended to detecting changes in images. In [11] the authors have used Independent Component Analysis (ICA) features for authenticating images by watermarking. A similar approach can be tried to extract block-wise changes.

To handle computational complexity tradeoff, in [12], a new oriented bounding box (OBB) has been proposed for extracting regional area descriptor and has been used for 3D point cloud registration when no coordinate references exist. These techniques can only capture global surface properties and cannot handle dataset with isometric or topological changes which spectral signatures can capture in a succinct manner. In this article, we propose a OBB and LBS based approach for detecting changes in 3D point cloud.

2 Proposed Work—Deformation Detection and Localization Framework

The proposed framework given in Fig. 1 consists of three key components. In the first component, the estimation of OBB is done on both individual point cloud of source and target data using covariance-based method and is partitioned into segments via sub-blocks. The salient partitioned blocks are fed to the second component to get the LBS spectral signatures. These signatures are fed into the third component where the interpolation of signatures is done to estimate the deformation and visualization of point cloud with deformation.

A. *Estimation of OBB and Partitioning*

This proposed approach involves estimation of OBB for both source M and target point cloud M' analogous to Chen et al. [12] based on covariance-based method and partitioning of cloud into sub-blocks.

The estimation of OBB is done as given below

1. Let p represent the number of vertices in a point cloud given as $X = (a, b, c)$
2. Calculate centroid of the point cloud

Fig. 1 Framework for
deformation estimation and
segmentation

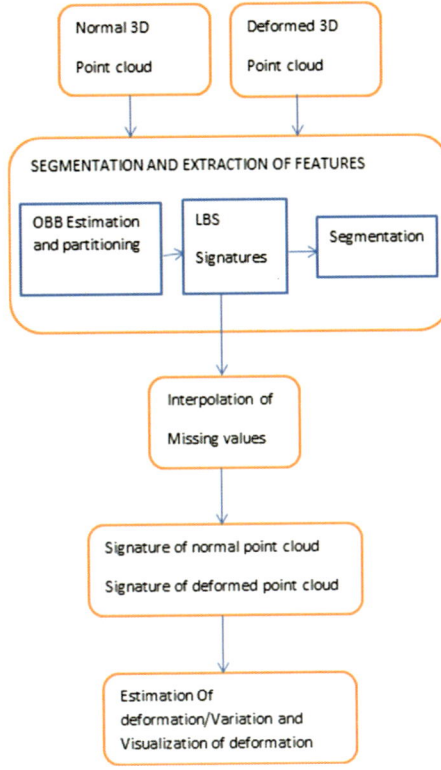

$$\mu = (\mu_a, \mu_b, \mu_c)$$

$$\mu_a = \sum_{j=0}^{p} a_j \mu_b = \sum_{j=0}^{p} b_j \mu_c = \sum_{j=0}^{p} c_j$$

3. Calculate the covariance matrix for the point cloud

$$\text{Covariance} = \frac{1}{p} X(X)^T$$

4. Calculate eigenvalues and eigenvectors for covariance matrix and sort them as
   $\lambda_1 \geq \lambda_2 \geq \lambda_3$ with their respective eigenvectors $v_1 \geq v_2 \geq v_3$
5. OBB represented by $c = (a_c, b_c, c_c)$ and cc_1, cc_2, cc_3 is calculated using the
   following equations

$$a'_j = v_{11}(a_j - \mu_a) + v_{12}(b_j - \mu_b) + v_{13}(c_j - \mu_c)$$

$$b'_j = v_{21}(a_j - \mu_a) + v_{22}(b_j - \mu_b) + v_{23}(c_j - \mu_c)$$

$$c'_j = v_{31}(a_j - \mu_a) + v_{32}(b_j - \mu_b) + v_{33}(c_j - \mu_c)$$

$$a'_{min} = \min\left\{a'_j, \forall j = 1, 2, \ldots p\right\}$$

$$b'_{min} = \min\left\{b'_j, \forall j = 1, 2, \ldots p\right\}$$

$$c'_{min} = \min\left\{c'_j, \forall j = 1, 2, \ldots p\right\}$$

$$a'_{max} = \max\left\{a'_j, \forall j = 1, 2, \ldots p\right\}$$

$$b'_{max} = \max\left\{b'_j, \forall j = 1, 2, \ldots p\right\}$$

$$c'_{max} = \max\left\{c'_j, \forall j = 1, 2, \ldots p\right\}$$

6. OBB parameters are computed as:

$$a_c = \mu_a + v_{11}a'_{min} + v_{21}b'_{min} + v_{31}c'_{min}$$
$$b_c = \mu_b + v_{12}a'_{min} + v_{22}b'_{min} + v_{32}c'_{min}$$
$$c_c = \mu_c + v_{13}a'_{min} + v_{23}b'_{min} + v_{33}c'_{min}$$
$$cc_1 = (a'_{max} - a'_{min})v_1 \quad cc_2 = (b'_{max} - b'_{min})v_2$$
$$cc_3 = (c'_{max} - c'_{min})v_3$$

7. Finding corners from the above parameters is tricky $c_1 = cc_1 - c$ and based on dataset it may be $c_1 = cc_2$

$$d_1 = cc_{1a} \sim \mu_a$$
$$d_2 = cc_{1b} \sim \mu_b$$
$$d_3 = cc_{1c} \sim \mu_c$$

8. Corners are computed using

$$c_2 = (c_{1a}, \mu_b + d_2, c_{1c})$$
$$c_3 = (c_{1a}, c_b, \mu_c + d_3)$$
$$c_4 = (c_{1a}, \mu_b + d_2, c_{1c} + d_3)$$
$$c_5 = (\mu_a + d_1, c_{1b}, c_{1c})$$
$$c_6 = (\mu_a + d_1, \mu_b + d_2, c_{1c})$$
$$c_7 = (\mu_a + d_1, c_{1b}, c_{1c} + d_3)$$
$$c_8 = (\mu_a + d_1, \mu_b + d_2, c_{1c} + d_3)$$

3   Experimental Results

This proposed algorithm has been implemented on sample point clouds for exper-imentation. Using the above steps, OBB corners for a sample cloud of aircraft and the results are shown in Fig. 2.

Let the corners of a sample cloud be $(0, 0, -200)$, $(2000, 0, -200)$, $(2000, 0, 200)$, $(0, 0, 200)$, $(0, 1000, -200)$, $(2000, 1000, -200)$, $(0, 1000, 200)$, $(2000, 1000, 200)$; then the partition of cloud can be done by calculating the midpoints of corners and the sub-blocks of partition is shown in Table 1. The horizontal partition of the cloud is done by taking the midpoint of z_{min} and z_{max} of aircraft corners.

After obtaining the OBB corners of aircraft from the covariance-based method, the midpoints of corners for partition are calculated and with the reference of Table 1, the horizontal and vertical partition is done and the results are shown in Fig. 3.

B.   *Laplace Beltrami Spectral Analysis*

The divergence of the gradient of manifold f is known as Laplace Beltrami [13]. A manifold is known as a topological space which is locally Euclidean. Generally, any object which is as "flat" as possible when a small space is taken is called as manifold. The Eigen function of Laplace Beltrami differential operator captures the surface global properties. In the beginning, for shape descriptors the Eigen values of this operator were used but as they are unable to differentiate iso-spectral shapes a new signature has been proposed based on nodal counts of Eigen functions. The cropped set of eigenvalues of Laplace Beltrami is called as "Shape-DNA [14]". Poten-tial applications of these Eigen functions, as exemplified include signal processing on surfaces, and geometry. The authors in [16] and [17] have discussed Laplace

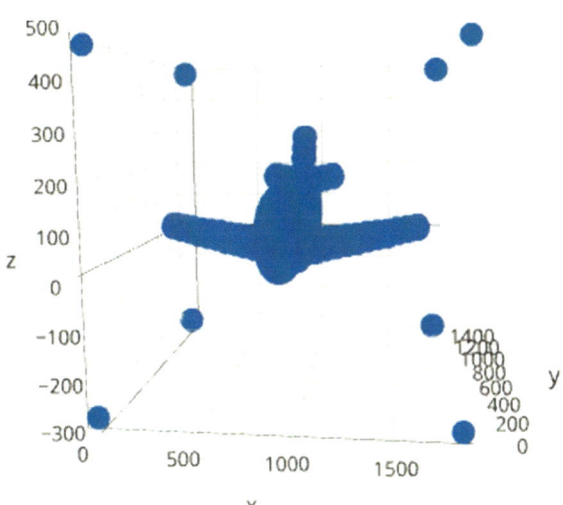

Fig. 2   OBB corners with aircraft cloud

Table 1 Partition block table sample H—Horizontal, V—Vertical, Q—Quad and O—Oct B—Block [12]

Block	x_{min}	x_{max}	y_{min}	y_{max}	z_{min}	z_{max}
HB_1	0	2000	0	1000	0	200
HB_2	0	2000	0	1000	−200	0
VB_1	0	1000	0	1000	−200	200
VB_2	1000	2000	0	1000	−200	200
QB_1	0	1000	0	1000	0	200
QB_2	1000	2000	0	1000	0	200
QB_3	0	1000	0	1000	−200	0
QB_4	1000	2000	0	1000	−200	0
OB_1	0	1000	0	500	0	200
OB_2	0	1000	500	1000	0	200
OB_3	1000	2000	0	500	0	200
OB_4	1000	2000	500	1000	0	200
OB_5	0	1000	0	500	−200	0
OB_6	0	1000	500	1000	−200	0
OB_7	1000	2000	0	500	−200	0
OB_8	1000	2000	500	1000	−200	0

(a)   (b)   (c)   (d)   (e)

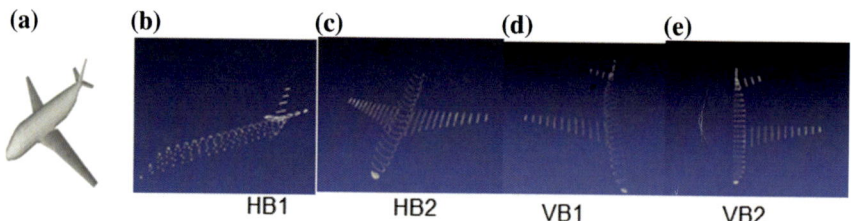

HB1   HB2   VB1   VB2

Fig. 3 **a** Aircraft model, **b** horizontal partitioned cloud HB1, **c** horizontal partitioned cloud HB2, **d** vertical partitioned cloud VB1, **e** vertical partitioned cloud VB2 (better viewed in color)

Beltami based shape extraction techniques for shape extraction and matching. These algorithms are a good basis for shape signature design.

We chose LBS because it is quite powerful due to its interesting properties such as invariant to scaling, rotation, and is also Isometric invariant with additional property as topology preserving capability. The principal direction of the surfaces of interest and geometrical information, such as localization of distension are provided by first and second Eigen function of Laplace Beltrami operator [15]. Isometric invariant holds since Eigen functions depend specifically on the gradient and divergence which are dependent on the Riemannian structure of the manifold. Similarly since Eigenfunctions are normalizable, scaling factor is also taken care of. Several variants of Beltrami based signatures are available in literature [13] processing, pose transfer, and parameterization. Another source of inspiration is where the eigenvalues of the same operator were used as a shape descriptor.

Let M be a normal manifold then Laplace Beltrami of M is given by [13]

$$\nabla_y f(y) = \text{div}(\text{grad}(f(y)))$$

The Laplacian can be calculated as the difference between the average of $f(y)$ and f on an infinitesimal sphere around f. For construction of shape descriptor several authors have used the Eigen functions of Laplace Beltrami. A general descriptor for this kind for Q-dimension is

$$f(y) = \sum_{i \geq 0} \tau(\lambda_i \varphi^2(y)) \approx \sum_{i=0}^{l} \tau(\lambda_i)\varphi^2(y)$$

where $y_{i=0}^{l} \in M$, λ is eigenvalue of Laplace Beltrami, φ is Eigen function of Laplace Beltrami, $\tau(\lambda_i)$ is a transfer f function computed using Eigen values. These descriptors are re-computed for every vertex of point cloud and are computed efficiently by using l (small number) of Laplacian Eigen values and Eigen functions. It may be observed that the descriptor changes for every change of transfer function.

Heat Kernel Signature (HKS): This is a special case of LBS [13] where

$$\tau(\lambda) = e^{-t_0 \lambda a^\tau}$$

$$\text{HKS}(y) = \sum_{i=0}^{l} e^{-t_0 \lambda a^{\tau_i}} \varphi^2(y)$$

and $\tau(\lambda)$ is a low-pass filter. The disadvantage of HKS by using a low-pass filter is its poor spatial localization.

Scale Invariant Heat Kernel Signature (SIHKS): As the major limitation of heat kernel signatures is its sensitivity to scale, in this work SIHKS [15] has been used as an alternate. For a given shape Y and its scaled version $Yo = \beta Y$, the new eigenvalues and Eigen functions will satisfy $\lambda_o = \beta^2 \lambda$ and $\emptyset_o = \beta\emptyset_o$. If a shape is scaled β then the time shift will be $s = 2\log_\alpha \beta$ and scaling of amplitude is given by β^2. In this proposed work, we remove β^2 by taking log of HKS and differentiate w.r.to τ then the resultant signature is computed as given below

$$\text{hks}(y) = \frac{\sum_{i=0}^{l} -t_0 \alpha^{\tau_i} \log \alpha e^{-t_0 \lambda \alpha^{\tau_i}} \varphi^2(y)}{\sum_{i=0}^{l} e^{-t_0 \lambda \alpha^{\tau_i}} \varphi^2(y)}$$

$$\text{SHKS}(\tau) = \text{hks}(\tau + 1) + \text{hks}(\tau)$$
$$\text{SIHKS} = |\text{FFT}(\text{SHKS}(\tau))|$$

Wave Kernel Signature (WKS): This is also a particular case o LBS [13] given by

(a)   **(b)**   **(c)**

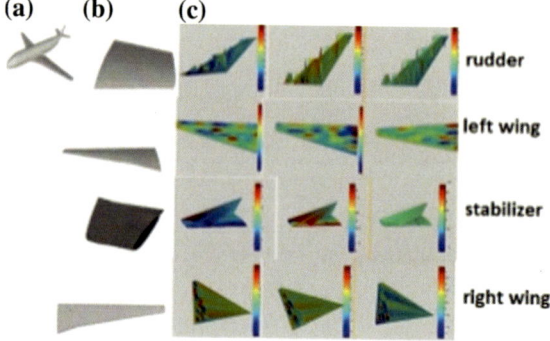

rudder

left wing

stabilizer

right wing

Fig. 4 Signatures of WKS for aircraft parts in the order of **a** first Eigen function, **b** second Eigen function and **c** fourteenth Eigen function

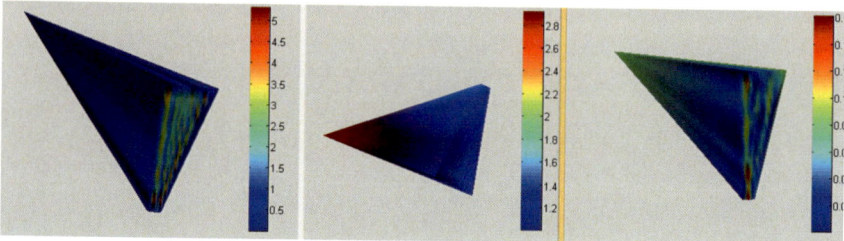

Fig. 5 Signatures for aircraft left wing for WKS, HKS and SIHKS respectively

$$\tau(\lambda) = e^{\dfrac{(\log v_{1i} - \log \lambda)^2}{2\sigma^2}}$$

$$\mathrm{WKS}(y) = \sum_{i=0}^{l} e^{-\dfrac{\left(\log v_{1_i} - \log \lambda\right)^2}{2\sigma^2}} \varphi^2(y)$$

where v_1 is the mean energy level, $v_1 = \mathrm{func}(\lambda)$ and σ is the variance of kernel. WKS shape descriptor is based on the behavior of quantum particle on a manifold that a particle is having an initial energy distribution of log normal with mean energy level. The probability of finding an energy particle of v_1 at y is measured by WKS (Figs. 4 and 5).

C. *Deformation Estimation*

The signatures of the normal and the deformed cloud are different in size hence the missing values of the deformed signature can be computed using linear interpolation. The signatures are then compared using Euclidean distance and the visualization of deformation can be done based on different colors. The percentage of deformation is calculated if k the number of deformed vertices present in the deformed cloud is known prior.

(a)   **(b)**

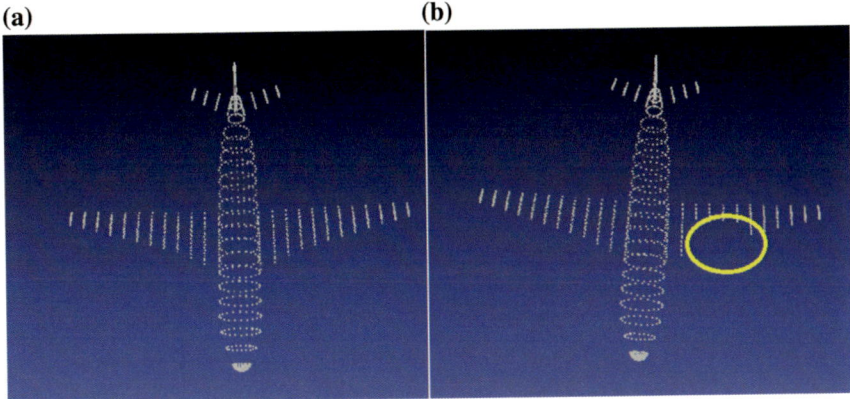

Fig. 6 Aircraft **a** normal, **b** deformed: Ground truth (better viewed in color)

Fig. 7 Deformation
visualized in red color
highlighted in yellow circle
(better viewed in color)

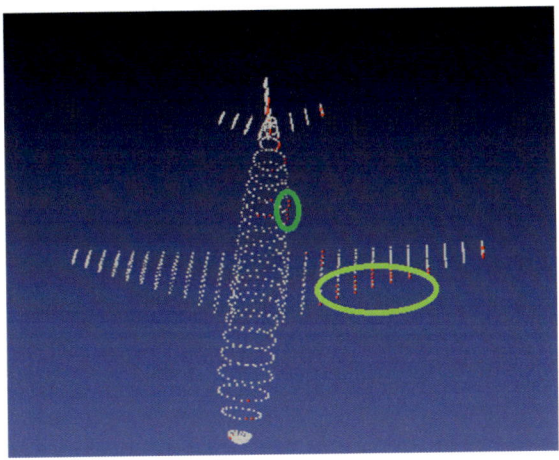

The images of Fig. 6 the aircraft cloud without deformation and with deformation (marked in circle) in right wing is shown. The estimated deformation is shown with red color (marked in yellow circle) and false positive is marked in green circle is shown in Fig. 7.

D. *Segmentation*

One of the applications of LBS signatures is segmentation. Here the segmentation of right wing of the aircraft is shown in the red color by taking the first Eigen function of WKS. The median of first Eigen function is calculated and thresholding is applied to get the segmented cloud.

In Fig. 8 the segmentation is shown in red color. The visualization of the cloud with segmented color can be seen in part "a" of Fig. 8 but the segmented wing is better when the segmentation is done for partitioned cloud.

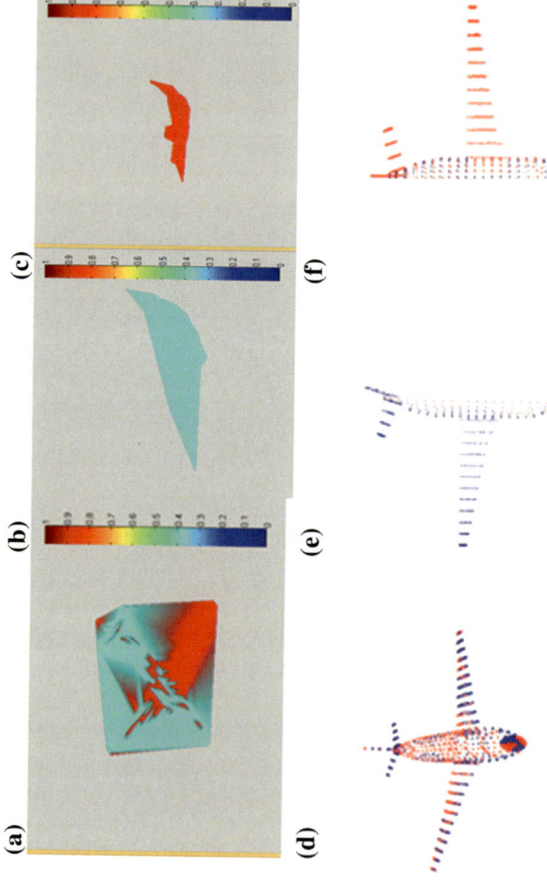

Fig. 8 **a** Wing segmentation (red color) of aircraft from WKS signatures, **b** OBB point cloud of the same segmentation of vertical bi–block 1, **c** the same segmentation of vertical bi–block 2 (**d**–**f**) point cloud representation in the same order

4   Conclusion and Future Work

In this work, we visualize the deformation of point cloud in red color along with segmentation. To get better estimation of deformation the interpolation can be bilinear and cubic. In future, this work can be extended by using various signatures and can match the clouds for deformation detection. In addition, this can be extended to calculate the correspondence between the clouds.

References

1. Andra S, Radke RJ, Roysam B, Al-Kofahi O (2005) Image change detection algorithms: a systematic survey. IEEE Trans Image Process 14(3):294–307
2. Wiselin Jiji G, Naveena Devi R (2015) Change detection techniques—a survey. Int J Comput Sci Appl (IJCSA) 5(2)
3. Konstantinos K (2014) Recent advances on 2D and 3D change detection in urban environments from remote sensing data. In: Computational approaches for urban environments, Geotechnologies and the environment. Springer, Cham, pp 237–272
4. Núñez P, Drews P, Bandera A, Rocha R, Campos M, Dias J (2010) Change detection in 3D environments based on gaussian mixture model and robust structural matching for autonomous robotic applications. In: IEEE/RSJ international conference on intelligent robots and systems (IROS). IEEE, pp 2633–2638
5. Vieira AW, Drews PL, Campos MF (2012) Efficient change detection in 3D environment for autonomous surveillance robots based on implicit volume. In: IEEE international conference on robotics and automation (ICRA). IEEE, pp 2999–3004
6. Monserrat O, Crosetto M (2008) Deformation measurement using terrestrial laser scanning data and least squares 3D surface matching. ISPRS J Photogram Remote Sens 63(1):142–154
7. Vaiapury K, Aksay A, Lin X, Izquierdo E, Papadopoulos C (2011) A vision based audit method and tool that compares a systems installation on a production aircraft to the original digital mock-up. SAE Int J Aerosp 4:880–892
8. Houda C, Thomas K, Pablo A, Peter R (2010) 3D change detection inside urban areas using different digital surface models. In: IAPRS, XXXVIII
9. Lu W, Vinutha K, Mayank B, Jayan E, Harpreet S, Denise P, Richard S (2014) Automatic 3D change detection for glaucoma diagnosis. In: WACV
10. Karthika R, Parameswaran L (2016) Study of Gabor wavelet for face recognition invariant to pose and orientation. In: Proceedings of the international conference on soft computing systems, Volume 397, 2016, of the series Advances in intelligent systems and computing, pp 501–509
11. Parameswaran L, Anbumani K (2008) Content-based watermarking for image authentication using independent component analysis. Informatica J 32:299–306
12. Chen L, Hoang D, Lin H, Nguyen T (2016) Innovative methodology for multi-view point cloud registration in robotic 3D object scanning and reconstruction. Appl Sci 6:132
13. Boscaini D, Masci J, Rodolà E, Bronstein MM, Cremers D (2016) Anisotropic diffusion descriptors. EUROGRAPHICS 35(2)
14. Levy B (2006) Laplace-Beltrami eigenfunctions towards an algorithm that "understands" geometry. In: Shape modeling and applications, international conference 2006, p 13
15. Kokkinos I, Bronstein MM (2010) Scale-invariant heat kernel signatures for non-rigid shape recognition. In: CVPR 2010
16. Arteaga RJ, Ruuth SJ (2015) Laplace-Beltrami spectra for shape comparison of surfaces in 3D using the closest point method. In: ICIP 2015
17. Buendia AE, Martinez LH, Vega RM, Torres JM, Crisóstomo ON (2015) Skeletons of 3D surfaces based on the Laplace-Beltrami operator eigen functions. Appl Math 6:414–420

A Study on Firefly Algorithm for Breast Cancer Classification

Harikumar Rajaguru and Sunil Kumar Prabhakar

Abstract One of the most serious and prominent cancers which affect more women in this world is breast cancer. Various kinds of breast cancer have been reported in medical literature that significantly differ in their capability to spread to other tissues in the human body. Plenty of risk factors have been traced through research though the exact reasons of breast cancer are not yet fully understood. Advances in medicine and technology help to improve the quality of life for breast cancer patients. A lot of contribution especially in the area of artificial intelligence and data mining is done to aid the diagnosis and classification of breast cancer. In order to provide assistance to the doctors, oncologists, and clinicians, neural networks and other optimization techniques serves as a great boon. Here, in this paper, firefly algorithm is utilized effectively as a powerful tool for the analysis and classification of breast cancer. Results show that an average classification accuracy of 98.52% is obtained when firefly algorithm is used for the classification of breast cancer.

1 Introduction

One of the cancers that develop from the tissue of the breast is known as breast cancer [1]. Breast cancer is a serious malignant tumor which arises from the cells of the breast. The significant signs and symptoms of breast cancer can include a red patch of skin, inverted nipples, oozing of fluid from nipples, skin dimpling, change in size and shape of the breast, and sometimes there could be a lump in the breast also [2]. If the breast cancer has spread, then the lymph nodes get swollen and there is a significant bone pain along with shortness of breath accompanied by fatigue [3]. The major factors and reasons for breast cancer in women are obesity and lack of nutritional diet. Lack of exercise, irregular menstruation, hormonal changes, and general lifestyle also contribute greatly to the occurrence of breast cancer. By the self-examination of the breasts using ultrasound testing, mammography, and biopsy, the diagnosis

H. Rajaguru (✉) · S. K. Prabhakar
Department of ECE, Bannari Amman Institute of Technology, Sathyamangalam, India
e-mail: harikumarrajaguru@gmail.com

© Springer Nature Switzerland AG 2019
D. Pandian et al. (eds.), *Proceedings of the International Conference on ISMAC in Computational Vision and Bio-Engineering 2018 (ISMAC-CVB)*, Lecture Notes in Computational Vision and Biomechanics 30,
https://doi.org/10.1007/978-3-030-00665-5_42

of breast cancer can be easily done [4]. Depending on the type of cancer based on its position and severity and stages (0–4), the treatment options such as radiation, surgery, and chemotherapy can be decided easily by the doctor. Some of the common types of breast cancers include lobular carcinoma in situ, medullary carcinoma, adenoid cystic carcinoma, inflammatory breast cancer, mucinous carcinoma, invasive ductal, and lobular carcinoma, ductal carcinoma in situ, etc. [5]. With the aid of data mining, artificial intelligence, and soft computing, the classification procedure of breast cancer has become a boon for clinicians. A few significant and relevant works done by researchers for breast cancer classification are discussed as follows. For the diagnosis of breast cancer, feature selection combined with Support Vector Machine (SVM) was employed by Akay. The breast cancer classification was analyzed based on Radial Basis Functions (RBF) and Gaussian Mixture Model (GMM) by Rajaguru and Prabhakar [6]. A swarm intelligence-based wavelet neural network methodology for breast cancer detection and classification was reported by Dheeba et al. [7]. By the comparison of backpropogation training algorithm, the breast cancer can be classified easily by Paulin [8]. With the advantages of Linear Discriminant Analysis (LDA) and Soft Discriminant Classifier (SDC), the classification of breast cancer was done by Prabhakar and Rajaguru [9]. The multiclassifiers were utilized on various datasets for the diagnosis of breast cancer by Salama et al. [10]. For the breast cancer diagnosis and classification, a fuzzy-genetic approach was used by Pena-Reyes and Sipper [11]. Rajaguru and Prabhakar proposed an Expectation–Maximization Based Logistic Regression for Classification of Breast Cancer [12]. In this paper, firefly algorithm is used for breast cancer classification. The pictorial illustration of the paper is shown in Fig. 1.

2 Materials and Methods

As the first step, the clinical values obtained from the hospital are converted into numerical values. Chi-square test is implemented to all the numerical values. The values obtained by means of chi-square test are compared with different TNM stages of classification. The different stages of breast cancer are classified with the aid of firefly algorithm in this paper. For this study, a total of 82 patients are considered from the Department of Oncology of Kuppuswamy Naidu Memorial Hospital (KNMH), Coimbatore, Tamil Nadu, India. A lot of consolidation was done from hospital charts, referral letters, radiation therapy reports, radiological study reports, operative reports, and pathological reports. All the cancer patients were categorized into four stages based on their severity level. General parameters like food habits, personal hygiene, family history cancer, marital status, menstrual phase and duration, menopause, abortions (if any), size of breast, lump size and location, level and position of breast nipple, abdominal analysis, etc. were analyzed thoroughly. According to the standard International Union Against Cancer (UICC), 20 patients are present in the first stage and 23 patients are present in the final stage. Stage two has 28 patients and stage three has 11 patients.

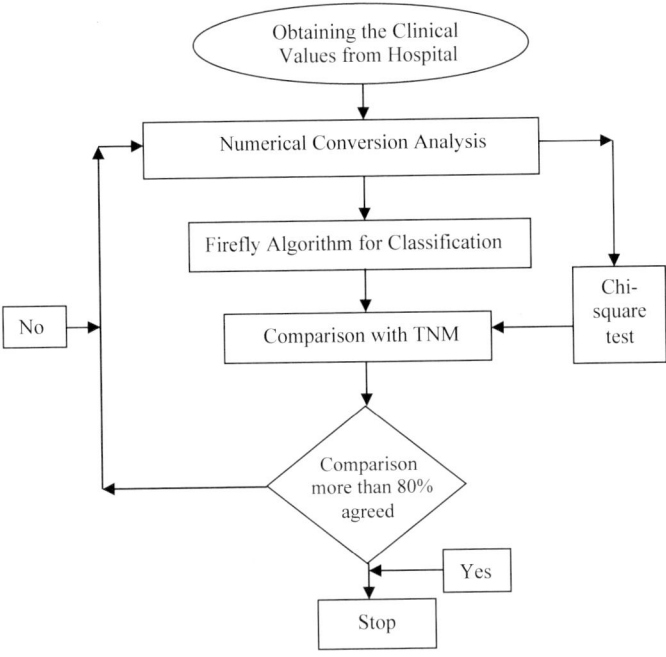

Fig. 1 Pictorial illustration of the work

2.1 Implementation of Chi-Square Test

A famous mathematical distribution which has a wide applicability in statistics oriented work is chi-square distribution [6]. To define this distribution, the Greek letter χ is used. The elements upon which this specific distribution is dependent are squared and so the symbol χ^2 is used to indicate the chi-square distribution. Every χ^2 distribution has an associated degree of freedom with it. The χ^2 statistic is pretty different from other statistics and has very little resemblance to the theoretical chi-square distribution. For the test of independence and for the goodness of fit test, the chi-square statistic is similar. For both the tests, all the categories into which the specific data is divided are utilized. The information obtained from the samples is the number of cases observed. These are known as the frequencies of occurrence for each and every category into which the information is grouped. In chi-square test, a statement is made by the null hypothesis dealing with the total number of cases to be expected in every category if this hypothesis is correct. For every category, the chi-square test is based on the difference between the observed and the expected values for all the categories.

The chi-square test is expressed as

Table 1 Performance measures of chi-square test for the 82 Breast Cancer Patients

	Performance of chi-square test		
	Minimum value	Maximum value	Average value
T1	3.84	19.52	11.7
T2	21.2	32.96	26.9
T3	30.72	40.85	35.5
T4	41.92	90.08	66

$$\chi^2 = \sum_i \frac{(Q_i - F_i)^2}{F_i}$$

where

Q_i denotes the observed number of cases in category i,
F_i denotes the expected number of cases in category i.

By the simple computation of difference calculation between the observed number of cases and the expected number of cases in every category, the chi-square statistic is used. This particular difference is squared and divided by the expected number of cases in a particular category. Then for all the different categories, these values are added and the total is referred to as chi-squared value. The null hypothesis bothers much about the distribution of data. For each chi-square test, the null and alternative hypothesis is stated as

$$H_0 : Q_i = F_i$$
$$H_1 : Q_i \neq F_i$$

If the null hypothesis claim is true, then the expected and observed values are pretty close to each other and $(Q_i - F_i)$ is pretty small for every category. Thus, when the null hypothesis claim is true, the chi-square statistic is small and when the null hypothesis claim is false, then the chi-square statistic is large. The degree of freedom is dependent on the total number of categories utilized in the calculation of the statistic. Thus, with the aid of chi-square distribution and chi-square statistic, one can determine easily whether the data is claimed or distributed. The performance analysis of chi-square test for the breast cancer patients is shown in Table 1.

3   Firefly Algorithm for Classification of Breast Cancer

The flashing phenomena of fireflies are simulated with the help of firefly algorithm and it was introduced by Yang [13]. In order to simulate the idea, the assumptions are made as follows. All the fireflies belong to the same gender. The fireflies which are brighter have a greater attraction than the fireflies which are less brighter. Between the fireflies if the distance increases, then there is a decrease in attraction. If none of the fireflies is brighter than a particular one, then the firefly will move randomly. The landscape of a particular objective function greatly influences the brightness of a firefly. For a specific maximization problem, the particular relation of the objective function and the brightness J is represented as $J(y) \alpha f(y)$. For the second assumption, the mathematical representation is as follows:

$$J(q) = (J_0)(e^{-\gamma q^2})$$

where

J_0 denotes the actual density of light and
γ is named as the light absorption coefficient.
   The attractiveness β is selected as follows:

$$\beta(q) = (\beta_0)\left(e^{-\gamma q^2}\right)$$

where q denotes the distance between the two fireflies and β_0 represents the actual attractiveness at $q = 0$. The distance q_{jk} between the 2 fireflies at respective positions y_j and y_k is calculated as

$$q_{jk} = \|y_j - y_k\| = \sqrt{\sum_{m=1}^{d}\left(y_{j,m} - y_{k,m}\right)}$$

where $y_{j,m}$ is the mth component of the spatial coordinate y_j of the jth firefly. The specific movement of a firefly 'j' which is attracted to a much brighter firefly 'k' is determined as

$$y_j = y_j + \alpha\left(rand - \frac{1}{2}\right) + \beta_0 e^{-\gamma q_{jk}^2}(y_k - y_j)$$

here, α denotes a random parameter in the interval [0, 1]. *Rand* denotes a random number and is derived from a Gaussian distribution which is uniform in nature in [0, 1].
   The Firefly Algorithm is given as follows:

(1) Objective function $f(y)$, $y = (y_1, \ldots, y_d)^T$
(2) A population of fireflies is initiated $y_j(j = 1, 2, \ldots, n)$
(3) The light intensity J_j at y_j is calculated by $f(y_j)$
(4) The light absorption coefficient γ is defined
(5) while ($t < Max\ Generation$)
(6) for $j = 1 : n$ all n fireflies
(7) for $k = 1 : n$ all n fireflies
(8) The distance $'r'$ is calculated between y_j and y_k using Cartesian distance Eq.

(9) if $(J_k > J_j)$
(10) Attractiveness varies with distance r through $\beta_0 e^{-\gamma q^2}$
(11) The firefly j is moved toward k in all the $'d'$ dimensions
(12) end if
(13) The new solutions are evaluated and the light intensity is updated
(14) end for k
(15) end for j
(16) The fireflies are ranked and the current best is found out
(17) end while

4 Results and Discussion

For the performance analysis of firefly algorithm as classifier for breast cancer classification , the standard benchmark parameters utilized for the analysis are sensitivity
and specificity measures, classification accuracy, Perfect Classification Measures,
False Alarm measures, Missed Classification Measures, and are given by the following mathematical formulae as:

$$PI = \left(\frac{PC - MC - FA}{PC} \right) \times 100$$

PC is observed as Perfect Classification, MC is indicated as Missed Classification, and FA is denoted as False Alarm. The Sensitivity, Specificity, and Accuracy
measures are mathematically expressed by the following formulae:

$$Sensitivity = \frac{PC}{PC+FA} \times 100$$

$$Specificity = \frac{PC}{PC+MC} \times 100$$

$$Accuracy = \frac{Sensitivity+Specificity}{2}$$

Table 2 depicts the performance analysis of firefly classifier for the various stages
of breast cancer. Table 3 indicates the detailed analysis of comparison of clinical
values with the firefly algorithm as classifier for the different stages of breast cancer
classification.

5 Conclusion

Breast cancer is one of the most dangerous and the second leading reason of cancer
death in women after lung cancer. In a most encouraging manner, the death rate from
breast cancer has declined to a greater extent due to the advancement in medicine

Table 2 Performance analysis of firefly classifier for the different stages of breast cancer

	PC	MC	FA	PI	Sensitivity	Specificity	Average
T1	100	0	0	100	100	100	100
T2	100	0	0	100	100	100	100
T3	100	0	0	100	100	100	100
T4	88.23	11.76	0	86.67	100	88.23	94.11
Average	97.05	2.94	0	96.66	100	97.05	98.52

Table 3 Consolidate analysis of clinical values with firefly classifier for the different breast cancer classification stages

S.No	Breast cancer stage	Classification accuracy (%) through clinical procedure	Classification accuracy (%) through firefly algorithm as a classifier
1	T1	98	100
2	T2	100	100
3	T3	97	100
4	T4	100	94.11

and technology along with awareness. When cells in the tissue of the breast mutate and keep reproducing, then this problem occurs. These abnormal cells generally form a cluster and become a tumor. Some of the major risk factors of the breast cancer have been identified as age, gender, breast cancer gene mutation, change in size and shape of breast, race/ethnicity, hormonal fluctuations, weight, alcohol consumption, pregnancy history, radiation exposure, etc. In this paper, the firefly algorithm is utilized effectively as a classifier for breast cancer classification for all the four different stages and it is compared with the classification done through clinical procedure too. The classification accuracy of stages T1, T2, and T3 is 100% and for T4 stage, the classification accuracy is 94.11%. The average classification accuracy for all the stages from (0-4) produced is 98.52% and average performance index is 96.66%, along with an average specificity and sensitivity of 97.05% and 100%, respectively. Future works aim to make modifications in firefly algorithm for obtaining a better classification accuracy.

References

1. Polat K, Günes S (2007) Breast cancer diagnosis using least square support vector machine. Digit Signal Process 17:694–701
2. Setiono R (2000) Generating concise and accurate classification rules for breast cancer diagnosis. Artif Intell Med 18:205–219
3. Übeyli ED (2007) Implementing automated diagnostic systems for breast cancer detection. Expert Syst Appl 33:1054–1062

4. Mert A, Kılıç NZ, Bilgili E, Akan A (2015). Breast cancer detection with reduced feature set. Comput Math Methods Med 1–11
5. Rajaguru H, Prabhakar SK (2017) Bayesian linear discriminant analysis for breast cancer classification. In: 2nd IEEE international conference on communication and electronics systems. Coimbatore, India, pp 19–20 October 2017
6. Rajaguru H, Prabhakar SK (2016) A comprehensive analysis on breast cancer classification with radial basis function and gaussian mixture model In: 16th international conference on biomedical engineering (ICBME). Singapore, 7–10 December 2016
7. Dheeba J, Singh NA, Selvi ST (2014) Computer aided detection of breast cancer on mammograms: a swarm intelligence optimized wavelet neural network approach. J Biomed Inf 49:45–52
8. Paulin F (2011) Classification of breast cancer by comparing backpropagation training algorithm. Int J Comput Sci Eng 3:327–332
9. Rajaguru H, Prabhakar SK (2017) Performance analysis of breast cancer classification with softmax discriminant classifier and linear discriminant analysis. In: International conference on biomedical and health informatics. Thessaloniki, Greece, pp 18–21 November 2017
10. Salama GI, Abdelhalim MB, Zeid MA (2012) Breast cancer diagnosis on three different data sets using multi-classifiers. Int J Comput Inf Technol 1:36–43
11. Pena-Reyes CA, Sipper M (1999) A fuzzy-genetic approach to breast cancer diagnosis. Artif Intell Med 17:131–155
12. Rajaguru H, Prabhakar SK (2017) Expectation maximization based logistic regression for breast cancer classification In: IEEE proceedings of the international conference on electronics, communication and aerospace technology (ICECA 2017). Coimbatore, India, pp 603–606
13. Yang X-S (2009) Firefly algorithms for multimodal optimization. In: Proceedings of the international symposium on stochastic algorithms, vol 5792. Springer, Berlin, pp 169–178

Fuzzy C-Means Clustering and Gaussian Mixture Model for Epilepsy Classification from EEG

Harikumar Rajaguru and Sunil Kumar Prabhakar

Abstract Due to various disorders in the functionality of the brain, epileptic seizures occur and it affects the patient's mental, physical and emotional health to a great extent. The prediction of epileptic seizures before the beginning of the onset is pretty useful for seizure prevention by medication. One of the major causes for epilepsy is molecular mutation which results in irregular behaviour of neurons. Though the exact reasons for epilepsy are not known, early diagnosis is very useful for the treatment of epilepsy. Various computational techniques and machine learning algorithms are utilized to classify epilepsy from Electroencephalography (EEG) signals. In this paper, Fuzzy C-Means (FCM) Clustering algorithm is used as a clustering technique initially and then the features obtained through it is classified with the help of Gaussian Mixture Model (GMM) used as a post-classification technique. Results report that an average classification accuracy of 97.64% along with an average performance index of 95.01% is obtained successfully.

1 Introduction

Due to the unexpected and sudden surge of electrical activity in the brain, recurrent seizures occur [1]. The sudden electrical activities cause a huge distribution in the messaging system below the brain cells. Therefore, epilepsy is quite a serious and yet common neurological disorder with seizures as its primary symptoms [2]. Depending on the individual, the seizures can have a wide range of severity on the patients. Some common symptoms of epilepsy are: the patient falls down suddenly for no reason, occurrence of a convulsion with no fever, confused memory, inappropriate repetitive moments, sudden bouts of chewing and blinking, short spells of blackout, peculiar change in seizures such as smell and sound, rapid jerking movements and body stiffness [3]. Some of the similar symptoms often misdiagnosed as epilepsy include fainting, high fever, cataplexy, narcolepsy, panic attacks, nightmares, etc.

H. Rajaguru (✉) · S. K. Prabhakar
Department of ECE, Bannari Amman Institute of Technology, Sathyamangalam, India
e-mail: harikumarrajaguru@gmail.com

© Springer Nature Switzerland AG 2019
D. Pandian et al. (eds.), *Proceedings of the International Conference on ISMAC in Computational Vision and Bio-Engineering 2018 (ISMAC-CVB)*, Lecture Notes in Computational Vision and Biomechanics 30,
https://doi.org/10.1007/978-3-030-00665-5_43

Epilepsy results when the messaging system in the brain is disrupted because of faulty electrical activity [4]. In many cases, the causes of occurrence of epilepsy are not known. Some patients developed epilepsy because of inherited genetic factors and epilepsy happens more often [5]. Some major factors that increase the risk of epilepsy is trauma in head, infectious diseases, serious brain conditions, prenatal injury, developmental disorders, etc. The various aspects of a person's life is affected terribly because of epilepsy which includes academic and work disturbances, unable to socialize well in the society and prohibits social development and interaction. EEG is the most famous physiological technique to record the electrical activity generated from the scalp of the brain [6]. Due to the synchronized activity of thousands of neurons, the electrical activity is generated and it is measured easily by EEG. EEG provides excellent time resolution and so it is widely used for the analysis of neurological disorders. A variety of works have been discussed in the literature regarding epilepsy classification from EEG signals. A clear understanding of the automated EEG analysis for epilepsy classification was done by Acharya et al. [7]. The epileptic seizure prediction in advance in a most efficient manner was done by Moghim and Corne [8]. With the aid of Adaboost Classifier and some important dimensionality reduction techniques, the epilepsy classification was performed with some interesting results by Prabhakar and Rajaguru [9]. Using Lyapunov exponent of the EEG signals, the characterization of epileptic seizures was done by Osowski et al. [10]. Based on computational intelligence techniques, a comparative study was made for epilepsy prediction and classification by Teixeira et al. [11]. The interpretation of Probabilistic Mixture Model (PMM) was discussed by Prabhakar and Rajaguru for the epilepsy classification from EEG signals [12]. The epileptic seizure prediction was performed with the relative spectral power features by Bandarabadi et al. [13]. Based on variational Bayesian GMM of zero-crossing intervals, the epileptic seizures were easily predicted in scalp EEG by Zandi et al. [14]. An efficient e-health system design for epilepsy classification was proposed by Prabhakar and Rajaguru [15]. In this work, Fuzzy C-Means Clustering and GMM algorithm were utilized well for epilepsy classification from EEG signals. The illustration of the work is shown in Fig. 1. The organization of the work is structured as follows. The materials and methods are explained in Sect. 2 followed by the usage of clustering technique in Sect. 3. Section 4 explains the classification using GMM algorithm followed by the results and discussion in Sect. 5 and ended with conclusion in Sect. 6.

2 Materials and Methods

The clinical EEG recordings of 20 epileptic patients who were admitted for their treatment from the Department of Neurology of Sri Ramakrishna Hospital, Coimbatore, India are taken for the analysis in this study [9]. The standard 10–20 International system was deployed to obtain the EEG recordings of the epileptic patients. A total of 16 channel electrodes were placed on the scalp of the epileptic patients and for various stages, the EEG recordings were easily measured. The recordings were measured

Fig. 1 Block diagram of the work

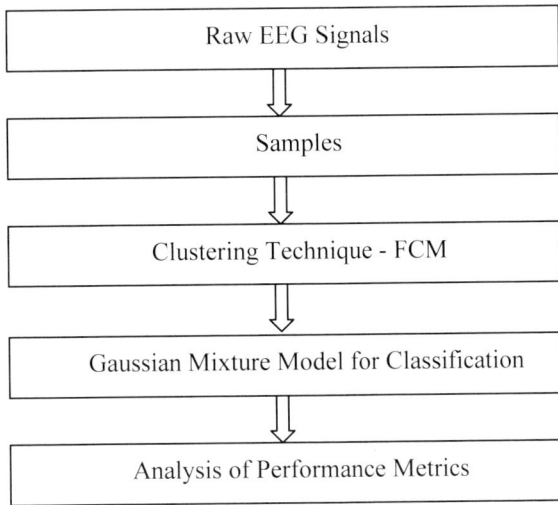

for a period of 55 min and it was obtained in European Data Format. Each channel was split into epochs and each epoch has the values pertaining to the instantaneous amplitude levels of the EEG signals.

3 Fuzzy C-Means Clustering Algorithm

To the raw EEG data and its samples, FCM algorithm is implemented to extract its features. Fuzzy Clustering algorithms have been of great utility to researchers and it is discussed and implemented in different areas [16]. FCM is one of the most commonly utilized algorithms. The total number of clusters used in this study 128. The dataset Z is divided into 'f' fuzzy clusters. This algorithm holds that every object belongs to a particular cluster but with a different membership degree. It implies that a cluster is considered as a fuzzy subset on the entire dataset.

Assuming that $Z = \{z_1, z_2, \ldots, z_n\}$ is a finite dataset, where $z_i = (z_{i1}, z_{i2}, \ldots, z_{iq})$ denote a q dimensional object and z_{ij} is the jth property of the ith object. $F = \{F_1, F_2, \ldots, F_f\}$ indicates the f clusters. $W = \{w_1, w_2, \ldots, w_f\}$ represents 'f' l-dimensional cluster centroid points, where $w = (w_{i1}, w_{i2}, \ldots, w_{iq})$. $Y = (Y_{iy})_{(n \times f)}$ is a fuzzy partition matrix and y_{ik} denotes the degree of membership of the ith object in the gth cluster, where

$$\sum\nolimits_{g=1}^{f} y_{ig} = 1, \forall_i = 1, \ldots, n$$

The objective function is nothing but the quadratic sum of the weighed distance from all the samples to the cluster centroid in every cluster

$$P_r(Y, W) = \sum_{g=1}^{f} \sum_{i=1}^{n} y_{ig}^r d_{ig}^2$$

where $d_{ig} = \|z_i - w_g\|$ depicts the Euclidean distance between the ith object and the gth cluster centroid. $r[r \in (1, \infty)]$ denotes the fuzziness index which controls the fuzziness of the memberships. The membership becomes fuzzier when the value of $'r'$ becomes higher. FCM is carried out as an iterative process aiding in the minimization of the objective function P_r, with the updation of both Y and W. The steps are as mentioned below:

Step: 1 The initial value of the total number of clusters $'f'$, fuzziness index r, the threshold ξ and the maximum iteration I_{max} are assigned.

Step: 2 In a random manner, the fuzzy partition $Y^{(0)}$ is initialized based on the constraints of the membership degree.

Step: 3 The $'f'$ cluster centroids $W^{(t)}$ at the t-step is computed.

Step: 4 The objective function $P_r^{(t)}$ is calculated. If $\left| P_r^{(t)} - P_r^{(t-1)} \right| < \xi$ or $t > I_{max}$, then stop otherwise continue to step 5

Step: 5 The $Y^{(t+1)}$ is calculated and returned to step 3. Finally, every object is arranged into one single cluster with accordance to the maximum degree of membership. The main advantages are that the algorithm is simple, has quick convergence capability and it is quite easy to extend.

The standard technique to determine the total number of clusters of FCM is as follows:

(1) The search range of the total number of clusters is set as the first step
(2) Run the FCM to generate the clustering results of various number of clusters
(3) Selection of a suitable clustering validity index
(4) Evaluate the clustering results
(5) Obtaining the optimal number of clusters based on the evaluation result.

The consolidated steps involved in the clustering procedure are as follows:

Step 1: Input the search range $[f_{min}, f_{max}]$
Step 2: For every integer $gn \in [f_{min}, f_{max}]$, run FCM and then the clustering validity index is calculated
Step 3: All the values of clustering validity index are compared. gn represents the optimal number of clusters f_{opt}
Step 4: The output is f_{opt}. It is the optimal value of clustering validity index and thus the clustering result is found out.

4   Gaussian Mixture Model as a Post Classifier

The cluster values are now fed inside the GMM classifier for epilepsy classification. Let $a \in \Re^d$ and G be the total number of components where every component has a prior probability d_i and probability density function with mean μ_i and covariance $\Sigma_i, i = 1, \ldots, G$.

A Gaussian Mixture Model is expressed as

$$\sum_{i=1}^{G} d_i \phi(a|\mu_i, \Sigma_i) = \sum_{i=1}^{G} d_i \frac{1}{\sqrt{(2\pi)^d |\Sigma_i|}} \exp\left(\frac{-(a - \mu_i)^t \Sigma_i^{-1} (a - \mu_i)}{2} \right)$$

where $\sum_{i=1}^{G} d_i = 1$

The likelihood function is expressed as

$$H(A|\theta) = \prod_{j=1}^{n} f\left(a_j|\theta\right)$$

and log-likelihood function is expressed as

$$l(A|\theta) = \sum_{j=1}^{n} \log\left(\sum_{i=1}^{G} d_i \phi\left(a_j|\mu_i, \Sigma_i\right) \right)$$

where $A = \left(a_1^t, \ldots, a_n^t\right)^t$, respectively.

Expectation Maximization was utilized for GMM [17]. The unobservable variable A in a particular space is indirectly observed through an observed variable Z in its respective sample space. Let $f(a|\theta)$ be the sampling density and it depends on the parameter $\theta \in \Omega$, therefore the corresponding family of sampling densities for Z is expressed as follows:

$$q(z|\theta) = \int_{a(z)} f(a|\theta) da$$

where $a(z)$ is a subset of A under the mapping $a \to z(a)$ from A to Z. The main objective of the EM algorithm is to find the value of θ so that $q(z|\theta)$ is minimized.

5   Results and Discussion

With FCM clustering and GMM algorithm, the parameters like Performance Index, Accuracy, Sensitivity and Specificity are computed in Table 1. The mathematical formulae for the parameters like Performance Index (PI), Sensitivity, Specificity and Accuracy are mathematically written as follows:

Table 1 Performance metrics analysis of FCM with GMM algorithm

Name of the parameter (%)	Average
PC	95.28
MC	0
FA	4.70
PI	95.01
Specificity	100
Sensitivity	95.28
Accuracy	97.64

$$PI = \left(\frac{PC - MC - FA}{PC} \right) \times 100$$

where

PC— Perfect Classification,

MC— Missed Classification and

FA— False Alarm. The Sensitivity, Specificity and Accuracy measures are expressed by the following equations:

$$Sensitivity = \frac{PC}{PC+FA} \times 100$$

$$Specificity = \frac{PC}{PC+MC} \times 100$$

$$Accuracy = \frac{Sensitivity+Specificity}{2}$$

6 Conclusion

Thus, epilepsy is a chronic neurological disorder causing recurrent and unprovoked seizures. If the seizure is stronger, then it causes uncontrollable muscle spasms and twitches and it lasts for a few minutes. The patients lose their consciousness or become quite confused if the seizure is stronger. EEG has a pivotal role to play in the diagnosis, classification and analysis of epilepsy classification. In this work, FCM is used as a clustering technique and later it is classified with the help of GMM. Results show that an average classification accuracy of about 97.64% is obtained, an average sensitivity rate of 95.28% is obtained, an average specificity rate of 100% is obtained and an average performance index of 95.01% is obtained. Future works aim to work with modified FCM and modified GMM algorithm for obtaining a better epilepsy classification rate.

Acknowledgements The authors are grateful to Dr. Asokan, Neurologist, Ramakrishna Hospital Coimbatore and Dr. B. Rajalakshmi, Diabetologist, Govt. Hospital Dindigul for providing the EEG signals.

References

1. Harikumar R, Kumar PS (2015) Fuzzy techniques and aggregation operators in classification of epilepsy risk levels for diabetic patients using eeg signals and cerebral blood flow. J Biomater Tissue Eng 5(4): 316–322

2. Rajaguru H, Prabhakar SK (2016) A framework for epilepsy classification using modified sparse representation classifiers and native bayesian classifier from EEG signals. J Med Imaging Health Inf

3. Prabhakar SK, Rajaguru H (2015) Analysis of centre tendency mode chaotic modeling for electroencephalography signals obtained from an epileptic patient. Adv Stud Theor Phys 9(4): 171–177, HIKARI Ltd., http://dx.doi.org/10.12988/astp.2015.5117

4. Harikumar R, Kumar PS (2015) Frequency behaviors of electroencephalography signals in epileptic patients from a wavelet thresholding perspective. Appl Math Sci 9(50): 2451–2457, HIKARI Ltd., http://dx.doi.org/10.12988/ams.2015.52135

5. Prabhakar SK, Rajaguru H (2016) Classification of epilepsy risk using variable thresholding based feature extraction technique and suitable post classifiers. Int J Simul Syst Sci Technol (IJSSST) 17(33): 28.1–28.8

6. Rajaguru H, Prabhakar SK (2017) Analysis of probabilistic neural networks with dimensionality reduction for epilepsy classification from EEG. Int J Mech Eng Technol

7. Acharya UR, Sree SV, Swapna G, Martis RJ, Suri JS (2013) Automated EEG analysis of epilepsy: a review. Knowl- Based Syst 45:147–165

8. Moghim N, Corne DW (2014) Predicting epileptic seizures in advance. PLoS ONE 9(6), Article ID e99334

9. Prabhakar SK, Rajaguru H (2017) Adaboost classifier with dimensionality reduction techniques for epilepsy classification from EEG. In: International conference on biomedical and health informatics. Thessaloniki, Greece, 18–21 November 2017

10. Osowski S, Swiderski B, Cichocki A, Rysz A (2007) Epileptic seizure characterization by Lyapunov exponent of EEG signal. COMPEL - Int J Comput Math Electr Electron Eng 26(5):1276–1287

11. Teixeira CA, Direito B, Bandarabadi M et al (2014) Epileptic seizure predictors based on computational intelligence techniques: a comparative study with 278 patients. Comput Methods Program Biomed 114(3):324–336

12. Prabhakar SK, Rajaguru H (2017) Conceptual analysis of epilepsy classification using probabilistic mixture models In: 5th IEEE winter international conference on brain-computer interface. South Korea, 9–11 January 2017

13. Bandarabadi M, Teixeira CA, Rasekhi J, Dourado A (2015) Epileptic seizure prediction using relative spectral power features. Clin Neurophysiol 126(2):237–248

14. Zandi AS, Tafreshi R, Javidan M, Dumont GA (2013) Predicting epileptic seizures in scalp EEG based on a variational bayesian gaussian mixture model of zero-crossing intervals. IEEE Trans Biomed Eng 60(5):1401–1413

15. Prabhakar SK, Rajaguru H (2016) Efficient wireless system for telemedicine application with reduced PAPR using QMF based PTS technique for epilepsy classification from EEG signals In: IFBME proceedings (Springer), international conference on advancements of medicine and health care through technology (MEDITECH), Romania, 12–15 October 2016

16. Saha I, Maulik U, Bandyopadhyay S (2009) A new differential evolution based fuzzy clustering for automatic cluster evolution. In: Proceedings of the IEEE international advance computing conference (IACC'09), IEEE, Patiala, India, 706–711 March 2009

17. Prabhakar SK, Rajaguru H (2016) Performance analysis of GMM classifier for classification of normal and abnormal segments in PPG signals In: 16th international conference on biomedical engineering (ICBME), Singapore, 7–10 December 2016

Analysis on Detection of Chronic Alcoholics from EEG Signal Segments—A Comparative Study Between Two Software Tools

Harikumar Rajaguru, Vigneshkumar Arunachalam
and Sunil Kumar Prabhakar

Abstract Alcohol consumption is vulnerable to the brain and has a high risk of brain damage and other neurobehavioral deficits. This paper primarily focuses on massive data generated from EEG signals and its characterization with respect to various states of the human brain under influence of alcohol. A single trial 64-channel EEG database is utilized for classification of alcoholic states for a single patient. Singular Value Decomposition (SVD) features of EEG segments are computed. Even though EEG signals are acquired from alcoholic patient some of the EEG signal segments resemble EEG segments of normal, alcoholic, and epileptic persons. Depending on the SVD values, EEG segments are labeled as normal, alcoholic, and epileptic and then classified through Hard Thresholding and K-means clustering techniques. The classification is done using two different softwares in this paper, namely, MATLAB and R studio and then the results are compared. The results show that MATLAB software classifies better than R studio software with comparatively highest classification accuracy of 83.5% which is obtained when Hard Thresholding method is utilized.

1 Introduction

Alcoholism is one of the most common reasons for a lot of complications happening in the human body. Consuming alcohol for a very long time leads to larger health diseases for sure. The long-term use of alcohol damages all human organs and has a great toll on the human body [1]. Alcohol abuse causes blurred vision, impaired memory, and has a lot of brain deficits [2]. Electroencephalography (EEG) is an inter-

H. Rajaguru (✉) · S. K. Prabhakar
Department of ECE, Bannari Amman Institute of Technology, Sathyamangalam, India
e-mail: harikumarrajaguru@gmail.com

V. Arunachalam
Department of CSE, Bannari Amman Institute of Technology, Sathyamangalam, India

© Springer Nature Switzerland AG 2019
D. Pandian et al. (eds.), *Proceedings of the International Conference on ISMAC in Computational Vision and Bio-Engineering 2018 (ISMAC-CVB)*, Lecture Notes in Computational Vision and Biomechanics 30,
https://doi.org/10.1007/978-3-030-00665-5_44

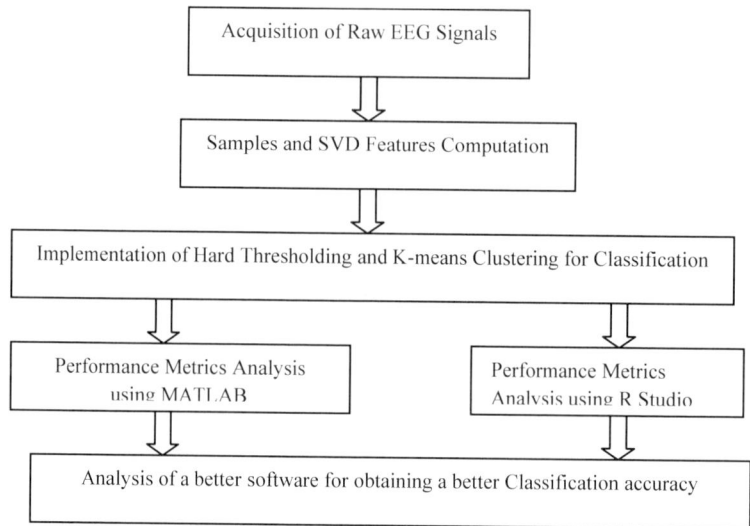

Fig. 1 Flow diagram for detection of chronic alcoholic segments from EEG signals through SVD labels using MATLAB and R studio

esting and vibrant electrophysiological monitoring technique utilized to record the electrical activity of the human brain [3]. EEG generally refers to the recording of the spontaneous electrical activity of the brain over a long/short specific period of time. In this work, the alcoholic EEG signal of a single patient is examined deeply with the help of two classification techniques and it is performed in two different softwares in order to analyze both the best software and the best classification technique.

Figure 1 depicts the flow diagram for the detection of chronic alcoholic segment from EEG signals through SVD labels. The alcoholic EEG data set is sampled and then mean, SVD features are calculated for each channels. These SVD values are used to label the channel through hard Thresholding and K-means clustering algorithms. The structure of the paper as follows: Sect. 1 introduces the manuscript. Section 2 discusses the materials and methods. Section 3 describes the classification procedure and the results are discussed in Sect. 4. Section 5 concludes the paper.

2   Materials and Methods

In this section, the acquisition of alcoholic EEG data acquisition and the feature extraction concept using SVD is discussed. Table 1 shows the details of the alcoholic EEG data [4].

Table 1 Details of the alcoholic EEG data

Total number of patients analyzed	1
Utilization of channel	Single trial 64 channel
Electrodes used	3
Signal sampling	256 samples/second
Analog to digital converter utilized	12 bit signed representation
Time required for every channel	10 s duration
Samples obtained per particular channel	2560
The total number of samples present for all the 64 signals	1,63,840
Total number of groups of samples	64

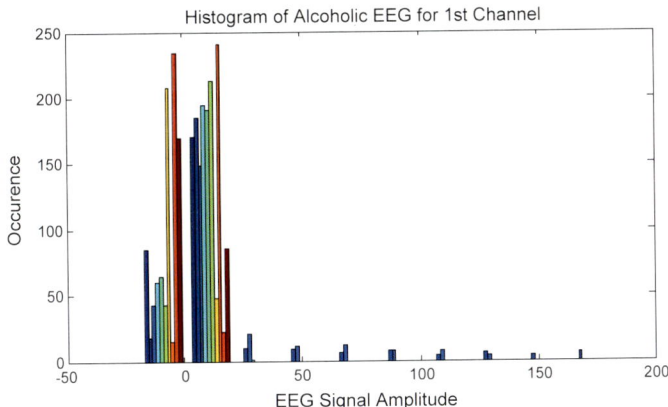

Fig. 2 Histogram of EEG signal amplitude for channel 1

2.1 Histogram Analysis

Histogram is a representation of repeated patterns based on their frequency of occurrence. It is observed from the analysis of amplitude components of EEG signals that nonlinear dynamics is responsible for this alcoholic state. The histogram of an alcoholic person for two different channels is depicted in Figs. 2 and 3. Figure 2 shows that the histograms for the first channel of alcoholic EEG signal which is skewed and non-Gaussian in nature, whereas the histogram of 64th channel of EEG signals as shown in Fig. 3 attained a nonlinear, non-Gaussian shape. Therefore, it is felt that the histogram may also be useful to differentiate the brain activity of an alcoholic patient on channel basis.

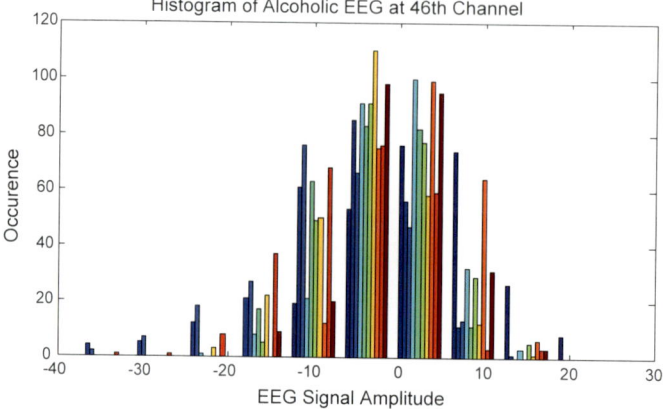

Fig. 3 Histogram of EEG signal amplitude for channel 64

2.2 Feature Extraction Using SVD

SVD plays a major role in most of the mathematical operations and is just a factorization of a real or complex matrix [5]. The SVD of a $l \times h$ matrix L is a factorization of the form $W \sum Y^*$, where W is a $l \times l$ unitary matrix (real or complex), $\sum$ is a $l \times h$ rectangular diagonal matrix and Y is a $h \times h$ unitary matrix (real or complex).

3 Classification Using Hard Thresholding and K-Means Clustering

The SVD features are classified with the help of Hard Thresholding classifier and K-means clustering Classifier to detect the risk of alcohol for the particular patient.

3.1 Hard Thresholding Classifier

Once the SVD features are computed, hard Thresholding classifier [6] is designed with the following conditions as mentioned in Table 2. The following representations are used.

μ_1 = the mean SVD of the entire normal segments of EEG signal
μ_2 = the mean SVD of the entire alcoholic segments of EEG signal
μ_3 = the mean SVD of the entire epileptic segments of EEG signal.

Based on the following constrains, the gold standard for the Thresholding classifier is arrived.

Table 2 Gold standard for thresholding classifier

Sl. no.	Mean SVD value of EEG segment	Clinical state
1	Up to μ_1	Normal
2	More than μ_1 and less than μ_3; with center at μ_2	Alcoholic
3	More than μ_2; with center at μ_3	Epileptic

Table 3 Classification through hard thresholding classifier for MATLAB and R studio

Platform classes	Alcoholic	Epileptic	Normal	Total
MATLAB	46	9	9	64
R studio	43	9	12	64

i. $\mu_1 < \mu_2 < \mu_3$;
ii. $(\mu_1 + \mu_2)/2 < \mu_2 < (\mu_2 + \mu_3)/2$;
iii. $(\mu_2 + \mu_3)/2 < \mu_3$.

Table 2 depicts the gold standard for Thresholding classifier based upon SVD values.

The algorithm is applied to a classifier and here they received pattern is a[n] = b[n]+d[n], where b(n) indicates an unknown pattern to be detected and d(n) indicates a white Gaussian noise. A lot of high-frequency features are present in the original pattern. The standard Thresholding techniques consist of both soft and hard Thresholding function.

Soft Thresholding is denoted as

$$Y(t) = \{a - sgn(a)T \text{ if } |a| \geq T;$$
$$0 \text{ if } |a| < T$$

Hard thresholding is denoted as

$$Y(t) = \{ \ a \text{ if } |a| \geq T;$$
$$0 \text{ if } |a| < T$$

Table 3 shows the classification effectiveness of the alcoholic EEG data obtained through a single trial 64 channel using Hard Thresholding classifier for both MATLAB and R studio.

Out of the 64 channels, 46 channels were fully alcoholic, 9 channels were false classified as epileptic, and 9 channels were misclassified as normal by the MATLAB. For the R studio out of the 64 channels, 43 channels were fully alcoholic, 9 channels were false classified as epileptic, and 12 channels were misclassified as normal.

Fig. 4 Flow chart K-means
clustering algorithm

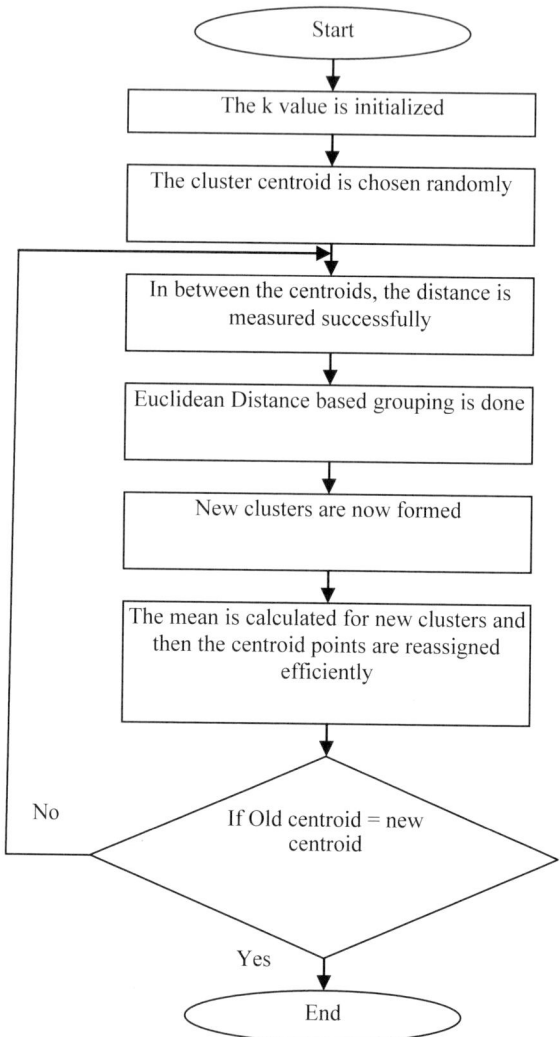

3.2 K-Means Clustering

K-means clustering algorithm belongs to partitioning methods and is used widely
because of its robustness, fast convergence, and simplicity [7]. Figure 4 shows the
flow chart for K-means algorithm.

The basic process can be explained as follows:

First, the initialization of the K cluster centers is done and is chosen in a random
manner.

Table 4 Classification through K-means clustering for the MATLAB and R studio

Platform classes	Alcoholic	Epileptic	Normal	Total
MATLAB	36	8	20	64
R studio	32	8	24	64

Second, each y_i is assigned to its nearest cluster center q_k by means of employing the Euclidean Distance (d).

$$KM(Y, Q) = \sum_{i=1}^{n} \min_{j \in \{1....k\}} \left\| y_i - q_j \right\|^2$$

$$d(f, g) = d(g, f) = \sqrt{\sum_{i=1}^{n} (g_i - f_i)^2}$$

Third, each cluster center q_k is updated as the mean of all y_i that belongs to it.

Finally, unless the cluster centers have obtained stability, the steps are repeated.

Table 4 shows the classification usefulness of the alcoholic EEG data obtained through a single trial 64 channel using K-means clustering for both MATLAB and R studio.

As in the case of K-means clustering both the MATLAB and R studio classifies the data set in a similar manner. Out of the 64 channels, 36 channels were correctly classified as alcoholic, 8 channels were false classified as epileptic, and 20 channels were misclassified as normal in MATLAB software. Out of the 64 channels, 32 channels were correctly classified as alcoholic, 8 channels were false classified as epileptic, and 24 channels were misclassified as normal in R Studio software.

4　Results and Discussion

This paper identifies a good classifier and a better platform for an alcoholic EEG data set using SVD values and two different types of software. The performance of the classifiers as in this case of Hard Thresholding and K-means clustering are evaluated against their benchmark parameters like Performance Index, Accuracy, Specificity, and Sensitivity and the average results are computed in Table 5. The mathematical formulae for the Performance Index (PI), Sensitivity, Specificity, and Accuracy are given as follows:

$$PI = [PC - MC - FA/PC] * 100$$

Table 5 Average performance of SVD features with hard thresholding and K-means clustering in MATLAB software

Parameters (%)	Hard threshold as a classifier	K-means clustering as a classifier
Perfect Classification PC	71.8	40.62
Missed Classification MC	14.1	46.87
False Alarm FA	14.1	12.5
Performance Index PI	60.7	46.15
Specificity	83.5	46.43
Sensitivity	83.5	76.47
Accuracy	83.5	61.45

Table 6 Average performance of SVD features with hard thresholding and K-means clustering in R studio software

Parameters (%)	Hard threshold as a classifier	K-means clustering as a classifier
Perfect Classification PC	67.18	56.25
Missed Classification MC	14.1	31.25
False Alarm FA	18.72	12.5
Performance Index PI	51.15	55.47
Specificity	78.2	81.8
Sensitivity	82.65	64.28
Accuracy	80.42	73.05

where Perfect Classification is indicated by PC, Missed Classification is denoted by MC and False Alarm is specified as FA.

The Sensitivity, Specificity, and Accuracy measures are calculated as follows:

$$\text{Sensitivity} = [PC/(PC + FA)] * 100$$
$$\text{Specificity} = [PC/(PC + MC)] * 100$$
$$\text{Accuracy} = (\text{Sensitivity} + \text{Specificity})/2$$

Table 5 shows the classification worthiness of the both classifiers in the MATLAB platform. As tabulated in the Table 5, the Hard Thresholding method better classifies the data set with an accuracy of 83.5% when compared to the accuracy of 61.45% in K-means clustering method.

Average Performance of SVD features with Hard Thresholding and K-means clustering in R is shown in Table 6. It is observed from Table 6 that once again Hard Thresholding classifier scores better in the R studio platform with an accuracy of 80.42% when compared to the accuracy of 73.05% for K-means clustering algorithm.

5   Conclusion

Identification of valuable features and a better classifier is discussed in this paper for alcoholic EEG Signals of a single patient. SVD features are calculated and classified by using K-means clustering and Hard Thresholding. Softwares like MATLAB and R studio results are utilized and the results are compared. The results show that MATLAB software classifies better than R studio software with the comparatively highest classification accuracy of 83.5% is obtained when Hard Thresholding method is utilized. Further research will be in the usage of more heuristic and meta-heuristic classifiers for the analysis on detection of chronic alcoholics from EEG data segments.

References

1. Zou Y, Miao D, Wang D (2010) Research on sample entropy of alcoholic and normal people. Chin J Biomed Eng 29:939–942
2. Wu D, Chen ZH, Feng RF, Li GY, Luan T (2010) Study on human brain after consuming alcohol based on EEG signal. In: Proceedings of the 2010 3rd IEEE international conference on computer science and information technology, vol 5
3. Prabhakar SK, Rajaguru H (2016) Comparison of fuzzy output optimization with expectation maximization algorithm and its modification for epilepsy classification. In: International conference on cognition and recognition (ICCR 2016), Mysore, India, 30–31 December 2016
4. Prabhakar SK, Rajaguru H (2017) Softmax discriminant classifier for detection of risk levels in alcoholic EEG signals. In: IEEE Proceedings of the international conference on computing methodologies and communication (ICCMC July 2017), Erode, India
5. Prabhakar SK, Rajaguru H (2015) Performance comparison of fuzzy mutual information as dimensionality reduction techniques and SRC, SVD and approximate entropy as post classifiers for the classification of epilepsy risk levels from EEG signals. In: Proceedings of 2015 IEEE student symposium in biomedical engineering and sciences (ISSBES), 4 November 2015, Universiti Teknologi Mara, Malaysia
6. Zhang Q, Rossel RA, Choi P (2006) Denoising of gamma-ray signals by interval-dependent thresholds of wavelet analysis. Meas Sci Technol 731–735
7. Prabhakar SK, Rajaguru H (2015) PCA and K-means clustering for classification of epilepsy risk levels from EEG signals–a comparative study between them. In: Proceedings of the international conference on intelligent informatics and biomedical sciences (ICIIBMS), 28–30 November 2015, Okinawa, Japan

A System for Plant Disease Classification and Severity Estimation Using Machine Learning Techniques

Anakha Krishnakumar and Athi Narayanan

Abstract In India, more than 80% of agrarian crops are produced by smallholder farmers. The reports point that almost half the yield loss is mainly due to pests and diseases. Unlike pests, diseases are more difficult to detect and treat. Numerous studies and researches have been put forward to identify the behaviour of different diseases. Traditionally, farmers use naked eye observation for detecting disease but one of the areas considered today is processing the images with machine learning concepts to assist the farmers technologically. This paper presents an image processing strategy to classify sort of disease in a cucumber plant and gives a severity measure of malady spots in the cucumber leaf caught under real field condition.

1 Introduction

India is a developing country and around 70% of the people depend on agriculture. Agricultural profitability on production is something in which economy particularly depends. This is the one reason that disease detection in plants accepts an indispensable part in cultivation. The quantity, quality and productivity of crops are very important and hence proper care should be taken to resolve serious effects on plants.

The methods in image processing aim to recognize the leaf ailments. Otsu thresholding, K-means are the different segmentation techniques to separate the region of interest. Support Vector Machine (SVM), LDA, Neural System (NN) are the disease classification techniques to identify the affected symptoms on plants. The past works have a couple of demerits as lack of accuracy, lack of severity measure and lack of generic model [1].

A. Krishnakumar (✉) · A. Narayanan (✉)
Department of Computer Science and Engineering,
Amrita Vishwa Vidyapeetham, Amritapuri, India
e-mail: anakhakmr@gmail.com

A. Narayanan
e-mail: mail2athi@gmail.com

© Springer Nature Switzerland AG 2019
D. Pandian et al. (eds.), *Proceedings of the International Conference on ISMAC in Computational Vision and Bio-Engineering 2018 (ISMAC-CVB)*, Lecture Notes in Computational Vision and Biomechanics 30,
https://doi.org/10.1007/978-3-030-00665-5_45

447

Fig. 1 Different types of affected leaf: **a** Alternaria leaf blight. **b** Angular leaf spot. **c** Bacterial leaf spot. **d** Bacterial Wilt. **e** Cercospora leaf spot. **f** Cucumber mosaic. **g** Target leaf spot. **h** Powdery mildew. **i** Downy mildew. **j** Phytophthora blight

Cucumis sativus is the scientific name of cucumber and is yearly creeping plant with sprawling vine, immense leaves and turning rings. Alternaria leaf blight, angular leaf spot, bacterial leaf spot, bacterial wilt, cercospora leaf spot, cucumber mosaic, target leaf spot, powdery mildew, downy mildew, phytophthora blight are the different kinds of diseases seen in cucumber leaf [2]. The following are mentioned in Fig. 1.

The paper is organized as follows: Sect. 2 specifies the related work. Section 3 introduces the proposed approach and Sect. 4 includes implementation and experimental results.
Section 5 concludes and specifies the future work.

2 Related Works

Lots of work in the field of detection of leaf diseases have been done by utilizing advanced image processing techniques. The methods include image acquisition, image preprocessing, image segmentation and feature extraction.

2.1 *Image Processing*

Initially, the technique of capturing the images in the RGB format are used for further processing and analysis purpose. The image taken under real field condition may contain inconsistency due to the lighting effects, which reduces the quality of picture. To deal with this situation, various preprocessing techniques are implemented such as mean filter, median filter and sharpening filter etc. To identify the distinctive portion in the image (the area of related pixels with equivalent properties), to discover the boundaries between the regions are the next important concerns. There are different strategies like K-means, Canny and Sobel division and ostu thresholding, etc. [3].

The preprocessing of image deals with removal of noise. There are different techniques to remove salt and pepper noise, Gaussian noise, Poisson noise, speckle noise, etc. [4]. The Fourier filtering transforms the digital image into frequency domain which implies the no. of frequency corresponding to no. of pixels in the spatial domain. Therefore, it is easy to examine the influence of particular pixels in an image and can be processed for further analysis. Sobel operator and Laplacian operator are used in edge detection. The edge detection aims to get the shape information of image where edge implies the sudden change in pixel values of an image [4]. The edge detection preprocessing which enhances the area of the image encloses edges and hence the sharpness of image increases, which makes the image clearer [5].

In the morphological preprocessing process the image is based on shape which mainly includes dilation and erosion operations. No of pixels added or removed from objects in the image depends on the size and shape of the structuring element [6]. Mainly concentrated to enable the selective enhancement of target structures. Gaussian filtering is used to blur the image to reduce the noise whereas in thresholding, it is used to classify the pixel values into different categories for later image analysis.

K-means is an unsupervised learning technique in which clusters with equivalent property are obtained. Initially, K-means picks k centroids and then reshape the clusters by assigning all data centre to the closest centroid of each cluster. In K-means we need to specify the number of clusters explicitly. But in otsu thresholding technique, all the possible threshold values are considered for choosing the one with minimum intraclass variance. The edges of diseased part are obtained by applying Canny and Sobel technique [7].

In order to get informative part from the segmented portion of the image, we predominantly choose shape, texture and colour features. These features are extracted mainly to identify the disease part from the image. Structural features give the information about size and shape of image whereas texture features deal with the intensity values [3].

2.2 Machine Learning

Machine learning algorithms are used to build the model which iteratively learns from the given information and forecast the future results. The area is broadly classified as supervised learning, unsupervised learning and reinforcement learning [7].

Support Vector Machine (SVM) classifies the data on the basis of decision planes modelled from the input samples. It can be linear classifier or nonlinear classifier depending upon the no of classes to be classified. In K-Nearest Neighbour classifier (K-NN), the training examples are classified in the view of nearest examples in the feature space [8]. Neural Network (NN) is another technique which uses the backpropagation system to classify the training examples but the system seems to be slower to train [7]. There are many kinds of classifiers used based on different features [9]. Effective results can be obtained from effective learning by training the model with large amount of dataset [10].

3 Proposed Method

3.1 System Design

The system mainly concentrates on early detection of plant disease. The basic diagrams for leaf disease detection are shown in Figs. 2, 3 and 4.

3.2 Disease Severity

Disease severity is the ratio of diseased area to total leaf area. The formula is given below:

$$DS = DA/LA \tag{1}$$

where

DS   Disease Severity

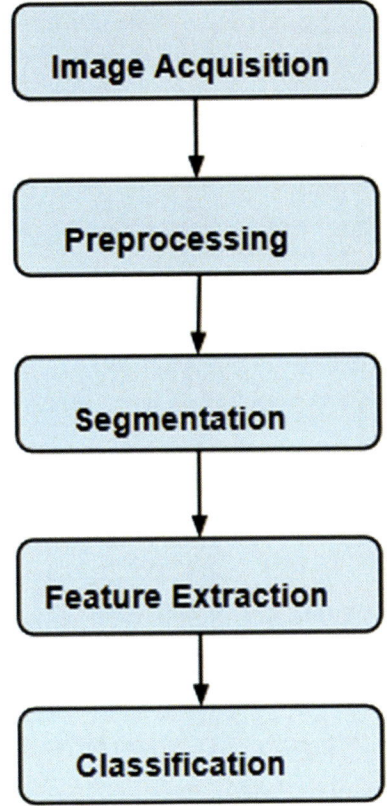

Fig. 2 System architecture

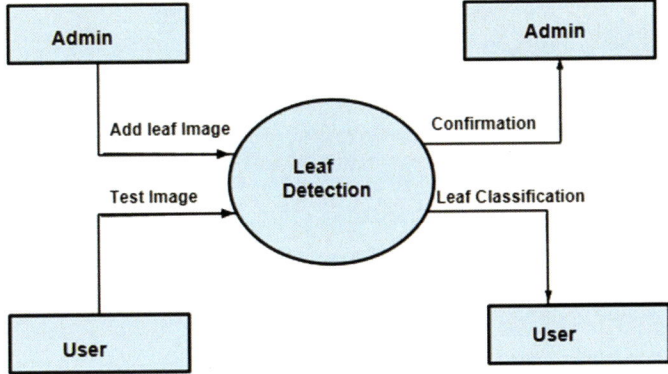

Fig. 3 Basic mechanism of the model

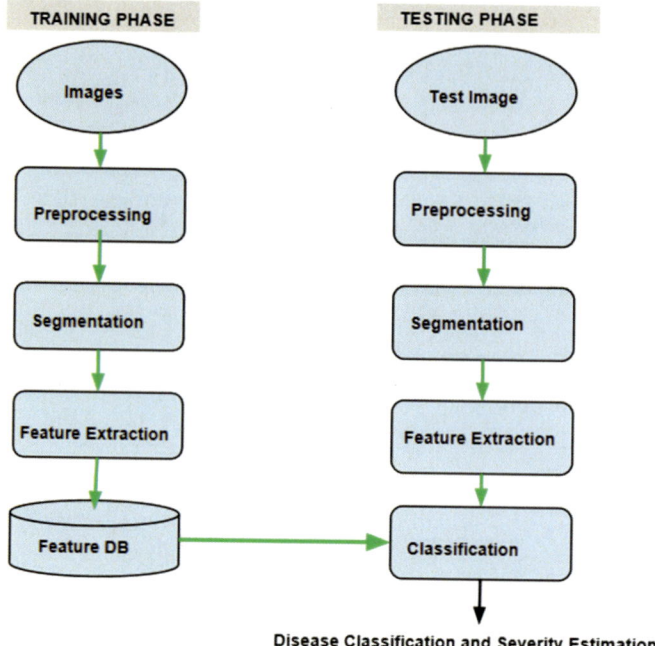

Fig. 4 Detailed workflow of model

DA   No. of Pixel constitute disease area
LA   No. of Pixel constitute normal leaf area.

3.3 Accuracy

The proportion of accurately recognized test images to the total number of test images and is given in the equation:

$$\%\text{Accuracy} = [\text{CRT}/\text{TI}] * 100 \qquad (2)$$

CRT   Correctly Recognized Test Images
TI   Total Test Images.

4   Implementation and Experimental Results

Classification includes training and testing phase. The extracted feature values and its respective target values are utilized to train the model. Later this trained classifier

Fig. 5 Input and preprocessed image

model is used to predict the future results. Performance of SVM classifier is 86% when 50% of images from dataset is used for testing and the remaining for training.

4.1 Image Preprocessing

First step is the enhancing of the original image. Figure 5 shows original image and processed image.

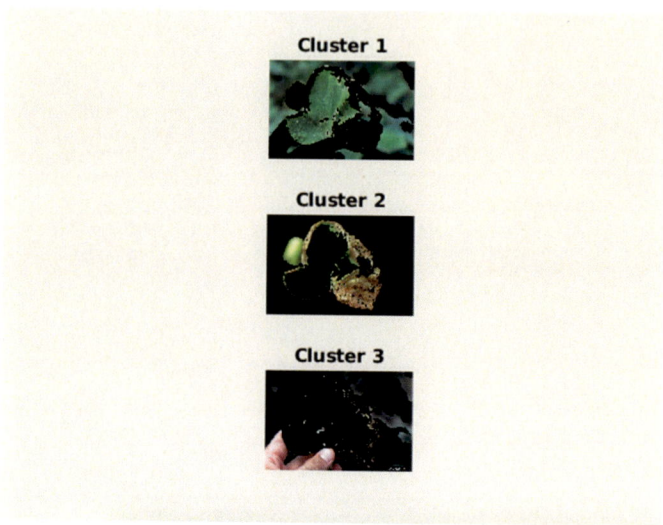

Fig. 6 Segmentation using K-means clustering

4.2 Segmentation

The enhanced image is then segmented into three clusters. Figure 6 shows three clusters formed using K-means clustering.

4.3 Disease Classification and Disease Measure

The features include colour and texture which are extracted from the segmented images. In this work, ten features are figured for the sectioned parts of single leaf image [2]. These features, aggregately known as feature vector, and are utilized to train the SVM classifier. As a result, classifying the sample input leaf image into various classes. Features can be extracted using texture method or HOG to identify the different kind leaf disease and further using SVM for classification [9]. Figure 7 demonstrating the classification results.

Brightness refers to overall lightness or darkness of the image. Increasing the pixel value refers to lightening and decreasing the pixel value refers to darkening. Contrast is actually related to brightness where increasing the contrast makes light area lighter and dark area darker. Mathematically contrast refers to difference between maximum and minimum pixel. Images with high contrast are easier to recognize. As a part of feature analysis, it is important to understand how correlated a pixel to its neighbours. This can be measured based on the correlation factor [5].

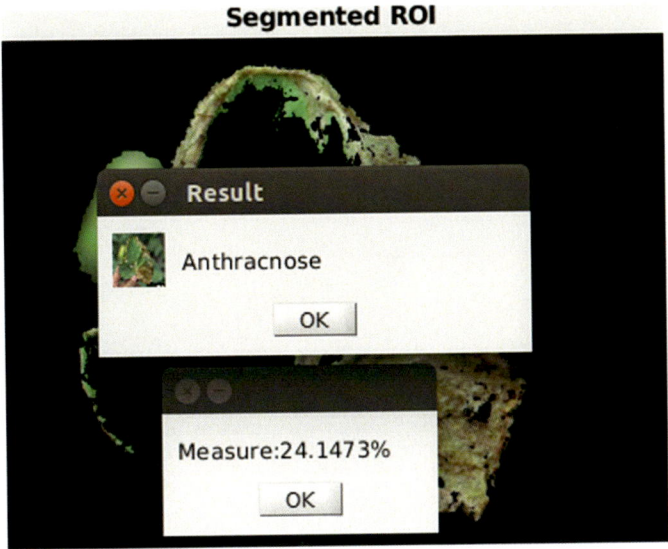

Fig. 7 Disease and its measure

We can analyze image based on pixel values since the image is composed of different ranges of pixels. Sometimes the values seem to be more similar or more random. To track these changes, there are different technical terms introduced. The measure of similarity of pixel refers to homogeneity and measure of randomness refers to the entropy. Uniformity of image refers to energy. Hence the homogeneity and energy are strongly correlated to each other. If the pixel seems to be more similar then the image will be more uniform and for a uniform image the value is set to 1. Therefore the image is a constant one and hence the contrast for constant image is 0.

Skewness is the measure of symmetry which implies the distribution on the left and right of the centre point is same. Negative value of skewness refers to skewed left implies left tail is long relative to right tail whereas positive value of skewness refers to skewed right and vice versa of skewed left. Kurtosis is the measure that deals whether the dataset is peaked or flat relative to normal distribution. The dataset with high kurtosis tends to have outliers whereas dataset with low kurtosis lack the outliers. The structural features provide information about the size and shape whereas texture features provides information based on the variation in intensities. The measure of image texture refers to IDM with 0.0 as highly textured and 1.0 as untextured [5] (Table 1).

Table 1 Image texture features

Features	Description
Contrast	Difference in brightness. Contrast is 0 for constant image
Correlation	How a pixel correlated to its neighbours. Range can be from -1 to $+1$
Energy	Measure of uniformity
Homogeneity	Measure of similarity of pixel
Entropy	Measure of randomness in pixel value
Variance	Indicates how each pixel varies from the mean pixel
Mean	Contribution of each pixel in the entire image
Standard deviation	Measure of deviation from mean
Skewness	Measure of symmetry where $-ve$ value of skewness indicates left tail is long relative to right tail and vice versa
Smoothness	Measure of blurring to reduce noise
Kurtosis	If the dataset with high kurtosis tend to have outliers else lack of outliers
IDM	Measure of image texture where 0.0 is highly textured and 1.0 is untextured

Table 2 Management techniques

Disease	Cause	Pesticides
Alternaria leaf spot	Fungus	Chlorothalonil or mancozeb or copper fungicide
Anthracnose	Fungus	Chlorothalonil or mancozeb
Powdery mildew	Fungus	Sulphur or chlorothalonil or horticultural oil + baking soda
Downy mildew	Oomycete	Chlorothalonil or mancozeb or copper fungicide
Cercospora leaf spot	Fungus	Chlorothalonil or mancozeb
Bacterial wilt	Bacterium	no chemical control

5 Conclusion and Future Work

In this paper, we mainly focus on disease classification and severity measure which helps the farmers technologically to identify the leaf disease at the earliest. Also, we can identify the stage of affected plant based on the measure and classification and required management technique can be applied. This gives proficient and exact plant disease detection, discovery and characterization, measure figuring strategy by utilizing MATLAB. As a part of future expansion, we can bring out computerized model with the help of embedded system which can naturally splash the fertilizer by system spraying mechanism effectively at appropriate time (Table 2).

References

1. Sladojevic S, Arsenovic M, Anderla A, Culibrk D, Stefanovic D et al (2016) Deep neural networks based recognition of plant diseases by leaf image classification. Comput Intell Neurosci. http://dx.doi.org/10.1155/2016/3289801, Article ID 3289801, 11 pp
2. Prajapati BS, Dabhi VK, Prajapati HB et al (2016) A survey on detection and classification of cotton leaf diseases. In: International conference on electrical, electronics, and optimization techniques (ICEEOT)
3. Khirade SD, Patil AB (2015) Plant disease detection using image processing. In: 2015 International conference on computing communication control and automation
4. Thangavel SK, Murthi M et al (2017) A semi automated system for smart harvesting of tea leaves. In: 2017 International conference on advanced computing and communication systems (ICACCS-2017) 06, 07 Jan 2017, Coimbatore, India
5. Padol PB, Yadav AA et al (2016) SVM classifier based grape leaf disease detection. In: 2016 Conference on advances in signal processing (CASP), Cummins College of Engineering for Women, Pune. 9–11 Jun 2016
6. Singh V, Misra AK et al (2017) detection of plant leaf diseases using image segmentation and soft computing techniques. Inf Process Agric 4:41–49
7. Pujari JD, Yakkundimath R, Byadgi AS et al (2016) SVM and ANN based classification of plant disease using feature reduction technique. Int J Interact Multimed Artif Intell 3(7)
8. Mohanty SP, Hughes DP, Salath M et al (2016) Using deep learning for image-based plant disease detection, vol 7, September 2016. Article 1419
9. Venkataraman D, Mangayarkarasi N et al (2017) Support vector machine based classification of medical plants using leaf features. In: 2017 International conference on advances in computing, communications and informatics (ICACCI)
10. Anjali Rani KA, Supriya P, Sarath TV et al (2017) Computer vision based segregation of carrot and curry leaf plants with weed identification in carrot field. In: Proceedings of the IEEE 2017 international conference on computing methodologies and communication (ICCMC)

A Clinical Data Analytic Metric for Medical Ontology Using Semantic Similarity

Suraiya Parveen and Ranjit Biswas

Abstract Ontology is a set of concepts in a domain that shows their properties and the relations between them. Medical domain Ontology is widely used and very popular in e-healthcare, medical information systems, etc. The most significant benefit that Ontology may bring to healthcare systems is its ability to support the indispensable integration of knowledge and data (Pisanelli et al, Proceedings biological and medical data analysis, 6th international symposium, 2005, [1]). Graph structure is very important tool for Foundation, Analysis, and Domain Knowledge. Ontology as a graphical model envisages the process of any system and present appropriate analysis (Pedrinaci, Ontology-based metrics computation for business process analysis, [2]). In this study, the knowledge provided by the Ontology is further explored to obtain the related concepts. An algorithm to compute the related concepts of Ontology is also proposed in a simplified manner using Boolean Matrix. The inferences from this study may serve to improve the diagnosis process in the field of Biomedical Intelligence and Clinical Data Analysis.

1 Introduction

The Ontology is intended to accomplish a common and shared knowledge that can provide interoperability between computer applications, web pages, etc. ICT is becoming key foundation to improve healthcare by storing and analysing huge amount of data. ICT ensures its availability to all stakeholders in real time [3]. Medical ontologies are big support to realize the opportunity. Ontology is the most basic component of semantic technologies. This makes knowledge sharing fast and convenient. Interoperability, reuse and sharing ability of Ontology are powerful aspects

S. Parveen (✉) · R. Biswas
Department of Computer Science & Engineering,
SEST, Jamia Hamdard, New Delhi, India
e-mail: husainsuraiya@gmail.com

R. Biswas
e-mail: rbiswas@yahoo.com

© Springer Nature Switzerland AG 2019
D. Pandian et al. (eds.), *Proceedings of the International Conference on ISMAC in Computational Vision and Bio-Engineering 2018 (ISMAC-CVB)*, Lecture Notes in Computational Vision and Biomechanics 30,
https://doi.org/10.1007/978-3-030-00665-5_46

to make semantic search possible [4]. The Ontology is visualized in the form of a graph structure. Graph forms the foundation for the design of many information processing systems such as transaction processing systems, decision support systems, project management systems, workflow systems, knowledge management, intelligent integration, search engine, and information retrieval [5]. The Ontology/schema is used to represent complex domain knowledge. Ontology represents knowledge of the domain with the help of "*IS-A*" relationship which is hierarchical [6]. In this study, Ontology is represented using Graph Structure in lieu Tree Structure. A graph represents Ontology more meritoriously in case of simple as well as complex domain. The nodes and link types are related through an Ontology graph [7].

A Graph is defined by a pair $G = \{V, E\}$ where $V = \{x1, x2, x3... xn\}$ is a finite set of agile vertices and E a collection of edges that happen to connect these vertices.

Ontology is knowledge representation and to retrieve relevant information from the ontology is very essential. Similarity measurement in Ontology has been proposed by many researchers [8–12]. Boolean Matrix representation of graph structure is exploited to find the semantic similarity in Ontology. The idea is to convert Ontology in Boolean matrix form [6], and then use graph techniques to find semantic similarity of concepts. Ontology has sometimes been very complex; so it is useful to represent it in some simplified form.

In Sect. 1, we introduce the ontology and its representation. In Sect. 2, the approach of APO ontology, developed for K4Care project [13] is outlined by giving an overview. Section 3 elaborates the methodology of research and give the algorithm based on the study. In Sect. 4, we give an evaluation of our algorithm. Results are given in Sect. 5. Section 6 conclude the paper.

2 Case Study of APO Ontology

We are using Actor Profile Ontology (APO) to study the conceptual relationship in the Healthcare Domain. It has 399 concepts. APO Ontology is developed in the EU K4Care Project "Knowledge-Based Home care Service for an aging Europe". This project takes care of health issues of senior citizens in Europe. This is a knowledge-based e-health care project. The K4CARE [13] summarized below is a simple web-based model to provide home care services. In Fig. 1 the base concepts of APO Ontology are represented as hierarchal structure. These are the minimum elements required to provide basic home care. APO summarizes the contents of these concepts.

- Entity: An Entity is somebody who can perform an action on the K4Care project.
- Actor: The Actor class represents somebody in the K4Care. Actors are people interacting with the HC system. The class Actor is divided into subclasses such as Stable Members (Composed of Family Doctors, Physicians in Charge, Head Nurses, Social Workers, and Nurses), Additional Care Givers (Specialist Physician, Social Operator, Continuous Care Provider, and Informal Care Giver) and Patients.

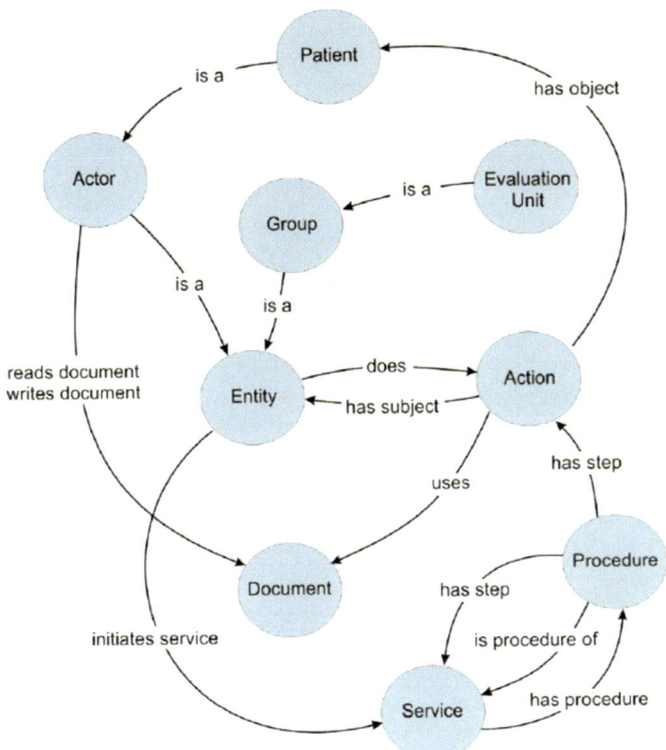

Fig. 1 APO base concept

- Evaluation Unit: This class represents a medical evaluation unit, a group that works together in the assessment and evaluation of patients. It is composed of a Family Doctor, a Physician Incharge, a Head Nurse, and a Social Worker.
- Services: The Service class represents the home care services. These services are classified into Access services, Patient Care services, and Information services.
- Procedure: This class represents a procedure associated with a service.
- Action: The class Action represents a set of simple actions that are executed inside one or more procedures.
- Document: This class represents the documents that are read or written during the execution of one action. These documents are stored in the EHR.

3 Methodology

In Ontology, we have considered only the taxonomical relation within a domain [6]. We require measuring semantic similarity within ontological concepts. Ontology is

Table 1 NCP matrix

	Patient	Actor	Entity	Action	Evaluation unit	Group	Document	Procedure	Service
Patient	#	1	0	1	0	0	0	0	0
Actor	1	#	1	0	0	0	1	0	0
Entity	0	1	#	1	0	1	0	0	1
Action	1	0	1	#	0	0	1	1	0
Evaluation unit	0	0	0	0	#	1	0	0	0
Group	0	0	1	0	1	#	0	0	0
Document	0	1	0	1	0	0	#	0	0
Procedure	0	0	0	1	0	0	0	#	1
Service	0	0	1	0	0	0	0	1	#

knowledge representation about a specific domain and presented in the form of the graph structure. Graphs are analytical tools which themselves give lots of information. In this paper, Ontology graph is manipulated with computation technique to find the sub-concept and super-concept of any particular concept (C_i). This is done for fast and simple retrieval of information from the Ontology.

This paper is an attempt to modify adjacency matrix representation of graph and using it for similarity measurement. Table 1 shows the Boolean matrix, which includes immediate successor and the immediate predecessor of the APO Ontology. This matrix leads to form a cluster of closely related concepts, which are interdependent on each other. The algorithm for the computation does not consider directions, but only adjacent pairs of a concept. We have named it **Neighbor Concepts Pair (NCP)**. To validate our results, we have used a statistical technique called Simple Matching Coefficient (SMC) [14].

3.1 Neighbor Concept Pair Matrix for APO Ontology

The aim of our work is to extract similar concepts within an Ontology. The above discussion about Ontology reveals that the predecessor and successor of any concept in the Ontology are related to each other being hierarchal. The following is the NCP matrix (n*n) of APO Ontology with nine concepts represented in Table 1. Each row contains 1 if the concept is adjacent (sub-concept/super-concept) with column concept, otherwise 0. Hash # symbol in the matrix shows same concept in row and column [15].

$$A_{ij} = 1, \text{ If there is an edge from } v_i \text{ to } v_j \text{ or } v_j \text{ to } v_i$$
$$0, \quad \text{otherwise.}$$

Table 2 Algorithm

Adjacency list

Input: Adjacency Matrix (m × n)

Output: Adjacent vertices of V$_i$

Step 1: Each Vertex maps to corresponding adjacent vertices (do not consider directions)

Step 2: Vertex List<adjacent vertices>

Step 3: To get all vertices adjacent to v$_1$,

Step 4: List<adjacent vertices> neighbors = adj.get(v$_1$)

 // input value 1means i is adjacent to j

 // input value 0 means i is not adjacent to j

Adjacency map (adjacency hash_table for concepts)

Each Vertex maps to a hashtable of adjacent vertices
Vertex (Vertex Edge)
To find out whether there's an edge from v$_1$ to v$_2$, call
adj.get(v$_1$).containsKey(v$_2$)
To get the edge from v$_1$ to v$_2$, call adj.get(v$_1$).get(v$_2$)

3.2 Algorithm for Extracting Adjacent Clusters from NCP Matrix

An algorithm has been designed on above explanation of neighboring concept pair. This will extract sub-concepts and super concepts of any concept (C_i) in an Ontology. The algorithm is given in Table 2.

3.3 Clusters of Similar Concepts Generated by NCP Matrix

Ontology may be hierarchal or cyclic. In any case, the concepts in the Ontology are related to each other. When we represent the Ontology using proposed NCP matrix, it gives adjacent neighbors (subclass and super-class) of each concept in the Ontology, i.e., similarity cluster for each concept in Ontology. We obtain following related concepts, shown in Table 3 using this technique.

The semantically similar cluster has all nine basic concepts of APO ontology and corresponding row of each concept contain its related concepts.

4 Validation of Proposed Approach

Simple Matching Coefficient [14] is used for validation of our work. As per above matrix, APO Ontology shows the relationship between patient, actor, and action. Actor has relationship with patient, entity, and document. The cluster of similar concepts is shown in Table 3. To verify the similar concepts obtained from the Ontology using the proposed method; we have used Simple Matching Coefficient, as defined below:

Simple Matching Coefficient:

Simple Matching Coefficient (symmetric attributes) = number of matches/number of attributes

$$= (M\,11 + M\,00)/(M\,01 + M\,10 + M\,11 + M\,00)$$

The range of SMC is 0.00–1.00. The value SMC above 0.5 in a concept pair mean concepts are quite similar or related. SMC values below 0.5 means less similar concept pair.

We have applied the proposed method on APO Ontology and calculated Simple Matching Coefficient for concept pair as shown in Tables 4, 5, 6, and 7.

Table 3 Semantically similar cluster for APO ontology

Vertex	AV1	AV2	AV3	AV4	AV5
Patient	Actor	Action			
Actor	Patient	Entity	Document		
Entity	Actor	Group	Action	Service	Action
Action	Entity	Procedure			
Evaluation unit	Group				
Group	Evaluation unit	Entity			
Document	Actor	Action			
Procedure	Service	Action			
Service	Entity	Procedure			

Table 4 Similarity calculation patient and actor (neighboring concepts)

Patient	1	1	0	1	0	0	0	0	0
Actor	1	1	1	0	0	0	1	0	0

Table 5 Similarity calculation patient and action (neighboring concept)

Patient	1	1	0	1	0	0	0	0	0
Action	1	0	1	1	0	0	1	1	0

Table 6 Similarity calculation patient and document (neighboring concept)

Patient	1	1	0	1	0	0	0	0	0
Document	0	1	0	1	0	0	1	0	0

Table 7 Similarity calculation actor and evaluation unit (non-neighboring concept): SMC value for concept pair with no adjacency

Actor	1	1	1	0	0	0	1	0	0
Evaluation unit	0	0	0	0	1	1	0	0	0

5 Results

Simple Matching Coefficient is applied to NCP Matrix to verify the concept similarity within the Ontology. A minor modification is done so that the SMC can be applied. Hash # in NCP Table shows the same concept, but to find SMC it is taken as 1.

$M\,01 = 2$ (the number of attributes where p was 0 and q was 1)
$M\,10 = 1$ (the number of attributes where p was 1 and q was 0)
$M\,00 = 4$ (the number of attributes where p was 0 and q was 0)
$M\,11 = 2$ (the number of attributes where p was 1 and q was 1)

$$SMC = (M\,11 + M\,00)/(M\,01 + M\,10 + M\,11 + M\,00)$$
$$= (2+4)/(1+1+0+4) = 6/9 = 0.66$$

$M\,01 = 3$ (the number of attributes where p was 0 and q was 1)
$M\,10 = 1$ (the number of attributes where p was 1 and q was 0)
$M\,00 = 3$ (the number of attributes where p was 0 and q was 0)
$M\,11 = 2$ (the number of attributes where p was 1 and q was 1)

$$SMC = (M\,11 + M\,00)/(M\,01 + M\,10 + M\,11 + M\,00)$$
$$= (2+3)/(3+1+3+2) = 5/9 = 0.55$$

$M\,01 = 0$ (the number of attributes where p was 0 and q was 1)
$M\,10 = 0$ (the number of attributes where p was 1 and q was 0)
$M\,00 = 5$ (the number of attributes where p was 0 and q was 0)
$M\,11 = 2$ (the number of attributes where p was 1 and q was 1)

$$SMC = (M\,11 + M\,00)/(M\,01 + M\,10 + M\,11 + M\,00)$$
$$= (2+5)/(1+1+2+5) = 7/9 = 0.77$$

M 01 = 2 (the number of attributes where p was 0 and q was 1)
M 10 = 4 (the number of attributes where p was 1 and q was 0)
M 00 = 3 (the number of attributes where p was 0 and q was 0)
M 11 = 0 (the number of attributes where p was 1 and q was 1)

$$SMC = (M\,11 + M\,00)/(M\,01 + M\,10 + M\,11 + M\,00)$$
$$= (0 + 3)/(2 + 4 + 3 + 0) = 3/9 = 0.33$$

The similarity results for APO Ontology fulfil the four criteria of similarity measures [16]. These measures are as follows:

(1) Non-negativity: Similarity value cannot be less than zero.
(2) Identity: Sim (A, A) = Sim (B, B) = 1
(3) Symmetry: If any attribute of A is similar to attribute of B, Then same attribute of B is similar to same attributes of A.
    Sim (A, B) = Sim (B, A)
(4) Uniqueness: Sim (A, B) = 1 → A = B

The range of semantic similarity is from zero to one.

The above similarity result clearly shows that neighboring concept pair have high similarity values. Hence, they are more similar as compared to non-neighboring concept pair.

6 Conclusion

ICT is becoming a very valuable asset for healthcare, Patient's Medical Records, Drug Records, Schedules, Diagnosis, Test Report, and many related issues are maintained and retrieved in real time. The accuracy, timely reminders, fast processes, decision support tool for diagnosis are most important in Healthcare; the proposed technique of finding semantically similar clusters in this work can be very useful for the same. It is crucial to avoid any human errors, missing information, related disease to help patients under the observation of many doctors. The work can be used as a decision support tool in healthcare.

References

1. Pisanelli DM, Pinciroli F, Masseroli M (2005) The Ontological lens: zooming in and out from genomic to clinical level. In: Proceedings biological and medical data analysis, 6th international symposium, Nov 2005
2. Pedrinaci D (2009) Ontology-based metrics computation for business process analysis, Knowledge Media Institute, The Open University

3. Riañoa D, Joan FR, López A, Sara VF, Patrizia E, Roberta M (2012) Annicchiaricod carlo caltagironede: an ontology-based personalization of health-care knowledge to support clinical decisions for chronically ill patients. J Biomed Inform 45(3):429–446

4. Pérez G, López F, Corcho O (2004) Ontological engineering, 2nd printing. Springer, Berlin. ISBN: 1-85233-551-3

5. Biswas R, Gaur D (2008) Fuzzy meta node fuzzy metagraph and its cluster analysis. J Comput Sci 4(11):922–927. ISSN: 1549-3636

6. Batet M, Valls A, Giber K (2008) Measuring similarities in ontology by means of Boolean metrices, Intelligent Technologies for Advanced Knowledge Acquisition, Department of Computer Engineering and Maths, Universitat Rovira i Virgili

7. Barthélemy M, Chow E, Eliassi-Rad T (2005) Knowledge representation issues in semantic graphs for relationship detection, U.S. Department of Energy by University of California Lawrence Livermore National Laboratory, UCRL-CONF-209845

8. Wu Z, Palmer M (1994) Verb semantics and lexical selection. In: 32nd annual meeting of the association for computational linguistics, pp 133–138

9. Leacock C, Chodorow M (1998) Combining local context and WordNet similarity for word sense identification. In: Fellbaum C (ed) WordNet: an electronic lexical database. MIT Press, Cambridge, pp 265–283

10. Al-Mubaid H, Nguyen HA (2006) New ontology-based semantic similarity measure for the biomedical domain. In: Proceedings of the IEEE conference on granular computing, GrC-2006, Atlanta, GA, 10–12 May 2006, pp 623–628

11. Pedersen T, Patwardhan S, Michelizzi J (2004) WordNet: similarity –measuring the relatedness of concepts. AAAI, pp 1024–1025

12. Zhang X, Jing L, Hu X, Ng M, Zhou X (2007) A comparative study of ontology based term similarity measures on PubMed document clustering. In: Advances in databases: concepts, systems and applications DASFAA, vol 4443, pp 115–126

13. K4CARE, www.k4care.net

14. Segaran T (2007) Programming collective intelligence: building smart Web 2.0 applications. Beijing, O'Reilly

15. Samantha D (2001) Classic data structure. PHI

16. Veltkamp RC, Latecki LJ (2006) Properties and performances of shape similarity measures

True Color Image Compression and Decompression Using Fusion of Three-Level Discrete Wavelet Transform—Discrete Cosine Transforms and Arithmetic Coding Technique

Trupti Baraskar and Vijay R. Mankar

Abstract In this research paper, we have done the implementation and analysis of true color image compression and decompression technique. The implemented paper divides the color image into RGB component then after applying three-level Discrete Wavelet Transform, RGB components are split into nine higher frequency sub-bands and one lower order sub-band. The lower frequency sub-band is compressed into T-Matrix using One Dimension Discrete Cosine Transform. At the same time, higher frequency sub-bands are compressed using scalar quantize and eliminate zero and store data algorithm are applied to remove zeros in sub-band matrixes. Last, the encoded mode adopted arithmetic encoding. This algorithm has use two level of quantization this show significance improve in performance of compression algorithm. The decompression process is reverse process of encoder. The decompression algorithm decoded high-frequency subbands using return zero matrix algorithm and recover low-frequency sub-bands and other sub-bands using applying inverse process.

1 Introduction

Now a days, the use of image in decompression algorithm. The visual lossless compression is a practice that was presented by the imaging industry and it is trying to tell their users that encoding error are not noticed and viewing by persons. Maximum computer graphic adapter cards practice a true color model for display. It uses 8 bits per channel. Most graphics file formats accumulate 24-bit image where three

T. Baraskar (✉)
Department of Electronic Engineering, SGBAU, Amravati, India
e-mail: trupti.baraskar@mitpune.edu.in; baraskartn@gmail.com

V. R. Mankar
Department of Electronics Engineering, Government Polytechnic, Amravati, India
e-mail: vr_mankar@rediffmail.com

© Springer Nature Switzerland AG 2019
D. Pandian et al. (eds.), *Proceedings of the International Conference on ISMAC in Computational Vision and Bio-Engineering 2018 (ISMAC-CVB)*, Lecture Notes in Computational Vision and Biomechanics 30,
https://doi.org/10.1007/978-3-030-00665-5_47

component red, blue, and green has 8 bits each. This produces probable 16 million colors. A true color array can be of less unit8, unit16, single or double. In a true color array, each color component is a value between 0 and 1. If pixel has color values are (0, 0, 0) is displayed as black and if color value are (1, 1, 1) is displayed as white. A pixel can represent three color components in three dimensions of the data array. An example is RGB (5, 2, 1).

1.1 Image Format Used to Store RGB Image

The visual information that can be stored in digital computer has two main classes. The vector class presents the image information that can be described by simple geometric function, i.e., segment, color, data bands, circumferences. The raster class deal with the basic element of image, i.e., pixel value. The raster class arranges the pixel value in a pre-defined order that is later used by compression schemes.

Generally, raster class contains a representation of a graphic stored as pixels at a fixed position with a fixed resolution. The popular available raster formats are GIF, JPEG, PNG, TIFF, and BMP. The three dimension voxel raster class are used in medical imaging such as MRI Scanners. Another commonly used format is PBM (Portable Bitmap). This classifies into three different image format; binary, grayscale, and color images. This format has a common structure with uncompressed size [1, 2].

The proposed work is organized as follows. First in Sect. 2, we describe the existing techniques for true color image compression and decompression. In Sect. 3, there is a description of block diagram for compression and decompression, Sect. 4 has display result with a brief discussion. Finally, Sect. 5 is dedicated to the conclusion of implemented work.

2 Related Work

The compression of the true color image has various surveyed image compression approaches. The true color image compression algorithms are categorized into lossy or lossless compression schemes. The outmoded lossless compression method; RLE (Run-Length encoding) method has given better compression measure on the specific data with a large number of redundant information. In the true color image, the data has continuous pixel for one color line and this continuity can break down for multiline of the same color. Thus, the traditional RLE cannot guarantee to get reliable compression measures [3]. Thus, a new approach combined the DCT transform with improving RLE coding is implemented to achieve better compression ratio within acceptable range of data loss [4]. The Discrete Wavelet Transform (DWT) gives a high degree of correlation between RGB planes of the true color image. By combining Genetic Algorithm (GA) with DWT, more suitable space planes are generated. This result is found to be superior in terms of quality of the reconstructed image [5].

The available lossless compression methods include Shannon–Fano coding, Huffman coding, LZW (Lempel Ziv Welch) coding, and arithmetic coding. The Lossy compression methods contains predictive coding, block truncation, transform, vector quantization, sub-band, and fractal coding. Out of all abovementioned methods, discrete cosine transform is often used in the realization of transform coding for the image pixel [6–8].

The literature survey is based on transform approaches used by available compression techniques like discrete cosine transform and discrete wavelet transform. A new lossy compression algorithm has two transforms (DCT and DWT) with T-matrix coding and optimal sub-band threshold by using SURE and neigh shrink. The method enhances compression performance of image without reducing the PSNR quality [9, 10]. The served hybrid scheme combing the DWT and DCT transform which gives high compression ratio. In this paper, simulation work shows constant improved performance as compared to the JPEG-based DCT and the Daubechies-based DWT [11].

3   Proposed Work

In this paper, we have used fusion of transforms for implementation of color image compression. Mostly, color image has integer value of the pixels. Thus, for the compression, a luminance–chrominance model is considered where luminance represents the intensity of the image and looks like a grayscale and the chrominance components represent color information in the image [12, 13]. The color components of an 8-bit RGB image are integers in the range between 0 and 255 rather than floating-point values in the range [0, 1]. The proposed method works on luminance–chrominance models using DWT–DCT Transform. In this proposed method, image format used to store RGB image is Portable Pix Map [14].

3.1   Block Diagram of Compression

This compression approach is split into two parts; first, a three-level DWT is applied to a color image (A), which splits the image into nine high-frequency sub-bands (HL3, LH3, HH3, HL2, LH2, HH2, HL1, LH1, HH1) and one low-frequency sub-band (LL3). The high-frequency sub-bands at level one, two and three are quantized and encoded directly by EZSD. Second, the low-frequency sub-band are compressed by DCT and quantized then encoded by arithmetic coding (Fig. 1).

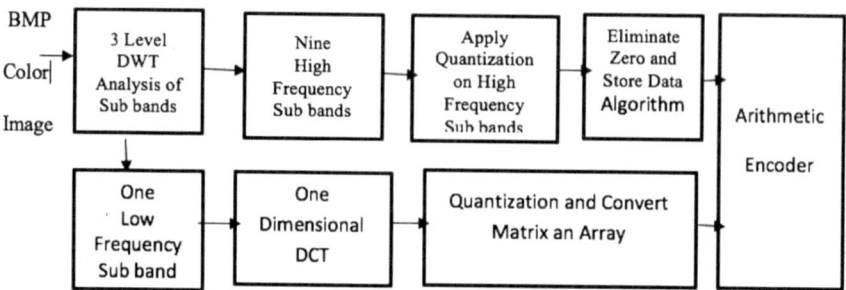

Fig. 1 The block diagram of compression

3.1.1 Three-Level DWT Decomposition

If the 2D analysis filter bank with Haar wavelet filter is applied on the digital image, the result is a first-level decomposition of Discrete Wavelet Transformed (DWT). The first level of decomposition has four parts, i.e., LL1, LH1, HL11 and HH1 sub-bands. It indicates the approximated image and the images with horizontal, vertical, and diagonal boundaries separately. The LL1 sub-band further to have a second and third level of DWT decomposition. The pixels within each sub-band represents the DWT coefficients corresponding to the sub-band. Mostly, the LL sub-band at the three level of decomposition has significant information as compared to all other sub-bands The approximated image has high-scale, low-frequency components and the detail of image has low-scale, high-frequency components. These results determine excellent energy compaction properties of DWT. This property can archive image compression. Normally, we have to transmit the image in limited bandwidth for compression and hence the DWT coefficients are to be quantized and efficiently encoded (Fig. 2).

3.1.2 Apply Quantization on Higher Frequency Sub-Bands

Quantization in an image compression is an example of lossy technique. Quantization is achieved by many-to-one mapping technique that replaces set of value into one value. There are two basic methods of quantization; SQ (scalar quantization) performs many-to-one mapping and VQ (vector quantization) replaces each block of input pixels with the index of a vector in the codebook. The decoder simply receives each index and looks up the corresponding vector in the codebook [15, 16]. In this proposed compression algorithm, we utilized high-frequency sub-bands as input vector to quantize using Scalar Quantization. Now Scalar Quantization performs many-to-one mapping on each value. The true value image contains standard quality factor and it varying [0.01–0.1] [17].

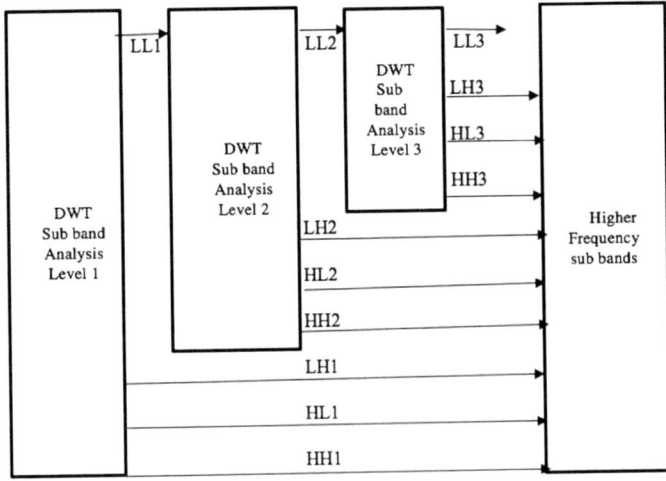

Fig. 2 The decomposition process of three-level DWT

Table 1 Evaluation of quantized level based on different quality value (Qv) and standard quality factor (Q)

Quantization level	If quality value (Qv) = standard quality factor (Q)	If quality value (Qv) > standard quality factor (Q)	If quality value (Qv) < standard quality factor (Q)
$Qm = Q \cdot 2^{m-1}$	$Qm = Qv \cdot 2^{m-1}$ if $Qv = 0.01$ and $Q = 0.01$	$Qm = Qv \cdot 2^{m-1}$ if $Qv = 0.05$ and $Q = 0.01$; $Qv = 0.1$ and $Q = 0.01$; $Qv = 0.2$ and $Q = 0.01$;	$Qm = Qv \cdot 2^{m-1}$ if $Qv = 0.005$ and $Q = 0.01$
$Q1 = Q$	0.01	0.05, 0.1, 0.2	0.005
$Q2 = 2Q$	0.02	0.1, 0.2, 0.4	0.01
$Q3 = 4Q$	0.04	0.2, 0.4, 0.8	0.02
$Q4 = 8Q$	0.08	0.4, 0.8, 1.6	0.04
$Q5 = 16Q$	0.16	0.8, 1.6, 3.2	0.08

If quantizes Value > 0.01 or 1% then quantized level of high-frequency sub-band are as following:

Table 1 shows the evaluation of the quantization level for different relation between Quality Value (Qv) = 0.01, 0.05, 0.1, 0.2, 0.05 where Standard Quality Factor = 0.01. Table 2 indicates that the output parameter of Quantization level depends on Quality Value. If $Qv \geq Q$; Image Quality low but we can achieve high compression ratio and vice versa (Fig. 3).

Table 2 Comparison of output parameter for selection of quality value (Qv)

S. No.	Condition for selection of quality value (Qv)	Output parameters	
		Image quality	Compression ratio
1	If quality value (Qv) > standard quality factor (Q)	Low	High
2	If quality value (Qv) < standard quality factor (Q)	High	Low
3	If quality value (Qv) = standard quality factor (Q)	Accept level	Accept level

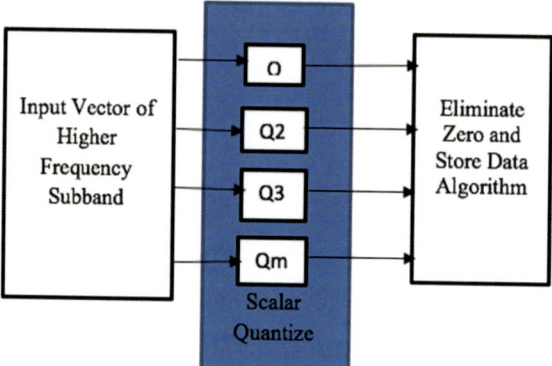

Fig. 3 The quantization process on higher frequency sub-bands

3.1.3 Apply Eliminate Zero and Store Data Algorithm on High-Frequency Sub-Bands

As level decomposition increases, sub-bands undergo partition and information details shift toward left corner (LL bands), meanwhile high-frequency sub-bands become more insignificant. Hence, the ignorance of this insignificant information is not perfect solution toward image quality. If quantized (If Quality Value (Qv) >= Standard Quality Factor [Q = 0.01]), high-frequency sub-bands have a rich number of zeroes, unnecessary coding of these zeroes make the algorithm complex and time-consuming. Therefore, EZSD is used to eliminate blocks of these zeroes and to store the blocks of nonzero data [15]. The EZSD algorithm is useful to increase compression ratio for high-frequency sub-bands. This algorithm is mainly applicable for eliminating zero and store any one nonzero data in a reduced array (Fig. 4).

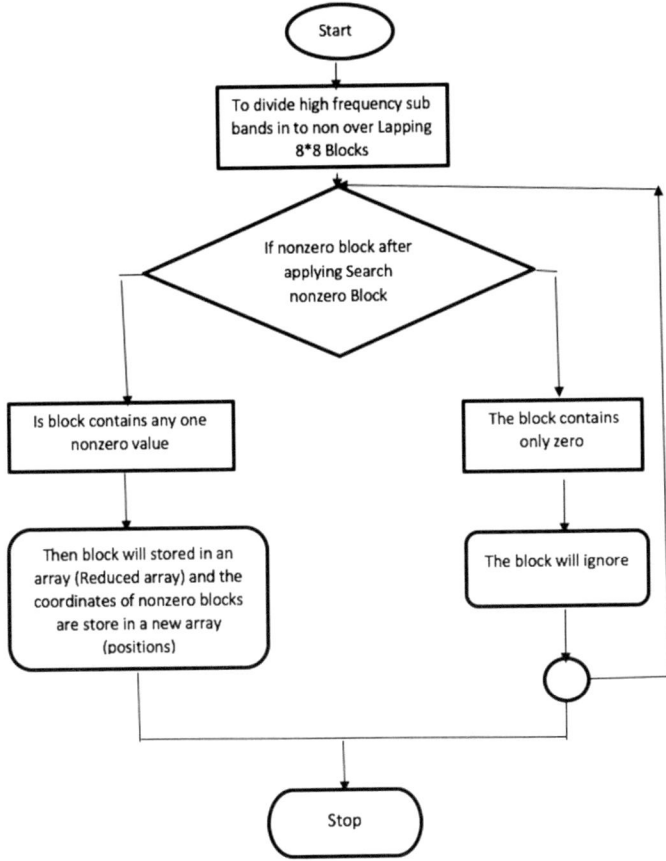

Fig. 4 The flow chart of eliminate zero and store data algorithm on high-frequency sub-bands

3.1.4 LL3 Sub Band Compression Using One-Dimensional DCT, Quantization Converted to an Array

The one-dimensional DCT and quantization process is applied to every row and column of LL3 sub-band array. The discrete cosine transforms (DCT) helps to separate the image into spectral sub-bands. Each sub-band has different importance (with respect to the visual quality of image). The equation for one dimension (N data items) DCT is as following (Fig. 5):

The equation of One-Dimensional DCT

$$X_c(k) = (1/N) \sum_{n=0}^{N-1} X_n \cos(k2\pi n/N) \quad \text{where } k = 0, 1, 2 \ldots N - 1$$

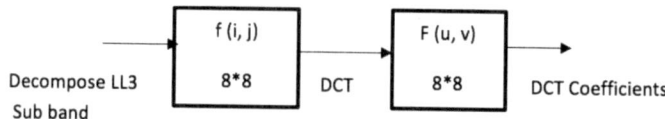

Fig. 5 One-dimensional DCT

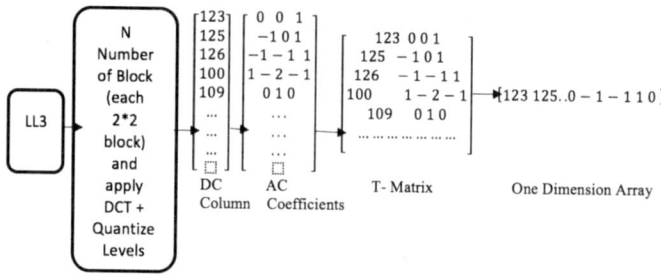

Fig. 6 The flow of implementation for quantizing levels and T-Matrix

The equation of One-Dimensional IDCT

$$X_c(k) = \sum_{n=0}^{N-1} C(u)X_n \cos(k2\pi n/N) \quad \text{where } k = 0, 1, 2 \ldots N - 1$$

$$C(u) = 1 \quad \text{for } u = 0 \text{ and } C(u) = 2 \quad \text{for } u = 1, 2, 3, \ldots N - 1$$

The LL3 sub-band coefficients are difficult to encode directly by arithmetic coding because of its properties of DCT-like De-correlation Property, Energy Compaction, and integer value. The proposed image compression algorithm using DCT and arithmetic encoding for true color image model are as follows: The LL3 sub-band is divided into 2 * 2 blocks of matrix.

1. Apply DCT on each 2 * 2 blocks of matrix.
2. The DCT produce DC and AC coefficients. Every DC coefficient is stored in the DC column matrix and, respectively, other three AC coefficients are stored in AC matrix.
3. The DC column is partitioned into 64 or 128 parts, then transformed value and stored in the matrix as row, called Transform Matrix (T-Matrix).
4. Then, scan T-Matrix column by column and convert them into one-dimensional array.
5. Apply compression process using arithmetic coding (Fig. 6).

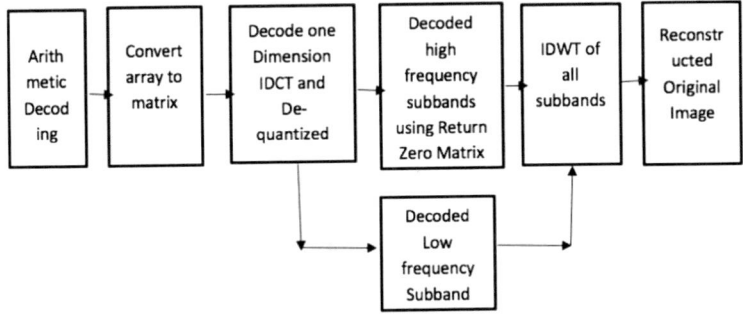

Fig. 7 Block diagram of decompression

3.2 Block Diagram of Decompression

See Fig. 7.

4 Result Analysis

The result shows that lower the MSE higher the PSNR and lower the SNR and better the compression ratio, out of five image4.bmp gives good result (Table 3).

Table 4 shows that if Quality value (Qv) > Quality Factor (Q) we get Higher PSNR and better compression ratio. Out of four third case gives good results (Figs. 8, 9, 10 and 11).

Table 3 Our compression and decompression methods result for five image inputs

Image name	Original image size (KB)	Compressed image size (KB)	Decompressed image size (KB)	PSNR	MSE	SNR	Compression ratio
image1.bmp	1352	79	Decompression not working	NA	NA	NA	17.11392405
image2.bmp	732	47	732	31.5364	59.182	26.4046	15.57446809
image3.bmp	769	38	732	35.5833	58.4021	29.7125	20.23684211
image4.bmp	2026	103	2026	30.1916	51.0853	22.3542	19.66990291
image5.bmp	769	35	769	35.992	60.5467	31.5013	21.97142857

Table 4 Comparison of image2.bmp, compression ratio for various quantization levels

	Original image size (KB)	Compressed image size (KB)	Decompressed image size (KB)	PSNR	MSE	SNR	Compression ratio
If quality Qv = Q; Qv = 0.01 and Q = 0.01	732	79	732	32.0296	59.182	26.8979	9.265822785
If quality Qv > Q; Qv = 0.05 and Q = 0.01	732	56	732	31.891	59.182	26.7592	13.07142857
If quality Qv > Q; Qv = 0.1 and Q = 0.01	732	47	732	31.5364	59.182	26.4046	15.57446809
If quality Qv < Q; Qv = 0.005 and Q = 0.01	732	63	Decompression not working	NA	NA	NA	11.61904762

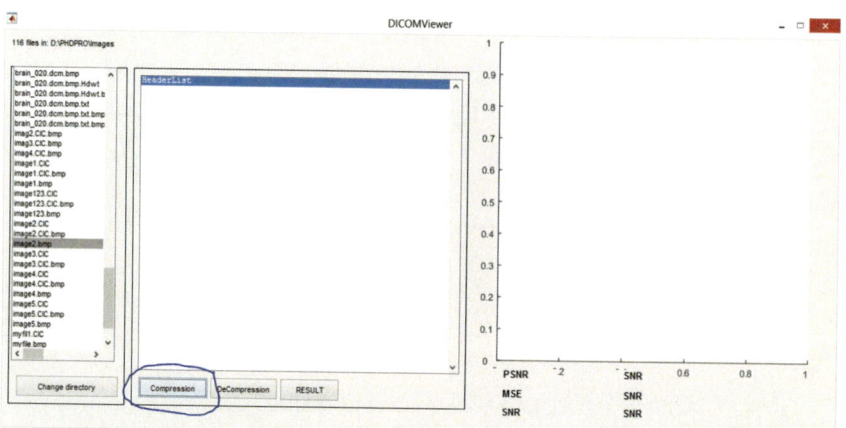

Fig. 8 Graphical user interface of DICOM viewer

5   Conclusion

This paper introduces new compression and decompression technique using fusion of three-level Discrete Wavelet Transform and Discrete Cosine Transforms with T-matrix coding. This implement helped to increases number of high-frequency coefficients and also to rises compression ratio. It also shows a method which is useful for quantization-level selection for different condition. This approach identifies if the quality value is greater than the standard quality factor, we can achieve higher compression and maintain image quality. The implemented method increase number

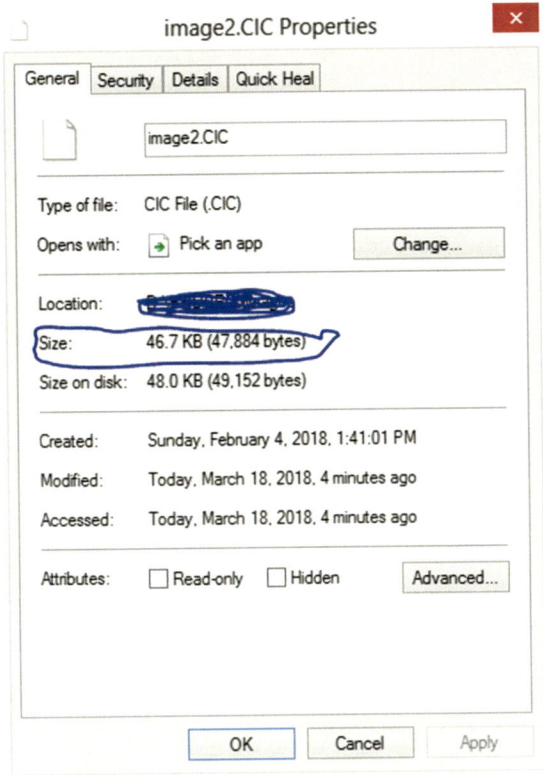

Fig. 9 Graphical user interface of image2.CIC with compression size 47.884 bytes

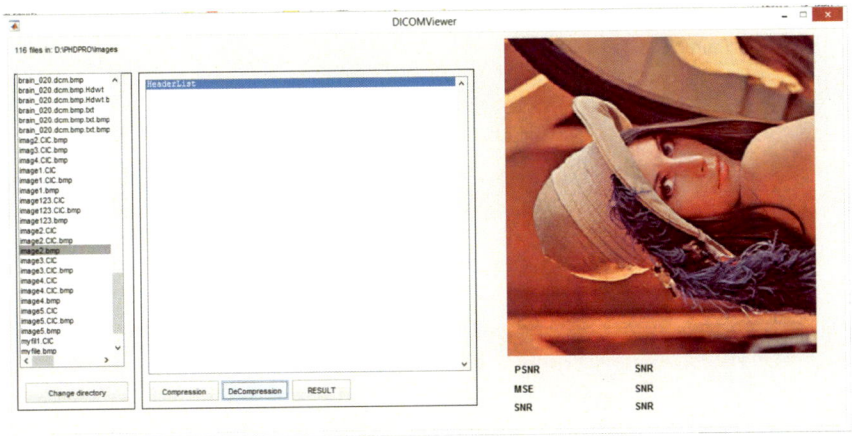

Fig. 10 Graphical user interface of image2.BMP for decompression

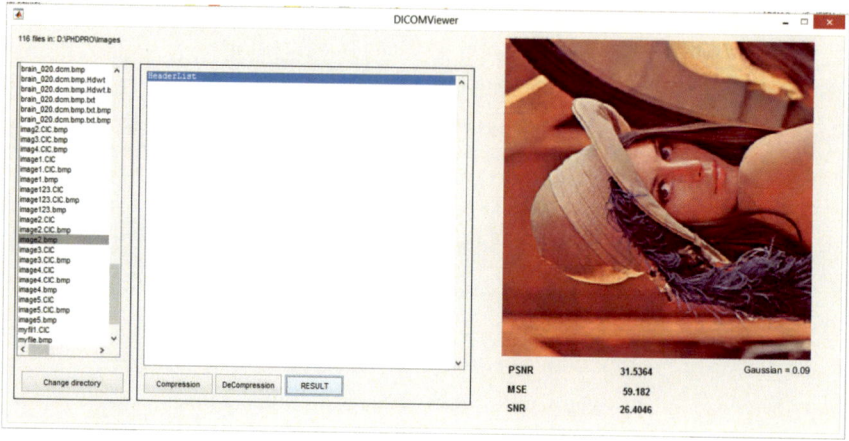

Fig. 11 Graphical user interface of image2.BMP for PSNR, MSE, SNR

of zeros in the transformation matrix. The EZSD algorithm used to remove zeros and at the same time it converts a high-frequency sub-bands into an array. Last, the arithmetic coding mapped entire sequence of symbols into a single codeword. It gives high encoding capacity.

Acknowledgements Authors thank Dr. S. V. Dudal, HOD, Department of Applied Electronics, SGBA University, Amravati and Maharashtra, India for providing all kind of facilities and support.

References

1. Malacara D (2011) Color vision and colorimetric: theory and applications. Press monograph, SPIE. http://books.google.com.br/books?id=xDU4YgEACAAJ
2. Frery AC, Perciano T (2013) Introduction to image processing using R. 21 Springer briefs in computer science. https://doi.org/10.1007/978-1-4471-4950-7-2
3. Gupta P, Bansal V (2015) The run length encoding for RGB images. Int J Converg Technol Manage 1(1). ISSN: 2455-7528
4. Xijun Y, Lei J (2015) A method of lossy compression for RGB565 format true color image. Int J Signal Process Image Process Pattern Recognit 8(5):279–288. http://dx.doi.org/10.14257/ijsip.2015.8.5.29. ISSN: 2005-4254
5. Boucetta A, Melkemi KE (2012) DWT based-approach for color image compression using genetic algorithm. In: International conference on image and signal processing, ICISP 2012: image and signal processing. LNCS 7340, pp 476–484
6. Sangwine SJ, Horne RE (1998) The colour image processing handbook, 1st edn. Chapman & Hall
7. Gonzalez RC, Woods RE (2001) Digital image processing. Addison Wesley Publishing Company, Reading
8. Sayood K (2000) Introduction to data compression, 2nd edn. Academic, Morgan Kaufman Publishers
9. Siddeq MM (2012) Using two level DWT with limited sequential search algorithm for image compression. J Signal Inf Process 3:51–62. https://doi.org/10.4236/jsip.2012

10. Naveen Kumar R, Jagadale BN, Sandeepa KS (2016) Use of optimal threshold and T-Matrix coding in discrete wavelet transform for Image compression. In: 3rd international conference on electronics and communication systems (ICECS 2016), IEEE Explore. 978-1-4673-7832-1/16
11. Buela Divya G, Krupa Swaroopa Rani K. Implementation of image compression using hybrid DWT–DCT algorithms, Int J Modern Trend Eng Res. ISSN: 2393-8161
12. Sangwine SJ, Horne RE (1998) The colour image processing handbook, 1st edn. Chapman & Hall
13. Sonka M, Halva V, Boyle T (1999) Image processing analysis and machine vision, 2nd edn. Brooks/Cole Publishing Company
14. Frery AC, Perciano T (2013) Color representation, introduction to image processing using R. 21 Springer briefs in computer science. https://doi.org/10.1007/978-1-4471-4950-7
15. Gersho A, Gray RM (1992) Vector quantization and signal compression, Boston. Kluwer Academic Publishers, MA
16. Suma S, Sridhar V (2014) A review of effective techniques of compression in medical image processing. Int J Comput Appl (0975–8887) 97(6)
17. Siddeq MM (2012) Using two level DWT with limited sequential search algorithm for image compression. J Signal Inf Process 3:51–62

Application of Neural Networks in Image Processing

Ishan Raina, Chirag Chandnani and Mani Roja Edinburgh

Abstract In today's world, the need for computers and their logics is critical for the development of any system. This paper is for those people who have little or no knowledge about Artificial Neural Networks (ANNs). Various advances have been made in creating intelligent systems, some enlivened by biological neural networks. Analysts from numerous logical orders are outlining artificial neural networks (ANNs) to take care of a variety of problems in design recognition, prediction, optimization, associative memory, and control. And will be discussing some of the algorithms of ANNs such as image segmentation using edge detection, image enhancement using multiscale retinex and technique combining sharpening and noise reduction and some practical applications such as diagnosing liver disease and face detection systems.

1 Introduction

ANNs are processing devices (paradigm or real equipment) that are loosely modeled after the neuronal structure of the mammalian cerebral cortex yet on considerably littler scales. A huge ANN may have hundreds or thousands of processor units, though a mammalian mind has billions of neurons with a relating increment in extent of their general connection and emanant conduct. Despite the fact that ANN analysts are for the most part not worried about whether their systems precisely take after organic frameworks, some have. For instance, specialists have precisely mimicked the capacity of the retina and displayed the eye rather well.

Neural systems are regularly composed in layers. Layers are comprised of various interconnected "hubs" which contain an "initiation work". Examples are introduced to the system by means of the "information layer", which conveys to at least one "hidden layers" the place the real preparing is done through an arrangement of weighted "associations". The shrouded layers at that point connect to a "yield layer" where the appropriate response is yield as demonstrated above. Modern computerized

I. Raina · C. Chandnani (✉) · M. R. Edinburgh
Thadomal Shahani Engineering College, Mumbai, India
e-mail: chandnanichirag512@gmail.com

© Springer Nature Switzerland AG 2019
D. Pandian et al. (eds.), *Proceedings of the International Conference on ISMAC in Computational Vision and Bio-Engineering 2018 (ISMAC-CVB)*, Lecture Notes in Computational Vision and Biomechanics 30,
https://doi.org/10.1007/978-3-030-00665-5_48

483

Table 1 Comparison between VN and biological computer [1]

	Von Neumann computer	Biological computer
Processor	Complex High speed One or a few	Simple Low speed A large number
Memory	Separate from a processor Localized Noncontent and addressable	Integrated into processor Distributed Content addressable
Computing	Centralized Sequential Stored programs	Distributed Parallel Self-learning
Expertise	Numerical and symbolic manipulations	Perceptual problems
Reliability	Very vulnerable	Robust
Operating environment	Well-defined Well-constrained	Poorly-defined Unconstrained

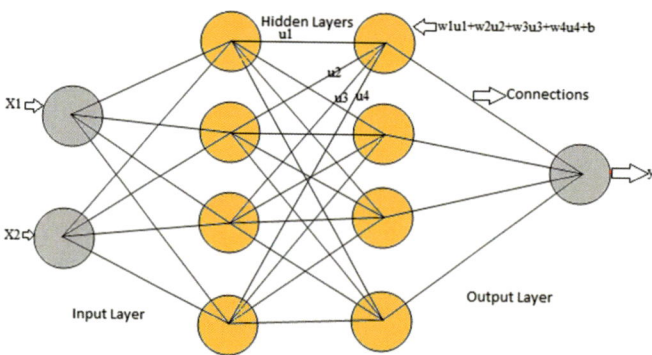

Fig. 1 The structure of a basic neural network

PCs beat people in the area of numeric calculation and related image control. In any case, people can easily take care of complex perceptual issues (like perceiving a man in a group from a minor look at his face) at such a rapid and degree as to predominate the world's fastest PC. Why is there such a surprising contrast in their execution? The organic neural framework design is totally not quite the same as the von Neumann engineering (see Table 1). This distinction essentially influences the sort of capacities each computational model can best perform (Fig. 1).

Various endeavors to create "savvy" programs in light of von Neumann's brought together engineering have not brought about broadly useful canny projects. Propelled by natural neural systems, ANNs are greatly parallel figuring frameworks comprising of an amazingly extensive number of straightforward processors with numerous interconnections. ANN models endeavor to utilize some "organizational" standards accepted to be utilized as a part of the human.

Moeled loosely on the human cerebrum, a neural net comprises of thousands or even a large number of basic preparing hubs that are thickly interconnected. The vast majority of the present neural nets are sorted out into layers of hubs, and they are "sustain forward," implying that information travels through them in just a single heading. An individual hub may be associated with a few hubs in the layer underneath it, from which it gets information, and a few hubs in the layer above it, to which it sends information.

To every one of its approaching associations, a hub will appoint a number known as a "weight." When the system is dynamic, the hub gets an alternate information thing—an alternate number—over every one of its associations and increases it by the related weight. It at that point includes the subsequent items together, yielding a solitary number. On the off chance that number is beneath a limit esteem, the hub passes no information to the following layer. On the off chance that the number surpasses the limit esteem, the hub "fires," which in the present neural nets by and large means sending the number—the entirety of the weighted information sources—along all its active associations.

At the point when a neural net is being prepared, the greater part of its weights and limits are at first set to arbitrary esteems. Preparing information is bolstered to the base layer—the information layer—and it goes through the succeeding layers, getting duplicated and included in complex courses, until the point that it at last arrives, fundamentally changed, at the yield layer. Amid preparing, the weights and limits are ceaselessly balanced until the point when preparing information with similar names reliably yield comparable yields.

The neural framework learns by altering its weights and slant (edge) iteratively to yield needed yield. These are moreover called free parameters. For making sense of how to happen, the neural framework is arranged first. The readiness is performed using described arrangement of standards generally called the training algorithm.

Popular Learning Algorithms used in Neural Network are as follows [1]:

Gradient Descent: This is the simplest training algorithm used in case of supervised training model. In case, the actual output is different from target output, the difference or error is find out. The gradient descent algorithm changes the weights of the network in such a manner to minimize this error.

Back Propagation: It is an extension of gradient-based delta learning rule. Here, after finding an error (the difference between desired and target), the error is propagated backward from output layer to the input layer via hidden layer. It is used in case of multilayer neural network.

1.1 Types of Learning Neural Network

Supervised Learning: In Supervised Learning, the training data as the initial input to the input layers and the required output is known to us. Weights are balanced until the point when output gives the required value.

Unsupervised Learning: The input information is utilized to prepare the system whose yield is known. The system characterizes the information and modifies the weight by feature extraction in input information.

Reinforcement Learning: Here the estimation of the yield is obscure, yet the system gives the feedback whether the yield is correct or off-base.

Offline Learning: The alteration of the weight vector and threshold is done simply after all the preparation set is displayed to the system.

Online Learning: The change of the weight and threshold is done. Subsequently, the every preparation sample is shown to the system.

1.2 Types of Data Sets Used in ANNs

Training Set: An arrangement of cases utilized for discovering that is to fit the parameters (i.e., weights) of the system. One Epoch contains one full preparing cycle on the training set.

Validation Set: An arrangement of cases used to tune the parameters (i.e., architecture) of the system. For instance to pick the quantity of hidden units in a neural network.

Test Set: An arrangement of cases utilized just to evaluate the performance of a completely determined system or to apply effectively in anticipating yield whose information is known.

1.3 Different Uses of ANNs*

Classification: A neural system can be prepared to characterize given pattern or informational index into predefined class. It utilizes feed-forward systems.

Prediction: A neural system can be prepared to create yields that are expected from given info.

Clustering: The Neural system can be utilized to recognize a distinct component of the information and classify them into various classes with no earlier learning of the information.

2 Applications

2.1 Image Enhancement Using Multiscale Retinex

The retinex is a human recognition based image processing algorithm which gives shading steadiness and dynamic range compression. There are two kinds of retinex algorithms to be specific single scale retinex (SSR) and multiscale retinex (MSR). The MSR can be minimalistically composed as

$$F_i(x, y) = \sum_{n=1}^{N} W_n * \{\log[S_i(x, y)] - \log[S_i(x, y) * M_n(x, y)]\}, \qquad (1)$$

where [2] where $F_i(x, y)$ is MSR and $S_i(x, y)$ is SSR the subscripts: $i \in R, G, B$ speak to the three color bands, N is the quantity of scales being utilized, and W_n are the weighting factors for the scales. The $M_n(x, y)$ are the encompass function given by

$$M_n(x, y) = K_n e^{[-(x^2+y^2)/\sigma_n^2]} \qquad (2)$$

Rahman et al. [2] where σ_n the will be the standard deviations of the Gaussian distribution that decide the scale. The greatness of the scale decides the sort of data that the retinex gives: littler scales giving more unique range compression, and bigger scales giving more shading steadiness. The I (are selected so that [2]).

$$\iint F(x, y) \mathrm{d}x \mathrm{d}y = 1 \qquad (3)$$

As expressed over, the MSR still experiences turning gray out of uniform zones much as the SSR did. The favorable position that the MSR has over the SSR is in the mix of scales which give both dynamic range compression and tonal version in the meantime. The general consequence of the use of the MSR is still more saturated than human perception, giving the last picture a washed-out appearance, however, it saves the vast majority of the detail in the scene. This turning gray of territories of consistent intensity happens in light of the fact that the retinex handling enhances each color band as a component of its encompass.

2.2 Image Enhancement Technique Combining Sharpening and Noise Reduction [3]

Another way to deal with contrast enhancement of picture information is displayed. The proposed strategy depends on a numerous yield framework that adopts fuzzy models with a specific end goal to keep the noise increase during the sharpening of the picture as points of interest. Key highlights of the proposed strategy are preferable execution over accessible techniques in the upgrade of pictures adulterated by Gaussian noise and no confounded tuning of fuzzy set parameters.

Enhancing the nature of sensor information is a key issue in picture based instrumentation. To be sure, preprocessing strategies can play an extremely important part in expanding the precision of consequent errands, for example, parameter estimation and object recognition. In this regard, contrast enhancement is frequently fundamental keeping in mind the end goal to highlight essential features embedded within the picture information. The improvement of noisy information, in any case, is an extremely basic process in light of the fact that the honing activity can altogether increase the noise. Distinctive systems have been proposed in the writing. They generally use the unsharp masking approach.

In the outstanding straight unsharp masking technique, a small amount of the high-pass filtered version of the input picture is added to the first information keeping in mind the end goal to get an enhanced picture. The reception of the high-pass channel, be that as it may, makes this basic strategy exceptionally sensitive to noise and frequently unacceptable for genuine applications. More powerful strategies utilize a nonlinear channel rather than the high-pass straight administrator with a specific end goal to accomplish a tradeoff between noise attenuation and edge enhancement. In this regard, an exceptionally intriguing class of nonlinear strategies is spoken to by polynomial unsharp masking methods, such as the Teager-based operator and the effective cubic unsharp masking strategy. In an alternate class of nonlinear methodologies, a weighted middle replaces the high-pass linear filter.

2.3 Image Segmentation Using Edge Detection

Image Segmentation is the way toward partitioning a computerized picture into numerous locales or sets of pixels. As a matter of fact, partitions are diverse objects in picture which have a similar texture or color. The outcome of image segmentation is an arrangement of areas that all in all cover the whole picture or an arrangement of forms extricated from the picture [4]. The majority of the pixels in a region are comparable with regard to some characteristic or computed property, such as color, intensity, or texture. Nearby areas are fundamentally different regarding the same attributes. Edge detection is a standout amongst the most as often as possible utilized procedures in computerized picture processing. The limits of object surfaces in a scene regularly prompt arranged confined changes in intensities of a picture, called edges. This perception joined with an accepted way of thinking that edge recognition is the initial phase in image segmentation and has powered a long inquiry for a decent edge detection paradigm to use in image processing. This search has constituted a foremost region of research in low-level vision and has prompted a constant flow of edge detection paradigm distributed in the image processing journals in the course of the most recent two decades. Indeed, even as of late, new edge detection paradigms are published every year.

Artificial neural systems (ANN) are generally connected for pattern recognition. Their handling potential and nonlinear attributes are utilized for clustering. Self-organization of Kohonen Feature Map (SOFM) network is an intense apparatus for grouping. Ji and Park proposed a paradigm for watershed segmentation in light of SOM [4]. This strategy finds the watershed segmentation of luminance part of color picture. The technique can be clarified as takes after. It comprises of two free neural systems one each for saturation and intensity planes. In edge location

(a)   **(b)**

Fig. 2 **a** Original image [4]; **b** segmented image [4]

process, Initialize the synaptic weights of the system, to little, unique, arbitrary numbers at iteration, draw an example output from the information set. Locate the best coordinating (winning) neuron at this iteration utilizing the minimum distance Euclidean criterion. Continue this procedure until the synaptic weights__ achieve their Steady state values. This results in Image Segmentation (Fig. 2).

2.4 Diagnose of Disease Using Ultrasound Liver Images

Ultrasound imaging methodology is very well known and most generally utilized methodology for envisioning and concentrate the liver for any disease conditions without making any agony or distress the patient. Ultrasound liver imaging is generally utilized because of its noninvasive nature and ease when contrasted with other imaging modalities. The diagnosis of different illnesses is performed based on different picture features, for example, the echogenicity, legion shape, and echotexture. Liver imaging is a standout amongst other strategies of early detection of liver infections and early recognition is imperative since it spares patients from advance afflictions, for example, enlarged stomach filled with ascites fluid, bleeding varices, and encephalopathy or sometimes jaundice [5]. Liver sickness conditions, for example, liver disease conditions such as fatty liver, cirrhosis, and hepatomegaly are known for delivering unmistakable reverberate designs during US imaging, however, the pictures are likewise known to be visually challenging for interpreting them because of their imaging artifacts and speckle noise. Because of it, the sonographers need to depend upon additional pathological tests.

In this method image acquired are first processed using preprocessing techniques like cropping, edge detection then image processing is done in which feature extrac-

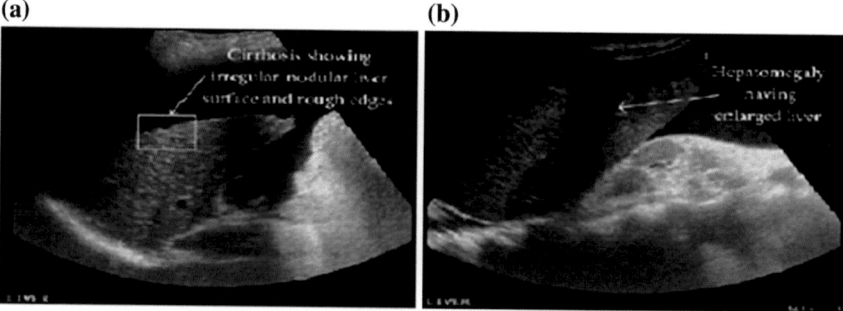

Fig. 3 **a** Cirrhosis liver [5]; **b** hepatomegaly liver [5]

tion is done after this final step is to select the feature which distinguishes the livers from each other. This is implemented using a ANN as explained below.

The backpropagation algorithm was picked because of the system's capacity to learn and store tremendous measures of mapping relations of information yield demonstrate without the requirement for earlier disclosure of mathematical conditions relating to these mapping relations. The algorithm likewise manages the system's weight and edge esteems so as to acquire minimum error sum of square. The outlined neural system classifier utilizes a two-layer feed-forward backpropagation network. Two-layer feed-forward network can be best characterized as a system with sigmoid hidden up and yield neurons.

To train the system, the input information and target information should be fed into the system. The system at that point partitions the input test information into three unique examples, which are training, validation, and testing samples. The training samples are utilized to prepare the system, and the system is balanced by its error. The validation samples are utilized to quantify organize generalization and to halt training when generalization stops improving. Testing samples are then used to give an autonomous measure of the system execution during and after training. In the event that the error of the system is still extensive, the system can be retrained back as to get more precise and proficient outcome. This is the way it is used to diagnose liver diseases using ANNs (Fig. 3).

2.5 Neural Network-Based Face Detection System

A neural network-based algorithm to identify upright, frontal perspectives of faces in grayscale images. The algorithm works by applying at least one neural networks straightforwardly to parts of the information picture and arbitrating their outcomes. Each system is prepared to yield the presence or absence of a face. The algorithms and training techniques are intended to be general, with little customization for faces.

This framework works in two phases.

- **Stage One: A Neural Network-Based Filter**

To detect faces anyplace in the input, the filter is connected at each area in the picture. To detect faces bigger than the window estimate, the information picture is over and over lessened in measure (by subsampling), and the filter is connected at each size. This filter must have some invariance to position and scale. The measure of invariance decides the quantity of scales and positions at which it must be connected.

- **Stage Two: Merging Overlapping Detections and Arbitration**

Two methodologies to enhance the unwavering quality of the detector: merging overlapping detections from a single network and arbitrating among multiple networks.

Most faces are detected at numerous adjacent positions or scales, while false discoveries frequently happen with less consistency. This perception prompts a heuristic which can dispense with numerous false detections. For every area and scale, the quantity of detections inside a predetermined neighborhood of that area can be tallied. In the event that the number is over a limit, at that point that area is delegated a face. The centroid of the close-by detections characterizes the area of the detections result, accordingly falling numerous detections. This is alluded to as "Thresholding" [6].

In the event that a specific area is effectively detected as a face, at that point all other detection areas which cover it are probably going to be error and can in this manner be eliminated. With the higher number of detections inside a small neighborhood and eliminate areas with less detections. This heuristic is called "overlap elimination" [6].

Arbitration Among Multiple Networks.

To additionally lessen the quantity of false positives, we can apply numerous systems and arbitrate between their yields to deliver an final decision. Each system is prepared in a comparable way, however with arbitrary beginning weights, irregular introductory nonface pictures, and permutations of the order of presentation of the scenery images. As will be found in the following segment, the detection and false-positive rates of the individual systems will be very close. Be that as it may, as a result of various preparing conditions and due to self-choice of negative preparing illustrations, the systems will have diverse predispositions and will make distinctive errors. One approach to consolidate two such pyramids is by ANDing them. This technique flags a detection just if the two systems identify a face at decisively a similar scale and position. Because of the diverse predispositions of the individual systems, they will once in a while concur on a error in detection of a face. This enables ANDing to take out most false detections from the given image in the input data.

3 Conclusion

In this paper, a brief information is mentioned on ANNs, types of ANN's that are currently being used in the world, different types of learning algorithm. The paper discusses mainly on different algorithms that are currently being used for various application of image processing such as image enhancement and Image Segmentation. Also, some real-life applications of these image processing techniques are mentioned in the article such as image processing for identification of a disease in the medical field and face detection algorithm.

References

1. Jain AK, Mao J, Mohiuddin KM. Artificial neural networks: a tutorial
2. Rahman Z, Jobson DJ, Woodell GA. Multi-scale retinex for color image enhancement
3. Russo F. An image enhancement technique combining sharpening and noise reduction
4. Senthilkumaran N, Rajesh R. Edge detection techniques for image segmentation
5. Systems Biomedicine Division, India Research Advisory Council, Haffkine Institute for Training Research and Testing, Parel, Mumbai, Maharashtra
6. Rowley HA, Baluja S, Kanade T. Neural network-based face detection

Region of Interest (ROI) Based Image Encryption with Sine Map and Lorenz System

Veeramalai Sankaradass, P. Murali and M. Tholkapiyan

Abstract In this research work, ROI-based grayscale image encryption with chaos is proposed. First, ROI areas are identified using Sobel edge detection operator and categorized into important and unimportant regions based on number of edges present in the particular block. Next, the important regions are encrypted with Lorenz system (both confusion and diffusion process) and unimportant regions are encrypted using Sine map. Finally, the entire image is shuffled by Lorenz system with new initial conditions to get final encrypted image. The significant advantage of this research work is that important and unimportant regions are encrypted separately with different chaos equation and system which increases the security of the image and also the end user can vary the important regions depends upon the requirements. The experimental results show that the proposed encryption approach provides better results for different cryptographic attacks.

1 Introduction

Security of data including image, video, text, audio and multimedia is gaining much attention due to vast development of communication technologies. During transmission over public networks as well as private networks, hackers and attackers may steal that data for various purposes such as modification, fabrication, etc. Therefore, data security has become an important issue during transmission as well as

V. Sankaradass · P. Murali (✉)
Department of Computer Science and Engineering, Vel Tech High Tech Dr. Rangarajan Dr.
Sakunthala Engineering College, Avadi, Chennai, India
e-mail: pmuraliphd@gmail.com

V. Sankaradass
e-mail: veera2000uk@gmail.com

M. Tholkapiyan
Department of Civil Engineering, Vel Tech High Tech Dr. Rangarajan Dr. Sakunthala
Engineering College, Avadi, Chennai, India
e-mail: m.tholkapiyan@gmail.com

© Springer Nature Switzerland AG 2019
D. Pandian et al. (eds.), *Proceedings of the International Conference on ISMAC in Computational Vision and Bio-Engineering 2018 (ISMAC-CVB)*, Lecture Notes in Computational Vision and Biomechanics 30,
https://doi.org/10.1007/978-3-030-00665-5_49

storage. Recently, lots of images are transmitted and uploaded to the Internet and image encryption is serious issue. Traditional encryption algorithms such as Data Encryption Standard (DES), Advanced Encryption Standard (AES), etc., are mainly developed for text data and not suitable for images due to bulk volume of data, high correlation among pixels and intrinsic features of image [1–5]. By contrast, chaos-based image encryption algorithms have attracted many researchers due to high sensitivity to initial conditions, ergodicity, pseudorandomness and periodicity [6–9].

2 Literature Survey

In this section, related works are presented. Nowadays, lots of image encryption schemes based on chaos are presented. Ye et al. [10] presented image encryption algorithm based on generalized Arnold's map. In this work, first, total circular function is used in the permutation stage and in the diffusion stage, double diffusion process is applied with keystream. Image encryption with combined confusion and diffusion is designed in [1]. The spatiotemporal chaos is utilized to shuffle the pixels as well as diffusion and also used for generating random numbers.

In [11], image encryption scheme based on logistic map and DNA is developed. First, 2-D logistic map is employed to shuffle the pixels in the image then, DNA encoding is applied to encode the shuffled pixels. Finally, DNA addition and subtraction are performed to obtain the ciphered image. Region of Interest (ROI) based image encryption is gaining popularity in recent days. In [12], selective image encryption based on orthogonal polynomials transformation (OPT) and chaos for low power devices are proposed. In this work, ROIs are identified with orthogonal polynomials and important regions are encrypted in bit level with chaos. Next, unimportant regions are encrypted in the OPT domain with chaos. Finally, square-wave SFC is applied to obtain the final encrypted image. Tanja et al. [13] presented combinational domain image encryption. First, they identified significant and insignificant areas in the image using Prewitt edge detection. Next, the identified significant areas are encrypted in the spatial domain with chaos and insignificant areas are encrypted in the wavelet domain. Finally, they applied reverse diffusion process to get the final encrypted image.

Motivated by existing image encryption methods, in this research work, we propose novel image encryption scheme based on ROI and chaos. First, the Sobel edge detection operator is applied to plain image and categorized into important and unimportant regions based on number of edges. Then, important region blocks are encrypted with Lorenz system and unimportant region blocks are encrypted with Sine map. Finally, overall shuffling is performed using random sequences from Lorenz system. The main feature of the proposed work is separate encryption of important and unimportant regions which increases the security of the image.

The remainder of this paper is organized as follows: In Sect. 3, details of Lorenz system and sine map are presented. Section 4 describes the proposed cryptosystem. The experiments, results and discussions carried out to show the effectiveness of the proposed system are presented in Sect. 5 and finally conclusions are drawn in Sect. 6.

3 Chaotic System and Map

In the proposed research work, Lorenz system and sine map are used. The Lorenz system is used to encrypt the important regions in the image as well as shuffling of entire image. The sine map is used to encrypt unimportant regions.

3.1 Lorenz System

The Lorenz system with x, y and z as state variables is mathematically defined as

$$\dot{x} = -a(x - y)$$
$$\dot{y} = -xz + bx - y$$
$$\dot{z} = xy - cz \tag{1}$$

where a, b and c are control parameters and when $a = 10$, $b = 28$ and $c = 8/3$, the system is in chaotic state.

3.2 Sine Map

In this proposed work, sine map is used to generate random values and hence to construct random window in the random artificial image. The 1D sine map, $x(i)$ is mathematically defined as

$$x(i) = \lambda * \text{sine}(\pi * (i - 1)) \tag{2}$$

where $\lambda = 0.99$ and $\pi = 3.14$.

Having described the mathematical preliminaries, the proposed cryptosystem is presented in the next section.

4 Proposed Cryptosystem

4.1 Proposed Encryption Process

The proposed encryption method starts with identification of ROI areas in the image. Generally, image contains various areas like edges, smooth regions, etc. The proposed research work applies Sobel edge detection operator on the input image of size $(M \times N)$ and obtains the edge detected binary image. Next, the binary image is divided into $(k \times k)$ non-overlapping blocks where $k < M, N$ and number of edges present in the each block is counted. Then, based on threshold value, T from the user, blocks with edges are separated as important and unimportant regions where number of edges are greater than in important regions and less than in unimportant regions. Also, the identified important and unimportant regions are termed as 1 and 0 that is act as secret key in the decryption side. After that, important blocks are encrypted with Lorenz system. Before encryption, the sine map is iterated and their values are digitized to 1, 2 and 3. Then, Lorenz system is iterated $(k \times k)$ times and generate the sequences. Based on sine map values, one of the sequence is selected from Lorenz system and sorted. Finally, index values are obtained from sorted sequence to shuffle the block. Next, exclusive-OR (XOR) operation is performed between generated sequence (digitized values) from Lorenz systems and shuffled values. The above process is repeated for all important blocks with new initial conditions.

Next, Similar to important blocks encryption, unimportant blocks are encrypted with sine map. Finally, Lorenz system is iterated $(M \times N)$ times and the index values are used to shuffle the entire image to produce the final encrypted image. The steps involved in the proposed encryption method is presented as an algorithm hereunder:

Algorithm

Input: Grayscale image of size $(M \times N)$
Output: Encrypted image of size $(M \times N)$
Begin
Step 1: Read plain image of size $(M \times N)$
Step 2: Apply Sobel edge detection operator on the plain image
Step 3: Divide the binary image obtained from Step 2 into $(k \times k)$ non-overlapping blocks where $k < M, N$
Step 4: Count the edge points cnt, in each block of the binary edge map
Step 5: Accept threshold T from user
Step 6: If the $cnt \geq T$, then the corresponding block is considered as important and termed as ROI; Else, it is considered as unimportant, non-ROI
Step 7: Identified blocks (important and unimportant) are numbered from left to right, 1 and 0
Step 8: Repeat the above steps for all blocks in the image
Step 9: Iterate the Sine map up to total number of important blocks and digitize the values to 1, 2 and 3

Step 9: Important Blocks Encryption
Step 9.1: Iterate Lorenz system $(k \times k)$ times with new initial conditions
Step 9.2: Based on digitized sequence, select one of the sequence from Lorenz system and sort the sequence into S_x and obtain the respective index I_x. It can be represented as

$$I_x[i] \leftarrow [S_x, X^l]$$

Step 9.3: Shuffle the block using I_x and obtain the shuffled blocks
Step 9.4: Consider the same sequence and digitize the values between 0 and 255
Step 9.5: Diffuse the shuffled block by performing Exclusive-OR (XOR) between values from Step 9.4 and shuffled block
Step 9.6: Repeat Step 9.1 to 9.5 for all important blocks
Step 10: Unimportant blocks encryption using Sine map that is similar to important blocks encryption
Step 11: Iterate Lorenz system $(M \times N)$ times with new initial conditions and select one of the sequence and obtain index values from sorted sequence
Step 12: Perform shuffling operation based on index values for the entire image and obtain the final encrypted image
End

4.2 *Proposed Decryption Process*

The main aim of the decryption process is to retrieve the original image from encrypted image. The proposed decryption process is the exact reverse of encryption process with same secret keys.

5 **Experiment and Security Analysis**

The proposed encryption scheme is experimented with more than 100 benchmark images of size (512×512). First, the edge detected image is divided into (32×32) non-overlapping blocks. Then, Sine map is iterated to generate values 1, 2 and 3 and the corresponding keys are $x = 0.1346$, $\lambda = 0.99$ and $\pi = 3.14$. For important blocks encryption, the lorenz system values are $a = 10$, $b = 28$, $c = \frac{8}{4}$, $x_1 = 10.0$, $y_1 = 15.0$ and $z_1 = 45.0$. Similarly, for unimportant blocks encryption, the sine map values are $x = 0.16746$, $\lambda = 0.99$ and $\pi = 3.14$. Finally, overall shuffling is performed with Lorenz system and the corresponding values are $a = 10$, $b = 28$, $c = \frac{8}{4}$, $x_1 = 10.0$, $y_1 = 17.0$ and $z_1 = 45.0$. The proposed encryption method encrypts the original mandrill image with above keys. Figure 1 shows the encryption and decryption results of Mandrill image.

  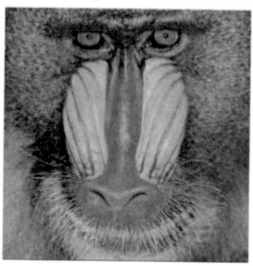

(a) Original Mandrill   (b) Encrypted Mandrill   (c) Decrypted Mandrill
        Image   Image   Image

Fig. 1 Encryption and decryption results of proposed technique

The proposed encryption technique is analyzed with various performance measures and the corresponding results are presented in the following subsections.

5.1 Key Sensitivity Analysis

The encrypted image is sensitive with secret keys values. To test the key sensitivity of the proposed cryptosystem, first, the original grey image is encrypted with above said key 1 ($x = 0.1346$, $\lambda = 0.99$, $\pi = 3.14$, $a = 10$, $b = 28$, $c = \frac{8}{4}$, $x_1 = 10.0$, $y_1 = 15.0$, $z_1 = 45.0$, $x = 0.16746$, $\lambda = 0.99$, $\pi = 3.14$, $a = 10$, $b = 28$, $c = \frac{8}{4}$, $x_1 = 10.0$, $y_1 = 17.0$ and $z_1 = 45.0$). Then, the original key 1 is modified as key 2. The modified key values are ($x = 0.1347$, $\lambda = 0.99$, $\pi = 3.14$, $a = 10$, $b = 28$, $c = \frac{8}{4}$, $x_1 = 10.0$, $y_1 = 15.0$, $z_1 = 45.0$, $x = 0.16746$, $\lambda = 0.99$, $\pi = 3.14$, $a = 10$, $b = 28$, $c = \frac{8}{4}$, $x_1 = 10.0$, $y_1 = 17.0$ and $z_1 = 45.0$). Again, the key 1 is slightly modified as key 3 and the corresponding values are ($x = 0.1346$, $\lambda = 0.99$, $\pi = 3.14$, $a = 10$, $b = 28$, $c = \frac{8}{4}$, $x_1 = 10.0$, $y_1 = 15.0$, $z_1 = 45.0$, $x = 0.16746$, $\lambda = 0.99$, $\pi = 3.14$, $a = 10$, $b = 28$, $c = \frac{8}{4}$, $x_1 = 10.0$, $y_1 = 17.000001$ and $z_1 = 45.0$). Now the encrypted image is decrypted with key 2 and key 3 and the corresponding results are shown in Fig. 2a, b.

From Fig. 2, the proposed encryption technique is highly sensitive to secret keys. So, small change in the key structure will not produce original image in the proposed method.

5.2 Histogram Analysis

Figure 3a, c shows the histogram of original Lena and Mandrill image. Similarly, Fig. 3b, d shows the corresponding encrypted Lena and Mandrill image, respectively.

(a) Decrypted Mandrill image   (b) Decrypted Mandrill image
with Key2   with Key3

Fig. 2 Decrypted images with wrong keys

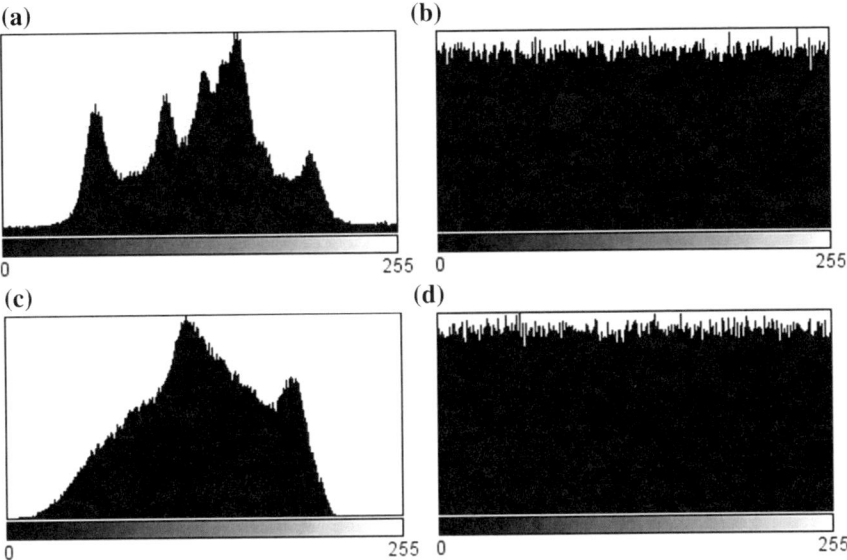

Fig. 3 Histogram analysis results **a** original Lena image, **b** encrypted Lena Image, **c** original Mandrill image, **d** encrypted Mandrill image

From the figures, histogram of encrypted images is flat and one cannot identify the different images. Therefore, the proposed encryption method is successful against statistical attack.

5.3 Differential Attack

To find out the relationship between original image and encrypted image, the attacker encrypts the original image before and after changing and comparing those images.

It is called differential attack and it is measured by the number of pixel rate change (NPCR) and the unified average changing intensity (UACI).

5.3.1 Number of Pixel Change Rate (NPCR)

The NPCR is used to measure the number of pixels indifference of a component between two images.

$$\text{NPCR} = \frac{\sum_{i,j} D(i, j)}{W \times H} \times 100\% \tag{3}$$

$$\text{where } D(i, j) = \begin{cases} 1, & c_1(i, j) \neq c_2(i, j) \\ 0, & \text{otherwise} \end{cases}$$

Here W and H are the width and height of the image, c_1 and c_2 are two cipher images with slightly modified keys, $c_1(i, j)$ and $c_2(i, j)$ are pixel values of c_1 and c_2 at position (i, j).

5.3.2 Unified Average Change Intensity (UACI)

UACI is used to measure the average intensity difference in component and can be defined as

$$\text{UACI} = \frac{1}{N} \left[\sum_{i,j} \frac{|c(i, j) - c'(i, j)|}{2^L - 2} \right] \times 100\% \tag{5}$$

The proposed cryptosystem is analyzed with NPCR and UACI and the values are presented in Table 1.

From Table 1, it shows that there is a small change in the original images will result in a significant change in the encrypted images. The table also implies that the proposed encryption technique successfully resists the differential attack.

Table 1 NPCR and UACI values between original Lena image and encrypted Lena image

Image	NPCR%	UACI%
Lena	99.87	31.59
Mandrill	99.64	31.68
House	99.39	31.62
Lake	99.44	31.57
Fruits	99.71	31.78

Table 2 Results of PSNR values obtained from proposed encryption scheme	Image	PSNR in dB
	Lena	10.67
	Mandrill	10.91
	House	9.78
	Lake	9.29
	Fruits	10.22

5.4 Peak Signal-to-Noise Ratio (PSNR)

The performance of the proposed cryptosystem is evaluated objectively via the PSNR between the original image and the encrypted image. The mathematical formula for computation of PSNR is

$$\text{PSNR} = 10 \log_{10}\left(R^2/\text{MSE}\right) \tag{6}$$

where R is maximum gray level and MSE is the mean square error given by

$$\text{MSE} = \left(\sum_{m,n}[I_1(i,\,j) - I_2(i,\,j)]^2/(m \times n)\right)$$

$I_1(i,\,j)$ and $I_2(i,\,j)$ represent the intensity of the original and encrypted image pixel positions at $(i,\,j)$, and m and n are the height and width of the image. The PSNR results of the proposed encryption scheme are tabulated in Table 2.

From Table 2, low PSNR values are produced for all images that means difficulty in retrieving the original image.

6 Conclusion

This research work presents ROI-based image encryption with Lorenz system and Sine map. First, the ROIs are identified using Sobel edge detection operator and important and unimportant regions are encrypted with Lorenz system and Sine map, respectively. The main advantage of the proposed encryption technique is encryption of ROI areas separately with different chaos which maximizes the security of the encrypted images. Extensive security analyses have been carried out to demonstrate the efficiency of the proposed system, suitable for practical applications.

References

1. Wang Y, Wong K-W, Liao X, Chen G (2011) A new chaos-based fast image encryption algorithm. Appl Soft Comput 11(1):514–522
2. Abdullah AH, Enayatifar R, Lee M (2012) A hybrid genetic algorithm and chaotic function model for image encryption. AEU Int J Electron Commun 66(10):806–816
3. Kalpana J, Murali P (2015) An improved color image encryption based on multiple DNA sequence operations with DNA synthetic image and chaos. Opt Int J Light Electron Opt 126(24):5703–5709
4. Afarin R, Mozaffari S (2013) Image encryption using genetic algorithm. In: 2013 8th Iranian conference on machine vision and image processing (MVIP), pp 441–445
5. Ayoup AM, Hussein AH, Attia MAA (2015) Efficient selective image encryption. Multimed Tools Appl
6. Babaei M (2013) A novel text and image encryption method based on chaos theory and DNA computing. Nat Comput 12(1):101–107
7. Bhatnagar G, Jonathan Wu QM (2012) Selective image encryption based on pixels of interest and singular value decomposition. Digit Signal Process 22(4):648–663
8. Bigdeli N, Farid Y, Afshar K (2012) A novel image encryption/decryption scheme based on chaotic neural networks. Eng Appl Artif Intell 25(4):753–765
9. Chai X, Gan Z, Yuan K, Chen Y, Liu X (2017) A novel image encryption scheme based on DNA sequence operations and chaotic systems. Neural Comput Appl
10. Ye G, Wong K-W (2012) An efficient chaotic image encryption algorithm based on a generalized Arnold map. Nonlinear Dyn 69(4):2079–2087
11. Wang X-Y, Zhang Y-Q, Zhao Y-Y (2015) A novel image encryption scheme based on 2-D logistic map and DNA sequence operations. Nonlinear Dyn 82(3):1269–1280
12. Krishnamoorthi R, Murali P (2017) A selective image encryption based on square-wave shuffling with orthogonal polynomials transformation suitable for mobile devices. Multimed Tools Appl 76(1):1217–1246
13. Taneja N, Raman B, Gupta I (2012) Combinational domain encryption for still visual data. Multimed Tools Appl 59(3):775–793

Chaos-Based Color Image Encryption with DNA Operations

P. Murali, Veeramalai Sankaradass and M. Tholkapiyan

Abstract In this paper, an efficient yet simple color image encryption is proposed based on DNA sequence operation and chaotic maps for Android-based mobile devices. We designed our encryption algorithm based on such constraints. The main idea of the proposed method is different kind of DNA encoding for RGB channels and then Arnold's cat map shuffling is performed in RGB channel in the combined manner. Then DNA addition is performed to diffuse the values between the RGB channels itself. Finally, DNA decoding is performed to form the color encrypted image. The promising feature of the proposed encryption algorithm is simple operations which comfortably suites for low processing power computing devices like Android mobile phones. The analysis and experiment's results show the efficiency of the proposed encryption method.

1 Introduction

Nowadays, the communication between the people is drastically increased when compared to last decade. The development of many electronic devices like mobile phones, smartphones, etc., made communication much easier. So the people can use the Internet within mobile devices at any time and anywhere. However, the secrecy of data is a major issue. To protect data while transmission, encryption plays an important role. Among various encryption methods, image encryption is totally

P. Murali (✉) · V. Sankaradass
Department of Computer Science and Engineering, Vel Tech High Tech Dr. Rangarajan
Dr. Sakunthala Engineering College, Avadi, Chennai, India
e-mail: pmuraliphd@gmail.com

V. Sankaradass
e-mail: veera2000uk@gmail.com

M. Tholkapiyan
Department of Civil Engineering, Vel Tech High Tech Dr. Rangarajan Dr. Sakunthala Engineering
College, Avadi, Chennai, India
e-mail: m.tholkapiyan@gmail.com

© Springer Nature Switzerland AG 2019
D. Pandian et al. (eds.), *Proceedings of the International Conference on ISMAC
in Computational Vision and Bio-Engineering 2018 (ISMAC-CVB)*, Lecture Notes
in Computational Vision and Biomechanics 30,
https://doi.org/10.1007/978-3-030-00665-5_50

different due to bulk capacity, redundancy, etc. The traditional encryption methods like DES, AES, etc., are unsuitable for multimedia encryption due to intrinsic nature of data [1–5]. So, the researchers developed various image encryption methods based on spatial domain and frequency domain. In these methods, the chaos-based image encryption is familiar due to high security, high sensitivity, large key space, etc. [3–5].

2   Literature Survey

In this section, related works are discussed. Recently, Kanso et al. has been proposed color image encryption algorithm, which uses 3D chaotic map [6]. In [7], chaotic neural networks based image encryption and decryption is introduced, which uses like Lorenz, Chua system, Lu system, Tent map, and Arnold's cat map. Liu et al. [8] presented bit level based color image encryption which Piecewise Linear Chaotic Map (PWLCM) and random numbers. The development of DNA computing, DNA cryptography is introduced and biological technology is used as a tool for implementing encryption, message hiding, etc. [9–12]. In [9, 10], initially, DNA encoding is applied to pixels, and diffusion is performed using DNA addition operation. Finally, DNA decoding is performed to construct color encrypted image. The main drawback is fixed DNA encoding and decoding. DNA subsequence operation is introduced in [11, 12].

The above-discussed DNA based image encryption algorithms have drawbacks of fixed DNA encoding and decoding rules which creates the loophole for attackers. In order to overcome the above shortcomings and motivated by DNA computing, we propose simple and efficient color image encryption scheme based on DNA sequence operations and chaotic maps in this paper. Also, we propose dynamic DNA encoding and decoding rules for RGB channels which increase the randomness and security of the encryption algorithm. The most important feature is diffusion operation, which is done between RGB channels. The remaining of this paper is organized as follows. Arnold's cat map and 1D Chebyshev map are introduced in Sect. 2. In Sect. 3, DNA sequences and DNA addition and subtraction are presented. In Sect. 4, the proposed encryption and decryption method is presented, and experiments and analysis are presented in Sect. 5. Finally, conclusion is drawn in Sect. 6.

3   Chaotic Maps

In our proposed algorithm, we use two kinds of chaotic maps: Arnold's cat map and 1D Chebyshev map. The cat map is used for shuffling purpose and Chebyshev map is used for generating the different kind of DNA encoding and decoding rules for RGB channels with different initial conditions.

3.1 Arnold's Cat Map

It is a two dimensional invertible map. It was discovered by Russian mathematician Vladimir. I. It is used to scramble the DNA encoded values in the proposed method. It is mathematically defined by

$$\begin{pmatrix} x_{n+1} \\ y_{n+1} \end{pmatrix} = \begin{pmatrix} 1 & p \\ q & pq+1 \end{pmatrix} \begin{pmatrix} x_n \\ y_n \end{pmatrix} \bmod 1 = A \begin{pmatrix} x_n \\ y_n \end{pmatrix} \bmod 1 \tag{1}$$

where $x \bmod 1$ refers to the fractional part of x for any real number x, p, and q are cat map control parameters. The map is area-preserving since $\det|A| = 1$. An interesting property of Arnold's cat map is it will return into a state which is very close to the initial state after a few iterations. The cat map parameter p and q acts as secret key while decryption process.

3.2 1D Chebyshev Map

The k_{th} degree 1D polynomial can be mathematically defined as

$$T_k(x) = \cos(k \arccos x) \tag{2}$$

where $k = 1, 2, 3, \ldots, n$ and $x \in [-1, 1]$.

Here, 1D Chebyshev map is used to generate the encoding and decoding rule for RGB channels with different initial condition and the mathematical formula is defined as

$$Rul_i = f((\cos(k \arccos x) * 1000))\%8 \tag{3}$$

where $i = 1, 2, 3, \ldots, (M \times 4N)$ and $f()$ denotes the integer function.

4 Chaotic Maps

The DNA sequence consists of four nucleic bases A (adenine), C (cytosine), G (guanine), T (thymine), and can be represented ad 00, 01, 10 and 10. There are 24 encoding rules, but only eight rules are satisfying the Watson–Crick complement rule. The sequence is shown in Table 1. In the four bases, A and T are complementary. G and C are complementary.

The DNA sequence is used to encode the pixels. The pixel consists of 8 bits, and it is separated into four parts each consists of 2 bits, and one of the DNA rules is used to encode the separated four parts. For example, the pixel value is 189, and

Table 1 DNA sequence

	1	2	3	4	5	6	7	8
A	00	00	01	01	10	10	11	11
T	11	11	10	10	01	01	00	00
G	01	10	00	11	00	11	01	10
C	10	01	11	00	11	00	10	01

the binary value is 10111101. It is separated into four parts like 10, 11, 11 and 01 and DNA encoded with rule 5 and the DNA sequence becomes TGGA. To get the original binary sequence, rule 5 is applied if we use different DNA rule, we can get wrong DNA sequence.

4.1 DNA Addition and Subtraction Operation

DNA addition and subtraction operation is performed in binary level. The DNA addition and subtraction rule table are shown in Tables 2 and 3. In image encryption, these two operations are performed while diffusing the pixel's values. In Encryption, the two pixels are added (DNA addition) and DNA subtraction is performed during the decryption process to retrieve the original pixel value.

For example, two DNA sequences [AGCT] and [TAGG] and we applied DNA addition rule as shown in Table 2, and we get the DNA sequence [TGTA]. Similarly, we can obtain the original sequence [AGCT] by subtracting [TAGG] from [TGTA]. In our proposed encryption algorithm, we adopted DNA addition operation during the encryption process and DNA subtraction operation during the decryption process.

Table 2 DNA addition

+	A	G	C	T
A	A	G	C	T
G	G	C	T	A
C	C	T	A	G
T	T	A	G	C

Table 3 DNA subtraction

−	A	G	C	T
A	A	T	G	C
G	G	A	T	C
C	C	G	A	T
T	T	C	G	A

5   Proposed Encryption Technique

First, the original color image with size $(M \times N)$ is separated into RGB channel and then converted into a bit format with size $(M \times 8N)$. Next, we applied DNA encoding operation on RGB channel using Eq. 3 with different initial condition for every channel. So every pixel is differently encoded unlike DNA encoding. Then the entire encoded RGB channel is combined in vertical direction, and we applied Arnold's cat map to scramble the encoded values and it is separated into three equal parts with size $(M \times 4N)$. DNA addition operation is performed between scrambled channels to diffuse the values furthermore. After that, DNA decoding operation is performed on the diffused channels with three different rules as described in Eq. 3, and every channel is converted to the pixel format. Finally, the entire three channels are combined to form the encrypted color image. These are presented as an algorithm in the following sections.

5.1   Proposed Encryption Algorithm

Input: Plain Image of size $(M \times N)$
  Output: Encrypted Image of size $(M \times N)$
  Begin
  Step 1   Read an original color image of size $(M \times N)$.
  Step 2   Convert color image into R, G, and B channel with size $(M \times N)$.
  Step 3   Convert R, G and B channels into bit format with size $(M \times 8N)$.
  Step 4   Apply DNA encoding on RGB channel as described in Eq. 3. It is denoted as R_{en}, G_{en}, and B_{en} as and the size of every channel becomes $(M \times 4N)$.
  Step 5   Combine the encoded channels R_{en}, G_{en} and B_{en} in vertical direction with size $(M \times 3(4N))$. It is denoted as RGB_{com}.
  Step 6   Initialize Arnold's cat map parameters p and q.
  Step 7   Shuffle the RGB_{com_s} by Arnold's cat map with n times. It is denoted as RGB_{com_s}.
  Step 8   Split the RGB_{com_s} into three separate RGB_{com} channels with size $(M \times 4N)$. It is named as R_{en_s}, G_{en_s} and B_{en_s}.
  Step 9   Perform the following DNA Addition operation on shuffled channels:
           $R' = R_{en\_S} + B_{en\_s}, G' = G_{en\_S} + B_{en\_s}, B' = B_{en\_S} + R'$
           where $+$ denotes the DNA addition operation.
  Step 10 Perform DNA decoding on channels R', G' and B' as described in Eq. 3. It is denoted as R_f, G_f and B_f with size $(M \times 8N)$.
  Step 11 Convert R_f, G_f and B_f into pixel format with size $(M \times N)$. It is denoted as R_{pix}, G_{pix} and B_{pix}.
  Step 12 Combine R_{pix}, G_{pix} and B_{pix} to get final encrypted color image.

  End

5.2 Proposed Decryption Algorithm

The decryption process is just reverse of the encryption algorithm. In encryption, we performed DNA addition operation to diffuse the pixel values but in decryption, DNA subtraction operation is applied to get back the original values. Here, we present only DNA subtraction formula:

$$B_{en\_s} = B' - R', G_{en\_s} = G' - B_{en\_s}, R_{en\_s} = R' - B_{en\_s}$$

where denotes the DNA addition operation.

6 Experiments and Security Analysis

The proposed color image encryption algorithm is experimented with more than 100 color images which include standard image processing images with size of (512×512).

6.1 Encryption and Decryption

The proposed encryption algorithm uses the following keys to obtain the encrypted image. The key values are 0.000001, 2, 0.000011, 3, 0.000111, 4, 50, 49, 2, 0.000002, 5, 0.000022, 6, 0.000222, and 3. The values 0.000001, 2, 0.000011, 3, 0.000111, 4 in the key structure denotes the 1D Chebyshev map initial values and k values for RGB channel to generate DNA encoding rules and the values 50, 49, and 2 denotes Arnold's cat map parameter and number of iteration. Finally, the values 0.000002, 5, 0.000022, 6, 0.000222, and 3 denotes the 1D Chebyshev map initial values and k values for RGB channel to generate DNA decoding rules. For representation purpose, we show only Lena image with size of (512×512). Figure 1a shows the original color Lena image, and the encrypted color image is shown in Fig. 1b. The cipher image is decrypted successfully with abovementioned key, and the decrypted image is shown in Fig. 1c.

6.2 Key Sensitivity Analysis

For good cryptosystem, the encrypted image is sensitive with secret key values. To test the key sensitivity analysis, first we encrypt the original color image with key 1 (0.000001, 2, 0.000011, 3, 0.000111, 4, 50, 49, 2, 0.000002, 5, 0.000022, 6, 0.000222, and 3) and then we slightly modify the key 1, and it is denoted as key

| (a) Original Lena image | (b) Encrypted Lena image | (c) Decrypted Lena image |

Fig. 1 Encryption and decryption results of the proposed algorithm

| (a) Decrypted Lena image with key2 | (b) Decrypted Lena image with key3 |

Fig. 2 Decrypted color images with wrong keys

2. The modified key values are (0.000011, 2, 0.000011, 3, 0.000111, 4, 50, 49, 2, 0.000002, 5, 0.000022, 6, 0.000222, and 3). Again, the key1 is slightly modified as key 3 and the corresponding values are (0.000001, 2, 0.000011, 3, 0.000112, 4, 50, 49, 2, 0.000002, 5, 0.000022, 6, 0.000222, and 3). Now, the encrypted image is decrypted with key2 and key3, and the results are presented in Fig. 2.

It is clear from the results that the minor modification in the key structure will give a completely different output. So the proposed encryption algorithm is highly sensitive with the secret key.

6.3 Histogram Analysis

The histogram of the image shows the how pixels are distributed. The pixel distribution of the proposed encryption algorithm is tested in terms of histogram analysis. The results are presented in Fig. 3.

Figure 3a–c shows the pixel distribution of original Lena image R, G, and B channel and Fig. 3d–f shows the encrypted Lena's image R, G and B channel. It is

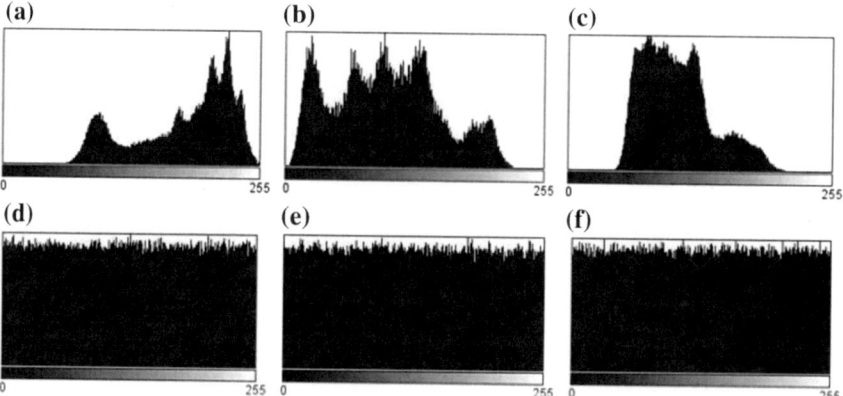

Fig. 3 Histogram analysis results, **a** original Lena image Red channel, **b** Green channel, **c** Blue channel, **d** encrypted Lena image Red channel, **e** Green channel, **f** Blue channel

evident that pixels in encrypted image are evenly arranged and does not give any information about the original image.

6.4 *Differential Attack*

The attacker usually slightly modifies the original image and then performs the encryption and compares the results after changing and before changing. It is called differential attack. It is measured by the number of pixel rate change (NPCR) and the unified average changing intensity (UACI).

6.4.1 Number of Pixel Change Rate (NPCR)

The NPCR is used to measure the number of pixels in difference of a component between two images.

$$\text{NPCR} = \frac{\sum_{i,j} D(i,j)}{W \times H} \times 100\% \tag{4}$$

where W and H is the width and height of the image and $D(i, j)$ is defined as

$$\text{where } D(i, j) = \begin{cases} 1, & c_1(i, j) \neq c_2(i, j) \\ 0, & \text{otherwise.} \end{cases} \tag{5}$$

Table 4 NPCR and UACI values between original Lena image and encrypted image

Image	NPCR (%)			UACI (%)		
	Red	Green	Blue	Red	Green	Blue
Lena	99.16	99.82	99.41	31.67	31.88	31.98
Pepper	99.29	99.11	99.67	31.44	31.24	31.67

6.4.2 Unified Average Change Intensity (UACI)

UACI is used to measure the average intensity difference in component and can be defined as

$$\text{UACI} = \frac{1}{N} \left[\sum_{i,j} \frac{|c(i, j) - c'(i, j)|}{2^L - 2} \right] \times 100\% \tag{6}$$

The proposed cryptosystem is analyzed with NPCR and UACI and the values are presented in Table 4.

From Table 4 shows clearly NPCR is achieved above 99% similarly UACI is achieved above 31%. So, the proposed encryption algorithm shows the extreme sensitivity when small changes in the original image and successfully resists the differential attack.

6.5 Correlation of Adjacent Pixels

For good cryptosystem, adjacent pixels in the encrypted image should be less correlated but in the original image, adjacent pixels are highly correlated. To test the correlation of adjacent pixels in the original image and cipher image, we randomly select 2000 pair's pixels in diagonal, vertical and horizontal direction. The following formula is used to calculate a correlation coefficient:

$$r_{xy} = \frac{\text{cov}(x, y)}{\sqrt{D(x)}\sqrt{D(y)}} \tag{7}$$

$$\text{where cov}(x, y) = \frac{1}{N} \sum_{i=1}^{N} (x_i - E(x))(y_i - E(y))$$

$$E(x) = \frac{1}{N} \sum_{i=1}^{N} x_i \quad D(x) = \frac{1}{N} \sum_{i=1}^{N} (x_i - E(x))^2$$

x and y are gray-level values of two adjacent pixels in the image.

Table 5 Correlation coefficients of adjacent pixels in the original Lena image and encrypted image

	Original image			Proposed method			Wei et al. [9]		
	Red	Green	Blue	Red	Green	Blue	Red	Green	Blue
H	0.9734	0.9631	0.9342	0.0018	0.0034	0.0072	0.0356	0.0763	0.0012
V	0.9621	0.9982	0.8912	−0.012	0.0034	0.0012	0.0127	0.0067	0.0098
D	0.9823	0.9102	0.9112	0.0046	−0.194	0.0041	0.0783	0.0562	0.0058

The results are presented in Table 5 and also we compared our results with Wei et al. [9].

From Table 5, the adjacent pixels in the original Lena image (Red, Green, and Blue channel) are highly correlated but in the encrypted image (Red, Green, and Blue channel), the adjacent pixels are less correlated and also better than [9]. So, our proposed encryption algorithm is successfully resisting the statistical attack.

7 Conclusion

In this paper, we have proposed simple color image encryption algorithm based on DNA operation and chaotic maps for Android environment. The Android environment is mainly for low processing devices like mobile phones, tablets, etc. So we have designed our proposed encryption algorithm to works in such environments. The main feature of the proposed algorithm is different encoding and decoding rules for every channel and diffusion operation, which was done between R, G, and B channels, not any values generated by chaotic equations. It reduces the execution time considerably. The experiment's results show the good stability against cryptographic attacks and also provide large key space. The proposed cryptosystem shows better results compared with Xiaopeng Wei et al. method. So it is suitable for any real-time applications.

References

1. Schneier B (1996) Applied cryptography-protocols, algorithms and source code, 2nd edn. Wiley, New York
2. Stallings W (2006) Cryptography and network security, 4th edn. Pearson Education
3. Matthews R (1989) On the derivation of a 'chaotic' encryption algorithm. Cryptologia 13(1):29–41
4. Li S, Chen G, Zheng X (2004) Chaos-based encryption for digital images and videos. In: Multimedia security handbook. CRC Press, LLC, Boca Raton, FL, USA, pp 133–67
5. Fridrich J (1998) Symmetric ciphers based on two-dimensional chaotic maps. Int J Bifurcat Chaos 8(6):1259–1284
6. Kanso A, Ghebleh M (2012) A novel image encryption algorithm based on a 3D chaotic map. Commun Nonlinear Sci Numer Simulat 17:2943–2959

7. Bigdeli N, Farid Y, Afshar K (2012) A novel image encryption/decryption scheme based on chaotic neural networks. Eng Appl Artif Intell 25:753–765
8. Liu H, Wang X (2011) Color image encryption using spatial bit-level permutation and high-dimension chaotic system. Opt Commun 284:3895–3903
9. Wei X, Guo L, Zhang Q, Zhang J, Lian S (2012) A novel color image encryption algorithm based on DNA sequence operation and hyper-chaotic system. J Syst Softw 85:290–299
10. Zhang Q, Guo L, Wei X (2010) Image encryption using DNA addition combining with chaotic maps. Math Comput Model 52:2028–2035
11. Zhang Q, Xue X, Wei X. A novel image encryption algorithm based on DNA subsequence operation. https://doi.org/10.1100/2012/286741
12. Zhang Q, Wei X (2013) A novel couple images encryption algorithm based on DNA subsequence operation and chaotic system. Optik 124:6276–6281

A Lucrative Sensor for Counting in the Limousine

R. Ajith Krishna, A. Ashwin and S. Malathi

Abstract In developing countries like India, the governments are in dire need of funds for all its projects. Hence it relies on a number of external factors such as exports, trading and FDI. On the other hand, the government also depends on internal factors such as tax from its citizens and indirectly from departments such as transportation. It is the responsibility of the government to provide transportation facilities like buses and trains to its citizens at a nominal charge and at the same time make a profit to run the government and its aided projects. To start from the lowest level, the sad truth is that the number of people travelling without tickets in buses is increasing at an alarming rate. This leads to a big loss for the transportation department and indirectly to the government which falls short of funds. When such losses are incurred due to defaulters in a large scale the government is forced to increase prices of other commodities to look at other sources of income from its citizens though hike in tax, interest rates, fuel price, VAT, etc. The proposed system reduces manpower by excluding checking inspector's role and their responsibilities. As we are all facing a lot of problems in identifying the number of passengers travelling without tickets the design of an automation system emerges with an in-built lucrative sensor that can count the number of passengers entering inside and leaving the limousine. Further, it can also detect the number of passengers who have taken tickets and who have not taken tickets, which will be handy to track down defaulter and make the bus ticketing system more efficient.

R. Ajith Krishna (✉) · A. Ashwin
Department of ECE, College of Engineering, Guindy, Chennai, India
e-mail: ajithkrishna1997@hotmail.com

A. Ashwin
e-mail: aashwin1@outlook.com

S. Malathi
Computer Science and Engineering, Panimalar Engineering College, Chennai, India
e-mail: malathi_raghu@hotmail.com

© Springer Nature Switzerland AG 2019
D. Pandian et al. (eds.), *Proceedings of the International Conference on ISMAC in Computational Vision and Bio-Engineering 2018 (ISMAC-CVB)*, Lecture Notes in Computational Vision and Biomechanics 30,
https://doi.org/10.1007/978-3-030-00665-5_51

1   Introduction

In many applications like safety, transportation security, etc., embedded system plays a pivotal role as it is reliable system which can be isolated from hacking. By making subtle changes in the software, a general embedded system with a standard input and output configuration can be made to perform in a totally different manner. Embedded processors can be classified into two main categories namely, Ordinary microprocessors (μP) that use separate integrated circuits for memory and peripherals whereas Microcontrollers (μC) containing on-chip peripherals which gives benefits like low power consumption, size and cost.

The main objective is to design an efficient system wherein the number of persons coming in and out of a room or inside a vehicle is displayed on a screen. The presence of visitors is sensed using IR sensing mechanism and the entire counting mechanism is done by a microcontroller. This function is implemented using a pair of infrared sensors. LCD display placed outside the room displays this value of person count.

Nowadays, people counting system finds wide use for security purpose [1]. However, many algorithms have been implemented for the system design but clear approach is not yet followed in using these algorithms based on their drawbacks employed in people tracking and counting system. For the evaluation purpose, the number of people with digital camera makes it hard for automatic analysis of large images from many camera due to the high computing overhead. Moreover, video technology does not work in dark environments. In order to overcome this, people counting sensors [2, 3] are used in the proposed system. People counters provide valuable data related to the amount of traffic that flows through facilities. Placing people counters not only at entrances but also at separate displays and exhibits will give you more in-depth data about the popularity of each display.

In people counting system, the sensors are placed at the entry door. Whenever a passenger enters through the bus, there is a breakage in the infrared beam of the sensor [4]. By using this methodology, the total number of people inside the bus will be calculated and the total revenue of the trip will be estimated. Since the counting is done through the electronic equipment (sensor), there will be no manual errors [5] in total count unless there is a problem in the system. Thus, people counting system in transportation provide greater efficiency for counting the people.

2   Literature Survey

Estimating crowd density in an RF-based Dynamic Environment [6] was attempted by Yuan et al.; proposing a Device-free Crowd Counting approach without objects carrying any assistive device, which overcome the problem of high cost and low-light environments conducted by pattern recognition technologies on video surveillance. Accurate instantaneous and cumulative counts especially in crowded scenes [7] were determined using an integer programming method for estimating the instantaneous count of pedestrians crossing a line of interest in a video sequence. The video is first converted into a temporal slice image through a line sampling process. Then using regression function the number of people is estimated in a set of overlapping sliding windows on the temporal slice image which map from local feature to a count.

A concept-based indexing is a challenging approach that permits users to access videos that are conceptually related to the information provided in a search query to recognize the presence of concepts in a video segment for video retrieval over cross domains [8]. This is performed using a flexible energy optimization-based fusion and multicue fusion method integrates both the likelihood predicted by classifiers and high-order contextual-temporal relationships revealed from annotations and pseudo labels. Estimating the size of in homogenous crowd was the problem addressed by Chan and Vasconcelos [9] which are composed of pedestrians travel in different directions, without using explicit object segmentation or tracking is proposed.

An innovative approach called Occlusion Handling [10] is introduced to cross camera people for counting that can adapt itself to a new environment bypassing the manual inspection. In the analysis, a method for motion detection called ViBe as Universal Background Subtraction algorithm for video sequences was explained by Barnich et al. [11] that stores for each pixel, a set of values taken in the past at same location or in neighborhood. First, background subtraction [12] results estimated the number of people moving only slightly in a complicated scene. Second, an Expectation Maximization (EM) based method has been implemented to find the individuals in a low-resolution scene. The representation of each person in the scene is done using a novel cluster model. The current state-of-the-art image processing methods for automatic-behaviour-recognition techniques [13] is described by the system to monitor human activities in haulage. In the year 2009, Embedded Vision Modules for Tracking and Counting People showed the algorithm implementation [14] for a field-programmable gate array (FPGA)-based design for people counting using a low-level head-detection method.

In case of people in crowd areas, the crowd counting was found out using Bayesian Poisson Regression [15, 16] approach and it is analyzed by initiating a prior distribution on the weights of the linear function algorithm. Based on the flow velocity field estimation model and offline learning the algorithm has unified both of the LOI and ROI problems [17] together. Multi-object tracking which considers object detection and space-time trajectory estimation as a coupled optimization problem [18] were implemented by Leibe and Schindler. This proposed system employs a Minimum Description Length hypothesis selection framework, making the system to recover from mismatches and temporarily lost tracks. Segmentation and Tracking using multiple cameras in a crowded environment proposed [19] handled the inter-object occlusion problem.

Privacy-preserving crowd monitoring [20] handled the in homogenous crowds in different directions and a Gaussian process (GP) is used to regress feature vectors to the number of people per segment. High accuracies obtained by using motion capture and surveillance which automatically track the articulations of people from a video sequence [21]. This creates ambiguity while examining the people in both frame and in estimation of their layout. Bayesian clustering algorithm [22] assessing the value of motion and the information content of unsupervised data-driven has been left out as the control. In the survey, an efficient technique for Pedestrian detection [23] by Leibe et al.; specifically address the task of detection in crowded scenes. It handles the problem of detecting people in crowd with severe overlaps. Stauffer and Grimson focused on motion tracking and showed how one can use observed motion to learn patterns of activity [24] in a site. In enclosed surroundings this system is used to track people alone, whereas both people and vehicles are tracked in open atmosphere. Also, the system is used even for finding the fish in a tank, ants on a floor, and remote control vehicles in a lab setting.

3 Proposed Work

In recent days, people counting system is being generally used for security purpose. Many algorithms have been designed and implemented, but still there lies a confusion in selecting the suitable algorithms considering the advantages and disadvantages. Moreover, video technology does not work in dark environments. In order to overcome this, people counting sensors are used in the proposed system.

Figure 1 shows the architecture diagram for the system Where the IR sensor is given as input to the port1 of the ATMEL microcontroller. When the conductor is issuing tickets to passenger, count value of issued tickets will be stored in the database at each bus stop. The system detects the frauds that are travelling without tickets by comparing the difference between count value of passengers inside the bus and count value of issued tickets at each stop. The components included in the system are ATMEL microcontroller represented as 'AT89S52' which is a typical 8051 microcontroller. The IR sensor is included in the system that produces or detects infrared radiation in order to sense some aspect of its environment.

ALGORITHM: COMMUTER SENSOR (p1, p2. . . pS). Given a Set of N Training Examples with their S types of Distance Descriptors.

Step 1: Initialize Ptc=0
Step 2: Initialize s=0
Step 3: while(s==x)
//varying the distance by adjusting the probe
Step 4: for i=1 to P
Step 5: if (! ir_in) //infrared beam interrupted at the entrance
Step6:Ptc++
Step 7: else if (! ir_out)
//infrared beam interrupted at the exit
Step 8: Ptc- -
Step 9: else Print Ptc
Step 10: end if
Step 11: end for
Step 12: end while
Step 13: return Pi count value

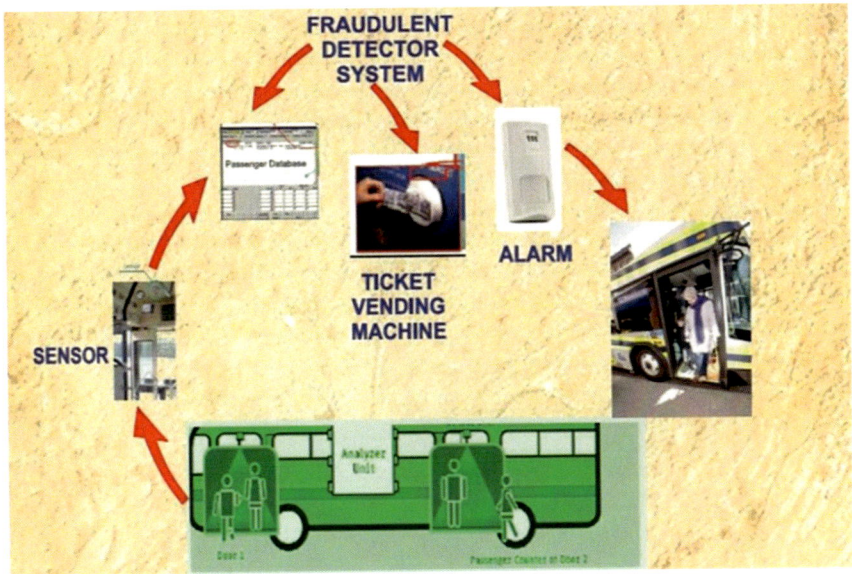

Fig. 1 Architecture framework for counting in the Limousine

A set (p1, p2, p3 … pS) of N training examples is specified as an input to the sensor with their types of distance descriptors, where p represents the number of people entered inside the bus. This parameter is considered for N number of samples and gets repeated until the limousine reaches its destination and Δs is the symbol of distance that can be attuned by varying their ranges.

ALGORITHM: TRAVELLING WITHOUT TICKETS combines (p1,p2, . . . ,pS) and (t1,t2, . . . ,tD).Given a Set of N Training examples with their S types of distance descriptors.

Step 1: Initialize place=source
Step 2: While (! place=destination)
Step 3: for i=1 to 5
            //iterate the loop for every 5 minutes
Step 4: for j=1 to T
            //loop iterates for generation of each ticket
Step 5:   if (! BS)
            //check whether bus stop has not arrived
Step 6:   wt=Ptc - tc
            //compute the travelers who haven't taken tickets
Step 7: end if
Step 8: end for
Step 9: end for
Step10: end while
Step11: return total wt value

A set (p1, p2, p3 … pS) of N training examples is specified as an input to the sensor with their types of distance descriptors, where p represents the number of people entered inside the bus. This parameter is considered for N number of samples and gets repeated until the limousine reaches its destination and s is the symbol of distance that can be attuned by varying their ranges. Initially, a parameter named Ptc represents the people total count to be zero then validates the while condition loop by checking whether s equals the varying distance x. The count of without tickets can be calculated for every stop till the destination place.

4   Results and Discussion

Labview finds its application in science and engineering field for creating custom applications which interact with real-world data or signals. Instead of sequential line by line operation, G is an efficient graphical dataflow language in which nodes (operations or functions) operate on data once it becomes available. Table 1 represents the estimation of number of passengers travelling without tickets is detected.

Figure 2 represents the plotted graph used for estimating the number of passengers travelling without tickets by indicating the blue bar for number of passengers entering inside the bus and red bar for the tickets issued for passenger entered inside the limousine.

Figures 3 and 4 depict the outcome of the passengers travelling from source station and final destination where the passengers without tickets are verified.

Table 1 Number of passengers without tickets

Bus stop	Time (min)	Iteration	
		No. of passengers	No. of tickets issued
s1	3	10	8
s2	2	8	6
s3	2.5	9	7

Fig. 2 Passengers without tickets based on time

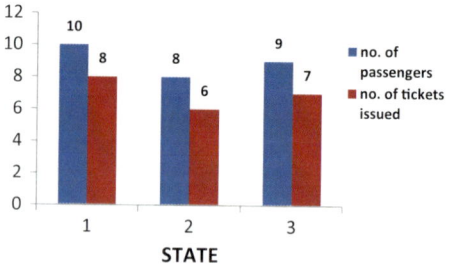

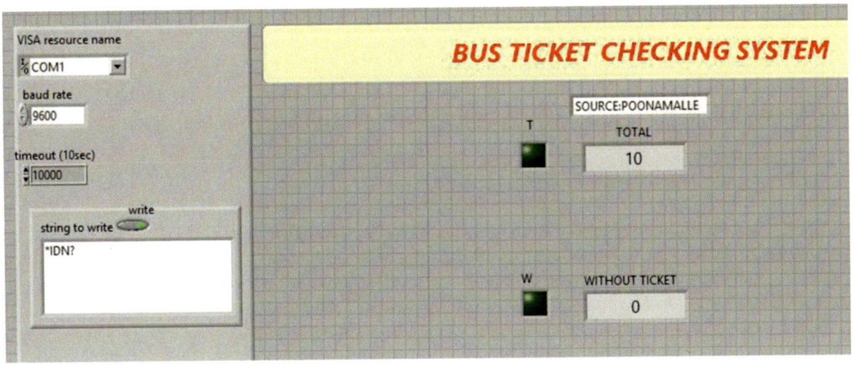

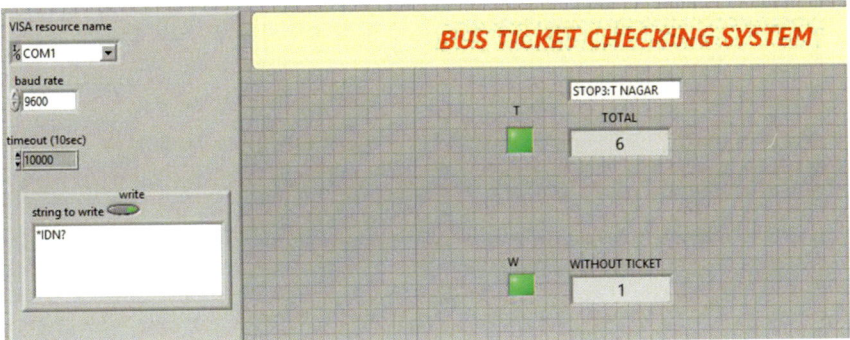

Fig. 3 Passengers at source station

Fig. 4 Passenger without tickets at the destination

5 Conclusion

Transportation department is facing numerous problems in identifying the number of passengers travelling without tickets, an automation system has been designed with an in-built commuter computational sensor for counting number of passengers entering and leaving and detecting number of passengers travelling with tickets and without tickets which reduces manpower by excluding checking inspector's role and their responsibilities and loss of transportation's fund. There is still much to improve the system in identifying passengers who are travelling without tickets by using a fixed camera inside that can work by integrating with ATMEL microcontroller combining the facial recognition techniques in image processing it will also feature a notification system called Rap Back, which will give investigators live updates of any given criminal's movements.

In a bid to modernize US detective work, an alternate replacement for a countrywide fingerprint database using the current system. This recognition can also be implemented by scanning the social security card like Adhaar card that contains details like in an individual file followed by linking them to personal and biographic data like name, home address, ID number, immigration status, age, race, etc. The other federal agencies also share the database which contains new records that reveals that this is capable of processing 55,000 direct photo enrollments and also conducting numerous searches like tens of thousands every day.

References

1. Leibe B, Schindler K, Van Gool L (2007) Coupled detection and trajectory estimation for multi-object tracking. In: Proceedings of IEEE international conference computer vision, 2007, pp 1–8
2. Yang DB, González-Bãnos HH, Guibas LJ (2003) Counting people in crowds with a real-time network of simple image sensors. IEEE Conf Comput Vis Pattern Recogn 2
3. Li Y, Huang C, Nevatia R (2009) Learning to associate: hybridboosted multi-target tracker for crowded scene. In: Proceedings of IEEE international conference
4. Liu X, Taommingli Song D, Zhang L, Bu J, Chen C (2014) Learning to track multiple targets. IEEE Trans Neural Netw 20
5. Muñoz-Salinas R, Medina-Carnicer R, Madrid-Cuevas FJ, Carmona-Poyato A (2009) Multi-camera people tracking using evidential filters. Elsevier Inc.
6. Yuan Y, Zhao J, Qiu C, Xi W (2013) Estimating crowd density in an RF-based dynamic environment. IEEE Sens J 13(10)
7. Ma Z, Chan AB (2013) Crossing the line: crowd counting by integer programming with local features. In: Proceedings of conference computer vision pattern recognition, pp 2539–2546
8. Weng M-F, Chuang Y-Y (2012) Cross-domain multicue fusion for concept-based video indexing. IEEE Trans Pattern Anal Mach Intell 34(10):1927–1941
9. Chan AB, Vasconcelos N (2012) Counting people with low-level features and Bayesian regression. IEEE Trans Image Process 21(4):2160–2177
10. Lin T-Y, Lin Y-Y, Weng M-F, Wang Y-C, Hsu Y-F, Liao H-YM (2011) Cross camera people counting with perspective estimation and occlusion handling. In: Proceedings of IEEE international workshop information forensics security, pp 1–6, Nov/Dec 2011
11. Barnich O, Van Droogenbroeck M (2011) vibe: a universal background subtraction algorithm for video sequences. IEEE Trans Image Process 20(6):1709–1724
12. Hou Y-L, Pang GKH (2011) People counting and human detection in a challenging situation. IEEE Trans Syst Man Cybern A Syst Humans 41(1):24–33
13. Candamo J, Shreve M, Goldgof DB, Sapper DB, Kasturi R (2010) Understanding transit scenes: a survey on human behavior-recognition algorithms. IEEE Trans Intell Transport Syst 11(1)
14. Vicente AG, Muñoz IB, Molina PJ, Galilea JLL (2009) Embedded vision modules for tracking and counting people. IEEE Trans Instrument 58(9)
15. Chan AB, Vasconcelos N (2009) Bayesian Poisson regression for crowd counting. In: Proceedings of IEEE international conference computer vision 545–551
16. Zhao T, Nevatia R (2003) Bayesian human segmentation in crowded situations. Proc IEEE Conf Comput Vis Pattern Recogn 2:459–466
17. Cong Y, Gong H, Zhu S-C, Tang Y (2009) Flow mosaicking: realtime pedestrian counting without scene-specific learning. In: Proceedings of IEEE conference computer vision pattern recognition, June 2009, pp 1093–1100
18. Leibe B, Schindler K, Cornelis N, Van Gool L (2008) Coupled object detection and tracking from static cameras and moving vehicles. IEEE Trans Pattern Anal Mach Intell 30(10)
19. Zhao T, Nevatia R, Wu B (2008) Segmentation and tracking of multiple humans in crowded environments. IEEE Trans Pattern Anal Mach Intell 30(7):1198–1211

20. Chan AB, Liang ZSJ, Vasconcelos N (2008) Privacy preserving crowd monitoring: counting people without people models or tracking. Proc IEEE Conf Comput Vis Pattern Recogn 1–7
21. Ramanan D, Forsyth DA, Zisserman A (2007) Tracking people by learning their appearance. IEEE Trans Pattern Anal Mach Intell 29(1):65–81
22. Brostow GJ, Cipolla R (2006) Unsupervised Bayesian detection of independent motion in crowds. Proc IEEE Conf Comput Vis Pattern Recognit 1:594–601
23. Leibe B, Seemann E, Schiele B (2005) Pedestrian detection in crowded Scenes. Proc Conf Comput Vis Pattern Recognit 878–885
24. Stauffer C, Grimson E (2000) Learning patterns of activity using real-time tracking. IEEE Trans Pattern Anal Mach Intell 22(8):747–757

Energy Expenditure Calculation with Physical Activity Recognition Using Genetic Algorithms

Y. Anand and P. P. Joby

Abstract Physical health is associated with physical activity, physical activity also ensures the wellbeing of the humans, physical activity is recognized using body worn sensors, and three Inertial Measurement units (IMU) are used to capture the data from the sensors. The activity recognition chain consists of Data Acquisition, Preprocessing, segmentation, Feature extraction, and Classification. Different levels of research are carried out on each stage. In feature selection genetic algorithms are used but the paper proposing the memetic algorithms an enhanced version of the genetic algorithms with local search in the each stage of genetic algorithm. This technique shall eliminate the chances of energy loss and consequently increase efficiency of the current system.

1 Introduction

The physical activity is solely related to the human health, but unfortunately advancements in transportation and technology made everything available at doorsteps and fingertips, reducing physical activity thereby reducing physical health. Several works were done in the fields of relating physical activity and physical health, works of Janssen and LeBlanc [1] stating that even physical activity can improve the health considerably. Next big challenge is to monitor and record the physical activity, so we could have qualitative analysis of the physical activity, this can be done using three IMUs (Inertial measurement units). Specialized systems can record these activities and can be used for the further processing, machine learning data set is from the available dataset is done using running data through Activity Recognition Chain

Y. Anand (✉)
Computer Science and Engineering, Mar Baselios Christian College of Engineering and Technology, Peermade, India
e-mail: michaelanand2020@gmail.com

P. P. Joby
Mar Baselios Christian College of Engineering and Technology, Peermade, India
e-mail: jobymone@gmail.com

© Springer Nature Switzerland AG 2019
D. Pandian et al. (eds.), *Proceedings of the International Conference on ISMAC in Computational Vision and Bio-Engineering 2018 (ISMAC-CVB)*, Lecture Notes in Computational Vision and Biomechanics 30,
https://doi.org/10.1007/978-3-030-00665-5_52

(ARC) [2]. Energy expenditure will be different for different activities based on MET value. Calculating accurate MET value with minimal use of sensors and is the idea explained in the paper.

2 Overview of Physical Activity and Health Relation

The importance of keeping physical health is much important for well-being and keeping life expectancy high, reduced physical activity in any age group will lead to the serious health issues. People are aware of it but lack motivation, giving them a quantitative detail with the aid of technology will aid in improving activity and increasing health. Different physical activity requires different level of energy and some intense activities requires only small time to burn more calories (energy spent in a activity) and light activities requires more time to do the same. Building a specific set of activities to burn more calories in a shorter time is the key aspect. Doing the activities for burning more energy will increase the health, thereby reduce the risk of many dangerous lifestyle disorders like cholesterol, blood pressure, etc. In order to find activities with more energy burning relation we need to know some value to calculate the energy. The collected data from the sensors will be converted into some useful features, which can be analyzed in an efficient manner.

3 Physical Activity and Energy Expenditure

MET or metabolic equivalent is amount of energy we spent in each activity, as per m. Jette et al. [3] work about MET, one MET is equal to amount of oxygen intake when our body being rested, taking this value as unity other activities MET can be calculated as the multiple of these, one MET means the 3.5 ml of O_2 (oxygen) one MET is equivalent to 17.5 W (watts) and one Watt is equal to 0.01435 kcal of energy Compendium of Physical Activities has provided the provided MET values [4] and Kozey et al. proposed work of calculating MET Value from the accelerometer outputs. Getting the Data from the sensors and processing it to find the values.

4 Physical Activity Recognition Using Sensors

The human physical activity recognition been in interest of the researchers in the early 20s, later many research has been done on the physical activity recognition using sensors, Some devices from gaming consoles are very well known for tracking and analyzing the human activity and also found some space in fitness applications, devices like these enhanced the computer human interaction. Main characteristics of human activity recognition systems are given in Bulling et al. [5].

4.1 The Activity Recognition Chain

The ARC (activity recognition chain) is the sequence of processing the signal, specific to machine learning techniques the activity recognition chain explains how the raw data from the sensors are converted to the useful data, Fig. 1 explains how dataset is generated with the help of the sensors.

Sensor Data Acquisition and Preprocessing The main objective is to capture the data from the sensors that are worn on subject's body, three colibri wireless sensors were used on this process, position of the IMU (inertial measurement units) [6].

- 1 IMU over the wrist on the dominant arm
- 1 IMU on the chest
- 1 IMU on the dominant side's ankle.

The sensors should have components like accelerometer, gyroscope, temperature sensor, magnetometer, etc., All these components were collected with the corresponding attributes from the body, 9 (8 men and 1 women aged 27.22 ± 3.31) subjects are requested to do certain activities.

Each IMU collects 17 columns of data in total which consists of

- 1 temperature (°C)
- 2–4 3D-acceleration data (ms^{-2}),
- 5–7 3D-acceleration data (ms^{-2})
- 8–10 3D-gyroscope data (rad/s)
- 11–13 3D-magnetometer data (μT)
- 14–17 orientation (invalid in this data collection).

When building a dataset that can be useful for the later processing we should also add some useful attributes like timestamp, activity id for uniquely identifying each activity.

Activity IDs:

- 1 lying
- 2 sitting
- 3 standing

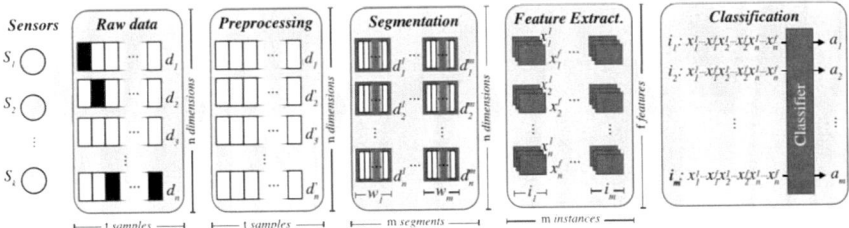

Fig. 1 Activity recognition chain

- 4 walking
- 5 running
- 6 cycling
- 7 Nordic walking
- 9 watching TV
- 10 computer work
- 11 car driving
- 12 ascending stairs
- 13 descending stairs
- 16 vacuum cleaning
- 17 ironing
- 18 folding laundry
- 19 house cleaning
- 20 playing soccer
- 24 rope jumping
- 0 other (transient activities).

HR (Heart Rate) is also included so that the bpm (beats per minutes) will be recorded by the heart rate monitor. All these data are synchronized and labeled data from all the sensors are merged into a single file. finally there will be 54 columns of data that contains the following data:

- 1 timestamp (s)
- 2 activity ID
- 3 heart rate (bpm)
- 4–20 IMU hand
- 21–37 IMU chest
- 38–54 IMU ankle.

Data Segmentation The data need to be segmented in order to processing, data collection is continuous and repeated over 0.01 s (seconds) so we have to specify a start time t_1 and end time t_2 and all the data collected from this time should be considered as a segment [5]. The whole data collected can be converted to some useful segments, a single segment can be defined as,

$$W_i = (t_1, t_2) \tag{1}$$

W is set of segments were

$$W = (w_1 \ldots w_m) \tag{2}$$

Segmenting the data can be done based on different methods, we use sliding window method. The size of sliding window is fixed as 512 (5.12 s), that is the sensor data of 5.12 s are grouped together.

Feature Extraction This stage consist of converting the segments into some useful features, features are brief abstract of the raw data, data collected can be represented as features like mean and variance, running the DFT (Discrete Fourier Transform) over the segmented data of every signal and dimension using the (FFT) Fast Fourier Transform [6] in the feature selection challenges arise due to various factors selecting the correct feature selection techniques this is directly linked to the accuracy of the whole data processing.

Classification This is the final process in ARC. In this data should be classified for each person and this data can be used for cross person classification done based on the feature extraction. A technique known as extra trees in machine learning library is used for the work Alejandro et al. [6] and obtained an accuracy of 97.45 in the final result.

5 Calculating the Energy Expenditure

To find the energy spend on each activity in terms of calories or joules, this could be very useful for the user to assess the performance every day. Calculating the energy expenditure is given as

$$\text{Total amount of calorie burned} = \text{MET} * \text{Body weight} * \text{time duration} \qquad (3)$$

Oxygen intake is directly related to the energy expenditure, for calculating MET, accelerometer data and heart rates can be used [4]. With the help of accelerometer data and wireless unit that measures breath-by-breath gas exchange. MET value can be calculated for each individual. This Data can be compared to user's previous data or other persons with cross person validation. More energy spent on the activity if it consumes more oxygen, the oxygen gas flow is key factor to compute the total energy expenditure.

6 Conclusion and Future Work

The physical activity is essential to maintain the physical health, new age of technology will aid us in tracking and improving our health by analyzing our physical activity, the different techniques used for acquisition and processing of data and how this data can be used to calculate the energy is explained. Future work on this paper would be by building a machine learning system can be used for learning the user movements and an artificial intelligence system can be used for the giving continuous suggestion to the user about the his/her physical activity and it can be programmed such a way that we could do a intense work in small amount of time.

References

1. Janssen I, LeBlanc AG (2010) Systematic review of the health benefits of physical activity and fitness in school-aged children and youth. Int J Behav Nutr Phys 7(1):40
2. Reiss A, Stricker D (2012) Introducing a new benchmarked dataset for activity monitoring. In: Proceeding of ISWC'12, pp 108–109
3. Jette M, Sidney K, Blumchen G (1990) Metabolic equivalents (METS) in exercise testing, exercise prescription, and evaluation of functional capacity. Clin Cardiol 13(8):555–565
4. Kozey SL, Lyden K et al (2010) Accelerometer output and MET values of common physical activities. US National Library of Medicine National Institutes of Health. https://www.ncbi.nlm.nih.gov/pmc/articles/PMC2924952/
5. Bulling A, Blanke U, Schiele B (2014) A tutorial on human activity recognition using body-worn inertial sensors. ACM Comput Surv 46(3):33:1–33:33
6. Baldominos A, Isasi P, Saez Y (2017) Feature selection for physical activity recognition using genetic algorithms. IEE Explore

Improved Intrinsic Image Decomposition Technique for Image Contrast Enhancement Using Back Propagation Algorithm

Harneet Kour and Harpreet Kaur

Abstract The technique of intrinsic image decomposition is based on the illumination value of the image. The histogram equalization value of the input image is calculated to increase the image contrast. In this research work, the back propagation algorithm is applied for the calculation of histogram equalization. The iterative process of back propagation is executed until error is reduced for the histogram equalization calculation. The simulation of the proposed modal is performed in MATLAB. The performance of proposed modal is compared in terms of PSNR and MSE.

1 Introduction

A mechanism in which an image is converted into digital form is known as image processing. In order to enhance the image or to extract important information present within, various operations are performed on that image. In order to provide certain benefits within applications, image processing has been growing in demand. Digital image processing techniques are now applicable within numerous other applications along with medicine and space programs [1]. For easy interpretation of X-rays and various other images available from different sources, various computer procedures are applied which help in enhancing the contrast or other features present within the image. Within most of the image processing applications, image enhancement is the most interesting area of research. The details are made more obvious and specific features that are of interest are highlighted with the help of applying image processing techniques [2]. The images that are ambiguous and uncertain can include different aspects within them. The qualities of an image are enhanced with respect to contrast, brightness properties, removal of noise and so on, by applying image

H. Kour
ECE Department, Chandigarh University, Gharuan, India

H. Kaur (✉)
CSE Department, Chandigarh University, Gharuan, India
e-mail: harpreet8307@gmail.com

© Springer Nature Switzerland AG 2019
D. Pandian et al. (eds.), *Proceedings of the International Conference on ISMAC in Computational Vision and Bio-Engineering 2018 (ISMAC-CVB)*, Lecture Notes in Computational Vision and Biomechanics 30,
https://doi.org/10.1007/978-3-030-00665-5_53

enhancement operation. An image might be blurred in some cases due to which the objects present within it might not be visible clearly. Thus, there is a need to enhance that image such that its contents can be visible. In order to provide solutions to this issue, contrast enhancement is one important research aspect [3]. There are several contrast enhancement techniques used among which some are explained below:

a. Histogram Equalization (HE): In order to enhance the contrast of an image, the most popular technique known is HE. On the basis of probability distribution of input gray levels, the available fray levels are mapped within this technique. The dynamic range of image's histogram is flattened and stretched through this technique due to which the complete contrast enhancement is achieved [4]. The image is treated globally through the HE technique which is a major advantage. The images that include within them both bright or both dark backgrounds can attain better results when this technique is used.

b. Contrast-Limited Adaptive Histogram Equalization (CLAHE): An adaptive contrast enhancement technique which is based on the adaptive histogram equalization which is enhancement to the traditional HE technique is known as CLACHE. Instead of computing the complete image, a specific section of an image which is known as tile is computed using various histograms in this technique. In order to redistribute the pixel values of an image, the contrast of each tile is improved. Further, for removing any artificially added boundaries, the bilinear interpolation is utilized in order to combine the neighboring tiles [5]. In order to prevent noise amplification that is present within an image, the contrast is ensured to be limited within the homogeneous regions. The local contrast of an image is thus enhanced and more specific details are achieved through this technique. Instead of focusing on the overall contrast of an image, the local contrast is only focused on through this method.

c. Morphological Enhancement: A new technique is generated in order to solve the various issues arising in this field by applying mathematical morphology to the image processing and analysis methods. On the basis of set theoretic concepts of shape, this approach is generated [6]. The objects that are present in an image are considered to be sets as per the morphology technique. The machine vision and recognition processes can easily use mathematic morphology since the objects and the features present in them can be identified on the basis of their shapes.

2 Literature Review

Wang et al. (2017) proposed an enhanced to the image contrast enhancement preprocessing approach such that the model parameters can be derived in automatic manner from the cell video sequences as per each frame [7]. This technique is derived in order to avoid issues that arise due to low image intensity contrast characteristics within cell images. Two image sequences were tested and comparisons were made against Viterbi-based segmentation and the proposed technique. As per the evaluations it was seen that there was around 37 and 33% of enhancement in accuracy through application of proposed technique.

Murinto et al. (2017) proposed a novel approach in order to enhance the leather image. On the basis of un-sharp masking algorithm, piecewise linear contrast stretch technique is utilized here [8]. Upon four different animal leather images of animals namely lizard, crocodile, cow and goat, the experiments are conducted here. Comparisons are made amongst other existing enhancement techniques and it is seen as per the simulation results that the proposed technique is better in comparison to proposed technique in terms of PSNR, MSE and SSIM values.

Nnolim (2017) proposed an enhanced algorithm for single underwater image enhancement. More generalized global contrast enhancement is utilized in the PDE-based framework in order to develop adaptive as well as completely automated algorithms [9]. The color correction, edge as well as contrast enhancement results are improved as per the results generated by applying proposed technique. The underwater images that include flat color histograms are very difficult to improve by utilizing traditional techniques can be enhanced by applying this proposed technique.

Fornes et al. (2013) proposed a novel technique in order to enhance and eliminate the show-through scheme present within gray-scale documents [10]. Only one side of the page is utilized by this method and there is no need to eliminate the framing borders. There is a need to tune only one parameter within this approach. As per the simulation results it is seen that the readability of documents is enhanced due to which the unreadable words can be recovered. Also, the ambiguities of the document can be solved and the amount of time that is required to read each document is also minimized by using this proposed technique.

Rosenberger (2012) proposed the utilization of Spartan 6 FGPA within the developed camera. In order to attain higher processing power, the properties of DSP 48 slices were utilized [11]. The frame rate of CCD sensor is the speed limitation for this application. Lots of additional sampling time is saved with the help of generating VCE image. A low uncertainty of optical measurement is achieved when a well illuminated image is required along with various sub-images. The results achieved show that this mechanism helps in eliminating previous problems and providing enhanced results.

In Fig. 1, the framework of proposed contrast enhancement method is shown. Initially, an input color image I is converted into HSV representation. Further, the proposed intrinsic decomposition model is used that decomposes the value (V) channel image into illumination (L) and reflectance (R) layers as secondary step. An adjusted L layer which is represented as L_a is generated when L layer is adjusted using Gamma mapping function which is the third step. In order to generate the enhanced V channel image which is represented here as V_e, the adjusted L_a is multiplied by reflectance layer R. In order to enhance the local contrast of V_e the contrast-limited adaptive histogram equalization (CLAHE) [12] is utilized due to the fact that the mapping function can be performed globally. $\hat{V}_e$ is used to denote the enhanced result. The technique of back propagation will be applied to calculate the histogram equalization. The technique of back propagation take input the pixel number of pixel value

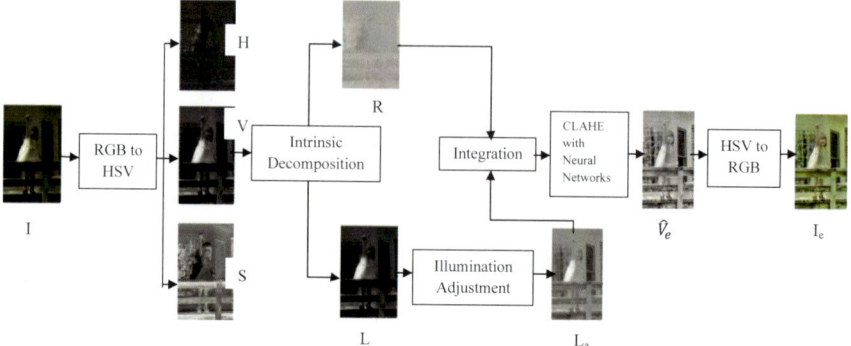

Fig. 1 The framework of the proposed contrast enhancement method

to generate the final equalization value. The formula is given in Eq. (1) to calculate actual value of equalization

$$\text{Actual equalization value} = \sum_{\substack{x=0 \\ w=0}}^{\substack{x=n \\ w=n}} x_n w_n + \text{bias} \tag{1}$$

To calculate the histogram equalization, we need to define the desired value. The final equalization will be at which error is least. The error will be calculated with the Eq. (2)

$$\text{Error} = \text{Desired Equalization value} - \text{Actual equalization value} \tag{2}$$

At the end, the final result I_e is generated by transforming the enhanced HSV image into RGB space. Intrinsic decomposition and illumination adjustment are the two major modules of this proposed mechanism which are explained further.

a. Intrinsic Image Decomposition

In order to compute the equation $V = L.R$, one image is decomposed into reflectance and illumination layers within the Intrinsic image decomposition. In the form of constraints that involve the neighboring pixels with same color to have similar reflectance and the neighboring pixels that have similar illumination, intrinsic decomposition is proposed here [13]. Thus, the intrinsic decomposition issue is formulated here within Eq. (3), as a minimization problem of the energy function. Here, vector form of V, L, and R are used which are v, l, and r.

$$\min_{l,r} \; E(l,r) = E_r(r) + \mu E_l(l) + \theta E_d(v;l,r) + \beta E_o(l,l_o),$$

$$\text{s.t.} \qquad\qquad\qquad\qquad 0 \le r \le 1, \tag{3}$$

Here, the weighting parameters are denoted by $\mu, \theta,$ and β. Upon the reflectance and illumination layers, μ and θ are regularizer. In order to ensure the reliability of decomposition, β terms is used. l_2 norm penalty which is $\left(\|v - l \cdot r\|_2^2\right)$ is utilized instead of previous equality constraint which was $v = l \cdot r$ within the intrinsic decomposition such that the noise can be tolerated. The noise might decompose otherwise. In order to constrain the scale of l, the final term $E_o = \left(\|l - l_o\|_2^2\right)$ is used. The chromatic normalization value which is $\sqrt{I_r^2 + I_g^2 + I_b^2}$ is set for l_o here.

The reflectance later is constrained to be piecewise constant as per the similarity of color by the $E_r(r)$ term. For instance, at pixel i, the reflectance value is denoted by r_i. The representation of this constraint is shown in Eq. (4).

$$E_r(r) = \sum_i \sum_{j \in N(i)} w_{ij} \| r_i - r_j \| 1 \tag{4}$$

Here, for pixel i, the neighborhood is $N(i)$. For pixel i and j, the similarity of chromatic value is measured by w_{ij}. The weight values are increased here such that the different amongst r_i and r_j can be penalized for the pixels that have similar colors. Thus, the weighting function is defined in Eq. (5) as:

$$w_{ij} = \exp\left(-\frac{\|f_i - f_j\|_2^2}{2\sigma^2}\right) \tag{5}$$

Here, the value of pixel i is represented by f_i which can be denoted as $f_i = [\tau l_i, a_i, b_i]^{\mathrm{T}}$. The values of τ and σ are constant. It is to be ensured that $\tau < 1$ in order to minimize the impact of illumination variations on the color similarity measurement.

The isotropic total variation is utilized in order to enforce that the illumination is smooth with the application of $E_l(l)$, and is defined by Eq. (6) as:

$$E_l(l) = \|D_x l\|_2^2 + \|D_y l\|_2^2 \tag{6}$$

Here, along the horizontal and vertical directions, D_x and D_y represent the matrix representation of derivative operators respectively.

b. Illumination Adjustment

In order to enhance the image details, the adjustment of illumination values is the next major task to be performed after calculating the reflectance and illumination layers. Through the results obtained it is seen that the by preserving the lightness order, all the dark areas are darkened. Within the bright areas however, the intensities are compressed such that there is minimization of variation of large intensities. Thus, as a result, within the bright areas, the details will be lost mainly within the Log and

(a) Input Image (b) Intrinsic image decomposition (c) Improved Intrinsic image
decomposition

Fig. 2 **a** Input image. **b** Intrinsic image decomposition. **c** Improved intrinsic image decomposition

Sigmoid functions [13]. Thus, Gamma function is utilized here for adjusting the illuminations which is represented in Eq. (7).

$$L_a = 255 \times (L/255)^{1/\gamma} \tag{7}$$

Here, for conducting experiments, the value of γ is set as 2.2.

3 Experimental Results and Analysis

The performance of proposed technique is evaluated in this section. The performance of proposed technique will be compared with the intrinsic image decomposition technique. The performance of the proposed modal can be viewed in both subjective and objective form. The proposed algorithm can be tested on the dark image for the contrast improvement.

As shown in Fig. 2a is the input image which is the dark image whose contrast needs to improve. The technique of intrinsic image decomposition is applied in Fig. 2b. The Improved Intrinsic image decomposition technique is applied in Fig. 2c which gives best result as compared to existing technique.

As shown in Fig. 3, the Intrinsic Image decomposition technique is compared with the proposed intrinsic image decomposition technique in terms of PSNR and MSE. The simulation results illustrated that proposed technique performs well in terms of all parameters.

4 Conclusion

In this paper, improvement in the intrinsic image decomposition technique is proposed. The intrinsic image decomposition technique is based on the illumination layer in which histogram equalization is calculated to image contrast. In this research work, the improvement in the intrinsic image decomposition technique is proposed

Fig. 3 Comparison of techniques

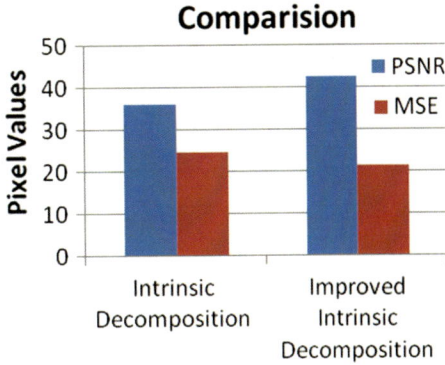

based on back propagation technique. The back propagation technique will be executed in the iterative manner until error gets reduced for the calculation of histogram equalization. The iteration at which error is least is defined as the final histogram equalization point. The simulation results shows that proposed technique performs well in terms of PSNR and MSE than existing intrinsic technique.

References

1. Cheng P, Cui A, Yang Y, Luo Y, Sun W (2017) Recognition and classification of coating film defects on automobile body based on image processing. In: 2017 10th International congress on image and signal processing, biomedical engineering and informatics (CISP-BMEI 2017)
2. Purohit AD, Khandare ST (2017) A survey on different color image segmentation techniques using multilevel thresholding. IJCSMC 6(4):267–273
3. Ritika, Kaur S (2013) Contrast enhancement techniques for images–a visual analysis. Int J Comput Appl 64(17)
4. Liang Z, Liu W, Yao R (2016) Contrast enhancement by nonlinear diffusion filtering. IEEE Trans Image Process 25(2):673–686
5. Li Y, Guo F, Tan RT, Brown MS (2014) A contrast enhancement framework with JPEG artifacts suppression. In: European conference on computer vision. Springer, pp 174–188
6. Yun S-H, Kim JH, Kim S (2011) Contrast enhancement using a weighted histogram equalization. In: 2011 IEEE international conference on consumer electronics (ICCE). IEEE, pp 203–204
7. Wang X, Cheng E, Burnett IS (2017) Improved cell segmentation with adaptive Bi-Gaussian mixture models for image contrast enhancement pre-processing. IEEE
8. Murinto, Winiarti S, Ismi DP, Prahara A (2017) Image enhancement using piecewise linear contrast stretch methods based on unsharp masking algorithms for leather image processing. In: 2017 3rd International conference on science in information technology (ICSITech)
9. Nnolim UA (2017) Improved partial differential equation-based enhancement for underwater images using local–global contrast operators and fuzzy homomorphic processes. IET Image Process 11(11):1059–1067
10. Fornes A, Otazu X, Llados J (2013) Show-through cancellation and image enhancement by multiresolution contrast processing. In: 2013 12th International conference on document analysis and recognition

11. Rosenberger M (2012) Virtual contrast enhancement intelligent illumination adjustment processing with field programmable gate array based camera systems for imaging applications enhancing contrast in multi AOI applications. IEEE
12. Reza AM (2004) Realization of the contrast limited adaptive histogram equalization (CLAHE) for real-time image enhancement. J VLSI Sig Process Syst Sig Image Video Technol 38(1):35–44
13. Yue H, Yang J, Sun X, Wu F, Hou C (2016) Contrast enhancement based on intrinsic image decomposition. IEEE Trans Image Process

Low-Power High-Speed Hybrid Multiplier Architectures for Image Processing Applications

U. Saravanakumar, P. Suresh and V. Karthikeyan

Abstract Multipliers play an imperative role in communication, signal and image processing, and embedded ASICs. Generally, multipliers are designed through various steps and they are occupying more area in the hardware, consumes more power, and causes an effect on performance. This paper is aimed to implement low-area and high-speed multiplier using various data compressors in partial product stages and tested for image processing. To suppress the vertical dimension of the partial product stage in multiplier, Sklansky adder is considered for the last stage and five hybrid multiplier architectures (HyMUL1–HyMUL5) have been implemented. For application verification, the two grayscale images are given as the inputs of the proposed multipliers and produce a new image which is an overlap of the two input images. The comparative analysis indicates that the proposed multiplier HyMUL2 consumed less area compared to other multipliers and its speed is also improved. The obtained new overlapped image using proposed multiplier HyMUL2 has high PSNR and low NMED.

1 Introduction

Most of the computing processes for real-time applications are implemented using digital building blocks such as adders, multiplier, comparators, etc., thus operating with a high degree of consistency and exactness. However, many applications including multimedia can accept a small number of errors and inaccuracy in computation but produce valid results. Since multimedia applications work with inaccurate mod-

U. Saravanakumar (✉) · P. Suresh · V. Karthikeyan
Department of ECE, Vel Tech Rangarajan Dr. Sagunthala R&D Institute of Science and Technology, Chennai 600062, India
e-mail: saran.usk@gmail.com

P. Suresh
e-mail: suresh3982@gmail.com

V. Karthikeyan
e-mail: vkarthikeyan652@gmail.com

© Springer Nature Switzerland AG 2019
D. Pandian et al. (eds.), *Proceedings of the International Conference on ISMAC in Computational Vision and Bio-Engineering 2018 (ISMAC-CVB)*, Lecture Notes in Computational Vision and Biomechanics 30,
https://doi.org/10.1007/978-3-030-00665-5_54

els, designers have approached several architectures for approximate computation to achieve the simplest design process and other hardware-related benefits like power consumption and area occupancy [1]. The key element in digital building blocks is data compressors and they have been studied in detail for approximate implementation [2]. Since the multipliers for handheld devices require more number of adders to minimize the height of partial product, the demand for low area multiplier is continuously increasing.

A multiplication process consists of generating the partial product's matrix, reducing the matrix to 2 rows followed by the final carry regarding the performance characteristics of a multiplier. Previously, partial product reduction has been achieved by using carry-save adders (CSA) consisting of rows of 3:2 compressors. Recently, a focus has been put on higher order reduction scheme mainly by using 4:2 compressors and it is constructed with 23-bit adder cells [3]. Several optimization approaches have been investigated and employed in [4–9] for 4-2 exact and approximate compressors [10]. In this work, we considered both the type of compressors and parameters like area and speed considered for VLSI implementation. Numerous multiplier architectures for approximate multiplication had been implemented in [11–14] and most of them used the truncation method. In [15], a new inaccurate multiplier architecture has been designed and validated for Arithmetic Data Value Speculation (ADVS). However, these works have involved in designing novel structures, investigating and implementing optimization methods for the reduction of partial product in multiplication without considering compressors. The metrics like PSNR and Normalized Mean Error Distance (NMED) are considered to evaluate the results of image multipliers. As like trade-off between various metrics in VLSI circuits, here the trade-off between precision and power is also quantitatively evaluated to find the suitability for multimedia applications.

This paper utilized two unique approximate 4-2 compressors and exact 5-2 compressors for partial product reduction in the multiplier. Then, these inaccurate data compressors are used in appropriate stages of Dadda multiplier and 4 novel architectures are designed for inexact multiplication. In new multiplier architectures, Sklansky Tree Adder (STA) is implemented at the last phase of addition to reduce the area and the overall latency.

2 Data Compressors

The fundamental concept of an n to 2 data compressor is, any number of input bits can be added to produce 2-bit outputs while separating sums and carries. This shows that all the input columns in the table can be added in parallel by omitting the result of an earlier column, generating 2-output adder with a small amount of delay which is independent of the input's size.

2.1 Exact 4 to 2 Compressor and 5 to 2 Compressor

Exact 4 to 2 Compressor (E42C): The conventional arithmetic circuits like CSA and parallel multiplication were designed to scale down the number of bits from n to 2; therefore, n-2 compressors have been generally implemented in computer arithmetic applications in computer arithmetic applications. Figure 1(a) demonstrates the basic version of E42C. An E42C is operated with 4 primary inputs from x_1 to x_4 plus one carry input C_{in} and produces 3 outputs such as C, C_{out}, and S. So simply, it is a 5-bit adder. Here, two carry outputs C and C_{out} are carrying equal weightages. The C_{out} depends on only primary inputs and it forms the C_{in} of the next column. Therefore, it clearly shows that C_{out} is independent of C_{in} [10].

$$\text{Sum} = x_1 \oplus x_2 \oplus x_3 \oplus x_4 \oplus C_{in} \tag{1}$$

$$C_{out} = (x_1 \oplus x_2)x_3 + (x_1 \oplus x_2)'C_{in} \tag{2}$$

$$C = (x_1 \oplus x_2 \oplus x_3 \oplus x_4)C_{in} + (x_1 \oplus x_2 \oplus x_3 \oplus x_4)'x_4 \tag{3}$$

5 to 2 Compressor (52C): A widely used compressor of significant importance is the 52C for high precision and high-speed multipliers and it is illustrated in Fig. 1(b). It is working with 5 primary inputs $(x_1$–$x_5)$ and 2 carry-in $(C_{i1}$ and $C_{i2})$ from the earlier step. For the 7 input bits, the 52C produces sum, carry, and 2-carry out bits $(C_{o1}$ and $C_{o2})$. The 52C generates an output of the same weight as the inputs, and three outputs, Carry, C_{o1}, and C_{o2} and weighted one binary bit order higher. The outputs are given to the neighboring compressor of higher significance. All the 52Cs of different designs abide by Eq. (4). A new set of 52C implementations with an adjournment of 4 XORs is presented in [16]. A modified set of output expressions satisfying Eq. (4) are used in their design and they are shown in equations from (5) to (8).

$$x_1 + x_2 + x_3 + x_4 + x_5 + C_{i1} + C_{i2} = \text{sum} + 2(\text{carry} + C_{o1} + C_{o2}) \tag{4}$$

$$\text{Sum} = x_1 \oplus x_2 \oplus x_3 \oplus x_4 \oplus x_5 \oplus C_{i1} \oplus C_{i2} \tag{5}$$

$$C_{o1} = x_1x_2 + (x_1 + x_2)x_3 \tag{6}$$

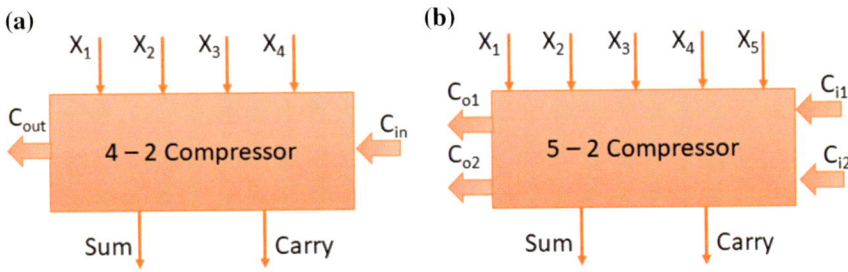

Fig. 1 Block diagram of **a** E42C, **b** 52C

$$C_{o2} = (x_4 \oplus x_5)C_{i1} + (x_4 \oplus x_5)x_3 \tag{7}$$

$$\text{Carry} = ((x_1 \oplus x_2 \oplus x_3)(x_4 \oplus x_5 \oplus C_{i1}))C_{i2}$$
$$+ (\text{not}(x_1 \oplus x_2 \oplus x_3)(x_4 \oplus x_5 \oplus C_{i1}))(x_1 \oplus x_2 \oplus x_3) \tag{8}$$

2.2 Approximate Compressors

In the truth table of an E42C, the carry output of 24 states out of 32 are having the same value of C_{in} and it helps for approximate design. In Approximate 4 to 2 Compressor I (A42CI), the carry is adjusted to C_{in} by changing the remaining 8 carry outputs. Here, carry output C is assumed as the higher weight of a binary bit and an incorrect value will generate a difference value of 2 in the output. For the input value 01001, the correct output is 010 that is equal to 2. By assuming the carry output C to C_{in}, the A42CI will generate the 000 at the output which is equal to 0. This difference value is inadequate for accurate computations but it can be paid by adjusting the C_{out} and sum values. In this case, we will consider the second half of Table 1. The sum value is simplified to 0 and it helps to minimize the difference between inexact and exact outputs. This kind of approach simplifies the design complexity with increased error rate and it is realized by comparing Eqs. (1–3) and (9–11).

$$C = C_{in} \tag{9}$$

$$\text{Sum} = C'_{in}(x_1 \oplus x_2)' + (x_3 \oplus x_4)' \tag{10}$$

$$C_{out} = x_1 \cdot x_2 + x_3 \cdot x_4 \tag{11}$$

Further to improve the performance of A42CI, the carry C and C_{out} equations are interchanged even they are having the same weight. This design is called as Approximate 4 to 2 Compressor II (A42CII). In this design, carry C uses the right-hand side of Eq. 6 and C_{out} is C_{in}. Initially, C_{in} value is 0 and C_{out} value is either 0 or 1. Then C_{out} and C_{in} value will be 0 in all stages so that they are discounted in our circuit realization. The A42CII has realized by the expressions in Eqs. 12 and 13.

$$\text{Sum} = (x_1 \oplus x_2 + x_3 \oplus x_4)' \tag{12}$$

$$C_{out} = (x_1 \cdot x_2)' + (x_3 \cdot x_4)' \tag{13}$$

For example, when all inputs are 1, the addition value is 4_{10}. However, the A42CII produces a 1 for the carry and sum. The base 10 representation of the outputs, in this case, is 3. This design has therefore 4 incorrect outputs out of 16 outputs and the difference value is presented in Table 2.

Table 1 Truth table for E42C and A42CI

C_{in}	Primary inputs				E42C			A42CI			
C_{in}	X_4	X_3	X_2	X_1	C_{out}	C	Sum	C_{out}	C	Sum	Diff.
0	0	0	0	0	0	0	0	0	0	1	1
0	0	0	0	1	0	0	1	0	0	1	0
0	0	0	1	0	0	0	1	0	0	1	0
0	0	0	1	1	1	0	0	0	0	1	−1
0	0	1	0	0	0	0	1	0	0	1	0
0	0	1	0	1	1	0	0	1	0	0	0
0	0	1	1	0	1	0	0	1	0	0	0
0	0	1	1	1	1	0	1	1	0	1	0
0	1	0	0	0	0	0	1	0	0	1	0
0	1	0	0	1	0	1	0	1	0	0	0
0	1	0	1	0	0	1	0	1	0	0	0
0	1	0	1	1	1	0	1	1	0	1	0
0	1	1	0	0	0	1	0	0	0	1	−1
0	1	1	0	1	1	0	1	1	0	1	0
0	1	1	1	0	1	0	1	1	0	1	0
0	1	1	1	1	1	1	0	1	0	1	−1
1	0	0	0	0	0	0	1	0	1	0	1
1	0	0	0	1	0	1	0	0	1	0	0
1	0	0	1	0	0	1	0	0	1	0	0
1	0	0	1	1	1	0	1	0	1	0	−1
1	0	1	0	0	0	1	0	0	1	0	0
1	0	1	0	1	1	0	1	1	1	0	1
1	0	1	1	0	1	0	1	1	1	0	1
1	0	1	1	1	1	1	0	1	1	0	0
1	1	0	0	0	0	1	0	0	1	0	0
1	1	0	0	1	0	1	1	1	1	0	1
1	1	0	1	0	0	1	1	1	1	0	1
1	1	0	1	1	1	1	0	1	1	0	0
1	1	1	0	0	0	1	1	0	1	0	−1
1	1	1	0	1	1	1	0	1	1	0	0
1	1	1	1	0	1	1	0	1	1	0	0
1	1	1	1	1	1	1	1	1	1	0	−1

Table 2 Truth table for A42CII

X_4	X_3	X_2	X_1	C	Sum	Difference
0	0	0	0	0	1	1
0	0	0	1	0	1	0
0	0	1	0	0	1	0
0	0	1	1	0	1	−1
0	1	0	0	0	1	0
0	1	0	1	1	0	0
0	1	1	0	1	0	0
0	1	1	1	1	1	0
1	0	0	0	0	1	0
1	0	0	1	1	0	0
1	0	1	0	1	0	0
1	0	1	1	1	1	0
1	1	0	0	0	1	−1
1	1	0	1	1	1	0
1	1	1	0	1	1	0
1	1	1	1	1	1	−1

3 Tree Structure Adders

The low-cost multipliers can be constructed by reducing the partial product sizes. The tree adders address the solution for power consumption and latency in arithmetic circuits. The tree adders are developed from the computation concept of Carry-Lookahead (CLA) adders and it is viewed as parallel CLA. These structures target high-performance applications. In this work, the Sklansky Tree Adder (STA) is considered because it reduces power consumption of adder circuits and increases speed than other tree adders in the earlier works. The binary trees of cells in the STA first generate all possible carry inputs at the same time and it follows a systematic configuration. The STA ultimately reduces the delay to logarithmic of number of bits (number of bits = N) by calculating the midway prefixes with the large group prefixes. It increases the number of fan-outs at each stage. In this adder, a binary tree of propagate and generate cells will first simultaneously generate all the carries and C_{in}. It constructs 2-, 4-, 8-bit adders and so on by adjoining 2 smaller adders (2-bit adders), respectively [17].

4 Hybrid Multipliers

This work considered Dadda multiplier because it requires a minimum number of adder stages for the summation of partial products. The design of Dadda multiplier

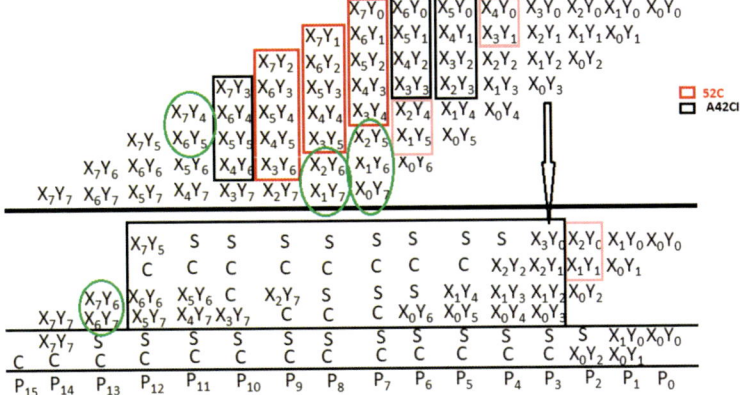

Fig. 2 Reduction structure of partial product in HyMUL 1

comprises three steps and they are followed. The partial product is generated in the first step and it is reduced to a height of 2 in the second step. At last step, these 2-rows are combined by using various addition algorithms. Initially, 8×8 unsigned Dadda multiplier is considered to evaluate the performance of compressors in the multiplier. In this work, all the partial products are generated using AND gates and inexact compressors discussed in this paper are employed at the addition portion to minimize the partial products. The last part is designed with STA structure to calculate the final product value. The various stages of HyMUL1 for $n = 8$ is shown in Fig. 2. The reduction part uses adders and compressors. In this paper, five cases are considered for designing the inexact multipliers and they are named as HyMUL 1, HyMUL 2, HyMUL 3, HyMUL 4 and HyMUL 5. The feature of each multiplier is given in Table 2. (These multipliers are called as approximate multiplier because the multiplier is designed with approximate compressors.) For understanding, HyMUL 1 is explained here with its feature. The hybrid multiplier 1 (HyMUL 1) is designed with the approximate compressor I and exact 5-2 compressor to minimize the partial product hight.

4.1 HyMUL 1

A42CI is used for all 4-variable additions and 52C is used for all 5-variable additions. The height of subsequent stages $= 2(X - i)$, where $2X$ is a nearest smaller integer to N; i ranges from 0 to $X-1$. The stage heights are stage $2 = 2X$, Stage $3 = 2X - 1$, Stage $4 = 2X - 2$... until final stage height is 2. This setup is achieved with the usage of 52C while maintaining stage height. The 8×8 hybrid multiplier 1 architecture is presented in Fig. 2. The maximum height of the partial products stage (Stage1) is 8 bits. The nearest 2 m integer smaller than 8(23) is 4, i.e., 2(2-0). So, the height

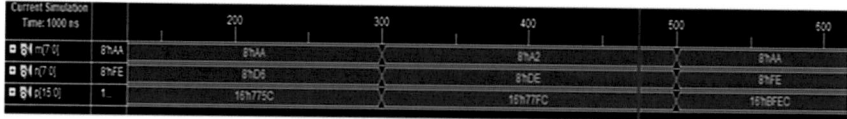

Fig. 3 Simulation result of HyMUL 2

Table 3 Area and delay comparison of various multipliers

Multipliers	Features	CPA		STA	
		Area	Delay (ps)	Area	Delay (ps)
Multiplier 1	A42CI in all columns	858	22.077	801	20.489
Multiplier 2	A42CII in all Columns	738	22.42	716	19.921
Multiplier 3	E42C in all columns	807	22.654	784	21.638
HyMUL 1	A42CI + 52C	840	21.739	817	19.95
HyMUL 2	A42CII + 52C	718	21.448	693	18.721
HyMUL 3	A42CI in LSB + 52C in MSB	810	21.934	802	20.351
HyMUL 4	A42CII in LSB + 52C in MSB	775	20.756	768	20.478 ˙
HyMUL 5	E42C + 52C	786	20.324	778	20.324

of Stage 2 should be 4 bits, which is maximum column (C) height. This hybrid multiplier 1 uses 14 approximate compressor I structures, 3 exact 5-2 compressor, 3 half adders, and 18 full adders. The usage of STA at the last portion of multipliers, the required number of 3-bit adders can be minimized.

5 Simulation Results

The multipliers have designed using VHDL and simulated with Xilinx ISE Simulator. Figure 3 demonstrates the simulation result of HyMUL 2 for 8×8. The area in terms of number transistors and delay (in ps) are tabulated and it is presented in Table 3. The HyMUL 2 has less transistor count compared to other multipliers. The transistor count is used in this paper as metric of area occupancy. And HyMUL 3 also consumes less hardware area with lower delay parameter. From Table 3, it is clearly understood that STA at the last portion of the proposed hybrid multipliers helps to reduce the latency during the computation process. Among all multipliers, the HyMUL 2 which is designed using A42CII and 52C in all columns of partial products has less delay as shown in Table 3.

Fig. 4 Example 1: multipliers for image processing

Fig. 5 Image multiplication results for Example 1

6 Hybrid Multipliers for Image Processing

The multipliers are applied to image processing applications like image manipulations or image overlapping. A multiplier for image processing can be designed to multiply the pixels of two images. This multiplication process is actually a pixel-by-pixel multiplication, thus two images are overlapped. Figure 4 shows an example of two input images and the resulting output image is provided. The image multiplication results of examples are given in Figs. 5 and 6.

6.1 Performance Analysis

PSNR: PSNR that is based on the Mean Squared Error (MSE) are computed to evaluate the quality of the output image. The equations for the MSE and PSNR are presented and discussed in [1]. Table 4 shows the comparison of PSNR of the output images generated by all the multipliers. The HyMULs are designed using various combinations of A42CI, A42CII, E42C, and 52C compressors and produces incorrect value. However, when applied to image processing applications, these multipliers

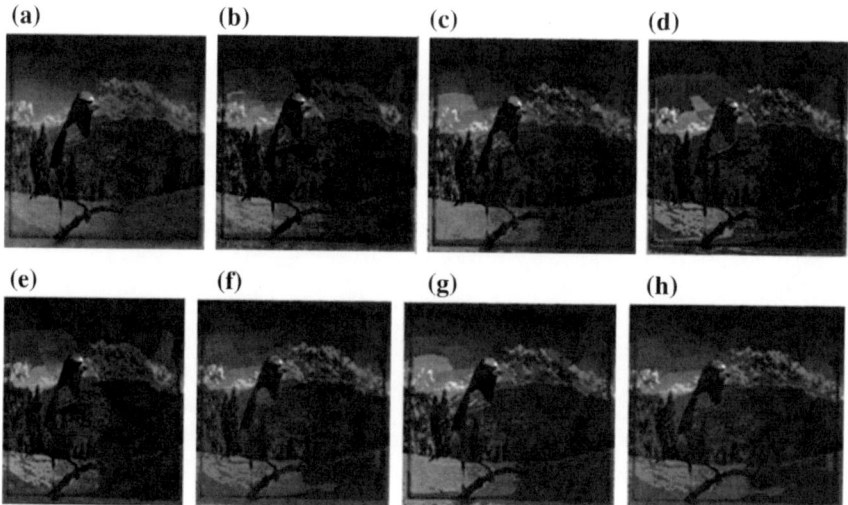

Fig. 6 Image multiplication results for Example 2 (for Figs. 5 and 6: **a** Multiplier 1, **b** Multiplier 2, **c** Multiplier 3, **d** HyMUL 1, **e** HyMUL 2, **f** HyMUL 3, **g** HyMUL 4, **h** HyMUL 5)

Table 4 PSNR and NMED comparison of all the multipliers with STA

Multipliers	PSNR	NMED
Multiplier 1	26.6688	3.0942
Multiplier 2	22.3255	7.4891
Multiplier 3	29.111	5.0204
HyMUL 1	45.2801	0.0757
HyMUL 2	49.7683	0.0709
HyMUL 3	38.8971	0.4601
HyMUL 4	41.7465	0.4263
HyMUL 5	37.9894	0.5728

results in improved peak signal-to-noise ratio (PSNR) value as shown in Table 4. The proposed multiplier HyMUL 2 of examples is found to have high PSNR value when compared to other multipliers and its value is around 50 dB which is applicable for most applications.

NMED: The Error Distance (ED) and NMED are considered to evaluate the performance of approximate arithmetic circuits for image processing applications. For multiplication, ED is the arithmetic difference between the exact product (M) and the approximate product (M'). The MED is computed by taking the average of EDs for a set of outputs (for the set of inputs). The error distance is measured between for various multipliers are listed in Table 4. The equations for ED, MED and NMED are presented in [1] and simulated using MATLAB.

7 Conclusions

The multipliers are considered as an important module of processors and DSPs so that various algorithms like convolution and filtering for image and signal processing can be implemented. The computer arithmetic supports various schemes and structures for both exact and inexact computation. In this work, five different schemes are presented in 8-bit Dadda multiplier using A42CI, A42CII, and 52C for only inexact computation. On the experimental results, HyMUL 2 occupied a smaller hardware area and its latency is considerably reduced than all other multipliers. The multipliers designed using the various combinations of A42CI, A42CII, E42C and 52C results in the inexact final product but when applied to image processing applications, image multipliers result in high PSNR and low NMED values compared to conventional or exact multipliers. And in future, higher order multiplier design with approximately 5-2 compressor can be designed to reduce the delay and transistor count further.

References

1. Momeni JH, Montuschi P, Lombardi F (2014) Design and analysis of approximate compressors for multiplication. IEEE Trans Comput 64(4):984–994
2. Han J, Orshansky M (2013) Approximate computing: an emerging paradigm for energy-efficient design. In: IEEE ETS, pp 1–6
3. Chang P, Ahmadi M (2009) High speed low power 4:2 compressor cell design. In: International symposium on signals, circuits and systems, Iasi, pp 1–4
4. Chang C, Gu J, Zhang M (2004) Ultra Low-voltage low-power CMOS 4-2 and 5-2 Compressors for fast arithmetic circuits. IEEE Trans Circuits Syst 51(10):1985–1997
5. Gu J, Chang CH (2003) Ultra low-voltage, low-power 4-2 compressor for high speed multiplications. In: 36th IEEE international symposium circuits systems, Bangkok, Thailand
6. Margala M, Durdle NG (1999) Low-power low-voltage 4-2 compressors for VLSI applications. In: IEEE Alessandro volta memorial workshop low-power design, pp 84–90
7. Parhami B (2010) Computer arithmetic: algorithms and hardware designs, 2nd edn. Oxford University Press, New York
8. Prasad K, Parhi KK (2001) Low-power 4-2 and 5-2 compressors. In: 35th Asilomar conference on signals, systems and computers, vol 1, pp 129–133
9. Ercegovac MD, Tomas L (2003) Digital arithmetic. Elsevier, Amsterdam
10. Ma J, Man K, Krilavicius T, Guan S, Jeong T (2011) Implementation of high performance multipliers based on approximate compressor design. In: International conference on electrical and control technologies (ECT), pp 96–100
11. Mahdiani HR, Ahmadi A, Fakhraie SM, Lucas C (2010) Bio-inspired imprecise computational blocks for efficient VLSI implementation of soft-computing applications. IEEE Trans Circ Syst I: Regul Pap 57(4):850–862
12. Schulte MJ, Swartzlander EE (1993) Truncated multiplication with correction constant. In: IEEE workshop on VLSI signal processing VI, pp 388–396
13. King EJ, Swartzlander EE (1998) Data dependent truncated scheme for parallel multiplication. In: 31st Asilomar conference on signals, circuits and systems, pp 1178–1182
14. Kulkarni P, Gupta P, Ercegovac MD (2011) Trading accuracy for power in a multiplier architecture. J Low Power Electron 7(4):490–501
15. Kelly D, Phillips B, Al-Sarawi S (2009) Approximate signed binary integer multipliers for arithmetic data value speculation. In: Conference on design and architectures for signal and image processing, pp 97–104

16. Menon R, Rdhakrishnan D (2006) High performance 5:2 compressor architectures. In: IEE Proceeding of circuits devices systems, vol 153, no 5
17. Sankar DR, Ali SA (2013) Design of Wallace tree multiplier by Sklansky adder. Int J Eng Res Appl 3(1):1036–1040

Intervertebral Disc Classification Using Deep Learning Technique

J. V. Shinde, Y. V. Joshi and R. R. Manthalkar

Abstract This paper describes the semiautomatic method for diagnosis of intervertebral disc degeneration according to Pfirrmann's five scale (1–5) grading system, which is used in the assessment of disc degeneration severity. Total 1123 discs are obtained after augmentation from 120 subject's T2-weighted lumbar scans. Manual classification into five grades is done by experts. Our method is extracting 59 features using Local Binary Pattern for texture analysis and 4096 features using pretrained CNN. 1×59 and 1×4096 feature vectors are fused to form 1×4155 feature vector to train our multiclass Support Vector Machine classifier. This feature level fusion method is able to achieve 80.40% accuracy. A Quantitative analysis is done using parameters, viz.,—Accuracy, Sensitivity, Specificity, Precision, Recall, F1 score, etc.

1 Introduction

Spinal degeneration [1–3] is commonly found in two anatomical parts called vertebra and intervertebral disc that results in mild to severe back pain. Degenerative lumbar spine diseases include disc herniation and Modic changes which are highly associated with lower back pain. Modic changes and different types of disc herniation [4] are clearly visible on Magnetic Resonance Imaging (MRI). In clinical evaluation disc degeneration, severity is assessed with mid-sagittal slice of T2-weighted images. It provides both internal chemical composition and structural integrity information

J. V. Shinde (✉)
Department of Computer Engineering, L.G.N. Sapkal College of Engineering, Nasik, India
e-mail: jv.shinde@rediffmail.com

Y. V. Joshi · R. R. Manthalkar
Department of Electronics & Telecommunication, S.G.G.S Institute of Engineering and Technology, Nanded, India
e-mail: yvjoshi@sggs.ac.in

R. R. Manthalkar
e-mail: rrmanthalkar@sggs.ac.in

© Springer Nature Switzerland AG 2019
D. Pandian et al. (eds.), *Proceedings of the International Conference on ISMAC in Computational Vision and Bio-Engineering 2018 (ISMAC-CVB)*, Lecture Notes in Computational Vision and Biomechanics 30, https://doi.org/10.1007/978-3-030-00665-5_55

(a) **(b)** **(c)**

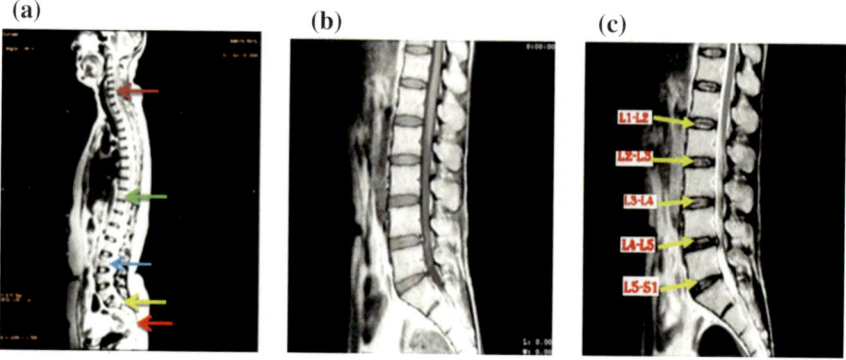

Fig. 1 **a** Whole spine image, Brown arrow cervical, Green arrow Thoracic, Blue arrow Lumbar, Yellow arrow Sacrum and Red arrow shows coccyx section, **b** mid-sagittal T1—weighted image and **c** T2-weighted image with 5 disc positions in between vertebras. MR Image source: Samarth Diagnostic center, Nasik (M.S), India

about disc tissue. Biochemical changes are reflected in MRI signal intensity and compared with normal levels.

In addition, T2 image provides useful information like disc shape, size, water content, erosion of vertebral endplates, wear and tear of the outer wall of disc, etc., which allows investigating various disc disorders like disc narrowing and disc herniation. Pfirrmann's five scale (1–5) grading system is very popular in Magnetic Resonance Imaging assessment [5].

Spine is made up of small bones called vertebra, muscles, ligaments, nerves, and intervertebral discs. The whole spinal column is divided into 5 sections namely cervical, thoracic, lumbar, sacrum, and tiny coccyx. The cervical section contains 7 vertebrae C1–C7 and discs, Thoracic contains 12 vertebrae T1–T12 and discs, lumbar section contains 5 vertebrae L1–L5 and 5 discs, sacrum shows fused vertebrae S1–S5 which articulates with the hip bone of pelvis, spinal column is terminated by tiny coccyx consist of fused vertebrae shown in Fig. 1.

Commonly used pulse sequences are T1, T2-weighted and T2-weighted with fat saturation (STIR) technique

T2-weighted mid-sagittal sequence images are used for assessing Intervertebral Discs (IVD), which provide better visualization to morphological- and tissue-related biochemical changes. These changes are reflected in the image in terms of variations in MR signal intensities and irregularities in the shape and size of the disc. Alterations in intervertebral disc height, signal intensity, and the distinction between center part nucleus pulposus and outer wall annulus fibrosus are major signs of disc degeneration. All these features are evaluated for determining disc degeneration severity. In the literature, various grading schemes have been proposed to assist clinicians [6, 7]. Figure 2 shows Pfirrmann's grading scheme (1–5).

In Grade-1 class, the disc looks normal and a clear difference can be observed in NP and AF. Also, the height of the disc seems to be normal. NP reflects in a

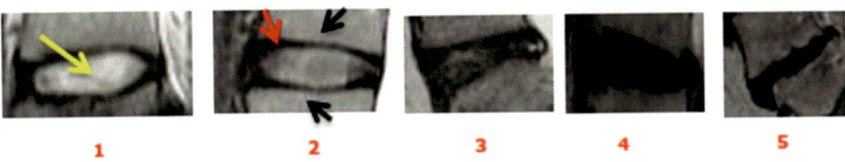

Fig. 2 1. Normal Disc 2. Normal Disc 3. Mild Degeneration 4. Moderate Degeneration 5. Severe Degeneration. Yellow arrow shows NP, Black arrows shows vertebra and Red arrow shows AF

brighter white shade. Grade-2 class indicates a normal disc with normal height but NP looks inhomogeneous with or without horizontal bands. Grade-3 class shows mild degeneration and NP is Inhomogeneous, Gray with slightly decreased disc height. There is an unclear distinction between NP and AF. If disc height is moderately decreased and the distinction between NP and AF is lost, NP looks inhomogeneous, Gray to black then disc falls into Grade-4 class. Grade-5 belongs to the class of severe degeneration, where disc height is totally collapsed and NP is black. Also, the distinction between NP and AF is completely lost.

2 Related Work

Intervertebral Disc Degeneration (IVDD) is becoming a universal health issue, which results in tremendous economic losses and damage to productivity every year. IVD degeneration is a common cause of lower back pain found commonly in almost every person in late adulthood to a varying degree.

For surgical removal of disc radiological scoring systems and histological systems are utilized to assess the progression of intervertebral disc degeneration. MRI helps to record signal variations in the disc while microstructural changes are less appreciated. MRI also helps to find the changes in the endplate of vertebra known as the Modic changes.

2.1 Lumbar Disc Degeneration Scoring Systems Utilized in Following Imaging Modalities

2.1.1 X-ray Imaging

There are totally six classification systems are developed and used for X-ray diagnosis, Kellgren's [8] 0–2 scale is based on movement, sclerosis, lipping, and spacing criteria. Gordon [9] proposed 1–4 scale scoring based on sclerosis, narrowing, osteophytes. Lane [10] has been developed 0–3 scale scoring method using criteria joint space narrowing, osteophytes and sclerosis. Mimura [11] developed 1–4 scale using disc height, osteophytes, endplate, and sclerosis. Taking into account the disc height, osteophytes, endplate sclerosis, Schmorl's nodes, and vacuum phenomenon Madan [12] developed a 0–3 scale classification system.

Thalgott [13] used A–F scale scoring using both X-ray and MRI on sclerosis and osteophytes criteria. Wilke [14] in (2006) proposed 0–3 scoring on height loss, osteophyte formation, and diffuse sclerosis.

2.1.2 Magnetic Resonance Imaging

Eight classification systems have been proposed for IVD degeneration using MRI. One of the strong radiological grading systems proposed by Pfirrmann et al. [5] is based on morphological alterations found in the disc. Pfirrmann's 1–5 grades are shown in Fig. 2. Schneiderman [15] formed 1–4 scales based on disc height, signal intensity, and patterns in MRI. Butler [16] proposed 1–4 scaling system based on criteria borders of annulus and nucleus, disc space, herniation, signal intensity. Tertti's [17] 1–3 scale is based on only signal intensity variations found in the sagittal plane.

Gunzburg [18] proposed 0–3 scale scoring observing only central part that is nucleus signal. Disc space narrowing and signal intensity are observed in all three planes namely Axial, Sagittal, and Coronal by Southern et al. [19] to propose a 1–4 scaling system. Askar [20] has proposed 1–4 grading using both axial and sagittal planes based on criteria like disc height reduction, radial tears, and annular's shape. Griffith's [21] 1–8 scoring system is based on rim lesions, annular tears, bulging, disc height and difference between NP and AF.

2.1.3 Histological Grading

Histological grading involves microscopic tissue level investigation. As per Boos et al. [22] criteria four parameters are evaluated at the microscopic level. Chondrocyte proliferation, Tears of clefts, granular changes and mucoid changes with 0–22 scale. Three more grading systems are presented: Gunzburg [18] proposed 0–3 scale scoring, Berlemann [23] proposed 1–4 scale scoring, and Weiler [24] proposed 0–15 scale scoring.

Many contributions have been made in literature for automatic diagnosis/classification of a number of spinal degeneration conditions like Modic changes and intervertebral disc degeneration. Methods proposed by Corso et al. [25] and Alomari et al. [26–28] deals with binary classification on the basis of absence or presence of abnormality like disc desiccation/degeneration taken in consideration the appearance and shape of disc. The various intensity and texture features are extracted from ROI and shape information are evaluated using ASM and GVF-snake-based segmentation algorithms. One of their contributions is using a probabilistic model for automatically labeling the disc and extracting ROI using ASM. Classification accuracy is compared using heterogeneous classifiers. But radiological grading is not performed.

Instead of using conventional T2 weighted MR images for classification of intervertebral discs, Watanabe et al. [29] proposed a system which is very useful in the detection of early degeneration in the disc using axial T2 mapping technique. Using human cadaveric disc specimens Southern et al. found a correlation between MRI and quantitative discomanometry in the assessment of lumbar disc degeneration [19].

Marcelo et al. [30] developed a semiautomatic method for diagnosis of intervertebral disc degeneration using T2-weighted lumbar images according to Pfirmann's five scale grading system. Each ROI is detected using a binary mask followed by Haralick feature extraction for texture analysis using gray level co-occurrence matrix. Disc classification is done using a multilayer perceptron artificial neural network with full attribute vectors.

Using Pfirmann's scoring system, automatic classification of intervertebral disc degeneration using T2-weighted images is performed by Castro-Mateos et al. [31]. The region of interest is segmented using an active contour model. Using shape and intensity features of intervertebral disc Neural Network is trained to achieve classification task.

Most of the spinal disease diagnosis studies have been primarily conducted with handcrafted features. A very few approaches utilize deep learning based frameworks for classification of IVD degeneration. The Pretrained Convolution Neural Networks (CNN) has been utilized mainly in medical diagnosis for pathology detection.

One successful example based on CNN is 'U-net' a medical image segmentation framework developed by Ronneberger et al. [32]. It works on few training image dataset and yields more precise segmentation.

Pfirmann's grading/scoring system is used by the method proposed by Jamaludin et al. [33] their scheme automatically predicts radiological scoring as well as disc narrowing and marrow changes from multiple slice lumbar spine MR images. Their CNN based framework is modified version of VGG-M known by name 'SpineNet'. İt is employed to detect, localize, and classify multiple abnormalities of spinal objects like vertebra and discs in T2-weighted MRI.

3 Proposed Methodology

We have proposed feature-level fusion approach for IVD classification based on Pfirmann's grading scale. The procedure of the proposed algorithm is described as follows (Fig. 3).

3.1 Image Preprocessing

T2-weighted images are manually cropped and five discs from lumbar sections are saved separately. Categorizations of those discs are done in five classes as per the

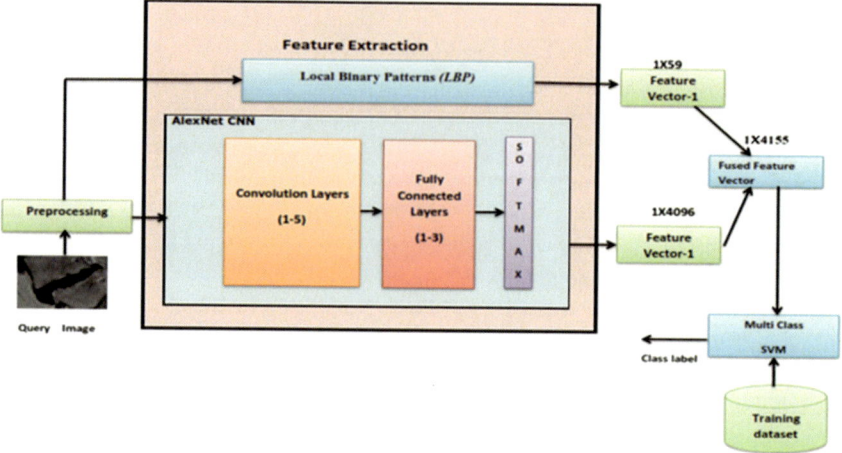

Fig. 3 Architecture diagram of classification process using feature level fusion

recommendation of experts. Images are preprocessed by using super-resolution [34] technique to enhance the low-resolution images.

3.2   *Local Binary Patterns*

Local binary pattern [35] is computationally simple and an efficient texture operator used on spatial domain images in texture analysis applications. It uses grayscale image for texture classification as it is very robust to monotonic grayscale variations, which exist due to illumination variations. When LBP operator is applied on an image a binary number is obtained as a result. LBP representations can be expressed by following Eqs. (1) and (2). Image pixel labels are provided after thresholding the 3 × 3 neighborhood with a center pixel value.

The value of the LBP code of pixel (x_c, y_c) is given by

$$\text{LBP}_{N,R} = \sum_{n=0}^{N-1} s_x(g_n - g_c)2^n \tag{1}$$

where N =no. of sampling points, R =radial distance, and g_c =center pixel

$$s_x(\text{diff}) = \begin{cases} 1, & (\text{diff}) \geq 1 \\ 0, & (\text{diff}) < 0 \end{cases} \tag{2}$$

Following Fig. 4 shows the LBP code computation process.

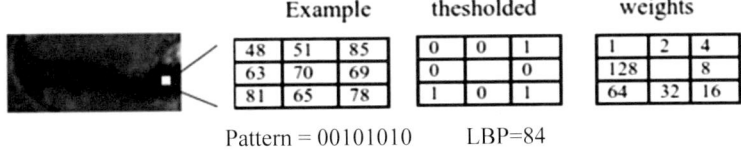

Fig. 4 Local binary pattern process

3.3 Convolution Neural Networks

Convolution Neural Networks (CNN) consist of two types of layers: convolution and pooling layers. In convolution layer entire image is convolved with the set of learned filters, f_n to get n corresponding feature maps which are then subsampled by average or max in pooling layer. It makes resulting feature invariant to small deformation and translation.

CNNs are used for improving the accuracy of image classification. It processes multiple array data. In last few years pretrained network models have gained popularity and widely used in medical imaging domain. 'AlexNet' [36] is one such pretrained model available for feature extraction. AlexNet contains 25×1 layer array with a series of five convolution layers, three fully connected layers followed by single softmax layer. A series of convolution layers are intermixed with rectified linear units (ReLU) and max-pooling layers. Classification is done by the final layer and its properties are depending on the classification task. This net can process RGB images that are $227 \times 227 \times 3$ in size.

3.4 Data Augmentation

In order to maximize the classification accuracy and to increase the number of samples available for the training. We have applied rotations by -2 and $+2$ degrees. Augmentation samples are used as extra samples in training. Finally, the number of class-wise samples is as follows.

Class-1 → 210 samples, Class-2 → 223 samples, Class-3 → 212 samples, Class-4 → 261 and Class-5 → 217 samples. In total, we have 1123 samples. From these 1123 samples, our system will select randomly at runtime 60% training and 40% testing samples. That results in overall 674 samples for training and 449 for testing.

3.5 Support Vector Machine Classifier

Support Vector Machine (SVM) is a supervised non-probabilistic binary linear classifier [37] widely used in classification and pattern recognition problems. To per-

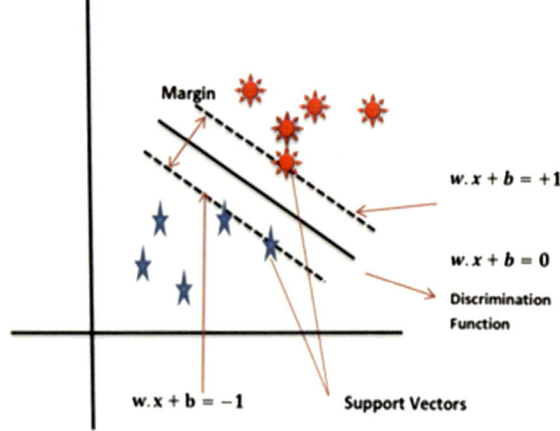

Fig. 5 Support Vector Machine for linear separable data

form classification SVM constructs decision boundaries called as hyperplanes in a multidimensional space that separates a set of objects belonging to different class memberships. The optimal hyperplane can be found by using feature points called support vectors (Fig. 5).

SVM training can be considered as the constrained optimization problem which minimizes the structural risk while maximizing the width of the margin.

$$\min(w, b) \quad \frac{1}{2} a^T a + C \sum_{i=1}^{N} \varepsilon_i \tag{3}$$

Subject to

$$y_i \left(W^T \varphi(x_i) + b \right) \geq 1 - \varepsilon_i, \quad \varepsilon_i \geq 0, \forall_i, \tag{4}$$

where b is the bias, C is the tradeoff parameter, ε_i variable measures the deviation of a data point from the ideal condition and $\varphi(.)$ is the feature vector. A Support Vector Machine classifier algorithm constructs a model that allocates new sample to one of the two variable regions. Classification of the test sample p is performed by

$$y = sgn \left(\sum_{sv=1}^{N_{sv}} \alpha_{sv} \, y_{sv} K(x_{sv}, p) \right) \tag{5}$$

Mapping of input samples into higher dimensional space is performed by the kernel function $K(x_{sv}, p)$.

3.6 Imaging Data Details

Mid-sagittal DICOM MR images are acquired using 1.5 T Siemens scanner from Samarth diagnostic center Nasik, India for 120 individuals of both sexes with age varying from 20 to 89 yrs. Imaging protocol as follows with parameters.

- T2-weighted: FRFSE SAG TR = 4220 ms, TE = 105 ms, Slice thickness = 4 mm, image matrix size is 512×512.
- T1-weighted: FRFSE SAG TR = 700 ms, TE = 114.34 ms, Slice thickness = 4 mm, image matrix size is 512×512.

T2-Weighted images are used for experimentation.

3.7 Implementation

Algorithm 1: Proposed Training Algorithm

Input: Train image, LBP block size.
Output: Fused feature vector for Train input image
Algorithm steps
Step 1: Image preprocessing (image resolution enhancement)
   1.1 Read the train image **I** of size 227x227x3
   1.2 Enhance resolution of **I** by applying Super resolution technique.
   1.3 Refer enhanced image as **I_Sr**
Step 2:
   Pass **I_Sr** image as an input to **CNN** to get feature vector **f1_cnn** of size **1x4096.**
Step 3: Computation of LBP feature vector
   J = Resize (**I**) to **128x256**
   3.1 Convert image **J** into blocks with block size **B_size.**
      for every pixel in the block **do**
   3.2 Perform computations to obtain LBP representation
   3.3 Computation of histogram **hist_i** for each block **i**
   3.4 All histograms are concatenated to obtain a feature vector **f2_lbp** of 1x59 size.
Step 4: Perform feature fusion to obtain fused feature vector.
   Concatenate **f1_cnn** & **f2_lbp** to yield fused feature vector **f_fused** of **1x4155** size.
Step 5. Repeat Step 1 to 4 for all training images
   Save feature vector in training database **F_train** matrix

Table 1 5 × 5 confusion matrix in the IVD classification

Correct class	Classified as				
	1	2	3	4	5
1	62	16	6	0	0
2	0	71	18	0	0
3	0	20	51	12	2
4	0	1	11	91	1
5	0	0	0	1	86

Algorithm 2: Proposed Testing Algorithm

Input: Test image sample
Output: class label for test image
Algorithm steps
Step 1: Perform step 1 through 4 of Algorithm 1 to obtain fused feature vector
   F_test.
Step 2: Multiclass Support Vector Machine classification
   2.1. Compare the **F_test** feature vector against feature vector matrix
   F_train.
   2.2. Find the probability of particular class using one against all model to
   find maximum probability.
   Class Label having maximum probability will be the output label.

----

4   Results and Discussion

In medical image classification many handcrafted feature descriptors are used, LBP is a texture descriptor having high discriminative power and invariant to grayscale and illumination changes, so widely used in medical image analysis.

While using deep learning methods very complex functions can be learned. In this work we propose feature level fusion process where 59 features are extracted from LBP and total 4096 features are extracted from CNN. By fusion, we got 1 × 4155 size feature vector. SVM training is done and classification is performed by the same classifier using one against all model. Parameters Accuracy, Sensitivity, Specificity, Precision, Recall, and F1-Score are reported in Table 2. The resultant confusion matrix is 5 × 5 is shown in Table 1.

Table 2 Quantitative Analysis Accuracy, Sensitivity, Specificity, Precision, Recall, and F1-score

Class	Accuracy	Sensitivity	Specificity	Precision	Recall	F1-score
Class-1	0.95	0.74	1.00	1.00	0.73	0.84
Class-2	0.87	0.80	0.89	0.65	0.79	0.72
Class-3	0.84	0.60	0.90	0.59	0.60	0.59
Class-4	0.94	0.88	0.96	0.87	0.88	0.87
Class-5	0.99	0.98	0.99	0.96	0.98	0.97

5 Summary and Conclusion

İn this paper, we present a feature fusion method for intervertebral disc degeneration classification based on Pfirrmann's 1–5 scale. This method takes advantages of both feature extraction methods from the spatial domain using LBP and CNN form neural network domain. Accuracy is calculated as 80.40%. We are working on the relatively small dataset, but deep networks provide good results on a huge dataset. Our future study will be on the huge dataset and will try different CNN models for comparisons. In addition, dimensionality reduction techniques can be applied to achieve maximum accuracy.

Acknowledgements We thank Dr. Hemant Borse, Consultant Radiologist, Samarth Diagnostic Center, Nasik and Dr. Rajesh Jawale, Consultant Radiologist, Wockhardt Hospital Nasik (M.S.) India who provided insight and expertise that greatly assisted the research. We are immensely grateful to Dr. Hemant Borse and his technical team for providing spine MR image dataset for research work.

References

1. Bindra S, Sinha AGK, Benjamin AI (2015) Epidemiology of low back pain in indian population: a review. Int J Basic Appl Med Sci 5(1):166–179. ISSN 2277–2103 (Online) 015
2. Fardon DF (2014) Lumbar disc nomenclature: version 2.0 Recommendations of the combined task forces of the North American Spine Society, the American Society of Spine Radiology and the American Society of Neuroradiology Review Article. Spine J 14:2525–2545
3. Nazeer M, Rao SM, Soni S, Ravinder M (2015) Lower back pain in South Indians: causative factors and preventive measures. Sch J App Med Sci 3(1D):234–243. ISSN 2347-954X
4. Modic MT, Ross JS (2007) Lumbar degenerative disk disease. Radiology 245:43–61
5. Pfirrmann CWA, Metzdorf A, Zanetti M, Hodler J (2001) Magnetic resonance classification of lumbar intervertebral degeneration. SPINE 26:1873–1878
6. Ketler A, Wilke HJ (2006) Review of existing grading systems for cervical or lumbar disc and facet joint degeneration. Eur Spine J 15:705–718
7. Bennekar LM, Heini PF, Anderson SE (2005) Correlation of radiographic and MRI parameters to morphological and biochemical assessment of intervertebral disc degeneration. Eur Spine J 14:27–35
8. Kellgren JH, Lawrence JS (1952) Rheumatism in miners Part II X-ray study. Br J Ind Med 9(3):197–207

9. Gordon SJ, Yang KH, Mayer PJ, Mace AH Jr, Kish VL, Radin EL (1991) Mechanism of disc rupture. A preliminary report Spine (Phila Pa 1976) 16(4):450–456

10. Lane NE, Nevitt MC, Genant HK, Hochberg MC (1993) Reliability of new indices of radiographic osteoarthritis of the hand and hip and lumbar disc degeneration. J Rheumatol 20(11):1911–1918

11. Mimura M, Panjabi M, Oxland T, Crisco J, Yamamoto I, Vasavada A (1994) Disc degeneration affects the multidirectional flexibility of the lumbar spine. Spine (Phila Pa 1976) 19(12):1371–1380

12. Madan SS, Rai A, Harley JM (2003) Interobserver error in interpretation of the radiographs for degeneration of the lumbar spine. Iowa Orthop J 23:51–56

13. Thalgott J, Albert T, Vaccaro A, et al (2004) A new classification system for degenerative disc disease of the lumbar spine based on magnetic resonance imaging, provocative discography, plain radiographs and anatomic considerations. Spine J 4(6 Suppl):S167

14. Wilke HJ, Rohlmann F, Neidlinger-Wilke C, Werner K, Claes L, Kettler A (2006) Validity and interobserver agreement of a new radiographic grading system for intervertebral disc degeneration: Part I. Lumbar spine. Eur Spine J 15(6):720–730

15. Schneiderman G, Flannigan B, Kingston S, Thomas J, Dillin WH, Watkins RG (1987) Magnetic resonance imaging in the diagnosis of disc degeneration: correlation with discography. Spine (Phila Pa 1976) 12(3):276–281

16. Butler D, Trafimow JH, Andersson GB, McNeill TW, Huckman MS (1990) Discs degenerate before facets. Spine (Phila Pa 1976) 15(2):111–113

17. Tertti M, Paajanen H, Laato M, Aho H, Komu M, Kormano M (1991) Disc degeneration in magnetic resonance imaging. A comparative biochemical, histologic, and radiologic study in cadaver spines. Spine (Phila Pa 1976) 16(6):629–634

18. Gunzburg R, Parkinson R, Moore R, Cantraine F, Hutton W, Vernon Roberts B, Fraser R (1992) A cadaveric study comparing discography, magnetic resonance imaging, histology, and mechanical behavior of the human lumbar disc. Spine 17(4):417–426

19. Southern EP, Fye MA, Panjabi MM, Patel TC, Cholewicki J (2000) Disc degeneration: a human cadaveric study correlating magnetic resonance imaging and quantitative discomanometry. Spine 25(17):2171–2175

20. Askar Z, Wardlaw D, Muthukumar T, Smith F, Kader D, Gibson S (2004) Correlation between intervertebral disc morphology and the results in patients undergoing Graf ligament stabilisation. Eur Spine J 13(8):714–718

21. Griffith JF, Wang YX, Antonio GE, Choi KC, Yu A, Ahuja AT, Leung PC (2007) Modified Pfirrmann grading system for lumbar intervertebral disc degeneration. Spine (Phila Pa 1976) 32(24):E708-E712

22. Boos N, Weissbach S, Rohrbach H, Weiler C, Spratt KF, Nerlich AG. Classification of age-related changes in lumbar intervertebral discs: 2002 volvo award in basic science. Spine. 2002; 27(23):2631-44.

23. Berlemann U, Gries NC, Moore RJ (1998) The relationship between height, shape and histological changes in early degeneration of the lower lumbar discs. Eur Spine J 7(3):212-217

24. Weiler C, Lopez-Ramos M, Mayer HM, Korge A, Siepe CJ, Wuertz K, Weiler V, Boos N, Nerlich AG (2011) Histological analysis of surgical lumbar intervertebral disc tissue provides evidence for an association between disc degeneration and increased body mass index. BMC Res Notes 4(1):497

25. Corso JJ, Alomari RS, Chaudhary V (2008) Lumbar disc localization and labeling with a probabilistic model on both pixel and object features. In: Proceeding of MICCAI, vol 5241 of LNCS Part 1. Springer, pp 202–210

26. Alomari RS, Corso JJ, Chaudhary V (2009) Abnormality detection in lumbar discs from clinical MR images with a probabilistic model. In: Proceeding of CARS

27. Alomari RS, Corso JJ, Chaudhari V, Dhillon G (2009) Desiccation diagnosis in lumbar discs from clinical MRI with a probabilistic model. In: Proceeding of ISBI'09, pp 546–549

28. Alomari RS, Corso JJ, Chaudhary V, Dhillon G (2009) Computer-aided diagnosis of lumbar disc pathology from clinical lower spine MRI. Int J Comput Assist Radiol Surg 5(3):287–293

29. Watanabe A, Benneker L, Boesch C, Obata T, Anderson S, Watanabe T (2007) Classification of intervertebral disk degeneration with axial T2 mapping. AJR 189(4):936–942
30. da Silva Barreiro M, Marcello H, Rangayyan R (2014) Semiautomatic classification of intervertebral disc degeneration in magnetic resonance images of the spine. In: 5th ISSNII-IEEE Biosignals and biorobotics for better and safer living (BRC) conference, pp 1–5
31. Castro-Mateos I, Hua R, Pozo JM, Lazary A, Frangi AF (2016) Intervertebral disc classification by its degree of degeneration from T2 weighted magnetic resonance images. Eur Spine J 25(9):2721–2727
32. Ronnerberger O, Fischer P, Brox T (2015) U-net: convolution networks for biomedical image segmentation. In: MICCAI, Springer, LNCS, vol 9351, pp 234–241
33. Jamaludin Amir, Kadir Timor, Zisserman Andrew (2017) SpineNet: automated classification and evidence visualization in spinal MRIs. Med Image Anal 41:63–73
34. Kim KI, Kwon Y (2010) Single-image super-resolution using sparse regression and natural image prior. IEEE Trans Pattern Anal Mach Intell 32(6):1127–1133
35. Ojala T, Pietikainen M, Maenpaa T (2002) Multiresolution gray-scale and rotation Invariant texture classification with local binary patterns. IEEE Trans Pattern Anal Mach Intell 24:971–987
36. Krizhevsky A, Sutskever I, Hington GE (2012) ImageNet classification with deep convolution neural networks. In: NIPS, pp 1106–1114
37. Cortes C, Vapnik V (1995) Support-vector networks. Mach Learn 20:273–297

Thermal Image Segmentation of Facial Thermograms Using K-Means Algorithm in Evaluation of Orofacial Pain

Nida Mir, U. Snekhalatha, Mehvish Khan and Yeshi Choden

Abstract The study aims at analyzing skin surface temperature, aided by the thermal camera, a supporting software, and application of k-means algorithm, and feature extraction in MATLAB to diagnose dental diseases, specifically, orofacial pain. The thermal camera is employed for capturing thermal images of the Left, Right, and Front profiles of all the subjects taken into account. MATLAB-based image segmentation using k-means algorithm, and feature extraction was carried out for control and test group data. The results obtained from the study depict that the mean temperature difference of maximum, minimum and average values of temperature recorded were found to be 1.09% in the front, 3.78% in the right, and 3.97% in the left facial regions between the normal subjects and abnormal diseased subjects. Of the regions examined using thermography, and subsequent feature extraction, the right and left sides show almost similar percentage differences, that is, of 3.78 and 3.97%. These findings point toward a clear, and significant rise of temperature, due to presence of infections, or ailments in the From this data it is safe to infer that, presence of infections, significantly increases the temperature of the region they are present in, and hence give an indication of possible application of thermography in dental disease detection.

N. Mir · U. Snekhalatha (✉) · M. Khan · Y. Choden
Department of Biomedical Engineering, SRM University, Kattankulathur, Chennai 603203,
Tamilnadu, India
e-mail: sneha_samuma@yahoo.co.in

N. Mir
e-mail: Nidamir33@yahoo.com

M. Khan
e-mail: Meh.pahtan13@gmail.com

Y. Choden
e-mail: wishfangle@gmail.com

© Springer Nature Switzerland AG 2019
D. Pandian et al. (eds.), *Proceedings of the International Conference on ISMAC in Computational Vision and Bio-Engineering 2018 (ISMAC-CVB)*, Lecture Notes in Computational Vision and Biomechanics 30,
https://doi.org/10.1007/978-3-030-00665-5_56

1   Introduction

Orofacial pain is a general term covering any pain usually used to describe symptoms of pain in the mouth, head and neck region. Facial and oral pain, both are
included in Orofacial pain. Out of this, pain within the mouth is called orofacial
pain, whereas Facial pain includes the pains anterior to ears, below the canthometal
line, and over the neck [1]. The most commonly reported symptom of orofacial pains
is toothache (57.6%) [2]. Prevalence of OFP was 26.94% which is more common
in age group of 35–44 yrs (10.3%), females (14%) and low-income group (9.05%)
[3]. These diseases might go unnoticed, or undiagnosed, leading to increased infection or severe problems. Common methods of diagnosis include clinical intra- and
extra-oral examination, radiographic method, occlusal examination, and fiber-optic
Transillumination, Electrical conductance, quantitative light-induced fluoroscope [4,
5]. Thermography can also be used for studying the physiology of skin temperature
of the subjects, and can serve as a non-invasive imaging modality, for detection of
dental disorders, according to the amount of heat, and consequently IR rays emitted
by the subject's body.

Gratt et al. diagnosed chronic orofacial pain and TMJ disorders by measuring the
skin temperature in facial region, using static area telethermography and classified
patients with orofacial pain and normal subjects attaining 92% accuracy [6] a few
researchers including Durnovo et al., and Pogrel et al. suggested the use of thermography as a potential diagnostic tool for the evaluation of diseases in maxillofacial
areas [7, 8].

The aim and objectives of the study were to analyze the skin surface temperature
distribution in dental disorders using thermal imaging and to segment the region of
interest using k-means algorithm and to perform statistical feature extraction for the
total population studied.

2   Methodology

2.1   Study Design and Population

The study was approved by an institutional ethical committee of SRM medical college
and Hospital research centre, consisted of a total of 10 normal subjects and 10 subjects
suffering from OFP. The institutional scientific committee of SRM medical college
and Hospital research centre had approved the study, and the participant's signed the
informed consent form.

2.2 Procedure

The subjects were made to sit in a room with the temperature at approximately 20 °C with humidity of 45–50%. Subjects with beard, or wearing ornaments or jewelry within the region of interest that could not be removed were excluded. The distance between the subjects face, and the camera (FLIR A300) was standardized at 1 m, and thermal images of the left, right, and front profiles of the face region of both the subject groups (normal and diseased) were captured. FLIR software was used for image analysis, including measurement of average skin temperature using rectangular and elliptical area tool for ROI selection and further processing was done using MATLAB R2012a (Math Works Inc., Natick, MA, USA).

2.3 Image Segmentation Algorithm

1. Images of left, right, and front profiles of test group and control group subjects were acquired using FLIR 300 thermal camera

    1. images captured were converted from RGB Color Space to $L * a * b *$ Color Space (L is lightness, varying from 0 to 100, 'a' refers to green-red component and 'b' refers to blue-yellow component)
    2. the image was then segmented into 16 different clusters to obtain the Hot Spot
    3. region separately, using k-means clustering
    4. images were then labeled according to the results obtained from k-means algorithm.

Table 1 The mean of maximum, minimum and average values of temperature recorded from the front, left and right profiles of the normal and diseased subjects are calculated and depicted in Table 1. A difference of 1.09% in the front, 3.78% in the right and 3.97% in the left facial regions between the normal subjects and abnormal diseased subjects was calculated. Of the regions examined using thermography, and subsequent feature extraction, the right and left sides show almost similar percentage differences, that is, of 3.78 and 3.97% respectively. The average temperature recorded in the region of interest, of front, left and right profiles, of the diseased patients is always higher in comparison to that of normal subjects, though the difference fluctuates, as sides differ.

   Table 2: The temperature values collected from front, left and right profiles of the normal and diseased subjects are recorded and are analyzed statistically. The mean, standard deviation, variance, Median, skewness, kurtosis and moment obtained from this analysis are depicted in this table. The values in diseased subjects are present, unlike normal cases, where they are repetitively zero.

   Figure 1 shows the right profile of the normal subject. Figure 1a displays the original thermogram to be processed by K-means clustering. Figure 1b, c displays

Table 1 Skin surface temperature measured at various facial regions using thermal imaging

Facial region		Temperature		
		Max	Min	Mean
Front	Normal	35.2875	25.25	31.7125
	Diseased	35.7625	26.35	32.0625
	% difference	1.33	4.26	1.09
Right	Normal	35.0625	25.3375	30.7625
	Diseased	36.1	26.2125	31.95
	% difference	2.91	3.394	3.78
Left	Normal	35.2125	26.1625	30.85
	Diseased	36.0875	26.4125	32.1
	% difference	2.45	0.951	3.97

Table 2 Feature extraction performed for the segmented image of normal and orofacial pain

Feature extracted parameters	Region	Mean	Std. dev	Variance	Median	Skewness	Kurtosis	Moment
Normal ($n = 10$)	Front	0	0	0	0	0	0	0
	Left	0	0	0	0	0	0	0
	Right	0	0	0	0	0	0	0
Orofacial pain ($n = 10$)	Front	56.7201	68.3811	4.69E+03	19.875	1.34949	4.24321	4.69E+03
	Left	56.1394	67.6715	4.59E+03	18.375	1.33023	4.16089	4.59E+03
	Right	57.0711	68.4343	4.70E+03	19.875	1.31145	4.08315	4.70E+03

the color-based segmented images using K-means clustering. Figure 1d displays the segmented image after thresholding. Similarly, Fig. 2 shows the right profile of the diseased subject. Figure 2a displays the original thermogram and Fig. 2b, d displays the various color-based segmented images using K-means clustering. Figure 2c displays the high-temperature regions due to inflammation or infection. Figure 2e displays the segmented image of the highest temperature regions after thresholding. Due to the absence of high-temperature zone in normal subjects, the regions of maximum temperature are not segmented contrary to diseased subjects.

3 Discussion

In the literature, the methods used for diagnosis of dental diseases include observation of symptoms, anatomical changes, and X-rays. This study, aims at establishing

(a) **(b)** **(c)** **(d)**

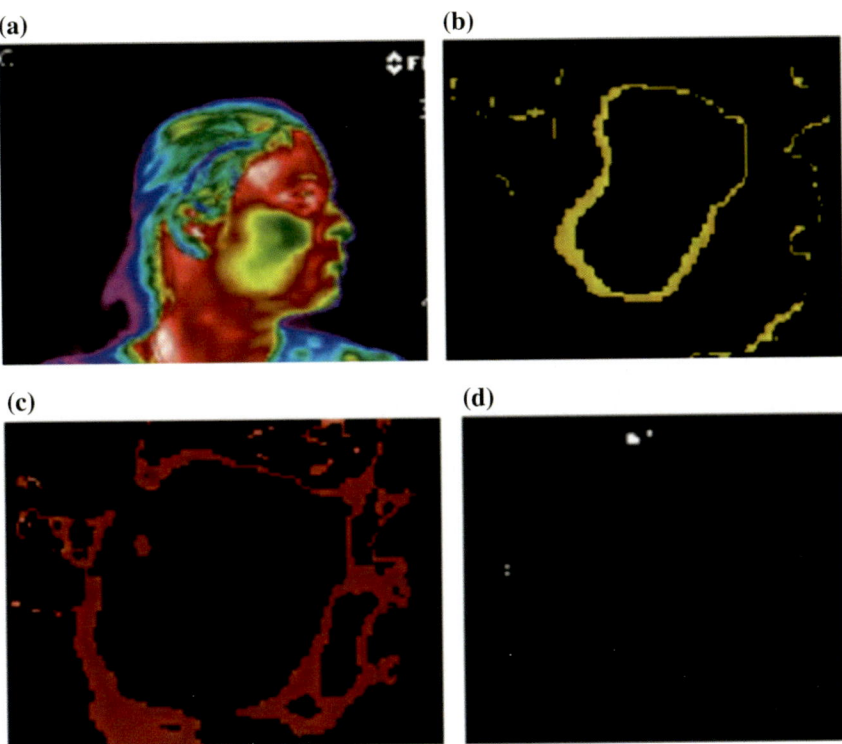

Fig. 1 Normal subject right profile **a** input thermal image, **b** and **c** random clusters obtained using the algorithm, **d** segmented image (absence of high temperature)

thermography as a means, or at least, an aid, in the detection of such diseases, which might otherwise go unnoticed, or might be difficult to rule out, or diagnose. A safe inference, that is, the temperature in case of, diseased patients is higher, can be made, and used as a support for the study. This is specially, advantageous for asymptomatic diseases, as detection of such abnormalities, still remains a matter of concern for the medical professionals.

Gratt et all selected a region of interest, in normal and diseased subjects and classified them into normal, hot and cold, on the basis of temperature recorded in that region [9, 10]. Another study conducted by Dibai Filho et al. suggested that, within the upper trapezius muscle region, infrared imaging was suitable for diagnosis of trigger points in the in clinical and research practices [11].

In a separate study conducted in 2014, by Haddad et al., it was significantly apparent that the temperature measured at anterior temporal muscle regions and masseter muscles in controls were higher than myogenic TMD volunteers, along with posing the suggestion of thermography being a possible tool in diagnosis or aid of other diagnostic methods [12].

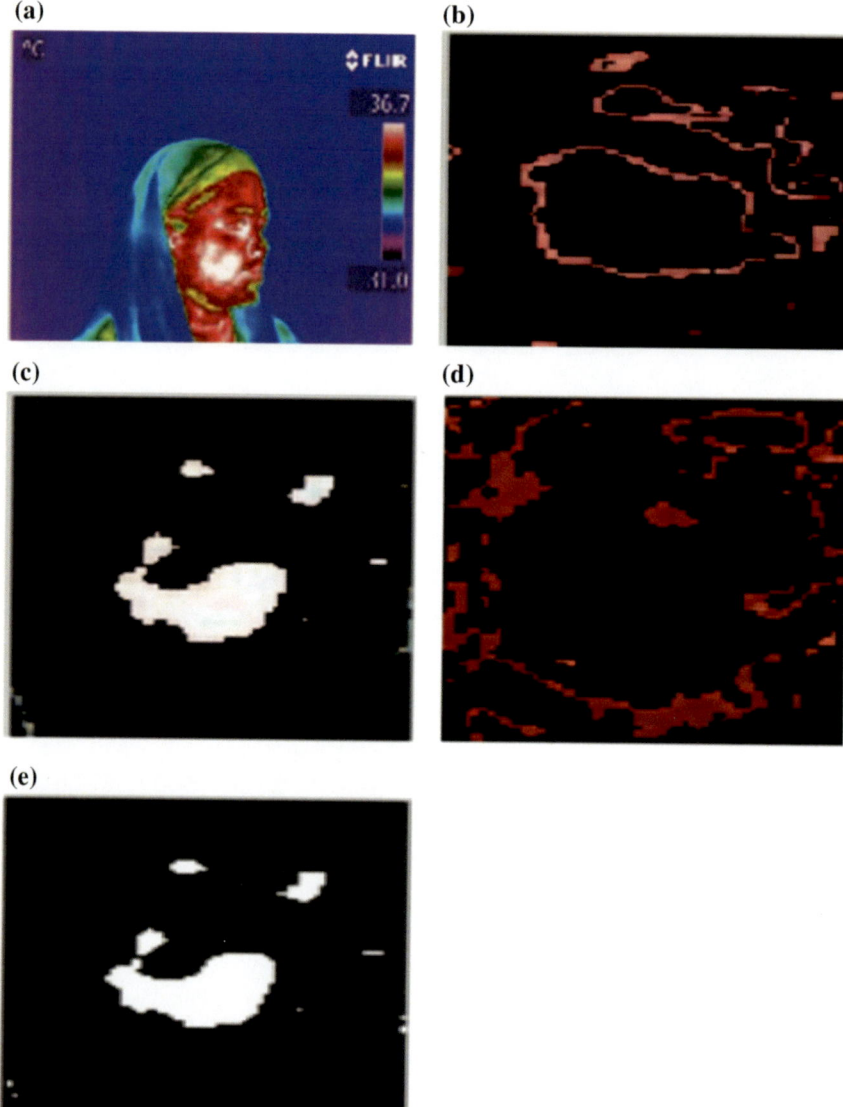

Fig. 2 Diseased subject right profile **a** input thermal image, **b–d** random clusters obtained using the algorithm, **e** segmented image showing high temperature alone

Also the studies, like that carried out by McBeth et al., Biagioni et al., etc., are in agreement with the current study, implying the possible use of thermography in future for better diagnosis [13–16].

4 Conclusion

Thermography can be used as a complementary diagnostic tool in dentistry with reference to the facial thermograms as there is a notable escalation of regional temperature. The mean of maximum, minimum, and average values of temperature recorded from the front, left, and right profiles of the normal and diseased subjects were calculated and subsequent feature extraction was implemented. The right and left profiles showed a comparable percentage difference (i.e., 3.78 and 3.97%). As inferred, thermographic findings differ with different subjects' side profiles, however, the average temperature of the diseased subjects are always higher than that of the normal subjects.

References

1. Macfarlane TV, Blinkhorn AS, Davies RM, Kincey J, Worthington HV (2002) Orofacial pain in the community: prevalence and associated impact. Commun Dent Oral Epidemiol 30:52–60
2. Oberoi SS, Hiremath SS, Yashoda R et al (2014) Prevalence of various orofacial pain symptoms and their overall impact on quality of life in a tertiary care hospital in India. J Maxillofacc Oral Surg 13:533. https://doi.org/10.1007/s12663-013-0576-6
3. Manjunath G, Prasad KVV (2012) Prevalence of Orofacial pain among urban adult population of Hubli—Dharwad and its impact on daily. J Indian Assoc Public Health Dent 10:7–18
4. Abesi F, Mirshekar A, Moudi E, Seyedmajidi M, Haghanifar S, Haghighat N, Bijani A (2012) Diagnostic Accuracy of digital and conventional radiography in the detection of non-cavitated approximal dental caries. Iran J Radiol 9(1):17–21
5. Gomez J (2015) Detection and diagnosis of the early caries lesion. BMC Oral Health 15(Suppl 1):S3
6. Gratt BM, Graff-Radford SB, Shetty V, Solberg WK, Sickles EA (1996) A six-year clinical assessment of electronic facial thermography. Dentomaxillofacial Radiol 25:247–255
7. Durnovo EA, Potekhina YP, Marochkina MS, Yanova NA, Sahakyan MY, Ryzhevsky DV (2014) Diagnostic capabilities of Infrared thermography in the examination of patients with diseases of maxillofacial area. Clin Med 6(2):61–65
8. Pogrel MA, Yen CK, Taylor RC (1989) Infrared thermography in oral and maxilla facial surgery. Oral Surg Oral Med Oral Pathol 67(2):126–131
9. Gratt BM, Sickles EA, Ross JB (1994) Thermographic characterization of an internal derangement of the temporomandibular joint. J Orofacial Pain 8:197–206
10. Gratt BM, Sickles EA, Wexler CA (1993) Thermographic characterization of osteoarthrosis of the temporomandibular joint. J Orofacial Pain 7:345–353
11. Dibai-Filho AV, Guirro EC, Ferreira VT, Brandino HE, Vaz MM, Guirro RR (2015) Reliability of different methodologies of infrared image analysis of myofascial trigger points in the upper trapezius muscle. Braz J Phys Ther 19:122–128
12. Haddad DS, Brioschi ML, Vardasca R, Weber M, Crosato EM, Arista ES (2014) Thermographic characterization of masticatory muscle regions in volunteers with and without myogenous temporomandibular disorder: preliminary results. Dentomaxillofacial Radiol 43:20130440
13. Gratt BM, Sickles EA (1995) Electronic facial thermography: an analysis of asymptomatic adult subjects. J Orofacial Pain 9:255–265
14. Biagioni PA, Longmore RB, McGimpsey JG, Lamey PJ (1996) Infrared thermography. Its role in dental research with particular reference to craniomandibular disorders. Dentomaxillofacial Radiol 25:119–124

15. Komoriyama M, Nomoto R, Tanaka R, Hosoya N, Gomi K, Iino F, Yashima A, Takayama Y, Tsuruta M, Tokiwa H, Kawasaki K, Arai T, Hosoi T, Hirashita A, Hirano S (2003) Application of Thermography in dentistry-visualization of temperature distribution on oral tissues. Dent Mater J 22(4):1–7

16. McBeth SA, Gratt BM (1996) A cross-sectional thermographic assessment of TMJ problems in orthodontic patients. Am J Orthod Dentofac Orthop 109:481–488

Analysis of Web Workload on QoS to Assist Capacity

K. Abirami, N. Harini, P. S. Vaidhyesh and Priyanka Kumar

Abstract Workload characterization is a well-established discipline, which finds its applications in performance evaluation of modern Internet services. With the high degree of popularity of the Internet, there is a huge variation in the intensity of workload and this opens up new challenging performance issues to be addressed. Internet Services are subject to huge variations in demand, with bursts coinciding with the times that the service has the most value. Apart from these flash crowds, sites are also subject to denial-of-service (DoS) attacks that can knock a service out of commission. The paper aims to study the effect of various workload distributions with the service architecture 'thread-per-connection' in use as a basis. The source model is structured as a sequence of activities with equal execution time requirement with an additional load time of page (loading embedded objects, images, etc.). The threads are allocated to the requests in the queue; leftover requests if any are denied service. The rejection rate is used as a criterion for evaluation of the performance of the system with a given capacity. The proposed model could form a basis for various system models to be integrated into the system and get its performance metrics (i.e. QoS) evaluated.

1 Introduction

Today Internet has emerged as the default platform for application development. Unfortunately, modern applications demand more complexity than traditional applications. As the Internet was not designed to suit the requirements of modern applications, the execution results in high frustration of users. This factor demands a research on how the existing infrastructure could be modified for efficient execution of modern web applications [1]. The complexity exhibited by applications are multifold (process, data, load, configuration, scale, etc.). With an intent to improve the

K. Abirami (✉) · N. Harini · P. S. Vaidhyesh · P. Kumar
Department of Computer Science and Engineering, Amrita School of Engineering, Amrita Vishwa Vidyapeetham, Coimbatore, India
e-mail: k_abirami@cb.amrita.edu

© Springer Nature Switzerland AG 2019
D. Pandian et al. (eds.), *Proceedings of the International Conference on ISMAC in Computational Vision and Bio-Engineering 2018 (ISMAC-CVB)*, Lecture Notes in Computational Vision and Biomechanics 30,
https://doi.org/10.1007/978-3-030-00665-5_57

performance of web applications a study on workload characterization is compelling. Many researchers have focused their study on understanding the characteristics and intensity of workloads. In this work, we discuss the role of workload models for resource assignment in the scenario of the e-commerce application. The impact of the workload on system properties and behavior is analyzed using a capacity planning model. The proposed system evaluates the Quality of Service (Qos) and Quality of Experience (QoE) perceived by the users for different workload distributions. These observations could aid in framing security mechanisms, recommendation engines, data distribution policies, etc.

When the system is scaled, the work also presents major findings from experimentation indicating performance implications. The rest of the paper is organized as follows: Sect. 2 presents a comprehensive overview of the literature on different workload distributions. Section 3 summarizes the characterization methodologies and related measurement process. Section 4 presents the results and analysis of experimentation and finally, Sect. 5 presents concluding remarks.

2 Literature Review

Many research work addresses the black box approach for the assessment of performance based on workloads. Rejection rates have a huge impact on the performance of the system [2]. Recent rates have a huge impact on the performance of the system [2]. Recent studies have also considered performance measurements based on user behavior patterns and businesses [3, 4]. The response time metric has been chosen in most of the research work for performance evaluation [5]. The Zipf law's applicability of web workloads is addressed by Levene et al. [6] and Menasce et al. [7]. Mi et al. [8] and Harini and Padmanabhan [9] discuss the need for stationary of arrival processes to study and characterize web load. Harini and Padmanabhan [9, 10] addresses the issue of the presence of malicious request in the incoming lot which needs to be weeded out before the commencement of processing. Workload management is a process of effective workload distribution to achieve optimal performance and productivity levels. Modelling workload distributions would aid one to understand the performance and the scalability of the system. Workload model of an application depicts how the application would perform in the given infrastructure. The performance is usually assessed using Service-Level Agreements (SLA). Little Theorem gives a relationship between the average number of users, arrival rate and average time, an end user spends in the system.

The theorem state that

$$L = \lambda N$$

where λ is the arrival rate and L is the effective arrival rate.

The only prerequisite being system should not preempt and must be stable. The arrival pattern of the request can be modeled based on different probability distributions like exponential, normal, binomial, Poisson, Zipfian.

2.1 Distributions

2.1.1 Exponential Distributions

Exponential distribution is a well-known concept in the theory of probability and statics. The distribution denotes the time between two events in processes where the events are continuous and occur independently. The key property of the distribution is memorylessness. This general exponential distribution is given by

$$f(x;\lambda) = \begin{cases} \lambda e^{-\lambda x} & x \geq 0 \\ 0 & x < 0 \end{cases}$$

where λ greater than 0 is the rate parameter. The distribution is well supported in the interval 0 to infinity. This distribution is mainly used to model service times rather than arrival patterns. These can have a strong effect on performance evaluation results.

2.1.2 Normal Distribution

Normal distribution is a very commonly used distribution to determine whether an observation falls between two extreme limits. This distribution is used to model random variables in natural and social sciences. The normal distribution is used in real-valued random variables, where the distributions are not available. The general normal distribution is given by

$$f\left(x|\mu, \sigma^2\right) = \frac{1}{\sigma}\psi\left(\frac{x - \mu}{\sigma}\right)$$

The standard normal deviate is given by Z where

$$Z = \frac{(X - \mu)}{\sigma}$$

These could be used to model the peak of arrivals and at a more concrete level, it can help one to identify results of random effects on workloads.

2.1.3 Poisson Distribution

The application of Poisson distribution in traffic problems is not new. A Poisson distribution is a probability distribution of a discrete random variable that represents the number of statistically independent events occurring within a unit of time or space. Time-based Poisson variables are more popular. Given the expected Value μ of the Poisson variable x the probability function is defined as the probability of observing k events in an interval is given by the equation

$$P(k \text{ events in interval}) = e^{-\lambda} \frac{\lambda^k}{k!}$$

where the average number of events per interval e is the number 2.71828 ... (Euler's number) the base of the natural logarithms k is any natural number, $k! = k \ (k \ 1) \ (k \ 2) \ 2 \ 1$ is the factorial of k. This could be used to model the rate of arrival of request patterns and it could be also used to measure the performance when requests are queued in the system.

2.1.4 Zipf Distribution

A Zipf distribution is sometimes referred to as zeta distribution. This is particularly used for modeling rare events. The probability density function for Zipf distribution is the nth raw moment is defined as the expected value of X_n:

$$m_n = E(X^n) = \frac{1}{\zeta(s)} \sum_{k=1}^{\infty} \frac{1}{k^{s-n}}.$$

The series on the right is just a series representation of the Riemann zeta function, but it only converges for values of s-n that are greater than unity. Thus:

$$m_n = \begin{cases} \frac{\zeta(s-n)}{\zeta(s)} & \text{for } n < s - 1 \\ \infty & \text{for } x \geq 0 \end{cases}$$

Note that the ratio of the zeta functions is well defined, even for $n > s-1$ because the series representation of the zeta function can be analytically continued. The Zipf distribution turns out to better describe varied human activities. It is a good model for popularity distribution. This does not change the fact that the moments are specified by the series itself, and are therefore undefined for large n.

2.1.5   Binomial Distribution

The Binomial distributions will have two outcomes, success or failure. The experiment can have n number of trials and the outcomes are independent. The general equation of the distribution is given by the following:

$$b(x;n, p) = \binom{n}{r} p^x (1 - p)^x$$

where,

n represents the number of trials, x represents the number of successes, p represents the probability of success in an individual trial.

The distribution could be used for modeling random arrival patterns, study effects of peak load, perform resource assignments, etc.

2.2   Summary of Findings

A special case of performance evaluation that deserves individual attention is capacity planning. Many research works propose different methodologies for setting up configurations that would provide desired performance. The required system capacity obviously depends on workload intensity, i.e., one needs more capacity to do more work. The relationship between capacity and workload is often not linear. Researchers have also stated that burst is an important attribute that contribute to capacity planning. Burst refers to large fluctuations in workload intensity.

A good characterization technique thus requires a clear understanding of burst characteristics. A combination of system model with workload characterization can enhance the performance of the system. Although individual schemes specific to Internet services have been explored in large, to the best of our knowledge a comprehensive study based on multiple workload distribution based analysis with its performance assessment has not been addressed to a greater extent.

2.3   Problem Statement

To build a system capable of characterizing the performance of a system model for varied workload distributions, which could aid in capacity planning, arriving at optimal configuration to improve QoS, assist in data movement with applied security features.

3 Proposed System

Though Internet services remain simple in the structure at the start they become more complex when functionalities of the service expand. The block diagram for service architecture used for experimentation is shown in Fig. 1.

3.1 Model Description

The basic service is taken as composed of n sequential activities. A single processor system with T threads is used as a basis for service. The service request is taken as a stationary random process having a selected distribution (Normal, Binomial, Exponential, Poisson, Zipf) with its associated parameters.

3.1.1 Request Arrival Pattern

A Service request is taken as a stationary random process with an associated mean and variance attributes. The proposed system considers discrete random arrival pattern. Each arrival is independent of the previous arrival. An arrival set is characterized with a number of incoming requests and a class type associated with it. The arrival capacity is not limited.

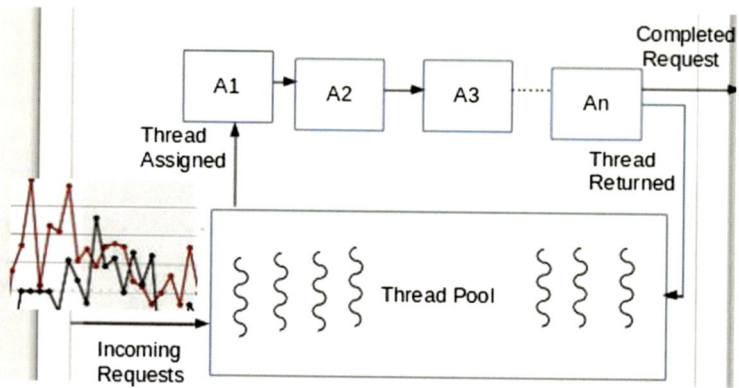

Fig. 1 Service architecture

3.1.2   Request Characteristics

A request is characterized by arrival time and service class. Each request is assumed to have the same number of tasks to be completed. Each request is assumed to have a load time that is dependent on the dynamic content of the webpage and an associated service time.

3.1.3   Service Scheme

At the beginning of every time slots, threads from the request that completed execution are returned to the free pool of threads. The incoming requests are assigned threads for until request list is exhausted or threads in the thread pool are exhausted. Once a thread is allocated to a request it remains associated with the request until completion of execution. There can be time slots when threads are free and those when request are dropped. Both do not happen simultaneously. An extensive simulation was carried out with different distributions. The variations in terms of request drops are modeled and presented in Sect. 4.

3.1.4   Service Scheme Algorithm

The algorithm for modeling arrival:

Step 1: Generate the total number of arrivals for each time slot based on the random number generated by the distribution parameters.

Step 2: Assign the arrivals randomly to n categories as category1, category2 ...categoryn with:

$$\sum_{k=1}^{n}(\text{category1} + \text{category2} + \cdots \text{categoryn})$$

It should be equal to a number of arrivals in the time slot category1 = rand()% number of arrivals.

3.1.5   Resource Allocation

Step 1: Initialize index and rejections as 0, Initialize T as maximum number of threads

Step 2: For the arrivals in timeslot allocate thread from the thread pool

Step 3: Update thread counter in the thread pool

Step 4: If not enough threads for allocation.

```
for index in value:
    if value[index]≥ rejection threshold:
        rejection threshold −=value[index]
    else
        rejections +=value[index]
return rejections
```

4   Results and Discussions

The effect of workloads based on different distributions was studied through extensive simulation process. The representative results for thread pool capacity 100 are
presented in Fig. 2. The simulation run duration in each case was selected in such a
way that all possible service request values appeared enough number of times to bring
out all behavioral characteristics. The rejection of requests for different distributions
is also presented in Fig. 2. To facilitate service differentiation, three categories of
arrivals were considered (this could be used for priority scheduling). Automated service history collection which enables culling intelligent information out of it is also
collected by the system.

4.1   Measuring Overhead in Dynamic Pages

Processing dynamic webpage requires additional page load time that includes loading
time of images, audio video links etc., To understand the effect of this additional time
a webpage with following specification (i.e.,) load time approximately 9.14 s etc.,
is presented in the table. Image loading time for a website is approximately 1–2 ms.
As an example, the website "www.amrita.edu", the total no of requests is 151. The
size of the page is 2.6 MB, load time is 9.14 s. Out of this 151 requests 94 (64.1%)
requests are given for image. The download time for the images is approximately
1–2 ms each.

   This paper clearly forms a useful contribution for assessing the impact of load on
the web server for a selected system configuration. Our observations indicated that
system model integration with the architecture can enable one to analyze the performance of an Internet service. High rejection rates indicate the need for an increase in
the capacity of the system. While doing so, one should ensure that the resources are
not underutilized. A low percentage of rejection rates with Zipf distribution indicate
the identical behavior of incoming requests. The rejection rate under the Normal distribution shows the random occurrences of peak load in the traffic. The rejection rate
under Binomial and Poisson distribution clearly demand scaling the system capacity.

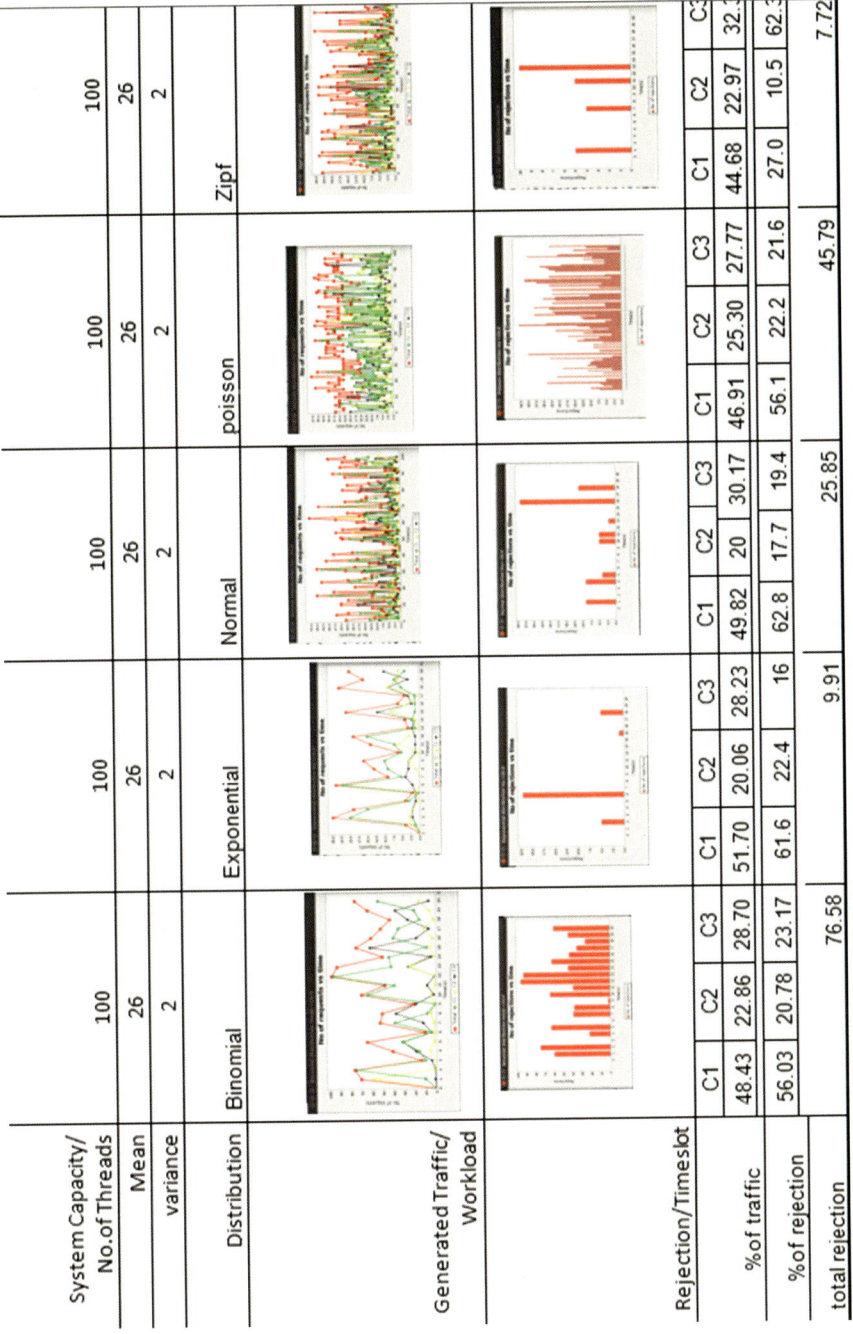

System Capacity/No.of Threads	100			100			100			100			100		
Mean	26			26			26			26			26		
variance	2			2			2			2			2		
Distribution	Binomial			Exponential			Normal			poisson			Zipf		
Generated Traffic/Workload															
Rejection/Timeslot															
	C1	C2	C3	C1	C2	C3	C1	C2	C3	C1	C2	C3	C1	C2	C3
%of traffic	48.43	22.86	28.70	51.70	20.06	28.23	49.82	20	30.17	46.91	25.30	27.77	44.68	22.97	32.3
%of rejection	56.03	20.78	23.17	61.6	22.4	16	62.8	17.7	19.4	56.1	22.2	21.6	27.0	10.5	62.3
total rejection	76.58			9.91			25.85			45.79			7.72		

Fig. 2 Table 1

5 Concluding Remarks

Modern web services have thrown up many unconventional challenges for monitoring QoS. Although ways for monitoring QoS parameters have been addressed extensively in the literature, the methodologies and techniques applied for creating workload models are strictly related to the objectives of the studies. With the aim of studying the effect of the movement of data in distributed systems in terms of response metric, the scheme proposed in the paper was implemented. The scheme enabled to understand the impact of different distributions on system performance (Completed vs. Rejected Services). Experimentation clearly revealed the effect of dynamic contracts in processing concurrent requests. Schemes like loading essential partial images rather than complete contents could be used to improve the QoS.

References

1. Calzarossa MC, Massari L, Tessera D (2016) Workload characterization: a survey. ACM Comput Surv (CSUR) 48(3):48
2. Galletta DF, Henry R, Mccoy S, Polak P (2004) Web site delays: how tolerant are users. J Assoc Inform Syst pp 1–28
3. Goncalves MA, Almeida JM, dos Santos LG, Laender AH, Almeida V (2010) On popularity in the blogosphere. IEEE Internet Comput 14(3):42–49
4. Gusella R (1991) Characterizing the variability of arrival processes with indexes of dispersion. IEEE J Sel Areas Commun 9(2):203–211
5. Gunther NJ (2001) Performance and scalability models for a hypergrowth e-commerce web site. In: Performance engineering, state of the art and current trends. Springer, London, UK, pp 267–282
6. Levene M, Borges J, Loizou G (2001) Zipfs law for web surfers. Knowl Inf Syst 3(1):120–129
7. Menasce D, Almeida V, Riedi R, Ribeiro F, Fonseca R, Meira W Jr (2000) In search of invariants for e-business workloads. In: EC 00: proceedings of the 2nd ACM conference on electronic commerce, New York, NY, USA. ACM, pp. 56–65
8. Mi N, Casale G, Cherkasova L, Smirni E (2008) Burstiness in multi-tier applications: symptoms, causes, and new models. In: Middleware 08Proceedings of the 9th ACM/IFIP/USENIX international conference on middleware, New York, NY, USA. Springer, New York, Inc., pp 265–286
9. Harini N, Padmanabhan TR (2012) A secured-concurrent available architecture for improving performance of web services. In: Communications in computer and information science, vol 292, no 1. Springer, pp 621–631
10. Harini N, Padmanabhan TR (2013) Admission control and request scheduling for secured-concurrent-available architecture. Int J Comput Appl 63(6):24–30

Content-Based Image Retrieval Using Hybrid Feature Extraction Techniques

B. Akshaya, S. Sruthi Sri, A. Niranjana Sathish, K. Shobika, R. Karthika and Latha Parameswaran

Abstract Images consist of visual components such as color, shape, and texture. These components stand as the primary basis with which images are distinguished. A content-based image retrieval system extracts these primary features of an image and checks the similarity of the extracted features with those of the image given by the user. A group of images similar to the query image fed is obtained as a result. This paper proposes a new methodology for image retrieval using the local descriptors of an image in combination with one another. HSV histogram, Color moments, Color auto correlogram, Histogram of Oriented Gradients, and Wavelet transform are used to form the feature descriptor. In this work, it is found that a combination of all these features produces promising results that supersede previous research. Supervised learning algorithm, SVM is used for classification of the images. Wang dataset is used to evaluate the proposed system.

B. Akshaya · S. Sruthi Sri · A. Niranjana Sathish · K. Shobika · R. Karthika (✉)
Department of Electronics and Communication Engineering, Amrita School of Engineering,
Amrita Vishwa Vidyapeetham, Coimbatore, India
e-mail: r_karthika@cb.amrita.edu

B. Akshaya
e-mail: akshayabalan96@gmail.com

S. Sruthi Sri
e-mail: shruthisridhar.10@gmail.com

A. Niranjana Sathish
e-mail: niranjana.sathish97@gmail.com

K. Shobika
e-mail: shobi2197@gmail.com

L. Parameswaran
Department of Computer Science and Engineering, Amrita School of Engineering, Amrita
Vishwa Vidyapeetham, Coimbatore, India
e-mail: p_latha@cb.amrita.edu

© Springer Nature Switzerland AG 2019
D. Pandian et al. (eds.), *Proceedings of the International Conference on ISMAC
in Computational Vision and Bio-Engineering 2018 (ISMAC-CVB)*, Lecture Notes
in Computational Vision and Biomechanics 30,
https://doi.org/10.1007/978-3-030-00665-5_58

1 Introduction

With a splurge of visual data, it is a vastly cumbersome task to scan tens of thousands of images manually. The content-based image retrieval system extracts the basic features of every image in a dataset and compares the same with those of the image provided by the user. General flow states that post the query matching, the system ranks the images in a descending order of similarity with the given input query and the output is all the images that are ranked highest. Every image will have three basic components: Shape, Color, and Texture. A CBIR relies on these extracted features, individually or combinations of them, to extract and run similarity algorithms on them.

While CBIR completely relies on the contents of the image itself, image retrieval using the metadata of images can also be done. There is textual data associated with each image and traditional methods of retrieval, such as retrieval using keywords, can be done. While the annotation process is time consuming and laborious, there is also an additional limitation with this system which pertains to lack of standardization. Meaning, no two users perceive an image in the same way. Since there is no standard way to perceive an image, multiple users will give multiple annotations which are not likely to match. This is will also result in large amounts of junk data which is undesirable. Thus, CBIR is more suitable for large amounts of visual data.

In this paper, a combination of color, texture, and shape features are extracted in order to maximize the accuracy of extraction and efficacy of output. A proposed method uses best of multiple techniques to improve quality of output and reduce error margins.

In order to enhance the results of a robust dataset (Wang dataset), the images are classified by Supervised Vector Machines (SVM) algorithm. SVM is the most efficient supervised learning algorithm for image recognition, face recognition, speech recognition, and face detection. It is reliable, accurate and is most efficient for binary classification.

2 Literature Survey

The properties of Hue, Saturation, and Intensity values color space is analyzed in [1]. The values are varied and the visual perception is studied. The saturation value is used to decide if the Hue or the Intensity of the pixel is closer to human perception.

Various CBIR tools are compared in [2]. From this comparison, it is observed that most of the systems use color and texture features. Shape feature is not as common. Layout feature is very rarely used. Retrieval techniques based on a single feature worked well only for a specific set of images.

One of the most commonly used color feature in CBIR system is color histogram, [3–5]. Color Histogram concentrates only on the proportion of the number of various types of colors in an image, but does not focus on the spatial location of the colors.

Noise is not handled efficiently by histograms because they are very sparse. To overcome the drawback of this feature, features such as color-correlogram and color moments are applied. Preprocessing the images will increase the accuracy.

In [6], it is observed that color and texture features are used. Support Vector Machine (SVM) and Euclidean distance are applied to retrieve similar images.

Different approaches of different combinations of color, shape and texture retrieval are compared in [7]. When color (color histogram) and texture features (standard wavelet) are combined, accuracy was enhanced but the feature set was inadequate. On merging color, texture and shape feature (Color moment, Gabor filter, Gradient Vector Flow), the strong feature set was created.

Ecosembles is formed by concatenating different combinations of weak feature views (color, shape, texture, etc.), which must be extracted from images. Histograms are extracted from a group of images with varying numbers of bins for each histogram. By examining the Intelligence, Surveillance, and Reconnaissance data, it is understood that Ecosembles performed slightly better than GIST descriptors coupled with SVM. This is observed in [8].

In [9], images are retrieved separately using the features like color Histogram, Gabor and wavelet transform for texture, and Shape information from Phase congruency (edge detection for any change in illumination and contrast in the image). A combination of these produced an accuracy of 96.4%.

A system for biometric security for CBIR is developed based on the extraction of Shape (moment invariant), Color (Histogram), and Texture (Gabor wavelet) in [10].

The color and texture features are concatenated where Wavelet-Based Color Histogram (WBCH) method is used. The precision of this proposed method is found to be better. The computational steps are reduced with the help of wavelet transform, thereby increasing the retrieval speed [11].

In [12], a relative study on several features like merged color histogram and Gabor transform is performed.

With only statistical entities of the first order such as mean and standard deviation, Gabor wavelet showed better classification results. This method is proven to be slightly superior to the co-occurrence matrix which is usually used for texture classification [13].

A CBIR system in which the features like HOG, SIFT, SURF, and color histogram are used to extract the features of the image and formed a collection of local feature vectors is observed in [14].

3   Proposed Work

The primary goal of the proposed work is to retrieve similar images from a database as a response to passing a single query image. Low-level features of images are used in this approach. The prime motive of this work is to obtain an efficient and less complex image retrieval system. To achieve this, a comparative study on image retrieval using different features is performed. Initially, images are retrieved with

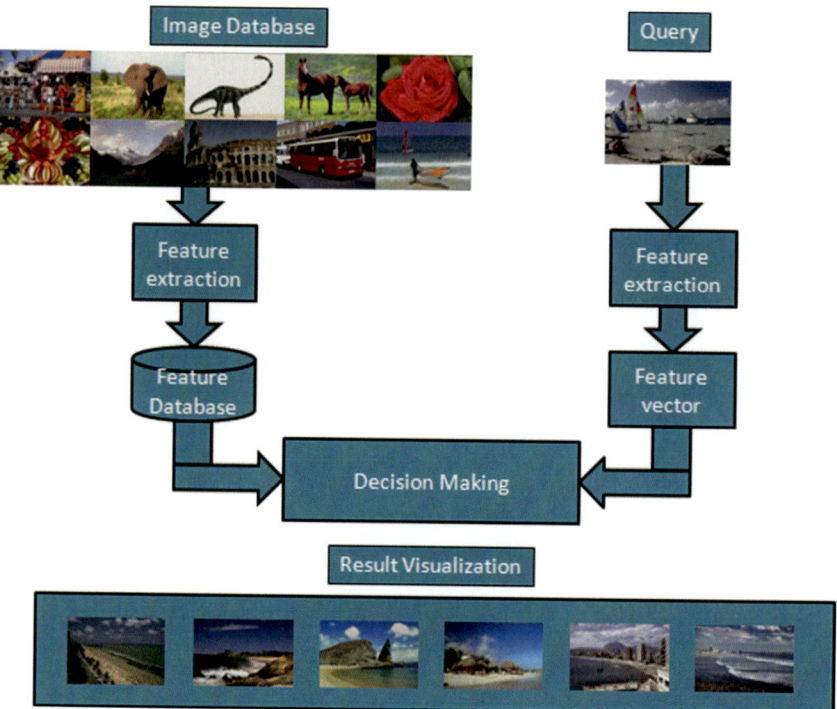

Fig. 1 CBIR system architecture

color, shape and texture features separately. Further, all three features are combined and results are compared.

In this approach, Support Vector Machine (SVM) algorithm is used for classification. The dataset used here is Wang dataset [15], which consists of 1000 images consisting of 10 different classes. Each class has 100 images. The classes include Africa, Rose, Beach, Monument, Bus, Horse, Food, Dinosaur, Scenery, and Elephant. The sizes of the images are either 384×256 or 256×384.

The schematic representation of the CBIR is shown in Fig. 1.

4 Feature Extraction

This is the most crucial stage in CBIR. The feature of each image is extracted and stored in a vector. The process is as described below.

4.1 Color Feature

HSV Histogram. Color Histogram is a schematic impression of the distribution of colors in an image. Although simple and straightforward, Color Histogram is sensitive to change in brightness and does not account for spatial information. Hence, images are converted from Red, Blue, and Green (RGB) toHue, Saturation, Value (HSV) color space. The histogram is computed for the HSV color model as it has an advantage that it separates the chrominance and luminance of an image. Each HSV component is quantized into (8 * 2 * 2) bins which gives a vector of 32 dimensions.

Color Auto Correlogram. A color auto correlogram, which is an indexed table of color pairs and their probabilities, includes spatial correlation of colors and is easy to compute. In color auto correlogram, color distribution is calculated as a function of the distance between two pixels. The image is quantized into 4 * 4 * 4 bins, which give a feature vector of 64 dimensions.

Color Moment. Color moment is the simplistic distribution of color in the image. Likeliness of two images can be compared with the help of color moments. The mean and standard deviation moments are calculated. Mean is the average color of the image and the standard deviation is calculated by taking the square root of the variance. Thus, the first two moments of each channel (RGB) is extracted to form a vector of six dimensions.

In order to compensate for the deficits of each technique, a combination of three techniques are used in order to leverage their best outcomes for color feature extraction.

Mean:

$$E_i = \sum_{j=1}^{N} \frac{1}{N} P_{ij} \tag{1}$$

N Number of pixels in the image
P_{ij} value of the jth pixel in the image at ith color channel.

Standard deviation:

$$\sigma = \sqrt{\left(\frac{1}{N} \sum_{j=1}^{N} (P_{ij} - E_i)^{\wedge}2 \right)} \tag{2}$$

E_i mean value for ith color channel of the image.

4.2 Shape Feature

Wavelets are a more general way to represent and analyze multiresolution images. Thus, shape feature extraction is accomplished using wavelet transform and Histogram of Oriented Gradients.

Histogram of Oriented Gradients (HOG). The HOG technique counts the number of occurrences of a specific orientation in each part of the image. It forms a feature vector of 1-N length, where N is the length of the HOG feature. The mean is used to form a feature vector.

Wavelet Transform. The wavelet transform is a multiresolution filtering technique that eliminates noise efficiently. The DWT (2 discrete wavelet transform) is used for detection of edges. Coiflet wavelet is applied with a 3 level decomposition and the mean and standard deviation is used to form a feature vector of 40 dimensions.

4.3 Texture Feature

Gabor Wavelet. Gabor Wavelet is used for the extraction of texture feature. The Gabor representation minimizes uncertainty in space and frequency dimensions and the micro-features extracted characterize texture information. Gabor wavelet filters are applied to each image spanning across four scales and six orientations. This produces a vector of 48 dimensions.

All features from the aforementioned steps are concatenated to form a feature vector of 192 dimensions.

A query image is an input for feature extraction and the feature vector is stored.

5 Classifier

5.1 Support Vector Machines (SVM)

Post the feature extraction process, all the images in the database are classified using SVM. It is a supervised learning algorithm that is used for classification and regression analysis. The approach used is "one-versus-one", where $\frac{n!}{(n-k)!k!}$ binary classifiers have to be trained for a k-way problem. It differentiates the samples of a pair of classes at a time. When a query image is given as input, a voting scheme is applied to all $\frac{n!}{(n-k)!k!}$ classifiers. The predicted output by the classifier is the class that gets the highest number of '+1' predictions.

6 Result Analysis

Various experiments were performed to show the efficiency of the proposed method. The system receives a single query image and returns 20 similar images from the database. The result is tested using test images from each class. Of 100 images in each class, 95 images are trained and 5 images are used for testing. For each query image, there exist 95 relevant images.

Performance evaluation can be done using numerous metrics. In this paper, precision and recall have been used for performance evaluation.

Precision

Precision is the ratio of retrieved relevant images to the total number of images retrieved.

$$\text{Precision} = \frac{\text{no. of relevant images retrieved}}{\text{total no. of images retrieved}} \tag{3}$$

Recall

Recall is the measure of how many number of truly relevant results are retrieved. A high recall implies that the algorithm has returned most of the relevant images.

$$\text{Recall} = \frac{\text{no. of relevant images retrieved}}{\text{no. of relevant images in the database}} \tag{4}$$

Table 1 shows the average values of the precision and recall of Wang dataset. It shows the retrieval performance when color, texture, and shape features are used in isolation. The average value is taken for each class. Table 2 shows the values based on a combination of color, texture, and shape feature.

Table 3 shows a comparison of the existing technique to the proposed technique. It can be seen that the color feature is more effective compared to shape and texture when the features are used in isolation. It is clear that the proposed system where the features are combined offers the highest value of precision and recall (Figs. 2, 3, 4 and 5).

The results demonstrate that the output obtained by using color, shape and texture features separately do not match with the query image whereas a combination of the three produced much accurate results. The performance measure was calculated based on precision and recall.

7 Conclusion

The goal of this paper is to retrieve images from a database with reliable accuracy by using multiple techniques in tandem with one another. The proposed work introduces an integrated approach to CBIR which helps retrieve similar images from a

Table 1 Retrieval result using color, shape, and texture separately

Class	Color		Shape		Texture	
	Precision	Recall	Precision	Recall	Precision	Recall
Africa	0.8	0.16	0.4	0.08	0.6	0.12
Beach	0.8	0.16	0.6	0.12	0.6	0.12
Monument	0.8	0.16	0.4	0.08	1	0.2
Bus	0.8	0.16	0.6	0.12	0.6	0.12
Dinosaur	0.8	0.16	1	0.2	1	0.2
Elephant	0.8	0.16	0.8	0.16	0.2	0.04
Rose	1	0.2	0.8	0.16	1	0.2
Horse	1	0.2	0.8	0.16	0.4	0.08
Mountain	0.8	0.16	0.6	0.12	0.6	0.12
Food	0.6	0.12	0.4	0.08	0.6	0.12
Mean	0.82	0.164	0.64	0.128	0.66	0.132

Table 2 Retrieval result combining color, shape, and texture

Class	Precision	Recall
Africa	0.8	0.16
Beach	0.8	0.16
Monument	0.8	0.16
Bus	1	0.21
Dinosaur	1	0.21
Elephant	0.8	0.16
Rose	1	0.21
Horse	1	0.21
Mountain	0.8	0.16
Food	0.8	0.16

Table 3 Comparison of the existing and proposed method

	Color	Shape	Texture	Color, shape and texture
Precision	0.82	0.64	0.66	0.88
Recall	0.164	0.128	0.132	0.18

Fig. 2 Precision and recall using color

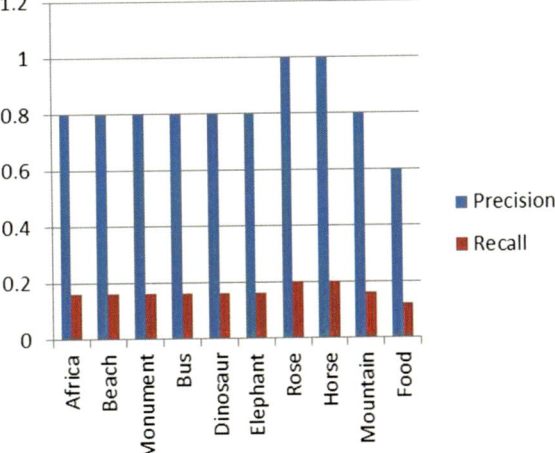

Fig. 3 Precision and recall using shape

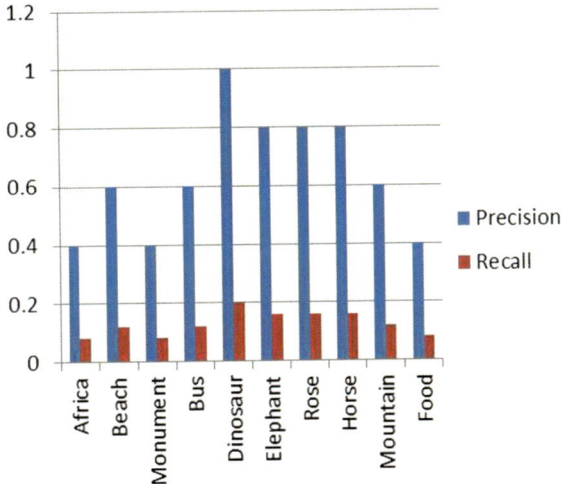

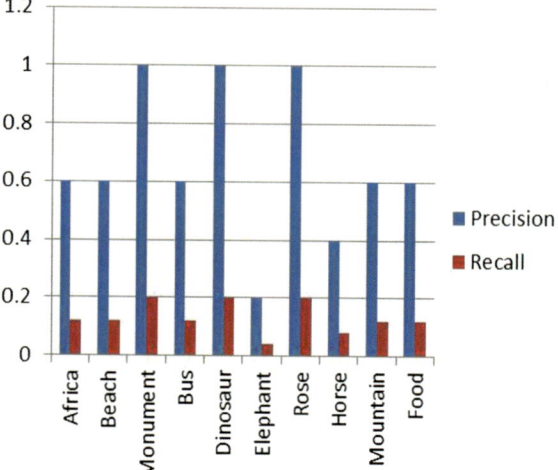

Fig. 4 Precision and recall using texture

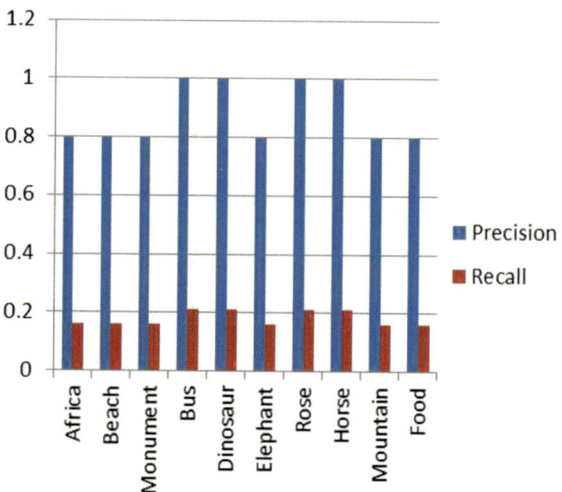

Fig. 5 Precision and recall combining all features

database using SVM. A comparative study on retrieval results was performed and the results using color, shape, and texture features in tandem with one another has given the precision value of 0.88 and recall value of 0.18 for the Wang dataset, which superseded previous research on CBIR. Future refinement of the work will involve research using a bag of words as a feature for larger datasets.

References

1. Sural S, Qian G, Pramanik S (2002) Segmentation and histogram generation using the HSV color space for image retrieval. In: IEEE international conference on image processing
2. Shirazi S, Khan NUA, Umar AI, Razzak MI, Naz S, AlHaqbani B (2016) Content-based image retrieval using texture color shape and region. Int J Adv Comput Sci Appl (IJACSA) 7(1):418–426
3. Iqbal Q, Aggarwal JK (2002) CIRES: a system for content-based retrieval in digital image libraries. In: Seventh international conference on control, automation, robotics, and vision (ICARCV), Singapore
4. Gonde AB, Maheshwari RP, Balasubramanian R (2013) Modified curvelet transform with vocabulary tree for content based image retrieval. Digit Signal Proc 23(1):142–150
5. Manjunath BS, Ohm JR, Vasudevan VV, Yamada A (2001) Color and texture descriptors. IEEE Trans Circ Syst Video Technol 11(6):703–715
6. Giri A, Meena YK (2014) Content based image retrieval using integration of color and texture features. Int J Adv Res Comput Eng Technol (IJARCET) 3(4)
7. Pandey D, Shivpratapkushwah A (2016) Review on CBIR with its advantages and disadvantages for low-level features. Int J Comput Sci Eng 4(7):161–167
8. Rosebrock A, Oates T, Caban J (2013) Ecosembles: a rapidly deployable image classification system using feature-views. In: 12th International conference on machine learning and applications
9. Prakash KSS, Sundaram RMD (2007) Combining novel features for content based image retrieval. In: EURASIP conference focused on speech and image processing
10. Aravind G, Andan HM, Singh T, Joseph G (2015) Development of biometric security system using CBIR and EER. In: IEEE international conference on communication and signal processing (ICCSP)
11. Singha M, Hemachandran K (2012) Content based image retrieval using color and texture. Signal Image Process Int J (SIPIJ) 3(1):39
12. Raghupathi G, Anand RS, Dewal ML (2010) Color and texture features for content based image retrieval. In: Second International conference on multimedia and content based image retrieval
13. Arivazhagan S, Ganesan L, Priyal SP (2006) Texture classification using Gabor wavelets based rotation invariant features. Pattern Recogn Lett 27(16):1976–1982
14. Bagyammal T, Parameswaran L (2015) Context based image retrieval using image features. Int J Adv Inf Eng Technol (IJAIET) 9(9)
15. Li J, Wang JZ (2003) Automatic linguistic indexing of pictures by a statistical modeling approach. IEEE Trans Pattern Anal Mach Intell 25(9):1075–1088. http://wang.ist.psu.edu/docs/related/

Review of Feature Extraction and Matching Methods for Drone Image Stitching

M. Dhana Lakshmi, P. Mirunalini, R. Priyadharsini and T. T. Mirnalinee

Abstract Image stitching is the process of combining multiple overlapping images of different views to produce a high-resolution image. The aerial perspective or top view of the terrestial scenes will not be available in the generic 2D images captured by optical cameras. Thus, stitching using 2D images will result in lack of information in top view. UAV (Unmanned Aerial Vehicle) captured drone images tend to have the high aerial perspective, 50–80% of overlapping of information between the images with full information about the scene. This work comprises of discussion about methods such as feature extraction and feature matching used for drone image stitching. In this paper, we compare the performance of three different feature extraction techniques such as SIFT (Scale-Invariant Feature Transform), SURF (Speeded-Up Robust Features), and ORB (ORiented FAST and rotated BRIEF) for detecting the key features. Then the detected features are matched using feature matching algorithms such as FLANN (Fast Library for Approximate Nearest Neighbors) and BF (Brute Force). All the matched key points may not be useful for creating panoramic image. Further, RANSAC (Random sample consensus) algorithm is applied to separate the inliers from the outlier set and interesting points are obtained to create a high-resolution image.

1 Introduction

Many technologies have been developed to produce high-resolution images with a wide view of the scene. However, they have limitation to capture the whole scene at an instance. Panoramic stitching is a technique used widely to overcome this problem. Image or panoramic stitching is the process of combining multiple overlapping images of different views to produce a high-resolution image. The process can be achieved in two ways: the direct pixel to pixel approach and feature-based approach [1]. The direct technique performs matches on each of the pixel to other pixel in

M. Dhana Lakshmi (✉) · P. Mirunalini · R. Priyadharsini · T. T. Mirnalinee
Department of Computer Science and Engineering, SSN College of Engineering, Chennai, India
e-mail: dhanalakshmi1605@cse.ssn.edu.in

© Springer Nature Switzerland AG 2019
D. Pandian et al. (eds.), *Proceedings of the International Conference on ISMAC in Computational Vision and Bio-Engineering 2018 (ISMAC-CVB)*, Lecture Notes in Computational Vision and Biomechanics 30,
https://doi.org/10.1007/978-3-030-00665-5_59

Fig. 1 2D images and 2D
drone images with respect to
different degree of rotation

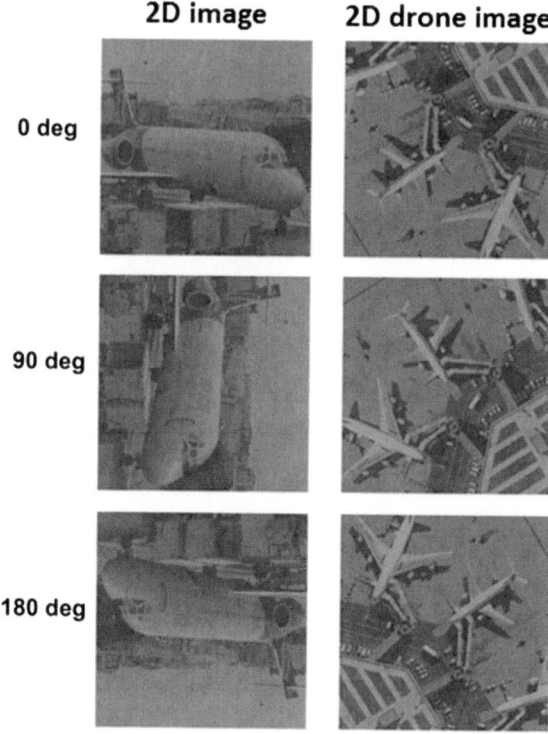

order to reduce the mismatches. The feature-based technique, extract the features and perform the matches on the extracted features. Terrestrial scenes (2D images) captured by the digital handheld cameras and smart phones are significantly affected by rotation and become much more difficult to recognize the scene of an image. As the degree of rotation for a terrestrial scene increases, the harder it becomes to identify the scene information as shown in Fig. 1. In stitching of the terrestial scenes, the aerial perspective or top view cannot be retained or captured as such with respect to 2D images. Thus, stitching using 2D images will result in lack of information in top view. When 2D images are used to generate an image stitching of a building, then the roof parts and other structures that are visible from an aerial perspective will not be captured.

But image stitching can also be done using drone images captured by Unmanned Aerial Vehicle (UAV). Drone images contain the high aerial perspective, overlapping of an image and complete information of a scene of an image than the 2D terrestrial images. These information helps the image stitching technique to build a high-quality image. The panoramic view of drone images obtained after stitching can be applied in the several fields such as movie industry, civil and mechanical industry.

2 Related Work

The image stitching has been implemented by many researchers in different ways. In [2] the authors reviewed the different feature-based image stitching methods such as SIFT, SURF, and ORB and also proposed a new method A-KAZE. It has been found among three methods that SURF-based visual odometry shows best accuracy for KITTI benchmark dataset. The proposed A-KAZE features demonstrated variation of motion estimation accuracy and computation efficiency. Parallel architecture for image fusion based on ORB feature identifier on a multicore DSP platform has been proposed in [3]. The methodology uses a position weighted image fusion algorithm to stitch the images. A panoramic image stitching technique for rotational images was proposed in [4], which used SIFT and SURF feature detector algorithms and blends the two images using DWT (Discrete Wavelet Transform) after obtaining the matches between the images. A comparison between different feature detector algorithm for the image stitching such as SIFT, SURF, ORB, FAST, Harris corner detector, FAST, MSER detector was done in [5].

3 Feature Extraction Techniques

Keypoints are the dominant features of an image. Features are contributed by the structures and the properties of an image such as color, texture, points, edges, objects, etc. Various feature extraction algorithm such SIFT, SURF and ORB were used to extract key features from drone images.

3.1 SIFT

The SIFT algorithm [6], is invariant to scaling and rotation of an image. It can also handle significant changes in illumination and efficient to run in real time. The SIFT algorithm can be achieved in four steps. It detects a scale space extrema, by generating the several octaves of the original image. Within an octave, images are progressively blurred using the Gaussian Blur operator. Two consecutive images in an octave are taken and one is subtracted from the other. Then the next consecutive pair is taken, and the process repeats. This is done for all octaves. The resultants are the approximation of Gaussian. Localization of keypoint is done by comparing neighboring pixels in the present scale, the successor scale and the predecessor scale. Keypoints can be rejected if they had a low contrast or if they were located on an edge. The assignment of orientation can be done by gradient directions and magnitudes around each keypoint. An image descriptor at each keypoint has been computed using a descriptor generator and stored as a descriptor [2, 7]. SIFT which is scale-invariant, extract the key features of image by resizing the image at different scales. So, all the important features have been extracted due to its invariant property. This property greatly enhances the degree of orientation and performs better in close range and aerial photography.

3.2 SURF

SURF algorithm is fast, robust feature detection and extraction algorithm. It approximates the Laplacian of Gaussian with Box filter and computes local extrema using second-order derivative. Implementation of Haar-like operators over an integral image can fasten the SURF in an efficient manner when compared to SIFT. For orientation assignment, it uses Haar wavelet by applying Gaussian weights for feature description. A keypoint may have connected neighbors which can be chosen and splitted into subparts. The wavelet responses are applied on each of the part to obtain feature descriptor. The features having same type of contrast can perform matching in faster rate [7, 8].

3.3 ORB

ORB is the combination of oriented FAST (Features from Accelerated Segment Test) and rotated BRIEF (Binary Robust Independent Elementary Fast) with some modification in order to enhance the performance of keypoint identification. FAST method is repeatedly applied to each layer of the pyramid in order to achieve the scale-invariant feature. The N keypoints which are computed based on the Harris corner measure are retained and uninteresting keypoints are eliminated. ORB adopts a rotation-aware variant of BRIEF. For detected keypoints, it finds patch centroids by image moments. The moments of a vector, links the keypoint's center to patch's centroid. The binary test pattern is rotated by the moments of a patch which allows feature to be in rotation-invariant form [8, 9].

4 Feature Matching Techniques

Keypoint matching is the process of finding correspondences between two images of the same scene or object. Drone images contain the 50–80% of overlapping between the images. Among the feature points extracted, the points that can be used for stitching are identified by the feature matching methods and the similar points from one image is mapped to points in the other image.

4.1 FLANN

FLANN is a library [10] of optimized algorithms that performs fast nearest neighbor search in high-dimensional features and large datasets. The FLANN uses randomized kd tree algorithm and does the priority search using k-means tree algorithm. Randomized kd-tree algorithm can search multiple trees in parallel by finding a point in the kd-tree which is nearest to a given input point [11]. The search can quickly eliminate the part of the search space by using the tree properties. Priority Search K-Means Tree Algorithm splits the data into M multiple regions and recursively partitioning each zone until the each of the leaf node has no more than M items. Then, picks up the initial centers in random manner [12].

4.2 Brute Force Matcher

BF matcher tries all possibilities and finds the best matches [10]. It takes the descriptor of a feature in an image and compared it using distance measure with all other features in the second image. The nearest point is represented as matched keypoints between the images. The steps involved in Brute force algorithm are as follows:

1. The distance between reference points (first image) and query points (second image) are calculated.
2. The calculated distances are sorted.
3. The k-smallest distances are selected as reference points.
4. The steps 1–3 is repeated for all query points.

FLANN considered to be faster since it compares between the nearest points of two images. But BF matcher compares every point in one image with all other point in another image.

5 Image Stitching Technique

An interest point (TP) is a specific location that is recognizable visually in the overlap area between two or more images. Interest points are also considered as inliers which is separated from the outliers to create an image matches.

RANSAC

RANSAC is used to separate matching keypoints (inliers) from non-matching keypoints (outliers) and create an image match. A part of data items is selected aimlessly from the input. An applicable model and the corresponding model parameters are computed using the items of this sample part. The algorithm checks the elements of the whole dataset whether it is consistent with the already represented model. A data element which does not fit the model is considered as outliers. The set of inliers obtained from the applicable model is called concord set. The RANSAC algorithm will run repeatedly until the obtained concord set in certain iteration has enough inliers (valid keypoints). This helps in generation of the interest points used to stitch the images. It also calculates the homography between the images in 3×3 matrix form. This homography matrix establishes the relationship between the two images in order to obtain the stitched image [1, 13–15].

6 Experimental Results

The sample overlapping drone images shown in Fig. 2 are taken from DJI building dataset. These images are captured by the UAV named AscTec Falcon 8 (Ascending Technologies) using Sony NEX-5 (RGB) camera.

The sample images consist of 50–80% of overlapping region. In this paper, feature-based image stitching have been done using Opencv. The methods used for the discussion are as follows:

  i. Feature extraction—SIFT, SURF and ORB
 ii. Feature matching—FLANN and BF
iii. Interest point generation—RANSAC.

A possible combination of feature extraction and feature matching methods with RANSAC has been performed and the results have been visually compared. The following combination of methods has been carried out and the results are shown below in Fig. 3.

(a) SIFT + FLANN + RANSAC
(b) SIFT + BRUTE FORCE + RANSAC
(c) SURF + FLANN + RANSAC
(d) SURF + BRUTE FORCE + RANSAC
(e) ORB + FLANN + RANSAC
(f) ORB + BF + RANSAC.

Fig. 2 Sample drone images

Fig. 3 Experimental results of various feature extraction and matching methods with RANSAC

7 Conclusions

In this paper, SIFT, SURF, and ORB algorithms are used for feature extraction as they efficiently detect the features in distinctive descriptor vector form. On the other hand, ORB extracts the features in binary string. Then feature matching algorithm FLANN and BF are performed on the extracted features. Though the features are matched between the two images of a scene, all matched features are not reliable that is they may not be an interest points used for stitching. RANSAC separates the valid inliers from outliers of matched features. It also calculates the homography between the images in 3×3 matrix form. This homography matrix establishes the relationship between the two images in order to obtain the stitched image. This work has justified that feature-based approach can be used for aerial drone images in image stitching which helps in full view of the scene.

References

1. Szeliski R (2006) Image alignment and stitching: a tutorial. Found Trends Comput Graph Vis 2(1):1–104
2. Chien HJ, Chuang CC, Chen CY, Klette R (2016) When to use what feature? SIFT, SURF, ORB, or A-KAZE features for monocular visual odometry. In: International conference on image and vision computing New Zealand
3. Wang G, Zhai Z, Xu B, Cheng Y (2017) A parallel method for aerial image stitching using ORB feature points. In: 2017 IEEE/ACIS 16th international conference on computer and information science
4. Bind VS, Muduli PR, Pati UC (2013) Robust technique for feature-based image mosaicing using image fusion
5. Adel E, Elmogy M, Elbakry M (2014) Real time image mosaicing system based on feature extraction techniques. In: 9th International conference computer engineering and systems (ICCES)
6. Lowe DG (2004) Distinctive image features from scale-invariant keypoints
7. Karami E, Prasad S, Shehata M (2017) Image matching using SIFT, SURF, BRIEF and ORB: performance comparison for distorted images
8. Rao T, Ikenaga T (2017) Quadrant segmentation and ring-like searching based FPGA implementation of ORB matching system for Full-HD video. In: 2017 Fifteenth IAPR international conference on machine vision applications, Nagoya University, Nagoya, Japan
9. Rublee E, Rabaud V, Konolige K, Bradski G (2011) ORB: an efficient alternative to SIFT or SURF. In: 2011 IEEE international conference on computer vision
10. Opencv.org. (2018) OpenCV | OpenCV. [online] Available at: http://opencv.org
11. Suju DA, Jose H (2017) FLANN: fast approximate nearest neighbour search algorithm for elucidating human-wildlife conflicts in forest areas. In: 4th International conference on signal processing, communications and networking
12. Ha Y-J, Kang H-D (2017) Evaluation of feature based image stitching algorithm using OpenCV. In: 10th International conference on (HSI) human system interactions
13. Steedly D (2005) Efficiently registering video into panoramic mosaics. Computer Science Department Faculty Publication Series 84
14. Tidke KS, Banarase SJ (2014) Review on image Mosaicing based on phase correlation and Harris Algorithm. Int J Adv Res Comput Sci Manage Stud 2(2)
15. Agarwal A, Jawahar CV, Narayanan PJ (2005) A survey of planar homography estimation techniques

An ANN-Based Detection of Obstructive Sleep Apnea from Simultaneous ECG and SpO$_2$ Recordings

Meghna Punjabi and Sapna Prabhu

Abstract Obstructive sleep apnea (OSA) is one of the most common sleep disorders characterized by a disruption of breathing during sleep. This disease, though common, goes undiagnosed in most cases because of the inconvenience, cost, and/or unavailability of opting for polysomnography (PSG) and a sleep analyst. Many researchers are working on devising an unsupervised, cost-effective, and convenient OSA detection methods which will aid the timely diagnosis of this sleep disorder. Commonly used signals to detect OSA are ECG, EEG, pulse oximetry (SpO$_2$), blood oxygen saturation (SaO$_2$), and heart rate variability (HRV). In this work, an attempt to detect the OSA using simultaneously acquired ECG and SpO$_2$ signals has been presented. Various features from the RR intervals of ECG, and a couple of features—-namely, CT90 and delta index—from the SpO$_2$, were extracted as indicators of OSA. The features were then fed to a trained artificial neural network (ANN) which classified the signals as OSA positive or OSA negative. The proposed technique boasts a very high accuracy of 98.3%, which is superior to other competing techniques reported so far.

1 Introduction

A balanced sleep is essential for normal functioning of the human body. Sleep disorders disrupt the ability to sleep well and can potentially cause physical, emotional, and psychological damage. Among the numerous sleeping disorders currently known, the most common ones are insomnia, narcolepsy, sleep apnea, and restless leg syndrome. Sleep apnea is characterized by disruption in a person's breathing during sleep, breathing might get shallow, or even stop for few seconds. The breaks during

M. Punjabi (✉) · S. Prabhu
Department of Electronics Engineering, Fr. Conceicao Rodrigues College of Engineering,
University of Mumbai, Bandra 400050, Mumbai, India
e-mail: meghna_1911@hotmail.com

S. Prabhu
e-mail: sapna@frcrce.ac.in

© Springer Nature Switzerland AG 2019
D. Pandian et al. (eds.), *Proceedings of the International Conference on ISMAC
in Computational Vision and Bio-Engineering 2018 (ISMAC-CVB)*, Lecture Notes
in Computational Vision and Biomechanics 30,
https://doi.org/10.1007/978-3-030-00665-5_60

breathing called apnea could be of different frequency and duration. If left untreated could cause repeated interruption while sleeping. There are three types of sleep apnea: obstructive (OSA), central (CSA), and a combination of the two called mixed. The most common form is the first type OSA found in 2–4% of middle-aged adults and 1–3% of children. OSA is caused by a collapse of the upper respiratory airway. The CSA occurs due to inhibited or absent respiratory drive. In general, the mixed type is found, and the occurrence of CSA only cases is quite rare. Although they are commonly present, diagnosing can be challenging. The reason behind undiagnosed cases in sleep apnea is due to the inconvenience, cost, and unavailability of testing. Current diagnosis for testing sleep apnea is done using polysomnography (PSG). PSG is a standard test for all sleeping disorders, the breath airflow, respiratory movements, oxygen saturation, body position, electroencephalogram (EEG), electrooculogram (EOG), electromyogram (EMG), and electrocardiogram (ECG).

There are different methods proposed for diagnosing sleep apnea till date. Widely used statistical parameters for detecting sleep apnea are nasal airflow, thorax and abdomen effort signals, acoustic speech signal, oxygen saturation, electrical activity of the brain (EEG), and electrical activity of the heart (ECG).

The detection technique for sleep apnea using thoracic and abdominal signals was reported by Ng et al. [13]. They used mean absolute amplitude analysis and showed that the combination of the thoracic and abdominal signal had the best overall and individual performance when compared to the separate performances of the thoracic and abdominal signal.

Goldshtein et al. [9] hypothesized that the patients with OSA would exhibit different speech signal properties than those without OSA. They had studied acoustic speech features of 93 subjects recorded using text-dependent speech protocol and a digital audio recorder before doing polysomnography study. With the help of a Gaussian mixture model-based system, they had developed a model and classified based on the features like vocal tract length and linear prediction coefficients. It was concluded that the acoustic features from speech signals during wakefulness can detect OSA patients with good specificity and sensitivity.

Another method of distinguishing OSA positive and negative patients was done using nocturnal oximetry which was superior to polysomnography with respect to its low cost and simplicity. The nocturnal oximetry assesses nonlinear analysis of blood oxygen saturation (SaO_2). The oximetric indices evaluated were cumulative time spent, oxygen desaturation, and delta index. The result of the nonlinear analysis of SaO_2 signals [4] from nocturnal oximetry suggested that it could lead to useful information in OSA diagnosis.

Another method of identifying sleep apnea episodes was reported by Lin et al. [11]. This method uses artificial neural networks to extract EEG signal characteristics of sleep apnea episodes. They used wavelet neural network as a sleep apnea identification system. They were able to achieve a sensitivity of approximately 69.6% and specificity of 44.4%.

There have been several methods for detection of sleep apnea through heart rate variability analysis. For example, Manrique et al. [16] used time–frequency distribution to extract dynamic features for detecting sleep apnea that are recorded from

ECG signals during sleep. They proposed a method that can be used as a simple diagnostic tool for sleep apnea with an accuracy of 92.7% in one-minute intervals. Another work on detecting OSA using ECG [17] in which analysis and annotations are based on spectral components of heart rate variability, frequency analysis performed using Fourier and wavelet transformation with appropriate application of Hiber transform were able to achieve a sensitivity of 90.8% and a specificity of 92.7% on the learning set. Mendez et al. [12] showed that a bivariate autoregressive model used to evaluate beat-by-beat power spectral density of HRV and R peak area had higher than 85% accuracy in the classification result. The model is based on extraction of signal ECG signal characteristics. An improved technique based on automated classification algorithm was explored in Almazaydeh et al. [1]. This technique processed short duration epochs of ECG data. The automated classification algorithm was based on support vector machines (SVM). The resultant automated classification system developed showed a high degree of accuracy approximately 96.5% in recognizing epochs of sleep disorder.

Xie et al. [18] explored various features in order to find an efficient alternative for polysomnography (PSG). They investigated real-time sleep apnea and hypopnea syndrome (SAHS) based on ECG and SpO_2 signals both separately and combined. They showed that the SpO_2 with the proposed features outperformed ECG in terms of diagnostic capability.

In this work, we are proposing a novel ANN-based technique for the detection of OSA using the features derived from simultaneously acquired ECG and SpO_2 signals, which is an improvement on the previously reported techniques.

2   Methods

An attempt to detect OSA using simultaneous ECG and SpO_2 signals acquired from the patients during the sleeping phase is presented in this work. In short, the acquired signals were divided into segments of 1 min duration. For each of those segments, the features obtained from the ECG and SpO_2 were combined to form a feature vector, which was given as the input to an ANN. The ANN is trained to classify the feature vector into OSA positive or OSA negative. Using the annotated data obtained from the PhysioNet Databank [15], the performance of the proposed OSA detection technique was evaluated. Each of the steps involved in the method is described in the subsections below.

2.1   Preparation of the Dataset

The "Apnea-ECG" database [15] of "PhysioNet" [8] contains recordings from 70 patients—which include a continuous digitized ECG signal, a set of apnea annotations derived by human experts on the basis of simultaneously recorded respiration

Fig. 1 Schematic diagram
of the OSA detection system

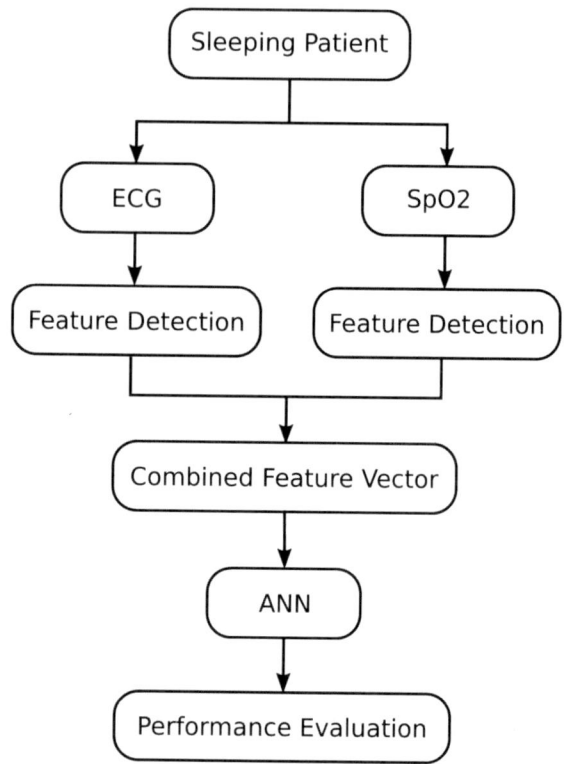

and related signals, and a set of machine-generated QRS annotations in which all
beats regardless of type have been labeled normal. Among those 70 recordings,
eight recordings (a01 through a04, b01, and c01 through c03) are accompanied by
four additional signals (Resp C and Resp A, chest and abdominal respiratory effort
signals obtained using inductance plethysmography; RespN, oronasal airflow measured using nasal thermistors; and SpO_2, oxygen saturation). As our technique makes
use of simultaneous ECG and SpO_2 recordings, we restrict our study to these eight
recordings.

The ECG and SpO_2 records are present in recording Channels 1 and 5, respectively. Annotations done by expert physicians are available for each of the shortlisted
records. These annotations are done at the beginning of every minute of the recording.
As the sleeping sessions are roughly 8 h long, there is around 500 min of recordings
available. Every minute of the recordings was treated as an individual signal segment
(Fig. 1).

The annotations which are present at the beginning and end of the segments were
used to label the segments as OSA positive and OSA negative. The following rules
were used:

1. If the segment is bounded by two "N" annotations, it represents a normal segment or N-Seg.
2. If the segment is bounded by two "A" annotations, it represents an apnea segment or A-Seg.
3. If the segment is bounded by one "N" annotation and one "A" annotation, the status of the patient is ambiguous during this segment and hence these segments were not used in the study.

Once the segments were identified, the key features of the segments which aid the detection of OSA were identified. The features used in this study are described in the following subsection.

2.2 Feature Detection

For every segment of sleep, there is a corresponding ECG and SpO$_2$ segment. The features of ECG and SpO$_2$ were separately evaluated.

The ECG features used were all based on the variations in RR interval observed during OSA [1, 5, 19]. As the R peaks of the ECG signals from Apnea-ECG database were not annotated, those were detected using the well-established Pan-Tompkins algorithm [14]. Once the R peak locations were identified, the RR intervals for the ECG segment were obtained. The features extracted from the series of RR intervals corresponding to the segment of interest are given below:

1. Mean RR interval,
2. Standard deviation (SD) of RR interval,
3. NN50 variant 1:The number of RR interval pairs in which the first interval exceeds the second by at least 50 ms,
4. NN50 variant 2: The number of RR interval pairs in which the second interval exceeds the first by at least 50 ms,
5. pNN50 variant 1: NN50v1 represented as a fraction of total number of RR intervals in the segment,
6. pNN50 variant 2: NN50v2 represented as a fraction of total number of RR intervals in the segment,
7. SDSD: Defined as "the standard deviation of the differences between the adjacent RR intervals" [1],
8. RMSSD: Defined as "the square root of the mean of the sum of the squares of differences between adjacent RR—intervals" [1],
9. Median of the RR intervals,
10. Interquartile range of the RR intervals,
11. MADV: Mean absolute deviation values, defined as "the mean of absolute values obtained by the subtraction of the mean RR-interval values from all the RR-interval values in an epoch" [1], and

12. Spectral Ratio: defined as the ratio of the area under the power spectrum of the RR interval between the frequency bands (0.02–0.1) and (0.01–0.02), relative to the Nyquest frequency [6].

The features from the SpO_2 segments suitable for OSA detection were defined in the earlier works [2, 3]. Two of those features, namely CT90 and delta index, were used for the work presented here. These features were defined as follows:

1. CT90: The cumulative time spent below 90% oxygen saturation level is quantified by the feature CT90. In [3], CT90 is defined as the percentage of time during which the SpO_2 value is below 90%.
2. Delta Index: The factor delta index quantifies the SpO_2 variability. This feature was originally defined in [10] and also used by recent research works such as [2] and [7]. The metric delta index is defined as the average of absolute differences of the mean SpO_2 between successive 12-s intervals [7].

2.3 Preparation of the Feature Vector

Once these features values were evaluated from the ECG and SpO_2 segments, they were combined to form a feature vector. The feature vector is a series of 14 values, each of which represents individual features—12 ECG features and 2 SpO_2 features—in order. It is assumed that this feature vector contains all the relevant information from the raw ECG and SpO_2 segments required for the detection of the OSA.

2.4 Detection of OSA Using ANN

The individual features extracted from the ECG and SpO_2 segments vary in their values for normal and apneatic ECG/SpO_2 records. The extent to which each of the features expresses their variation sensitive to the presence or absence of OSA is different for different features. A machine learning algorithm was designed to classify the sleep segment as Normal or Apnea using a neural network classifier which learns the appropriate weights of each of the features by making use of a training set.

For the application presented in this work, the minimum number of neurons needed in the input is the length of the feature vector, i.e., 14. The number of output layer neurons is also fixed, which is unity, as a Boolean output (0 or 1) is required which can be provided by a single neuron. Our choice in deciding the network topology lies in the hidden layer, where the number of layers and the number of neurons in each layer can be decided by the user. We have chosen for one hidden layer with 10 neurons. The Neural Fit Tool (NFTOOL) of the NN toolbox available in the Matlab platform was used to create the neural network design.

The individual neuron in the network was assigned a log-sigmoid (logsig) transfer function. The inputs to the neuron are scaled using a weight assigned to the neuron. The process of optimizing these weights so that the network produces desirable outputs for all input vectors is the training process. The Levenberg–Marquardt algorithm (LMA) was used as the training function, which computes the optimized neuron weights which produce the least mean square error (LMS error) between the NN outputs and the TARGETS, where TARGETS are the desired output for each vector in the training set.

Once the training is completed, the network can be used to detect the sleep apnea from a pair of ECG and SpO$_2$ segments by providing the 14-element feature vector as the input. The logsig transfer function of the output neuron provides a real number between 0 and 1. Hence a rounding off operation is carried out on the ANN output for the ease of apnea detection, which makes the output Boolean, representing OSA positive and OSA negative detection.

2.5 Performance Evaluation

Multiple test sets were prepared separately from different patients. Individual samples in these datasets were taken through the OSA detection procedure explained in the previous section. From the ground truth annotations available from the database and the outputs obtained from the trained ANN, the performance indicators such as sensitivity, specificity, and accuracy were measured. The confusion matrix which displays detailed performance indicators were also obtained for each of the datasets. The results thus obtained were then compared with the results from the similar work found in the literature.

3 Results

Eight datasets, one each from individual patients, were prepared for the training of the ANN. Each dataset consists of multiple ECG—SpO$_2$ segment pairs of 1 min duration. Each segment pair was considered as one training sample. The number of samples was limited to 200 from each patient, which were randomly chosen from the entire recording. The details of the datasets used are given in Table 1. A combined dataset containing the pool of all patient-specific datasets was used to train the ANN.

The samples present in the above data pool were randomly grouped into three categories in the following manner:

1. Training Set: 70% of the entire samples are chosen to represent the Training Set. This was the set used to train the ANN parameters. For each segment-pair in the Training Set, a feature vector was constructed as mentioned in the methods. These feature vectors were given as the INPUTS to the ANN along with the

Table 1 Details of the data sets which were used to train the ANN in detecting the presence of OSA

Dataset	Patient	Total segments	OSA +ve	OSA −ve
Set-1	a01er	150	135	15
Set-2	a02er	150	75	75
Set-3	a03er	200	100	100
Set-4	a04er	200	165	35
Set-5	b01er	63	13	50
Set-6	c01er	200	0	200
Set-7	c02er	200	0	200
Set-8	c02er	200	0	200

corresponding annotations as TARGETS which indicate the desired output of the network.

2. Validation Set: 15% of the samples were chosen to represent the Validation Set, which is used for cross-validation. There is a possibility that the NN overfit the training set and generate 100% accuracy for that training set. A validation set was used for avoiding this overfit. If the ANN gets overfit to the training data, it gives a very low accuracy for the validation dataset. In such cases, the training process was repeated with a different set of values for the NN weights. This process was repeated until a similar performance was obtained for training dataset and the validation dataset.

3. Test Set: The final 15% of the samples were used as test data. After the entire training process, these data will be used to evaluate the efficiency of the apnea detection.

The training process was repeated until a maximum accuracy of 94.5% was obtained for the Test Set. In order to compare the performance of our technique to previously published techniques, a test dataset used in the work of Almazaydeh et al. [1] was chosen for detailed evaluation of the proposed technique. Almazaydeh dataset was taken from recording a03er. For regular (OSA −ve) samples, the recording stretch. The confusion matrix obtained by the proposed method. A test data set containing 58 samples was used. TP, TN, FP, and FN represent True Positives, True Negatives, False Positives, and False Negatives, respectively. From 2:27:00.000 to 2:57:00.000 was used. For apnea (OSA +ve) samples, the stretch from 3:06:00.000 to 3:36:00:000 was used. As our samples are of 1 min duration, we have total 58 samples—29 each for OSA +ve and OSA −ve samples. For this dataset, the proposed technique was able to yield 98.3% accuracy with 100% specificity and 96.6% sensitivity (Fig. 2) which is superior to the accuracy achieved by any other technique reported in the literature so far. Comparison of the proposed technique with similar prominent works published earlier are shown in Table 2.

Fig. 2 The confusion matrix obtained by the proposed method. A test data set containing 58 samples was used. TP, TN, FP, and FN represent True Positives, True Negatives, False Positives, and False Negatives, respectively

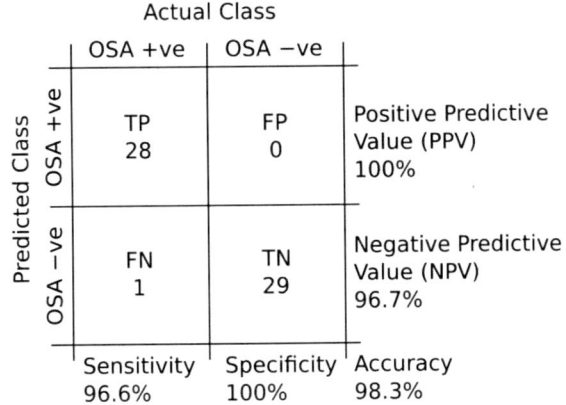

Table 2 Comparison of performances in OSA detection approaches

Method by	Refs.	Signal(s) used	Performance (%)		
			Se	Sp	Acc.
Schrader et al.	[17]	Fourier and wavelet transformation of HRV	90.8	NA	NA
Chazal et al.	[5]	Measure of minutes of sleep disordered respiration	NA	NA	91
Lin et al.	[11]	EEG	69.6	44.4	NA
Alvarez et al.	[3]	SaO$_2$	90.1	82.9	NA
Mendez et al.	[12]	Bivariate autoregressive model of HRV	NA	NA	85
Alvarez et al.	[4]	SaO$_2$ and EEG	91	83.3	88.5
Manrique et al.	[16]	ECG	NA	NA	92.7
Yilmaz et al.	[19]	RR interval-based classification	NA	NA	89
Xie et al.	[18]	SpO$_2$ and ECG	79.7	85.9	84.4
Almazaydeh et al.	[1]	ECG	92.9	100	96.5
Proposed		SpO$_2$ and ECG	96.6	100	98.3

4 Conclusions

This work proposes an ovel method for the detection of OSA using the features detected from simultaneously acquired ECG and SpO$_2$ recordings from apnea patients during sleep. A machine learning algorithm based on ANN was designed and trained to detect OSA using the shortlisted features. The performance analysis of the technique shows that it is superior to similar algorithms designed so far for unsupervised OSA detection.

The two signals used in the proposed techniques are ECG and SpO$_2$—two non-invasive, cheap, and commonly available instruments in hospitals and affordable in home. The patients can easily acquire these signals and, with the help of the proposed technique, can get the diagnosis done for OSA. In future, this technique can be incorporated into a real-time ECG + SpO$_2$ acquisition system and aid the process of sleep analysis.

Acknowledgements We thank our institute, Fr. Conceicao Rodrigues College of Engineering, for providing all possible supports to this work.

References

1. Almazaydeh L, Elleithy K, Faezipour M (2012) Obstructive sleep apnea detection using SVM-based classification of ECG signal features. In: 2012 Annual international conference of the IEEE engineering in medicine and biology society (EMBC). IEEE, pp 4938–4941
2. Almazaydeh L, Faezipour M, Elleithy K (2012) A neural network system for detection of obstructive sleep apnea through SpO$_2$ signal. Editorial Preface 3(5)
3. Alvarez D, Hornero R, Abasolo D, Del Campo F, Zamarron C (2006) Nonlinear characteristics of blood oxygen saturation from nocturnal oximetry for obstructive sleep apnoea detection. Physiol Meas 27(4):399
4. Alvarez D, Hornero R, Marcos JV, del Campo F, Lopez M (2009) Spectral analysis of electroencephalogram and oximetric signals in obstructive sleep apnea diagnosis. In: Annual international conference of the IEEE engineering in medicine and biology society. EMBC 2009. IEEE, pp 400–403
5. de Chazal P, Penzel T, Heneghan C (2004) Automated detection of obstructive sleep apnoea at different time scales using the electrocardiogram. Physiol Meas 25(4):967
6. Drinnan M, Allen J, Langley P, Murray A (2000) Detection of sleep apnoea from frequency analysis of heart rate variability. In: Computers in Cardiology 2000. IEEE, pp 259–262
7. Garde A, Dehkordi P, Wensley D, Ansermino JM, Dumonf GA (2015) Using oximetry dynamics to screen for sleep disordered breathing at varying thresholds of severity. In: 2015 23rd European signal processing conference (EUSIPCO). IEEE, pp 439–443
8. Goldberger AL, Amaral LA, Glass L, Hausdorff JM, Ivanov PC, Mark RG, Mietus JE, Moody GB, Peng CK, Stanley HE (2000) Physiobank, physiotoolkit, and physionet. Circulation 101(23):e215–e220
9. Goldshtein E, Tarasiuk A, Zigel Y (2011) Automatic detection of obstructive sleep apnea using speech signals. IEEE Trans Biomed Eng 58(5):1373–1382
10. Levy P, Pepin JL, Deschaux-Blanc C, Paramelle B, Brambilla C (1996) Accuracy of oximetry for detection of respiratory disturbances in sleep apnea syndrome. Chest 109(2):395–399

11. Lin R, Lee RG, Tseng CL, Zhou HK, Chao CF, Jiang JA (2006) A new approach for identifying sleep apnea syndrome using wavelet transform and neural networks. Biomed Eng Appl Basis Commun 18(03):138–143

12. Mendez MO, Ruini DD, Villantieri OP, Matteucci M, Penzel T, Cerutti S, Bianchi AM (2007) Detection of sleep apnea from surface ECG based on features extracted by an autoregressive model. In: 29th Annual international conference of the IEEE engineering in medicine and biology society. EMBS 2007. IEEE, pp 6105–6108

13. Ng AS, Chung JW, Gohel MD, Yu WW, Fan KL, Wong TK (2008) Evaluation of the performance of using mean absolute amplitude analysis of thoracic and abdominal signals for immediate indication of sleep apnoea events. J Clin Nurs 17(17):2360–2366

14. Pan J, Tompkins WJ (1985) A real-time QRS detection algorithm. IEEE Trans Biomed Eng 32(3):230–236

15. Penzel T, Moody GB, Mark RG, Goldberger AL, Peter JH (2000) The apnea-ECG database. In: Computers in cardiology 2000. IEEE, pp 255–258 (2000)

16. Quiceno-Manrique A, Alonso-Hernandez J, Travieso-Gonzalez C, Ferrer-Ballester M, Castellanos-Dominguez G (2009) Detection of obstructive sleep apnea in ECG recordings using time-frequency distributions and dynamic features. In: Annual international conference of the IEEE engineering in medicine and biology society. EMBC 2009. IEEE, pp 5559–5562

17. Schrader M, Zywietz C, Von Einem V, Widiger B, Joseph G (2000) Detection of sleep apnea in single channel ECGs from the PhysioNet data base. In: Computers in Cardiology 2000. IEEE, pp 263–266 (2000)

18. Xie B, Minn H (2012) Real-time sleep apnea detection by classifier combination. IEEE Trans Inf Technol Biomed 16(3):469–477

19. Yilmaz B, Asyali MH, Arikan E, Yetkin S, Ozgen F (2010) Sleep stage and obstructive apneaic epoch classification using single-lead ECG. Biomed Eng Online 9(1):39

Design of an Image Skeletonization Based Algorithm for Overcrowd Detection in Smart Building

R. Manjusha and Latha Parameswaran

Abstract Crowd analysis has found its significance in varied applications from security purposes to commercial use. This proposed algorithm aims at contour extraction from skeleton of the foreground image for identifying and counting people and for providing crowd alert in the given scene. The proposed algorithm is also compared with other conventional algorithms like HoG with SVM classifier, Haar cascade and Morphological Operator. Experimental results show that the proposed method aids better crowd analysis than the other three algorithms on varied datasets with varied illumination and varied concentration of people.

1 Introduction

Video analytics has gained its prevalence over recent years due to its flexibility and reduction in cost for the overall system. People counting and crowd analysis are active areas of research with the rapid increase in surveillance videos. Crowd alert is inevitable in public scenarios which demands crowd-induced disasters. Several disasters like the recent stampede at Mumbai's Elphinstone Road station on September 29, 2017, in which 22 were dead and several were injured, happens due to lack of proper crowd analysis and crowd alert. Crowd alert in smart building aids in providing emergency evacuation, analyzing abnormal crowd at unusual time, etc.

This paper focuses on counting the number of people to estimate and provide an alert if it is overcrowded, using skeleton-based contour extraction and compare it with three algorithms like HOG with SVM, Haar Cascade, and Morphological operator based algorithm. The paper is organized as follows: Sect. 2 discusses the existing algorithms for people counting and crowd analysis; Sect. 3 elaborates on the

R. Manjusha (✉) · L. Parameswaran
Department of Computer Science and Engineering, Amrita School of Engineering, Amrita
Vishwa Vidyapeetham, Coimbatore, India
e-mail: r_manjusha@cb.amrita.edu

L. Parameswaran
e-mail: p_latha@cb.amrita.edu

© Springer Nature Switzerland AG 2019
D. Pandian et al. (eds.), *Proceedings of the International Conference on ISMAC
in Computational Vision and Bio-Engineering 2018 (ISMAC-CVB)*, Lecture Notes
in Computational Vision and Biomechanics 30,
https://doi.org/10.1007/978-3-030-00665-5_61

proposed method and the three conventional methods with which the proposed work is compared; Sect. 4 provides the comparison and analysis of the proposed method with the other three algorithms.

2 Related Works

Many authors have published work on people counting. A brief literature is presented here.

In [1], Scale-Invariant Feature Transform (SIFT) was used for blob analysis. Complex wavelet transform (CWT) and dimensionality reduced Optical flow was applied to the detected blobs for estimating the movement direction and speed of the key points. If the result of CWT and Optical flow contains the blob, it is treated as a person. The people map generated was then applied with Kalman filter for tracking and thus occlusion was dealt. In [2] authors used head shoulder region for identification of people; Histogram of Gradients (HoG) combined with Completed Local Binary Pattern (CLBP) features which were provided to Support Vector Machine (SVM) for identifying people.

A hybrid face detector [3] a combination of Normalized Pixel Difference (NPD), Haar classifier, and Haar classifier for profile face was used. If all the three detectors, NDP which detects faces at different scales and sizes, Haar cascade and Haar cascade for profile faces which works well for profiled and rotated faces identifies a region as a face then it is considered to be a face. False positives are further eliminated based on the area of the bounding box. In [4] the camera was placed overhead and people counting has been done based on the headcount. ViBe algorithm has been used for foreground extraction followed by closing and opening for removal of noise and holes in the extracted foreground. Local Binary Pattern (LBP) based Adaboost classifier has been used for detecting human heads. Head tracking has been further done based on mean shift algorithm. In [5] authors used Mixed Gaussian background model (MGM) for extraction of foreground, followed by dilation and erosion for removal of noise and blur. Histograms of Gradient (HoG) features were extracted and Principal Component Analysis (PCA) has been applied on HoG features to reduce the dimension of the feature vector. Support Vector Regression (SVR) has been used for training and identification of people.

In [6] authors have developed a Frame differential method along with dilation and erosion has been used for extracting foreground of each frame. Each connected component in the resulting foreground was treated as a group. Each frame has been divided into blocks and the area, the number of edge pixels and number of SURF points of each group falling in each block has been estimated. Support Vector Regression algorithm has been used for training and for predicting the number of people in each frame. A contracted graph with an entry point of people as a source (S) and the exit point of people as a sink (T) for a sequence of frames in a video has been constructed. The prediction of the regression algorithm has been improvised by using an integer quadratic programming model on the contracted graph.

Lidar has been placed vertically in [7] for collecting the 2D points which indicated the contours of each frame. These 2D points obtained from various frames were superimposed to identify contour over time. RANdom SAmple Consensus (RANSAC) algorithm has been further used for registration of these images which in turn creates a point cloud. Since the pedestrians too close to each other may be grouped into one cloud, the head shoulder information has been used for isolating people. SVM has been further used for detection and classification of people. Perng et al. [8] aim at counting the number of people in the bus at a point of time, based on the entry and exit count at each stop. Background subtraction intending to identify the moving objects as foreground has been done based on frame differencing. Erosion and dilation have been done to remove noise and broken edges. Connected component analysis based on area, aspect ratio, and velocity was used as a means to identify people. A virtual counting line and a counting zone have been set for tracking and counting the number of people who enter and exit the bus.

In [9] the authors focus on counting the number of people from ATM surveillance video. Static background has been subtracted from each image frame and based on a threshold moving target has been identified. Gaussian Model has been further used for extracting foreground pixels from the thresholded image. The resulting image was eroded and then dilated using a 2×7 structuring element. A rectangle was drawn for each contour and to identify people close to each other, the height and width of the rectangle have been considered. SURF (Speeded-Up Robust Features) has been used by [10] for identifying the interest points on blob-like structures throughout the image. From the set of interest points, those which corresponds to moving objects indicates a person. Adaptive Rood Pattern Search (ARPS) algorithm has been used for identifying those motion vectors. Training and recognition of people have been done using linear regression. In [11] an algorithm using head-based people counting has been implemented using Statistically Effective Multi-scale Block Local Binary Pattern (SEMB-LBP) features and Adaboost classifier. The region of interest which was 0.7–1.3 times larger than the human head has been initialized for counting the heads of people. A modified version of compressive tracking (CT) tracker based on model matching was used for tracking the detected head in the region of interest. In [12] saliency map had been generated based on the color and luminance information of the image. Object segmentation had been done based on dynamic mode decomposition. In [13] based on facial features like eyes, nose, and mouth of the person and using K-Nearest Neighbor classifier people identification had been done. A summary of literature helps to infer that people counting and overcrowding identification is still an open research.

3 Proposed Work

The proposed algorithm counts the number of people and with a user-defined threshold identifies overcrowd, if any, in each frame of a video using skeleton-based contour extraction. The proposed algorithm is compared with the standard existing algorithms

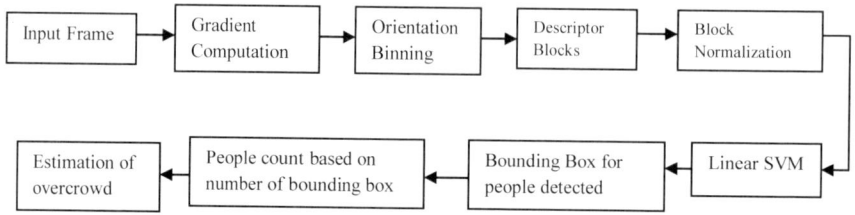

Fig. 1 Architecture of HOG with SVM

like HoG with SVM, Haar cascade, and morphological operator based algorithm. Existing algorithms and by the proposed algorithm for people counting overcrowding identification is presented here.

3.1 HOG with SVM

Histogram of oriented gradients (HOG) descriptor, divides the image (each frame in a video) into cells and histogram of gradient directions are computed for every pixel in each cell. The local histogram intensities of cells in a block are further normalized. Figure 1 shows the architecture of HOG descriptor with SVM algorithm [14]:

Algorithm:

a. Gradient Computation:

The first step involved is the computation of Gradients for each pixel in the frame. Gradients are computed using the standard derivative mask $[-1\ 0\ 1]$, $[-1\ 0\ 1]^T$ in one direction or both horizontal and vertical direction.

b. Orientation Binning:

The next step is to calculate the cell histograms. Histogram contains channels ranging from $0°$ to $180°$ which corresponds to unsigned gradients. Depending on the gradient computed for each pixel, the corresponding histogram channels vote is updated.

c. Descriptor Blocks:

Cells are further grouped into larger blocks that are spatially connected which are then normalized. The normalized cell histograms are concatenated to form the HOG descriptor for each block.

d. Block Normalization:

The blocks are normalized using L1 or L2 normalization. This Normalized HOG descriptor is provided to SVM classifier for differentiating humans from non-humans.

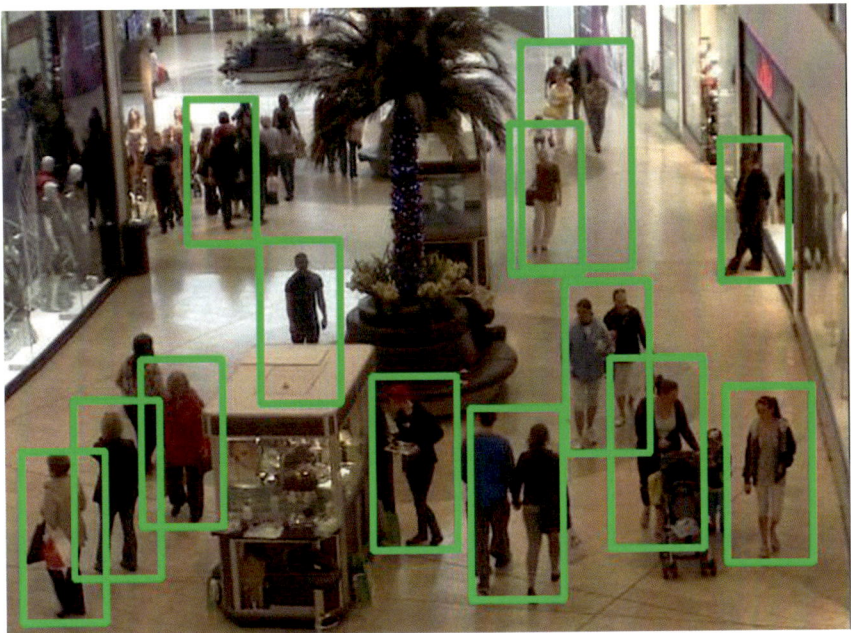

Fig. 2 HOG with SVM on Mall dataset

e. SVM Classifier:

HOG Descriptors is further used for detecting human which aids in people counting by providing those features to SVM (Support Vector Machine). HOG features are extracted both from positive and negative samples in the dataset. SVM is trained with all the samples and from each input video frame, the humans are detected based on the HOG descriptor of that frame and the trained samples. A rectangular bounding box is drawn around each detected human and is further used for counting the number of people in each frame. Based on user-defined threshold overcrowd is alerted.

Figures 2 and 3 shows the result of HOG with SVM on Mall dataset and Dataset from a surveillance camera of our University.

3.2 Haar Cascade

Haar feature-based cascade classifiers [15] represent each frame of a video as an intermediate representation known as an integral image. Integral image is of the same size as the original image. The sum of all pixels on the up-left region of the original image constitutes each element in an integral image. $Sum = I(C) + I(A) - I(B) - I(D)$ for a particular position as shown in Fig. 4.

Fig. 3 HOG with SVM on dataset from surveillance camera of our university

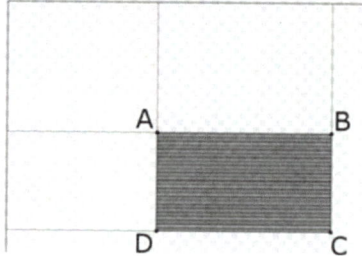

Fig. 4 Integral image

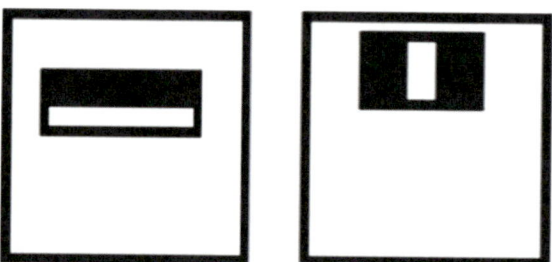

Fig. 5 Best features for face detection

From the integral image the features shown in Fig. 5 are extracted which suits best for detecting faces. The first feature is based on the fact that region of eyes is mostly darker than the cheeks and the nose region. The second feature is supported by the fact that eyes are darker than the nose bridge.

These features are extracted from the training images and an Adaboost classifier which is a combination of several weak classifiers classifies the face and the non-face

Fig. 6 Haar cascade on Mall dataset

Fig. 7 Haar cascade on dataset from surveillance camera of our campus

images. A cascade of classifiers has been used where only if the first classifier returns a positive result, the image shall be passed on to the second classifier, otherwise discarded and so on, thus minimizing false negatives.

Using the pre-trained Haar classifier for upper body, people are detected, which aids in people counting and overcrowd detection. Figure 6 shows the people detected from the Mall dataset and Fig. 7 shows the people detection on the dataset from the surveillance camera of our university.

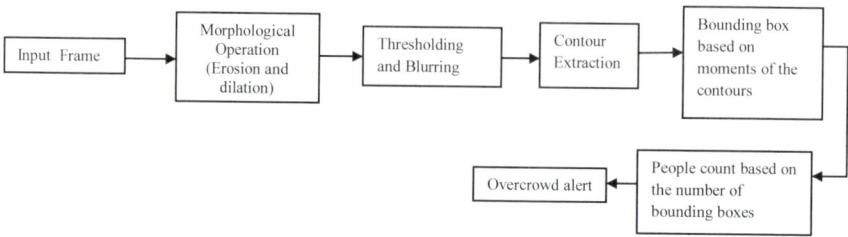

Fig. 8 Architecture of morphological operator based people counting

Fig. 9 Morphological operator based people counting on Mall dataset

3.3 *Morphological Operators*

Gaussian Mixture Model is used to separate the background from each frame of the video. On the foreground image of each frame, a series of erosion followed by a series of dilation are applied. Resultant images are threshold followed by median blurring. Contours are extracted from the blurred image by compressing the horizontal, vertical, and diagonal segments and leaving behind only the endpoints based on chain codes and only the outer contours are extracted. Based on the moments extracted from the contours, the rectangular bounding box is drawn on each region identified to be the person in the image. Crowd alert is provided based on the user-defined threshold. Figure 8 shows the architecture of morphological operators based people counting.

Figure 9 shows the result of morphological operator based people counting on Mall dataset and Fig. 10 shows the result of morphological operator based people counting on surveillance dataset of our university.

Fig. 10 Morphological-based people counting on dataset from surveillance camera of our campus

3.4 Proposed Skeletonization Based People Counting

Each frame of the video is applied with a Mixture of Gaussian Model for background separation. The extracted foreground image is converted to binary and then skeletonization is done on the binary image. An opening morphological operator which is erosion followed by dilation, where, erosion aids in removing smaller elements and dilation helps in restoring the shape of the remaining skeleton is then applied on the skeleton. The resultant skeleton is further smoothened and edges are detected using a canny edge detector. From the resulting skeleton, contours are extracted based on the chain codes by compressing the horizontal, vertical, and diagonal segments and leaving behind only the endpoints. Moments are calculated for the extracted contours and a bounding box is drawn on each person detected. The number of bounding box corresponds to the number of people in each frame. Based on the user-defined threshold, overcrowding is estimated. Figure 11 shows the architecture of people counting based on skeletonization.

Figure 12a shows the skeleton of the 1997th frame of the Mall dataset. Figure 12b shows the Skeletonization-based people counting algorithm applied on the video and the result on the 1997th frame of the Mall dataset.

Figure 13a shows the skeleton of 223rd frame and Fig. 13b shows the result of Skeletonization-based people counting algorithm applied on the same frame of the video from the surveillance camera of our university.

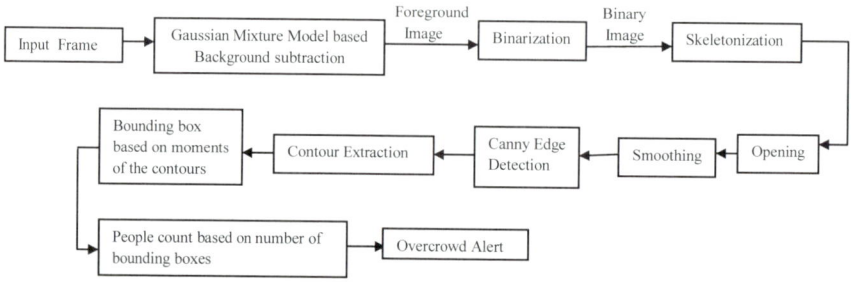

Fig. 11 Architecture of people counting based on skeletonization

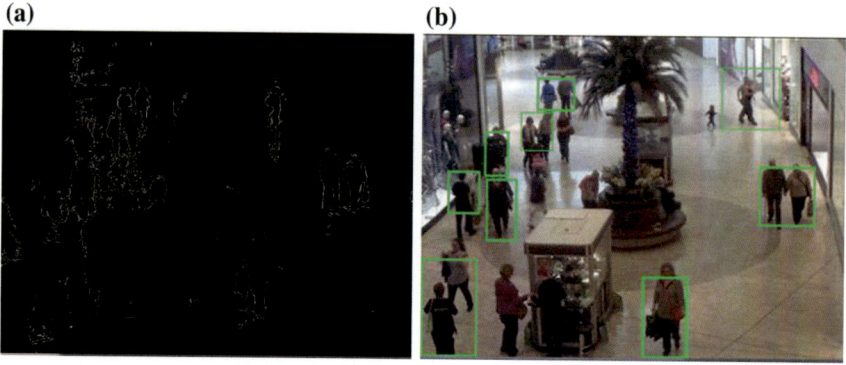

Fig. 12 **a** Skeleton. **b** Skeletonization-based algorithm on Mall dataset

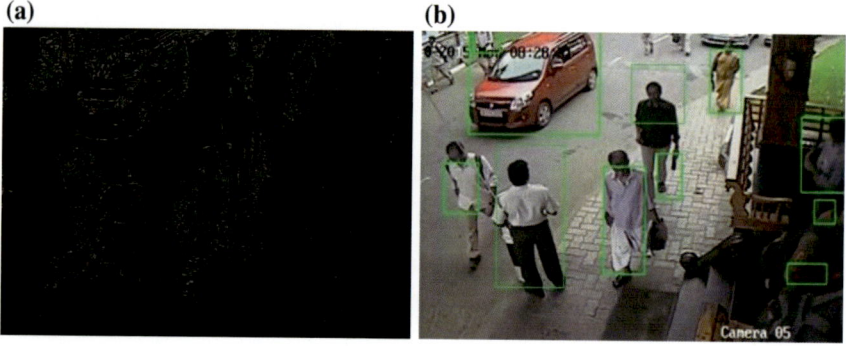

Fig. 13 **a** Skeleton. **b** Skeletonization-based algorithm applied on 223rd frame of the video from surveillance camera of our campus

4 Experimental Results

Open source Computer Vision with C++ has been used for implementing the over-crowd detection algorithms. All the algorithms were tested on two different datasets, Mall dataset [16, 17] and the surveillance dataset from our campus. Each dataset has videos captured under different illumination conditions and with varying concentration and occlusion of people in it.

All the algorithms were tested on 35 videos in Mall Dataset and 40 videos in surveillance dataset from our campus. The videos ranged from 100 frames to 6304 frames. For a video from mall dataset with 100 frames, based on the ground truth of each frame and the number of people identified by each algorithm in each frame, the graphs plotted are shown in Fig. 14a–d.

Graphs show that compared to HoG with SVM, Haar Cascade, and Morphological Operator based algorithms, the proposed algorithm gives people count very close to the ground truth which is calculated manually.

Based on the confusion matrix which depicts the true positive, false positive and false negative, the precision and recall are calculated. True positive stands for people being detected correctly, false positive where people are detected erroneously, false negative indicates non-detection of people by the algorithm in spite of their presence.

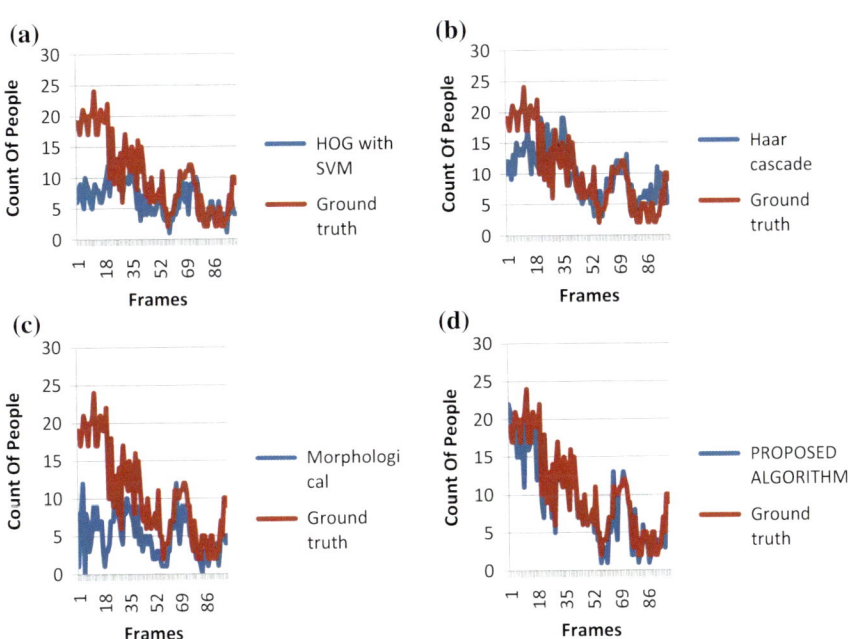

Fig. 14 a HoG with SVM. **b** Haar cascade. **c** Based on morphological operator. **d** Proposed algorithm

Table 1 Confusion matrix for Mall dataset video with 600 frames (Ground truth: 6186)

Mall dataset	Algorithm	True positive	False positive	False negative
	HOG with SVM	1602	204	2088
	Haar cascade	3834	842	1218
	Morphological operator based algorithm	222	43	3204
	Proposed algorithm	5142	84	438

Table 2 Confusion matrix for a video from surveillance camera with 812 frames (Ground truth: 8203)

Dataset from the surveillance camera of our campus	Algorithm	True positive	False positive	False negative
	HOG with SVM	2143	271	2684
	Haar cascade	5105	1015	1724
	Morphological operator based algorithm	305	62	4282
	Proposed algorithm	6976	114	574

True negative is not considered for calculating precision and recall as true negative is an indication of people not being detected when people are not there, which is not a prominent indication required for counting the number of people.

Table 1 indicates the confusion matrix for Mall dataset tested for HoG with SVM, Haar cascade, Morphological based, and the proposed Skeletonization-based people counting algorithms. A video which comprised of 600 frames was considered and the ground truth was calculated manually from each frame of the video which counted up to 6186 people.

Table 2 is the confusion matrix for the video from the surveillance camera of our campus tested for HoG with SVM, Haar cascade, Morphological-based, and the proposed Skeletonization-based people counting algorithms. A video which consisted of 812 frames was considered and the ground truth was calculated manually from each frame of the video which counted up to 8203 people.

Table 3 is generated based on the values from Tables 1 and 4 are generated based on the values from Table 2. Precision, recall, and accuracy are calculated. Precision is calculated as

$$Precision = \frac{True\ positive}{True\ Positive + False\ positive}$$

Table 3 Precision, recall, and accuracy for the confusion matrix in Table 1

Mall dataset	Algorithm	Precision	Recall	Accuracy
	HOG with SVM	0.887	0.434	0.411
	Haar cascade	0.819	0.75	0.65
	Morphological operator based algorithm	0.83	0.065	0.064
	Proposed algorithm	0.984	0.922	0.907

Table 4 Precision, recall, and accuracy for the confusion matrix in Table 2

Dataset from the surveillance camera of our campus	Algorithm	Precision	Recall	Accuracy
	HOG with SVM	0.888	0.444	0.42
	Haar cascade	0.83	0.747	0.66
	Morphological operator based algorithm	0.83	0.07	0.071
	Proposed algorithm	0.99	0.924	0.91

Recall is calculated as

$$Recall = \frac{True\ Positive}{True\ positive + False\ Negative}$$

Accuracy is calculated as

$$Accuracy = \frac{True\ Positive}{True\ positive + False\ positive + False\ Negative}$$

From the metrics evaluated it is observed that for both the datasets the proposed skeletonization based algorithm gave better precision, recall, and accuracy compared to the other three algorithms. Even though HoG with SVM, Haar cascade and morphological operator based algorithms seemed to provide better precision but failed to provide good accuracy due to the increase in the false negatives reported. Hence the proposed algorithm outperformed for all the other three algorithms on all the varied videos with which it was tested.

5 Conclusion

Crowd alert based on people counting was done using Skeletonization-based algorithm and has been compared with three algorithms like HoG with SVM, Haar Cascade, and Morphological-based algorithms. The confusion matrix and the graphical plot prove that the proposed skeletonization-based algorithm works better than the other algorithms tested upon. The test was done on varied datasets of different sizes and under varied illumination conditions and varied concentration and occlusion of people. Future scope of this work is to improvise on occlusions which are complex.

Acknowledgements The proposed work is tested on three different datasets. (i) Mall dataset [16, 17]; (ii) Dataset from surveillance camera of Amrita Vishwa Vidyapeetham, the university with which we work and where the proposed work is carried on; (iii) User defined video, which is video of myself on which the proposed work was tested upon.

References

1. Sağun MAK, Bolat B (2017) A novel approach for people counting and tracking from crowd video. In: IEEE international conference on INnovations in Intelligent SysTems and Applications (INISTA). IEEE, pp 277–281
2. Hassan MA, Pardiansyah I, Malik AS, Faye I, Rasheed W (2016) Enhanced people counting system based head-shoulder detection in dense crowd scenario. In: 6th International conference on intelligent and advanced systems (ICIAS). IEEE, pp 1–6
3. Neethu A, Bijlani K (2016) People count estimation using hybrid face detection method. In: International conference on information science (ICIS). IEEE, pp 144–148
4. Li B, Zhang J, Zhang Z, Xu Y (2014) A people counting method based on head detection and tracking. In: International conference on smart computing (SMARTCOMP). IEEE, pp 136–141
5. Wang Y, Lian H, Chen P, Lu Z (2014) Counting people with support vector regression. In: 10th International conference on natural computation (ICNC). IEEE, pp 139–143
6. Gao L, Wang Y, Wang J (2016) People counting with block histogram features and network flow constraints. In: International congress on image and signal processing, biomedical engineering and informatics (CISP-BMEI). IEEE, pp 515–520
7. Chen Z, Yuan W, Yang M, Wang C, Wang B (2016) SVM based people counting method in the corridor scene using a single-layer laser scanner. In: 19th International conference on intelligent transportation systems (ITSC). IEEE, pp 2632–2637
8. Perng JW, Wang TY, Hsu YW, Wu BF (2016) The design and implementation of a vision-based people counting system in buses. In: International conference on system science and engineering (ICSSE). IEEE, pp 1–3
9. Baozhu Z, Qiuyu Z, Yufeng X (2016) People counting system based on improved Gaussian background model. In: International conference on smart and sustainable city and big data (ICSSC). IEEE, pp 118–122
10. Riachi S, Karam W, Greige H (2014) An improved real-time method for counting people in crowded scenes based on a statistical approach. In: 11th International conference on informatics in control, automation and robotics (ICINCO), vol 2. IEEE, pp 203–212
11. Cai Z, Yu ZL, Liu H, Zhang K (2014) Counting people in crowded scenes by video analyzing. In: 9th Conference on industrial electronics and applications (ICIEA). IEEE, pp 1841–1845
12. Sikha OK, Sachin Kumar S, Soman KP (2017) Salient region detection and object segmentation in color images using dynamic mode decomposition. J Comput Sci 25:351–366

13. Pranav V, Manjusha R, Parameswaran L (2018) Design of an algorithm for people identification using facial descriptors. Lecture Notes in computational vision and biomechanics, vol 28. Springer, pp 1117–1128
14. Dalal N, Triggs B (2005) Histograms of oriented gradients for human detection. In: IEEE Computer society conference on computer vision and pattern recognition, CVPR, vol 1. IEEE, pp 886–893
15. Viola P, Jones M (2001) Rapid object detection using a boosted cascade of simple features. In: Proceedings of the 2001 IEEE computer society conference on computer vision and pattern recognition, CVPR, vol 1. IEEE, pp I–I
16. Loy CC, Chen K, Gong S, Xiang T (2011) Feature mining for localised crowd counting. In: British machine vision conference (BMVC), vol 1, no 2
17. Loy CC, Chen K, Gong S, Xiang T (2013) Crowd counting and profiling: methodology and evaluation. In: Modeling, simulation and visual analysis of crowds. Springer, pp 347–382

Performance Comparison of Pre-trained Deep Neural Networks for Automated Glaucoma Detection

Manas Sushil, G. Suguna, R. Lavanya and M. Nirmala Devi

Abstract This paper addresses automated glaucoma detection system using pre-trained convolutional neural networks (CNNs). CNNs, a class of deep neural networks (DNNs), extract features of high-level abstractions from the fundus images, thereby eliminating the need for hand-crafted features which are prone to inaccuracies in segmenting landmark regions and require excessive involvement of experts for annotating these landmarks. This work investigates the applicability of pre-trained CNNs for glaucoma diagnosis, which is preferred when the dataset size is small. Further, pre-trained networks have the advantage of the quick model building. The proposed system has been validated on the High-Resolution (HRF), which is a publicly available benchmark database. Results demonstrate that among other pre-trained CNNs, VGG16 network is more suitable for glaucoma diagnosis.

1 Introduction

Many ocular diseases progress asymptomatically, ultimately resulting in vision impairment [1, 2]. Glaucoma is a prominent cause of blindness, with noticeable vision loss occurring only in advanced stages. It has no cure but early diagnosis can slow progression rate. Worldwide, glaucoma count in the age group of 40–80 years was estimated to be 64.3 million in 2013 and is projected as 76 million in 2020 [3]. Glaucoma affects the optic nerve and is associated with the improper draining

M. Sushil · G. Suguna · R. Lavanya (✉) · M. Nirmala Devi
Department of Electronics and Communication Engineering, Amrita School of Engineering,
Amrita Vishwa Vidyapeetham, Coimbatore, India
e-mail: r_lavanya@cb.amrita.edu

M. Sushil
e-mail: 66manas@gmail.com

G. Suguna
e-mail: g_suguna@cb.amrita.edu

M. Nirmala Devi
e-mail: m_nirmala@cb.amrita.edu

© Springer Nature Switzerland AG 2019
D. Pandian et al. (eds.), *Proceedings of the International Conference on ISMAC in Computational Vision and Bio-Engineering 2018 (ISMAC-CVB)*, Lecture Notes in Computational Vision and Biomechanics 30,
https://doi.org/10.1007/978-3-030-00665-5_62

of aqueous humor through the drainage angle in the eye. Symptoms include high intraocular pressure (IOP), degeneration of optic nerve fibers, and irreversible vision impairment. Glaucoma diagnosis involves assessing medical history, visual field loss, IOP, as well as Optical Nerve Head (ONH) defects and Retinal Nerve Fiber Layer (RNFL) defects by analyzing retinal images [4].

Digital Fundus Photography (DFP) is a retinal imaging modality that is most widely used in clinical practice for glaucoma screening. It is a quick, simple, cheap, and non-invasive technique that captures a large retinal field. However, manual diagnosis using fundus is subject to ambiguity and is time-consuming.

The advances in image processing and machine learning have paved way for automated diagnosis systems, which can aid time-effective and objective diagnosis. Many researchers have proposed automated systems for diagnosing glaucoma in fundus images. The approaches employed are classified as segmentation-based and non-segmentation-based techniques. In the former approach, also called structural analysis, retinal structures of interest are detected to extract clinically relevant features which ophthalmologists normally use for diagnosis. On the contrary, the non-segmentation-based approach is used to extract high level, non-clinical information from the images, which normally cannot be perceived by human experts [5].

2   Related Work

The segmentation approach in automated glaucoma diagnosis involves localization and extraction of anatomical structures for analyzing defects in the ONH and RNFL. The clinical features of importance that characterize the ONH defects include the Cup-to-Disc Ratio (CDR), ISNT rule and notching. Among these parameters, CDR is widely used in clinical practice; nevertheless, it is susceptible to large inter-individual variability. RNFL defects serve as an early indicator of Glaucoma. However, the detection differentiating RNFL striations and vasculature structure is quite challenging.

Major steps involved in automated glaucoma diagnosis for determination of CDR include pre-processing of the fundus image, ONH localization, OD segmentation, and OC segmentation. Nayak et al. [6] employed neural network for classifying the images as normal or glaucoma based on CDR. An accuracy of 90% was obtained for glaucoma diagnosis. The accuracies of OD and OC segmentation in fundus images face several challenges and have a cascaded effect on the accuracy of glaucoma diagnosis. The segmentation accuracy can be affected by many factors that include low contrast and blood vessel occlusion. Little work has been reported on the analysis of RNFL defects using 2-D retinal images, demonstrating preliminary results.

Some approaches using non-clinical features that are independent of segmenting the retinal structures, such as Textural analysis, Fourier and Wavelet transforms have also been proposed for glaucoma diagnosis [7, 8]. Chen et al. [9] employed convolutional neural network (CNN) to capture hidden, discriminative patterns related to glaucoma resulting in AUC of 0.887.

3 Deep Neural Networks

The conventional segmentation approach based on pre-defined hand-crafted clinical features is not optimal, as the characteristics of the underlying problem may be patient specific. Recent developments in the field of deep learning (DL) offer powerful tools to intelligent image analysis. DL is a promising machine learning technique that is capable of characterizing hidden complex features by applying neural network architectures with multiple hidden layers to solve complex problems.

Deep neural networks (DNN) exhibit feature hierarchy and learn features of increasing complexity and abstraction with increasing network depth. A DNN can discover latent structures within the raw data automatically, eliminating the need for manual labor and high-quality expert knowledge to generate hand-crafted features. Further, it is independent of the cascading effect of segmentation inaccuracy associated with extraction of hand-crafted features [10].

CNN is a branch of deep learning that can visualize and classify visual imagery and is being successfully employed to tackle complex image recognition tasks. CNNs have demonstrated to perform well for medical image analysis and interpretation [9, 11].

4 Methodology

The proposed methodology for automated glaucoma diagnosis using deep learning is illustrated in Fig. 1.

Various steps involved in the proposed work are explained below.

4.1 Data Augmentation

This is a strategy used to increase the images in the dataset. This is performed by combinations of flipping the images with 90° and 270° rotation, left–right rotation, and top–bottom rotation.

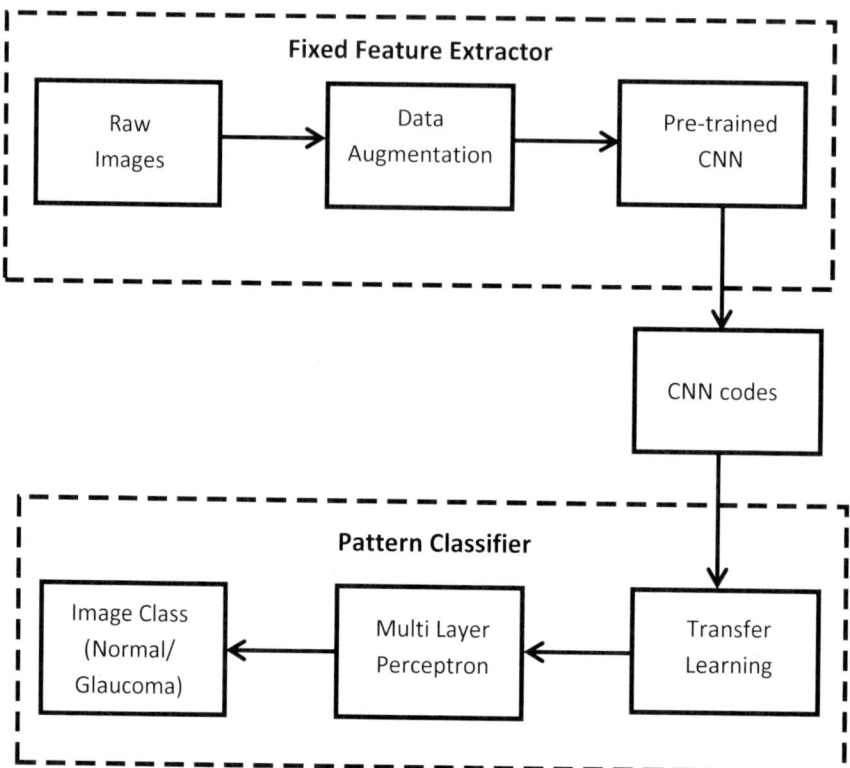

Fig. 1 Proposed methodology

4.2 Pre-trained CNN

In practice, random initialization for training a CNN with is rare as a sufficiently large dataset is not available in most cases. Instead, a CNN pre-trained on large datasets are used as an initial network or a fixed feature extractor. This approach is called transfer learning.

In this work, VGG16 architecture [12], a network pre-trained on ImageNet [13] has been employed. ImageNet is a large image database containing 1.2 million images with 1000 categories. VGG16 is a very deep convolution neural network containing 16–19 weight layers as shown in Table 1. Due to this, VGG16 has shown significant improvements over other networks as reported in many studies. At ImageNet challenge 2014, VGG16 secured the first place in localization task and the second place in the classification task.

Table 1 Configuration of VGG16 net[a]

VGG Configuration		
16 weight layers	16 weight layer	19 weight layer
C	D	E
Input image		
CL3-64 cnn3-64	CL3-64 CL 3-64	CL3-64 CL3-64
Maxpool		
CL3-128 CL3-128	CL3-128 CL3-128	CL3-128 CL3-128
Maxpool		
CL3-256 CL3-256 CL1-256	CL3-256 CL3 -256 CL3-256	CL3-256 CL3-256 CL3-256 CL3-256
Maxpool		
CL3-512 CL3-512 CL1-512	CL3-512 CL3-512 CL3-512	CL3-512 CL3-512 CL3-512 CL3-512
Maxpool		
CL3-512 CL3-512 CL1-512	CL3-512 CL3-512 CL3-512	CL3-512 CL3-512 CL3-512 CL3-512
Maxpool		
FCL-4096		
FCL-4096		
FCL-1000		
Soft-max		

[a]Covolutional layer parameters—"CL<filter size>-<number of filters>"
FCL fully connected layer

4.3 Transfer Learning

Two different approaches are widely employed for transfer learning. Generally, when large datasets are involved, the weights of all the layers of pre-trained network layers are fine-tuned using back propagation, as there is no concern of overfitting. However, for small datasets, such as the one employed in this work, fine-tuning will result in overfitting. As an alternate strategy, the final layer of the network is removed, with the other layers serving as a fixed feature extractor. These features are called CNN codes. These codes are used to train a much simpler classifier using the data set under consideration.

5 Results

The performance of pre-trained CNNs in diagnosing glaucoma is evaluated using High-Resolution Fundus (HRF) [14] database which comprises 15 normal fundus

Table 2 Comparison of the performance of pre-trained networks

	VGG16	InceptionV3	ResNet50
Loss	0.07	0.4	0.4
Precision	1	0.48	0.4
Recall	1	0.48	0.08
Accuracy	1	0.6	0.6

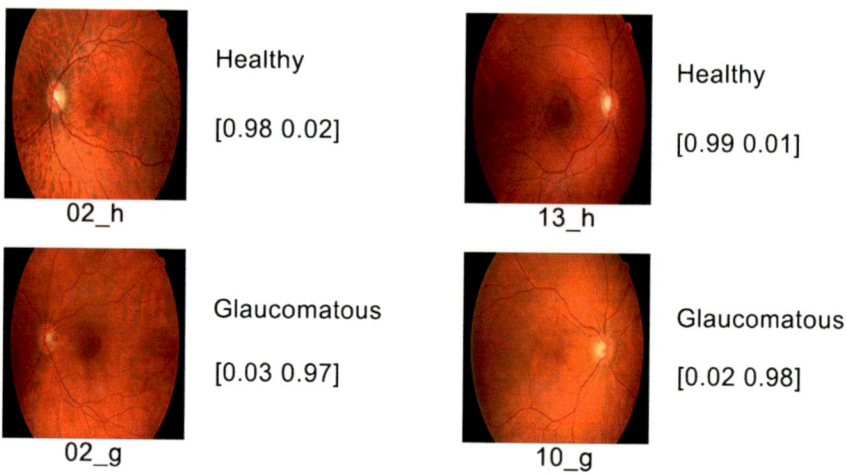

Fig. 2 Softmax outputs of sample HRF images

images and an equal number of glaucomatous patients. Data augmentation was performed to increase the number images to 120, comprising 60 healthy cases and 60 glaucoma cases. Thirty images corresponding to healthy cases and glaucoma cases each used for training. A similar distribution was used for testing. Transfer learning was performed with VGG16, and the results were compared with other pre-trained CNNs including InceptionV3 and ResNet50. As HRF is a relatively small database, the pre-trained networks were used as fixed feature extractors with fully connected layers being replaced by a Multilayer Perceptron (MLP). The MLP employed consists of one hidden layer followed by the output layer. The hidden layer contains 1024 neurons with "Relu" activation and a dropout of 0.5 to avoid overfitting. Output layer consists of two neurons with "Softmax" activation. Fundus image of glaucomatous patients and healthy patients were labeled as [0 1] and [1 0], respectively. Performance of the networks was measured using mean squared error.

In Table 2, the performance of the three pre-trained CNNs is compared. It is observed that VGG16 yields a categorical accuracy of 100% when compared to 60% accuracy achieved by the other two networks. Further, the loss associated with VGG16 is trivial and equal to 0.07, while it is 0.4 each for the other two networks. Sample images from the HRF database with the respective softmax outputs of VGG16 are depicted in Fig. 2.

6 Conclusion

The suitability of pre-trained CNNs for automated glaucoma diagnosis has been investigated in this work. Promising results have been achieved with VGG16 architecture for the High-Resolution Fundus (HRF) database. Future work involves the extension of this work to larger databases using fine-tuning approach for transfer learning.

References

1. Dharani V, Lavanya R (2017) Improved microaneurysm detection in fundus images for diagnosis of diabetic retinopathy. In: International symposium on signal processing and intelligent recognition systems. Springer, pp 185–198 (2017)
2. Sharma A, Subramaniam SD, Ramachandran KI, Lakshmikanthan C, Krishna S, Sundaramoorthy SK (2016) Smartphone-based fundus camera device (MII Ret Cam) and technique with ability to image peripheral retina. Eur J Ophthalmol 26(2):142–144
3. Richter GM, Anne LC (2016) Minimally invasive glaucoma surgery: current status and future prospects. Clin Ophthalmol (Auckland, NZ) 10:189–206
4. Zhang Z, Srivastava R, Liu H, Chen X, Duan D, Wong WK, Kwoh CK, Wong TY, Liu Y (2014) A survey on computer aided diagnosis for ocular diseases. BMC Med Inf Decis Making 14(1):80
5. Haleem MS, Han L, van Hemert J, Li B (2013) Automatic extraction of retinal features from colour retinal images for glaucoma diagnosis: a review. Comput Med Imaging Graph 37(7):581–596
6. Nayak J, Acharya R, Bhat PS, Shetty N, Lim TC (2009) Automated diagnosis of glaucoma using digital fundus images. J Med Syst 33(5):337
7. Bock R, Meier J, Nyúl LG, Hornegger J, Michelson G (2010) Glaucoma risk index: automated glaucoma detection from color fundus images. Med Image Anal 14(3):471–481
8. Dua S, Acharya UR, Chowriappa P, Sree SV (2010) Wavelet-based energy features for glaucomatous image classification. IEEE Trans Inf Technol Biomed 16(1):80–87
9. Chen X, Xu Y, Wong DWK, Wong TY, Liu J (2015) Glaucoma detection based on deep convolutional neural network. IEEE EMBC 2015:715–718
10. Schmidhuber J (2015) Deep learning in neural networks: an overview. Neural Networks 61:85–117
11. Gulshan V, Peng L, Coram M, Stumpe MC, Wu D, Narayanaswamy A, Kim R (2016) Development and validation of a deep learning algorithm for detection of diabetic retinopathy in retinal fundus photographs. JAMA 316(22):2402–2410
12. Simonyan K, Zisserman A (2014) Very deep convolutional networks for large-scale image recognition. arXiv preprint arXiv:1409.1556
13. Deng J, Dong W, Socher R, Li LJ, Li K, Fei-Fei L (2015) Imagenet: a large-scale hierarchical image database. IEEE CVPR 2009:248–255
14. Odstrcilik J, Kolar R, Budai A, Hornegger J, Jan J, Gazarek J, Angelopoulou E (2013) Retinal vessel segmentation by improved matched filtering: evaluation on a new high-resolution fundus image database. IET Image Proc 7(4):373–383

Investigation on Land Cover Mapping of Large RS Imagery Using Fuzzy Based Maximum Likelihood Classifier

B. R. Shivakumar and S. V. Rajashekararadhya

Abstract The success of a large number of real-world applications such as mapping, forestry, and change detection depends on the effectiveness with which land cover classes are extracted from Remotely Sensed (RS) imagery. Application of Fuzzy theory in remote sensing has been of great interest in the remote sensing fraternity particularly when the data are inherently Fuzzy. In this paper, a Fuzzy theory based Maximum Likelihood Classifier (MLC) is discussed. The study aims at amplifying the classification accuracy of large heterogeneous multispectral remote sensor data characterized by the overlapping of spectral classes and mixed pixels. Landsat 8 multispectral data of North Canara District was collected from USGS website and is considered for the research. Seven land use land cover classes were identified over the study area. The study also aims at achieving classification results with a confidence level of 95% with ±4% error margin. The conducted research attains the predicted classification accuracy and proves to be a valuable technique for classification of large heterogeneous RS multispectral imagery.

1 Introduction

Classification of multispectral imagery has formed itself as one of the most sought-after technique for information extraction. The process of image classification, in the context or remote sensing, has been broadly classified into hard and soft classification techniques. In hard classification, a pixel is assumed to be an indecomposable part of the image and belongs to just one of the defined land cover class. However, in the real world due to the presence of mixed pixels (mixels), which form the salient feature of heterogeneous study areas, hard classifiers are shown to produce poor

B. R. Shivakumar (✉)
NMAM Institute of Technology, Nitte, India
e-mail: shivkumarbr@gmail.com

S. V. Rajashekararadhya
Kalpataru Institute of Technology, Tiptur, India
e-mail: svraradhya@gmail.com

© Springer Nature Switzerland AG 2019
D. Pandian et al. (eds.), *Proceedings of the International Conference on ISMAC in Computational Vision and Bio-Engineering 2018 (ISMAC-CVB)*, Lecture Notes in Computational Vision and Biomechanics 30,
https://doi.org/10.1007/978-3-030-00665-5_63

639

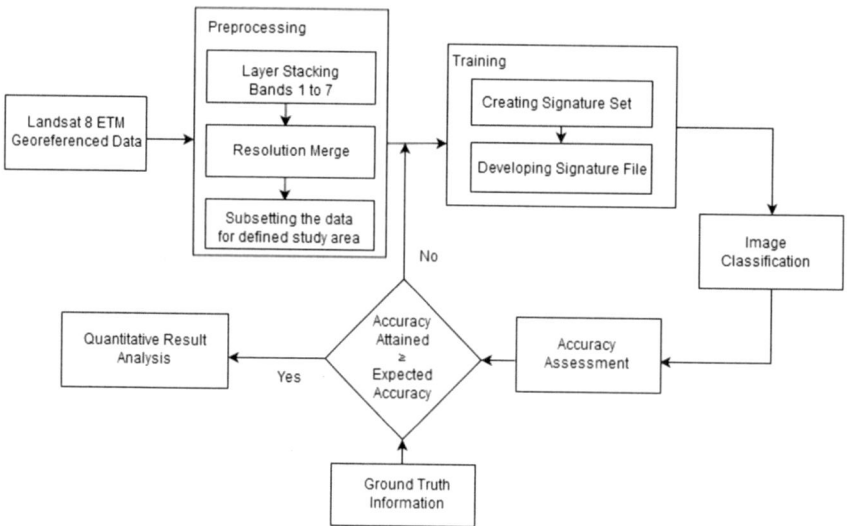

Fig. 1 General flow of the methodology

results as the ground data is imprecise [1–3]. Soft or Fuzzy classification comes in handy under these eventualities and has been known for its capability to extract a lot of helpful data from heterogeneous RS data [2, 4]. A large number of studies have been conducted for classifying RS data using Fuzzy logic [2, 5–8]. Though a large number of researchers have conferred their study towards remotely sensed image classification using different classifiers, it has still remained a challenging task within the remote sensing fraternity [1, 9–12]. Hence, it has become necessary to research and produce new classification techniques for obtaining better classification results.

Zadeh's concept of Fuzzy set theory has provided some very useful options for operating with heterogeneous datasets [13]. Using Fuzzy set concepts a pixel in an image is described with a membership function that links the pixel a real number from 0 to 1. This membership function is treated as the probability of the pixel belonging to each class [1, 13]. This principle is shown to provide an efficient solution for mixed pixel issue [3, 8].

The objective of this study is to perform LULC mapping of the considered study area using Fuzzy theory based MLC. The goal of classification is to obtain results with a confidence level of 95% with ±4% error margin. The methodology followed is illustrated in Fig. 1.

The rest of the paper is organized as follows. Section 2 discusses the data and study area used. Section 3 discusses the classification method employed. Section 4 presents the results obtained during the study. Section 5 presents the conclusions drawn from the results.

2 Materials and Methodology

Land Satellite (Landsat) 8 data was accessed by U.S. Geological Survey (USGS) website. Figure 2 indicates the study area considered and it envelops the North Canara District in Karnataka, India. The Western Ghats form the main geographic feature of the region and runs from North to South. The average rainfall on the coastal part is 3000 mm (120 in) and is as high as 5000 mm (200 in) in the west-facing slopes of Western Ghats. East facing ridge of the Western Ghats is the rain shadow region and receives, on an average, only 1000 mm (39 in) rainfall annually [14]. High rainfall in the region supports lavish forests that coat over 70% of the coverage area. The north part of the Western Ghats forms Moist Deciduous Forests ranging from 250 to 1000 m in elevation. Above 1000 m elevation are the Evergreen rain forests [15]. The study area also has chunks of degraded scrub jungles and savanna. The beach region is characterized by coconut plantations and screw pine. The study area is recognized as a coastal agro-climatic zone by the government of India. The overall spatial resolution of the image is 15 m. This data was acquired on May 18 2016, which is the mid-summer season and is free from clouds. Layer stacking of first seven bands produces a low-resolution multispectral (MS) imagery of 30 m. Resolution merge technique is used to merge the 30 m MS image with higher spatial resolution band 8 to obtain high resolution 15 m multispectral image. This image is then used for further processing.

2.1 Feature Extraction/Signature Collection

There are several methods of collecting the training data, including; in situ data collection, on-screen selection of polygonal data, and/or on-screen seeding of training data [1]. By employing an on-screen selection of polygonal data technique, seven land use land cover classes were identified over the study area; (i) Water Body, (ii) Kharif, (iii) Built Up, (iv) Scrub Land, (v) Double Crop/Horticultural Plantations, (vi) Moist Deciduous Forest, and (vii) Evergreen Forest. Spectral signatures were collected for each class with at least 300 sample pixels per class. The study considered both pure and mixed pixels for training the algorithm.

The considered study area exhibits the characteristics of a heterogeneous dataset with more than two classes severely overlapping one another. Class separability was measured in terms of Euclidean distance considering two classes at a time. The Euclidean distance was redefined to measure the severity of land cover overlapping and is used as the spectral similarity index (SSI). Highest class separability was measured between Water Body and Kharif classes and was considered as the reference for other classes. Classes with a SSI value of less than 0.4 were determined to be severely overlapping one another at certain locations. SSI measurements for all possible class pairs are shown in Table 1.

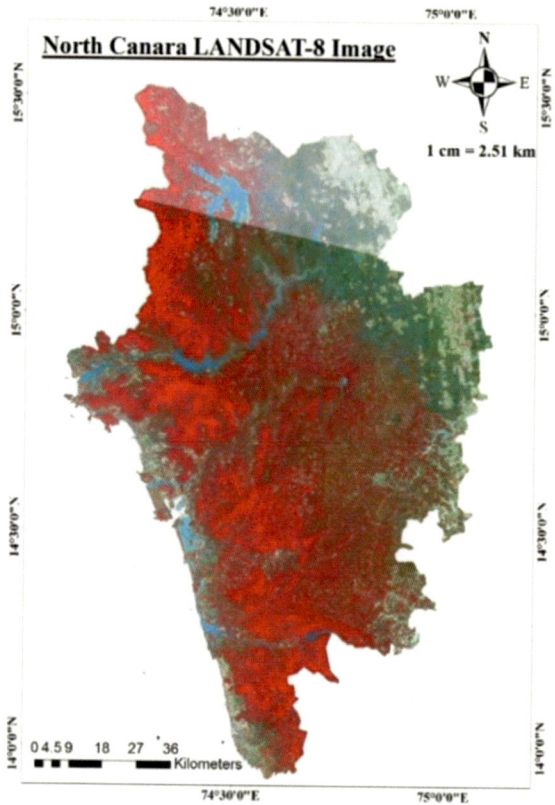

Fig. 2 15 m spatial resolution map of the study area [16]

3   Fuzzy Theory Based Maximum Likelihood Classifier

The Fuzzy theory based ML classifier discussed in this paper works in three stages;
(i) Fuzzification of the data, (ii) Classification using Fuzzy theory based MLC, and
(iii) Defuzzification.

3.1   Fuzzification

In the Fuzzification step, each pixel on the input image is converted into a pixel
measurement vector, x, of membership grades. The Fuzzy membership function for
any x must lie in the range 0–1, they should all add up to unity, and should be positive
values. These characteristics are listed in (1), (2), and (3) [1].

Table 1 Class separability in terms of spectral similarity index

Class Pair	SSI	Class pair	SSI	Class pair	SSI
1:2	1	2:4	0.441	3:7	0.963
1:3	0.879	2:5	0.796	4:5	0.544
1:4	0.628	2:6	0.486	4:6	**0.061**
1:5	0.666	2:7	0.968	4:7	0.649
1:6	0.598	3:4	0.528	5:6	0.518
1:7	0.668	3:5	0.774	5:7	**0.210**
2:3	**0.338**	3:6	0.562	6:7	0.612

1: Water Body, 2: Kharif, 3: Built Up, 4: Scrub Land, 5: Double Crop/Horticultural Plantations, 6: Moist Deciduous Forest, 7: Evergreen Forest
Bold characters represent the spectral similarity index for the most severely overlapping class pairs. The class pairs having spectral similarity index (SSI) value of 0.3 and lesser are the most severely overlapping class pairs. It is highlighted to make it easy for the reader to identify the severely overlapping class pairs easily

$$0 \leq f_{F_i}(x) \leq 1 \qquad (1)$$

$$\sum_{x \in X} f_{F_i}(x) > 0 \qquad (2)$$

$$\sum_{i=1}^{m} f_{F_i}(x) = 1, \qquad (3)$$

where, F_i is one of the spectral classes, X represents all pixels in the dataset, x is a pixel measurement vector, m is the number of classes, and f_{F_i} is the membership function of the Fuzzy set $F_i (1 \leq i \leq m)$. All the membership function values are recorded as a Fuzzy partition matrix

$$\begin{bmatrix} f_{F_1}(x_1) & f_{F_1}(x_2) & \cdots & f_{F_1}(x_n) \\ f_2(x_1) & f_{F_2}(x_2) & \cdots & f_{F_2}(x_n) \\ \vdots & \vdots & \cdots & \vdots \\ f_{F_m}(x_1) & f_{F_m}(x_2) & \cdots & f_{F_m}(x_n) \end{bmatrix}, \qquad (4)$$

where, n represents the total number of pixels, and x_i is the ith pixel's measurement vector [1].

3.2 Classification

Fuzzy based classification involves the use of Fuzzy mean and covariance matrices. For class c, Fuzzy mean is computed as

$$\mu_c^* = \frac{\sum_{i=1}^{n} f_c(x_i)x_i}{\sum_{i=1}^{n} f_c(x_i)}, \tag{5}$$

where, x_i is a sample pixel measurement vector $(1 \le i \le n)$, f_c is the membership function of class c, and n is the total number of sample pixel measurement vectors [1].

The Fuzzy covariance matrix V_c^* is computed as;

$$V_c^* = \frac{\sum_{i=1}^{n} f_c(x_i)(x_i - \mu_c^*)(x_i - \mu_c^*)^T}{\sum_{i=1}^{n} f_c(x_i)} \tag{6}$$

These mean and covariance values replace the conventional mean and covariance matrix in the classical MLC algorithm. This will convert a classical MLC algorithm into a Fuzzy based soft classification algorithm [1].

Fuzzy set theory solely provides membership functions for every pixel over the defined number of classes and requires a parametric rule for assigning those pixels to relevant classes. Parametric rules such as Maximum Likelihood, Mahalanobis Distance et al. may be used in the process. This study involves the utilization of Maximum Likelihood classifier as the parametric rule.

Fuzzy based Maximum Likelihood classifier uses Fuzzy mean and Covariance matrices replacing the conventional mean and covariance matrices. For an n-band multispectral image the likelihood function for a pixel belonging to class k is given by [1],

$$p^*(x|w_i) = \frac{1}{(2\pi)^{n/2}|V_k^*|^{1/2}} \exp\left[-\frac{1}{2}(x - \mu_k^*)V_k^{*-1}(x - \mu_k^*)^T\right], \tag{7}$$

where, x is one of the brightness values on the x-axis, μ_k^* is the Fuzzy mean as in (5), V_k^* is the Fuzzy covariance matrix as in (6), and $(x - \mu_k^*)^T$ is the transpose of vector $(x - \mu_k^*)$. Similarly, $p^*(x|w_i)$ is calculated for each pixel for all classes. A membership function then enables the algorithm to decide to which class the corresponding pixel is to be assigned. For Maximum Likelihood classifier, the membership function can be defined as [1];

$$f_c(x) = \frac{p^*(x|w_k)}{\sum_{i=1}^{m} p^*(x|w_k)} \tag{8}$$

The membership grades of a pixel vector depend on x's position in the vector space. $f_c(x)$ increases exponentially with the decrease of $(x - \mu_k^*)V_k^{*-1}(x - \mu_k^*)^T$, i.e., the Mahalanobis distance between x and class k. The factor $\sum_{i=1}^{m} p^*(x|w_k)$ is a normalization factor [1]. Applying this type of Fuzzy logic creates a membership grade matrix for each pixel.

3.3 Defuzzification Using Fuzzy Convolution

Fuzzy convolution technique is used to convert the n-layer output of classification into a map like structure. It creates the map by computing the total weighted inverse distance of all the classes in a window of pixels. The process first computes the total inverse distance summed over the entire set of Fuzzy classification layers for each class. It then assigns the center pixel to the class for which the value $T[k]$ is largest. The total inverse distance, $T[k]$, can be computed using [17]:

$$T[k] = \sum_{i=0}^{s} \sum_{j=0}^{s} \sum_{l=0}^{n} \frac{w_{ij}}{D_{ijl}[k]} \tag{9}$$

where, i is the row index of window, j is the column index of window, s is the size of window, l is the layer index of fuzzy layers used, n is the number of fuzzy layers used, W is the weight table for window, k is the class value, $D[k]$ is the distance file value for class k, and $T[k]$ is the total weighted distance of window for class k [17]. This study considers a 5×5 size window given by

$$\begin{bmatrix} 0.500 & 0.605 & 0.646 & 0.605 & 0.500 \\ 0.605 & 0.750 & 0.823 & 0.750 & 0.605 \\ 0.646 & 0.823 & 1.000 & 0.823 & 0.646 \\ 0.605 & 0.750 & 0.823 & 0.750 & 0.605 \\ 0.500 & 0.605 & 0.646 & 0.605 & 0.500 \end{bmatrix} \tag{10}$$

4 Results and Discussion

Figure 3 illustrates the Fuzzy topology based maximum likelihood classification map of the study area. Table 2 shows the results obtained after accuracy assessment. User's Accuracy and Kappa value are considered as the pivotal parameters in judging the classification process [18]. The classifier extracted dominant classes (Evergreen Forest and Deciduous Forest) very efficiently. Less dominant classes (Double Crop and Built Up) are extracted very poorly by the classifier. Fuzzy topology based maximum likelihood classifier has shown significant improvement in classification performance [12]. An overall Kappa value of 0.7870 indicates an excellent performance from the classifier [18].

To illustrate the usefulness of using the inverse weighted distance of all classes from the weight windows for pixels for assigning pixels to class values, Table 3 indicates the inverse weighted distance, $T[k]$, for 10 pixels randomly selected from the study area. A hard classified map is created by assigning a pixel to the class for which the distance measure, $T[k]$, is maximum.

Table 2 Results of Fuzzy topology based maximum likelihood classification

Class name	Reference totals	Classified totals	Number correct	Producer's accuracy (%)	User's accuracy (%)	Kappa value (k_{hat})
Evergreen Forest	534	538	502	94.01	93.31	0.8564
Scrub Land	59	70	40	67.80	57.14	0.5446
Moist Deciduous Forest	235	214	205	87.23	95.79	0.9450
Built Up	5	14	3	60.00	21.43	0.2103
Double Crop/Horticultural Plantations	57	80	37	64.91	46.25	0.4300
Water Body	31	16	16	51.61	100.00	1.0000
Kharif	79	68	59	74.68	86.76	0.8563
Total	1000	1000	862			
Overall classification accuracy = 86.2%						
Overall Kappa Statistic = 0.7870						

Table 3 Total inverse weighted distance of all classes for selected pixels

Pixel no.	Weighted inverse distance ($T[k]$) from							Class assigned
	EGF[a]	SL[b]	MDF[c]	BU[d]	DC/HP[e]	WB[f]	KH[g]	
1	**0.1988**	0.1199	0.1159	0.0654	0.1664	0.0026	0.0463	**EGF[a]**
2	0.1243	**0.1714**	0.1593	0.0821	0.1359	0.0011	0.1041	**SL[b]**
3	0.1426	0.1592	**0.1701**	0.0949	0.1348	0.0012	0.0815	**MDF[c]**
4	0.0639	0.1140	0.0576	**0.1655**	0.1443	0.0022	0.1345	**BU[d]**
5	0.0635	0.1561	0.0989	0.1361	**0.1709**	0.0019	0.1431	**DC/HP[e]**
6	0.0468	0.0587	0.0432	0.0951	0.0768	**0.1552**	0.0701	**WB[f]**
7	0.0148	0.1188	0.1634	0.1366	0.0540	0.0011	**0.1748**	**KH[g]**
8	0.0295	0.1161	0.0541	0.1059	0.0920	0.0007	**0.1939**	**KH[g]**
9	0.0215	0.0761	0.0314	**0.1742**	0.0582	0.0025	0.1235	**BU[d]**
10	0.1791	0.0758	0.0568	0.0326	**0.2013**	0.0018	0.0229	**DC/HP[e]**

EGF[a]: Evergreen Forest, SL[b]: Scrub Land, MDF[c]: Moist Deciduous Forest, BU[d]: Built Up, DC/HP[e]: Double Crop/Horticultural Plantations, WB[f]: Water Body, KH[g]: Kharif
Bold characters represent the spectral similarity index for the most severely overlapping class pairs. The class pairs having spectral similarity index (SSI) value of 0.3 and lesser are the most severely overlapping class pairs. It is highlighted to make it easy for the reader to identify the severely overlapping class pairs easily

Fig. 3 Fuzzy topology based maximum likelihood classified map of the study area

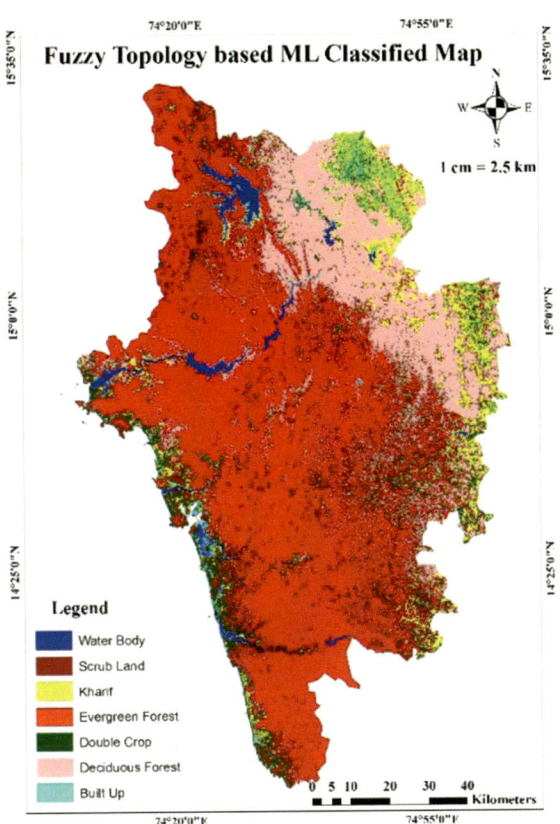

4.1 Placing Confidence Limits on Assessed Accuracy

A straightforward statistical approach may be used to express the interval within which the true map accuracy lies, say, with 95% certainty. It is possible to use the normal distribution to obtain this interval by the expression [2];

$$-Z_{\alpha/2} < \frac{x - nP}{\sqrt{nP(1-P)}} < Z_{\alpha/2}, \tag{11}$$

where, n is the number of testing pixels, $x(= np)$ is the number of correctly labelled pixels, P is the thematic map accuracy in percentage, $p(= x/n)$ is the proportion of pixels correctly classified, $Z_{\alpha/2}$ is the value of the normal distribution beyond which on both tails α of the population is excluded [2].

From Eq. (11), the estimate of the thematic map accuracy, P, estimated by the proportion of pixels that are correctly classified in the testing set, at the 95% confidence level, for large values of n and x, and for reasonable accuracies are [2],

$$\frac{x \pm 1.960\sqrt{\frac{x(n-x)}{n}} + 0.960}{n} = p \pm \frac{1.960}{n}\sqrt{\frac{x(n-x)}{n}} + 0.960 \tag{12}$$

For 1000 testing pixels and a minimum of 80% of expected accuracy, it is expected to have, at least, 800 pixels to be correctly classified. From (12), the bounds on the estimated map accuracy are $P = p \pm 0.039$ or, in percentage terms, the map accuracy is approximated to be between 82.1 and 89.9%.

5 Conclusion

In this paper, a novel method for classification of large RS imagery through embedding Fuzzy theory into classical Maximum Likelihood Classifier (MLC) is discussed. A pixel is the basic building block of an image and is indecomposable in hard classification applications. The study proves that a pixel can be used as a decomposable unit in image classification. The results obtained indicate that Fuzzy theory based MLC permits obtaining accurate results for large heterogeneous study areas in the presence of mixed pixels and spectrally overlapping classes. The study conjointly confirms that one can use mixed pixels as training data and yet achieve good results. The objective of obtaining classification results with confidence level of 95% with $\pm 4\%$ error margin is achieved. Hence, it can be concluded that the discussed Fuzzy topology based MLC handles mixed pixel issue more successfully. However, more investigation is needed on the classification performance of some classes, such as Built Up and Double Crop/Horticultural Plantations. Future scope of the work involves exploring the information richness provided by embedding Fuzzy theory into MLC for other sensor data.

References

1. Jensen JR (2000) Introductory digital image processing: a remote sensing perspective. Prentice-Hall Inc., New Jersey
2. Zhang J, Foody GM (1998) A fuzzy classification of sub-urban land cover from remotely sensed imagery. Int J Remote Sens 19:2721–2738
3. Wang F (1990) Fuzzy supervised classification of remote sensing images. IEEE Trans Geosci Remote Sens 28:194–201
4. Ji M, Jensen JR (1996) Fuzzy training in supervised digital image processing. Geogr Inf Sci 2:1–11
5. Wang Y, Jamshidi M (2004) Fuzzy logic applied in remote sensing image classification. Syst Man Cybern 2004(7):6378–6382
6. Melgani F, Al Hashemy BAR (2000) Taha SMR (2000) An explicit fuzzy supervised classification method for multispectral remote sensing images. IEEE Trans Geosci Remote Sens 38(1):287–295
7. Droj G (2007) The applicability of fuzzy theory in remote sensing image classification. Stud Univ Babes, Bolyai, Inform LII, 89–96

8. Wang F (1990) Improving remote-sensing image-analysis through fuzzy information representation. Photogramm Eng Remote Sens 56:1163–1169
9. Lu D, Weng Q (2007) A survey of image classification methods and techniques for improving classification performance. Int J Remote Sens 28:823–870
10. Mas JF, Flores JJ (2008) The application of artificial neural networks to the analysis of remotely sensed data. Int J Remote Sens 29:617–663
11. Maselli F, Conese C, De Filippis T, Romani M (1995) Integration of ancillary data into a maximum-likelihood classifier with nonparametric priors. ISPRS J Photogramm Remote Sens 50:2–11
12. Shivakumar BR, Rajashekararadhya SV (2017) Spectral similarity for evaluating classification performance of traditional classifiers. In: 2017 International conference on wireless communications signal processing and networking (WiSPNET). Chennai, pp 1999–2004
13. Zadeh LA (1965) Fuzzy sets. Inf Control 8:338–353
14. Directorate of Census Operations Karnataka: District census handbook Uttara Kannada (2014)
15. Pascal JP (1986) Explanatory booklet on forest map of South India: Belgaum-Dharwar-Panaji, Shimoga, Mercara-Mysore. In: Institut Francais de Pondichery, p 88. Travaux de la Section Scientifique et Technique. Hors Serie N 18
16. USGS: Earth explorer, https://earthexplorer.usgs.gov/
17. Pouncey R, Swanson K, Hart K (1999) ERDAS field guide
18. Richards JA (2013) Remote sensing digital image analysis. Springer

Gaussian Membership Function and Type II Fuzzy Sets Based Approach for Edge Enhancement of Malaria Parasites in Microscopic Blood Images

Golla Madhu

Abstract This research presents a three-stage approach. In the first stage, the original image transformed into grayscale image, then normalizes grayscale image using min-max normalization, which performs a linear conversion on the original image data. The second stage calculates the Gaussian membership function on the normalized grayscale image then measure lower membership values and upper membership values using a threshold value. In addition, computed a novel membership function with Hamacher t-conorm using lower and upper membership values on given images. Finally, the median filter applied on these images to obtain edge enhanced microscopic images. The current study is conducted on the microscopic blood images of the malaria parasites. The experimental results compared with Prewitt filter, Sobel edge filter, and rank-ordered filter. The proposed approach is consistent and coherent in all microscopic malaria parasite images with four stages, with average entropy 0.90215 and 59.69% PSNR values, respectively.

1 Introduction

Medical imaging analysis has been increasing rapidly with the advent of imaging technologies and it plays a significant role in diagnosing a disease and for observing patient's health disorders and giving a competent treatment [1]. Most of the medical images such as blood images, magnetic resonance imaging (MRI), ultrasonic scanning, and X-rays consist with low contrast images [2]. In addition, these images are not clear because of their blur and uncertainties in edges or boundaries of the images. To handle these problems, we require a preprocessing technique that obtains the quality and clear edge images. Hence, edge enhancement technique plays a vital role in medical imaging analysis. In addition, it brings out the relevant features, which are not accurately detectable and interrupt redundant information [3].

G. Madhu (✉)
Department of Information Technology, VNR Vignana Jyothi Institute
of Engineering and Technology, Hyderabad 500090, Telangana, India
e-mail: madhu_g@vnrvjiet.in

© Springer Nature Switzerland AG 2019
D. Pandian et al. (eds.), *Proceedings of the International Conference on ISMAC in Computational Vision and Bio-Engineering 2018 (ISMAC-CVB)*, Lecture Notes in Computational Vision and Biomechanics 30,
https://doi.org/10.1007/978-3-030-00665-5_64

Malaria appears worldwide in more than 100 countries in the tropical and sub-tropical regions, and it becomes one of the prominent infection diseases, with an outcome of more than 2.3 billion deaths every year. In India, 80% of malaria cases occur among 20% of the population. In the year 2017, about 673,474 positive cases were found and of which 439,934 are of Plasmodium falciparum cases, out of a total 89,297,252 blood smears examined, until September 2017 [Source http://nvbdcp.gov.in/malaria3.html]. Therefore, parasite analysis in thick blood smears microscopic images became the gold standard for the diagnosis of malaria. Then, it requires the better edge enhancement technique for representation of blood cell recognition of the microscopic blood images that represents preprocessing step in image process-ing. Edge enhancement increase or enhance the cell edge or boundaries of the blood image [3]. One of the popular techniques also known as a median filter that preserves the edges or boundaries of the images while computing the variation with respect to the middle pixel component of the filter, then edges are enhanced on the image [3]. Fuzzy sets used to characterize uncertainty in the form of membership func-tion. This research, focus on the image obtained through microscopic blood images of malaria and use these images as input images. Then, generated grayscale image from input image and then computed Gaussian membership function on grayscale image using interval type II fuzzy sets, which represents lower membership values and upper membership values. Using these membership values, a novel member-ship function is computed by utilizing Hamacher t-conorm functionalities. Then, it applied the median filter on these images to obtain edge enhanced images that signif-icantly represent better quality of microscopic blood images. This technique is useful for medical experts to make easy decisions and predictions for further diagnosis of the patients.

2   Related Works

The aim of edge enhancement technique is to increase the quality of an image from the original image in order to highlight significant features of the image [4]. However, edge enhancement plays a significant role in imaging analysis that improves the edge contrast of an image [5]. In the literature, many researchers proposed and established several edge detection techniques, for example, Kirsch [6], Robert edge detection [7], canny edge detection [8], LOG [9], Sobel [10], and Prewitt [11]. However, this approach uses the notion of spatial variance filter using the local gradient approach on image pixels and search for local maxima to establish the stage edges. An advantage of these approaches is to process the image data, which is comparatively less in time to identify the edges and their positioning that computationally optimized. Nevertheless, they are very sensitive to noise data and missed data or to the false edges detected. Mokhtarian and Suomela [12] proposed an innovative approach for image edge detection using curvature scale-space illustration. Genming and Bouzong [13] developed a 5 × 5 kernel for finding the edges in an image using a threshold value. Conversely, their limitations are inadaptability of two regions with the fluctuating grayscale image due to their fixed threshold value. Fürhapter et al. [14] demonstrated an optical-based edge contrast enhancement in light microscopic images. Chaira [15] proposed an Attanassov's intuitionistic fuzzy sets for image edge detection. Marr and Hildreth [16, 17] suggested edge detection using a Gaussian filter and this approach has been a very popular technique, an earlier Canny liberated his detection. In this research, initially pointed out that the deviation of image intensity, and it happens at different levels of the image. Santis et al. [18] suggested a statistical approach for edge detection approach using the linear stochastic based signal technique, which obtained from the physical image description.

Fuzzy sets are used to perform analytical and logical approaches using rough approximations that are slightly crisp values which have been hired for edge detection significantly to reduce the uncertainty of the medical applications [19, 20]. Khamy et al. [20] recommended an improved fuzzy Sobel filter for edge detection as well as fuzzy reasoning methods, this method used to discover the image edges. Lu et al. [21] suggested a fuzzy set-based neural network (NN) approach for the edge enhancement then detection by improving the misplaced edges and discarding the false edges. Wu et al. [22] suggested a fast multilevel based fuzzy approach for image edge detection, which perform the robust and accurate detection of edges. Bustince et al. [23], proposed an interval valued-based fuzzy approach for edge detection of the image, then it is applied on common images but not on cellular images. Barrenechea et al. [24] discussed an interval-valued based fuzzy relation for the generation of the fuzzy edges on image data. Chaira [25] proposed an intuitionistic-based fuzzy set are used to edge enhancement of medical image data. Chaira [26] proposed a rank-order-based filter for edge enhancement of cellular images using an interval type II fuzzy sets and whole variation of the image of the edge enhancement that makes cell objects additional bulging with respect to background. Talari Tirupal et al. presented Sugeno's intuitionistic-based fuzzy set on medical image data which converted into Sugeno's intuitionistic based fuzzy image.

3   Interval Type II Fuzzy Sets

This research, presents basic notions of interval type II fuzzy sets and some important mathematical definitions and precise terms that will successfully communicate to proposed works [27]. In addition, interval type II fuzzy sets are differentiating from the type I fuzzy sets via adding uncertainty. Prof. Zadeh (1975) identified there was a difficulty with the type I fuzzy sets when he developed a fuzzy II and higher types [28]. The type I fuzzy set has commonly used in various real-world applications.

However, complex issues usually involve higher degree of uncertainty that cannot handle by the type I fuzzy set. To overcome such an issues, proposed an extension of the type I fuzzy set similarly known as the interval type II fuzzy sets [29]. In the interval-based type I fuzzy set membership, represent a type of vagueness or uncertainties. A sample of the basic Gaussian membership function is shown in Fig. 1a. Therefore, the type I fuzz set A is represented through $\mu_A(x)$, where

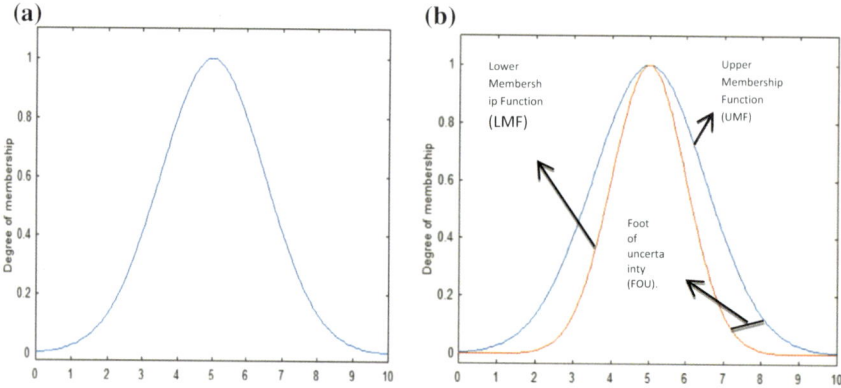

Fig. 1 **a** Represents the type I fuzzy set membership function, **b** Represents the interval type II fuzzy membership function also shown lower membership function (LMF), upper membership function (UMF) and foot of uncertainty (FOU)

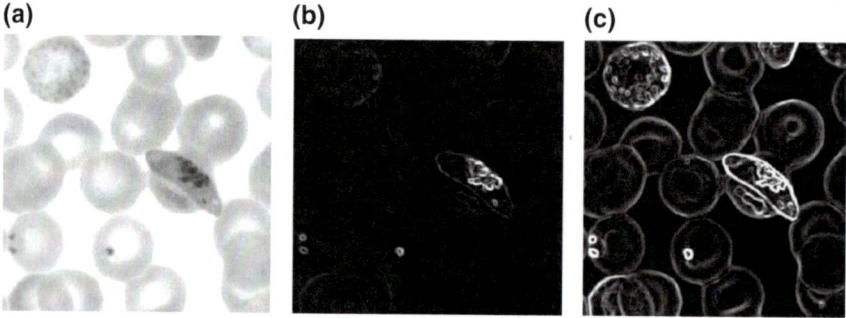

Fig. 2 **a** Representation of blood image with malaria parasite with gametocyte stage, **b** Corresponding edge enhanced results with $\lambda = 0.47$, **c** Edge enhanced results with for $\lambda = 1.37$

$A = \{x, \mu_A(x_{i,j}) | x \in X\}$ in [0,1]. Similarly, an interval type II fuzzy sets requires different types of membership grade functions that describe the uncertainty and each membership lies between [0,1]. Interval type II fuzzy set, defined the primary membership function, which represents ambiguity of boundary regions in the form of lower and upper. The uncertainty represented by a region is also known as foot of uncertainty (FOU) shown in Fig. 2. This represents the contour of Interval type II fuzzy set (shown Fig. 2) [30, 31].

Definition The interval type II fuzzy set [31] is defined in terms of membership function, which represents $\mu_{\tilde{A}}(x_{i,j}, a)$ where $x_{i,j} \in X$ and $a \in Y \subseteq [0, 1]$.

$$A = \{((x_{i,j}, a), \mu_{\tilde{A}}(x_{i,j}, a)) | \forall x_{i,j} \in X, \forall a \in Y \subseteq [0, 1] \tag{1}$$

In the Eq. (1), $\mu_{\tilde{A}}(x_{i,j}, a)$ represent the type II membership function and it belongs to $0 \leq \mu_{\tilde{A}}(x_{i,j}, a) \leq 1$.

To outline the interval type II fuzzy sets, describe an interval type I fuzzy set, and then assign a lower membership and an upper membership function that belongs to each pixel element and to (re)build the FOU [32]. A more concrete description for interval type II fuzzy set represented is as follows:

$$\tilde{A} = \left\{ \left(x_{i,j}, \mu_L(x_{i,j}), \mu_U(x_{i,j}) \right) \middle| \text{for all } x_{i,j} \in X, \mu_L(x_{i,j}) \leq \mu(x_{i,j}) \leq \mu_U(x_{i,j}) \right.$$
$$\text{where } \mu(x_{i,j}) \subseteq [0, 1] \tag{2}$$

In the Eq. (2), $\mu_L(x_{i,j})$ and $\mu_U(x_{i,j})$ represents lower membership and upper membership functions, respectively. In our research, these membership functions are represented as follows:

$$\mu_{\text{Lower}}(x_{i,j}) = \left[\mu(x_{i,j}) \right]^{1/\alpha} \tag{3}$$

$$\mu_{\text{Upper}}(x_{i,j}) = \left[\mu(x_{i,j}) \right]^{\alpha} \tag{4}$$

Here, the α value belongs to $0 \leq \alpha \leq 1$, this study, we used $\alpha = 0.37$ because of $\alpha \gg 2$ that is not significant representation for image data [32].

4 Proposed Methodology

The method of edge enhancement is mainly partitioned into three-stage approach that relates the Gaussian membership function using interval type II fuzzy sets. In the first stage, the image is transformed into grayscale image, and then normalized the image using min-max normalization, which performs a linear conversion on the original image data. The second stage calculates the Gaussian membership function on a normalized grayscale image and then measure lower membership values and upper membership values using a threshold value (α). In addition, calculate a novel membership function with Hamacher t-conorm using these lower and upper membership values on microscopic images. Finally, the median filter is applied on these images to obtain edge enhanced microscopic images. This research is conducted on the microscopic blood images of the malaria parasite to enhance the edges of the blood cells. The description of the proposed method is shown in algorithm-1.

In this research, the malaria blood images consist of four different stages like, ring forms, schizonts, trophozoites, and gametocyte stages. Thin peripheral blood smear samples were collected from Vivekananda Hospital, Begumpet, Hyderabad, Telangana 500016, India. Initially, the RGB images are transformed into the grayscale image A with the size M × N. Min-max normalization is to perform the linear transformation on the grayscale image using the following formula.

$$\mu' = \frac{x - \min_A}{\max_A - \min_A} (n\_\max_A - n\_\min_A) + n\_\min_A \tag{5}$$

where x is the gray values of the image that belongs to $[0, L-1]$. $\max_A$ & $\min_A$ are the, maximum and minimum values of the gray level of the image and $n\_\max_A$ & $n\_\min_A$ values are produced from the range of the image based on its class. This normalization formula preserves the relationships among the original image pixel values. Various types of type II fuzzy set based membership functions are proposed and applied in the literature such as Gaussian, triangular, sigmoid and trapezoidal, etc. In this work, we used Gaussian membership function $\mu_{\tilde{A}}(x_{i,j})$ which is defined as follows:

$$\mu_{\tilde{A}}(x_{i,j}) = \exp\left(-\frac{1}{2}\left(\frac{\vec{x}_n - \vec{m}_j}{\sigma_j}\right)^2\right) \tag{6}$$

where $\vec{x}_n = [x_{n1}, x_{n2}, \ldots, x_{nm}]^T$ is vector which is made by the pixel values of the nth sample of image. $\vec{m}_j = [x_{j1}, x_{j2}, \ldots, x_{jm}]^T$ is mean of the pixel values of the grayscale images denotes as $\vec{m}_j = \frac{1}{N}\sum_{\vec{x}_n \in S_j} \vec{x}_n$ and σ_j standard deviation that takes on values in lower membership and upper membership values. These lower and upper membership values are represented based on the type II fuzzy sets that are calculated using Eqs. (3) and (4). Fuzzy linguistic hedge with $\alpha \leq 1$ and its reciprocal value are drawn to the FOU, our study used $\alpha = 0.37$ that produce the lower membership and upper membership values, respectively. In addition, fuzzy linguistic hedges are utilized to adapt the membership values.

Utilizing the functionalities of generalized "Hamacher t-conorm" [33], we proposed a novel membership function as follows:

$$\text{Mem}(\mu_L(x_{i,j}), \mu_U(x_{i,j})) = \frac{\lambda(\mu_L(x_{i,j}) + \mu_U(x_{i,j})) + \mu_L(x_{i,j}) * \mu_U(x_{i,j})(1-2\lambda)}{(1+\lambda) + \mu_L(x_{i,j}) * \mu_U(x_{i,j})(1-\lambda)} \quad \text{for } \lambda \geq 0 \tag{7}$$

where $\mu_L(x_{i,j})$, $\mu_U(x_{i,j})$ represents the lower membership values, upper membership values of image dataset that are calculated from Eqs. (1) to (4). Using the results, a new fuzzy type image is generated which did not include an exact edges or non-edges of the image dataset. Therefore, we applied median filter (size 3×3) on this novel image dataset and then image edges are highlighted or enhanced. This study, a median filter technique is utilized to eliminate the impulse noise in microscopic images and this will not produce any new pixel value in the image when marks the edges when compared to averaging filters.

5 Experimentations and Results

The presentation of the proposed algorithm was assessed on microscopic blood smears of malaria parasite images of different stages such as ring forms, schizonts, trophozoites, and gametocyte stages. These microscopy images are prepared routine laboratory conditions of the hospital, Hyderabad, India under the supervision of pathologist. In addition, some malaria sample images are collected from differ-

---

Algorithm-1 A New Filtering Algorithm for Edge Enhancement

Input: Microscopic malaria blood image dataset $A = (m, n)$.

Output: Malaria parasite edge enhanced data image.

Step 1: Collect microscopic malaria blood images infected with Plasmodium falciparum of different stages such as ring forms, schizont, trophozoite, and gametocyte stages.

Step 2: Select a microscopic images and converted into grayscale images. Identify and eliminate the saturation and hue related information in given image while retentive the luminance.

Step 3: Min-max normalization technique used to perform the linear transformation on grayscale image using formula discussed in Sect. 3 in Eq. (5).

Step 4: Compute the membership function using Eq. (6) on normalized image dataset using step-3.

Step 5: Calculate lower and upper membership using Eqs. (3) and (4) with $\alpha = 0.37$ from step-4.

Step 6: Measure the new membership function using Eq. (7) with the help of step-5. A novel fuzzy image is generated which is not included an exact edge or non-edge image dataset.

Step 7: Apply median filter on this new membership function from derived from **Step-6**, and then edges of the new image are highlighted or enhanced.

Step 8: Repeat the Steps (2)–Step (8) for each image in data image.

Step 9: Finally, malaria parasite edge enhanced image obtained.

Step 10: Stop.

---

ent websites through online. In Fig. 2(a–c) presented a sample implementation of proposed algorithm on microscopic blood image. It clearly, we observe that with $0 \le \lambda \le 1.87$ edges are enhanced of the microscopic images which visually represent the better results other than $\lambda = 2$.

The parameter α is usually determined heuristically that belongs to [0,1], in this experiment we chosen $\alpha = 0.37 \in [0, 1]$. Gaussian membership values may be utilized by taking the standard deviation of the membership functions, i.e., lower and upper membership function results presented in Fig. 3.

From these outcomes, clearly, we observed that the microscopic blood images are infected parasites and edges are accurately enhanced with proposed method. Even though Sobel edge detection method and rank-ordered filter method, the edges are not clearly visible.

Figure 3a is grayscale of Plasmodium falciparum with schizonte stage image. Figure 3b shows type II lower membership values in terms of ambiguity and Fig. 3c presents type II upper membership values in terms of vagueness of image blood cells. Using the membership functions, the grayscale values is converted into fuzzy values. Figure 3d shows the Sobel edge operator approach, but this is not clear visual quality of malaria parasite edges in the image. Figure 3f represents the rank-ordered filtered method, which shows that the edges are not clearly highlighted. Figure 3e shows that edge boundaries are much more conspicuous using the proposed method. In terms of the assessment of malaria diagnosis, it is also important an estimation of the life stage of malaria parasite that infected erythrocyte.

Figure 4 shows illustrations sample of Plasmodium falciparum with ring stage, which can be confused with parasites in early stages. However, ring stage is difficult to segment because of overlapping edges with chromatin dots. Figure 4b, d parasite not clearly visible with Prewitt and rank-ordered filtered methods. However, in Fig. 4c

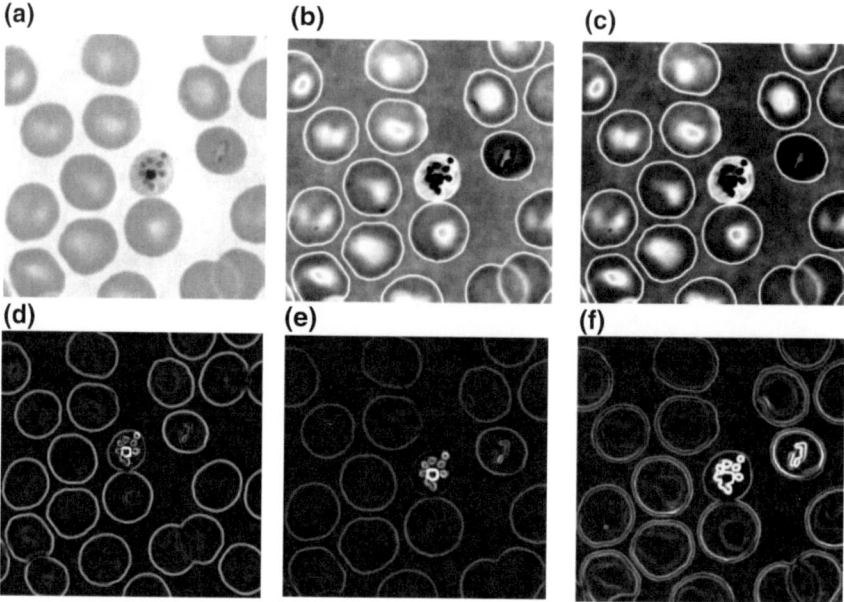

Fig. 3 **a** Original *P. falciparum* the schizonte image, **b** Gaussian lower membership function, **c** Gaussian upper membership function, **d** Sobel edge method, **e** Rank-ordered filtered method, **f** Proposed method

shows the results of Sobel filter method with clear edges only infected regions with blurry structures, which is quite better than Prewitt and rank-ordered filtered methods. Figure 4e demonstrated the outcomes of the proposed where image clearly enhanced edges that are much better than other methods with original image. Figure 5a indicates an image of *P. falciparum* of gametocyte stages, which is not clearly visible parasite.

Figure 5b illustrates the test outcomes of Prewitt filter method does not show the clear edges which shown some edge are broken also few were not clearly visible. Figure 5c illustrates the test results of using Sobel edge filter method that had slightly better than the Prewitt filter method. Figure 5d shows the test results of rank-ordered filter method, and it is observed that this approach does not enhance all the edges of the image and it enhanced on the infected gametocyte cell image. Figure 5e displays the results of proposed method, which performs well than Prewitt filter, Sobel edge filter, and rank-ordered filter methods. Finally, it concludes that the proposed approach which makes superior visual quality of edge enhancement among the existing methods.

Figure 6a shows an image of *P. falciparum* with trophozoite stage stages, which is not clearly visible parasites. Figure 6b shows the test results of Prewitt filter method which does not show the clear blood cell edges which shown some edges are broken also some cell edges are not clearly visible. Figure 6c shows the test results of using Sobel edge filter method that had slightly better edge enhancement than the Prewitt

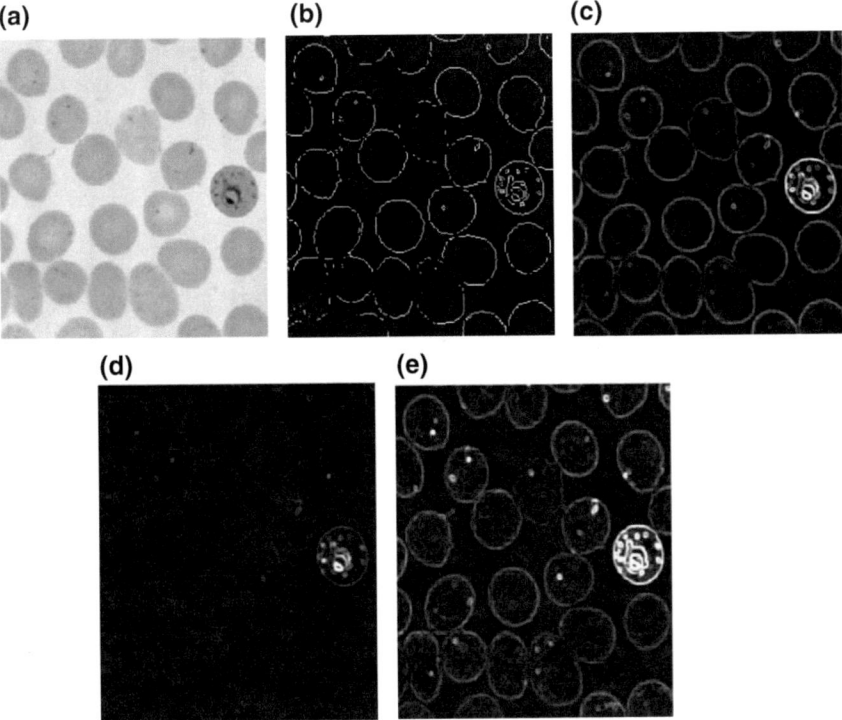

Fig. 4 **a** Represent the *P. falciparum* image with ringform stage, **b** Edge enhancement image based on Prewitt method, **c** Edge enhancement image based on Sobel edge method, **d** Edge enhancement image based on rank-ordered filter method, **e** Edge enhancement image of our method

filter method. Figure 6d shows the test results of rank-ordered filter method, it is observed that this approach also does not enhance all the edges of the blood image and it enhanced on the infected gametocyte cell image. Figure 6e represents the test results of suggested method, which performs better than Prewitt filter, Sobel edge filter, and rank-ordered filters. Finally, it determined that the suggested technique performs better visual quality on edge enhancement among all other methods.

5.1 *Quantitative Performance Analysis*

After visual valuation, quantitative performance analysis is also mandatory to compute the edge enhanced medical images. There are many methods to estimate the quantitative performance measure such as Shannon's entropy, Peak signal-to-noise ratio (PSNR), index of fuzziness, mean square error rate, and other methods. In this study, Shannon's entropy and PSNR is used to estimate the quality of the blood image. The Shannon's entropy is presented as follows:

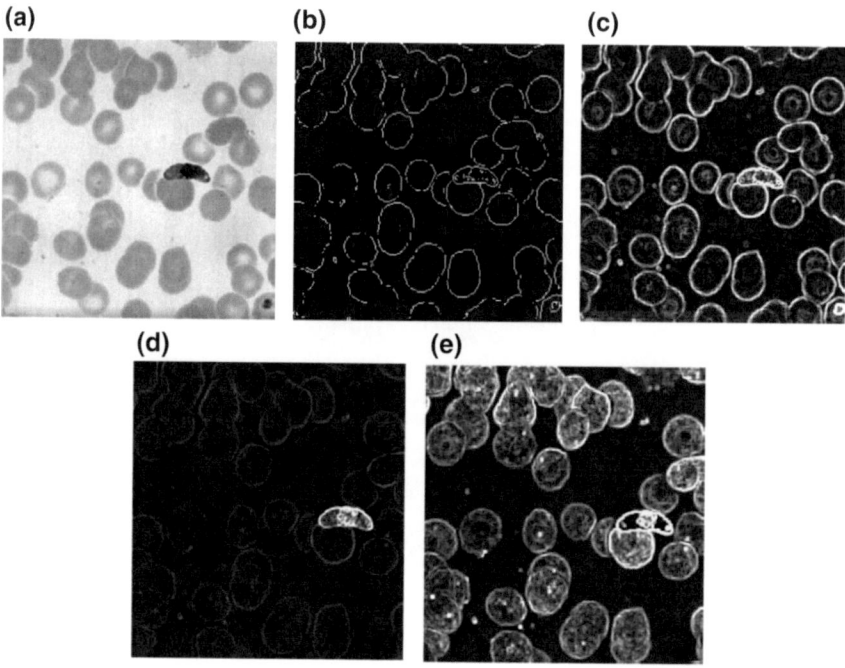

Fig. 5 **a** Represents the original *P. falciparum* with gametocyte stage, **b** Edge enhancement image based on Prewitt method, **c** Edge enhancement image based on Sobel edge method, **d** Edge enhancement image based on rank-ordered filter method, **e** Edge enhancement image based on proposed method

$$\text{Entropy}(X) = \frac{1}{(M \times N)} \sum_i \sum_j \mu_{\tilde{A}}(x_{ij}) \times \log\left[\mu_{\tilde{A}}(x_{ij})\right]$$
$$- \{1 - (\mu_{\tilde{A}}(x_{ij})) \times \log(\mu_{\tilde{A}}(x_{ij}))\} \tag{8}$$

where $M \times N$ represents the size of the image and $\mu_{\tilde{A}}(x_{ij})$ represent the membership function values of (i, j) the pixel values. PSNR is one of the standard quality measures in image analysis. In this experimentation, the PSNR value for grayscale image is defined by

$$\text{PSNR} = 20 * \log_{10}(A) - 10 * \log_{10}\left(A - \hat{A}\right)/255 \tag{9}$$

where $\hat{A}$ represent the edge enhanced image and A represents the original grayscale image.

Table 1 represents the entropy of microscopic blood image of malaria with different stages of malaria parasite. From this table, it is clearly observed that Prewitt filter shows the average entropy of 0.783425, Sobel edge filter demonstrate average entropy of 0.868975, and rank-ordered filter shows average entropy of 0.8081,

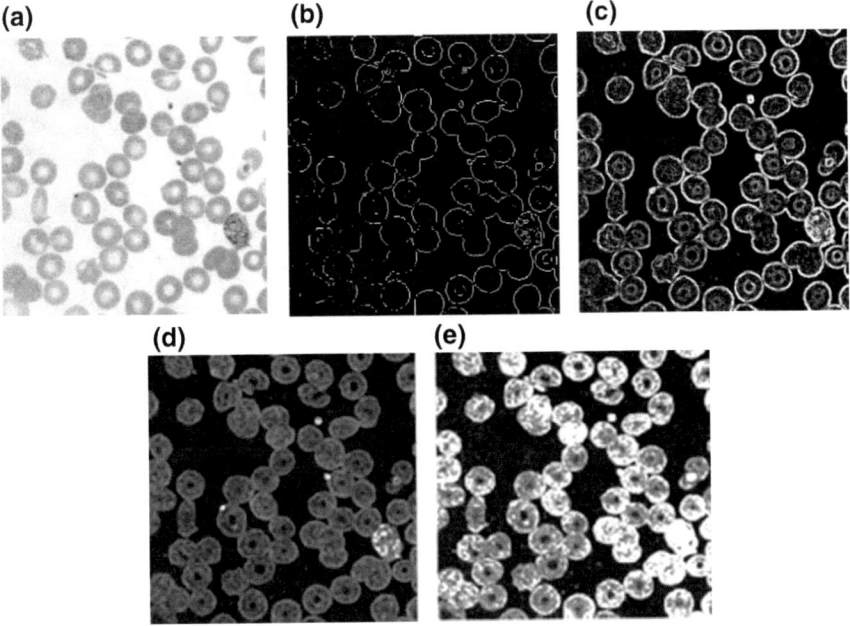

Fig. 6 **a** Represent the original *P. falciparum* with Trophozoite stage, **b** Edge image based on Prewitt method, **c** Edge image based on Sobel edge method, **d** Edge image based on rank-ordered filtered method, **e** Edge image based on proposed method

Table 1 Shannon's entropy values obtained while computing the quality of the malaria image dataset for different stages

Entropy measure				
Malaria dataset with different stages				
Microscopic images	Prewitt method	Sobel method	Rank-ordered filter	Proposed method
Ring forms	0.8682	0.8870	0.9235	0.9332
Gametocyte	0.7077	0.8447	0.8796	0.8912
Shizont	0.8409	0.8709	0.7578	0.8894
Trophozoites	0.7169	0.8733	0.6715	0.8948
Average	0.783425	0.868975	0.8081	0.90215

and our approach shows the average entropy of 0.90215. However, the suggested approach is significantly improved with higher entropy values when compared with other methods.

From Table 2, it clearly says that Prewitt method shows higher PSNR value for gametocyte stage with 56.3382 and lower PSNR value for trophozoites 54.1041. Sobel edge method shows higher PSNR value for ring form stage with 55.9305 and lower PSNR value tophozoites with 49.0710, rank-ordered filter method shows higher

Table 2 PSNR values obtained while computing the quality of the malaria image dataset for different stages

PSNR measure				
Malaria dataset with different stages				
Microscopic images	Prewitt method	Sobel method	Rank-ordered filter	Proposed method
Ring forms	55.2071	55.9305	54.3999	55.9601
Gametocyte	56.3382	55.3186	59.0502	59.6952
Shizont	55.0194	53.1093	57.8757	58.5438
Trophozoites	54.1041	49.0710	54.6634	54.7890

PSNR value for gametocyte with 59.0502 with lower PSNR value 54.3999 and the proposed method shows higher PSNR value for gametocyte stage with 59.6952 and lower PSNR value for trophozoites with 54.7890. However, the proposed method has higher PSNR values when compared with other edge enhanced methods.

6 Conclusions

In this study, we have demonstrated three-stage edge enhancement approach for microscopic blood images. In first stage, we converted microscopic blood image into grayscale image and normalized the grayscale image then computed Gaussian membership values with type II fuzzy sets. Second stage computes the lower membership and upper membership values of grayscale image. Using these membership values calculated a new membership function with Hamacher t-conorm on microscopic parasite images. Final stage, median filter applied on these images to obtain edge enhanced microscopic malaria parasite images. Quantitatively and qualitatively, results compared with other edge enhancement methods. It was found that the proposed approach is consistent and coherent in all microscopic malaria images with four stages, with average entropy 0.90215 and 59.69% PSNR values, respectively. It is clearly observed that the suggested technique is intelligent to determine blur and dark microscopic blood images. It can be used not only in microscopic images, but also in any other boundary edge enchantment of the image. This research is more capable for medical imaging analysis that makes sense for physicians decision for further diagnosis.

Acknowledgements The author acknowledge DRDO-DRL, Tezpur, Assam, India for providing financial support to carry out this work (Task No. DRDO-DRLT-P1-2015/Task-64).

References

1. Hassanpour H, Samadiani N, Salehi SM (2015) Using morphological transforms to enhance the contrast of medical images. Egypt J Radiol Nucl Med 46(2):481–489
2. Firoz R, Ali MS, Khan MNU, Hossain MK, Islam MK, Shahinuzzaman M (2016) Medical image enhancement using morphological transformation. J Data Anal Inf Process 4:1–12
3. Chaira T (2012) A rank ordered filter for medical image edge enhancement and detection using intuitionistic fuzzy set. Appl Soft Comput 12(4):1259–1266
4. Kumar A, Shaik F (2016) Image processing in diabetic related causes, forensic and medical bioinformatics. https://doi.org/10.1007/978-981-287-624-9_2
5. Russo F (2002) An image enhancement technique combining sharpening and noise reduction. IEEE Trans Instrum Meas 51(4):824–828
6. Kirsch RA (1971) Computer determination of the constituent structure of biological images. Comput Biomed Res 4(3):315–328
7. Rosenfeld A (1981) The max roberts operator is a Hueckel-Type edge detector. IEEE Trans Pattern Anal Mach Intell 3(1):101–103
8. Canny J (1986) A computational approach to edge detection. IEEE Trans Pattern Anal Mach Intell 8(6):679–698
9. Ulupinar F, Medioni G (1990) Refining edges detected by a LoG operator. Comput Vis Graph Image Process 51(3):275–298
10. Zhang J-Y, Chen Y, Huang X (2009) Edge detection of images based on improved Sobel operator and genetic algorithms. In: IASP 2009 international conference image analysis and signal processing, pp 31–35
11. Lei Y, Dewei Z, Xiaoyu W, Hui L, Jun Z (2011) An improved Prewitt algorithm for edge detection based on noised image. In: 4th International congress on image and signal processing (CISP), Shanghai, China. IEEE, pp 1197–1200, 15–17 Oct 2011
12. Mokhtarian F, Suomela R (1998) Robust image corner detection through curvature scale space. IEEE Trans Pattern Anal Mach Intell 20(12):1376–1381
13. Genming C, Baozong Y (1998) A new edge detector with thinning and noise resisting abilities. J Electron 6(4):314–319
14. Fürhapter S, Jesacher A, Bernet S, Ritsch-Marte M (2005) Spiral phase contrast imaging in microscopy. Opt Express 13(3):689–694
15. Chaira T (2008) A new measure on intuitionistic fuzzy set and its application to edge detection. Appl Soft Comput 8:919–927
16. Marr D, Hildreth E (1980) Theory of edge detection. Proc R Soc London B 207(1167):187–217
17. Kasturi R, Jain RC (eds) (1991) Computer vision: principles. IEEE Computer Society Press, Los Alamitos, CA
18. Santis AD, Sinisgalli C (1999) A Bayesian approach to edge detection in noisy images. IEEE Trans Circ Syst I Fundam Theory Appl 46(6):686–699
19. Kuo YH et al (1997) A new fuzzy edge detection method for image enhancement. In: 6th IEEE international conference on fuzzy systems, Barcelona, Spain. IEEE, pp 1069–1074, 1–5 Jul 1997
20. Khamy E et al (2000) A modified fuzzy Sobel edge detector. In: 17th National radio science conference, pp C32/1–C32/9
21. Lu S, Wang Z, Shen J (2003) Neuro-fuzzy synergism to the intelligent system for edge detection and enhancement. Elsevier J Pattern Recogn 36:2395–2409
22. Wu J, Yin Z, Xiong Y (2007) The fast multilevel fuzzy edge detection of blurry images. IEEE Signal Process Lett 14(5):344–347
23. Bustince H et al (2009) Interval valued fuzzy sets constructed from matrices: application to edge detection. Fuzzy Sets Syst 160:1819–1840
24. Barrenchea E et al (2011) Construction of interval valued fuzzy relation with application to generation of fuzzy edge images. IEEE Trans Fuzzy Syst 19(5):819–830
25. Chaira T (2012) A rank ordered filter for medical image edge enhancement and detection using intuitionistic fuzzy set. Appl Soft Comput 12(4):1259–1266

26. Chaira T (2015) Rank-ordered filter for edge enhancement of cellular images using interval type II fuzzy set. J Med Imaging 2(4):044005
27. Tirupal T, Mohan BC, Kumar SS (2017) Multimodal medical image fusion based on Sugeno's intuitionistic fuzzy sets. ETRI J 39(2):173–180
28. Mendel JM, Bob John RI (2002) Footprint of uncertainty and its importance to type-2 fuzzy sets. In: Proceeding of 6th IASTED international conference artificial intelligence and soft computing, Banff, Canada, pp 587–592
29. Mendel JM (2017) Type-2 fuzzy sets. In: Uncertain rule-based fuzzy systems, pp 259–306
30. Mendel JM, Bob John RI (2002) Type-2 fuzzy sets made simple. IEEE Trans Fuzzy Syst 10(2):117–127
31. Mendel JM (2001) Uncertain rule-based fuzzy logic systems. Prentice-Hall, Englewood Cliffs, NJ
32. Tizhoosh HR (2005) Image thresholding using type II fuzzy sets. Pattern Recogn 38(12):2363–2372
33. Roychowdhury S, Wang BH (1998) On generalized Hamacher families of triangular operators. Int J Approximate Reasoning 19(3–4):419–439

Implementation of Virtual Trial Room for Shopping Websites Using Image Processing

Niket Zagade, Akshay Bhondave, Raghao Asawa, Abhishek Raut and B. C. Julme

Abstract In today's world, the use of e-commerce websites has increased a lot. Online shopping gives us more information and choice about different products under various categories which are readily available. The customer just has to choose between the different products, purchase them, and the product comes to the customer's doorstep. Thus, many people like to buy many things online. Clothing is one of such categories that people can buy online. But for shopping clothes, this scenario is a little different. There is a major problem that people do not know how the clothes would actually look on them and so many people avoid buying clothes online. Sometimes the customers even send back the clothes they buy as it does not look good on them. This is why a virtual trial room has to be developed so that people do not have to wait to try on the clothes physically after it is delivered. They can try clothes virtually on the virtual trial room.

1 Introduction

In current scenario, online shopping is a big boon and people show quiet a lot interest in it. Many people buy clothes online from the e-commerce websites, but there is always a challenge while buying them. The customer who wants to buy apparel does

N. Zagade (✉) · A. Bhondave · R. Asawa · A. Raut · B. C. Julme
Department of Computer Engineering, PVG's College of Engineering and Technology, Pune, India
e-mail: niketzagade198@gmail.com

A. Bhondave
e-mail: akshaybhondave26@gmail.com

R. Asawa
e-mail: raghaosa@gmail.com

A. Raut
e-mail: abhishekraut695@gmail.com

B. C. Julme
e-mail: bcj_comp@pvgcoet.ac.in

© Springer Nature Switzerland AG 2019
D. Pandian et al. (eds.), *Proceedings of the International Conference on ISMAC in Computational Vision and Bio-Engineering 2018 (ISMAC-CVB)*, Lecture Notes in Computational Vision and Biomechanics 30,
https://doi.org/10.1007/978-3-030-00665-5_65

not know how it would actually look upon them. This is a major reason why some people do not buy clothes online. Some people even buy some apparel online and then later return it as it does not look good on them.

For this reasons, a virtual dressing room will be a great useful tool for many online sellers. In this software, people would know how the clothes would look on them. Shopping websites want to make the shopping process easy for the customers. However, while shopping the clothes people always have that last doubt of how the dress would look on them and so they rather end up going to the stores physically so that they can actually try the clothes before buying it.

In our project we are developing a virtual trial room for the above problem. This project will show the customers how the clothes would look on them virtually, i.e., without actually trying them. The customer just has to select the clothes that they want to try and the software will fit it virtually over their image and show how it looks upon them.

This software would be used by the shopping websites to help improve their relationship with the customer. By using this software customers would feel satisfied before buying any clothes online.

2 Literature Survey

Previously many researches have been done on virtual trial room. This work can be classified into 2 groups: (1) 2D image based and (2) 3D model based.

For 2D image based approach there is a website named Awasaba [1] which gives Internet-based-/web-based interface for virtualized clothes over a stationary model. Srinivasan [2] implemented virtual fitting room in which human silhouette extracted then wrapping of shirt and virtual fitting is done. On the basis of body contour of customer and model, yamada [3] has proposed a model for garment image reshaping and getting more reliable fitting over actual customer body.

3D model based approaches are mainly implemented when virtual humans are used as model. It consists of deformation transfer, garment shape estimation, wrinkle estimation, etc. [4, 5] For garment shape estimation [6], CNN (Convolutional Neural Network) has been used to estimate 3D vertex displacement from template mesh.

Deformation transfer copies the deformation exhibited by a source mesh onto different target mesh. Despite these progressive methodologies, customer experiences difficulties because photo-realistic rendering of 3D garments and virtual human model is not so efficient. Additionally the processing cost of such 3D model is comparatively higher than 2D image based methods. 2D based methods have an advantage of collecting data and photo-realistic rendering are easier than 3D model based approach.

Some methodologies use cloning and dress people using website [7]. In a body and garment creation method for an Internet-based virtual fitting room by Protopsaltou [8], they built a compelling, interactive, and highly realistic virtual shop. Here customers can check how the garments would look on the human body by virtually fitting the clothes over the animated bodies. They present a straight-forward garment virtualization technique. In Made-to-Measure Technologies [9], they provide a web application to give easy and fast access to and manipulate garments to facilitate the characteristics like design, pattern derivation, and sizing of garments.

In paper by Tong [10], using Kinect sensor modeling method, human body 3D modeling is done. To get the human body model three steps are followed, first with the help of anthropometric parameters model is parameterized, using the PCL library the point cloud data is processed and matched and then realistic human body model is obtained.

Ehara and Saito [11] has used a database of marker attach t-shirt images in different poses. Zhau et al. prepared animated garment in different poses and done real-time virtual trial fitting by superimposing garment over customer using Kinect [12]. But using a Kinect is not feasible for every user in the world because they cannot carry Kinect everywhere for now and it is costly. Sometimes Kinect gives distorted result because of environment condition/light disturbance/illumination.

3 Proposed Work

We are developing a java application using opencv for image processing. In our project, we are implementing a virtual trial room where first we will click/take a picture of customer according to our predefined pose. The customer has to stand accordingly inside the outlined structure given in the window. We are implementing haar classifier on that image to detect full body, upper body, and lower body of customer. After that we are imposing the garments on the customer's body according to the detected parts.

The system we are developing is for the e-commerce shopping websites where the customers can select the garments they want to try. These garments will be fed to java application as input.

The software will impose garments over the image of customer's body. The use of Haar classifier is during the superimposing of garments over the body of customer. Haar classifier detects body of customer and draws a rectangular contour around body. There are three kinds in haar classifier: (1) Full body (2) Upper body (3) Lower body.

According to contour drawn by haar classifier, we will give relative position to the garments. There can be small number of false alarms in this algorithm but it is reliable for still images [13]. So we are providing a manual interface to overcome the anomalies if any occurs. The manual interface can move the superimposed garment. Customers can do it to get the proper fitting as they desire.

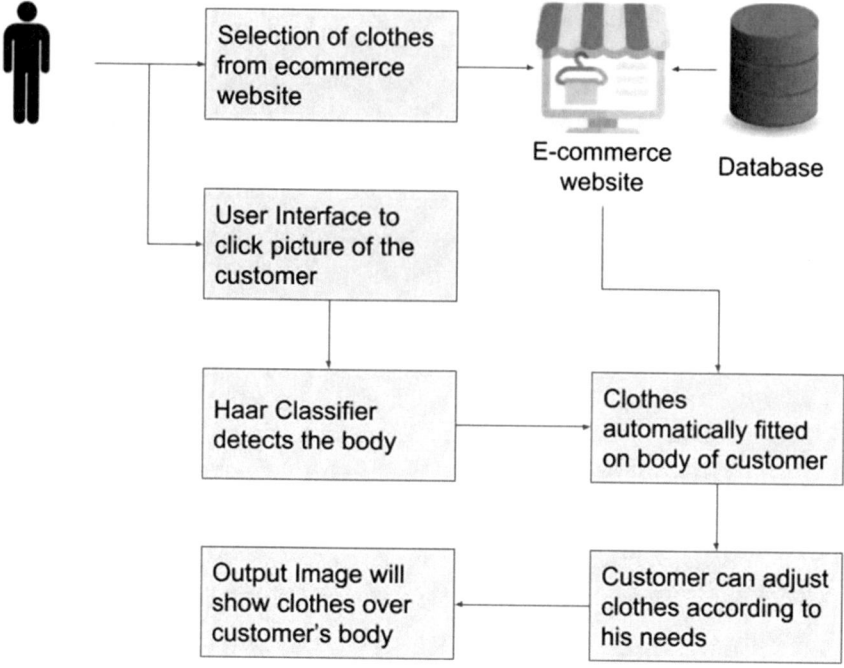

4 Conclusion

Several methods are used to improve the process of shopping clothes online. This software will definitely improve the customer–seller relationship and help the online sellers to expand their business. The processing speed will be better than that of Kinect and as we use webcams, the common man can easily use it. It is definitely reliable and will help both sellers and customers in their respective constraints.

References

1. Corporation, "Awaseba," www.awaseba.com/
2. Srinivasan K, Vivek S (2017) Implementation of virtual fitting room using image processing. In: IEEE International conference on computer, communication and signal processing (ICCCSP-2017)
3. Yamada H, Hirose M, Kanamori Y, Mitani J, Fukui Y (2014) Image-based virtual fitting system with garment image reshaping. In: 2014 International Conference on Cyberworlds (CW)
4. Sumner RW, Popović J (2004) Deformation transfer for triangle meshes. ACM Trans Graph (Proceeding of SIGGRAPH 2004) 23(3):399–405
5. Wang H, Hecht F, Ramamoorthi R, O'Brien J (2010) Example-based wrinkle synthesis for clothing animation. In: ACM SIGGRAPH 2010 Papers, pp 107:1–107:8

6. Guan P, Reiss L, Hirshberg DA, Weiss A, Black MJ (2012) DRAPE: DRessing Any PErson. ACM Trans Graph 31(4):35:1–35:10
7. Cordier F, Lee W, Seo H, Magnenat-Thalmann N (2001) From 2D photos of yourself to virtual try-on dress on the web. Springer, Berlin, pp 31–46
8. Protopsaltou D, Luible C, Arevalo-Poizat M, Magnenat-Thalmann N (2002) A body and garment creation method for an internet based virtual fitting room. In: Proceeding of computer graphics international 2002 (CGI '02). Springer, Berlin, pp 105–122
9. Cordier F, Seo H, Magnenat-Thalmann N (2003) Made-to-measure technologies for an online clothing store. IEEE Comput Graph Appl 23(1):38–48
10. Tong J, Zhou J, Liu L, Pan Z, Yan H (2012) Scanning 3D full human bodies using kinects. IEEE Trans Vis Comput Graph (Proceeding of IEEE Virtual Reality) 18(4):643–650
11. Ehara J, Saito H (2006) Texture overlay for virtual clothing based on PCA of silhouettes. In: Proceedings of the 5th IEEE and ACM international symposium on mixed and augmented reality, ser. ISMAR '06. IEEE Computer Society, pp 139–142
12. Zhou Z, Shu B, Zhuo S, Deng X, Tan P, Lin S (2012) Image-based clothes animation for virtual fitting. In: SIGGRAPH Asia 2012 Technical Briefs, pp 33:1–33:4
13. Kruppa H, Castrillon-Santana M, Schiele B (2003) Fast and robust face finding via local context. In: Joint IEEE international workshop on visual surveillance and performance evaluation of tracking and surveillance

Response Analysis of Eulerian Video Magnification

S. Ramya Marie and J. Anudev

Abstract The human eye has a very high optical resolution making it one of the most astonishing curiosities of the world. However, its spatial resolution is not good enough to capture everything happening around it, and it can miss out minor details, which can be termed as hidden movements. The hidden movements can be due to the extremely high speed of the visual, very small movements, long-term physical process, etc. Eulerian video magnification is a spatiotemporal video processing algorithm that can reveal hidden details that are otherwise hidden to naked eyes. In this process, a standard video sequence is spatially decomposed, and temporal filtering of the frames is done. The data so obtained as output can be used in many fields such as biomedical instrumentation, remote surveillance, etc. Here, an analysis has been done on Eulerian Video Magnification (EVM), for different video resolutions to understand its reliability.

1 Introduction

- The world around us is very dynamic and is in the process of continuous transition. The human eye is a highly complex organ which has evolved to react to a particular range of light and pressure. It can provide a three-dimensional, moving image, normally coloured in daylight. A normal human eye has a resolution of about 576 megapixels [1]. An evolutionary trait for an eye is to constantly keep collecting visual data of its surroundings and keeping track of threats, it does so by having a peripheral vision that gives it an initial impression or context before we focus on something. The peripheral vision helps us decide where to concentrate and helps us find the region of interest for our focused vision [2], such as the spotting of a

S. Ramya Marie · J. Anudev (✉)
Department of Electrical and Electronics Engineering, Amrita School of Engineering,
Amrita Vishwa Vidyapeetham, Amritapuri, Kollam, India
e-mail: anudevj@am.amrita.edu

S. Ramya Marie
e-mail: ramyamaries@am.students.amrita.edu

© Springer Nature Switzerland AG 2019
D. Pandian et al. (eds.), *Proceedings of the International Conference on ISMAC in Computational Vision and Bio-Engineering 2018 (ISMAC-CVB)*, Lecture Notes in Computational Vision and Biomechanics 30,
https://doi.org/10.1007/978-3-030-00665-5_66

lurking predator. This has helped human beings in survival as a species; however, the human eye has compromised on its spatiotemporal sensitivity, and this has caused many subtle changes to go unnoticed in everyday life, such pulse rate of a person, breathing rate, sag and sway of a bridge, subtle colour changes of our skin, etc. [3]. These visuals are not really necessary for normal everyday life, but these subtle motions have great potential which can be tapped in biomedical imaging [4, 5], E-health systems [6], telemedicine [7], remote sensing, predicting of natural disasters such as avalanche, etc. The rapid improvement in the fields of biomedical imaging can be attributed to the efforts aimed at making instrumentation systems capable of sensing multiple bio-signals, non-contact type [8], accurate such as retinal imaging [9], safer for vulnerable population [10], etc.

- It is very hard to notice these small changes using a human eye; however, a computer technique can be used for visualizing such subtle changes in colour and motion variations in videos by making the variations caused by these subtle changes larger. This can be done by amplifying motions in the videos which are otherwise moved only by a hundredth of a pixel to become magnified enough to span many pixels [3]. This process is called Video Magnification. Some common video magnification algorithms are the linear Eulerian [11], phase-based [12], and Riesz [13] algorithms. This is similar to how a microscope amplifies an optical image. Using video magnification algorithm, subtle colour changes, motions, etc., can be obtained from a video that seems static. These subtle signals can be quantitatively analysed for obtaining other parameters such heart rate, respiratory rate, reconstruct sound from an object by measuring the vibrations of the object on a high-speed video, etc.

- The video magnification algorithm to be used for these purposes need to be efficient and robust. Eulerian Video Magnification (EVM) is a linear video magnification technique and is both robust and efficient. It gives better results than the motion magnification [14] where a small motion is amplified by computing per-pixel motion vector and then displacing the pixel value by magnified motion vectors. This technique yields very good results and has very high reliability; however, it is computationally expensive and any error in motion analysis would generate artefacts in the outputs which are motion magnified. Thus, this technique with high potential risk gives extremely deviant output in case of any unexpected errors.

- The most basic version of the spatiotemporal video processing has been considered here for analysing the reliability of this algorithm. It identifies the intensity variations over time for each pixel and amplifies them. The technique is used to identify subtle colour changes by measuring the colour intensity of a particular pixel.

2 Methodology

2.1 Experimental Set-up and Procedure

The experiment is aimed at studying the reliability of Eulerian video magnifications in analysis or study of subtle colour changes that are usually hidden to human eyes in daily life. Three standard optical videos are considered here. They are of 144p, 180p and 192p in resolution, respectively. These optical videos can be considered as static optical videos with no visible changes that can be detected by a naked human eye.

The static video considered here was that of a human cheek which had no visible changes that could be observed. This subject was chosen so as to illustrate the potential of EVM in the field of biomedical instrumentation. The subject of the video was made to sit still, and the video was taken for a duration of 15 s. This process was done using three cameras of different resolutions at the same time. Human cheek is subject to constant colour changes due to the perfusion of blood through the face. The human face is redder when the heart pumps blood and is more of a shade of yellow than red when the heart contracts. This is because the blood rushes through the face [15] as the heart compresses, causing the face to gain a colour shade from the colour of blood and when the heart contracts the colour of the skin is more prominent. However, this colour changes are very subtle and hence not noticeable by human eyes.

For this experiment, the subject is advised to be at rest for a duration starting from at least 15 min prior to which the video was taken. The cameras are mounted on a tripod stand; this is for preventing any error due to external vibrations. It is would be better if the frame rate of the camera is high.

2.2 Eulerian Video Magnification (EVM)

The basic idea of colour magnification in Eulerian video magnification is to amplify variation of colour values at any point or spatial location (pixel) in a temporal frequency band, which suits a certain phenomenon [16]. A common example of EVM has been illustrated below. Here, a particular band of the temporal frequencies for skin colour has been amplified. This provides the variation of redness as the blood flows through the face. These variations have been in turn used to obtain human heart rate.

The technique used here to extract and reveal the required signal is localized spatial pooling and bandpass filtering. The whole process is schematically shown in Fig. 1. To obtain a decent quality of the output video in a limited time period, before analysing, down-sample the videos and filter it using a spatial low-pass filter. This reduces the noise and to boosts the subtle changes in the video. Then this video is decomposed into different bands of wavenumbers k ($k = 2\pi\lambda$).

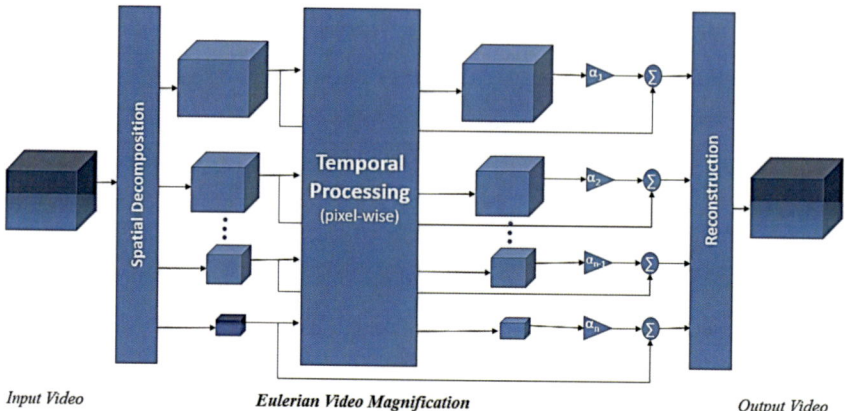

Input Video   Eulerian Video Magnification   Output Video

Fig. 1 Schematic diagram of EVM

Spatial processing is to be then followed by temporal processing, which is to be performed on each band of wavenumbers. Applying a bandpass temporal filter to each band of wavenumbers helps in extracting (pass) the motions that suit the frequency bands of the observed phenomenon. Magnification of motion of different passed wavenumbers is different because of two reasons. First, the signal-to-noise ratios corresponding to every wavenumber could be different. Second, some wavenumbers considered might not hold true for the linear approximation used in motion magnification. In the second case, the magnification is reduced to suppress artefacts. After the magnification is completed, these magnified bandpassed signals are added to the original signal to observe the effect of motion/colour magnification.

The process can be simply put as:

(1) The video is first decomposed into different bands of wavenumbers.
(2) The same temporal filter is then applied to all bands of wavenumbers, to reveal the time interval of the motion and the motion of each band of wavenumbers.
(3) Then a passband filter is applied to pass the bands of wavenumbers that suit the time interval of the observed phenomenon.
(4) The filtered bands of wavenumbers are then amplified by a given factor α, and then added back to the original signal and collapsed to generate the output video.

2.3 Filter and Amplification Criteria

The principle of Eulerian video magnification depends heavily on its filters and amplification factors. Procedures for selecting the filters and amplification parameters [16] are:

(1) Initially, a temporal bandpass filter that suits the phenomenon under observation was selected, to extract the desired motions or signals. The choice of temporal filter depends on the signal to be extracted; for example, temporal filters with broad passbands are used for broad but subtle motion magnification, narrow passband filters are used for colour amplification, etc.

(2) The user then selects the amplification factor α and a wavenumber cut off (specified by spatial wavelength, α) beyond which an attenuated version of α is used. α can be forced zero for all wavelength less than the spatial wavelength.

3   Simulation Results of Eulerian Video Magnification

The simulation was done using MATLAB. Here, a pre-recorded video was used. The output video from EVM was saved as a sequence of frames and was stored framewise into the hard disk as shown in Fig. 2. This output video was then further used for plotting mean light intensity of a primary colour to that of the number of frames in the video. The aforesaid video was 15 s long and had 104 frames. At first, the 144p video was considered, and a graph was plotted using data from EVM output of the video, which described the relationship between the intensity of light for a particular pixel with respect to the frame.

The graph shown in Fig. 3 draws comparisons between the output of EVM for the three different resolutions.

The output of the Eulerian video magnification for a static video consistently gave the same value as output irrespective of the number of times the video was processed over and over again. Also, for different resolutions, it has been observed that the Eulerian video magnification output tend to follow the input and didn't show much deviation from each other.

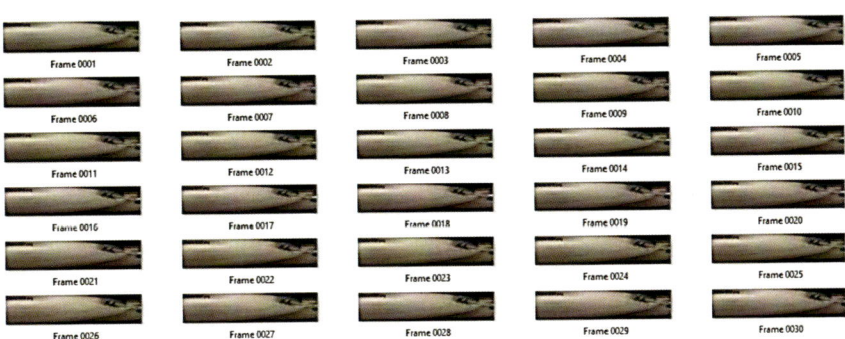

Fig. 2 Frames of the output video of Eulerian video magnification

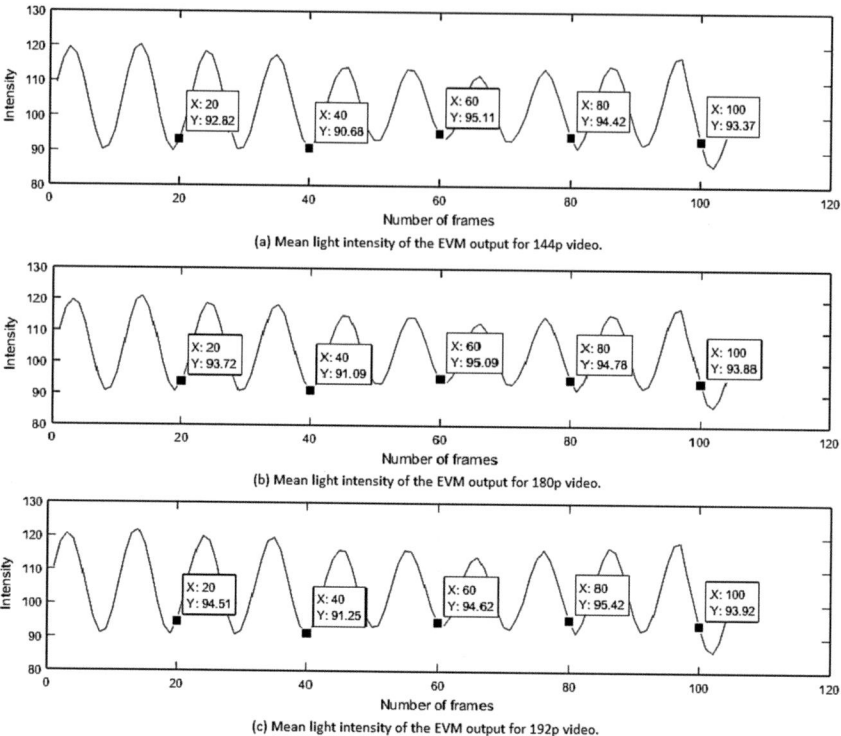

Fig. 3 Plots of mean intensity of a primary colour against the number of frames of the output of EVM for different video resolutions

4 Comparison of EVM Outputs for Different Resolutions

The light intensity of a primary colour pixel of Eulerian video magnification output for different resolutions has been plotted here. Comparisons have been drawn between light intensity at 144p and 180p in Table 1 and light intensity at 180p and 192p in Table 2. The difference between the light intensity for a particular frame for the outputs of the EVM has been compared and their difference has been found. It has been found that the magnitude of the difference between the resolutions with each other is always within the range of ±1. This difference is marginal and won't make much difference when it comes to major instrumentation systems when calculating error percentages, and they can be reduced by proper digital filtering and can still well within engineering standards [17, 18].

Table 1 Comparison between the light intensity of a particular primary colour pixel for EVM output of 144p and 180p videos

Frame	Light intensity		Deviation (±)
	144p	180p	
20	92.82	93.72	0.9
40	90.68	91.09	0.41
60	95.11	95.09	−0.02
80	94.42	94.78	0.36
100	93.37	93.88	0.51

Table 2 Comparison between the light intensity of a particular primary colour pixel for EVM output of 180p and 192p videos

Frame	Light intensity		Deviation (±)
	180p	192p	
20	93.72	94.51	0.71
40	91.09	91.25	0.16
60	95.09	94.62	−0.47
80	94.78	95.42	0.68
100	93.88	93.92	0.04

5   Conclusion

It has been observed that the graphs showed the same values irrespective of the number of times the program has been repeated. This shows that the algorithm fulfils the major instrumentation criterion such as repeatability, reproducibility, accuracy, reliability, etc. It has great fidelity of input to output ratio, and hence irrespective of the camera, the results are bound to be proportional. Upon using thermal imaging, the problem involving external factors such as light intensity, fluctuations due to fluctuations in the inconsistent light source can be avoided. In the future, this algorithm can find applications in biomedical instrumentation systems for measuring subtle processing in the human body such as heart rate, clots, respiratory rates, etc.

Acknowledgements The authors would like to express their gratitude to Dr. Ravikumar Pandi, Project Coordinator, Assistant Professor, Dept. of Electrical and Electronics Engineering, Amrita Vishwa Vidyapeetham, Amritapuri for his continuous support and motivation, and Chairperson Dr. Manjula G. Nair, Dept. of Electrical and Electronics Engineering, Amrita Vishwa Vidyapeetham, Amritapuri for providing all opportunities and facilities for the fulfilment of this work.
The authors also express their gratitude to the panel of reviewers who helped in reviewing the work and helped in organizing the contents.

References

1. Clarkvision photography [Online]. Available: http://www.clarkvision.com/articles/eye-resolution.html
2. The complexities of the human eye from the blind spot and macula to focused and peripheral vision. [Online]. Available: https://www.zeiss.co.in/vision-care/enin/better-vision/understanding-vision/eye-and-vision/the-complexities-of-the-human-eye-from-the-blind-spot-and-macula-to-focused-and-peripheral-vision.html
3. Wadhwa N, Wu H-Y, Davis A, Rubinstein M, Shih E, Mysore GJ, Chen JG, Buyukozturk O, Guttag JV, Freeman WT et al (2016) Eulerian video magnification and analysis. Commun ACM 60(1):87–95
4. Bennett SL, GoubranR, Knoefel F (2016) Adaptive Eulerian video magnification methods to extract heart rate from thermal video, pp 1–5
5. Menon HP, Narayanankutty KA (2016) MRI/CT image fusion using gabor texture features, vol 530, pp 47–60
6. Huang A, Xie L (2015) Healthinfo engineering: technology perspectives from evidence-based mhealth study in we-care project. Int J E-Health Med Commun 6(1):22–35
7. Liu L, Stroulia E, Nikolaidis I, Miguel-Cruz A, Rincon AR (2016) Smart homes and home health monitoring technologies for older adults: a systematic review. Int J Med Inf 91:44–59
8. Nilakant KR., Menon HP, Vikram K (2017) A survey on advanced segmentation techniques for brain MRI image segmentation. Int J Adv Sci Eng Inf Technol (Insight Society) 7(4):1448–1456
9. Menon HP, Gayathri V (2017) Vasculature detection from retinal color fundus images using linear prediction residual algorithm. Int J Pure Appl Math (Academic Press) 114(12):171–178
10. Vadivelu S, Ganesan S, Murthy OR, Dhall A (2016) Thermal imaging based elderly fall detection
11. Wu H-Y, Rubinstein M, Shih E, Guttag J, Durand F, Freeman W (2012) Eulerian video magnification for revealing subtle changes in the world
12. Wadhwa N, Rubinstein M, Durand F, Freeman WT (2013) Phase-based video motion processing. ACM Trans Graph (Proceedings SIGGRAPH 2013) 32(4)
13. Wadhwa N, Rubinstein M, Durand F, Freeman WT (2014) Riesz pyramids for fast phase-based video magnification. In: 2014 IEEE international conference on computational photography (ICCP). IEEE, pp 1–10
14. Liu C, Torralba A, Freeman WT, Durand F, Adelson EH (2005) Motion magnification. ACM Trans Graphics (TOG) 24(3):519–526
15. Edwards DJ, Cattell M (1930) The action of compression on the contraction of heart muscle. Am J Physiol Legacy Content 93(1):90–96
16. Brecelj T (2013) Eulerian video magnification. In: University of Ljubljana Faculty of Mathematics and Physics, pp 1–15
17. Instrumentation error calculation and setpoint determination. Engineering Standard ES-002, pp 1–14 (1994)
18. Rajevencelta J, Kumar CS, Cattell M (2016) Improving the performance of multi-parameter patient monitors using feature mapping and decision fusion. In: Region 10 conference (TENCON). IEEE, pp 1515–1518

Face Authentication and IOT-Based Automobile Security and Driver Surveillance System

Mahesh R. Pawar and Imdad Rizvi

Abstract Automobile industry is one of the largest and fastest growing industry and the actual reason behind it is, up-growing men to vehicle ratio. A lot of new vehicles are coming in the market and people are using them by spending substantial amount of money. This increasing ratio of man to vehicle is squeezing the crimes regarding vehicle robbery and accidents even though there are lots of safety features readily available. Hence this paper proposes simple low cost solution, based on strong biometric mechanism that involves face authentication. The system that uses night vision camera to capture the face of person seating on the driver's seat and some sensors to provide his surveillance in the accidental situations. This system also gives us instant alert with latest captured image of vehicle's interior on email.

1 Introduction

Automobile industry has tremendously grown up in the past years and still it is growing. It has become one of the largest as well as fastest growing industry and there is huge potential in it. Plenty of companies every year introduces new cars with many exciting and innovative features for luxury, security and also for comfort. The automobile industry is getting upgraded with the help of automation and moving towards new era. Many companies have already introduced their automated electric cars. If we observe that, even after having these features, there are still cases of vehicle thefts and misuse of vehicles has not reduced but increased. People die due to lack of surveillance systems, Many criminals walk free because of lack of proper evidence against them. Government, transport authority. The police are trying their best to overcome these crimes, Hence it is a time to upgrade the vehicle by itself with

M. R. Pawar (✉) · I. Rizvi
Department of Electronics and Telecommunication Engineering,
Terna Engineering College, Navi Mumbai, India
e-mail: maheshrpawar9@gmail.com

I. Rizvi
Electrical Engineering Division, Higher Colleges of Technology, Sharjah, UAE

© Springer Nature Switzerland AG 2019
D. Pandian et al. (eds.), *Proceedings of the International Conference on ISMAC in Computational Vision and Bio-Engineering 2018 (ISMAC-CVB)*, Lecture Notes in Computational Vision and Biomechanics 30,
https://doi.org/10.1007/978-3-030-00665-5_67

simple, low cost, easily available and trustworthy mechanism, which will actually help to give guaranteed security and surveillance of vehicle.

2 Literature Review

2.1 Tracking Method

If we consider last few years of work on automobile security, many techniques and methods are evolved. One among them is GPS (Global Positioning System) [1, 2] sensor which gives position of automobile in altitude and longitude. Then this data is transferred over air to owner or cop. This wireless transmission is mostly done with IP based modules like GPRS (General Packet Radio Service) to make IoT-based system or using the GSM (Global System for Mobile Communication) module [3, 4]. Figure 1 can give brief idea about it [5, 6].

2.2 Biometric Authentication

The biometric authentication-based methods are like, use of finger scanner, retina scanner, voice recognition, authentication, and giving commands [7–9], then one of the most acceptable method, that is face authentication. With above biometric methods they also implement security alarm or cutting off fuel-supply, locking doors, sending images to other side or owner. All transmission of data is mostly made by the SMS (Short Message Service) or MMS (Multimedia Messaging service) through GSM or GPRS modules [10–16].

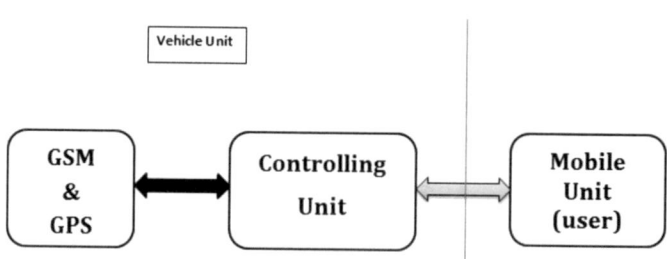

Fig. 1 Simplified block diagram GSM and GPS based system

3 Proposed Solution

Low cost with high reliability is primary objective of proposed system. The solution is conceptually much simple. Authors intention is to provide low cost, less complex, highly reliable system and At the same time it must be user friendly so that to a lay man can handle it. The person trying to access the vehicle must get identified first and then should get authorized. The following embedded system block diagram in Fig. 2 is the proposed idea with increased simplicity and functionality.

The proposed system requires 5 V/2 A power supply. Camera is used which acts like a transducer and it will take image and will provide suitable form of image in electric signal. This data will be send to processor for further processing and operation. We have used 8MP night vision Noir camera having maximum resolution of "3280 × 2464". Still we are taking pictures with resolution of "1024 × 768". Sensor detects the vibrations and makes output low or high, as it directly connected by wire to the controlling unit as it is shown in Fig. 2.

Controlling unit plays vital role, which takes all inputs and also process it. Raspberry pi 3b is used as development board. Vehicle control is an internal part of vehicle which include the key component of the vehicle, that is an ignition system. It will be handled by the controlling unit. Owner's device means it could be the mobile or Computer from where user can access an email account.

Raspbian stretch Operating System through noobs as per standard procedure is used. Python language is used for programming and VI editor for writing as well as editing code are used. OpenCV software is considered for the image processing. For face Recognition there are 3 algorithm-based functions made available in OpenCV. These are "Eigenfaces" second one "Fisherfaces" and third one is "Local Binary Patterns Histograms" (LBPH).

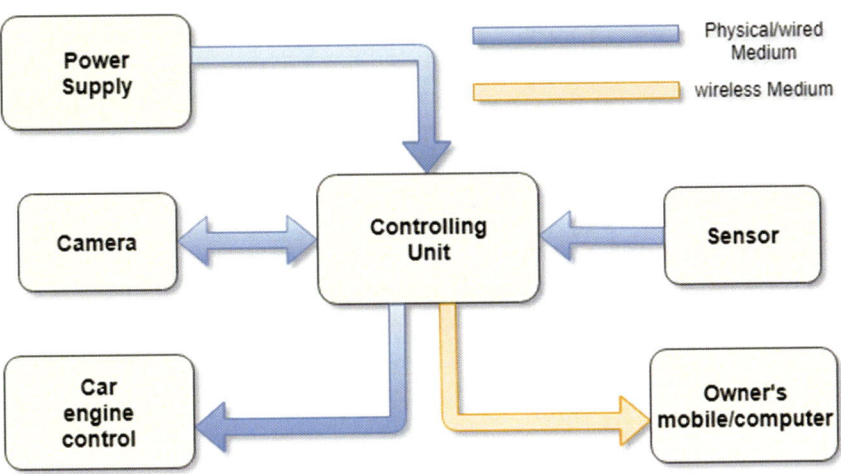

Fig. 2 Simplified block diagram of system

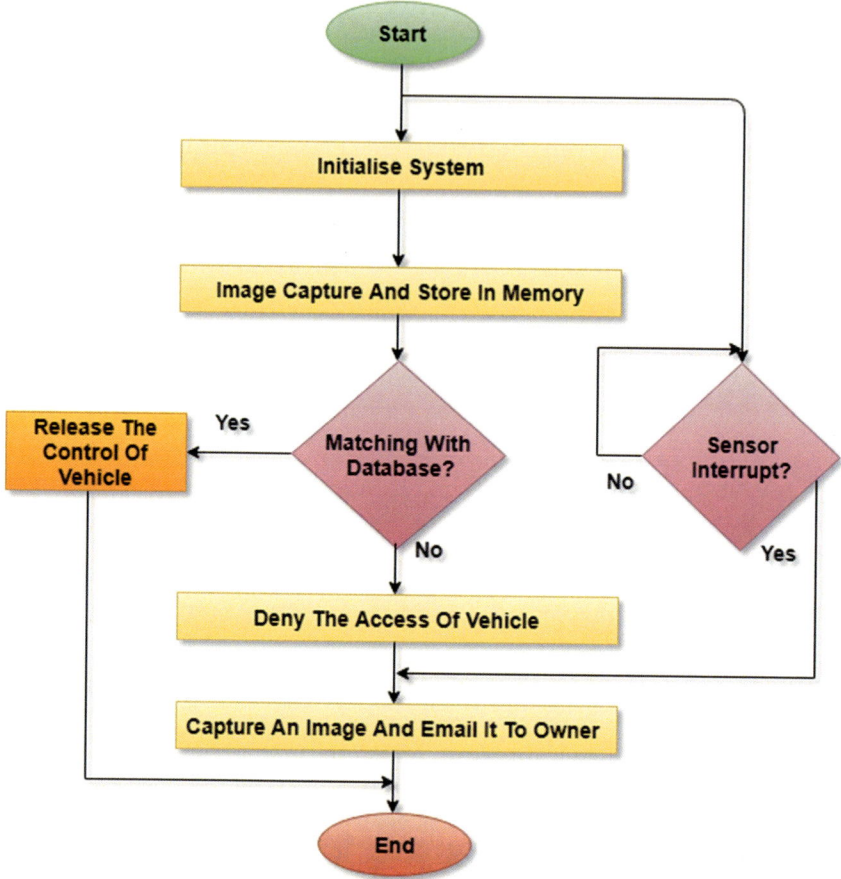

Fig. 3 Flowchart of proposed system

"Fisherfaces" and "Eigenface's" finds a mathematical relation and description of the most prior dominant property of our training image database as a whole. LBPH analyzes every face in training image database individually and independently.

The driver's image is captured via camera. Controller will do the comparison of recent image captured by the camera and the pre-stored images of authorized person as shown in Fig. 3. If the image matches, then controller will release the access of vehicle for authorized person, else controller denies the access of vehicle and it will send the image of unidentified person through an email to an authorized person. The vibration sensor will be continuously monitored for the vibrations. If any sudden vibrations are detected at any instant then camera will capture the image and send it to owner, on his email ID.

4 Result Analysis

Proposed system uses LBPH face recognition as it is simple and gives better result in different angle of views conditions. Proposed system also uses night vision camera which results in excellent face matching results in darkness too.

Figure 4 shows the database of vehicle owner. There could be multiple authorized individuals as per owners wish. Figure 4 indicates database of two authorized individuals.

Figure 5 shows face detection as well as authentication. Proposed system has successfully identified the authorized persons with their names.

Figure 6 shows that the proposed system has successfully identified an unknown person whose image was not in the database. Then it sends an email to an owner containing alert message with the image of the unauthorised person.

Similarly, the camera takes the picture after sensing heavy vibrations by vibration sensor. As soon as the proposed system senses the vibrations instantly sends the alert

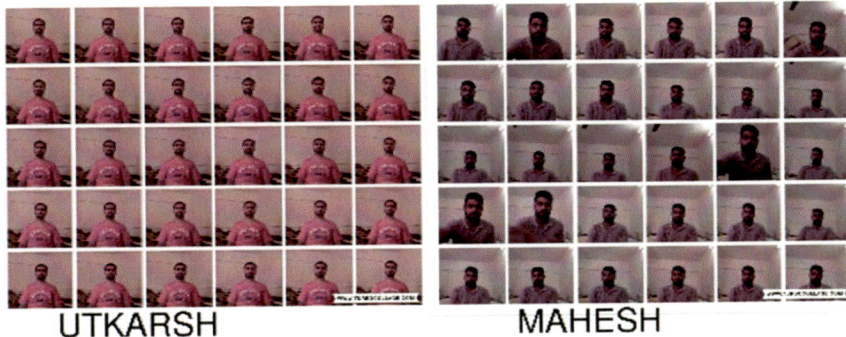

UTKARSH   **MAHESH**

Fig. 4 Database of Mahesh and Utkarsh

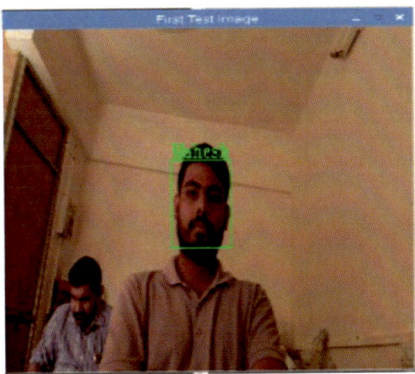

Fig. 5 Known person detected

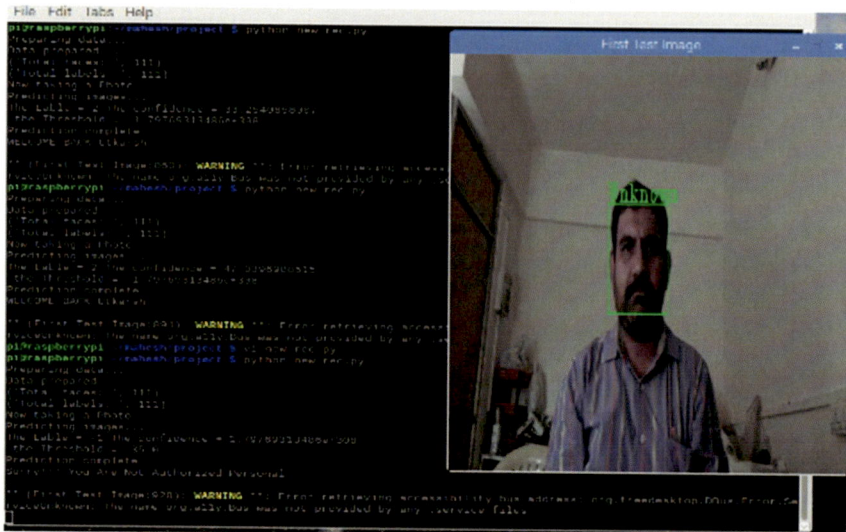

Fig. 6 Unknown person detected

Fig. 7 Complete hardware setup

message with picture of the interior of the vehicle. The actual setup containing all the hardware is shown in Fig. 7.

5 Conclusion

The proposed system easily distinguishes known and unknown person. The use of night vision camera gives excellent results even at the night time. Developed system completes the procedure within few seconds. The owner of car can easily park his vehicle anywhere as he gets instant security alert even for vibrations in vehicle.

The proposed system gives alert at low cost as well as evidence of unauthorised person. It also provides driver surveillance. This system works at very low power like on portable power bank used for the mobile charging. Hence it works like a silent watchman for an automobile.

Acknowledgements The authors would like to take the opportunity to thank all those individuals who have contributed or helped in some way in the preparation of this text. The authors would like to take the opportunity to thank all those individuals who have contributed or helped in some way in the preparation of this text. Special thanks to Mr. Vijay Kamble for their guidance, technical support and for providing me platform. I thank to Utkarsh chatte and again to Mr. Vijay Kamble for permission to use their faces for this research.
In my department, I thank to Dr. Imdad Rizvi for their academic support. For grammar and language support I thank to Prof. Geetanjali yampalle. I would also like to thank to Mandakini Chede and Apurav Patil for encouraging me.

References

1. Liu Z, Zhang A, Li S (2013) Vehicle anti-theft tracking system based on Internet of things. In: IEEE international conference on vehicular_electronics and safety (ICVES), Dongguan, pp 48–52. https://doi.org/10.1109/icves.2013.6619601
2. Shruthi K, Ramaprasad P, Ray R, Naik MA, Pansari S (2015) Design of anti-theft vehicle tracking system with a smartphone application. In: 2015 international conference on information processing (ICIP), Pune, pp 755–760. https://doi.org/10.1109/infop.2015.7489483
3. Lita I, Cioc IB, Visan DA (2006) A new approach of automobile localization system using GPS and GSM/GPRS transmission. In: 29th international spring seminar on electronics technology, St. Marienthal, pp 115–119. https://doi.org/10.1109/isse.2006.365369
4. Bavya R, Mohanamurali R (2014) Next generation auto theft prevention and tracking system for land vehicles. In: 2014 international conference on information communication and embedded systems (ICICES), Chennai, pp 1–5. https://doi.org/10.1109/icices.2014.7033987
5. Hameed SA, Khalifa O, Ershad M, Zahudi F, Sheyaa B, Asender W (2010) Car monitoring, alerting and tracking model: enhancement with mobility and database facilities. In: 2010 international conference on computer and communication engineering (ICCCE), Kuala Lumpur, pp 1–5. https://doi.org/10.1109/iccce.2010.5556796
6. Ajaz S, Asim M, Ozair M, Ahmed M, Siddiqui M, Mushtaq Z (2005) Autonomous vehicle monitoring & tracking system. In: 2005 student conference on engineering sciences and technology, SCONEST 2005, Karachi. https://doi.org/10.1109/sconest.2005.4382882
7. Bagavathy P, Dhaya R, Devakumar T (2011) Real time car theft decline system using ARM processor. In: 3rd international conference on advances in recent technologies in communication and computing (ARTCom 2011), Bangalore, pp 101–105. https://doi.org/10.1049/ic.2011.0059
8. Muji SZM, Wahab MHA, Zin MAM, Ayob J (2008) Simulation of smart card interface with PIC for vehicle security system. In: International conference on computer and communication engineering 2008, ICCCE 2008, pp 878–882
9. Ahilan A, James EAK (2011) Design and implementation of real time car theft detection in FPGA. In: 2011 third international conference on advanced computing, Chennai, pp 353–358. https://doi.org/10.1109/icoac.2011.6165201
10. Padmapriya S, KalaJames EA (2012) Real time smart car lock security system using face detection and recognition. In: 2012 international conference on computer communication and informatics (ICCCI), Coimbatore, pp 1–6. https://doi.org/10.1109/iccci.2012.6158802

11. Sreedevi P, Nair BSS (2011) Image processing based real time vehicle theft detection and prevention system. In: 2011 international conference on process automation, control and computing (PACC), Coimbatore, pp 1–6. https://doi.org/10.1109/pacc.2011.5979056
12. Sasikumar S, Ganesan R (2014) Facial and bio-signal fusion based driver alertness system using dynamic bayesian network. In: 2014 international conference on green computing communication and electrical engineering (ICGCCEE), Coimbatore, pp 1–5. https://doi.org/10.1109/icgccee.2014.6922268s
13. Sasikumar S, Ganesan R (2014) Facial and bio-signal fusion based driver alertness system using dynamic bayesian network. In: International conference on green computing communication and electrical engineering (ICGCCEE), Coimbatore, pp 1–5. https://doi.org/10.1109/icgccee.2014.6922268
14. Rizvi I, Chawda D (2009) Simulation of antilock braking system. In: International conference on emerging trends in software & networking technologies (ETSNT'09), Noida, India
15. Kolli A, Fasih F, Machot A, Kyamakyac K (2011) Non-intrusive car driver's emotion recognition using thermal camera. In: Proceedings of the joint INDS'11 & ISTET'11, Klagenfurt, pp 1–5. https://doi.org/10.1109/inds.2011.6024802
16. Saifullah A, Khawaja H, Arsalan, Maryam, Anum (2010) Keyless car entry through face recognition using FPGA. In: 2010 international conference on future information technology and management engineering (FITME), Changzhou, pp 224–227. https://doi.org/10.1109/fitme.2010.5654862

Highly Repeatable Feature Point Detection in Images Using Laplacian Graph Centrality

P. N. Pournami and V. K. Govindan

Abstract Image registration is an indispensible task required in many image processing applications, which geometrically aligns multiple images of a scene, with differences caused due to time, viewpoint or by heterogeneous sensors. Feature-based registration algorithms are more robust to handle complex geometrical and intensity distortions when compared to area-based techniques. A set of appropriate geometrically invariant features forms the cornerstone for a feature-based registration framework. Feature point or interest point detectors extract salient structures such as points, lines, curves, regions, edges, or objects from the images. A novel interest point detector is presented in this paper. This algorithm computes interest points in a grayscale image by utilizing a graph centrality measure derived from a local image network. This approach exhibits superior repeatability in images where large photometric and geometric variations are present. The practical utility of this highly repeatable feature detector is evident from the simulation results.

1 Introduction

The problem of image registration (IR) involves techniques for matching or alignment of several images of an object or scene taken by different sensing devices in different orientations, at different scales and times. IR has variety of applications in computer vision problems such as change detection, motion analysis, image matching, object recognition, and image fusion. The availability of diverse categories of data from many application areas has led to numerous research attempts during the recent past decades. For any image registration system, the following necessary aspects must be defined: feature space, transformation model, similarity metric, and an optimization method.

P. N. Pournami (✉) · V. K. Govindan
National Insitute of Technology Calicut, Calicut, Kerala, India
e-mail: pournamipn@nitc.ac.in

V. K. Govindan
e-mail: govi.kunnumal@gmail.com

© Springer Nature Switzerland AG 2019
D. Pandian et al. (eds.), *Proceedings of the International Conference on ISMAC in Computational Vision and Bio-Engineering 2018 (ISMAC-CVB)*, Lecture Notes in Computational Vision and Biomechanics 30,
https://doi.org/10.1007/978-3-030-00665-5_68

687

The feature space represents the information in the images before carrying out the matching process. The classes of transformations that are used to align the input images form the transformation model or search space. The choice of the transformation model from the search space is determined by the search strategy employed in the optimization process. The best choice is determined based on the similarity computed using the chosen similarity metric. The search process for obtaining the appropriate transformation parameters continues until a best acceptable match is determined between the reference and the float image [1]. The cases when structural information in images is predominant or significant, image representations using features in a concise form leading to compressed representation of data. The consequent reduction in search space and search time is the major advantage. Such representations allow faster registration of complex and even distorted images. These methods are faster when compared to pixel-based approaches. The choice of robust discriminative features, their extraction, and robust feature matching techniques are the major factors determining the success of this type of registration approaches. Feature points can either be automatically extracted or be manually selected. The robust features are invariant under rotation, translation, and scales changes. Use of discriminative features provides accurate and faster matching between float and reference images.

2 Feature-Based Image Registration

Feature-based registration algorithms are more robust to handle complex geometrical and intensity distortions when compared to area-based techniques. They can significantly reduce the execution time as less number of features is used for calculating the mapping between the input images. Feature point-based registrations effectively reflect the image structure information and they provide clear solutions to image registration. A set of appropriate geometrically invariant features forms the cornerstone for a feature-based registration framework. Because of these facts, the development of powerful feature-based registration methods has been the topic of current investigation [1].

Feature point detectors, descriptors and techniques for matching the derived features are the building blocks of the aforesaid algorithms. Feature-point or interest point detectors extract salient structures such as points, lines, curves, regions, edges or objects from the images. The mapping between the reference and the oat images is computed using the translation parameters, rotation angles, and the scaling parameters. This requires an acceptable number of matching control points from the input images, which in turn impose certain properties known as affine invariance on the extracted interest points. Such a detector will definitely improve the efficiency of a feature-based registration system when associated with a stable point descriptor. In recent years, many robust algorithms have been formulated in the category of feature-based approaches of image registration. Being the backbone of such techniques, various feature detectors have also been proposed by researchers. The following section discusses about feature point detectors.

2.1 Feature or Interest Point Detector

Feature detectors are algorithms for identifying features or points of interest in images. Points of interest are feature points that are to be detected robustly. They should be computed easily. The immediate neighborhood of a feature point has sufficient information content so that a meaningful point descriptor can be attached. Feature points should be stable under large photometric and geometric variations in the image. Feature based systems represent the current active direction of research in image registration technologies, which is the final issue investigated in this thesis. Image registration methods based on interest points or features achieve fast and high image registration.

2.2 Review of Recent Works

Zitova and Flusser [1] directed their efforts toward classifying image registration techniques according to the essential ideas, which reveals that an image registration process has the following components—detection and extraction of features, feature matching, transform model estimation, finally resampling and transformation of images. Each of these steps puts significant complications in image registration, starting from the selection of appropriate features for the given task. The features should be spread over the images and neighborhoods of feature points are rich in information content. The features usually are invariant to geometrical variations in the images and to any image degradation such as noise, blur, illumination changes, etc. Once, we get matching features in the input images, the correspondence can be established between them. This helps to derive a possible transformation, which aligns the float image with the reference image.

Harris Corner Detector, SUSAN detector, scale invariant feature transform (SIFT) detector, etc., are the major first-generation interest point detectors. The basic formulation of these detectors does not accommodate affine invariance. Later, many scientists tried to incorporate modifications to these detectors so that they become invariant to translation, rotation, scaling, camera viewpoint changes, illumination changes, noise, etc. Feature point detectors not affected by affine transformation are termed affine invariant detectors. These points are repeatable even after large distortions introduced by affine transformation [2–5]. This, in turn, motivated people to solve many computer vision applications that need feature-based matching such as object recognition, image registration, image retrieval from large databases, symmetry detection, texture recognition, etc.

One major work in this class is by Mikolajczyk and Schmid [2] where they proposed detector for interest points that are invariant under affine transformation. The algorithm is insensitive to translation, scale and shape of the immediate point neighborhood. The authors report that the proposed approach for interest point detection permits region matching even with large changes in viewpoints. Miao and Jiang [6] extracted interest points out of a nonlinear rank-order Laplacian of Gaussian filter. This algorithm can be used to extract regional structures in images where most pixels in the regions are darker or brighter than their corresponding surroundings. This permits detection of abrupt variations in illumination and geometric changes. Authors state that the approach is capable of detecting interest points efficiently with respect to repeatability and discriminative nature. The detected points of interest were applied to face recognition problem on five standard databases and depicted high recognition rates.

Criado et al. reported a graph centrality based feature point detector in [7]. Global and local centrality measures are employed. This method constructs a complex network out of the image regions created with morphological watershed operations. Though the performance of this technique is comparable with that of standard Harris and Stephens's detector, the authors report that the capability is low at over-segmented regions of images. Post-processing techniques can definitely improve the scope so that interest points are detected more sharply. A wavelet-based interest point detector for visible points is proposed in [8]. Annular color histogram and texture histogram were used to establish the affine invariance of this detector. This algorithm exhibited high precision value for an image retrieval application over a data set of 1000 images.

In 2008, Saydam et al. [9] narrated a point of interest detector with remarkable feature localization making use of non-sub-sampled Contourlet Transform. Both the local and global detectors outperformed many popular algorithms in the event of scale, viewpoint and rotational changes. This feature detector is suitable for computer vision and object recognition applications. Gevrekci and Gunturk [10] devised a feature point detector, capable of handling large photometric changes in the input images. By using the interest point detector along with a contrast-stretching operator, termed as contrast invariant feature transform, this algorithm enhances the Harris corner detector. Huge computational complexity prevents this technique to be used in practical applications.

Maver [11] developed a feature point locating technique utilizing the saliency of various regions in the image, computed at different scales. The authors reported good performances in image matching and object recognition. But this algorithm failed in many cases because it is not affine adaptable with respect to local regions. Lee and Chen [12] exploited low-level histograms computed from the images for finding interest points. Since this technique is invariant to blur and illumination changes, it could be effectively applied to image matching and action classification problems. A comparative study of four feature point detectors was performed by Martins and Carvalho [13]. In noisy images, the proposed detector showed stability when applied for image matching.

In 2010, Xie et al. [14] designed a feature point detector with scale invariance property. Harris detector is applied to detect feature points in low scale-space and these points are filtered using Hessian determinant to locate points in higher scale-space. This technique exhibited real-time response in large-scale data. Li et al. [15] presented a ranking scheme for interest point detection to identify stable local interest points. Repeatability of the interest points extracted is employed in this method. By limiting the number of stable interest points, this algorithm gives a noticeable performance in image retrieval problem.

Zukal et al. [16] reported a corroborative study of KLT detector, FAST, and Harris-Laplace detector in terms of information content. There are interest point detection algorithms which were designed specifically for image registration applications. Wen and Sheng [17] presented a framework for feature-based image registration under noisy imaging conditions, using local SIFT operator and cross-correlation information. A robust image registration algorithm combining SURF detector and FREAK descriptor is reported by Yanhai et al. in 2015 [18]. Ma et al. [19] suggested an image registration technique for synthetic aperture radar (SAR) images using modified SIFT. A registration technique for super-resolution image creation is reported by Nasir et al. [20].

The literature survey carried out opens a scope for devising an interest point detector based on local image properties. Local features strengthen image registration algorithms by handling image deformities such as rotation, scale changes, illumination changes, etc. Graphs give a very good representation of any local detail in an image. In this chapter, we present an algorithm to detect highly repeatable feature points from an image using a graph centrality measure. Utilizing graph theory, a novel interest point detector having higher repeatability, is proposed, which can be employed for extracting stable feature points to assist an image registration system.

3 Proposed Algorithm

Graph centrality measures the importance of nodes or vertices in a graph. For weighted graphs, a number of centrality measures are developed such as degree centrality, betweenness centrality, eigenvalue centrality and coseness centrality, etc. A highly efficient centrality measure known as Laplacian Centrality is presented in [21]. This metric was efficiently applied in social network analysis [22]. Image processing has also been benefitted abundantly by adopting theoretical concepts from graph theory. Graphs provide a unified representation for an image to study the global and local structural properties in detail. This, in turn, attracted researchers to devise procedures to solve many computer vision tasks utilizing graph theoretical concepts.

Fig. 1 A sample network
where u and v are two
random nodes

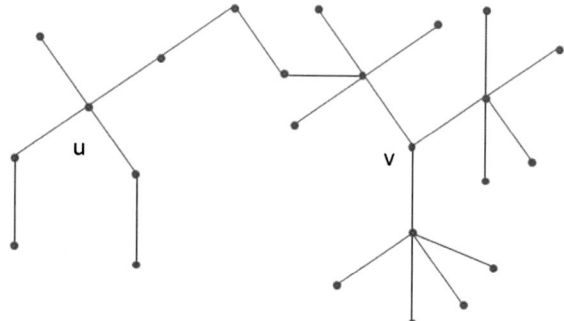

Consider a weighted, undirected graph or network $G=(V;E)$, where V is the finite set of nodes and E is the finite set of edges. Let $d_G(v)$ and $N(v)$ denote the degree of a node v and the finite set of nodes in the neighborhood of v, respectively. Laplacian centrality LC for the node v in a graph/network G can be calculated [22] as in Eq. (1).

$$LC(v) = d_G^2(v) + d_G(v) + 2 * \sum_{v_i \in N(v)} d_G(v_i) \qquad (1)$$

In the sample network shown in Fig. 1, let us mark two nodes u and v. Even though the node u has more number of neighbor nodes than v, the Laplacian centrality makes v a more important node (central node) than u. The Laplacian centrality values $LC(u)$ and $LC(v)$ for nodes u and v are 36 and 42, respectively. This is primarily because the immediate neighbors of v have more connectivity than those of u.

Pournami and Govindan [23] explains the process of identifying the central most nodes in the weighted graphs created out of local image patches. Laplacian centrality is an intelligible measure, which can be computed in linear time. Equation 1 provides an easier way of calculating Laplacian centrality from the adjacency matrix of the network [22].

Now the validness and robustness of this centrality are to be measured in cases where the images undergo various geometric transformations, using a standard dataset. This step is important because, to be incorporated into an image registration system, the feature points detected should be stable under all image distortions. Repeatability [24] is a measure of the geometric stability of interest points among images of the same object (or scene) captured under different imaging conditions. Let I_1 and I_2 be two images of the same scene. Also let $N(I_1)$ and $N(I_2)$ be the number of interest points extracted from the two images I_1 and I_2, respectively. Then repeatability, R, is defined as in Eq. (2).

$$R = \frac{N(I_1) \cap N(I_2)}{\min(N(I_1), N(I_2))} \qquad (2)$$

Fig. 2 Sample images from Oxford dataset

4 Experimental Results and Discussion

To demonstrate the robustness of this centrality measure, extensive experiments were designed with the algorithm written in MATLAB R2015b. The simulations were run on an Intel Core i5 (1.3 GHz) processor with 4 GB RAM and the operating system was 64-bit Mac OSX Yosemite. The robustness of this centrality measure is verified using Oxford dataset [25], where the images vary by angles of rotation, zoom or viewpoint changes, etc. Figure 2 shows sample input images from Oxford dataset.

As the first step, interest points are detected on the input images. Now, to verify the repeatability of the detected points, correspondence was established between matching points in the input images. Figures 3 and 4 show the sample output on Oxford dataset images.

Extensive experiments were conducted on more images from the Oxford dataset. The input images vary by angle of rotation, viewpoint changes, translation, illumination changes, blurring levels, etc. The repeatability of the proposed detector is calculated in each of these cases and reported in Table 1. Repeatability for speeded up robust features (SURF) points for these images is also added for comparison.

Fig. 3 Feature point correspondances on images from Oxford dataset

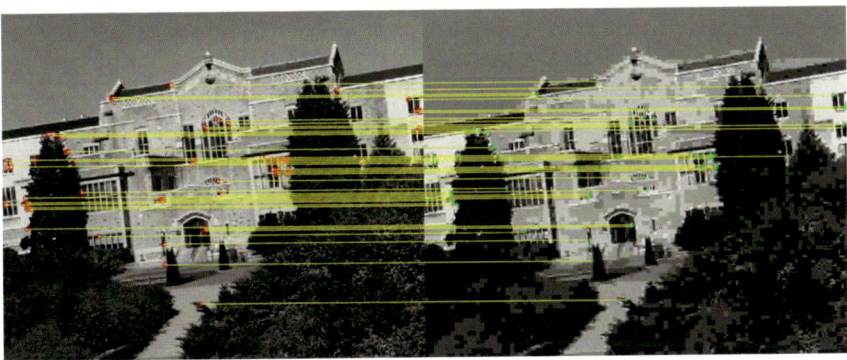

Fig. 4 Feature point correspondances on images from Oxford dataset

Table 1 Repeatability

Dataset	Nature of transformation	Proposed method	SURF
Cars (6 images)	Illumination changes	**97.98**	72.12
Boat (6 images)	Zoom, rotation	**91.76**	44.24
Building (6 images)	JPEG compression	**100**	89.98
Bikes (6 images)	Blurring	**84.26**	48.56
Graffiti (6 images)	Viewpoint changes	**96.26**	6.15
VanGogh (16 images)	Rotation	**99.29**	77.13

Repeatability values are higher for the proposed detector

5 Conclusion

The research aim of the study presented here was to test the repeatability of an efficient feature detector designed based on image geometry. We investigated the validness and robustness of this centrality measure under various geometric and photometric transformations, using standard datasets. The results obtained are highly reliable, which provide substantial evidence of the application of this centrality measure's utility computer vision applications such as image registration, image retrieval, object detection, etc. Parallel computation of node centrality values makes this approach faster.

References

1. Zitova B, Flusser J (2003) Image registration methods: a survey. Image Vis Comput 21(11):977–1000
2. Mikolajczyk K, Schmid C (2002) An affine invariant interest point detector. In: Proceedings of the 7th European conference on computer vision-part I, ECCV'02. Springer, London, pp 128–142. ISBN 3-540-43745-2
3. Lowe DG (1999) Object recognition from local scale-invariant features. In: Proceedings of the international conference on computer vision (ICCV'99), vol 2. IEEE Computer Society, USA, p 1150. ISBN 0-7695-0164-8
4. Lowe DG (2004) Distinctive image features from scale-invariant key points. Int J Comput Vis 60(2):91–110. ISSN 0920-5691
5. Mikolajczyk K, Schmid C (2004) Scale and affine invariant interest point detectors. Int J Comput Vis 60(1):63–86. ISSN 0920-5691
6. Miao Z, Jiang X (2013) Interest point detection using rank order log filter. Pattern Recognit 46(11):2890–2901
7. Criado R, Romance M, Sfanchez A (2012) Interest point detection in images using complex network analysis. J Comput Appl Math 236(12):2975–2980. ISSN 0377-0427
8. Ding G, Dai Q, Xu W, Yang F (2005) Affine-invariant image retrieval based on wavelet interest points. In: 2005 IEEE 7th workshop on multimedia signal processing, pp 1–4
9. Saydam SR, Rube IAE, Shoukry AA (2008) Contourlet based interest points detector. In: 2008 20th IEEE international conference on tools with artificial intelligence, vol 2, pp 509–513

10. Gevrekci M, Gunturk BK (2008) Reliable interest point detection under large illumination variations. In: 2008 15th IEEE international conference on image processing, pp 869–872

11. Maver J (2009) Self-similarity and points of interest. IEEE Trans Pattern Anal Mach Intell 32:1211–1226. ISSN 0162-8828

12. Lee WT, Chen HT (2009) Histogram-based interest point detectors. In: 2009 IEEE conference on computer vision and pattern recognition, pp 1590–1596

13. Martins P, de Carvalho P (2009) On interest point detection under a landmark-based medical image registration context. In: 2009 16th IEEE international conference on image processing (ICIP). IEEE, pp 2529–2532

14. Xie H, Gao K, Zhang Y, Li J, Liu Y (2010) GPU-based fast scale invariant interest point detector. In: 2010 IEEE international conference on acoustics, speech and signal processing, pp 2494–2497

15. Li B, Xiao R, Li Z, Cai R, Lu BL, Zhang L (2011) Rank-sift: learning to rank repeatable local interest points. CVPR 2011:1737–1744

16. Zukal M, Cika P (2012) Corner detectors: evaluation of information content. In: 2012 35th international conference on telecommunications and signal processing, pp 763–767

17. Wen H, Sheng XY (2011) An improved SIFT operator-based image registration using cross-correlation information. In: 2011 4th international congress on image and signal processing (CISP), vol 2. IEEE, pp 869–873

18. Yanhai W, Cheng Z, Jing W, Nan W (2015) Image registration method based on surf and freak. In: 2015 IEEE international conference on signal processing, communications and computing (ICSPCC), pp 1–4

19. Ma W, Wen Z, Wu Y, Jiao L, Gong M, Zheng Y, Liu L (2017) Remote sensing image registration with modified sift and enhanced feature matching. IEEE Geosci Remote Sens Lett 14(1):3–7. ISSN 1545-598X

20. Nasir H, Stankovic V, Marshall S (2010) Image registration for super resolution using scale invariant feature transform, belief propagation and random sampling consensus. In: 2010 18th European signal processing conference, pp 299–303

21. Qi X, Fuller E, Wu Q, Wu Y, Zhang CQ (2012) Laplacian centrality: a new centrality measure for weighted networks. Inf Sci 194:240–253

22. Qi X, Duval RD, Christensen K, Fuller E, Spahiu A, Wu Q, Wu Y, Tang W, Zhang C et al (2013) Terrorist networks, network energy and node removal: a new measure of centrality based on Laplacian energy. Soc Netw 2(01):19

23. P. N. Pournami and V. K. Govindan, "Interest point detection based on Laplacian energy of local image network," 2017 International Conference on Wireless Communications, Signal Processing and Networking (WiSPNET), Chennai, 2017, pp. 58–62

24. Schmid C, Mohr R, Bauckhage C (2000) Evaluation of interest point detectors. Int J Comput Vis 37(2):151–172

25. Oxford Dataset. http://www.robots.ox.ac.uk/~vgg/research/affine/

A Survey on Face Recognition in Video Surveillance

V. D. Ambeth Kumar, S. Ramya, H. Divakar and G. Kumutha Rajeswari

Abstract In today's world, enormous amount of threats arises due to terrorists, criminals, thieves and also illegal access of the data from the unwanted person, etc. This leads to a lot of challenges in our daily life. With the increase in threat globally the need to deploy reliable surveillance is to increase. Video surveillance is considered to be the major breakthrough in monitoring and security. In video surveillance, the facial recognition furthermore enhances the security and defense progressively. By face recognition the probe person can be recognized more accurately, efficiently and with short time. Various methods, approaches, algorithms were available for face recognition from surveillance video. The main objective of the paper is to discuss and analyze about the various facial recognition techniques.

1 Introduction

The concept of face recognition is to detect the particular character from the video or from the database and is used for security purposes. The face gives a rich source of information about a person and it is a most acceptable biometric. Face recognition, voice recognition, retinal scanning, fingerprint are some of the emerging biometric. In these, the method of spotting an individual by face recognition is far more accurate and faster than any others. Significant advances have occurred over the most recent couple of decades in face recognition. There is a significant attention and active research in this field. The human face tends to change significantly and quickly in

V. D. Ambeth Kumar (✉) · S. Ramya · H. Divakar · G. Kumutha Rajeswari
Computer Science and Engineering, Panimalar Engineering College, Chennai, India
e-mail: vdambethkumar@gmail.com

S. Ramya
e-mail: ramyasesha05@gmail.com

H. Divakar
e-mail: divakarvj06@gmail.com

G. Kumutha Rajeswari
e-mail: kums210819952@gmail.com

© Springer Nature Switzerland AG 2019
D. Pandian et al. (eds.), *Proceedings of the International Conference on ISMAC in Computational Vision and Bio-Engineering 2018 (ISMAC-CVB)*, Lecture Notes in Computational Vision and Biomechanics 30,
https://doi.org/10.1007/978-3-030-00665-5_69

Fig. 1 Stage of face
recognition algorithm

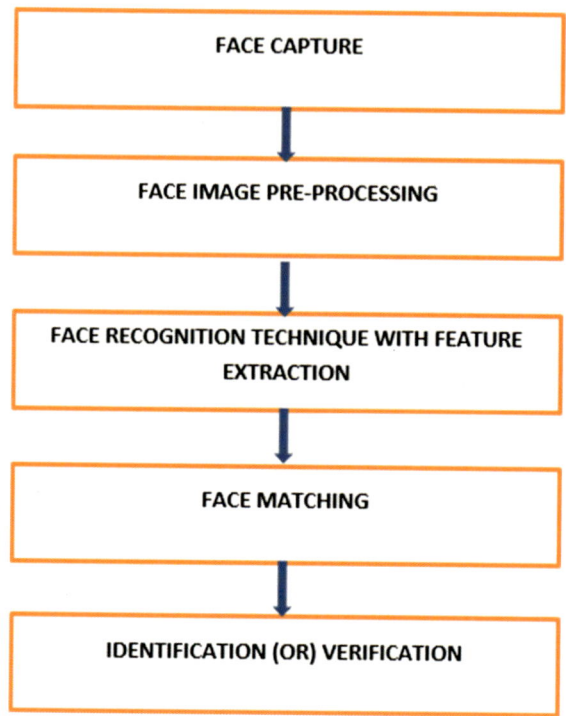

time, so it is considered as a complicated and dynamic structure. Face recognition is regarded to be challenging when there is variability in information due to pose variation, lighting conditions and different components. Face recognition should address these major challenges. Pose variation is considered to be a difficult problem because all faces seem to be similar with two eyes, mouth, nose and other features present in the same location.

The main aim of face recognition is to detect and verify a person from videos by one-to-many matching that compares the query face with the stored database faces [1]. The face recognition scheme may fail when there is a large difference in the query face and the stored database faces, e.g., makeup changes [2]. Face detection, feature extraction and face recognition are said to be the three stages of a face recognition system [3]. Face location is the way toward deciding the presence of a face in a video outline. Once a face is detected, face recognition is performed by isolating the face region and feature extraction is do carried out. Face verification and face identification are the two important stages of face recognition problem [4, 5].

The image preprocessing, feature extraction from face and the face matching are considered to be the main stages of the face recognition technique. In Fig. 1 these three main stages of face recognition algorithms are shown. Image normalization and enhancement are done during the image preprocessing. Also, the features of the face are extracted during this stage.

2 Face Detection Techniques

2.1 *Conditional Random Fields*

The Face Detection is done by dividing the face into various local regions optimally and integrating its multiple features and dependencies. These are demonstrated by a CRF structure. Multiple relationships modeling [6] are used for face detection in CRF.

2.2 *VGG-Face Algorithm*

Faces can be accurately recognized in spite of the difference in illumination using VGG-face algorithm [7]. Here, VGG-face performance is trialed with 8 NIST face recognition benchmarks. This algorithm performs exceptionally well in most difficult benchmarks.

2.3 *Convolution Neural Networks*

The CNN approach concentrates on the issue of face recognition in extreme pose variations. Different stances of particular models and rendered face images are utilized to handle posture variation [8]. This provides remarkably better performance. The CNN approach combined with a manifold based track comparison strategy can also be used for low-resolution video face recognition.

2.4 *Gabor Feature-Based Local Generic Representation (G-LGR)*

The G-LGR approach uses sparse representation properties for face recognition. This approach makes the recognition effectively, even when just a single probe subject for every class is feasible [9]. Feature extractions, such as uniform local binary patterns are not achieved by this technique.

2.5 *Feature-Augmented and Feature Transformation*

The technique of feature-augmented and feature transformation is combined in a hierarchical framework [10]. This multi-level component learning is utilized for confront acknowledgment under cosmetics changes. This is a robust method which is found to be superior to others. A level-wise change is utilized to limit the cosmetics and non-cosmetics confront contrasts.

2.6 Discriminant Correlation Analysis

DCA is one of the low-determination face detection strategies [11]. The low-resolution probe image is matched with high-resolution gallery face images. This computation is very much efficient and it can be applied to challenging real-time applications.

2.7 Heterogeneous Face Recognition

The heterogeneous face recognition focuses on the infrared-to-visible face matching scenario [12]. The main advantage of this technique is face recognition can be made possible in fog and haze and in low-light conditions. It uses an active illuminator which is not observable to the human eye. This method is purely passive and offers a truly covert surveillance capability.

2.8 Weber-Face and Singular Value Decomposition

This method Weber-Face and Singular Value Decomposition (SVD) is utilized to enhance the exactness of acknowledgment in a face acknowledgment framework. This technique can be utilized as a part of face acknowledgment in changing illumination [13]. At first, the face is represented by Weber–face (WF) method, and then the probe image is applied to singular value decomposition (SVD) method. Then, the SVD matrices value of test image is combined to adjust the illumination which is finally encoded by LBP/LTP descriptor.

2.9 Stacked Supervised Auto-Encoders (SSAE)

The SSAE method performs face identification task even though provided with only single sample per person (SSPP) in the gallery. This technique is additionally utilized for face detection under variety in posture, different facial expression and also in various illuminations [14]. Even though, the strategies combining with multiple samples per person probes (MSPPP) and SSAE will overcome the SSPP method.

2.10 Kernelized Locality-Sensitive Group Sparsity Representation

In paper [15], Face detection is performed by the joint sparse representation method. This sparse representation is the combination of group sparsity and locality-sensitive constraints. This improves the performance of recognition even in variation of pose. This KLS-GSRE method performs better than other sparse representation methods.

2.11 Discriminative Multimanifold Analysis for Face Recognition

Mostly the face recognition is done effectively by multiple samples per person (MSPP). But in this paper [16], a discriminative multimanifold analysis (DMMA) method is used for effectively to recognize face with single sample per person (SSPP).

2.12 Supervised Auto Encoder

The Supervised Auto encoder method should likewise be possible by utilizing single sample per person. This technique can extract facial features in various illumination changes, expression and occlusion, pose differences and facilitates face recognition [17]. Face verification can also be done by using this supervised autoencoder (Table 1).

3 Features of Human Face

The Description of the features of the face that are used in the face recognition system is discussed here. Some of the feature experts of the face are eye, chin, cheek, hair, jawline, mouth, nose, face outline etc. In these eyebrows, eyes, mouth, nose are considered to be the internal features while chin, face outline and hair are considered as external features. Although a system which uses an individual feature expert is good, the system which integrates these experts provides an even better result.

3.1 Eye

The Eye is the most widely used and attractive feature in face recognition. The features developed to detect eye is the variance projection function [18]. Eye detection

Table 1 Face recognition techniques

Name	Authors	Issues addressed	Methodologies used	Merits and demerits
A cross benchmark assessment of a deep convolutional neural network for face recognition	P. Jonathon Phillips	Face recognition in variable illumination	The VGG-face algorithm was trailed on eight NIST face recognition benchmarks	This algorithm performs exceptionally well in most difficult benchmarks It may not be possible to achieve optimal performance for all scenarios
Pose-aware face recognition in the wild	Iacopo Masi, Stephen Rawls, G´erard Medioni PremNataraja	Focus on the problem of extreme pose variations	Convolutional neural networks (CNNs)	Remarkably better performance
Gabor feature-based local generic representation for face recognition with single sample per person	Taher Khadhraoui	The primary problem tended to here is that if only one training subject per class is available	Gabor feature-based local generic representation (G-LGR) Virtual samples of each probe are considered and the new sample generic of a gallery set is used in order to generate the intrapersonal variations of different individuals	Have optimal localization properties in both spatial and frequency domains
Multi-level feature learning for face recognition under makeup changes	Zhenzhu Zheng and Chandra Kambhamettu	Propose a hierarchical framework to solve the problem of face recognition with cosmetic changes	Combine both strategies of feature augmentation and feature transformation in a unified way	A good representation is obtained by assembling multi-level representation. It is robust across makeup variations

(continued)

Table 1 (continued)

Name	Authors	Issues addressed	Methodologies used	Merits and demerits
Low-resolution face recognition in surveillance systems using discriminant correlation analysis	Mohammad Haghighat and Mohamed Abdel-Mottaleb	DCA analyzes the correlation of the features in high-resolution and low-resolution images and aims to find Projections that maximize the pair-wise correlations	A low-resolution face detection method based on Discriminant Correlation Analysis (DCA)	It has a very less computational Complexity. It can be used for real-time processing of several faces in a crowded image
Heterogeneous face recognition: recent advances in infrared-to-visible matching	Shuowen Hu1, Nathaniel Short, Benjamin S. Riggan, Matthew Chasse, M. Saquib Sarfraz	Matching between facial images acquired from different sensing modalities	Heterogeneous face recognition	Imaging through fog and haze, and in low-light conditions with an active illuminator not observable to the human eye heterogeneous face recognition is much more challenging
Five principles for crowd-source experiments in face recognition	Alice J. O'Toole, P. Jonathon Phillips	Crowd sourcing and human face identification on large databases. and to enhance the quality and accuracy of crowd-sourced data	Deep learning algorithms that consist of multi-layered neural networks	Accuracy and stability in face detection. It is critical to achieve human-sourced data that are meaningful and stable
Automatic face recognition.	Nawaf Yousef Al Mudhahka	The aim of semantic face recognition is to retrieve a suspect from a database of subjects using a human description of the suspect's face soft biometrics	Comparative soft biometrics is used due their ability for recognition and retrieval in constrained and unconstrained environments	Gives retrieval accuracy. Reduce dependency on human annotators. Bridges the semantic gap between humans and machines

(continued)

Table 1   (continued)

Name	Authors	Issues addressed	Methodologies used	Merits and demerits
Heterogeneous face recognition by margin-based cross-modality metric learning	Jing Huo, Yang Gao	Heterogeneous face recognition deals with matching face images from different modalities or sources	Margin-based cross-modality metric learning	Minimize intrapersonal cross-modality distances. Effective and superior
face recognition for movie character and actor discrimination based on similarity scores	Remigiusz Baran, Filip Rudzinski	A novel face detection approach dedicated to discriminate between motion picture characters and actors is presented in the paper	A bunching strategy in light of likeness scores ascertained by chose confront include descriptors is the key Element of this approach	Pretty high accuracy of face clustering up to 98%. Effectiveness strongly Depends on the accuracy of the eye localization process

is generally based on the eye motion and shape. The other method of face recognition using eye is Eye template matching [19]. The size of the image plays an important role in the calculation time. This method is considered to be efficient in terms of fast computation and also independent of the operating system platforms.

3.2   Eyebrows

Eyebrows are considered to be one of the prominent features in face recognition. The role of eyebrow in face recognition is considered to be as influential as that of the eyes. In the paper [20] eyebrows are used for detecting facial attractiveness and also to identify sexual dimorphism. An improved understanding of this system can contribute to the improvement of the artificial systems.

3.3   Skin Colors and Shape Information

The face recognition process is considered to be complex due to variation in illumination and background, difference in visual angle and facial expression. Once the skin color is evaluated, the shape information is used to locate the exact face. In the paper [21], the face candidates are located based on their color and shape information. Hue and saturation values are extracted and computed to find the best fit ellipse for each region. In the paper [22], facial regions are detected based on color and shape information. Here shapes are characterized by oval shape and HSV information is used for color localization. This is done by morphological operation and minima localization to intensity images.

3.4 Nose

A 3D nose tip localization method is used for face recognition. KNN AURA algorithm [23] is used to identify the nose tip, which is considered to be the facial feature to detect the faces. The identification rate of this method is 99.96% with much robustness and effectiveness.

The other method of face recognition using nose is by matching multiple overlapping regions around the nose [24]. This method is used for face recognition during varied facial expressions. This paper is considered to propose the first approach to solve expression variation problem.

4 Conclusion

In this paper, we have presented a review on face recognition. Different face recognition technique and the feature experts used for face recognition are discussed here. The three stages of the face recognition algorithm are image preprocessing, feature extraction and template matching. Obviously more accurate survey can be done by analyzing even more face recognition methods.

References

1. Patil SA, Deore PJ (2013) face recognition: a survey. Inform Eng Int J (IEIJ) 31–41
2. Zheng Z, Kambhamettu C (2017) Multi-level feature learning for face recognition under makeup changes. In: 12th International conference on automatic face and gesture recognition. IEEE, pp 918–923
3. Chihaoui M, Elkefi A, Bellil W, Ben Amar C (2016) A survey of 2D face recognition techniques. Computers 1–28
4. Meethongjan K, Mohamad D (2007) A summary of literature review: face recognition. In: Postgraduate annual research seminar, pp 1–12
5. Vijayakumari V (2013) face recognition techniques: a survey. World J Comput Appl Technol 41–50
6. Pang L, Ngo C-W (2015) Unsupervised celebrity face naming in web videos. IEEE Trans Multimed 17(6):854–856
7. Jonathon Phillips P (2017) A cross benchmark assessment of a deep convolutional neural network for face recognition. In: IEEE 12th international conference on automatic face and gesture recognition, pp 705–710

8. Masi I, Rawls S, Medioni G, Natarajan P (2015) Pose-aware face recognition in the wild. In: CVPR, pp 4838–4868
9. Khadhraoui T (2017) Gabor-feature based local generic representation for face recognition with single sample per person. IEEE, pp 157–160
10. Zheng Z, Kambhamettu C (2017) Multi-level feature learning for face recognition under makeup changes. In: IEEE 12th international conference on automatic face and gesture recognition, pp 918–933
11. Haghighat M, Abdel-Mottaleb M (2017) Low resolution face recognition in surveillance systems using discriminant correlation analysis. IEEE, pp 912–917
12. Hu S, Short N, Riggan BS, Chasse M, Sarfraz MS (2017) Heterogeneous face recognition: recent advances in infrared-to-visible matching. IEEE, pp 883–890
13. Tran C-K, Tseng C-D, Lee T-F (2016) Improving the face recognition accuracy under varying illumination conditions for local binary patterns and local ternary patterns based on weberface and singular value decomposition, pp 5–9
14. Vega PJS, Feitosa RQ, Quirita VHA, Happ PN (2016) Single sample face recognition from video via stacked supervised auto-encoder, pp 96–103
15. Tan S, Sun X, Chan W, Qu L, Shao L (2017) Robust face recognition with kernelized locality-sensitive group sparsity representation. IEEE, pp 1–8
16. Lu J, Tan Y-P, Wang G (2013) Discriminative multimanifold analysis for face recognition from a single training sample per person 1:39–51. IEEE
17. Gao S, Zhang Y, Jia K, Lu J, Zhang Y (2015) Single sample face recognition via learning deep supervised autoencoders. IEEE, pp 2108–2118
18. Feng GC, Yuen PC (1988) Variance projection function and its application to eye detection for human face recognition. Pattern Recognit Lett 19:899–906
19. Nikkhouy E, Abusham EEA (2011) Facial features detection using eyes-nose template. IJCSNS Int J Comput Sci Netw Secur 87–91
20. Sadrô J, Jarudi I, Sinha P (2003) The role of eyebrows in face recognition perception, pp 285–293
21. Sobotka K, Pitas I (1999) Extraction of facial regions and features using color and shape information. IEEE, pp 421–425
22. Sobottka K, Pitas I (1996) Face localization and facial feature extraction based on shape and color information. IEEE, pp 483–486
23. Ju Q (2013) Robust binary neural networks based 3D face detection and accurate face registration. Int J Comput Intell Syst 669–683
24. Chang KI, Bowyer KW, Flynn PJ (2006) Multiple nose region matching for 3D face recognition under varying facial expression. IEEE, pp 1695–1700

Driver's Drowsiness Detection Using Image Processing

Prajakta Gilbile, Pradnya Bhore, Amruta Kadam and Kshama Balbudhe

Abstract There are some causes of car accidents due to driver error which includes drunkenness, fatigue and drowsiness. Hence, the system is needed which will alert driver before he/she falls asleep and number of accidents can be reduced. In the proposed system, a camera continuously captures movement of the driver. To determine whether a driver is feeling drowsy or not the head position, eye closing duration and eye blink rate are used. Using this information, the drowsiness level is determined. As per the drowsiness level the alarm is generated. A night vision camera is used to handle different light conditions.

1 Introduction

Almost all the statistics have identified driver's drowsiness as a high priority vehicle safety issue [1]. Drowsiness refers to feeling sleepy, tired or being unable to keep eyes open. Due to drowsiness, the driver canot concentrate while driving, eye blink rate is decreased or increased and unable to keep eyes open. Fall-asleep crashes are very serious in terms of injury severity and may result in death. Drowsiness affects mental alertness and decrease an individual's capability to handle a vehicle safely. A driver is unable to predict when he or she will have an uncontrolled sleep onset. The advancement in technologies develops interest in driver's safety and comfort, increase traffic flow and reduce accidents. This paper introduces an alerting process when the driver falls asleep. It calculates the level of drowsiness depending on the

P. Gilbile (✉) · P. Bhore · A. Kadam · K. Balbudhe
Department of Information Technology, PVG's COET, Pune, India
e-mail: prajakta.u.gilbile136@gmail.com

P. Bhore
e-mail: pradnyab208@gmail.com

A. Kadam
e-mail: amrutakadam999@gmail.com

K. Balbudhe
e-mail: ksb_it@pvgcoet.ac.in

© Springer Nature Switzerland AG 2019
D. Pandian et al. (eds.), *Proceedings of the International Conference on ISMAC in Computational Vision and Bio-Engineering 2018 (ISMAC-CVB)*, Lecture Notes in Computational Vision and Biomechanics 30,
https://doi.org/10.1007/978-3-030-00665-5_70

head position, eye blinking rate. If the level exceeds the limit from a threshold, then the alarm is generated. Different sound alarms such as 'Take a break' or 'Have a coffee', etc., must be given at particular level of drowsiness. So that before falling asleep driver will get alert.

2 Related Work

Khunisuth et al. [2] discussed paper in which the drowsiness is detected based on various factors such as titling of head, blinking of eyes and eye blink rate [3]. Image is captured through camera then localization of head is done [4]. It is followed by localization of eyes and then titling of head angle is detected. Different templates such as both eyes closed, right eye closed, left eye closed, both eyes open are used to detect eyes. A drawback of this paper is that it is not steady in all light conditions. The accuracy of 99.59% is achieved for only stable light conditions. In the proposed system, the night vision camera is used to handle different light conditions.

Ahmed et al. [5] discussed paper in which drowsiness is detected based on only eyes. First, the location of driver's eyes is located and then it is decided whether the eyes are open or not. The captured image is in binary form through which location of edges of face are detected. This ultimately gives the location of eyes. Various frames are captured. If the driver's eyes found to be close in five successive frames, then the system assumes that drives is feeling sleepy and alert gets generated accordingly. The drawback of this paper is that, it may happen that some people have habit of blinking of eyes more than that the normal rate. The result which will get generated will be ultimately wrong. Only eyes cannot give accurate output to determine drowsiness.

Tadesse et al. [6] discussed paper which mainly focuses on the facial expressions for drowsiness detection. Many previous papers are specifically focus on eye closure and blinking of the driver. The facial expressions of driver are analysed through Hidden Markov Model (HMM-based dynamic modelling). Every time presetting of window size to fixed value is needed for different parameters.

Abtahi et al. [7] discussed a paper in which drowsiness is detected on basis of yawing of driver. A camera detects the face and eyes of the driver. After detection of eye, mouth is detected and then successively yawning is detected. On basis of only yawing, alert is generated. Yawing is detected in two steps. In first step yawn component is detected and in second step mouth location is used to verify the validity of detected component. This paper uses number of algorithms so they are insensitive to changes in light conditions, skin types. Various verification techniques are used to reduce false rate. The limitation of this paper is that it detects the drowsiness on basis of only yawning and uses number algorithms.

Saini et al. [8] discussed a way to find drowsiness which uses ECG and EEG, Steering Wheel Movement (SWM), Local Binary Pattern (LBP), and Optical Detection. This paper also discusses about eye blink-based technique, head nodding, and yawning-based technique.

3   Proposed Work

The main concept of 'Driver's Drowsiness Detection' is to capture a driver's face using a camera and accurately calculate the level of drowsiness. The proposed system consists of a camera pointing at the driver. A camera continuously captures images of driver. There are main five stages of processing: The first stage is to capture image using camera. Second stage is localization of head and check head position. Third stage is calculation of eye blink rate. Fourth stage is calculation of eye closing duration and fifth stage is to generate the alert. At different levels of drowsiness, different alerts will get generated.

3.1   Capture Images from Camera

The camera is used for continuously capturing video. The camera is placed such a way so that it will capture all head movements and eye movements of driver. Due to different light conditions the noise can be introduce in image. To handle different light conditions including night time, the night vision camera can be used. Camera will give whole picture of driver with background details but we are interested only in head position area. JMyron and OpenCV library is used to capture image and finding out area of interest. The captured image is RGB image and it is transformed into grey scale image for processing. As image is stored temporarily and after processing it is discarded, it does not take large space.

3.2   Find Head Position

For finding head position (Area of Interest) 'Haar-Cascade' algorithm is used [9]. JMyron library will remove unnecessary portion (background details) from captured image and give the area in which head is present. On which 'Haar-Cascade' algorithm is applied. As mentioned in algorithm JMyron library returns a vertex point of frame (ROI) in which head is present. Using this point, height and width of frame, the centre point of frame is calculated which is ultimately the centre point of head. The standard centre point is already defined to indicate position of head when driver is not feeling drowsy. The centre point of new frame is compared with the standard centre point and the difference between these two points is calculated. If the difference is greater than threshold value then alarm is generated. By using the difference, a level of drowsiness can be determined. If the difference is greater then, the drowsiness level is high and if low then, the drowsiness level is low. The threshold value is approximately 100 or 150 pixels.

3.3 Eye Blink Rate Calculation

After localization of eye, eye blink rate is calculated. The average eye blink rate of human beings is approximately fifteen to twenty times in one minute. For comparing with current eye blink rate two values for threshold are set.

High_threshold_binkrate = 25
Low_threshold_binkrate = 10

System captures eye blink rate for one minute. If current eye blink rate is greater than high_threshold_value or less than low_threshold_value then alarm is generated. Template matching is not used to detect eyes open or not. If eye blink rate is too fast or low then, drowsiness level is high.

3.4 Eye Closing Time Calculation

System captures eye closing time by checking successive frame. If eye closing time is greater than the threshold value, alarm is generated. The approximate threshold value is 3s.

3.5 Alarm Generation

This module is responsible to alert driver. Depending upon level of drowsiness different alarms are generated. This can be buzzer or voice message such as 'You need a break', 'Take a cup of coffee' etc. Another way is to include hardware like vibrator to alert driver.

4 Algorithm

Let P (xF, yF) = (0,0) be the vertex point of the captured frame. Let P (standX, standY) be the standard centre point for head position. Let fRect be the Region of Interest in which head is present.

Algorithm Drowsiness detection ()
{
Apply Haar Cascade algorithm to get vertex point P (xSelect, ySelect) of fRect.

$$currCX = ((xF + xSelect + fRect.x()) + fRect.width()/2) \qquad (1)$$

$$currCY = ((yF + ySelect + fRect.y()) + fRect.height()/2) \qquad (2)$$

Fig. 1 Terminologies used in algorithm

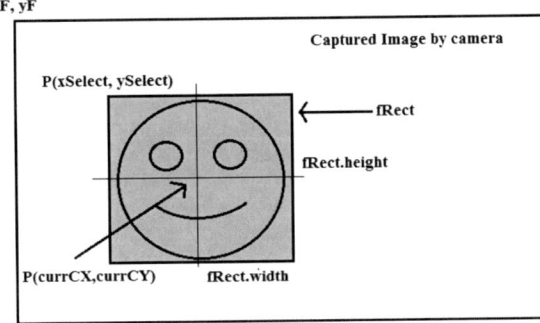

Find the distance between P (standX, standY) and P (currCX, currCY)

$$Distance = \sqrt{(currCX - standX) * (currCX - standX) + (currCY - standY) * (currCY - standY)} \tag{3}$$

If (Distance > threshold value of head position)
Generate alarm.

$$wF = f Rect.width() \tag{4}$$
$$hF = f Rect.height() \tag{5}$$

Select (0, 0, wF/2, hF) and (wF/2, 0, wF/2, hF) as Rect1 for left eye and Rect2 for right eye, respectively.

Apply Haar Cascade algorithm on Rect1 and Rect2 to detect eyes.

If (Rect1 == null && Rect2 == null)

Blink_count++;

If (Blink_count > high threshold value of eye blink rate ‖ Blink_count < low threshold value of eye blink rate)

Generate the alarm.

If (Rect1 == null && Rect2 == null) for continuous 3 s
Generate the alarm.

} (Fig. 1)

5   Conclusion

In this paper, the way to identify level of drowsiness is specified and according to that different alarms can be given to driver. The night vision camera is used to handle different light conditions. The algorithm to find head position is specified. In algorithm, the drowsiness level is determined depending upon head position, eye blink rate and eye closing duration. Proposed system detects drowsiness level and helps a driver to stay awake while driving. This system will reduce car accidents about great extent.

References

1. Ahmad R, Borole JN (2015) Drowsy driver identification using eye blink detection. Int J Comput Sci Inf Technol. 6(1):270–274
2. Khunpisuth O, Chotchinasri T, Koschakosai V, Hnoohom N (2016) Driver drowsiness detection using eye-closeness detection In: Signal-Image Technology & Internet-Based Systems (SITIS), 2016 12th International Conference on, pp. 661–668. IEEE, 2016
3. Parmar SH, Jajal M, Brijbhan YP (2014) Drowsy driver warning system using image processing. Int J Eng Dev Res, IJEDR1303017
4. Kuo Y-C, Hsu W-L (2010) Real-time drowsiness detection system for intelligent vehicles. Proceedings of the 5th Symposium on Smart Life Science and Technology (Part 1)
5. Ahmed J, Li J-P, Khan SA, Shaikh RA (2015) Eye behavior based drowsiness detection system In: Wavelet Active Media Technology and Information Processing (ICCWAMTIP) 2015 12th International Computer Conference on, pp. 268–272. IEEE, 2015
6. Tadesse E, Sheng W, Liu M (2014) Driver drowsiness detection through hmm based dynamic modelling In: Robotics and Automation (ICRA) 2014 IEEE international conference on robotics and automation (ICRA), pp. 4003–4008. IEEE, 2014
7. Abtahi S, Hariri B, Shirmohammadi S (2011) Driver drowsiness monitoring based on yawning detection In: Instrumentation and Measurement Technology Conference (I2MTC), pp. 1–4. IEEE, 2011
8. Saini V, Saini R (2014) Driver drowsiness detection system and techniques: a review. Int J Comput Sci Inf Technol. 5(3):4245–4249
9. Pamnani R, Siddiqui F, Gajara D, Gupta A, Pandya K Driver drowsiness detection using haar classifier and template matching. Int J Adv Res Eng Technol 3(IV), April ISSN 2320–6802
10. Nguyen TP, Chew MT, Demidenko S (2015) Eye tracking system to detect driver drowsiness In: Automation, Robotics and Applications (ICARA), 2015 6th International Conference on, pp. 472–477. IEEE, 2015
11. Assari MA, Rahmati M (2011) Driver drowsiness detection using face expression recognition In: Signal and Image Processing Applications (ICSIPA), 2011 IEEE International Conference on, pp. 337–341. IEEE, 2011
12. Flores MJ, Armingol JM, de la Escalera A (2010) Real-time warning system for driver drowsiness detection using visual information. J Intell Robot Syst 59(2):103–125

Identification of Optimum Image Capturing Technique for Corneal Segmentation—A Survey

H. James Deva Koresh and Shanty Chacko

Abstract Segmentation of corneal layers plays an important role in diagnosing the corneal disease and it also helpful for planning the refractive surgeries when measuring the corneal layer thicknesses. There are several imaging instruments available for capturing the image of the cornea which can be used for segmentation and thickness measurement of corneal layers. The work describes the various available imaging instruments along with the necessity of imaging the cornea. It also suggests a reliable imaging method for capturing the cornea for segmentation by doing a comparative study. It will be helpful for identifying the best image capture method for implementing the image processing technique for further segmentation and measurement flow from the image.

1 Introduction

The refractive surgeries like laser-assisted in situ keratomileusis (LASIK) and photorefractive keratectomy (PRK) which are the methods used to correct minor and modular refractive errors till -8.00 for myopia and $<+4.00$ for hyperopia. The corneal segments consist of Epithelium, Bowman's membrane, Stroma, Descemet's membrane, and the Endothelium. The typical corneal thickness of a human eye is $550\,\mu m$ at the center and it gradually increases towards the marginal area [1, 2]. The ideal thickness of the corneal layer may vary due to changes in oxygen consumption of cornea, acidosis and swelling. So, the accuracy in evaluation of corneal thickness is necessary to diagnose such problems to provide proper treatment [3, 4].

H. James Deva Koresh (✉)
Department of Electronics and Communication Engineering,
Karunya Institute of Technology and Sciences, Coimbatore, India
e-mail: jamesdeva@karunya.edu.in

S. Chacko
Department of Electrical and Electronics Engineering,
Karunya Institute of Technology and Sciences, Coimbatore, India
e-mail: shanty@karunya.edu

© Springer Nature Switzerland AG 2019
D. Pandian et al. (eds.), *Proceedings of the International Conference on ISMAC in Computational Vision and Bio-Engineering 2018 (ISMAC-CVB)*, Lecture Notes in Computational Vision and Biomechanics 30,
https://doi.org/10.1007/978-3-030-00665-5_71

The epithelium is the outermost layer of the cornea which covers less than 10% of the central corneal thickness (CCT), the thickness is almost around 50 μm for a healthy eye. The tissues in the epithelium layer have the capability of replacing itself after a damaged (in LASIK) or removal (in PRK). The regeneration time of epithelium layer is less than a week depends upon the changes. The Bowman's layer is a non-regenerative connecting tissue of 8–14 μm lies between epithelium and stroma. The Bowman's layer is ablated in the photorefractive keratectomy (LASIK or PRK). The stroma is a deeper layer that covers 90% thickness of cornea with a thickness of 500 μm. The stroma is contrary to epithelium consists of permanent tissues which will remain permanently after a reshape. The Descemet's membrane of 15 μm depth is the next layer to stroma which separates the stroma layer from endothelium. This membrane gets increases from 5 μm in children to 15 μm of an adult. The endothelium is a single-layer cell of 5 μm thickness and it is the innermost layer of the cornea [2].

In LASIK the epithelium layer is damaged in a circular manner to create a flap with the help of microkeratome or femtosecond laser to access further layers and the flap is laid back to position after the procedure to heal [5]. For PRK the epithelium layer is removed to access further layers. In both LASIK and PRK a cool ultraviolet light beam called laser is passed over the stroma to reshape it permanently. For myopia the laser beam is used to flatten the stroma layer and it makes steeper for hyperopia [5]. The corneal thickness must not be less than 470 μm for LASIK and PRK procedures and the final corneal thickness should be more than 400 μm after the procedures to have a healthy cornea [6]. The ablation depth of the stroma layer varies depends upon the refractive error of an eye and the percentage of tissues to be altered (PTA) should not exceed 40% of the CCT [6]. In refractive surgery the thinnest cornea may results in ectasia, it is an abnormal bulging over the surface of the cornea [7]. Hence it is must to measure the CCT of an eye before the photorefractive keratectomy (LASIK and PRK) to analyze the PTA.

2   Related Work

There are several techniques are available for estimating the corneal thickness. Pachymeter is a common name for devices which are used to measure the corneal thickness and the procedure for estimating the corneal thickness is called as pachymetry. In general, the pachymeter devices are broadly classified as contact and non-contact types.

2.1   Ultrasound Pachymetry

The Ultrasound Pachymetry (UP) is a simple, cost-effective and handy device used in medical field, for estimating the CCT. The UP is a contact-type pachymeter has an ultrasound probe which is to be kept directly over the corneal apex to make an ultrasound waves. The probe helps us to find out the CCT with the help of receiving the reflected sound from the surfaces of cornea. The reflected sounds are further taken into the connected device through the probe as multiple readings for the estimation of average CCT and standard deviation among the readings. The user also has the accessible to view each measured values individually and can delete the value which seems to be too irrelevant to other readings. As the UP is a contact-type device, it requires anesthetic drops for the cornea to avoid itchiness. Due to the contact of the probe in direct to the cornea, there is also chance for damage and abrasion over the cornea. Hence, this method requires trained practitioner for utilizing this device and there are also chances for variation in readings due to misalignment of the probe by the practitioner. This misalignment may happen due to keeping the probe just closer to the central cornea, it will result in showing higher corneal thickness than actual. Also, the ultrasound probe should be kept 90° perpendicular to the central cornea. Small changes in keeping the probe may also lead to a false estimation of CCT [8]. From this device, the intraocular pressure (IOP) is also measured as it has a direct relationship with the CCT. The ultrasound pachymetry is also called as the perimeter.

2.2 Orbscan

The Orbscan corneal topography is a quick and non-contact method of estimating corneal thickness along with anterior and posterior surface evaluation. This scan provides the overall health of a cornea including corneal shape and the result of this procedure will be 3D maps of the surface of an eye [9, 10]. The Orbscan is working on the principle of slit-scanning technology and video keratography estimates the thickness by measuring the reflections of rays from the corneal surface [11]. The Orbscan delivers a pair of vertical scans at 45° angle from both sides of the cornea to make it as slits. The slits are analyzed further with the help of digital video camera to make a topography of the eye in 3D maps [12]. As it is a non-contact method it does not require anesthetic drops during the procedure.

2.3 Optical Coherence Tomography

Optical Coherence Tomography (OCT) is another non-invasive method that can be used for imaging the corneal and retinal segments with the help of infrared light. The OCT is used to examine 2–3 mm of subsurface tissues in retina and cornea [13, 14]. The OCT performs the scanning of the corneal area with infrared spots produced by Michelson-type interferometer. The OCT was used to analyze the retinal layer by passing the infrared spot over the eye with the wavelength of 1310 nm. The cross-sectional images of the corneal layer can be viewed in this process by passing the infrared spots with 820 nm wavelength over the corneal surface in multiple axes [11].

The image also has the cross-sectional view of anterior and posterior segments of an eye. The anterior segment imaging is helpful for analyzing the space available for implanting the intraocular lens (IOL) to correct ew32 the high refractive error of an eye from the range of >-8.00 for myopia (nearsightedness) and $>+4.00$ for hyperopia (farsightedness) [15]. The cross-section image of cornea provides the curvature and thickness of central to peripheral corneal layer, which is also useful in diagnostic corneal edema and ocular hypertension.

2.4 Pentacam

The Pentacam is a device used for imaging the corneal surface and anterior segments of an eye. The Pentacam is a fast imaging instrument requires only 2 s for capturing; this is achieved with the help of scheimpflug camera by rotating it for 180° on the optical axis along with a monochromatic slit light source. It will take around 50 slit images consists of 500 assessing points for further estimation of CCT and each corneal layers [1, 16]. The Pentacam is also known as Scheimpflug Tomography.

3 Literature Survey

Martínez-Albert et al. [17] conducted the repeatability test on OCT in measuring the CCT. The test was repeated three times on 30 healthy individuals on their right eye. The result indicates that there is better correlation between the reading measured on each time by the OCT and it provides the intraclass correlation coefficient (ICC) of 0.984–0.988 [17]. Otchere et al. [16] assessed the CCT from 20 keratoconus individuals with OCT and Pentacam for analyzing their repeatability and reproducibility. The CCT was measured twice in a day from each instrument and repeated the test after 48 h in same manner. The assessed CCT values explores that both the devices give good repeatability on all the surfaces over the cornea but the reproducibility among the two devices are not good enough [16]. Ramesh et al. [18] compared the CCT measurements taken from 120 normal corneas by UP and OCT. The readings from OCT were measured twice and 25 times from UP. The values are average for taking comparison and the result shows that the values observed by OCT are slightly high always comparing to the values of UP. Upon the high repeatability the OCT stands better than UP [18].

Antonios et al. [19] assessed the CCT for 44 eyes of Keratoconus in before and after the corneal cross-linking (CXL) procedure in the time span of 14, 30, 90, 180, and 360 days. The CCT has been assessed with UP, OCT, and Scheimpflug Tomography (Pentacam). It is observed from the mean CCT assessment that the OCT has made similar measurement obtained by the UP throughout the assessment period [19]. Xu et al. [20] made the reliability test on Pentacam by means of measuring the corneal thickness for 60 Healthy eyes, 60 post-LASIK eyes, and 27 Keratoconus eyes.

The repeatability and reproducibility of the Pentacam are analyzed by allowing two trained operators to measure corneal thickness for three times of an eye. The CCT shows best repeatability and reproducibility than the peripheral corneal thickness and it also stated that measuring the peripheral cornea thickness through Pentacam should be taken care of unhealthy cornea and eyes [20].

Temstet et al. [21] made a test for finding the thickness of the outermost layer of the cornea called epithelium with the help OCT to measure its detection ability over Orbscan and Pentacam. There are 145 eyes consists of healthy, fruste keratoconus, moderate keratoconus, and severe keratoconus were undergone for the test and it explores that the subjects with severe keratoconus are having thinnest epithelium layer and it was found perfectly with the help of OCT than other methods [21]. Kuerten et al. [22] performed comparison study on OCT, Pentacam, and UP by analyzing the repeatability and reproducibility of estimating the CCT among control and post corneal edema. The estimation was done by two individual investigators randomly. The test started Pentacam, OCT and finally, with UP, the UP has the highest nature of user dependence for analyzing the CCT [22]. Randleman et al. [23] estimated the CCT among 100 eyes for comparison of performance between OCT, Orbscan, Pentacam, and UP. The test result indicates that there are significant variations among the results observed by all the instruments and stated that the observed values from any of the instrument should not be applicable for another instrument in comparison.

Lin et al. [24] evaluated the repeatability of OCT with UP by estimating the CCT for 51 patients with glaucoma, suspected glaucoma, and cataract with two different examiners. First, the CCT measurement was taken with OCT and then with UP, each device is repeated for two times of estimation. The repeatability test was done by analyzing the observed CCT value in the intraclass correlation of coefficient (ICC) between UP and OCT. The result indicates high repeatability range of ≥ 0.978 and for 2 mm from the center, the ICC increased to ≥ 0.999 [24]. Yazici et al. [25] made the CCT test for 100 eyes with the single operator through OCT, Orbscan and Pentacam. The comparative study shows that OCT measured the values of CCT thinner to other two methods but the variations among these methods are small and it does not affect the decision for photorefractive keratectomy [25]. Doors et al. [26] did a comparative study of CCT measurement with 66 healthy eyes and 42 eyes of after intraocular lens implantation (IOL). The comparative study takes place between the non-contact methods of OCT, Orbscan, and Pentacam. The conclusion indicates that there is a substantial variation among the values calculated by all three methods and the study recommends to not use these devices interchangeably for analysis [26].

Cheng et al. [27] measured the CCT with UP, Orbscan, and OCT for 68 eyes of 34 myopia individuals after 6 months of their LASIK surgery. The measured CCT were compared by the linear regression method, paired sample t-test, Bland and Altman plot and the result shows a closed correlation between OCT and UP comparing with Orbscan [27]. Kim et al. [28] done a relative test on CCT for 155 individuals with the clinically healthy eye. Two measurements from OCT and UP were taken and averaged for one eye randomly from a patient. The reproductive difference in CCT is significantly present on both UP and OCT and the test suggests that these two methods should not be interchanged in clinical practice for analysis and measurements [28].

Li et al. [29] estimated the value of CCT for 70 eyes with OCT, UP, and Orbscan. The OCT underestimates the CCT among the other two methods and the OCT has a better agreement than Orbscan with UP. As the CCT measurement is highly correlated, it is just to avoid taking measurement interchangeably in clinical practices [29]. Ho et al. [30] evaluated the CCT for 103 post-LASIK individuals with OCT, Orbscan, Pentacam, and UP. The evaluation has taken place after the period of six months from the LASIK procedure and the comparative evaluation specifies that OCT and Pentacam attained thinner values than the rest of the methods but it has a great correlation between all four methods [30]. Zhao et al. [31] tested the CCT with OCT and UP to 285 normal individuals. The observed CCT values between OCT and UP had a very close relationship of Pearson correlation coefficient of 0.98, $P < 0.001$. The Bland Altman analysis shows that the CCT of UP is comparatively high by $16.5 \pm 11.7\ \mu m$ [31].

Haque et al. [32] assessed the CCT along with epithelium and stromal thickness of 20 keratoconus and 20 with normal individuals with the help of OCT, Orbscan, and UP. The assessed values of OCT and Orbscan have closer association than UP. The UP estimated the CCT as higher in both central apexes of the cornea. It also indicates that the keratoconus subjects have thinner cornea layers than the normal individuals [32]. Leung et al. [33] performed the CCT test for 50 healthy individuals with OCT and then with UP, the correlation and Bland Altman comparison indicate a high correlation between the OCT and UP with Pearson coefficient of 0.934 and in Bland Altman plot the OCT results are in higher range continuously by mean of $23\ \mu m$. The OCT is agreed with UP with minor adjustment factor [33]. Li et al. [34] analyzed the CCT of LASIK undergoing patients of 21 individuals. The analysis was taken by UP and OCT procedures and the UP shows the higher value of CCT in Pre-Lasik phase and lower value than OCT on Post-Lasik phase. The OCT and UP techniques are highly correlated by Pearson correlation of 0.98 in Post-LASIK and 0.97 for Pre-LASIK [34].

Lackner et al. [35] conducted and compared the CCT measurement with 30 healthy eyes for twice the time with two independent observers. The comparative study was done by taking standard deviation among the values and it shows that Pentacam shows the values closer to UP than Orbscan and the reproducibility of Pentacam is also comparatively high [35]. Fishman et al. [36] made an assessment of CCT with OCT, Orbscan, and UP for 22 eyes. The result of the assessment states that the OCT is a good reproductive and accurate non-invasive method for estimating the CCT. The correlation of CCT between of OCT is 0.984 and UP is 0.981 where the Orbscan has the correlation of 0.942 [36]. Wong et al. [37] had performed the CCT test with OCT, Orbscan, and UP for 74 eyes. The test was first carried out with the non-contact methods (OCT and Orbscan) following to the UP. There are five readings were taken for non-contact methods and three for UP. The average values were comparatively analyzed using correlation, linear regression, and one-way analysis of variance methods and the result shows that OCT is more accurate than Orbscan and UP [37].

4   Discussion

Table 1 explores the CCT values measured by various experts with different imaging instruments in healthy and unhealthy eyes. It also indicates that the OCT method is an optimum technique for capturing the image of cornea. Hence the OCT is further compared individually with Pentacam, Orbscan, and UP to get a better solution.

4.1   Pentacam Versus OCT

Lackner et al. states that in comparing the values of non-invasive method with invasive method, the Pentacam values are very closer to the UP and the reproducibility of Pentacam stands better than the other non-invasive method. Otchere et al. made a test to find the repeatability of Pentacam and OCT by measuring the reading twice in two days span. The result indicates that the Pentacam is not the only device with better repeatability as stated by Lacknet et al. It also indicates that OCT has the good repeatability with lesser range difference than Pentacam. Ho et al. result explores that the correlation between the non-invasive and invasive methods are compromising to each other with higher values but the observed values of CCT from Pentacam and OCT are comparatively lesser than the other methods. Antonios et al. prove that the CCT values obtained by OCT and UP have better similarity than the Pentacam by readings observed from the individuals after multiple times. Xu et al. warns to take more care in measuring the peripheral thickness of cornea through Pentacam for the operated cornea and illness cornea.

4.2   Orbscan Versus OCT

Li et al. observed that the difference between the values of CCT are lesser in Orbscan with UP than OCT. Also the difference is very small in Orbscan with UP. The difference among the values observed by OCT is comparatively higher when comparing with UP and Orbscan and the difference with both the devices are almost equal with certain variations in microns. Cheng et al. suggest that both Orbscan and OCT has the ability to measure thinner corneas accurately than UP. The obtained values from Orbscan and OCT are having better correlation together. Fishman et al. state that the OCT is an accurate method in measuring the values of CCT. The study shows that the range values obtained from several individuals are higher in Orbscan than OCT in comparing with UP.

Table 1 Comparison of CCT values measured by various experts with different imaging instruments in control and uncontrolled eyes

Author name	Measured parameters
Martínez-Albert et al. [17]	Mean CCT + SD in microns 520 ± 30.59
Otchere and Sorbara [16]	Mean and SD of CCT in microns OCT = 484.97 |±| 43.14; range: 484.84–486.09 Pentacam = 478.86 |±| 45.31; range: 477.20–480.53
Ramesh et al. [18]	Mean and SD of CCT in the right eye in microns OCT = 516.28 ± 29.76; UP = 532.42 ± 29.71 Mean and SD of CCT in the left eye in microns OCT = 515.82 ± 29.88; UP = 532.36 ± 29.83
Antonios et al. [19]	Mean CCT in microns OCT = 470.02; UP = 469.79; Pentacam = 466.66
Xu et al. [20]	Coefficient of variation of Central and Peripheral corneal thickness in all 3 states, Healthy = ≤ 1.3%; Post-LASIK = ≤ 1.6%; Keratoconus = ≤ 1.6%
Temstet et al. [21]	Epithelium thickness in microns for Form fruste keratoconus = 52.8 + 3.3; Moderate keratoconus = 49.5 + 5.2; Severe keratoconus = 47.9 + 4.9; Control = 53.0 + 3.1
Kuerten et al. [22]	CCT from investigator 1 Control Pentacam = 561.5 ± 38.64; OCT = 552.8 ± 38.73; UP = 547.6 ± 39.81 Post corneal edema Pentacam = 615.9 ± 58.02; OCT = 608.8 ± 65.67; UP = 601.4 ± 63.77 CCT from Investigator 2 Control Pentacam = 560.7 ± 38.56; OCT = 553.1 ± 38.39; UP = 554.4 ± 4 Post corneal edema Pentacam = 615.1 ± 60.17; OCT = 606.9 ± 64.41; UP = 614.5 ± 70.91
Randleman et al. [23]	Mean and SD of CCT in microns OCT = 550.5 ± 32.7; Orbscan = 570.9 ± 36.1; Pentacam = 552.8 ± 33.8; UP = 563.9 ± 36.1
Lin et al. [24]	Corneal thickness with standard deviation measured by examiner 1 at Central = 554.43 (30.42); vertex = 540.22 (30.00); central 2 mm = 540.86 (29.81) Corneal thickness with standard deviation measured by examiner 2 at Central = 551.39 (29.85); vertex = 541.24 (29.70); central 2 mm = 540.04 (29.54)
Yazici et al. [25]	OCT = 529 ± 30.5 μm; Orbscan = 554 ± 32.7 μm; Pentacam = 552 ± 29.3 μm
Doors et al. [26]	For healthy eyes, OCT and Orbscan ($P = 0.422$) For IOL eyes, Orbscan and Pentacam ($P = 0.214$)

(continued)

Table 1 (continued)

Author name	Measured parameters
Cheng et al. [27]	Average CCT of post-LASIK under $UP = 436.65 \pm 43.82$ μm; Orbscan $= 422.84 \pm 51.04$ μm; OCT $= 422.26 \pm 42.46$ μm
Kim et al. [28]	Mean CCT + SD in microns $UP = 523.3 \pm 33.5$; OCT $= 499.0 \pm 32.0$ CCT range in microns $UP = 422–653$; OCT $= 428–613$
Li et al. [29]	$UP = 553.5 \pm 30.26$ μm; Orbscan $= 553.22 \pm 25.47$ μm; OCT $= 538.79 \pm 26.22$ μm
Ho et al. [30]	The mean CCT values $UP = 438.2 \pm 41.18$ μm; Orbscan $= 435.17 \pm 49.63$ μm; Pentacam $= 430.66 \pm 40.23$ μm and OCT $= 426.56 \pm 41.6$ μm
Zhao et al. [31]	Pearson coefficient $= 0.93$, $P < 0.001$ Bland Altman $= 16.5 \pm 11.7$ μm
Haque et al. [32]	CCT of Keratoconus $UP = 494.2 \pm 50.0$ μm; Orbscan $= 438.6 \pm 47.7$ μm; OCT $= 433.5 \pm 39.7$ μm
Leung et al. [33]	Pearson coefficient $= 0.934$ Bland Altmon $= 23$ μm
Li et al. [34]	Before LASIK mean CCT + SD in microns $UP = 546.9 \pm 29.4$; OCT $= 553.3 \pm 33.0$ After LASIK mean CCT + SD in microns $UP = 498 \pm 46.6$; OCT $= 513.7 \pm 44.5$
Lackner et al. [35]	Pentacam $= 542 \pm 29$ μm; Orbscan $= 576 \pm 37$ μm; OCT $= 530 \pm 34$ μm; $UP = 552 \pm 32$ μm
Fishman et al. [36]	Mean CCT + SD in microns $UP = 545.1 \pm 37.6$; OCT $= 545.1 \pm 36.8$; Orbscan $= 573.9 \pm 51.4$ CCT range in microns $UP = 479–609$; OCT $= 481–614$; Orbscan $= 512–656$
Wong et al. [37]	Orbscan $= 555.96 \pm 32.41$ μm; $UP = 555.11 \pm 35.30$ μm; OCT $= 523.2 \pm 33.54$ μm

4.3　Ultrasound Pachymetry Versus OCT

Zhao et al. found that the CCT values observed from OCT constantly lesser than UP with substantial correlation. Kim et al. also found that there is always a significant difference between the values obtained by OCT in comparing with UP but Leung et al. prove that the OCT is an agreeable device with UP with minor adjustment factor. Li et al. result indicates the high correlation between OCT and UP on pre- and post-LASIK individuals. Also Fishman et al. suggest OCT method than Orbscan for its better accuracy. The OCT also has the ability to measures thinner corneas Temstet et al. Upon the repeatability of OCT device Ramesh et al., Martínez-Albert et al., Otchere et al. state that there is no massive change in the obtained readings and in comparing with Orbscan the OCT has the better repeatability. Otchere et al.

also mention the OCT as a device with good repeatability as like Pentacam but the range of values obtained by OCT is similar to the UP than Pentacam.

5 Conclusion

The values of CCT obtained by OCT, Orbscan, and UP can be equal with correlation factor of 32 μm [17]. The differences between the values observed by the non-invasive devices are not going to affect the refractive surgeries [18] as the differences are very few microns range. Also if there are differences in the readings among all the devices the correlation among them is in higher range [21]. For analyzing the CCT the non-invasive methods are not to be interchanged and compared to each other for assessment on before and after refractive surgery [1, 23]. Among the all three non-invasive methods the OCT have the high correlation with UP also the accuracy and reliability is also high in OCT when measuring the thickness in central and apex of the cornea [33]. The OCT is correlated with UP but the UP is an invasive method so the readings are user-dependent [34]. So the OCT is the optimum image capturing technique for segmentation of the corneal layer and thickness measurement in central and peripheral area of cornea with high repeatability.

References

1. Rabsilber TM, Khoramnia R, Auffarth GU (2006) Anterior chamber measurements using Pentacam rotating Scheimpflug camera. J Cataract Refract Surg 32(3):456–459
2. Gary H (2017) Cornea of the eye—definition and detailed illustration. All about vision. www.allaboutvision.com/resources/cornea.htm
3. DelMonte DW, Kim T (2011) Anatomy and physiology of the cornea. J Cataract Refract Surg 37(3):588–598
4. Eichel J, Mishra A, Fieguth P, Clausi D, Bizheva K (2009) A novel algorithm for extraction of the layers of the cornea. In: Canadian conference on computer and robot vision, CRV'09, pp 313–320. IEEE
5. Alila Medical Media (2016) LASIK or PRK? Which is right for me? Animation. Filmed [Jan 2016]. YouTube video, 03:50, Posted [Jan 2016]. https://www.youtube.com/watch?v=dKANhIU7Sxk
6. Sinjab Academy (2015) Concepts in refractive surgery—part 2. Filmed [Oct 2015]. YouTube video, 22:10, Posted [Oct 2015]. https://www.youtube.com/watch?v=E2unBK5FkLI
7. Mazzotta C, Raiskup F, Baiocchi S, Scarcelli G, Friedman MD, Traversi C (2017) ACXL beyond Keratoconus: post-LASIK ectasia, post-RK ectasia and pellucid marginal degeneration. In: Management of early progressive corneal ectasia. Springer, Cham, pp 169–196
8. Paul T, Lim M, Starr CE, Lloyd HO, Jackson Coleman D, Silverman RH (2008) Central corneal thickness measured by the Orbscan II system, contact ultrasound pachymetry, and the Artemis 2 system. J Cataract Refract Surg 34(11):1906–1912
9. Liu Z, Huang AJ, Pflugfelder SC (1999) Evaluation of corneal thickness and topography in normal eyes using the Orbscan corneal topography system. Br J Ophthalmol 83(7):774–778
10. Matsuda J, Hieda O, Kinoshita S (2008) Comparison of central corneal thickness measurements by Orbscan II and Pentacam after corneal refractive surgery. Jpn J Ophthalmol 52(4):245

11. Colling AJ (2010) A comparison of three methods of measuring central corneal thickness in normal and thinned corneas. Ph.D. dissertation, The Ohio State University
12. Oliveira Cristina M, Ribeiro Celina, Franco Sandra (2011) Corneal imaging with slit-scanning and Scheimpflug imaging techniques. Clin Exp Optom 94(1):33–42
13. Huang D, Swanson EA, Lin CP, Schuman JS, Stinson WG, Chang W, Hee MR, Flotte T, Gregory K, Puliafito CA (1991) Optical coherence tomography. Science 254(5035):1178–1181
14. Fujimoto JG, Drexler W, Schuman JS, Hitzenberger CK (2009) Optical coherence tomography (OCT) in ophthalmology: introduction. Opt Express 17(5):3978–3979
15. Hurmeric V, Yoo SH, Mutlu FM (2012) Optical coherence tomography in cornea and refractive surgery. Expert Rev Ophthalmol 7(3):241–250
16. Otchere H, Sorbara L (2017) Repeatability of topographic corneal thickness in keratoconus comparing Visante™ OCT and Oculus Pentacam HR® topographer. Contact Lens Anter Eye 40(4):217–223
17. Martínez-Albert N, Esteve-Taboada JJ, Montés-Micó R (2018) Repeatability assessment of anterior segment biometric measurements under accommodative and nonaccommodative conditions using an anterior segment OCT. Graefe's Archive Clin Exp Ophthalmol 256(1):113–123
18. Ramesh PV, Jha KN, Srikanth K (2017) Comparison of central corneal thickness using anterior segment optical coherence tomography versus ultrasound pachymetry. J Clin Diagn Res JCDR 11(8):NC08
19. Antonios R, Abdul Fattah M, Maalouf F, Abiad B, Awwad ST (2016) Central corneal thickness after cross-linking using high-definition optical coherence tomography, ultrasound, and dual scheimpflug tomography: a comparative study over one year. Am J Ophthalmol 167:38–47
20. Xu Z, Peng M, Jiang J, Yang C, Zhu W, Fan L, Shen M (2016) Reliability of Pentacam HR thickness maps of the entire cornea in normal, post-laser in situ Keratomileusis, and keratoconus eyes. Am J Ophthalmol 162:74–82
21. Temstet C, Sandali O, Bouheraoua N, Hamiche T, Galan A, El Sanharawi M, Basli E, Laroche L, Borderie V (2015) Corneal epithelial thickness mapping using Fourier-domain optical coherence tomography for detection of form fruste keratoconus. J Cataract Refract Surg 41(4):812–820
22. Kuerten D, Plange N, Koch EC, Koutsonas A, Walter P, Fuest M (2015) Central corneal thickness determination in corneal edema using ultrasound pachymetry, a Scheimpflug camera, and anterior segment OCT. Graefe's Archive Clin Exp Ophthalmol 253(7):1105–1109
23. Randleman JB, Lynn MJ, Perez-Straziota CE, Weissman HM, Kim SW (2015) Comparison of central and peripheral corneal thickness measurements with scanning-slit, Scheimpflug and Fourier-domain ocular coherence tomography. Br J Ophthalmol 99(9):1176–1181
24. Lin C-W, Wang T-H, Huang Y-H, Huang J-Y (2013) Agreement and repeatability of central corneal thickness measurements made by ultrasound pachymetry and anterior segment optical coherence tomography. Taiwan J Ophthalmol 3(3):98–102
25. Yazici AT, Bozkurt E, Alagoz C, Alagoz N, Pekel G, Kaya V, Yilmaz OF (2010) Central corneal thickness, anterior chamber depth, and pupil diameter measurements using Visante OCT, Orbscan, and Pentacam. J Refract Surg 26(2):127–133
26. Doors M, Cruysberg LPJ, Berendschot TTJM, de Brabander J, Verbakel F, Webers CAB, Nuijts RMMA (2009) Comparison of central corneal thickness and anterior chamber depth measurements using three imaging technologies in normal eyes and after phakic intraocular lens implantation. Graefe's Archive Clin Exp Ophthalmol 247(8):1139–1146
27. Cheng ACK, Rao SK, Lau S, Lam DSC, Leung CKS (2008) Central corneal thickness measurements by ultrasound, Orbscan II, and Visante OCT after LASIK for myopia. J Refract Surg 24(4):361–365
28. Kim HY, Budenz DL, Lee PS, Feuer WJ, Barton Keith (2008) Comparison of central corneal thickness using anterior segment optical coherence tomography vs ultrasound pachymetry. Am J Ophthalmol 145(2):228–232
29. Li EYM, Mohamed S, Leung CKS, Rao SK, Cheng ACK, Cheung CYL, Lam DSC (2007) Agreement among 3 methods to measure corneal thickness: ultrasound pachymetry, Orbscan II, and Visante anterior segment optical coherence tomography. Ophthalmology 114(10):1842–1847

30. Ho T, Cheng ACK, Rao SK, Lau S, Leung CKS, Lam DSC (2007) Central corneal thickness measurements using Orbscan II, Visante, ultrasound, and Pentacam pachymetry after laser in situ keratomileusis for myopia. J Cataract Refract Surg 33(7):1177–1182

31. Zhao PS, Wong TY, Wong W-L, Saw S-M, Aung T (2007) Comparison of central corneal thickness measurements by visante anterior segment optical coherence tomography with ultrasound pachymetry. A J Ophthalmol 143(6):1047–1049

32. Haque S, Simpson T, Jones L (2006) Corneal and epithelial thickness in keratoconus: a comparison of ultrasonic pachymetry, Orbscan II, and optical coherence tomography. J Refract Surg 22(5):486–493

33. Leung DYL, Lam DKT, Yeung BYM, Lam DSC (2006) Comparison between central corneal thickness measurements by ultrasound pachymetry and optical coherence tomography. Clin Exp Ophthalmol 34(8):751–754

34. Li Y, Shekhar R, Huang D (2006) Corneal pachymetry mapping with high-speed optical coherence tomography. Ophthalmology 113(5):792–799

35. Lackner B, Schmidinger G, Pieh S, Funovics MA, Skorpik C (2005) Repeatability and reproducibility of central corneal thickness measurement with Pentacam, Orbscan, and ultrasound. Optom Vis Sci 82(10):892–899

36. Fishman GR, Pons ME, Seedor JA, Liebmann JM, Ritch R (2005) Assessment of central corneal thickness using optical coherence tomography. J Cataract Refract Surg 31(4):707–711

37. Wong ACM, Wong CC, Yuen NSY, Hui SP (2002) Correlational study of central corneal thickness measurements on Hong Kong Chinese using optical coherence tomography, Orbscan and ultrasound pachymetry. Eye 16(6):715

Hybrid SIFT Feature Extraction Approach for Indian Sign Language Recognition System Based on CNN

Abhishek Dudhal, Heramb Mathkar, Abhishek Jain, Omkar Kadam
and Mahesh Shirole

Abstract Indian sign language (ISL) is one of the most used sign languages in the Indian subcontinent. This research aims at developing a simple Indian sign language recognition system based on convolutional neural network (CNN). The proposed system needs webcam and laptop and hence can be used anywhere. CNN is used for image classification. Scale invariant feature transformation (SIFT) is hybridized with adaptive thresholding and Gaussian blur image smoothing for feature extraction. Due to unavailability of ISL dataset, a dataset of 5000 images, 100 images each for 50 gestures, has been created. The system is implemented and tested using python-based library Keras. The proposed CNN with hybrid SIFT implementation achieves 92.78% accuracy, whereas the accuracy of 91.84% was achieved for CNN with adaptive thresholding.

1 Introduction

Sign language is an efficient and natural way of communication for the hearing-impaired and verbally impaired community. Around 2.7 million of India's total population is hearing-impaired and 98% of this population uses Indian sign language as the primary language for communication. However, hearing-impaired community

A. Dudhal (✉) · H. Mathkar · A. Jain · O. Kadam · M. Shirole
Department of Computer Engineering and Information Technology, Veermata Jijabai
Technological Institute, Mumbai, India
e-mail: addudhal_b14@it.vjti.ac.in

H. Mathkar
e-mail: hkmathkar_b14@it.vjti.ac.in

A. Jain
e-mail: amjain_b14@it.vjti.ac.in

O. Kadam
e-mail: oskadam_b14@it.vjti.ac.in

M. Shirole
e-mail: mrshirole@it.vjti.ac.in

© Springer Nature Switzerland AG 2019
D. Pandian et al. (eds.), *Proceedings of the International Conference on ISMAC
in Computational Vision and Bio-Engineering 2018 (ISMAC-CVB)*, Lecture Notes
in Computational Vision and Biomechanics 30,
https://doi.org/10.1007/978-3-030-00665-5_72

experiences difficulties while communicating with people who lack knowledge of sign language. A human translator is required to translate sign language into speech. This solution is translator dependent and it fails in absence of a translator. A differently abled person can be empowered with a computer-based system for translation. A computer-based system can be trained to recognize ISL efficiently, thereby providing high availability, ease of use, and efficient navigation and trade to differently abled persons.

For Indian sign language recognition, the research work undertaken ranges from introducing a smart glove to monitor movements of fingers and hand to image processing that analyzes hand gestures captured. Heera et al. [1] introduced sensors incorporated glove-based approach to convert ISL into speech with the help of Bluetooth module and an Android smartphone. Ekbote et al. [2] proposed a method for ISL recognition using artificial neural network (ANN) [3] and support vector machine (SVM) [4] classifiers. Authors have used a self-created dataset for ISL, 0–9 numbers, which is very limited. Histogram of oriented gradients (HOG) [5] and ANN-based approach proposed by them was able to achieve 99% accuracy. Beena et al. [6] proposed a CNN [7] based ASL recognition system. They used ASL dataset with 33,000 images for 24 alphabets and 0–9 numbers and accuracy of 94.6774% was achieved. Pigou et al. [8] were able to classify 20 Italian gestures using CNN with validation accuracy of 91.7%. They used Microsoft Kinect to capture gestures. Microsoft Kinect is able to capture depth feature. Depth feature aids significantly in image classification.

The contemporary research focused on the numbers, alphabets, limited words, and single-handed gestures. In contrast, this paper aims to help the hearing-impaired community by developing a simple computer vision-based system, which works on 50 ISL words including numbers and double handed gestures.

This paper is composed of six sections; Sect. 2 discusses the basic concepts used in the paper. Section 3 discusses the proposed system. Results of the experiment are discussed in Sect. 4. A complete conclusion is drawn in Sect. 5. Section 6 highlights the future aspects of the paper.

2 Basic Concepts

2.1 Adaptive Thresholding

Image binarization can be achieved with the help of adaptive thresholding. Image binarization is a method of separation of pixel intensity in two groups. Setting black as foreground and white as background or vice versa. Image binarization and thresholding is an effective way to separate an object from the background. In adaptive thresholding, a threshold value is set such that pixel intensity below that threshold will be treated as zero while pixel intensity greater than the threshold will be treated as one. Equation 1 shows the formula for adaptive thresholding.

$$b(x, y) = \begin{cases} 0, & I(x, y) \leq T(x, y) \\ 1, & I(x, y) > T(x, y) \end{cases} \tag{1}$$

where $T(x, y)$ is the threshold, $b(x, y)$ is the binarized image, and $I(x, y)$ is the intensity of pixel at (x, y).

2.2 Image Smoothing Using Gaussian Blur

Image smoothing is a technique of removing noise from digital images. Smoothing can remove noise without losing important features from the image. Gaussian blur filter [9] uses a Gaussian function [10] to calculate transformation. Equation 2 represents Gaussian blur operator.

$$G(x, y, \sigma) = \frac{1}{2\pi\sigma^2} e^{-\frac{(x^2+y^2)}{2\sigma^2}} \tag{2}$$

2.3 Key-Point Generation Using Scale Invariant Feature Transform

SIFT [11] is scale, rotation, viewpoint, illumination invariant algorithm. The key-point generation involves three steps. The first step is the generation of scale space. In the second step, Laplacian of Gaussians (LoG) [12] is generated while in the final step, key points are calculated.

Scale-Space Generation In the scale-space generation step, the original image is taken, and progressively blurred out images are generated. Then, the original image is resized to half and blurred out images are generated again. Images of the same size form an octave. The number of scales and octaves depend on the user. Blurring can be thought of as a convolution of the Gaussian operator and the image.

$$L(x, y, \sigma) = G(x, y, \sigma) * I(x, y) \tag{3}$$

where L is the obtained Blurred image, G is the Gaussian blur operator, I is the image, (x, y) are the coordinates in the image, σ is scale parameter, and $*$ is convolution operator in (x, y). This operator applies Gaussian blur G onto the image I. Gaussian blur function is represented by Eq. 2. Let the amount of the blurring in one image be σ and then, the amount of blurring in the next image will be $k\sigma$, where k is a constant chosen by the user.

Laplacian of Gaussian (LoG) Approximation In the generation of LoG [12] step, an image is taken and blurred a little using Gaussian blur. Then, second-order derivatives are calculated on it (Laplacian). This process is computationally expensive. So to calculate LoG quickly, scale space obtained in the previous step is used. Difference between two consecutive scales is calculated. These differences of Gaussian images are approximately equal to LoG.

Finding Key Points In this step, each and every pixel is iterated, and all of its neighbors are checked. The check is done within the current image, one above and one below it. The pixel is marked as an approximate key point if it is greatest or least of all 26 neighbors. The minima or maxima lies somewhere between the pixels. So, subpixel value is calculated mathematically using Taylor expansion [13] of the image around the approximate key point.

2.4 CNN

CNN is a deep learning neural network. CNN is specialized for images, audios, videos, and speech processing. It is designed to learn features with the help of filters and hence requires very little data preprocessing and feature extraction. CNN is composed of one input layer, one or multiple convolutional layers, one or multiple max-pooling layers, a fully connected layer, and an output layer. Input layer accepts input which is passed to next layer. The convolutional layer is responsible to apply convolution operation to the input data. This layer works as eyes of CNN and looks for specific features useful for classification. The Filters are also known as the kernels. Max-pooling layer is useful for reducing parameters size and hence processing time. Fully connected layer acts as a classifier. Output layer gives an output vector consisting of probability for different classes. Each neuron uses activation function for mathematical processing of data. CNN has mechanism of dropout which is used to avoid overfitting.

2.5 Confusion Matrix

The confusion matrix is a matrix which is used to summarize the performance of the classifier. For a good classifier, it is a sparse matrix and can be represented in the form of a graph. Actual class is represented on X-axis while predicted class is represented on the Y-axis. Label to point (X, Y) represents a number of the example for which actual class is X and predicted is Y. When $X = Y$, then it is treated as accurate classification. Hence, (X, X) are treated as correctly classified examples. The confusion matrix is used to analyze the results of CNN with hybrid SIFT implementation, later in this research.

Precision Precision is a fraction of correctly identified examples to the number of examples for which that particular class is predicted as positive. Equation 4 specifies the formula for precision in the multiclass classifier.

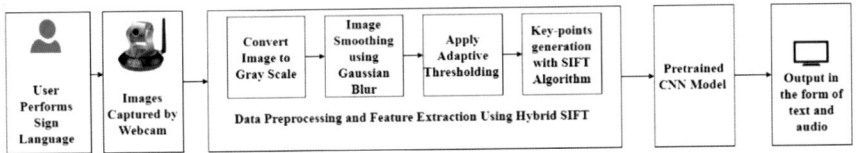

Fig. 1 Proposed system for ISL classification

$$\text{Precision}_i = \frac{\text{CF}_{ii}}{\sum_1^n \text{CF}_{ij}} \tag{4}$$

where CF_{ii} is (i, i)th entry in confusion matrix, CF_{ij} is (i, j)th entry in confusion matrix, and n is the total number of classes.

Recall A recall is a fraction of correctly identified examples to the number of examples available for that class. Equation 5 specifies the formula for recall in a multiclass classifier.

$$\text{Recall}_i = \frac{\text{CF}_{ii}}{\sum_1^n \text{CF}_{ji}} \tag{5}$$

where CF_{ii} is (i, i)th entry in confusion matrix, CF_{ji} is (j, i)th entry in confusion matrix, and n is total number of classes.

3 Proposed System

Figure 1 shows a flow diagram of a proposed system for ISL classification. Gestures performed by the user are captured by a webcam. The captured gesture is preprocessed and features are extracted using hybrid SIFT. Hybrid SIFT is discussed in Sect. 3.1. The preprocessed gesture is fed to a pretrained CNN model. The CNN model used for this research is explained in Sect. 3.2.

3.1 Data Preprocessing and Feature Extraction

Though CNN has the ability to work without any feature extraction and data preprocessing, data preprocessing is used for reducing computational power required and for better performance of the model. This paper presents a hybrid SIFT approach for data preprocessing and feature extraction.

Hybrid SIFT As discussed in Sect. 2.3, SIFT calculates key points. Data preprocessing before applying SIFT on the image can reduce noise and can help in better key-points generation. The key-point generation step of the SIFT is hybridized with

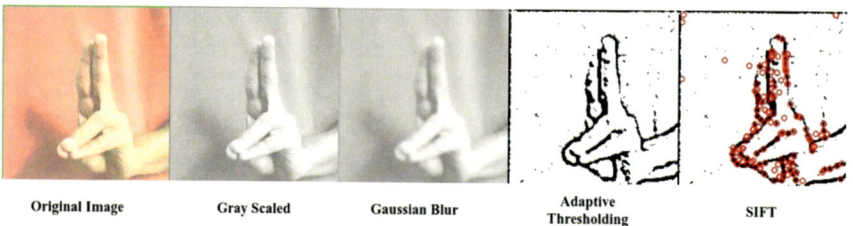

| Original Image | Gray Scaled | Gaussian Blur | Adaptive Thresholding | SIFT |

Fig. 2 Application of SIFT on preprocessed image (hybrid SIFT)

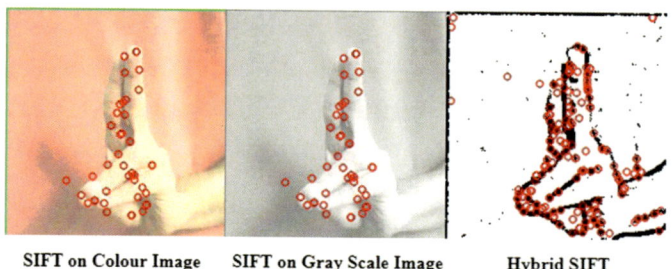

| SIFT on Colour Image | SIFT on Gray Scale Image | Hybrid SIFT |

Fig. 3 Key-point calculation on original image with SIFT, grayscaled image with SIFT and hybrid SIFT

Gaussian blurring and adaptive thresholding. Steps involved in performing hybrid SIFT are shown in Fig. 2. In the first step, the image is captured and resized to 200×200 pixels. The second step involves converting the resized original image to grayscaled image. In the third step as discussed in Sect. 2.2, the grayscaled image is smoothened using a Gaussian blur filter. In next step, the smoothened grayscaled image is binarized using adaptive thresholding. Finally, key-point generation step of SIFT algorithm is applied to the binarized image. Figure 3 shows a comparison of key-point calculation using SIFT algorithm on a simple grayscaled image and hybrid SIFT algorithm. Figure 3 suggests that SIFT applied on the adaptive thresholded image gives better key points than when applied directly to the original image.

3.2 Architecture of CNN Model

This paper presents a self-designed CNN model for gesture recognition. Figure 4 shows the code snippet of the proposed CNN model. Proposed CNN model has 10 convolutional layers. Convolutional layers are divided into five groups. Each group contains two convolutional layers. For the first group, the number of filters is kept 32, for second group 64, for third group 128, for fourth group 256, and for the final group, it is kept 512. Each convolutional layer has a filter size of 3×3. Max-pooling

```
model = Sequential()
model.add(Conv2D(32, (3, 3), padding='valid', activation='relu', input_shape=input_shape))
model.add(Conv2D(32, (3, 3), activation='relu'))
model.add(MaxPooling2D(pool_size=(2, 2)))
model.add(Dropout(0.25))
model.add(Conv2D(64, (3, 3), padding='valid', activation='relu'))
model.add(Conv2D(64, (3, 3), activation='relu'))
model.add(MaxPooling2D(pool_size=(2, 2)))
model.add(Dropout(0.25))
model.add(Conv2D(128, (3, 3), padding='valid', activation='relu'))
model.add(Conv2D(128, (3, 3), activation='relu'))
model.add(MaxPooling2D(pool_size=(2, 2)))
model.add(Dropout(0.25))
model.add(Conv2D(256, (3, 3), padding='valid', activation='relu'))
model.add(Conv2D(256, (3, 3), activation='relu'))
model.add(MaxPooling2D(pool_size=(2, 2)))
model.add(Dropout(0.25))
model.add(Conv2D(512, (3, 3), padding='valid', activation='relu'))
model.add(Conv2D(512, (3, 3), activation='relu'))
model.add(MaxPooling2D(pool_size=(2, 2)))
model.add(Dropout(0.25))
model.add(Flatten())
model.add(Dense(512, activation='relu'))
model.add(Dropout(0.25))
model.add(Dense(no_classes, activation='softmax'))
model.compile(loss='categorical_crossentropy', optimizer='adadelta', metrics=['accuracy'])
```

Fig. 4 Code snippet for CNN model

layer with pooling window of 2×2 is applied after each group of the convolutional layer. After each max-pooling layer, to avoid overfitting dropout ratio of 0.25 is used. Each convolution layer uses rectified linear unit (RELU) as the activation function [14]. Optimizer for training was kept fixed as Adadelta [15]. The model uses Softmax [16] as a classifier. To calculate the performance of model, the loss function is used as cross-entropy [17]. The CNN model is coded by using python-based library Keras [18].

4   Results

The proposed CNN model is applied to self-created dataset of ISL with 50 signs and 100 images per sign. CNN model is trained with an incremental increase of epochs approach. Batch size for the model training is kept fixed at 64. Starting from 10 epochs, epochs were increased by 5 till model converged with constant validation accuracy. With increments in the number of epoch approach, proposed model converged with constant validation accuracy at 25 epochs. The model was trained till 50 epochs. Section 4.1 describes the system requirement for the proposed system, Sect. 4.2 elaborates data acquisition process. The accuracy of the proposed CNN model is discussed in Sect. 4.3 which shows a confusion matrix generated for the proposed model and dataset. Finally, we discuss the comparison of the proposed system with related research.

4.1 System Requirements

The proposed system tried to keep user interaction with the system using desktop application. System requirements are as follows:

- 4 GB Ram
- 1 GB Free Space
- Web Cam.

4.2 Data Acquisition

The standard dataset for Indian sign language is not available. Two sets of datasets each of 5000 images were created. One was used for training and validation purpose while other for the testing purpose. Dataset was created with the help of 5 MP webcam attached to a laptop. Dictionary of 50 most used signs is created by taking help of Deaf and Dumb School, Mumbai. Each dataset contains total 5000 images, where for each sign 100 images of 200×200 pixel are captured. Dataset is created with the help of 20 people. The age group varies from 16 to 50. Around 70% of the people were in the age group of 20–25. Both males and females have participated in dataset creation.

4.3 Accuracy of Proposed CNN Model

Table 1 summarizes accuracy matrix of the proposed model. For CNN with adaptive thresholding, training accuracy (TA) achieved is 97.58% and validation accuracy (VA) is 91.84%. For CNN with hybrid SIFT, training accuracy increases to 98.83% and validation accuracy increases to 92.72%. It shows that CNN with hybrid SIFT performs better than CNN with only adaptive thresholding. Figure 5 shows the accuracy graph of the proposed model. In the accuracy graph, accuracy is represented as dependent variable on Y-axis and number of epochs as an independent variable on X-axis. Accuracy graph lists accuracy for training dataset by blue line and accuracy on validation dataset by the orange line. Accuracy graph shown in Fig. 5 suggests that early stopping of the model at around 25 epoch can avoid overfitting issue.

Table 1 Accuracy matrix for proposed CNN model	CNN with adaptive thresholding		CNN with hybrid SIFT	
	TA	VA	TA	VA
	97.58	91.84	98.83	92.72

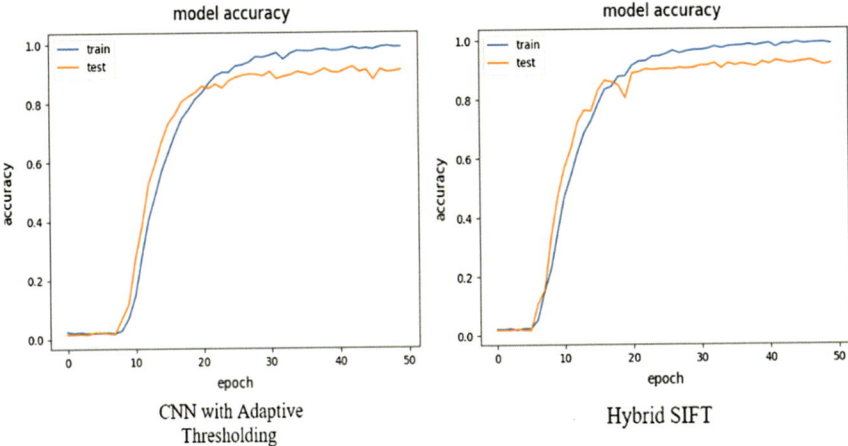

Fig. 5 Model accuracy

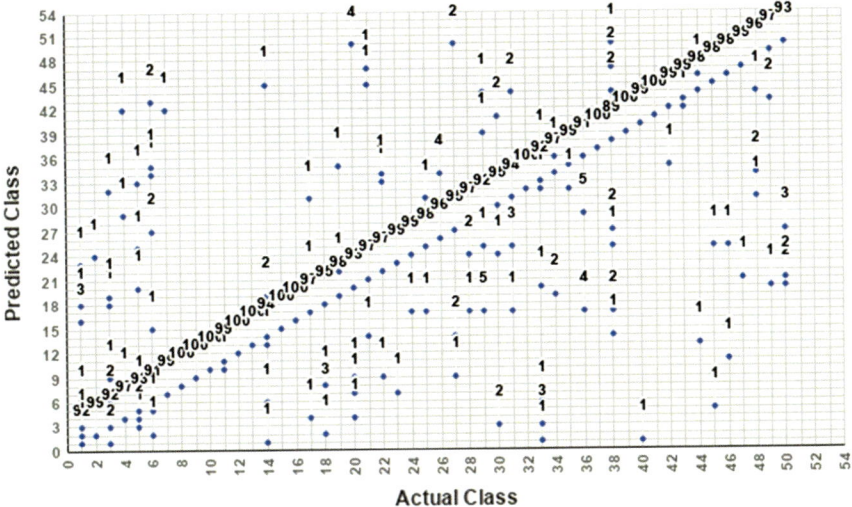

Fig. 6 Confusion matrix obtained by running trained CNN model on testing dataset

4.4 Confusion Matrix of Proposed System

The hybrid SIFT algorithm is first applied on the dataset and then CNN model is trained by using the same dataset. As discussed in Sect. 2.5, the confusion matrix is created by running trained CNN model on the testing dataset. Figure 6 shows the confusion matrix in a graphical format. The Y-axis of the graph represents predicted class while X-axis represents the actual class. Confusion matrix obtained for the proposed model is of sparse nature.

Table 2 Precision and recall obtained by running trained CNN model on testing dataset

Sign	Precision	Recall	Sign	Precision	Recall	Sign	Precision	Recall
House	94.8454	92	High	97.9381	95	Rose	97.0588	99
A board	97.0588	99	How many	95.1456	98	See	98.913	91
All Gone	92.9293	92	I	94.898	93	Seven	99.0099	100
Baby	96.0396	97	Man	96.0396	97	Short	100	89
Beside	97.8947	93	Marry	98.9796	97	Six	99.0099	100
Book	91.9192	91	Meat	99	99	Superior	100	99
Bowl	97.0588	99	Medal	96.1165	99	Ten	98.0392	100
Bridge	98.0392	100	Mid Day	92.4528	98	Thick	97.0588	99
Camp	96.1538	100	Middle	100	96	Thin	96.1165	99
Cartridge	99.0099	100	Money	93.1373	95	Three	94.2308	98
Eight	99	99	Moon	100	97	Tobacco	98	98
Five	100	100	Mother	93.8775	92	Two	98.9899	98
Fond	98.0392	100	Nine	100	95	Up	97.0588	99
Four	95.9184	94	One	96.9072	94	Watch	100	96
Friend	99.0099	100	Opposite	97.0874	100	Write	100	97
Glove	97.0874	100	Prisoner	97.8723	92	You	93	93
Hang	86.6071	97	Ring	92.381	97			

Precision and recall were calculated for multiclass using formula shown in Sect. 2.5. Table 2 summarizes precision and recall in percent for proposed model and 50 signs available in the dataset. Highest precision obtained is 100% while least precision is 86.6071%. Highest recall obtained is 100% while least recall is 89%.

5 Conclusion

The sensor incorporated glove-based systems which are most common in sign language recognition are usually costly and difficult to use; in contrast, image classification-based system proposed in this paper is much cheaper and easier to use. CNN is an important and efficient algorithm for image classification. This paper proposed a system for Indian sign language recognition using classification by CNN and feature extraction by hybrid SIFT. CNN is robust and stable such that it requires very little image preprocessing. But image processing using hybrid SIFT improves the performance of CNN classifier. The model proposed in this paper has achieved validation accuracy of 92.78% for CNN with hybrid SIFT approach while 91.84% accuracy was achieved for CNN with adaptive thresholding approach. The system proposed in this paper can work just with a laptop and web camera and hence can be used with mobility by the hearing-impaired community. The system proposed in this paper can also be used for learning Indian sign language.

6　Future Work

Dataset used in this paper contains 50 Indian signs. Further work can be done to increase the number of signs as well as images per sign. This paper considers only static Indian signs. In the future, CNN can also be implemented for motion-based Indian signs. The proposed system can be extended to work with handheld mobile devices by optimizing memory and power requirement.

Acknowledgements　We would like to thank Principal of Deaf and Dumb School, Mumbai for her help in understanding ISL gestures. We would like to thank teachers and students of Deaf and Dumb School for their help in the creation of ISL dataset.

References

1. Heera SY et al (2017) Talking hands—an Indian sign language to speech translating gloves. In: 2017 International conference on innovative mechanisms for industry applications (ICIMIA)
2. Ekbote J et al (2017) Indian sign language recognition using ANN and SVM classifiers. In: 2017 International conference on innovations in information, embedded and communication systems (ICIIECS)
3. Duch W (2005) Artificial neural network biological inspirations. In: ICANN 2005: 15th international conference, Warsaw, Poland, September 11–15, 2005, proceedings, Pt. 1. Springer, Heidelberg
4. Qi X et al (2017) Data classification with support vector machine and generalized support vector machine
5. Vo T et al (2015) Tensor decomposition and application in image classification with histogram of oriented gradients. Neurocomputing 165:38–45
6. Beena MV, Agnisarman Namboodiri MN (2017) Automatic sign language finger spelling using convolution neural network: analysis. Int J Pure Appl Math 117(20)
7. Aghdam HH, Heravi EJ (2017) Convolutional neural networks. In: Guide to convolutional neural networks, pp 85–130
8. Pigou L et al (2015) Sign language recognition using convolutional neural networks. Computer vision—ECCV 2014 workshops lecture notes in computer science, pp 572–578
9. Gaussian Smoothing (2008) Wolfram Demonstrations Project
10. Gaussian function. Springer Reference
11. Lowe DG (2004) Distinctive image features from scale-invariant keypoints. Int J Comput Vision 60(2):91–110
12. Laplacian of Gaussian Filtering (2008) Wolfram Demonstrations Project
13. Berz M (1997) From Taylor series to Taylor models
14. Ramachandran P et al (2017) Searching for activation functions. In: ICLR 2018 conference

15. Zeiler MD (2012) ADADELTA: an adaptive learning rate method
16. Pellegrini T (2015) Comparing SVM, Softmax, and shallow neural networks for eating condition classification. Interspeech 2015
17. Dahal P, Classification and loss evaluation—Softmax and cross entropy loss. https://deepnotes.io/softmax-crossentropy
18. Keras: The Python Deep Learning library. https://keras.io/

A Contemporary Framework and Novel Method for Segmentation of Brain MRI

A. Jagan

Abstract Brain magnetic resonance imaging plays a vital role in medical image processing for detection of brain tumor, therapy response evaluation, brain tumor diagnosis, and treatment selection. Contrast-enhanced T1C 3D brain magnetic resonance imaging is extensively used brain-imaging modalities. Automated segmentation and detection of brain tumors in contrast-enhanced T1C 3D brain magnetic resonance imaging is a very complicated task due to high disparity in shape, size, and appearance of brain tumor. Nevertheless, so many analyses have been conducted in the similar research work, it remains a very complicated task for automated segmentation and detection of brain tumor in magnetic resonance imaging, and enhancing segmentation accuracy of brain tumor is still continuing field. The main aim of this research work is to develop a fully automated segmentation framework for segmentation and detection of the brain tumor that is allied with contrast-enhanced T1C 3D brain magnetic resonance imaging. Consequently, this research work deals about development of segmentation framework for detection of tumor in brain 3D MR images. The proposed segmentation framework ingrates the most established fuzzy C means clustering method and improved Expectation Maximization (EM) method. An anisotropic filter is used for preprocessing and subsequently, it is employed to the fuzzy C means clustering method, improved Expectation Maximization (EM), and proposed method for superior segmentation and detection of tumor. The performance result of proposed framework is evaluated on 10 patients' simulated medical clinical dataset. The performance results of the proposed framework exhibited better results as compared with presented methods.

1 Introduction

The advancement in the segmentation and detection of brain tumor in magnetic resonance imaging is evolving. The similar research has exposed the various methods

A. Jagan (✉)
CSE Department, BVRIT, Narsapur, India
e-mail: jagan.amgoth@bvrit.ac.in

© Springer Nature Switzerland AG 2019 739
D. Pandian et al. (eds.), *Proceedings of the International Conference on ISMAC in Computational Vision and Bio-Engineering 2018 (ISMAC-CVB)*, Lecture Notes in Computational Vision and Biomechanics 30,
https://doi.org/10.1007/978-3-030-00665-5_73

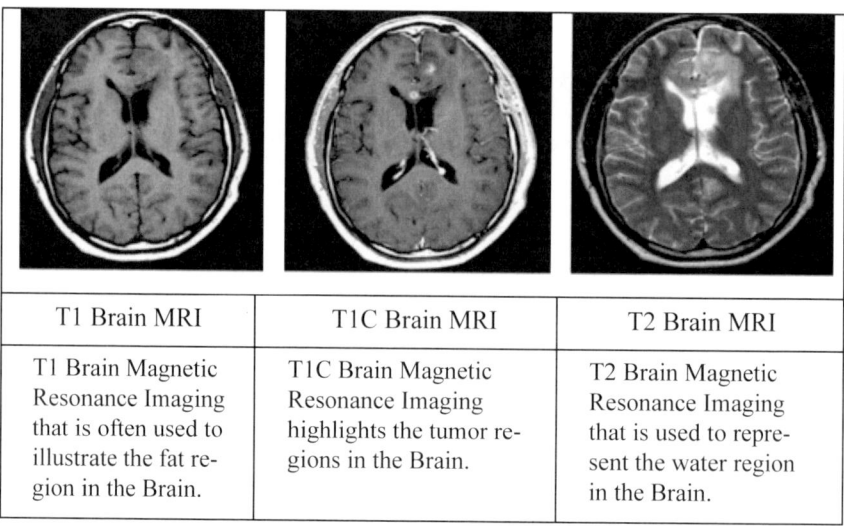

T1 Brain MRI	T1C Brain MRI	T2 Brain MRI
T1 Brain Magnetic Resonance Imaging that is often used to illustrate the fat region in the Brain.	T1C Brain Magnetic Resonance Imaging highlights the tumor regions in the Brain.	T2 Brain Magnetic Resonance Imaging that is used to represent the water region in the Brain.

Fig. 1 Magnetic resonance imaging series

and potential techniques to the brain image analysis and detection of tumor in 3D brain magnetic resonance imaging. The MRI images present the noninvasive and broad range of visualization of brain internal anatomical structure and analysis of the brain. The major research direction is focused on development of the segmentation accuracy for detection of tumor in brain magnetic resonance imaging. Hence, this research work deals about the analysis of contrast-enhanced T1C 3D brain magnetic resonance imaging with the intricate and superior quality became the most challenge job for magnetic resonance imaging analysis experts [1, 2]. Furthermore, because of the human intervention, the analysis of brain magnetic resonance imaging is bound to be invalid. Further, magnetic resonance imaging by manual analyses takes lots of time and very limited to discriminate of tumor in contrast-enhanced T1C 3D brain magnetic resonance imaging as compared with the modern automatic methods [1, 3, 4].

The usually used technique for analysis of brain magnetic resonance imaging is segmentation method that is most popular for imaging of medical applications especially brain magnetic resonance imaging. The limitations recognized in the research study exposes for assorted segmentation methods are inadequate in generating improved segmentation accuracy and more often than not focused on segmentation and detection of the brain tumor. The present research work focused on the incomparable segmentation accuracy results [4–6]. The brain magnetic resonance imaging has diverse MRI series such as T1 brain magnetic resonance imaging, contrast-enhanced T1C 3D brain MR imaging, and T2 brain magnetic resonance imaging as shown in Fig. 1.

This paper is focused on the segmentation of contrast-enhanced T1C 3D brain magnetic resonance imaging. The usefulness of T1C series has been estimated in various ways in the central nervous system, i.e., infarction, head injuries, subarachnoid hemorrhage, and multiple sclerosis. Contrast-enhanced T1C brain magnetic resonance imaging provides superior anatomical details had better than other brain MR imaging. In preprocessing stage, the anisotropic filtering [7, 8] is used to smooth edges of brain magnetic resonance imaging while preserving edges and it also used to improve the quality of input magnetic resonance imaging without losing the boundaries. The applications of anisotropic filter have fully extended, and it is now used for denoising of medical images and medical image enhancement, etc. The anisotropic filter is focused on input magnetic resonance imaging variance and standard deviation [7]. Fuzzy C means clustering method [9–13]. In this work, we analyze the accuracy of well-known methods, i.e., improved Expectation Maximization (EM) [5, 14–16] and Proposed method for segmentation and detection of brain disorder in contrast-enhanced T1C 3D brain magnetic resonance imaging. Hereafter the rest of research work is organized to obtain better segmentation accuracy and detection of tumor in contrast-enhanced T1C 3D brain magnetic resonance imaging.

The rest of the research work is organized as follows: Sect. 2 in this research work presents the Proposed Methodology and approach for enhancement of segmentation results, in Sect. 3 discuss in detail about applications and the evaluation of results which is executed on multiple contrast-enhanced T1C magnetic resonance imaging datasets and in Sect. 4 discuss the conclusions of the research work that is presented in this paper.

2 Proposed Methodology

The major objective of this proposed research work is to develop an automated segmentation framework for the detection of brain tumor in contrast-enhanced T1C 3D brain magnetic resonance imaging. In this research work, initially, the entire brain imaging is segmented with fuzzy C means clustering method and the improved Expectation Maximization (EM). These segmentation results are optimally integrated with proposed method to improve the segmentation accuracy. The main idea behind this work is to integrate the most established segmentation methods, i.e., fuzzy C means clustering method and the improved Expectation Maximization (EM). Consequently, the proposed framework is equipped with the anisotropic filter in the preprocessing stage to improve the quality of input brain MR image and for the better segmentation and detection of brain tumor. The block diagram of the proposed research work is shown in Fig. 2.

(a) FCMC Method = {CONTRAST-ENHANCED T1C $3D_1\_FCMCSeg$, CONTRAST-ENHANCED T1C $3D_2\_FCMCSeg$, CONTRAST-ENHANCED T1C $3D_3\_FCMCSeg$, ..., CONTRAST-ENHANCED T1C $3D_n\_FCMCSeg$}.

(b) EM Method = {CONTRAST-ENHANCED T1C $3D_1\_EMSeg$, CONTRAST-ENHANCED T1C $3D_2\_EMSeg$, CONTRAST-ENHANCED T1C $3D_3\_EMSeg$, ..., CONTRAST-ENHANCED T1C $3D_n\_EMSeg$}.

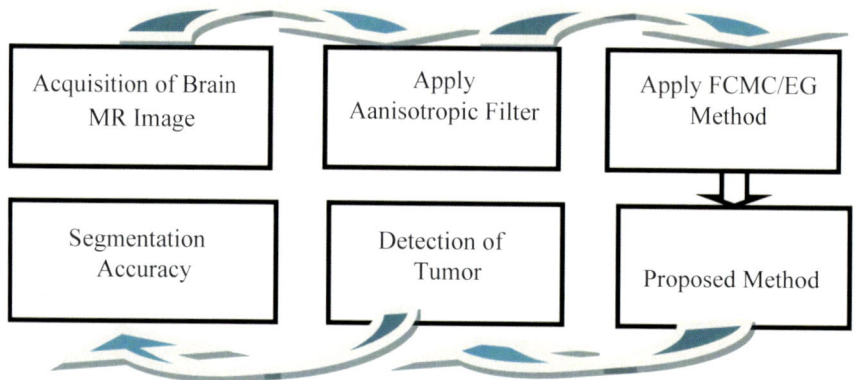

Fig. 2 Block diagram of the proposed work

(c) Proposed Method = {CONTRAST-ENHANCED T1C 3D$_1$_AMCSeg, CONTRAST-ENHANCED T1C 3D$_2$_AMCSeg, CONTRAST-ENHANCED T1C 3D$_3$_AMCSeg, ..., CONTRAST-ENHANCED T1C 3D$_n$_AMCSeg}.

(d) Segmentation Results (Tumor Detection) of the proposed Method = {Tumor$_1$_SegImage, Tumor$_2$_SegImage, Tumor$_3$_SegImage, ..., Tumor$_n$_SegImage}.

(e) Ground truth(Images) Dataset = {CONTRAST-ENHANCED T1C 3D$_1$_Truth, CONTRAST-ENHANCED T1C 3D$_2$_Truth, CONTRAST-ENHANCED T1C 3D$_3$_Truth, ..., CONTRAST-ENHANCED T1C 3D$_n$_Truth}.

The proposed procedure for automated segmentation for detection of tumor in contrast-enhanced T1C 3D brain magnetic resonance imaging:

Input: MRI Dataset – {CONTRAST-ENHANCED T1C 3D$_1$, CONTRAST-ENHANCED T1C 3D$_2$, CONTRAST-ENHANCED T1C 3D$_3$, ..., CONTRAST-ENHANCED T1C 3D$_n$}.

Output: Brain tumor detection in Contrast-Enhanced T1C 3D MR Image – {Tumor$_1$_SegImage, Tumor$_2$_SegImage, Tumor$_3$_SegImage, ..., Tumor$_n$_SegImage}.

Step 1   Preprocessing: The Anisotropic Filter (AF) is applied on input Original input MRI Dataset {CONTRAST-ENHANCED T1C 3D$_1$, CONTRAST-ENHANCED T1C 3D$_2$, CONTRAST-ENHANCED T1C 3D$_3$, ..., CONTRAST-ENHANCED T1C 3D$_n$} to enhance and improve the quality input MRI.

Step 2   Initial Segmentation: Primarily Improved Expectation Maximization (EM) method and fuzzy C means clustering method were functional on preprocessed brain MR images which is {CONTRAST-ENHANCED T1C 3D$_1$_PreP, CONTRAST-ENHANCED T1C 3D$_2$_PreP, CONTRAST-ENHANCED T1C 3D$_3$_PreP, ..., CONTRAST-ENHANCED T1C 3D$_n$_PreP}.

    a. Segmented results of FCMC is denoted as {CONTRAST-ENHANCED T1C $3D_1$_FCMCSeg, CONTRAST-ENHANCED T1C $3D_2$_FCMCSeg, CONTRAST-ENHANCED T1C $3D_3$_FCMCSeg, ..., CONTRAST-ENHANCED T1C $3D_n$_FCMCSeg}.

    b. Segmented results of EM method is denoted as {CONTRAST-ENHANCED T1C $3D_1$_EMSeg, CONTRAST-ENHANCED T1C $3D_2$_EMSeg, CONTRAST-ENHANCED T1C $3D_3$_EMSeg, ..., CONTRAST-ENHANCED T1C $3D_n$_EMSeg}.

Step 3 Proposed Integration Method is optimally integrating the segmentation results of Step 2 (a) and (b).

The outcome of optimal integration is denoted as {CONTRAST-ENHANCED T1C $3D_1$_AMCSeg, CONTRAST-ENHANCED T1C $3D_2$_AMCSeg, CONTRAST-ENHANCED T1C $3D_3$_AMCSeg, ..., CONTRAST-ENHANCED T1C $3D_n$_AMCSeg}.

Step 4 Detection of the Tumor: The proposed method is denoted as {CONTRAST-ENHANCED T1C $3D_1$_AMCSeg, CONTRAST-ENHANCED T1C $3D_2$_AMCSeg, CONTRAST-ENHANCED T1C $3D_3$_AMCSeg, ..., CONTRAST-ENHANCED T1C $3D_n$_AMCSeg}.

Step 5 The segmentation accuracy: Segmented contrast-enhanced T1C 3D MR image is compared with corresponding ground truth image for the performance evaluation of the proposed method, and it is evaluated over of improved Expectation Maximization (EM), fuzzy C means clustering method by using standard performance metrics, i.e., Segmentation Accuracy.

3 Results and Discussions

In this research work presents the MATLAB implementation of the proposed framework that is evaluated the performance of the current improved Expectation Maximization (EM) method, fuzzy C means clustering method, and proposed method on contrast-enhanced T1C 3D Brain MR Images as shown in Fig. 3.

The relative performance of the presented methods, i.e., fuzzy C means clustering method, improved Expectation Maximization (EM) method and proposed method have been demonstrated on BRATS simulated contrast-enhanced T1C 3D Brain MR Image datasets. The BRATS datasets that hold the simulated 3D brain MR images with corresponding ground truth image for analysis of segmentation accuracy as shown in Fig. 3. The visual segmentation result to the detection of tumor in contrast-enhanced T1C 3D Brain MR Image using improved Expectation Maximization (EM) method, fuzzy C means clustering method, and proposed method as shown in Table 1 and Fig. 3.

The relative performance of the presented methods, i.e., fuzzy C means clustering method, improved Expectation Maximization (EM) method, and proposed method

Fig. 3 T1C brain magnetic resonance imaging with segmented tumor regions

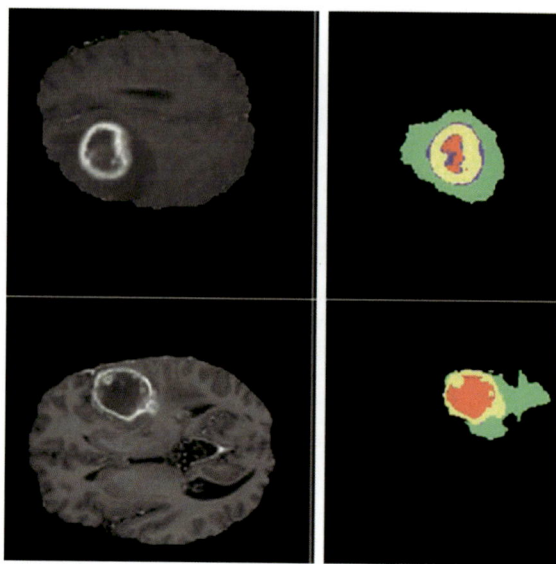

Table 1 Analysis of segmentation accuracy

Input 3D MR image	EG method (%)	FCMC method (%)	Proposed method (%)
Patient-1	92.31	96.22	97.27
Patient-2	96.81	96.56	97.29
Patient-3	93.31	96.91	97.7
Patient-4	93.31	95.22	96.27
Patient-5	97.11	95.56	97.79
Patient-6	94.38	95.31	96.71
Patient-7	98.31	97.13	98.91
Patient-8	97.19	97.21	97.48
Patient-9	96.31	98.13	98.78
Patient-10	96.99	96.21	97.78

have been demonstrated on BRATS simulated contrast-enhanced T1C 3D Brain MR Image datasets as shown in Fig. 4.

The proposed method achieved 97.98% of the average segmentation accuracy on 10 patients' contrast-enhanced T1C 3D Brain MR Image dataset that is superior compared with current methods and Comparison of average segmentation accuracy as shown in Fig. 5.

The research work clearly demonstrates the improved performance of proposed method, and it is compared to the current method on contrast-enhanced T1C 3D brain magnetic resonance imaging dataset.

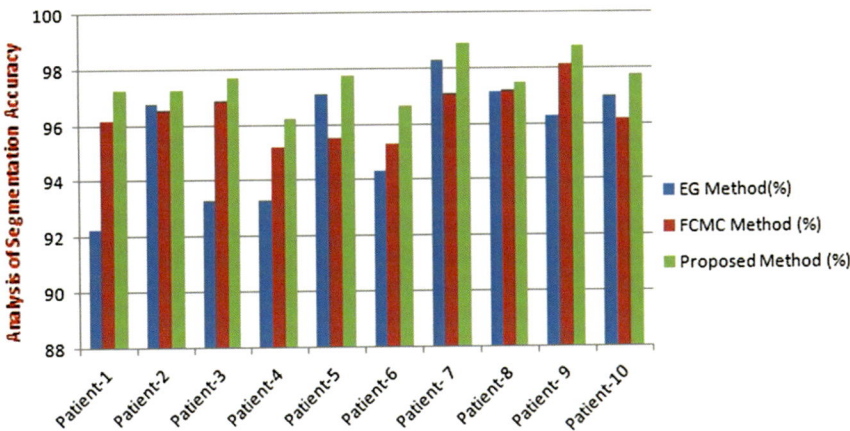

Fig. 4 Comparison of segmentation accuracy to proposed method, the improved Expectation Maximization (EM), and fuzzy C means clustering method

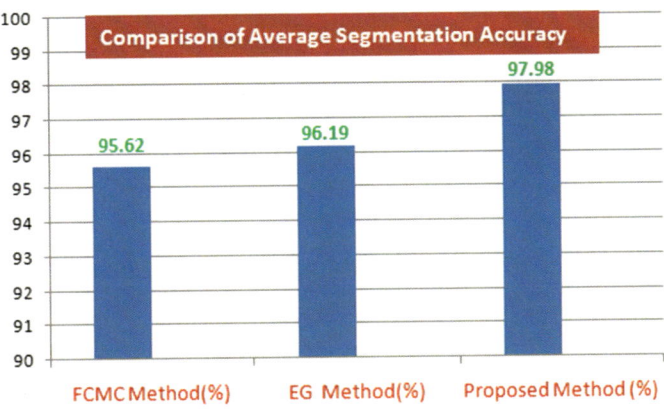

Fig. 5 Comparison of average segmentation accuracy

4 Conclusion

In this research paper, the wide-range of analysis has been done on numerous segmentation techniques and methods. The most popular contrast-enhanced T1C 3D brain magnetic resonance imaging modalities have been investigated for clinical tasks and detection of tumor. The proposed research work offered a contemporary fully automated segmentation framework for segmentation and detection of the brain tumor in contrast-enhanced T1C 3D brain magnetic resonance imaging which is successfully executed on BRATS simulated contrast-enhanced T1C 3D brain magnetic resonance imaging datasets and demonstrated the superior segmentation accuracy. The proposed segmentation method is evaluated on 10 patients' contrast-enhanced

T1C 3D brain magnetic resonance imaging dataset. The presented research work also demonstrated the relative analyses of various presented methods, i.e., fuzzy C means clustering method, improved Expectation Maximization (EM) method, and proposed method. The comparative study concludes that proposed method demonstrated improved segmentation accuracy than existing automated segmentation methods.

References

1. Bauer S, Wiest R, Nolte L-P, Reyes M (2013) A survey of MRI-based medical image analysis for brain tumor studies. Phys Med Biol 58(13):R97–R129
2. Weizman L (2012) Automatic segmentation, internal classification, and follow-up of optic pathway gliomas in MRI. Med Image Anal 16(1):177–188
3. Hameeteman K (2011) Evaluation framework for carotid bifurcation lumen segmentation and stenosis grading. Med Image Anal 15(4):477–488
4. Ahmed S, Iftekharuddin KM, Vossough A (2011) Efficacy of texture, shape, and intensity feature fusion for posterior-fossa tumor segmentation in MRI. IEEE Trans Inf Technol Biomed 15(2):206–213
5. Gooya A, Biros G, Davatzikos C (2011) Deformable registration of glioma images using EM algorithm and diffusion reaction modeling. IEEE Trans Med Imaging 30(2):375–390
6. Liu J, Li M, Wang J, Wu F, Liu T, Pan Y (2014) A survey of MRI-based brain tumor segmentation methods. Tsinghua Sci Technol 19(6):578–595
7. Palma CA, Cappabianco FAM, Ide JS, Miranda PAV (2014) Anisotropic diffusion filtering operation and limitations—magnetic resonance imaging evaluation. In: Preprints of the 19th world congress the international federation of automatic control, Cape Town, South Africa, 24–29 Aug 2014
8. Krissian K, Aja-Fernández S (2009) Noise-driven anisotropic diffusion filtering of MRI. IEEE Trans Image Process Arch 18(10):2265–2274
9. Bhima K, Jagan A (2017) An improved method for automatic segmentation and accurate detection of brain tumor in multimodal MRI. Int J Image Graph Signal Process (IJIGSP) 9(5):1–8. https://doi.org/10.5815/ijigsp.2017.05.01
10. Roerdink JBTM, Meijster A (2000) The fuzzy C means clustering method transform: definitions, algorithms and parallelization strategies. Fundam Inform 41:187–228
11. Li G (2010) Improved fuzzy C means clustering method segmentation with optimal scale based on ordered dither halftone and mutual information. In: 3rd IEEE international conference on computer science and information technology (ICCSIT), 9–11 July 2011, pp 296–300
12. Lin G-C, Wang W-J, Kang C-C, Wang C-M (2012) Multispectral MR images segmentation based on fuzzy knowledge and modified seeded region growing. Magn Reson Imaging 30(2):230–246
13. Gupta JMP, Shringirishi MM (2013) Implementation of brain tumor segmentation in brain MR images using k-means clustering and fuzzy c-means algorithm. Int J Comput Technol 5(1):54–59
14. Bhima K, Jagan A (2016) Novel technique for detection of anomalies in brain MR images. In: The international conference on frontiers of intelligent computing: theory and applications (FICTA-2016). Advances in intelligent systems and computing (AISC) Series. Springer, 16–17 Sept 2016
15. Stille M, Kleine M, Hagele J, Barkhausen J, Buzug TM (2016) Augmented likelihood image reconstruction. IEEE Trans Med Imaging 35(1):158–173

16. Nageswara Reddy P, Mohan Rao CPVNJ, Satyanarayana C (2016) Optimal segmentation framework for detection of brain anomalies. Int J Eng Manuf 6:26–37 (Published Online November 2016 in MECS)

Anatomical Segmentation of Human Brain MRI Using Morphological Masks

J. Mohamed Asharudeen and Hema P. Menon

Abstract Segmenting the anatomical parts of the human brain from MRI is a challenging task in medical image analysis. There is no evident scale for the distribution of intensity over a region in medical images. Region growing is performed to generate a mask to segment the anatomical parts from MRI. A new tailored version of dilation is used in renovating the segmentation mask. This custom-made dilation differs from typical dilation in computation. The structuring element size is fixed, and the anchor point norm is changed from usual dilation. The neighborhood evaluation is made only for certain pixels that are satisfied by the proposed constraints; thus, estimation is not made throughout the image. The computation of the classical dilation is reduced with the proposed custom-made dilation.

1 Introduction

Segmentation of a region in an image finds the homogenous similarity in the pixel intensities. The significant region of the image can be analyzed. Segmentation of brain anatomical parts is considered as one of the most critical steps in medical image analysis and it helps to examine the region with individual considerations. Since the regions of the brain image are difficult to observe from the gray level image segmentation. Equalization of the histogram over the image is made to make prominent regional differences. The image we have taken to examining the anatomical parts of the brain is the gray level image obtained from MRI Multiple Sclerosis Database (MRI MS DB) [1–4]. MRI determines the healthiness of the tissues present in the organs and growth of every tissue and also to measure the abnormal growth of tissues. MRI is used mostly in analyzing the tumors, brain injuries, and infec-

J. Mohamed Asharudeen (✉) · H. P. Menon
Department of Computer Science and Engineering, Amrita School of Engineering, Amrita
Vishwa Vidyapeetham, Coimbatore, India
e-mail: cb.en.p2cvi16002@cb.students.amrita.edu

H. P. Menon
e-mail: p_hema@cb.amrita.edu

© Springer Nature Switzerland AG 2019
D. Pandian et al. (eds.), *Proceedings of the International Conference on ISMAC
in Computational Vision and Bio-Engineering 2018 (ISMAC-CVB)*, Lecture Notes
in Computational Vision and Biomechanics 30,
https://doi.org/10.1007/978-3-030-00665-5_74

tions in the brain. MRI provides a detailed image of the brain. Brain lesions can be easily identified through MRI, as the appearance of the lesions will not be the same as a normal tissue. The interpretations of the brain parts are observed with color distributions in image processing. The hectic situation is there is no such scale for quantifying lesion intensity. The intensity is interpreted only through direct visual comparison with surrounding tissues.

2 Related Work

Hasan et al. [5] modified the existing morphological operations for 2D hole filling using dilation operation. The processing time of the algorithm reduced to one-third. Border image initial algorithm is modified by considering more starting points. This is considering the faster version of the existing hole-filling algorithm. Despotovic et al. [6] discussed the most popular methods used for brain parts segmentation in MR image, and comparison about them is also highlighted.

Guan et al. [7] proposed a novel model by integrating the Markov theory, the clustering theory, and normal curve model. Then the brain tissue has been extracted by the threshold segmentation. The work is compared with Markov algorithm and also, the algorithm of a 2D histogram with fuzzy clustering. Yazdani et al. [8] proposed a new technique for region dividing by combining the merits of region-based segmentation and the histogram-based image segmentation. This paper focused on the segmentation of the gray matter, white matter, and cerebrospinal fluid in brain MR images.

Somasundaram and Kalaiselvi [9] proposed an approach for filling the holes automatically in the binary masks based on run length encoded data. They claimed that the work is the best suitable holes-filling module for brain image segmentation. Piekar et al. [10] introduced the contrast agent for making the tumor visible. The region growing segmentation is employed. The region growing sometimes detects the healthy region as tumor boundaries. These overflow conditions are removed by using eight-neighborhood connectivity and specifies the importance of choosing the better homogeneity criteria for choosing the seed point.

Javadpour and Mohammadi [11] proposed a novel algorithm that uses regional growth methods and the seed points selections were automated by genetic algorithm for MRI segmentation. The proposed method claims that segmentation error can be reduced. Here, the standard deviation criterion is utilized to check the similarity. Mohd Saadd et al. [12] described the brain lesion segmentation. Lesion region is obtained by region splitting and merging. Automatic seed selection is performed using histogram thresholding that determines the optimal threshold value. Deng et al. [13] demonstrated the way to overcome the noises and threshold selection issues in tumor segmentation. An anisotropic diffusion filter is used to preserve the edges within which the proposed method chooses the mean–variance.

Roman-Alonso et al. [14] developed a tool called Data List Management Library used for segmentation of brain image volumes. They compared the proposed tool with other classical parallel approaches that are made with master–slave and static data distribution. Kapur et al. [15] presented a hybrid method for brain tissue segmentation from magnetic resonance images. They combined three existing techniques from the computer vision literature: expectation/maximization segmentation, binary mathematical morphology, and active contour models. They implemented the method on IBM's supercomputer Power Visualisation System and its results are validated by neuroanatomical experts. Park [16] presented the concept of mathematical morphological operations utilized for handling noises in tracking the objects.

3 Proposed Work

Region growing is one of the popular approaches of image segmentation [17]. It starts from the user interest to which region need to be segmented in an image. It plays a vital role in medical image analysis. The image we have taken to examine the anatomical parts of the brain is the gray level image obtained from MRI Multiple Sclerosis Database (MRI MS DB) [1–4]. The algorithm starts budding from a seed pixel. This method checks for the homogeneity criteria around the seed. Once the neighborhood connectivity is checked, the next search for the seed pixel is started around the previous results to avoid the region's growth along the dominant direction. The pixels traced by the above-described process will determine the region wants to get segmented. The flowchart for the segmentation technique is shown in Fig. 1.

There are many challenges faced during this segmentation work such as noises [18, 19]. Noises interrupt the growth along with the chosen region of importance. There are many chances of discontinuity of the segmented parts. Analysis of the partially segmented region may result in wrong diagnostics. Distributions of the colors are made uniform as a result of histogram equalization. The precise tissue layer is also getting segmented through this method. Every subtle layer of the cerebellum folds can be segmented. The segmentation process is made by region growing methodology with better precision and the binary mask is developed as its result.

3.1 Morphological Operations

Mathematical morphology in image processing is popular in handling noises, extracting the features in the images. These functions are working under the concept of set

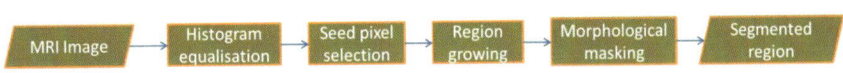

Fig. 1 Flowchart for brain parts segmentation

theory. The fundamental mathematical morphological operations are dilation and erosion.

These two primitive operations play a vital role in generating other morphological algorithms. The structuring element is a significant part in defining the shape of the object that required to be retained from the given image. The structuring element can be changed to any shape depends on the application.

Here in this proposed work, the structuring element is designed in an arrangement that it can be used for making the masks for the region to be segmented from human brain MRI. The typical dilation is applied in an inexplicable fashion for the fragmentary mask traced from the region growing technique. The dilation is tailored to two naïve operations, namely: waxing and waning. These operations are personalized to work with binary images.

Waxing: A structuring element B of size 3×3 has the anchor point in its origin and the waxing operation on the binary image A is performed only if the anchor point of the kernel has the intensity value 1. Here, the representation of waxing operation is chosen as $\oplus^+$. The anchor point of the kernel is checked and the individual neighborhood is changed only if the condition matched for waxing. The anchor point of the kernel is holding the intensity value as 1 and the conditions of the neighborhood hold the intensity value as 0 then replace them as 1.

$$A \oplus^+ B = \left\{ x \in E | (\hat{B})_x \cap A \neq \emptyset, x_B = 1 \right\} \tag{1}$$

Let E be the Euclidean space and waxing is applied for $x_B = 1$, x_B be the anchor point of the kernel.

Waning: A structuring element B of size 3×3 has the anchor point in its origin and the waning operation on the binary image A is performed only if the anchor point of the kernel has the intensity value 0. Here, the representation of waning operation is chosen as $\oplus^-$. The anchor point of the kernel is checked and the individual neighborhood is changed if the condition is matched for waning. The anchor point of the kernel is holding the intensity value as 0 and the conditions of the neighborhood hold the intensity value as 1 then replace them as 0.

$$A \oplus^- B = \left\{ x \in E | (\hat{B})_x \cap A \neq \emptyset, x_B = 0 \right\} \tag{2}$$

Let E be the Euclidean space and waxing is applied for $x_B = 0$, x_B be the anchor point of the kernel. The design of the structuring element is shown in Table 1.

These structuring elements are combined to form a 3×3 kernel. The red mark in Table 1 shows the anchor point of the individual structuring element. If the neighboring condition is satisfied the origin is changed its value. The combination of these structuring elements is shown in Fig. 2.

Figure 2 shows the designed kernel for waxing and waning as well as the existing kernel structures. The difference can be clearly visible from the red marks. The typical kernel structures will check for the neighborhood conditions and the anchor point located at the origin gets modified. The proposed kernel for waxing and waning

Table 1 (A) Horizontal structuring elements, (B) diagonal structuring elements, and (C) vertical structuring elements of waning and waxing kernels

Structuring element	Waning	Waxing
(A)		
(B)		
(C)		

Fig. 2 **a** Waning kernel, **b** waxing kernel, and **c** existing 3×3 morphological kernel

will check the anchor point only then the neighborhood condition is checked. The modification is not made for the anchor point of the kernel. Instead, those pixels satisfy the neighborhood conditions. These two conventional dilation operations are named as waxing and waning.

The result of the region growing technique is a binary image. This binary image creates a checkerboard pattern as a result of four-neighborhood connectivity comparison. The four-neighborhood connectivity pattern for this mask construction used is "cross" arrangement as shown in Fig. 3. This mask undergoes waxing operation to fulfill the mask shape and it has been used to segment the region from MRI.

Fig. 3 Four-neighborhood connectivity "cross" design

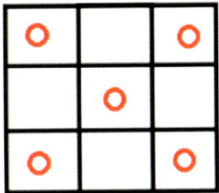

3.2 Methodology

The proposed method uses a region growing technique to identify the structural parts of the brain. Once the histogram equalization is performed, the region of interest mask is prepared from the region growing technique. The mask is outlined with holes. The holes appeared due to four-neighborhood connectivity comparison. The results of segmentation are also made using eight-neighborhood connectivity as claimed by [8]. The eight-connectivity dominates the region's growth beyond the region of interest as shown in Fig. 5. The shape of the mask is treated with waxing operation. The final mask has been utilized to segment the region.

Let the structuring element of waxing is superimposed over the binary image. Check for the waxing constraints. Replacement of the pixel value is made in its neighborhood. Loop throughout the image. Waxing is used for objects shape expansion and waning is used for objects shape shrinking. The algorithm appends one-pixel width around the boundary of the object. The proposed work utilizes the four connectivity estimation. Waxing, a new method proposed to fulfill the shape of the region growing resultant mask in this proposal.

Let $I(x, y)$ be the given MRI image, and the corresponding mask for the ROI is $M(x, y)$.

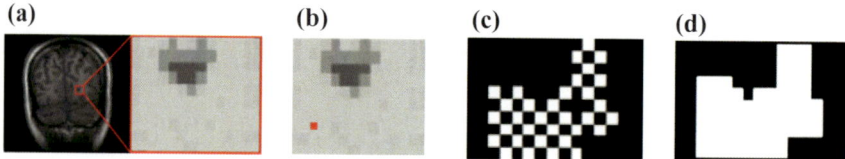

Fig. 4 **a** Region zoomed is examined with region growing, **b** red mark represents the seed pixel, **c** result of region growing technique, and **d** waxing output

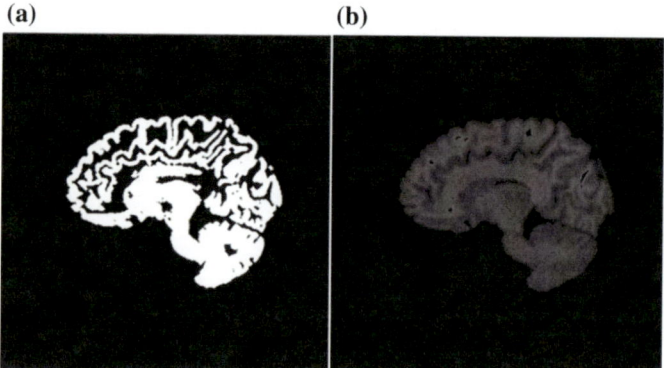

Fig. 5 Result of eight-neighborhood connectivity: **a** mask output and **b** segmented brain

$$M(x, y) = \begin{cases} 1, & N_4\{\nabla I(x, y)\} < T \\ 0, & \text{Otherwise} \end{cases}$$

Mask $M(x, y)$ undergoes *waxing* operation for filling the holes. With this mask, the region is segmented from the image $I(x, y)$.

$$M(x, y) \cap I(x, y) = R(x, y)$$

Let $R(x, y)$ be the segmented region.

Brain parts segmentation involves a region growing technique along with waxing operation as follows:

Step 1: Perform histogram equalization on the given image.

Step 2: Set the gradient value within the range of 0.00 and 0.01.

Step 3: Choose the region from the image, to get segmented by selecting any point in the region needs to be analyzed.

Step 4: Prepare the mask by selecting a seed pixel.

Step 5: Perform four-neighborhood connectivity evaluation from the selected seed pixel that is present in the mask.

Step 6: Mark those pixels that satisfy the constraints as white pixels in the mask.

Step 7: Loop through step 5 until the condition fails.

Step 8: Mask undergoes waxing and the result is shown in Fig. 4d.

Step 9: Segmentation of the region in the image is retrieved by the obtained mask.

4　Results and Analysis

The region growing is employed in brain segmentation for extracting the coherent intensity section and a preprocessing technique is also used for better localization of the region of interest. The eight-neighborhood connectivity is used, the region growing is dominated instead of the specified region of interest. The corpus callosum is selected as the region for segmenting from the given image. The result shows the entire brain from the raw MRI image as shown in Fig. 5. Eventually, the four-neighborhood connectivity is used for segmenting the definite region in the image. These results develop a mask for segmenting the region of interest from the original image. The created mask has some imperfect shape that is altered using waxing operation and the final mask is prepared. This mask will get superimposed over the original image for segmenting the definite region from the given image. The region growing process for corpus callosum from the sagittal slice is shown a successive growth step process in Fig. 6. The region growing algorithm using four-neighborhood connectivity is tested only using brain here. It can also be tested with some other organs for region segmentation.

The major difference between the standard dilation and these waxing and waning operations is: the origin differs as shown in Fig. 2, and the kernel matching is performed for replacing the anchor point of the kernel that satisfies the criteria. The neighborhood comparison is made throughout the image in the existing operation, but the proposed operation will perform the neighboring comparison only for the pixel that matches the anchor point of the kernel. The dimension is fixed as 3×3.

The result of the waxing and typical dilation will produce the same output as shown in Fig. 7 for the chosen structuring element. The major difference with these operations is the neighborhood pixel comparison shown in Table 2. The comparison of time complexity is shown in Table 3. This supports the claim that the proposed function works faster than the existing dilation operation. The proposed function has been verified with three categories of brain MRI.

Table 4 shows the anatomical parts of the brain from MRI of different slices. The result shows the partially generated mask and its zoomed view. The parts have been segmented through this method [20, 21] are evaluated for finding the similarity between the manually segmented and the proposed semiautomated segmented region of brain parts.

The values shown in Table 5 are almost nearer to 1 which represents the segmentation through this approach nearly matches the manual segments. The similarity

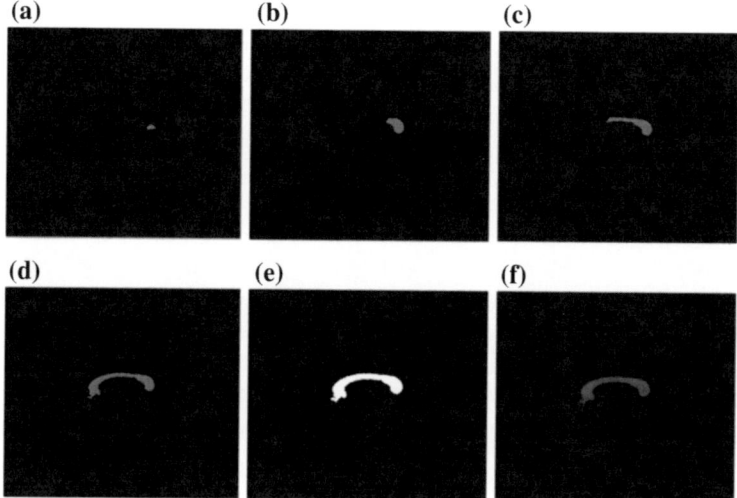

Fig. 6 a–d Successive region growths of corpus callosum, **e** final mask, and **f** segmented corpus callosum

Table 2 Neighborhood pixel comparison of waxing and dilation operation

Eight-neighborhood comparison	Waxing	Dilation
Total no. of pixels	80 pixels	192 pixels

measures [22–25] used here are SSIM and QIM. The structural similarity index is a method to compare the structure, contrast between the two images is calculated as in (3). Quality index measure compares contrast distortion, and luminance distortion between the images is calculated as in (4).

$$SSIM = \frac{(2\mu_x\mu_y + c1)(2\sigma_{xy} + c2)}{(\mu_x^2 + \mu_y^2 + c1)(\sigma_x^2 + \sigma_y^2 + c2)} \tag{3}$$

$$QIM = \frac{4\sigma_{xy}\bar{x}\bar{y}}{(\sigma_x^2 + \sigma_y^2)(\bar{x}^2 + \bar{y}^2)} \tag{4}$$

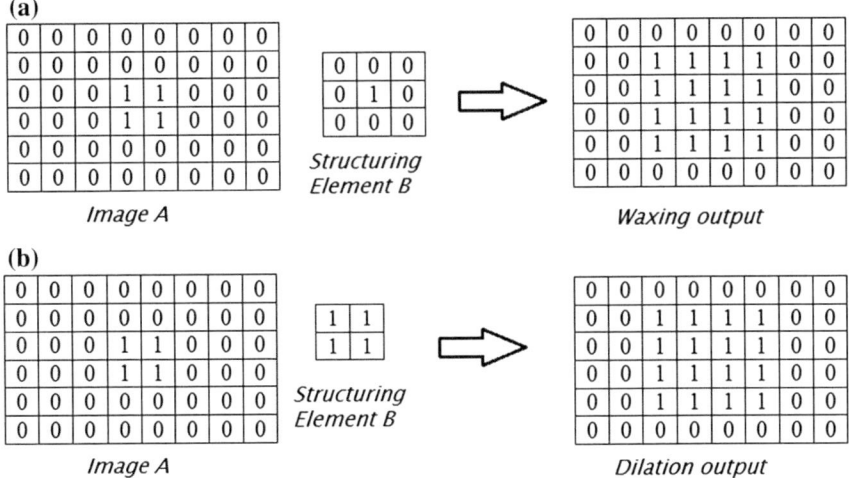

Fig. 7 **a** Waxing operation and **b** dilation operation

Table 3 Time taken for waxing and dilation operation

Masks developed from region growing	Time taken for waxing operation (s)	Time taken for dilation operation (s)
Sagittal_Corpus_Callosum	0.065	0.926
Sagittal_Arbor_vitae_of_cerebellum	0.073	0.997
Axial_Cerebellar_Tonsil	0.081	1.168
Axial_White_Matter	0.160	2.169
Coronal_Hippocampus	0.081	0.957
Coronal_White_Matter	0.074	1.003

Table 4 Brain MRI segmentation results

Image slice	MRI image	Histogram equalised image	Region segmented from the brain	Zoomed view of the region segmented	Name of the region
Sagittal slice					Corpus callosum
Sagittal slice					Arbor vitae of cerebellum
Axial slice					Cerebellar tonsil
Axial slice					White matter
Coronal Slice					Hippocampus
Coronal Slice					White matter

5　Conclusion

The anatomical parts of the brain are segmented using the region growing methodology. The tailored mask is computed efficiently. The distinct way of dilation is employed in constructing the mask. This operation is also clarified with the existing dilation operation. The similarity measures are also shown the segmented results are almost similar to the manually segmented results. The segmented results show the better path for diagnosis of the anatomical parts of the brain. The region growing has

Table 5 Similarity measures between manual segmentation and proposed segmentation

Image Slice	Manual segmentation	Proposed segmentation	SSIM	QIM
Sagittal slice	Corpus collusum		0.9597	0.9418
Sagittal slice	Pons		0.9554	0.9373
Sagittal slice	Septum pellucidum		0.9612	0.9482
Axial slice	Spinal cord		0.9663	0.9441
Coronal slice	Thalamus		0.9667	0.9492

its own limitations in tracing the region of interest. If the histogram equalization hindered the region boundary identification, then the segmentation gives unpredicted results not appropriate for diagnostics. Preprocessing has to develop the vibrant boundary localization.

References

1. Loizou CP, Murray V, Pattichis MS, Seimenis I, Pantziaris M, Pattichis CS (2011) Multi-scale amplitude modulation-frequency modulation (AM-FM) texture analysis of multiple sclerosis in brain MRI images. IEEE Trans Inf Technol Biomed 15(1):119–129

2. Loizou CP, Kyriacou EC, Seimenis I, Pantziaris M, Petroudi S, Karaolis M, Pattichis CS (2013) Brain white matter lesion classification in multiple sclerosis subjects for the prognosis of future disability. Intell Decis Technol J (IDT) 7:3–10
3. Loizou CP, Pantziaris M, Pattichis CS, Seimenis I (2013) Brain MRI Image normalization in texture analysis of multiple sclerosis. J Biomed Graph Comput 3(1):20–34
4. Loizou CP, Petroudi S, Seimenis I, Pantziaris M, Pattichis CS (2015) Quantitative texture analysis of brain white matter lesions derived from T2-weighted MR images in MS patients with clinically isolated syndrome. J Neuroradiol 42(2):99–114
5. Hasan MM, Mishra PK (2012) Improving morphology operation for 2D hole filling algorithm. Int J Image Process (IJIP) 6(1):1–12
6. Despotovic I, Goossens B, Philips W (2015) MRI segmentation of the human brain: challenges, methods and applications. Comput Math Methods Med 1–23. Article ID 450341
7. Guan Y-H, Lv L, Duan R, Ji Y-H (2009) The brain segmentation by Markov field and normal distribution curve. In: Image signal processing, pp 1–5
8. Yazdani S, Yusof R, Karimian A, Mitsukira Y, Hematian A (2016) Automatic region-based brain classification of MRI-T1 Data. PLoS ONE 11(4):e0151326
9. Somasundaram K, Kalaiselvi T (2010) A method for filling holes in objects of medical images using region labeling and run length encoding schemes. In: National conference on image processing (NCIMP), pp 110–115
10. Piekar E, Szwarc P, Sobotnicki A, Momot M (2013) Application of region growing method to brain tumor segmentation-preliminary results. J Med Inform Technol 22:153–160
11. Javadpour A, Mohammadi A (2016) Improving brain magnetic resonance image (MRI) segmentation via a novel algorithm based on genetic and region growth. J Biomed Phys Eng 6(2):95–108
12. Mohd Saadd N, Abu-Bakar SAR, Muda S, Mokji M, Abdullah AR (2012) Automated region growing for segmentation of brain lesion in diffusion-weighted MRI. In: Proceedings of the international multi conference of engineers and computer scientists, vol. I, pp. 14–16, 674–677, March 2012
13. Deng W, Xiao W, Deng H, Liu J (2010) MRI tumor segmentation with region growing method based on the gradients and variances along and inside of the boundary curve. In: 3rd international conference on biomedical engineering and informatics, vol 1, pp 393–396
14. Roman-Alonso G, Jimenez-Alaniz JR, Buenabad-Chavez J, Castro-Garcia MA, Vargas-Rodriguez AH (2007) Segmentation of brain image volumes using the data list management library. In: 29th annual international conference of the IEEE engineering in medicine and biology society, pp 2085–2088
15. Kapur T, Grimson WEL, Wells WM III, Kikinis R (1996) Segmentation of brain tissue from magnetic resonance images. Med Image Anal 1(2):109–127
16. Park H. A method of controlling mouse movement using a real time camera. Department of Computer Science, Brown University
17. Pohle R, Toennies KD (2001) A new approach for model-based adaptive region growing in medical image analysis. In: International conference on computer analysis of images and patterns, pp 238–246
18. Hiralal R, Menon HP (2016) A survey of brain MRI segmentation methods and issues involved. In: International symposium on intelligent systems technologies and applications (ISTA'16), Advances in intelligent systems and computing
19. Nilakant R, Menon HP, Vikram K (2017) A survey on advanced segmentation techniques for brain MRI image segmentation. Int J Adv Sci Eng Inf Technol 7(4)
20. Bhuvana D, Bhagavathi Sivakumar P (2014) Brain tumor detection and classification in MRI images using probabilistic neural networks. In: Proceedings of the second international conference on emerging research in computing, information, communication and applications (ERCICA-14). Elsevier, pp 796–801
21. Renjini H, Bhagavathi Sivakumar P (2013) Comparison of automatic and interactive image segmentation methods. Int J Eng Res Technol (IJERT) 2(6):3162–3170

22. Wang Z, Bovik AC (2005) Structural approaches to image quality assessment. In: The handbook of image and video processing, 2nd edn. Academic Press, New York, pp 961–974

23. Wang Z, Bovik AC, Sheikh HR, Simoncelli EP (2004) Image quality assessment: from error visibility to structural similarity. IEEE Trans Image Process 13(4):600–612

24. Wang Z, Bovik AC (2003) Objective video quality assessment. In: The handbook of video databases: design and applications. CRC Press, Boca Raton, pp 1041–1078

25. Wang Z, Bovik AC, Lu L (2002) Why is image quality assessment so difficult? In: IEEE international conference on acoustics, speech, and signal processing

Digital Image Restoration Using NL Means with Robust Edge Preservation Technique

Bhagyashree V. Lad, Bhumika A. Neole and K. M. Bhurchandi

Abstract We present a novel approach for image denoising with spatial domain edge preservation method to yield denoised image, from an image which is distorted by additive white Gaussian noise without loss of detail information of an image. Denoising of noise corrupted images while preserving its attributes like edges and fine details is an extreme challenge. During image acquisition and transmission often noise gets added in the image which degrades the image quality. Algorithms like NL means and BM3D have been very successful in this aspect. The weight assigned to the fellow pixels for calculating replacement for a noisy pixels have a pivotal role in these algorithms. In NL means, the weight assigned to a pixel depends only on its relative magnitude with respect to its neighborhood. In this paper we are proposing a robust mechanism for weight calculation which also considers the orientation of a pixel. Using this mechanism we got improvement in PSNR and visual quality of the de-noised image as compared to the NL means method and also got improvement in structural similarity as compared to the BM3D. From the experimental results it can be seen that our proposed algorithm is superior in comparison to the NL means and BM3D technique of image denoising, considering the factors like PSNR, SSIM and accordingly conclusions are drawn.

B. V. Lad (✉) · B. A. Neole
Department of Electronics and Communication Engineering, Shri Ramdeobaba College of Engineering and Management, Nagpur 440013, India
e-mail: ladbv@rknec.edu

B. A. Neole
e-mail: neoleba@rknec.edu

K. M. Bhurchandi
Department of Electronics and Communication Engineering, VNIT, Nagpur 440010, India
e-mail: bhurchandikm@ece.vnit.ac.in

© Springer Nature Switzerland AG 2019
D. Pandian et al. (eds.), *Proceedings of the International Conference on ISMAC in Computational Vision and Bio-Engineering 2018 (ISMAC-CVB)*, Lecture Notes in Computational Vision and Biomechanics 30,
https://doi.org/10.1007/978-3-030-00665-5_75

1 Introduction

Noise is an unsystematic and indeterminate variation in the intensity of the image. All the images contain certain amount of noise. An image captures a moment forever. Today there is almost no area of technical endeavor that is not impacted in some way by digital images and hence their processing. There are many fields where decisions are made solely based on the digital images as inputs. Then it becomes very important that these images are noise-free and pure. Thus this is the challenge that we need to obtain best possible approximation of original image from available degraded image. Many a times the noise gets added in an image while capturing it by the camera. Images are also contaminated by the noise while capturing the images through Charged Coupled devices (CCD), CMOS Sensors and Contact Image Sensors (CIS). Some signals required for TV to broadcast the data which is transmitted over cable or received at the antenna and in case of digital images the signal is the light which hits the camera sensor. In this transmission process the noise gets added in the signals. This noise appears as random signals which degrade the image quality. Thus, it is a great challenge for the researchers today to remove noise from images. There are various image denoising algorithms approaches which have been employed such as Perona and Malik proposed a scale space-based approach [1], Nobert Wiener presented an adaptive filter [2], furthermore several versions of wavelets [3–5] and spatial filters [6–9] are used and yielded interesting result to remove the noise. Every algorithm has certain merits and demerits.

There is a general belief that noise is a high-frequency component and the signal is a low-frequency component. An image itself has high-frequency attributes like edges which are sharp change in intensity and fine details which may get misinterpreted as high-frequency noise component by the de-noising algorithms and thus ruined.

An image can be considered as noisy if its pixel values do not match with the pixel values in some reference original image. In a way all the digital images are noisy as the process of quantization itself introduces the quantization noise. Further, noise may be introduced during any stage like while acquiring an image or maybe while transmitting it. The nature of this noise can be modeled by various noise models like salt and pepper noise, speckle noise, Gaussian noise, periodic noise etc. Out of these, generally, the independently and individually distributed additive white Gaussian noise can be considered as the better approximation to the practical noise.

Many of the methods used for denoising work poorly in case of recovery of edges. The edges become blurred because of averaging. Mostly, the noise present in digital images is additive in nature in which the uniform power is distributed in whole bandwidth and the noise generally has the Gaussian probability distribution. Such a noise is generally called as Additive White Gaussian Noise (AWGN). Because of poor image acquisition techniques and poor transmission through noisy communication channel the White Gaussian noise gets introduced in an image. In most of the image denoising algorithms the artificially distorted images by White Gaussian noise are considered to get the test results [3–7]. It can be seen that Gaussian filter even though has desirable features, the satisfactory results are not obtained because of edge detection techniques involved in the denoising algorithm [8].

Images are also affected by salt and pepper noise for which median filter is to be used for removal of such noise. Median Filter is an influential nonlinear filter which is one of best and simple method in spatial domain. But it is efficient only for low noise densities and yield blurred images at high noise densities though it is an easy method to perform image smoothing. Median filter reduces the intensity variation between the neighboring pixels. In median filtering technique we replace the pixel value of an image with the median of all the neighboring pixel values instead of replacing it with the mean of neighboring pixels. To calculate the median value, all the pixel values are sorted in the ascending order first and the pixel is replaced by the middle pixel value of the sorted sequence. If the sorted sequence consists of even number of pixels then the desired pixel is replaced by the average of two middle pixel values in the sequence. The median filter yields best result when percentage of impulse noise present in an image is less than 0.1%. When the impulse noise is present in large amount then the median filter does not give best result because it tends to remove image details at the time of removing noise such as lines and the corners and thus degrades the performance in case of signal-dependent noise.

The most popular and well-known Wiener filter can also be used for removal of image degradations. The Weiner filter, which is based on the statistical approach, filters the noise which has degraded the signal. The designing of all the filters is mostly based on desired frequency response, but the designing of Weiner filter takes the different approach. Weiner filter is designed to reduce the mean square error to the most extent. So, this filter more effectively reduces the noise and various degrading functions.

De-noising techniques can be implemented in spatial domain or in frequency domain. The non-local means and the BM3D are two very successful de-noising methods. The non-local means is a pure spatial method while the BM3D is implemented partly in spatial and partly in frequency domain.

An isotropic Gaussian kernel is used for denoising in NL means algorithm. Kernels provide NL means with more similar patches, in which maximum similar patches are obtained near the edges where the contrast is very high. Depending upon the patch size two drawbacks arise. By choosing the small patch size noise remains in homogeneous parts of image, on the other side, choosing the large patch size leads the noise to halo near the edges. All these situations happen when the patches that are similar to the current patch are rare. Some attempts have been made to remove noise halo, but a better technique can be used which is based on shaped and directional patches. Number of iterations are required to implement this technique so computational time increases and finally SURE criterion is used to choose the best kernel for each part of image. Although this method improves NL means performance by eliminating the noise halo, but it gives rise to some drawback. The first drawback is selection of limited number of predefined kernels which leads to difficulty in denoising the images with different structures. Also an image consists of a large number of different structures and textures which requires a huge number of predefined kernels. The second drawback is the algorithm becomes computationally inefficient because of big group of kernels which leads to numerous iterations. And the third drawback is due to lack of robustness of SURE. These drawbacks can be eliminated by using Adaptive NL means method.

The BM3D algorithm presents the redundant successive computation of transform of similar overlapping image patches and their so-called 3D transform computations followed by collaborative filtering and inverse 3D transform. Averaging of the overlapping redundant inverse transformed image patches is used to have a final pixel estimate. This is followed by a Wiener filter.

In this paper we are modifying subtly the weight calculation mechanism used in the non-local means algorithm. For finding a replacement for a noisy pixel, in non-local means, the idea is to assign weights to other pixels depending on the similarity of their neighborhood in terms of magnitude of the pixels with the neighborhood of the pixel which is under consideration. We propose dependence of the weight assigned to a pixel by considering adaptive kernel which is not only based on the magnitude of the pixels in its neighborhood but also on the orientation of that pixel.

The paper is divided in 5 sections. Section 1 briefs the noise model considered. Section 2 explains the non-local means algorithm, whereas the concept of determining orientation of a pixel is given in Sect. 3. Section 4 demonstrates the modified algorithm and finally in Sect. 5, the results are shown and finally the conclusion is drawn.

2   Methods and Materials

2.1   Noise Model

Let X be the original image of size $m \times n$ and let N represent the additive white Gaussian noise matrix of the same size. Then the corrupted image can be given by Y as

$$Y = X + N \tag{1}$$

Since the noise model is additive, the value at every point in the corrupted image is simply the addition of the corresponding values in the original image and the noise matrix. The values in this nose model have a Gaussian distribution [14].

2.2   Non-local Means

Antoni Buades presented the algorithm for noise removal that is non-local means filter which has received a lot of attention from image processing community [9, 10]. Non-local filtering considers all pixels for finding replacement of one pixel. Weights are assigned to all the pixels depending on their similarity with pixel to be replaced that is NL means uses the self-similarity concept. In this algorithm each pixel is represented by the weighted average of all the pixels in the neighborhood of the image. To compute similarity between two neighborhoods the weighted sum of the squares of difference between the two neighborhoods is calculated. The non-local means algorithm considers a small patch around a pixel. It doesn't consider only the central patch. The non-local means filter gives more weight to the pixels which are there in the center of neighborhood and less weight to those pixels which are near to the edges.

Thus taking the base of this NL means algorithm, in this paper we have further improved the image quality and compared the results with different filtering techniques and tried to bring the results close to the results obtained by the state of the art techniques. The image qualities are compared on the basis PSNR (Peak Signal to Noise Ratio) and SSIM (Structural Similarity Index Measure).

In NL-means algorithm the pixel intensity is represented as a sum of weighted Gaussians of each neighboring pixel. This algorithm is computationally very heavy due to the involved exponential computations in the neighborhood of all pixels. Also calculation of the weights is computationally heavy. It also blurs the images considerably if exact weights are not selected [11, 12].

To determine the weights of the filters for filtering the noisy images, non-local means filter uses all the possible self-predictions and self-similarities which the image imparts provided the image contains large amount of self-similarity. Because of the self-similarity in an image the pixels are highly correlated and because of the independent and identical distribution property of the noise, the averaging of such highly correlated pixels suppresses the noise which yields to the pixel which is nearly identical to its original value. For a given noisy image $v = \{v(i)|i \in I\}$, where I is a set of the whole pixels in the image, and i is an arbitrary pixel in I. The pixel i is processed by the non-local means filtering and the formula can be given by

$$NL[v](i) = \sum_{j \in I} w(i, j)v(j) \tag{2}$$

Where $w(i, j)$ is the weight between i and j, and it satisfies $0 \le w(i, j) \le 1$,

$$\sum_{j \in I} w(i, j) = 1 \tag{3}$$

By using this formula we can get the intensity of pixel i after denoising. Each pixel is expressed as the weighted average of all the pixels in an image.

2.3 BM3D

Another method which is most popularly used for image denoising is BM3D. In BM3D 3D arrays are formed by collecting the similar 2D image blocks. These 3D data arrays are called as groups. The filtering process used for these 3D groups is collaborative filtering. Collaborative filter gives the finest details of the grouped blocks and it also preserves all the features of an individual block. After this process of filtering, the blocks are returned to their original position [13].

BM3D, i.e. Block matching 3D generally deals with the improved thinly scattered data representation in the transform domain. And this improvement in the sparsity can be achieved by the formation of 3D data arrays by grouping the nearly identical 2D fragments in an image. For processing these 3D data arrays a special technique of the Collaborative filtering is used. This technique consist of the steps like, performing 3D transformation on 3D data array called as groups, then carrying out the shrinkage of the transformed coefficients and finally retrieving the image by applying inverse 3D transform to the obtained coefficients. Thus, a 3D estimate of the group is obtained which is nothing but the combination of filtered 2D image blocks. As the grouped fragments are highly similar, this technique gives the sparse representation of the true signal. And so the noise can be disarticulated by using the shrinkage technique. So the collaborative filtering divulges the sharp and fine details in conjunct with the groups. And also it preserves all the unique features of each individual grouped fragment. The grouping is done here so that the higher dimensional filters can be used for each group, which then utilizes the similarity between each data groups so that the true signal can be estimated for each of the data array.

In BM3D, block matching is used to perform the grouping of data and then the collaborative filtering is used which is achieved by shrinkage of 3D transform domain coefficients. In general the image group fragments are fixed sized square blocks. The generalized BM3D algorithm can be represented as follows. The reference blocks are extracted from the input noisy image and the following process is applied to each sub-block.

- Collect the blocks in a stack which are similar to the reference block to form a 3D data array, i.e. groups. This observation of the similar blocks to the reference one is called as block matching.
- Carry out the collaborative filtering of the group and put the estimated 2D data of all the grouped blocks to their original locations.
- Because of the overlapping of the obtained blocks during processing, there can be multiple estimates for each pixel. So average of all these estimates belong to one pixel and replace the original pixel value by this aggregate. In this way, the whole image can be estimated.

3 Proposed Algorithm

In this work, we have modified subtly the non-local means algorithm by using the concept of block matching from the BM3D technique. For finding a replacement of a noisy pixel in non-local means, the idea is to assign weights to other pixels depending on the similarity of their neighborhood with the neighborhood of the pixel which we are currently processing. Here, the similarity of the neighborhoods is computed in terms of their weighted averages. Instead of using weighted averages if the neighborhoods are matched as blocks, we obtain good results than classical non-local means algorithm.

The concept of block matching improves the performance of non-local means. In NL means the weight of a pixel (candidate pixel) depends only on how similar is its weighted average with respect to the weighted average of the pixel (central pixel) whose estimate is being calculated. More closely their weighted averages, more is the weight or dominance of the candidate pixel in calculating an estimate. If along with the weighted average, the similarity in the orientation of the candidate and the central pixel is also considered for calculation of weight, the estimated value of the central pixel gets even more close to its true value. In non-local means based on edge detection, the weights are calculated considering the similarity in both the weighted averages and the orientation of the candidate and the central pixel.

To enumerate the similarity between two neighborhoods we have to take the weighted sum of squares of difference between the two neighborhoods. The non-local means filter gives more weight to those pixels which are there in the center of the neighborhood and less weight to those pixels which are near the edges.

Here, the replacement value of a pixel completely depends on the other pixels in that image. The more similar a pixel is to the pixel for which replacement is to be found out, the more its weight is. Estimation of a corrupted pixel is affected by the pixels having higher weights.

Consider a noisy pixel p. For this noisy pixel, all the other pixels in the corrupted image will be considered for determining an estimate. Let the other pixels be denoted by qi, where $i = 1$ to $N - 1$, $N =$ total number of pixels in the given image. Let weights be denoted by $f(p, qi)$, i.e. weight of $q1 = f(p, q1)$, weight of $q2 = f(p, q2)$ and so on. Then, the estimate for pixel p denoted by $\hat{p}$ is given by the expression

$$\hat{p} = (q1 * f(p, q1) + q2 * f(p, q2) + q3 * f(p, q3) + \ldots)/c_p \qquad (4)$$

$$c_p = \text{normalizing factor} = \sum f(p, qi) \qquad (5)$$

3.1 Calculation of Weight

For calculation of $f(p, q1)$, i.e. weight for pixel $q1$ w.r.t p, we need to consider a neighborhood of $q1$ of size $n \times n$ where n is some odd number greater than or equal to 3, preferably 3. Let hq denote the average of neighborhood again of size $n \times n$ centered at $q1$ and hp denote the average of neighborhood of p, then

$$f(p, q1) = e^{\frac{-(hp-hq)^2}{sf}} \tag{6}$$

where, sf corresponds to smoothing factor, e.g. $sf = (10 * \text{sigma})^2$ where, sigma represents the standard deviation of the noise model added. Once we have calculated weights for all pixels, we can calculate the normalizing factor c_p and hence an estimate for pixel p. But computationally this algorithm is very complex and heavy. We can make it computationally efficient by incorporating the following steps:

- Instead of considering all the remaining pixels to find out replacement for a particular pixel, define a search window symmetric about the pixel which we have considered. In this way, only those pixels which are there within the search window of the given pixel and not all pixels of the image will influence the value of its estimate.
- Consider only those pixels about whom the search window fits properly. This will obviously leave the boundary pixels unchanged which can be omitted from the image later. Also within a search area the candidate pixels would be those about whom the $n \times n$ window fits.

These steps are rather rational as one expects a pixel most likely to be similar to a small block of pixels around it which forms its search area.

3.2 Algorithm

- Consider the noisy image.
- Define the window size and the search area.
- Start from the pixel about which the search area fits properly.
- Within the search area of current pixel (say p), start from such a pixel (say q) about which the window fits properly.
- To calculate weight for q ($f(p, q)$), we need hp = weighted average of window of pixel p, hq = weighted average of window of pixel q, rx = orientation of window of p along x-direction, ry = orientation of window of p along y-direction, ed = overall orientation of window of p, $rx1$ = orientation of window of q along x-direction, $ry1$ = orientation of window of q along y-direction and $ed1$ = overall orientation of window of q.
- $f(p, q) = e^{\left(-\frac{(hp-hq)^2}{(\text{smoothning factor}\times\text{sigma})^2} - \frac{(ed-ed1)^2}{(\text{smoothning factor}\times\text{sigma})^2}\right)}$
- Similarly, find the weights for the other pixels in the search area of p.
- Calculate a normalizing factor, $c_p = \sum_i f(p, q_i)$
- Estimation of pixel $p = \frac{\sum_i q_i \times f(p, q_i)}{cp}$.
- Repeat this procedure of all such valid p pixels.

Edge detection has been carried out by Sobel gradient operators along X-direction and Y-direction [14]. After Edge detection Wiener filter is also applied to denoised image to minimize the mean square error between estimated random process and desired process. So the experimental results demonstrate that our algorithms performance is superior to the NL-means and comparable with BM3D in terms of PSNR and SSIM.

4   Experimental Results

Table 1 presents the de-noising performance of the above algorithm along with NL means, BM3D methods. Zero mean Gaussian noise with standard deviation values of 10, 20, 25, 30, 35, 50, 80 and 100 have been introduced in the original gray level images of Cameraman of size 256×256, Lena and Boat of size 512×512. Peak signal to noise ratio (PSNR) in dB is chosen as the performance measuring parameter. More the value of PSNR more the removal of noise is. More value of PSNR indicates better results of the algorithm. It can be clearly observed that the PSNR values of the images de-noised by the proposed algorithm are greater than the PSNR values of the same set of images when de-noised by NL means.

As the standard deviation of the noise increased, the performance of the existing methods lowered as expected but the performance of the proposed algorithm still had an edge over that of NL means. In our proposed algorithm we experimented with the size of search area from 7×7 till 21×21 for a particular standard deviation of noise. It was found that for a given standard deviation, the PSNR first increased as the search area was widened but then after the search area reached to an extent, PSNR kept on dropping gradually.

On an average the PSNR increased until the search area was increased to 11×11. Beyond this point the PSNR mostly decreased. So the optimum search area according to the experimentation should be 11×11 with window size of 3×3. Visually the difference between the outputs of the two algorithms is indiscernible but mathematically, our proposed algorithm performs better than the NL means. Below table shows the denoised comparison of classical non-local means filtering, BM3D and proposed method.

In Table 1, consider denoising of corrupted image of Cameraman with Gaussian noise having standard deviation 10 which has PSNR of 28.13 dB. NL means improved the PSNR by 4.54 dB, making it 32.67 dB. BM3D improved the PSNR by 5.95 dB, making it 34.08 dB. Our algorithm improved it by 5.41 dB, making it 33.54 dB. As the standard deviation was increased, the PSNR of the corrupted image kept on decreasing. But the amount by which the PSNR of denoised image was lifted kept on increasing. This implies that more the noise gets added to the original image, the more noise is also removed by these algorithms and their performance thus increases. When the standard deviation was 50, the corrupted image's PSNR was 14.15 dB. BM3D uplifted this PSNR by 11.69 dB(against only 5.95 dB when standard deviation was 10), NL means improved the PSNR by 10.19 dB making it 24.34 dB and proposed

Table 1 De-noising performance of the proposed algorithm along with the performance of NL means and BM3D on gray images

Images	Sigma/PSNR	Non-local means		BM3D		Proposed algorithm	
Cameraman (256 × 256)		PSNR	SSIM	PSNR	SSIM	PSNR	SSIM
	10/28.13	32.67	0.90	34.08	0.93	33.54	0.92
	20/22.08	29.32	0.82	30.39	0.87	29.85	0.85
	25/20.15	27.35	0.77	29.45	0.85	28.96	0.82
	30/18.55	26.72	0.74	28.64	0.83	28.08	0.80
	35/17.55	26.09	0.71	27.93	0.82	27.30	0.79
	50/14.14	24.34	0.61	25.84	0.78	25.23	0.75
	80/10.04	22.2	0.48	24.05	0.71	23.73	0.69
	100/8.12	20.59	0.39	22.81	0.67	22.17	0.64
Lena (512 × 512)	10/28.13	33.28	0.85	35.93	0.96	35.13	0.95
	20/22.11	30.98	0.78	33.05	0.94	32.50	0.92
	25/20.16	29.84	0.80	32.08	0.92	31.64	0.90
	30/18.58	29.28	0.72	31.26	0.91	30.82	0.87
	35/17.22	28.49	0.69	30.47	0.89	29.98	0.84
	50/14.14	26.72	0.60	28.86	0.86	28.21	0.82
	80/10.06	24.32	0.47	27.02	0.80	26.53	0.78
	100/8.14	22.46	0.37	25.57	0.76	24.95	0.74
Boat (512 × 512)	10/28.13	32.25	0.84	33.92	0.96	33.17	0.95
	20/22.11	28.68	0.77	30.88	0.92	30.16	0.91
	25/20.16	27.44	0.71	29.91	0.90	29.30	0.89
	30/18.58	26.56	0.67	29.12	0.88	28.82	0.86
	35/17.22	25.80	0.61	28.43	0.86	27.90	0.84
	50/14.14	24.27	0.57	26.64	0.81	26.06	0.77
	80/10.06	22.62	0.45	24.74	0.72	24.29	0.69
	100/8.14	21.94	0.40	23.74	0.68	23.15	0.65

algorithm outperformed NL means by improving the PSNR by 11.08 dB making it to 25.23 dB.

In case of Lena, though the performance of BM3D has been the best but the PSNR obtained from the proposed algorithm is majorly more than that obtained from NL means. So, here our proposed algorithm clearly outperforms NL means and nearer to BM3D.

Figure 1 shows the comparative denoised visual results of Lena and Cameraman corrupted with Gaussian noise of standard deviation 25 by NL means, BM3D and proposed algorithm. Figures 2a, b represents the original zoomed image of Lena eye and the respective image added with zero mean Gaussian noise with standard

| (a) Original | (b) Noisy ($\sigma = 25$) | (c) NL means | (d) BM3D | (e) Proposed |

Fig. 1 **a** Original images of Lena and Cameraman, **b** Respective noisy images with $\sigma = 25$, **c, d**, **e** denoised images by NL means, BM3D, proposed algorithm respectively

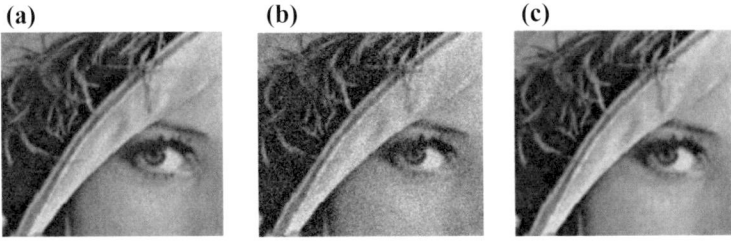

(a)   **(b)**   **(c)**

Fig. 2 **a** Original image of Lena eye . **b** Respective noisy image with $\sigma = 25$ and **c** denoised image of Lena eye by proposed algorithm

deviation 25 and Fig. 2c shows the quantitative results of the proposed algorithm. As it is visually difficult to observe the actual denoising for the whole image here, so we are presenting the results on zoomed portion of Lena eye. It can be seen that the PSNR of Lena eye is increased from 20.1 to 34.25 dB using the proposed algorithm with respect to noisy image.

5 Conclusion

A novel denoising technique for minimizing Gaussian noise using NL means and spatial domain edge preservation is presented in this work. The proposed algorithm gives good denoising performance as well as preservation of fine details and sharp edges of an image. The proposed algorithm yields better PSNR for low and moderate magnitudes of zero mean Gaussian noise. The recursive application of the algorithm is time-consuming like any other denoising algorithm. The algorithm is showing remarkable denoising capability of spatial domain edge preservation compared to other contemporary algorithms.

References

1. Perona P, Malik J (1990) Scale-space and edge detection using anisotropic diffusion. IEEE Trans Pattern Anal Mach Intell 12(7):629–639
2. Wiener N (1949) The extrapolation, interpolation and smoothing of stationary time series with engineering applications. Wiley, New York
3. Yaroslavsky LP (1985) Digital picture processing—an introduction. Springer, New York
4. Coifman RR, Donoho DL (1995) Translation-invariant de-noising. In: Antoniadis A, Oppenheim G (eds) Wavelets and statistics. Springer, New York, pp 125–150
5. Starck J, Candes EJ, Donoho L (2002) The curvelet transform for image denoising. IEEE Trans Image Proc 11(6):670–684
6. Smith SM, Brady JM (1997) SUSAN-A new approach to low level image processing. Int J Comput Vis 23(1):45–78
7. Buades A, Coll B, Morel JM (2005) A non-local algorithm for image denoising. In: Proceedings IEEE Computer society conference on computer vision and pattern recognition, San vol 2, pp. 60–65, Diego, USA
8. Liu Y-L, Wang J, Chen X (2008) A robust and fast non-local means algorithm for image denoising. J Comput Sci Technol 23(2):270–279
9. Buades A, Coll B, Morel J (2004) Technical report on image denoising methods: 2004–15, CMLA
10. Buades A, Coll B, Morel JM (2005) A review of image denoising algorithms, with a new one. Multiscale Model 4(2):490–530
11. Deledalle CA, Duval V, Salmon J (2011) Non-local methods with shape-adaptive patches (NLM-SAP). J Math Imag Vis 1–18
12. Tasdizen T (2009) Principal neighborhood dictionaries for nonlocal mean image denoising. IEEE Trans Image Proc 18:2649–2660
13. Dabov K, Foi A, Katkovnik V, Egiazarian K (2007) Image denoising by sparse 3-D transform-domain collaborative filtering. IEEE Trans image process 16(8):2080–2095
14. Gonzales RC, Woods RE (2012) Digital image processing, 3rd edn. Pearson Education Inc, pp 350–404

Speech Recognition Using Novel Diatonic Frequency Cepstral Coefficients and Hybrid Neuro Fuzzy Classifier

Himgauri Kondhalkar and Prachi Mukherji

Abstract Speech recognition is the ability of the machine to identify spoken words and classify them into appropriate category. First stage in the process of speech recognition is the extraction of appropriate features from the recorded words. We propose a novel algorithm for feature extraction using diatonic frequency cepstral coefficients. Diatonic frequencies are derived from a musical scale called as diatonic scale. The scale is based on harmonics of sound and models nonlinear behavior of human auditory filter. After feature extraction, the next classification stage uses a hybrid classifier using artificial neural network and fuzzy logic. If the difference between prediction values available at the output of the neural network is less, the classifier matches wrong patterns. Proposed algorithm overcomes this drawback using fuzzy logic. Proposed hybrid classifier improves the recognition rate significantly over existing classifiers. Test bed used in the experimentation focuses on Marathi language. It is the native language spoken in the state of Maharashtra.

1 Introduction

Spoken numeral recognition can be used for real-life applications like automated teller machines, lifts for visually sighted people. It can also be used for hands-free dialing systems for phones, search engines, etc. An efficient algorithm for both feature extraction stage and modeling stage of spoken word recognition is proposed in this work. Characteristics of sound are loudness, pitch, and quality. The speech signal can be represented using different types of parametric representations based on these characteristics. The existing techniques for the same are Mel frequency cepstral coefficients (MFCC) and Bark frequency cepstral coefficients (BFCC). Mel scale is a

H. Kondhalkar (✉)
Sinhgad College of Engineering, Pune, India
e-mail: gouri.ghule@viit.ac.in

P. Mukherji
Cummins College of Engineering, Pune, India
e-mail: prachi.mukherji@cumminscollege.in

© Springer Nature Switzerland AG 2019
D. Pandian et al. (eds.), *Proceedings of the International Conference on ISMAC in Computational Vision and Bio-Engineering 2018 (ISMAC-CVB)*, Lecture Notes in Computational Vision and Biomechanics 30,
https://doi.org/10.1007/978-3-030-00665-5_76

perceptual scale based on "pitch" measurement of the signal. The cutoff frequencies used in Mel scale model the nonlinear perception of frequencies in the human auditory system [1]. The maximum sampling frequency suggested for Mel scale is 14,000 Hz. Human hearing range is divided into 24 nonoverlapping critical bands, called as critical bandwidth. Each critical bandwidth represents one bark in the Bark scale [2]. The 24 center frequencies used in Bark scale are fixed. Bark scale is based on "loudness" measurement of speech signal [3]. Maximum sampling frequency suggested for Bark scale is 15,500 Hz. For frequencies higher than above values, downsampling of the speech signal needs to be done which leads to aliasing effect [4]. To avoid this effect, cutoff frequencies are formulated based on diatonic scale. The proposed diatonic frequency cepstral coefficients (DFCC) algorithm allows to extend the range of sampling frequencies to higher values above 15,500 Hz. Some speech recognition applications require sampling frequencies higher than 15,500 Hz like sampling frequency used for recording music is 44,000 Hz. A diatonic scale is based on the "quality" of sound. The proposed DFCC algorithm improved the accuracy and precision of classification over existing methods considerably. The feature vector produced by diatonic scale is used for classification of spoken words using feed forward neural network (NN). NN provides prediction accuracy which can be misleading if prediction values of the different classes are almost equal. As a result, the network misclassifies the pattern to the highest prediction accuracy class though it may not be the correct class. Hence, fuzzification can be used at the output of NN [5]. Application of fuzzy inference system at the output of NN tries to overcome abovementioned problem by reclassifying only the confusing patterns using proposed hybrid classifier. The overall accuracy as well as precision of the entire system is improved using hybrid neuro fuzzy (NF) classifier. The paper is organized as follows: review of related work in Sect. 2, methodology developed by authors in Sect. 3, results and discussion in Sect. 4 followed by conclusion in Sect. 5.

2 Related Work

Isolated word speech recognition in native languages spoken in different parts of India is an ongoing topic of research. An audio corpus of Marathi language containing 28,420 isolated words and 17,470 sentences is prepared by Santosh Gaikwad et al. It is available for experimentation in speech recognition system [6]. Recognition of Marathi numerals has been effectively implemented using Mel frequency cepstral coefficients (MFCC) and dynamic time warping (DTW) [7]. Speaker independent Urdu speech recognition system for district names of Pakistan has been studied and analyzed. The result proved that the system performs better if it is accent independent [8]. DNN–HMM hybrid model is developed for Chinese–English mix database to promote multilingual research [9]. The maximum word error rate achieved for Chinese database is 48.38%. Neuro fuzzy classification approach is used for recognizing sign language for hearing and speech impaired people. Combination of neural network and fuzzy logic resulted in 78% classification accuracy for the system [10].

Neuro fuzzy quantification is used for customer emotional state recognition. Multiple adaptive neuro fuzzy inference system with 95.29% classification accuracy is been developed [11]. Gujarati language speech corpora is prepared for speech recognition system. HMM is used as a classification algorithm. The system gives 95.9% word recognition rate [12]. Hierarchical speech recognition system is developed for Hindi language. The system aims at phoneme recognition [13]. MFCC and linear predictive cepstral coefficients techniques are used with HMM-based classification for Kannada language isolated spoken words [14]. The developed system was evaluated with 79% accuracy. Speech recognition is done for mobile applications in Tamil language using MFCC and dynamic time warping template matching [15]. Comparative study of different speech recognition techniques has been done by Gaikwad et al. One can choose an appropriate feature extraction technique with its merits and demerits [16]. Spoken Arabic digit recognition is done using MFCC and DTW techniques [17]. An average accuracy of 87% without normalization of feature vector was obtained.

3 Methodology

A novel algorithm to generate diatonic frequency cepstral coefficients using a musical scale called as diatonic scale is proposed in this work. The scale is based on quality of sound. Quality distinguishes two sounds of same pitch and loudness. Sound produced by musical instruments consists of series of tones with different frequencies called as overtones. Fundamental tone and series of overtones are in the ratio 1:2:3 called as harmonics. A diatonic scale has a series of eight notes between starting note and last note in an octave. It uses 8 frequencies starting on the tonic fundamental frequency, $F0$, and ending one octave above that at $2F0$. The scale then repeats, with 8 frequencies from $2F0$ to $4F0$, each being exactly twice the frequency considered as base octave. The next octave uses frequencies exactly four times those in the base octave. Input to the diatonic scale is the fundamental frequency $F0$ based on what the remaining octave frequencies are calculated. This property of the diatonic scale is used to generate cutoff frequencies in the feature extraction stage. Table 1 gives the ratio with $F0$ in the first octave span as well as the interval between consecutive notes. Human ear is most sensitive to frequencies between 300 Hz and 3 kHz. Hence, first octave starting with fundamental frequency 300 Hz is repeated to get diatonic scale frequencies within abovementioned range. Figure 1 shows the diatonic scale curve plotted with number of diatonic frequencies on x scale and corresponding frequency value in Hz on y scale.

Table 1 First octave used in diatonic scale

Ratio with $F0$	1	9/8	5/4	4/3	3/2	5/3	15/8	2
Interval	9/8	10/9	16/15	9/8	10/9	9/8	16/15	–

Fig. 1 Diatonic scale curve

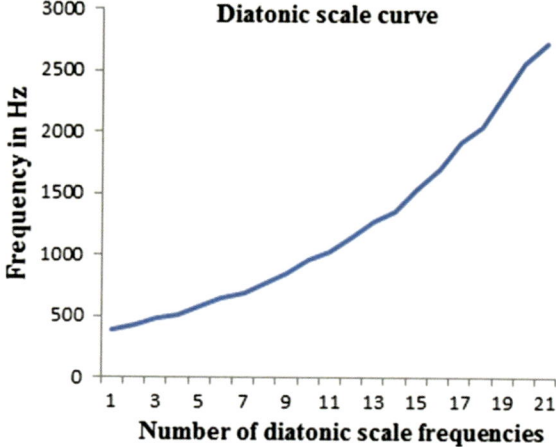

The process of speech recognition starting from input speech to output recognized word is illustrated in Fig. 2. In this section, feature extraction from the recorded signal, training, and speech recognition stages is described in detail.

3.1 Feature Extraction Using DFCC

Speaker utters a word which is recorded through microphone. This speech signal recorded has voiced part, unvoiced part, and silence portion. The voiced portion of the signal is needed to be separated out for further processing. Preprocessing of the speech signal includes voiced part detection and preemphasis.

Voiced Part Detection Let y be the input speech signal with length L and sampling frequency f_s.

$$y = \{y_0, y_1, y_2, \ldots, y_{n-1}\} \tag{1}$$

Speech processing requires nonstationary speech signal to be divided into number of frames denoted as N_f. When speech signals with different length undergo framing, unequal number of frames is generated. To maintain constant length of feature vector generated from these signals, zero padding is done to the frames. This affects the patterns generated by feature extraction algorithm reducing the recognition rate. To

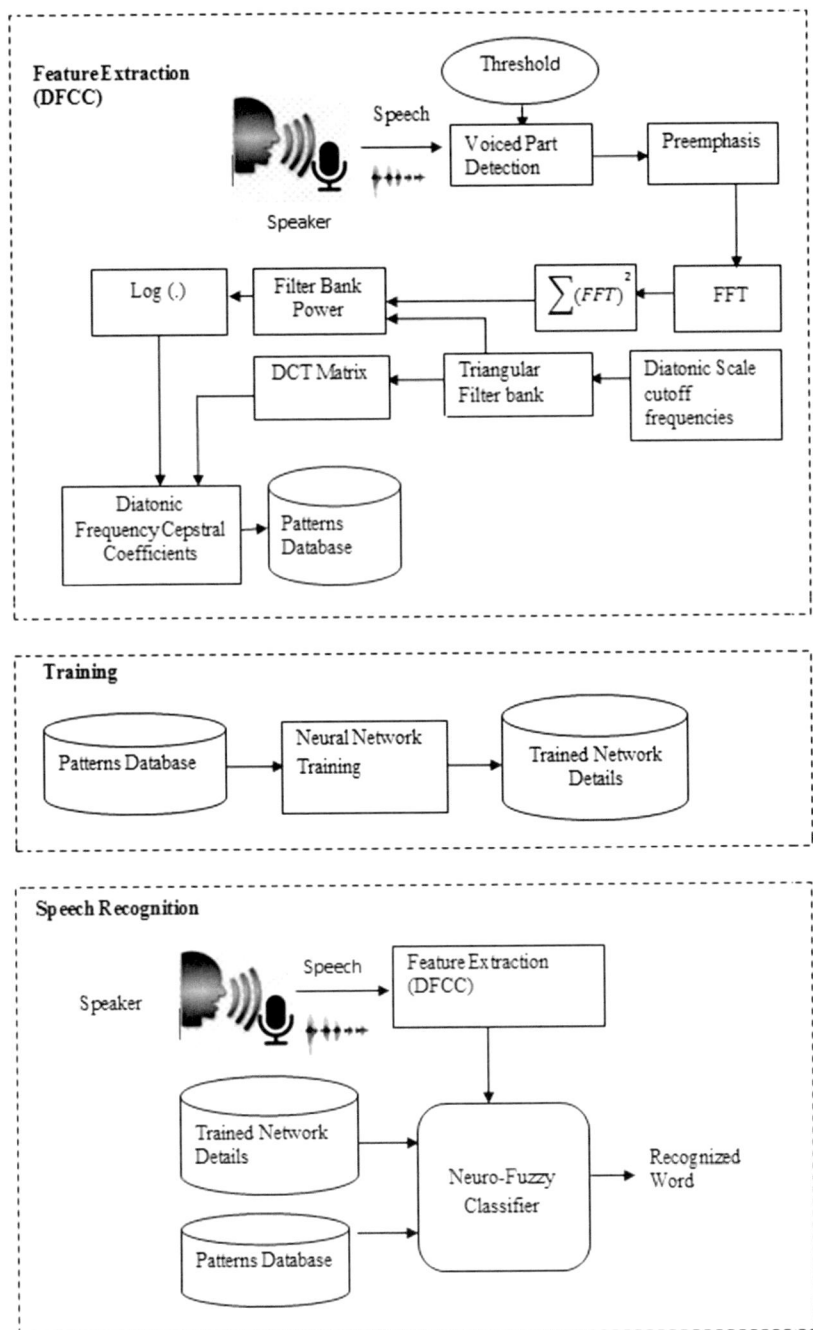

Fig. 2 Speech recognition using DFCC feature extraction and hybrid NF classifier

overcome this problem, the proposed algorithm uses variable length frames where
the frame length depends on the length of the original signal.

Irrespective of the length of the recorded signal, equal number of frames is generated using Eq. (2) in the proposed work.

$$N_f = \frac{(L - l)}{\text{step}} + 1, \quad l = 0.02 f_s, \text{step} = 0.01 f_s \tag{2}$$

Distortions are added due to number of edges created by frames. Hanning window
is applied to each frame to reduce the effect of distortions.

$$w(i) = 0.5\left(1 - \cos\left(\frac{2\pi i}{l - 1}\right)\right) \quad 1 \leq i \leq l \tag{3}$$

Since short time energy (STE) of the voiced part of a speech signal is high [18],
we calculate STE per frame.

$$\text{STE}_l = \sum_{i=1}^{l} (y_i * w_i)^2 \tag{4}$$

If STE is greater than a threshold value, it is considered to be the voiced part of
the speech sample.

$$y_1(n) = y(n), \quad \text{STE}_l \geq \lambda \tag{5}$$

where $y_1(n)$ is the voiced part of the signal.

Preemphasis Preemphasis filter is applied to the speech signal to spectrally flatten
the signal. This filtering helps to keep high-frequency components between the samples n and $n - 1$. Equation (6) represents the preemphasis of speech signal $y_1(n)$.
The value of α is chosen as 0.97, $y_2(n)$ is preemphasized speech signal.

$$y_2(n) = y_1(n) - \alpha(y_1(n - 1)) \tag{6}$$

Diatonic scale cutoff frequencies To generate the cutoff frequencies, diatonic scale
octave is repeated till maximum available frequency of the speech signal. Since
the sampling frequency used in the proposed work is 16 kHz, we need to repeat
the diatonic octave till maximum frequency of 8000 Hz. Table 2 shows 31 cutoff
frequencies thus obtained using diatonic scale. The ratio of frequencies with $F0$ is
repeated as shown in Table 1 for each consecutive octave.

Table 2 Diatonic scale cut off frequencies in Hz

384	426.66	480	512	572	640	682.66	768
853.33	960	1024	1152	1280	1365.33	1536	1706.66
1920	2048	2304	2560	2730.66	3072	3413.33	3840
4096	4608	5120	5461.33	6144	6826.66	7680	

Fig. 3 DFCC triangular filter bank

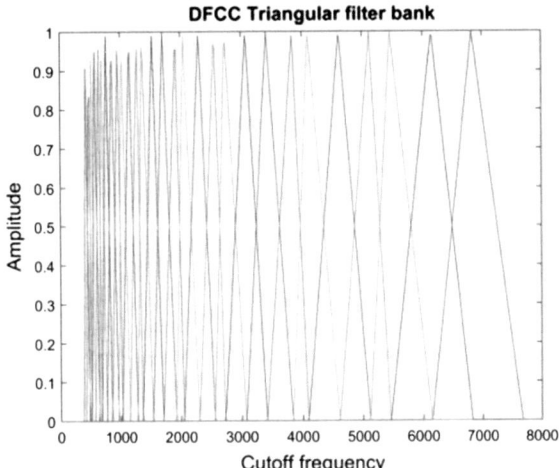

Next step in the generation of filterbank is calculating upslope and downslope coefficients for the filter bank channels. M is the number of filterbank channels. The value of M used in the proposed work is 29.

Upslope coefficients

$$k = f \geq \text{cf}(m) \text{ and } f \leq \text{cf}(m+1), \quad 1 \leq m \leq M$$

$$H(m,k) = \frac{(f(k) - \text{cf}(m))}{(\text{cf}(m+1) - \text{cf}(m))}, \quad 1 \leq k \leq K \tag{7}$$

Downslope coefficients

$$k = f \geq \text{cf}(m+1) \text{ and } f \leq \text{cf}(m+2), \quad 1 \leq m \leq M$$

$$H(m,k) = \frac{(\text{cf}(m+2) - f(k))}{(\text{cf}(m+2) - \text{cf}(m+1))}, \quad 1 \leq k \leq K \tag{8}$$

where cf is the cutoff frequency for each triangular filter bank, $K = n\text{FFT}/2$

Figure 3 displays the triangular filter bank generated using diatonic frequencies. It represents 29 filter banks placed between 300 and 8000 Hz. Discrete Fourier transform $F(u)$ with length N is calculated for each frame. Here $y_{2\text{frame}}$ is the preemphasized speech signal in a frame.

$$F(u) = \sum_{z=0}^{N} y_{2\text{frame}}(z) e^{\frac{-j2\pi zu}{N}} \tag{9}$$

Magnitude spectrum

$$|F(u)| = \sqrt{\text{real}(F(u))^2 + \text{imag}(F(u))^2} \tag{10}$$

Power spectrum

$$P(u) = |F(u)|^2 \tag{11}$$

Triangular filter bank $H(M, K)$ is applied to power spectrum to generate filter bank power vector fbp(M).

$$\text{fbp}(M) = \sum_{k=1}^{k} H(M, k) P(k, 1), \quad K = n\text{FFT}/2 \tag{12}$$

A discrete cosine transform matrix is created for each filter bank with number of cepstral coefficients as one of the dimensions.

$$\text{dctm}(p, m) = w(p) \sum_{m=1}^{M} \cos\left(\frac{\pi}{2M}(2m - 1)(p - 1)\right), \quad 1 \le p \le n\text{cep}$$

$$\text{where} \quad w(p) = \begin{cases} \frac{1}{\sqrt{N}} & p = 1 \\ \sqrt{\frac{2}{N}} & 2 \le p \le M \end{cases}$$

$n\text{cep}$ = number of cepstral coefficients $\tag{13}$

Filter bank power matrix is applied to discrete cosine transform matrix (dctm) to get cepstral coefficients vector, CC.

$$\text{CC}(p, j) = \sum_{m=1}^{M} (\text{dctm}(p, m) l\text{fbp}(m, j))$$

$$\text{where } 1 \le p \le n\text{cep}, \quad 1 \le j \le N_f,$$

$$l\text{fbp} = \ln(\text{fbp}) \tag{14}$$

N_f is the number of frames in which the speech signal is divided. The diatonic scale cepstral coefficients vector "CC" generated in Eq. 14 is stored in a patterns database which is input to the next training stage.

3.2   Training

The feature vector from the feature extraction stage is used for training neural network. It is a supervised training model that simulates the complex behavior of human brain. Training yields the class identification of each sample where we know the number of desired classes. It calculates the matching scores between the input pattern and sample pattern of a class. The proposed algorithm uses NN training. Test bed used for training NN and classification is recorded by us and named as Spoken Marathi Numeral Dataset (SMND) [19]. It consists of 7500 utterances of Marathi numerals from 0 ("Shunya") to 9 ("Nau"). 75 Speakers in the age group 18–55 with different gender and dialect pronounced the digits in succession. Recording was done in Mono mode with professional microphone and stored in.wav file format. Sampling frequency used for recording is 16 kHz. Closed room with provisions to minimize reverberation of sound was used for recording. 50% of these are used for training the network. The number of input neurons is 120. The number of hidden neurons used is 27 and the number of output neurons is 10 depending upon desired 10 classes at the output representing 10 numerals available in SMND. At the output of this stage, trained network details with best matching scores are obtained.

3.3   Speech Recognition

A hybrid classifier is used in this stage. This is a two-step process. The information retrieved from trained network details is used to determine desired output and calculate the error. This stage selects the class with best matching score and generates output which is closest to the input pattern. Input to this stage is the patterns database generated from the feature extraction stage and trained network details generated by training stage. Standalone NN when used for classification of isolated words produces prediction variables with almost similar values. In such cases, if the prediction variable of the wrong class is larger with minor difference than the actual class, NN produces wrong output. To overcome this problem, the prediction variables produced by NN are passed to fuzzy controller. The job of fuzzy controller is to improve the recognition accuracy by correctly classifying above patterns. Fuzzy logic is a part of fuzzy inference system that integrates human decision making in the form of IF-THEN rules [20]. Combination of fuzzy sets that represent some sort of uncertainty and IF-THEN rules determine the correct output class. Figure 4 represents the NF classifier architecture. This hybrid system has NN in the first step. NN classifier has inputs I_1 to I_m, input neurons a_1 to a_m, hidden neurons h_1 to h_n and NN output classes c_1 to c_k. The NN output is further given as input to fuzzy controller. The fuzzy inference system provides patterns to be reclassified to the next k nearest neighbors (k-NN) classifier. k-NN is a simple machine learning algorithm that uses a distance function to classify a new case. k-NN generates the correct classified word at the output. The role of fuzzy controller is to decide which misclassified

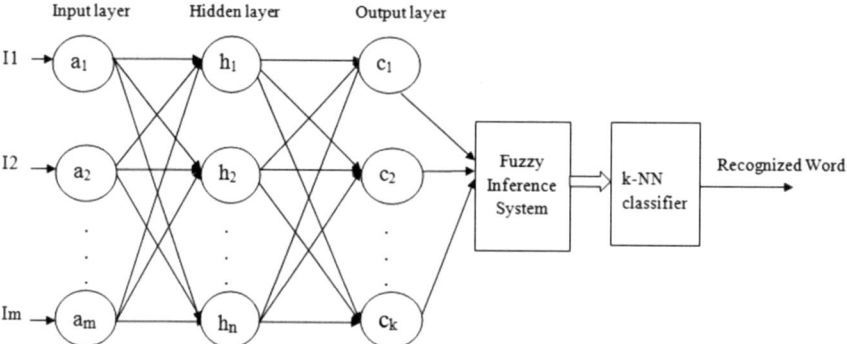

Fig. 4 Hybrid NF classifier architecture

patterns in the output of NN have very close prediction variables. These will be sent to next classifier. This improves the recognition rate of the system. It reduces overall computational complexity as few selected patterns are sent for reclassification. The proposed algorithm uses a triangular membership function in the fuzzy controller in order to define fuzzy rules using linguistic variables. This step is necessary to decide whether a pattern has to be sent to next classifier. We have used Euclidean distance measure as a similarity measure in k-NN classifier. k-NN classifies the pattern into correct class in majority cases. The proposed two-step hybrid classifier including NF as the first step followed by k-NN as the second step improves accuracy as well as precision of classification significantly over standalone NN classifier.

4 Results and Discussion

NN training of the patterns generated from SMND database is done using proposed DFCC feature extraction method. The dataset has utterances of ten Marathi digits from 0 to 9. It has 750 samples of each digit contributing to 7500 samples. Out of these 50% samples are used for training and 50% are used for testing of classifier. Thus, number of training and testing samples is 3750 each. Performance of the proposed DFCC method with an existing feature extraction method MFCC is compared. Table 3 represents recognition accuracy of ten digits for MFCC and DFCC using NN classifier. DFCC shows significant improvement over existing MFCC algorithm for all the ten digits. Figure 5 shows the plot of the recognition accuracy for MFCC and DFCC. From the results plotted, it can be concluded that the accuracy of digits 4, 5, 6, and 7 is lesser as compared to remaining digits. To improve overall accuracy, number of true positives for these digits should be increased which is achieved by implementing a hybrid classifier. We consider a case of digit 1 misclassified as digit 3. The prediction values at the output of NN classifier for the two digits are 0.4697 and 0.5729, respectively. NN misclassifies such patterns where the difference between

Table 3 Classification accuracy in % for ten digits using NN classifier

Digit category	0	1	2	3	4	5	6	7	8	9	Avg.
MFCC	96.3	94.8	93.6	96.8	93.3	92.3	92.2	88.7	91.1	96.4	93.5
DFCC	97.1	95.8	97.8	96.3	93.7	93.0	92.3	90.0	98.1	97.3	95.2

Table 4 Classification accuracy of DFCC algorithm for different datasets using NF classifier

Database	Accuracy (%)	Precision (%)
SMND	99.23	96.19
FSDD	99.00	95.56

Table 5 Summary of experimental results

Feature extraction	Accuracy (%)		
	SVM	NN	NF
MFCC	89.66	93.59	94.63
DFCC	89.77	95.15	99.23

prediction variables is less. Fuzzy controller assigns linguistic labels to these values as "High" as per triangular membership function. Since both of them have same linguistic label, these patterns are sent to k-NN classifier. In step 2 of the classification, k-NN classifies these patterns into actual classes. Figure 6 represents the comparative results of NN classifier and proposed NF classifier. It represents improvement by NF classifier in the number of true positives. True positives are the number of correct classified digits. From the figure, we can conclude that the proposed NF classifier improves classification accuracy for all the ten digits. We have tested the proposed method on another database comprising of English numerals. The database is available online for downloading known as free spoken digits dataset (FSDD). It has 2000 utterances of ten digits from 0 to 9 recorded by 4 speakers. We have used 1400 samples for training and 600 samples for testing purpose. Table 4 shows the comparative results for both the datasets. DFCC is used for feature extraction and classification is done with NF classifier. Accuracy and precision values are comparable for both the databases. Accuracy is the ratio of number of correct predictions and total number of patterns in the dataset. Precision is the ratio of number of correct positive predictions to the total number of positive predictions. Thus, the proposed algorithms can recognize words from different languages effectively. Table 5 shows the summary of the experimental results. MFCC and DFCC feature extraction techniques are compared for different classifiers. Classifiers used are NN, support vector machine (SVM) and proposed NF classifier. SVM is an existing supervised learning algorithm used for comparative analysis of the proposed hybrid classifier in our experimentation. The results are compared on the basis of average classification accuracy values. From the summary of results, we can conclude that both the proposed algorithms DFCC feature extraction and hybrid classifier outperform the existing techniques.

Fig. 5 Comparison of
MFCC and DFCC
classification accuracy using
NN classifier

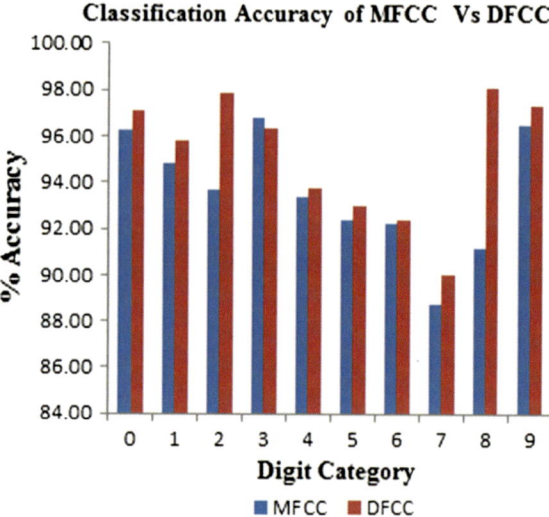

Fig. 6 Comparison of the
number of true positives
generated by NN classifier
and proposed NF classifier
using DFCC feature
extraction

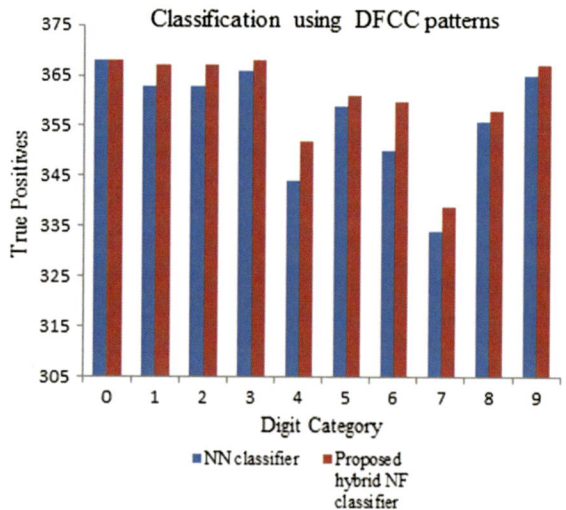

5 Conclusion

We have developed a language independent isolated word speech recognition system. The proposed system contributes at two stages. First, we have proposed a feature extraction method DFCC based on diatonic scale frequencies. This algorithm gives an overall classification accuracy 95.20% whereas existing MFCC technique has 93.50% accuracy. DFCC pattern generation can be used for a number of applications like speech recognition, music classification since diatonic scale cut off frequencies can be generated for any sampling frequency value. In the second stage of proposed work, we have designed a hybrid NF classifier to recognize patterns generated by DFCC algorithm. The combination of DFCC along with NF classifier generates an accuracy 99.23% which is significantly higher than existing SVM and NN classification methods. The NF classifier has a novel and unique implementation in the proposed work. This implementation ensures improved recognition rate for the speech recognition system irrespective of gender, age, and dialect variations of speakers.

References

1. Gupta D, Bansal P, Choudhary K (2018) The state of the art of feature extraction techniques in speech recognition. In: Agrawal S, Devi A, Wason R, Bansal P (eds) Speech and language processing for human-machine communications, vol 664. Advances in intelligent systems and computing. Springer, Singapore, pp 197–207
2. Lin Y, Abdulla WH (2015) Principles of psychoacoustics. Audio watermark. Springer, Cham, pp 15–49
3. Shanon BJ, Paliwal KK (2003) A comparative study of filter bank spacing for speech recognition. In: Microelectronic engineering research conference, Brisbane, pp 1–3
4. Hsieh SH, Lu CS, Pei SC (2013) Sparse fast fourier transform by downsampling. In: IEEE International conference on acoustics, Vancouver, pp 5637–5641
5. Bhavsar H, Trivedi J (2018) Image based sign language recognition using neuro fuzzy approach. Int J Sci Res Comput Sci, Eng Inform Technol, IJSRCSEIT 3:487–491
6. Gaikwad S, Gawali B, Mehrotra S (2013) Creation of Marathi speech corpus for automatic speech recognition. In: Conference on Asian spoken language research and evaluation (O-COCOSDA/CASLRE), Gurgaon, pp 1–5
7. Gedam YK, Magare SS, Dabhade AC, Deshmukh RR (2014) Development of automatic speech recognition of Marathi numerals. Int J Eng Innovative Technol (IJEIT) 3:198–203
8. Qasim M, Nawaz S, Hussain S, Habib T (2016) Urdu speech recognition system for district names of Pakistan. In: Conference of the oriental chapter of international committee for coordination and standardization of speech databases and assessment technique, Bali, pp 28–32
9. Wang D, Tang Z, Tang D, Chen Q (2016) A Chinese-English Mixlingual database and a speech recognition baseline. In: Conference of the oriental chapter of international committee for coordination and standardization of speech databases and assessment technique, Bali, pp 84–88
10. Li W, Hu X, Gravina R, Fortino G (2017) A neuro-fuzzy fatigue tracking and classification system for wheelchair users. IEEE Access 5:19420–19431
11. Diago L, Kitaoka T, Hagiwara I, Kambayashi T (2011) Neuro-fuzzy quantification of personal perceptions of facial images based on a limited dataset. IEEE Trans Neural Networks 22:2422–2432
12. Tailor JH, Shah DB (2018) HMM based light weight speech recognition system for gujarati language. In: Mishra D, Nayak M, Joshi A (eds) Information and communication technology for sustainable development. Lecture notes in networks and systems, vol 10. Springer, Singapore
13. Samudravijaya K, Ahuja R, Bondale N, Jose T, Krishnan S, Poddar P, Raveendran R (1998) A feature based hierarchical speech recognition system for Hindi. Sadhana. 23:313–340
14. Sneha V, Hardhika G, JeevaPriya K, Gupta D (2018) Isolated Kannada speech recognition using HTK-A detailed approach. In: Saeed K, Chaki N, Pati B, Bakshi S, Mohapatra D (eds) Process in advanced computing and intelligent engineering. Advances in intelligent systems and computing, vol 564. Springer, Singapore
15. Dalmiya CP, Dharun VS, Rajesh KP, (2013) An efficient method for tamil speech recognition using MFCC and DTW mobile applications. In: IEEE conference on information and communication technologies, Jeju Island, pp 1263–1268

16. Gaikwad S, Gawali B, Yannawar P (2010) A review on speech recognition technique. Int J Comput App 3:16–24
17. Ganoun A, Almerhag I (2012) Performance analysis of spoken arabic digits recognition techniques. J Electron Sci Technol 10:153–157
18. Jalil M, Butt FA, Malik A (2013) Short time energy, magnitude, zero crossing rate and autocorrelation measurement for discriminating voiced and unvoiced segments of speech signals. In: The international conference on technological advances in electrical, electronics and computer engineering (TAEECE), Konya, pp 208–212
19. Kondhalkar H, Mukherji P (2017) A database of Marathi numerals for speech data mining. Int J Adv Res Sci Eng 6:395–399
20. Bai Y, Wang D (2006) Fundamentals of fuzzy logic control-fuzzy sets, fuzzy rules and defuzzifications. In: Bai Y, Zhuang H, Wang D (eds) Advanced fuzzy logic technologies in industrial applications, advances in industrial control. Springer, London, pp 17–36

Performance Analysis of Fuzzy Rough Assisted Classification and Segmentation of Paper ECG Using Mutual Information and Dependency Metric

Archana Ratnaparkhi, Dattatraya Bormane and Rajesh Ghongade

Abstract The paper aims at the development of fuzzy rough set-based dimensionality reduction for discrimination of electrocardiogram into six classes. ECG acquired by the offline method is in the form of coloured strips. Morphological features are estimated using eigenvalues of Hessian matrix in order to enhance the characteristic points, which are seen as peaks in ECG images. Binarization of the image is carried out using a threshold that maximizes entropy for appropriate extraction of the fiducial features from the background. Various image processing algorithms enhance the image which is utilized for feature extraction. The dataset produced comprises the feature vector consisting of 79 features and 1 decision class for 6 classes of ECG. Extensive analysis of dimensionality reduction has been done to have relevant and nonredundant attributes. Fuzzy rough domain has been explored to take into account the extreme variability and vagueness in the ECG. Optimal feature set is subjected to fuzzification using Gaussian membership function. Further, fuzzy rough set concepts help in defining a consistent rule set to obtain the appropriate decision class. Classification accuracy of unfuzzified dataset is compared with the fuzzified dataset. Semantics of the data are well preserved using fuzzy rough sets and are seen from the performance metrics like accuracy, sensitivity and specificity. The proposed model is named as Fuzzy Rough ECG Image Classifier (FREIC) which can be deployed easily for clinical use as well as experimental use.

A. Ratnaparkhi (✉)
AISSMS-IOIT, SPPU, Pune, India
e-mail: archana.ratnaparkhi@gmail.com

D. Bormane
AISSMS COE, SPPU, Pune, India

R. Ghongade
BVCOE, Bharti Vidyapeeth, Pune, India

© Springer Nature Switzerland AG 2019
D. Pandian et al. (eds.), *Proceedings of the International Conference on ISMAC in Computational Vision and Bio-Engineering 2018 (ISMAC-CVB)*, Lecture Notes in Computational Vision and Biomechanics 30,
https://doi.org/10.1007/978-3-030-00665-5_77

1 Introduction

Arrhythmias cause a change in the normal rhythm of the heart which when sustained for a long time may result in irreparable damage to the heart as they actually cause a change in the normal rhythm of heart rate [1]. Thousands of beats are recorded by the Holter device carried out by the patients and immediate detection of the abnormality from such a huge database after preprocessing becomes a herculean task. 12-lead ECG system provides the information on complete ECG health from 12 different views. The ECG rhythm strips record the ECG which is analysed by the physician. Advancement in the field of Biomedical Image Processing has greatly assisted in feature extraction and dimensionality reduction that immensely contributes in quick and faster assessment of the patient. Dimensionality reduction to achieve appropriate feature extraction is generally a fundamental step in data processing. Removal of irrelevant features or redundant features enhances the discriminatory or predictive power of the algorithm. Selection of number of attributes which effectively constitute to increase the classification accuracy is possible due to rough set-based feature extraction. Rough set theory deals with datasets which are discrete in nature. Continuous-valued attributes need to undergo discretization which might add to information loss and further loss in classification accuracy. The previous approaches in ECG preprocessing and delineation has been widely done using Pan-Tompkins algorithm. However, the temporal information in ECG seems to get distorted by the use of filtering techniques like bandpass filtering [2], Kalman filtering and ensemble averaging [3] filtering [4]. Delineation of ECG has been done using various approaches like envelope detection and slope criteria for various segments in ECG that is seen in [5, 6]. The second-order differentiation and wavelet transform for extraction of the characteristic points is done by Duangsoithong and Windeatt [7] and Chan [8]. One can find the application of artificial neural networks for ECG delineation and classification [9]. Adaptive filtering and dynamic time warping has also been worked upon by Thakor and Zhu [3] and Laguna et al. [4]. Discrete wavelet transform has been widely used for the detection of QRS complex and other features [10]. For detecting congenital long QT syndrome DWT has been effectively used by Chevalier et al. [11]. Hybrid neural network approach in classifying the ECG into several classes has been done by Dokur and Olmez [12] using Daubechies wavelet (db2) [13]. The paper proposes a fuzzy rough assisted ECG classification of the ECG images that are obtained from the ECG rhythm strips.

1.1 Preliminaries

1.1.1 Rough Set Theory

Rough sets theory, a mathematical tool for representing datasets in an approximate manner, has been a primary tool to deal with imprecise and incomplete information.

It was originally proposed by Professor Pawlak in 1982 [1]. The original theory was put forth in such a simplified manner that it encouraged the researchers to apply this theory in varied fields ranging from image processing to higher dimensional signal processing systems and intelligent fault diagnosis systems [5]. Basically, the Rough Set Theory (RST) comprises of generation of reduced set of attributes called reducts which are used to form rules that classify a test dataset. A unique advantage of RST is that it deals with data and needs no other information about the dataset. Fuzzy set theory, rough set theory and probability theory have been an attraction for researchers by virtue of their immense capability to deal with uncertainty in the information. The structure of rough sets with abstract algebra is done in [8–10]. Rough sets have been effectively combined with soft computing tools in [5–10, 13, 14]. The study of evolutionary algorithms combined with RST has been done. Particle swarm organization, genetic algorithms and ant colony optimizations have been used in [15]. Thus, RST mainly focuses on attribute reduction and rule generation to be further processed by intelligent systems. However, as per the literature, RST suffers due to the crisp representation of the data. An acceptable range of variation in the input values is allowed in fuzzy domain. Hence with the advent of Fuzzy rough set as done by Jensen and Shen [15], representation of the biosignals has greatly simplified.

1.1.2 Fuzzy Rough Set Theory

A hybrid fuzzy rough set model first proposed by Dubois and Prade [16] was later extended and/or modified by many authors, and was applied successfully in various domains, most notably machine learning [17]. The upper and lower approximation sets in rough set theory have been fuzzified in fuzzy rough domain. However, certain principles as suggested by Jensen and Cornelis [17] need to be followed. And they have been stated as follows The set A may be generalized to a fuzzy set in X, allowing that objects can belong to a given concept to varying degrees. Instead of assessing objects indiscernibility, we may measure their approximate equality. As a result, objects are categorized into classes, or granules, with soft boundaries based on their similarity to one another. As such, abrupt transitions between classes are replaced by gradual ones, allowing that an element can belong (to varying degrees) to more than one class. The concept of belongingness to more than one class is able to efficiently handle the variability in ECG. The next section explains the use of fuzzy rough set theory applied to ECG feature vector which is extracted from the preprocessed ECG image (Fig. 1).

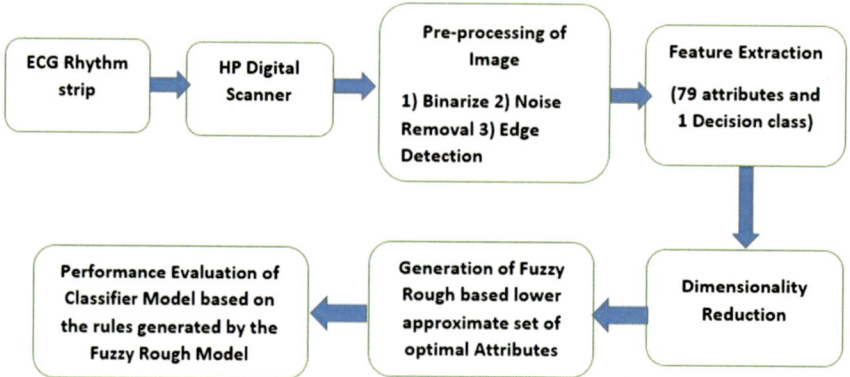

Fig. 1 Methodology

1.2 Methodology

1.2.1 Acquisition and Preprocessing

As given in the figure, ECG rhythm strip as recorded by the 12-lead ECG acquisition system is considered for further processing. This strip is scanned by HP digital scanner to get a good quality and clear image. Such a scanned image is stored in TIFF format and preprocessed using ImageJ toolbox. The preprocessing basically involves binarizing the image to enhance the ECG morphology. Further noise filtering is done using Gaussian filter which not only filters the image but also smooths it.

1.2.2 Feature Extraction and Dimensionality Reduction

The filtered binary image is subjected to various image processing tools that enhance the ECG morphology. The Sobel Edge detector highlights the edges in the ECG which helps in better feature vector formation. The features extracted include various parameters of the Hessian matrix. Second-order local image intensity variations around the selected pixel are described by the Hessian matrix. An orthonormal coordinate system is generated due to eigenvector decomposition which is perfectly aligned with the second-order structure of the image. A feature vector consisting of 79 attributes is further subjected to dimensionality reduction. Various standard attribute evaluators that are used in the literature have been tested.

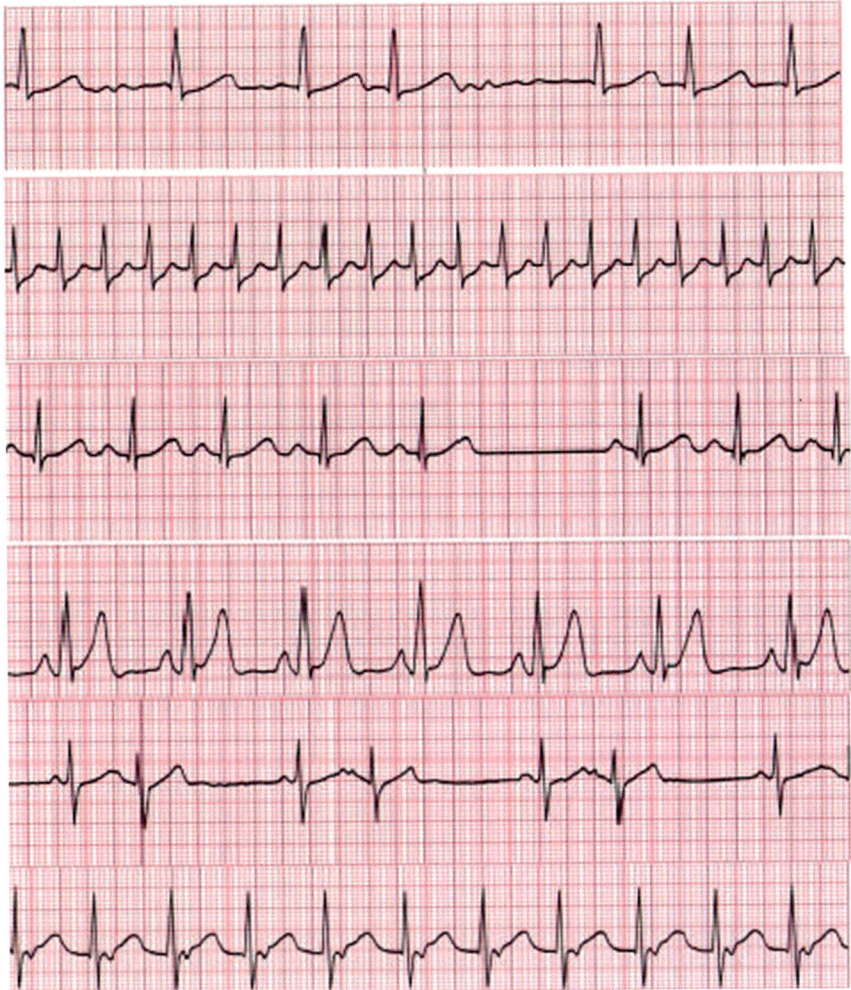

Fig. 2 ECG rhythm strip

1.3 Fuzzy Rough-Based Rule Generation and Classifier Evaluation

The optimal attributes generated are used to generate fuzzy rough-based lower approximate sets. Fuzzification of the inputs helps in incorporating the uncertainty, vagueness and extreme variability in the ECG. The novelty in Fuzzy Rough ECG Image Classifier (FREIC) lies in the use of fuzzy rough set-based rule induction using optimal attributes. According to [10], the QUICK-REDUCT algorithm is as follows:

$Quickreduct\,(C, D)$

$Input1 : C, the$ set of conditional attributes;

$Input2 :$ D,the set of decision features

$Output$: R,the feature subset

1.R ← { }

2.while $\gamma_R(D) \neq \gamma_C(D)$

3.$T \leftarrow R$

4.for each x ∈ (C - R)

   $if\ \gamma_{R \cup x}(D) > \gamma_T(D)$

   $T \leftarrow R \cup \{x\}$

5.$R \leftarrow T$

6. Re$turn$ R

The above fuzzy rough attributes evaluator helps in finding the optimal attribute set. The proposed FREIC classifier is compared with standard classifiers. Performance metrics for the same are discussed in the next section.

1.4   Experimentation and Results

The ECG strips obtained from 12-lead ECG acquisition system is scanned using the digital scanner. The figures below show the images obtained after scanning. These strips show six classes of Arrhythmia. The binary ECG image enhances the ECG morphology and assists in appropriate edge detection which is seen in the next Fig. 3. These preprocessed images are used for feature extraction using ImageJ toolbox. Hessian of the image is taken which sharply highlights the characteristic points in ECG. 79 such attributes further are subjected to Dimensionality Reduction. The dimensionality reduction using fuzzy rough metric is comparable to the reduction obtained using correlation-based method. The proposed model FREIC based on fuzzy rough metric looks for attributes which are relevant and nonredundant. Such attributes are then evaluated on the basis of their dependency metric (dependency of the decision class on a certain attribute). Thus, eventually dual dimensionality reduction occurs. The classification accuracy obtained using tenfold cross-validation on standard clas-

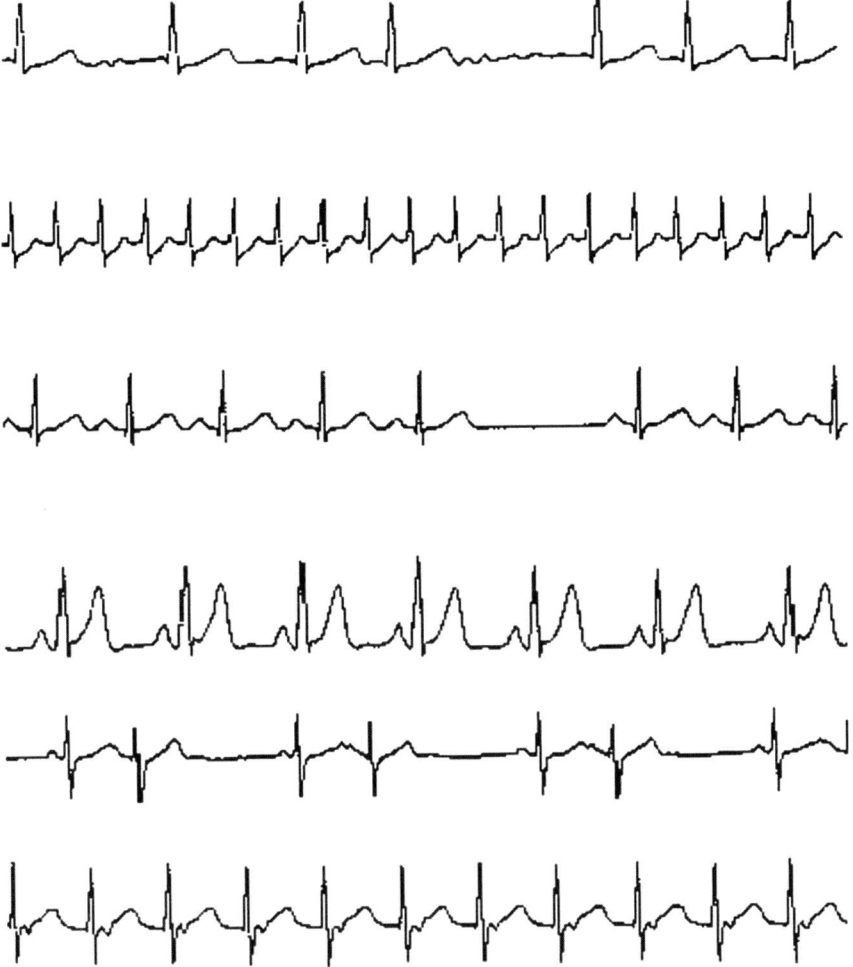

Fig. 3 Binary image of the ECG strip

sifiers like SVM, MLP, Random Forest, Rough sets and Fuzzy classifier is shown in the graph below. Comprehensive comparison of the classifiers is done. The proposed Fuzzy Rough ECG Image classifier reduces the number of attributes to nine which is considerably low as compared to SVM or MLP-based classification. Comparative analysis of the accuracies clearly shows the efficacy of rough sets and fuzzy rough sets-based classifier as they outperform the traditional methods in terms of optimal attribute selection as well as accuracy as is seen in ROC curves (Figs. 2, 4, 5, 6, 7, 8, 9 and 10).

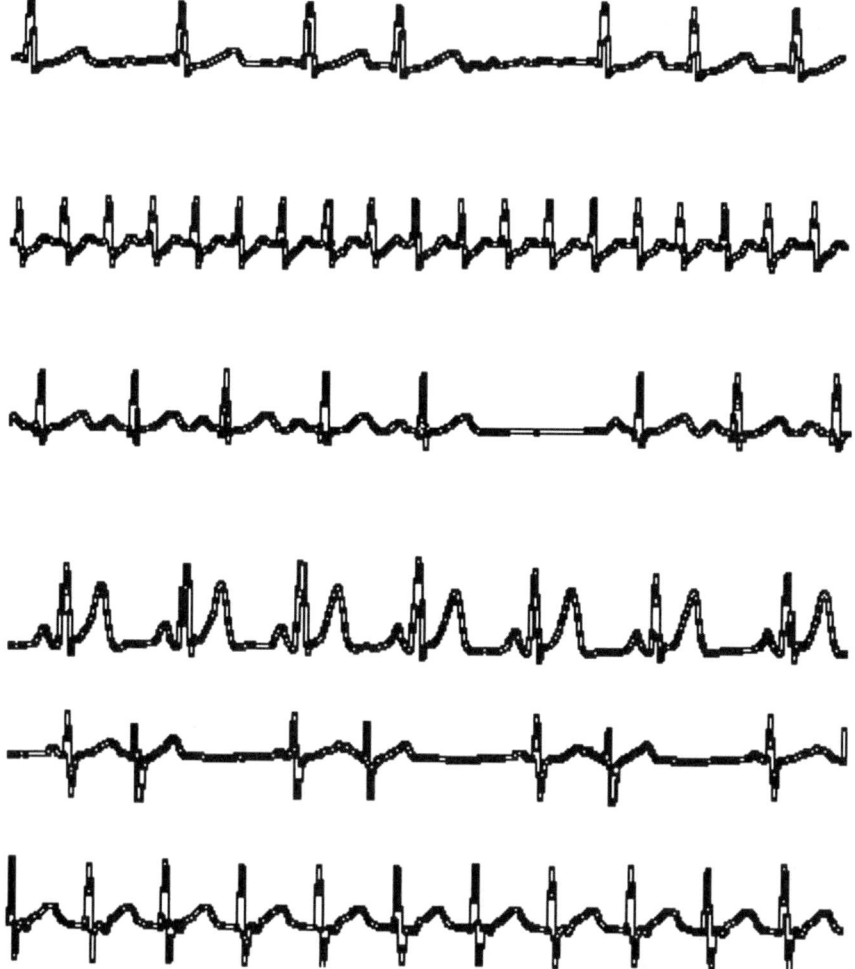

Fig. 4 Binary image of the edge detected ECG strip

1.5 Conclusion

The proposed Fuzzy Rough ECG Image classifier exploits the concepts of lower and upper approximate sets from fuzzy rough set theory for ECG rhythm strips which is completely novel. The algorithm not only outperforms other traditional classifiers in terms of accuracy by a comfortable margin, but also competes with much-proven correlation-based attribute evaluators to achieve an optimal attribute set. Additional resilience has been achieved by the image processing tools that effectively enhance the morphology of the ECG signal and assist in achieving high accuracy.

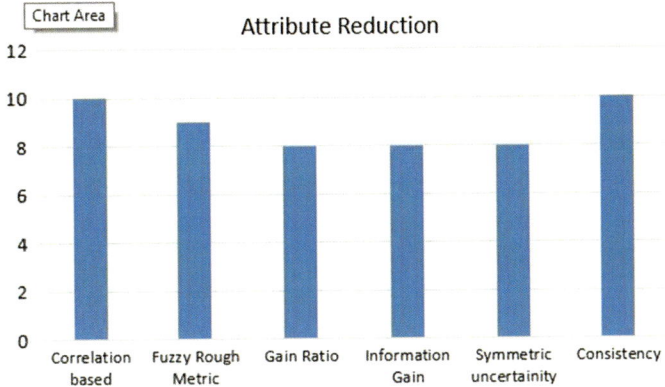

Fig. 5 Attribute reduction using standard attribute evaluators

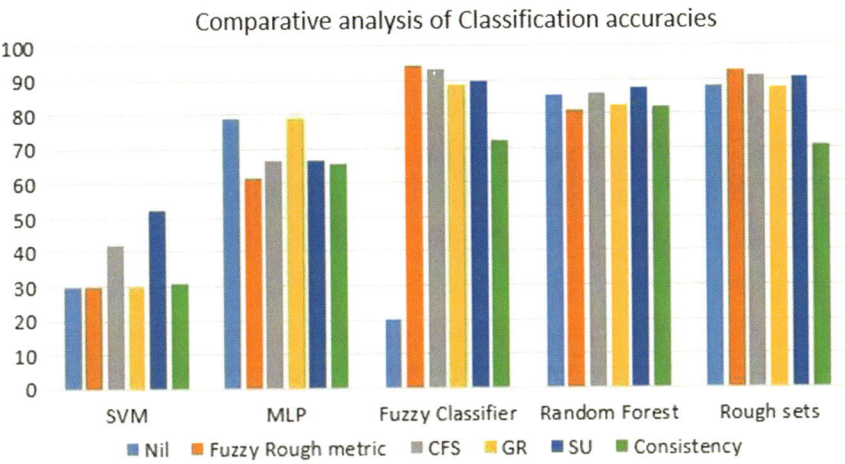

Fig. 6 Comparative analysis of classification accuracies

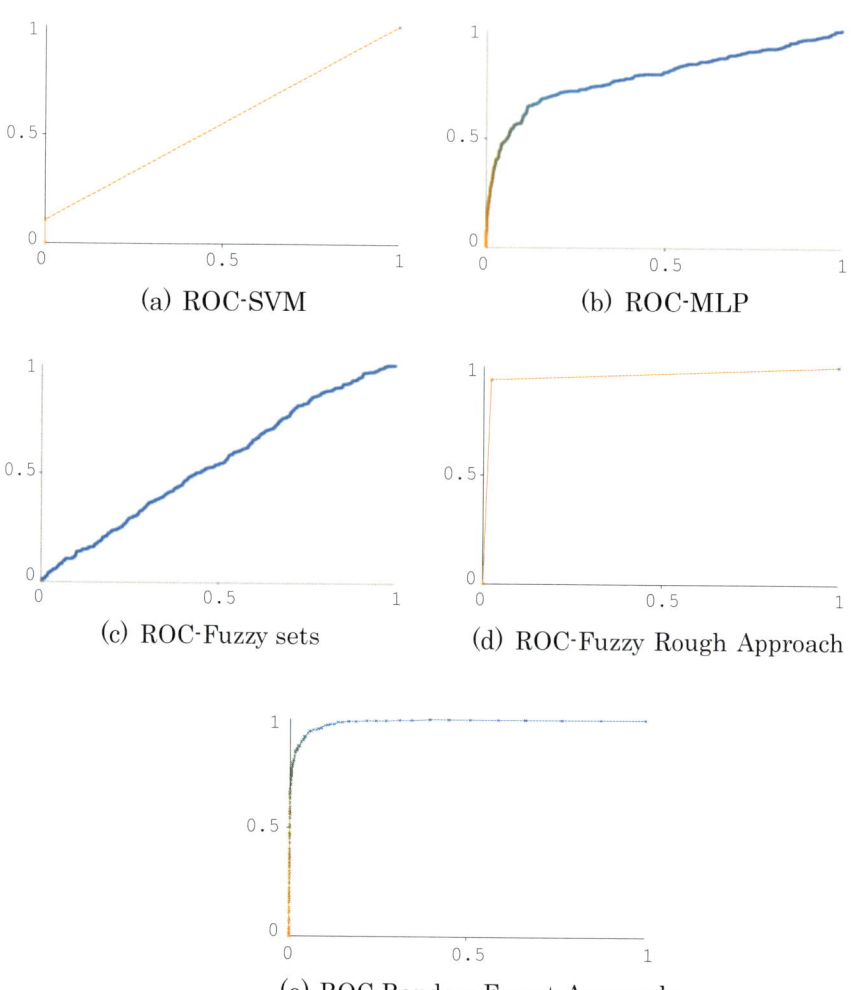

(a) ROC-SVM (b) ROC-MLP

(c) ROC-Fuzzy sets (d) ROC-Fuzzy Rough Approach

(e) ROC-Random Forest Approach

Fig. 7 ROC curves: classifiers

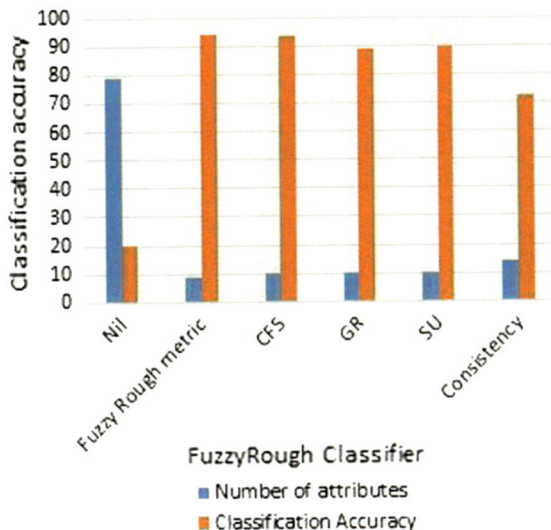

Fig. 8 SVM

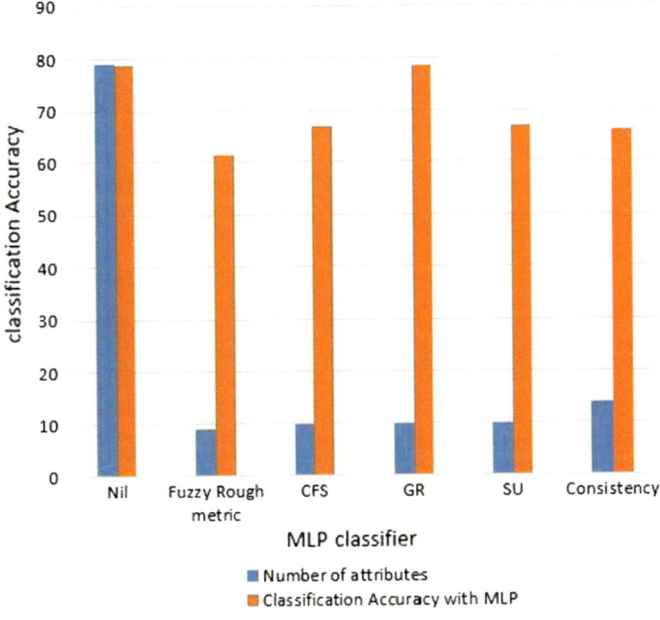

Fig. 9 MLP

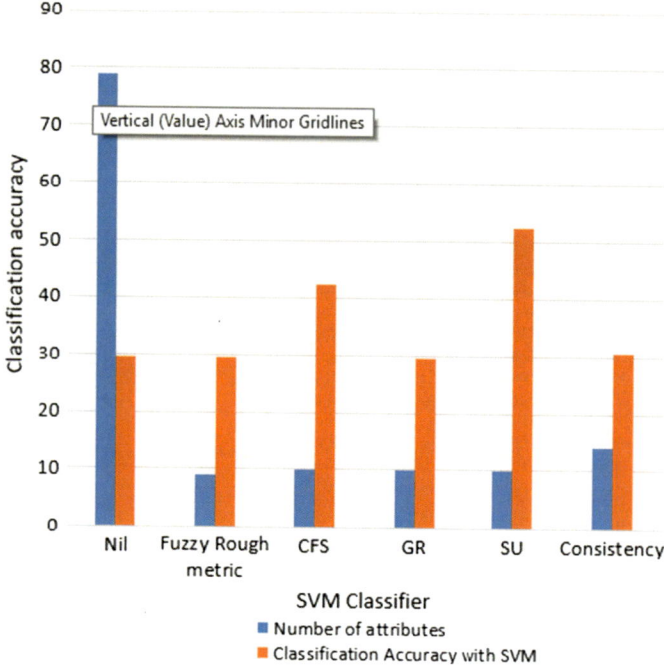

Fig. 10 Fuzzy rough sets

References

1. Tan WW, Foo CL, Chua TW (2007) Type-2 fuzzy system for ecg arrhythmic classification. In: IEEE International Fuzzy systems conference, 2007. FUZZ-IEEE 2007. IEEE, pp 1–6
2. Christov II, Daskalov IK (1999) Filtering of electromyogram artifacts from the electrocardiogram. Med Eng Phys 21(10):731–736
3. Thakor NV, Zhu Y-S (1991) Applications of adaptive filtering to ecg analysis: noise cancellation and arrhythmia detection. IEEE Trans Biomed Eng 38(8):785–794
4. Laguna P, Jané R, Meste O, Poon PW, Caminal P, Rix H, Thakor NV (1992) Adaptive filter for event-related bioelectric signals using an impulse correlated reference input: comparison with signal averaging techniques. IEEE Trans Biomed Eng 39(10):1032–1044
5. Zhang Q, Xie Q, Wang G (2016) A survey on rough set theory and its applications. CAAI Trans Intell Technol 1
6. Polkowski L (2002) Topology of rough sets. In Rough Sets. Springer, Berlin, pp 331–360
7. Duangsoithong R, Windeatt T (2011) Hybrid correlation and causal feature selection for ensemble classifiers. In: Ensembles in machine learning applications. Springer, Berlin, pp 97–115
8. Chan YT (2012) Wavelet basics. Springer Science & Business Media
9. Ghongade R et al (2007) A brief performance evaluation of ecg feature extraction techniques for artificial neural network based classification. In TENCON 2007-2007 IEEE region 10 conference. IEEE, pp 1–4
10. Martínez JP, Almeida R, Olmos S, Rocha AP, Laguna P (2004) A wavelet-based ecg delineator: evaluation on standard databases. IEEE Trans Biomed Eng 51(4):570–581
11. Chevalier P et al (2001) Non-invasive testing of acquired long QT syndrome: evidence for multiple arrhythmogenic substrates. Cardiovasc Res 50(2):386–398

12. Dokur Z, Ölmez T (2001) ECG beat classification by a novel hybrid neural network. Comput Methods Programs Biomed 66(2–3):167–181
13. Addison PS (2005) Wavelet transforms and the ecg: a review. Physiol Measur 26(5):R155
14. Mitra S, Mitra M, Chaudhuri BB (2006) A rough-set-based inference engine for ecg classification. IEEE Trans Instrum Measur 55(6):2198–2206
15. Jensen R, Shen Q (2007) Fuzzy-rough sets assisted attribute selection. IEEE Trans Fuzzy Syst 15(1):73–89
16. Dubois D, Prade H (1990) Rough fuzzy sets and fuzzy rough sets. Int J Gen Syst 17(2–3):191–209
17. Jensen R, Cornelis C (2011) Fuzzy-rough nearest neighbour classification and prediction. Theoret Comput Sci 412(42):5871–5884

A Mixed Reality Workspace Using Telepresence System

Viken Parikh and Mansi Khara

Abstract Even after years of advancement from telephone technology to video conferencing, it eventually does not provide the environment of real human interaction. It has been a challenge ever since. The system we propose here would provide an environment in the user's reality, where the user can interact with the person on the other side of the conversation with the help of a wearable mixed reality headset. The user will get a feeling of coexistence with objects and people who are far away. This advancement in technology will lead to a world, where people can connect with others who are only virtually present. This paper will present the technical system for teleporting virtual objects and virtual participants in the real-world environment. It will present a system where 3D audio and real-time communication will help build a more realistic environment with the presence of virtual participants and objects in 3D space.

1 Introduction

Telepresence is the hypothetical movement of material objects from one place to another without physically traversing the distance between them. By merging Mixed Reality and Telepresence, we can get a real world experience with virtual objects where people can communicate and interact with each other without being physically present. In this era, traveling seems to be more and more difficult, telepresence is an emerging research area and companies are moving forward to make this possible. It will seem astonishing to see how people travel across the globe virtually. This will save both time and money and will prove to be the next generation of technology. In this paper, we provide an approach to regenerate objects in a 3D space and display it to a user using light-field technology in Mixed Reality headset. The objects will

V. Parikh (✉) · M. Khara
K. J. Somaiya College of Engineering, Vidyavihar, Mumbai, India
e-mail: viken.parikh@somaiya.edu

M. Khara
e-mail: mansi.khara@somaiya.edu

© Springer Nature Switzerland AG 2019
D. Pandian et al. (eds.), *Proceedings of the International Conference on ISMAC in Computational Vision and Bio-Engineering 2018 (ISMAC-CVB)*, Lecture Notes in Computational Vision and Biomechanics 30,
https://doi.org/10.1007/978-3-030-00665-5_78

803

be captured with the help of 8 trinocular cameras where the RGB images along with the depths of these images will be captured, and the 3D model will be reconstructed. The color RGB image will provide color to the 3D model such that it appears to be the same as the original object. Finally, the object will be displayed on the other side to the person wearing the headset. The applications of the proposed model range from Virtual Meetings, where people can conduct important meetings virtually and can meet their loved ones from across the globe. This can also be useful in Virtual Tourism, where one can roam the world by sitting at just one place and capture memories, by storing the virtual meetings and regenerating them when needed. This can also be used in the field of Gaming, where playing a game will feel seamless, for example, the player playing the battle actually on the battlefield.

The paper is organized as follows, Sect. 2 shows the basic terminologies used by us throughout the paper. Section 3 gives a broader view of similar work done in the past, Sect. 4 describes the actual methodology used, applications are described under Sect. 5, Sect. 6 concludes our paper, and Sect. 7 described the work that can be continued in the future.

2 Basic Terminologies Used

2.1 Telepresence

It is the use of virtual reality technology for remote control of machinery or for apparent participation in distant events.

2.2 NIR

Near-infrared, a region of the infrared spectrum of light used for near-infrared spectroscopy. It is used for multi-depth mapping in imaging. For more details, refer [1].

2.3 Light-Field Technology for Imaging

Conventional images record the projection of light rays on a 2D plane as compared to light fields, which describe the distribution of light rays in a free space, consisting of their position, angle, and radiance. Light-field technology uses this idea to capture these image data for application in mixed reality. For more details, refer [2].

2.4 Binaural Recording

It is a method of recording sound using two microphones, arranged with the intent to create a 3D stereo sound sensation for the listener. It will perceive the listener of actually being in the room with the performers or instruments.

3 Related Work

The major differences and advantages what MR brings as compared to AR and VR are that, in AR, the position of the objects displayed to the user is not relative to the user's position in the 3D space. The user will be displayed data on their screen as it is, on top of the real-world environment. While VR creates a separate digital space, where the user can only interact with 3D objects and data that have been digitally made. MR provides the advantage over both of these technologies by providing relative positioning of objects on top of the real-world environment with the help of smart consumer glasses or displays.

There has been a lot of research work in this field. But the current research lacks the development of a system that can be used for teleconferencing with the help of mixed reality in real-time communication with efficient results.

HoloLens provides an AR technology but lacks the light-field technology to enable Mixed Reality. Many of the VR headsets have some related health problems like dizziness and headache due to low frame rates. They also lack the efficiency to place virtual world objects in the real world. Locomotion is restricted by the presence of limited object manipulation option in 3D space.

These technical smart video headsets and glasses are highly expensive and require continuous high bandwidth network usage and processing power. Hence compression, low-power processing, and immediate transmission of data are a few of the objectives while designing a mixed reality system.

For telepresence of human models, the need for continuous tracking is required at a high frame rate to provide a continuous communication experience to the participants.

There have been attempts but, none being able to demonstrate a mixed reality workspace and hence the need for research on this topic.

4 Methodology

Our main aim here is to create a two-way interaction between two participants in mixed reality. This project will allow a local participant and a distant participant to interact with each other and also interact with other virtual objects in a 3D space. The main importance here is a low latency model with the highest possible quality of the regenerated objects in 3D space.

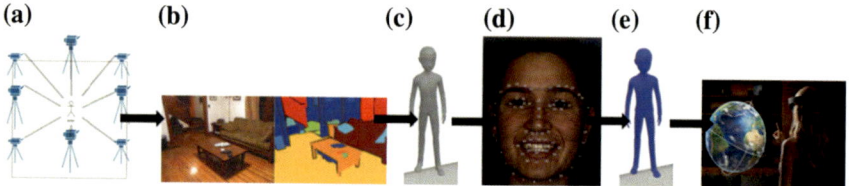

Fig. 1 **a** Capturing image. **b** Depth capturing and preprocessing. **c** 3D reconstruction. **d** Face and body tracking [8]. **e** Merging the colors with the 3D model. **f** Display in mixed reality to the remote participant

The model shown in Fig. 1 is a quick overview of how our proposed model would work. This overview provides a workflow of how the physical setup will enable interaction with virtual objects in the real-world environment.

4.1 Capturing Images

The participant and user will be in a room with a setup of 8 capturing trinocular cameras. The 8 trinocular cameras will be arranged as shown in Fig. 2a. The edge cameras are arranged higher as compared to the corner cameras. This would help create a better 3D model while reconstruction with the help of 8 images from 8 different positions. Each trinocular camera will be arranged as shown in Fig. 2b, which will consist of two Near-Infrared (NIR) cameras and one color camera.

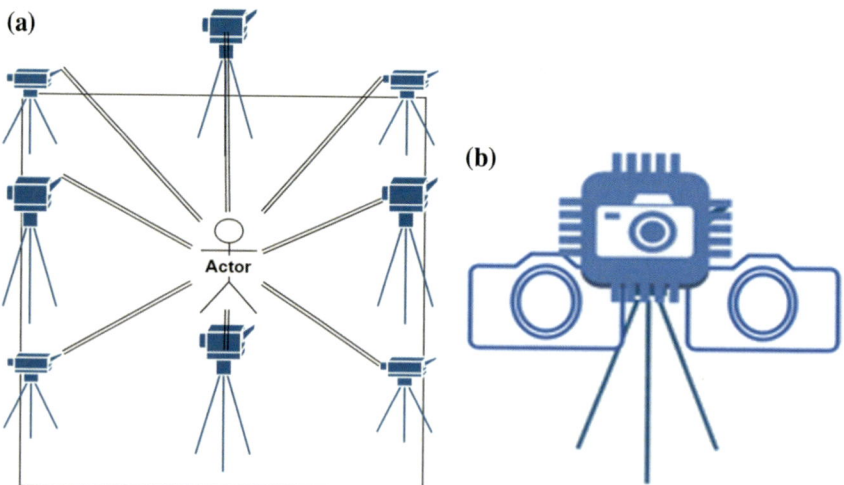

Fig. 2 **a** Arrangement of 8 trinocular cameras. **b** Trinocular camera setup with 2 NIR cameras and 1 color image capturing camera

- The distance of the object and position in the real world will be estimated with the help of the Depth cameras [3]. The first image captured is an RGB image and the second image is an NIR Image. There are several options available to capture depth images. We would propose Orbbec Astra Mini camera, which can be used for NIR image capturing. Kinect can also be used.
- For the RGB images, high-quality digital cameras will be used.
- All the cameras would be synchronized running at 60 fps.

4.2 Creating Virtual Objects in 3D Space

Figure 3 demonstrates how a 3D object can be placed along with real-world objects in the virtual space.

We use Autodesk Maya application to create 3D objects. The libraries already have a rich collection of redeveloped 3D objects. Other tools like Blender, 3ds Max can also be used. We use these objects and import them into Unity3d. Unity3d provides an environment which can be replicated to be similar to that of the real-world environment. Now the objects created can be placed in the 3D space of Unity3d, which would eventually be placed in the same virtual space as that of the models of human participants. We will use object recognition to detect a type of object, and based on object pairs created, we will place the objects in the real-world space. The light and placements of the objects can be trained to be placed in the places where it is most likely to be based on object placements pairs. For example, if an object is a human model, it is most likely to be placed on a chair, sofa, or floor based on his position and posture. A fruit basket object is most likely to be placed on a table.

(a)   **(b)**

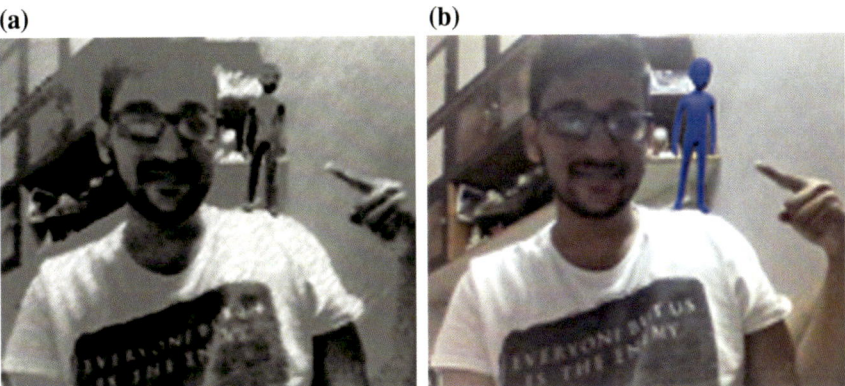

Fig. 3 **a** Image of virtual object silhouette to be added on top of the real world using mixed reality display. **b** Finally recreated 3D object added to the mixed reality display. Viken Parikh, K J Somaiya College of Engineering

4.3   Generating Silhouettes to Create a 3D Model of Real-World Objects

With the help of 2D images, we create silhouettes. These silhouettes help in image compression and also help us to extract the areas of interest like human models from the images. With the help of silhouettes, we can get the actual shape of the object, and a reverse algorithm can help us reconstruct that shape [4]. As we have multiple images, it will help us to create a better 3D model of the object.

4.4   3D Model Reconstruction

With the help of segmentation from the silhouettes, we get the objects to be reconstructed for our 3D model. Using temporal consistency, we can separate and create a more consistent model of our object. A statistical function can be used which contains objects of the same class with different appearances to be applied on the geometrical object captured from the real-world data. This mapping of the 3D object with common appearance statistics will provide more room for compression and better accuracy for low latency platforms (Fig. 4).

- Color Texturing or Matting
  After fusing depth data, a polygonal 3D model is extracted from its implicit volumetric representation (TSDF) with marching cubes, and then, this model is textured using the 8-input RGB images [5]. The approach presented for color texturing can classify unobserved views which form ghostly artifacts with traditional approaches.
- Human and Face tracking
  There are two types of human tracking, whole-body tracking and face tracking. Whole-body tracking corresponds to tracking a general tracking rather than small facial movements or gestures. Figure 5b shows whole-body tracking.

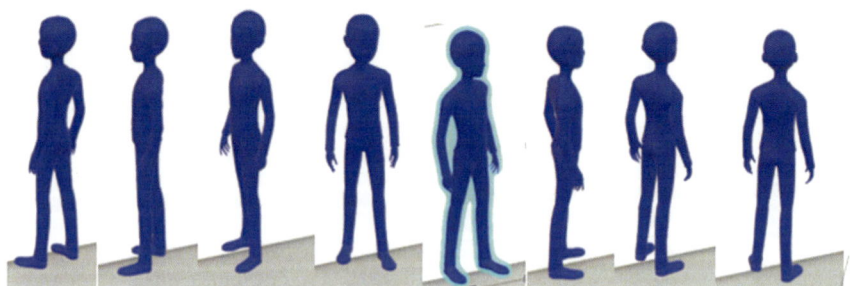

Fig. 4   3D object created used 8 different images from 8 different angles captured

Fig. 5 **a** Face tracking. **b** Whole-body tracking [8]

(a)

(b)

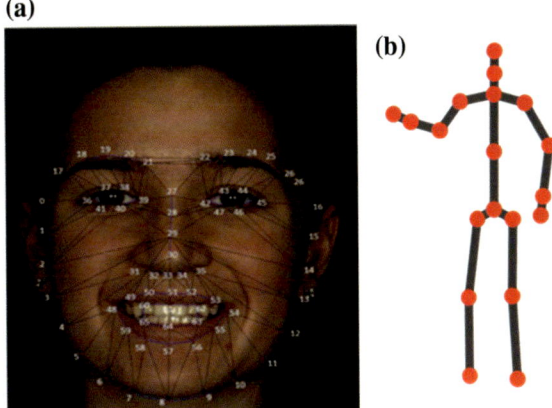

On the other hand, facial tracking is dedicated to tracking these small behaviors. For objects with minute details like face, a posture which changes quickly with respect to time, we apply face tracking for rigid and nonrigid facial tracking. Figure 5a shows face tracking. For more read [6].

4.5 3D Audio

Until a few years back, audio was recorded by using two methods: mono and stereo. Mono uses a single microphone to pick up a sound, while stereo uses two, spaced apart from each other.

The binaural audio setup is a simulation of how we receive sound in real life. That creates an experience, a user would if they were actually there.

Binaural recording is similar to the stereo method of recording. It records by placing two microphones in ear-like cavities on either side of a stand or dummy head. Though similar, it is better than stereo, as the microphones are placed on a person's ears, or inside ear-like imitations on a dummy's head. As the dummy head recreates the density and shape of a human head, these microphones capture and process sound exactly as it would be heard by human ears to preserve interaural cues.

In order for the user to have a complete experience of the real-world environment, along with the virtual one, the user cannot wear a headset or earpieces. Therefore, the speakers of the 3D audio would be placed just about the ears in order for a user to have a complete virtual and real-world experience (Fig. 6).

Fig. 6 A binaural audio
recording device for
capturing 3D audio

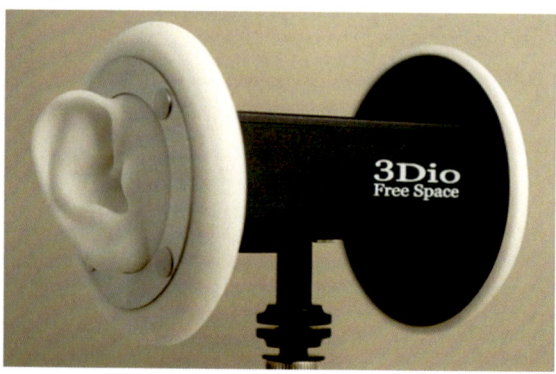

4.6 Compression

Compression is important for rendering such huge data to a distant location. There
are several ways of performing this compression, some of them include mesh compression and point cloud compression.

- Mesh Geometry
  Mesh compression was used to compress the polygonal model before sending
  over the Internet by serializing vertices. Two approaches were tested, 3Dzip for
  transmission over TCP and FFmpeg for UDP [7].

4.7 Processing and Transmission

- Parallelization across CPU or GPU
  First is the location of the reconstruction algorithm. In the model-based approach,
  it is executed on the CPU and in the render-based approach, on the GPU. This is
  because the former contains much branching, whereas, the latter is undertaking
  the same parallel instructions on different data [7].
- Transmission over the internet
  Based on the accuracy required by the user and bandwidth available, the number of
  participants, part of the mixed reality telepresence system can vary. For a system
  that can provide for a high bandwidth, it can support the telepresence for 5–10
  users in the same reality simultaneously.

4.8 Display in Mixed Reality

- Light-Field Technology
  Light Field is the term given to light rays traveling in all the directions in a given volume of space. Unlike conventional images, which record the 2D projection of the light rays by angularly integrating the rays at each pixel and missing out on important information, a light field describes the distribution of light rays in free space. This allows capturing richer information from the world. This is done using a micro-lens array, which is placed between the main lens and the camera sensor. Light-Field Technology solves one of the biggest challenges, which is enabling virtual objects to look real from both far and near distances.
- Gesture-Controlled Mixed reality headsets
  We propose a mixed reality headset that uses light-field technology to display 3D objects in the real world [2]. This will allow users to adjust the focus of near and far objects and have a better experience of the real world. The headsets will display 3D objects, which look very similar to real-world objects with low latency.

  With the advancements of light-field technology, the MR headset will be able to adjust the position of the object with respect to the real-world environment. The user will have a provision to change the position of the object in 3D space with the help of gesture patterns. By default, the objects will be placed in 3D space based on object pairs.

  These holograms of 3D objects on the top of real-world environments enable the user to have an experience of being with multiple users along with an ability to manipulate their positions and interact with them.

5 Applications

There are several other applications of teleportation. Some of them include the following:

- A new innovation in the field of Telecommunications
  From the traditional voice calls to video calls, technology is becoming more and more advanced. The concept of telepresence will give birth to people being virtually present at some other place, in 3D. Figure 7a is an example of the same.
- Gaming
  While gaming is still restricted to 2D or 3D with limited functionality, using Virtual Reality. Telepresence can be a whole new experience with actual objects inside the game can be seen in your surroundings, different players can play the game together by being virtually present. This innovation in teleportation will surely make gaming a wonderful experience.

- Creating memories

  Traditionally, the only way to store memories, be it some good holiday, or some important event was with the help of taking photographs. Telepresence will help us store the memories in 3D, resize them, replay them whenever one wishes to.
- Virtual Tourism

  People can teleport virtually from their place to a tourist location. This will save their time as well as money. Figure 7b shows a woman traveling to a distant location from her home by wearing a mixed reality headset.

6 Conclusion

Thus, we present a powerful object reconstruction technique, which can be used to map 3D objects in real time and present the same at a different location. The process needs high-quality cameras, mixed reality wearable headsets, and software tools to map and reconstruct the objects. We have mentioned the potential applications of our model and how this technology can enhance the future generation.

7 Future Work

New approaches can be considered, for capturing better quality pictures and reducing the latencies to regenerate them at the destination. A large number of users can be accommodated per mixed reality teleconferencing workspace by either increasing the bandwidth available, increasing the processing power and by improving the compression algorithm. Further, low-cost mixed reality displays like Aryzon can be made with additional features like 3D audio and gesture control so that a larger group of users can benefit from it.

(a)   **(b)**

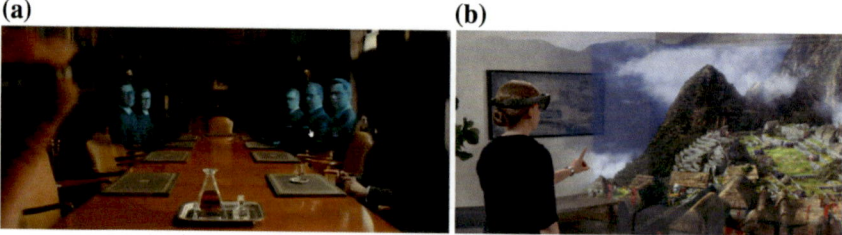

Fig. 7 **a** People sitting in a conference virtually. **b** A woman enjoying a tourist location with the help of telepresence

Acknowledgements The paper, its research, and its content is ethically ours and any of the other content used has been correctly referenced to its source. This is to clarify the ethical compliance of our paper to be published in Springer.

References

1. Ng R (2006) Digital light field photography
2. https://www.technologyreview.com/s/610458/vr-is-still-a-novelty-but-googles-light-field-technology-could-make-it-serious-art/
3. Xiong Y (2014) Automatic 3D human modeling: an initial stage towards 2-way inside interaction in mixed reality
4. Speakman S (2013) Dynamic pattern detection with temporal consistency and connectivity constraints
5. Orts-Escolano S et al (2016) Holoportation: virtual 3D teleportation in real-time
6. Mangold K, Shaw JA, Vollmer M (2013) The physics of near-infrared photography. Eur J Phys 34. IOP Publishing Ltd
7. Roberts DJ et al (2015) withyou—an experimental end-to-end telepresence system using video-based reconstruction. IEEE J Sel Top Signal Process 9
8. https://www.cl.cam.ac.uk/research/rainbow/projects/clmz/labels/indices.jpg

Detection of Static and Dynamic Abnormal Activities in Crowded Areas Using Hybrid Clustering

M. R. Sumalatha, P. Lakshmi Harika, J. Aravind, S. Dhaarani and P. Rajavi

Abstract In Computer Vision, to monitor the activities, behavior, and other changing information, surveillance is used. Surveillance is used by government organizations, private companies for security purposes. In video processing, anomaly is generally considered as a rarely occurring event. In a crowded area, it is impossible to monitor the occasionally moving objects and each person's behavior. The main objective is to design a framework that detects the occasionally moving objects and abnormal human activities in the video. Histogram of Oriented Gradient feature extraction is used for the detection of an occasionally moving object and abnormality detection involves in computing the motion map by the flow of motion vectors in a scene that detects the change in movement. The experimental analysis demonstrates the effectiveness of this approach which is efficient to run in real time achieves 96% performance, however, for effective validation of the system is tested with standard UMN datasets and own datasets.

1 Introduction

In current world video surveillance plays a vital role. It is a process of monitoring video clips for the behavior, activities, or other suspicious information for the purpose of managing or protecting the society. In a congested area, it is impossible to monitor each person's behavior. The behavior of each person varies. Some behave normally, while few behave abnormally. Therefore, it is a tedious process to monitor suspicious behavior. Arousing, fighting, rioting, jaywalkers, wrong way, etc., is considered as some of the suspicious behavior. Surveillance system has become an important aspect of security and a necessity to keep proper checking. With this fast growing technological world, the use of video surveillance is also increased. Now

M. R. Sumalatha (✉) · P. Lakshmi Harika (✉) · J. Aravind · S. Dhaarani · P. Rajavi
Department of Information Technology, MIT Campus, Anna University, Chennai, India
e-mail: sumalatha.ramachandran@gmail.com

P. Lakshmi Harika
e-mail: lakshmi.harika558@gmail.com

© Springer Nature Switzerland AG 2019
D. Pandian et al. (eds.), *Proceedings of the International Conference on ISMAC in Computational Vision and Bio-Engineering 2018 (ISMAC-CVB)*, Lecture Notes in Computational Vision and Biomechanics 30,
https://doi.org/10.1007/978-3-030-00665-5_79

a day, video surveillance is used in almost all areas. The main aim is designing the model to detect the abnormality in the human behavior either static or dynamic and to recognize the crowd event in the video sequence based on features extracted and the movement of motion vectors. Human anomaly detection in crowded areas is a great challenge. The existing framework has certain limitations like slow processing time, needs high storage space, cannot find occluded objects, need previous results as input.

2 Related Works

Lin et al. [1] designed a framework for unattended object detection. Immobile foreground is extracted and it is used to detect the unattended objects by backtracking the owner trajectories.

Lin et al. [2] framed a method which used grab cut segmentation [3] for unattended object detection. Background for detection is developed using dual Gaussian Mixture learning model [4]. But the work is limited to the complex environments and mainly focused on increasing efficiency.

Jadhav et al. [5] devised an algorithm for unattended object detection. By subtracting the frame from the background, the objects in the foreground are obtained. Then, using the feature of objects like shape, height, color, and size the unattended objects are identified and classified.

Jaouedi et al. [6] proposed a method to recognize the human actions by inspecting their behavior. In their work, they tracked the moving humans by using Kalman filtering [7]. Then their actions are determined by using their motion vectors and K nearest neighbor method is used for classification.

Kumar et al. [8] developed an approach for recognition of the activities of human in video sequence. This is done by finding the optical flow [9] of the edges of humans. But their work is based only on the known data set.

Zhu et al. [10] framed an approach to human behavior analysis by using the hybrid model. In this the behavior is inspected by extracting global silhouette feature and optical flow method is used to estimate their motions. These are the limitations in existing survey and it has been overcome in the proposed system.

3 Proposed System

The proposed system detects occasionally moving objects and uncommon human activities in video frames. The architecture for detection of occasionally moving objects and uncommon human activities is given in Fig. 1.

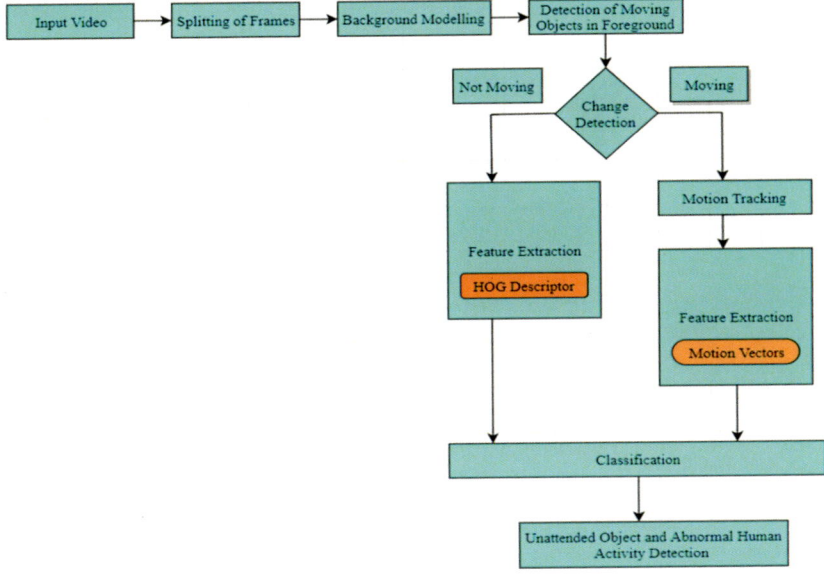

Fig. 1 Architecture for activity detection system

3.1 Preprocessing

The input to the system is a video file. The input video is split into frames and they are processed in a successive manner. Frames are sequences of images. The input video is preprocessed to remove noise and for the conversion of RGB frame to grayscale frame. The detection of occasionally moving objects and uncommon human activities is done by comparing the frames of video with background frame.

3.2 Background Modeling

Modeling of background is used to obtain accurate background from input video file. Gaussian Mixture model in Eq. (1) is used to model the background and the background is generated with first few frames.

Gaussian distribution

$$B = \arg\min\left(\sum_{K=1}^{b} w_k > T\right), \tag{1}$$

where B is a background model, T is a threshold, w_k is a weight factor of pixels.

Initially, video frames are given as an input to the system. A background model is initialized for all pixels in the background frame. The value of pixels in consecutive frames is checked in a successive manner to know whether they are pixels of background.

3.3 Occasionally Moving Static Object Detection

In video frames, subtraction between the images is a straightforward approach to detect the changes that occur between the images. The proposed system uses Histogram of Gradients (HOG) [11] to detect the changes between the frames. HOG is an image feature descriptor as well as a Human Feature descriptor. HOG descriptor is implemented by dividing of image into small cells and the computation of the histogram of gradient directions is performed in each cell. Support Vector Machine (SVM) [12], classifier is well trained to recognize HOG descriptors. HOG is not influenced by noise, clutter, and illumination changes.

Algorithm 1: Occasionally moving object detection
Input: Video Frames F^t (A Sequence of Images) $t = 1, 2... N$ (N = Total Frames)
    Output: Detection of Occasionally moving objects

Step 1:   Generate Background Model with first few frames
Step 2:   Get Background and use it as Reference Image (F^0)
Step 3:   Extract HOG^0 in F^0 where HOG is a Histogram of Gradients
Step 4:   For each frame $t = 1, 2... N$

      4.1:   Extract HOG^t of all frames
      4.2:   Calculate $Corr^t i, j$ & $Diff^t i, j$ for current image F^t and reference image
           F^0 by using Equ. (2) and Equ. (3)

$$Corr_{i,j}^t = \frac{HOG_{i,j}^0 \cdot HOG_{i,j}^t}{\left\| HOG_{i,j}^0 \right\| \cdot \left\| HOG_{i,j}^t \right\|} \tag{2}$$

$$Diff_{i,j}^t = \left| F^t - F^0 \right|, \tag{3}$$

      Where Corr is Correlation and Diff is the difference
      4.3:   Extract the Area and Perimeter of Detected object to draw Bounding
           Box
      4.4:   Get Detection Binary Map $Binary_{(i,j)}^t$

$$Binary_{i,j}^t = \begin{cases} 1, \text{ if } Corr_{i,j}^t < TCorr \,\&\&\, Diff_{i,j}^t > TDiff \\ 0, \text{ Otherwise} \end{cases}, \tag{4}$$

Where Binary$^t_{(i,j)}$ is the Detection Binary Map value of (i, j) Pixel in frame F^t

Tcorr and TDiff are thresholds of Correlation and Difference maps

4.5: For each object detected in a frame, check the object in consecutive frames using (4.4)

4.6: If the detected object is static in consecutive long successive frames, then classify it as an occasionally moving object

Step 5: End For

3.4 Dynamic Uncommon Behavior Detection

The Proposed System detects the global abnormal human activities that occur across the frame by using Optical flow approach. Dense Optical flow algorithm [13] is used to calculate the Optical flow of frames to obtain the motion vectors. It is represented by (r, θ), where "r" represents the magnitude of each pixel and "θ" represents the deviation angle of each pixel from Eqs. (5) and (6)

$$r = \text{sqrt}\big((x_2 - x_1)^2 + (y_2 - y_1)^2\big) \tag{5}$$

$$\theta = \tan^{-1}((y_2 - y_1)/(x_2 - x_1)) \tag{6}$$

In which (x_1, y_1) and (x_2, y_2) are initial and terminal coordinates of each pixel and r is the magnitude and (is the deviation angle of each pixel. Each frame is divided into A by B blocks in a uniform manner and find the Optical flow of all blocks in a successive manner.

Algorithm 2: Uncommon Human Activity Detection
Input: Video Frames f (A Sequence of Images)
    Output: Detection of Uncommon Human activity

Step 1: For each frame f, do Preprocessing

Step 2: For each frame f, Estimate the motion values of each pixel value (i, j) by using Optical Flow

$$\text{Optical Flow} = (r, \theta) \tag{7}$$

from Eqs. (5) and (6)
Where r- magnitude, θ-deviation angle of pixels

Step 3: Divide each and every frame f into A by B Blocks in a uniform manner

Step 4: Find Optical Flow for all blocks by using Eq. (8)

$$O_i = 1/P \sum_i Of_i^j \tag{8}$$

Where O_i is the Optical Flow of ith block, P is the Number of Pixels in a block, Of_i^j is the Optical Flow of jth Pixel in i^{th} block

Step 5: Generate Motion Vector Map of all blocks in a successive manner

    5.1: Assume M^i (j ϵK) to Zero at the start of processing of each frame //K-Set of blocks

    5.2: for all i $\in$ K do

        5.2.1: $Th_d = \|O_i\| * BS$; // Th_d -threshold, O_i -Optical Flow of ith block, BS-Block Size

        5.2.2: $Fv_i/2 = O_i + \pi/2$;

        5.2.3: $-Fv_i/2 = O_i + - \pi/2$; // Fv_i - Field view of Object i

    5.3: for all j ϵ K do

        5.3.1: if i $\neq$ j then

            5.3.1.1: Find the Euclidean Distance ED (i,j) between O_i and O_j

        5.3.2: if D (i,j) < Th_d then

            5.3.2.1: Calculate the angle ϕ_{ij} between O_i and O_j

        5.3.3: if $-Fv_i/2 < \phi_{ij} < Fv_i/2$ then

            5.3.3.1: Assign A $= (-ED (i,j) / \|O_i\|)$

            5.3.3.2: $M^j(O_i) = M^j(O_i) + \exp(A)$ (9)

        5.3.4: end if for all

    5.4: end for all

Step 6: Move TRAINING and TESTING Phase

3.5 Hybrid Clustering

The proposed framework uses hybrid combined clustering because it overcomes the disadvantages of hierarchical [14] and *k* means [15] approaches, which makes use of both features. Here the process starts by observing the movement of object vectors in blocks, and then each object is considered as initial cluster. The process goes on observing the features and spatial distance between the blocks to form clusters. It computes the vector matrix by matching the feature set and nearness in distances between the centroid of the clusters. The centroid value of the cluster is given in Eq. (10)

$$Ck = \sum_{i=1}^{n} x(k)/n \qquad (10)$$

These are the advantages to construct the optimal number of clusters. By using the patterns of normal crowd activity, uncommon crowd activity is identified in testing phase.

Training and Testing Phase

In Training phase, the patterns of normal crowd behavior are identified by using hybrid clustering and they are saved as pattern blocks. Here each frame is divided into non colliding mega motion vector blocks N. Then compute Z_N by using Eq. (11)

$$Z_N = \sum H^j, \tag{11}$$

where Z_N represents the Motion Vector map value of Mega Blocks
H^j represents Motion Vector map value of blocks.

Perform the extraction of Space-Time features for each mega block to perform hybrid clustering. Perform hybrid clustering to get pattern blocks and set the centers as pattern blocks. Pattern blocks represent the behavior pattern of normal crowd activity.

In Testing, the Minimum Distance Matrix has to be calculated over all blocks is given below Eq. (12)

$$E(i, j) = \min \left\| f^{(i,j)} - w_k^{(i,j)} \right\|^2, \tag{12}$$

where E is the Minimum Distance Matrix, $f^{(i,j)}$ is the feature vector, $w_k^{(i,j)}$ is the Pattern blocks. Then highest value is observed among all the values, if the value is higher than a threshold (Th_d) then classifies as abnormal frame.

4 Experimental Results

The Proposed system is developed using OpenCV-Python and the performance of the designed system is evaluated based on Pets [16], ABODA [17], UMN [18] and real time data sets. Pets and ABODA datasets are used for occasionally moving object detection and UMN dataset is used for uncommon human activity detection. The sample results are shown in Figs. 2 and 3 shows the occasionally moving object in indoor and outdoor scenario and abnormal human activity scenes. The performance is evaluated by considering the positive frames and negative frames taken from each dataset. The standard metrics of the classifier are calculated is shown in Tables 1 and 2 and resultant graph with existing system in Figs. 4 and 5.

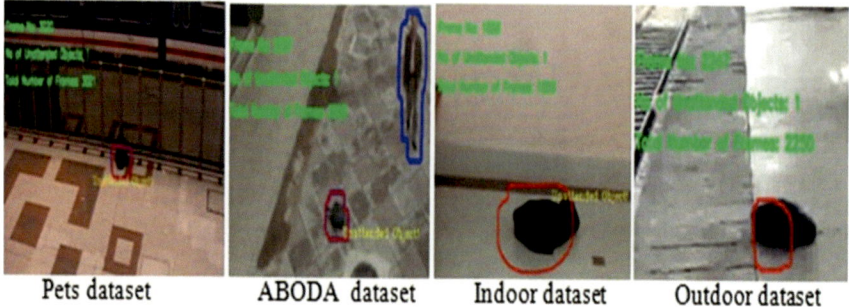

| Pets dataset | ABODA dataset | Indoor dataset | Outdoor dataset |

Fig. 2 The sample results for occasionally moving object with various datasets

| UMN dataset | UMN dataset | Violence dataset | College dataset |

Fig. 3 The sample results for uncommon human activity with various datasets

Table 1 The performance of the system—occasionally moving object detection

Algorithm	Sensitivity	Specificity	False alarm	Accuracy	Misclassify
Existing [19]	0.887	0.92	0.08	0.92	0.083
Proposed	0.89	0.932	0.068	0.95	0.05

Table 2 The performance of the system—uncommon human activity detection

Algorithm	Sensitivity	Specificity	False alarm	Accuracy	Misclassify
Existing [20]	0.904	0.907	0.093	0.933	0.066
Proposed	0.936	0.932	0.068	0.961	0.039

5 Conclusion

The detection of abnormality in the specific crowded areas is a difficult task. The main aim is designing the model to detect the abnormality in the human behavior either static or dynamic and to recognize the crowd event in the video sequence based on features extracted and the movement of motion vector pattern blocks. In this a new

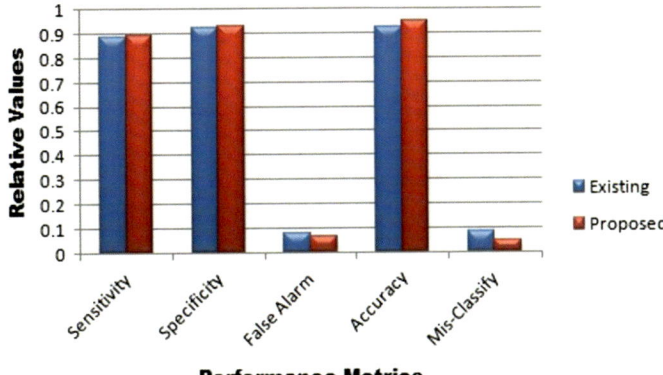

Fig. 4 Performance graphs of occasionally moving object detection

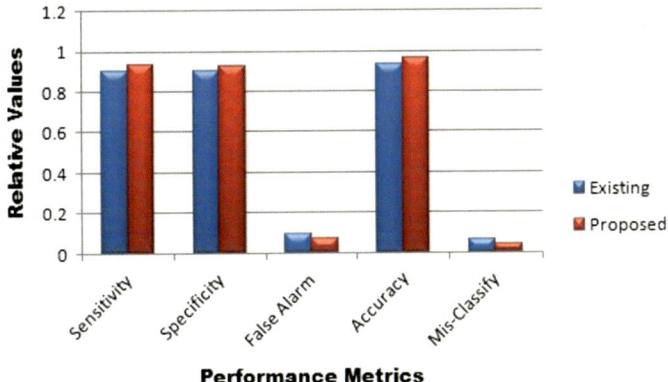

Fig. 5 Performance graphs of uncommon human activity detection

model is designed to identify peculiarity in crowd behavior involves in finding the displacement of motion vectors to form pattern blocks and HOG feature extraction and uses the hybrid clustering approach to construct the optimum number of clusters. It has been tested using standard PETS, ADOBA, and UMN dataset which showed accuracy of 96% in which processing time is lesser than the existing approaches. False alarm and misclassify is minimized. This work also tested with real time video shows it efficiency. In future the work is to extend various methods for different applications and to deal with the different events of multiple behaviors.

References

1. Lin K, Chen S-C, Chen C-S, Lin D-T, Hung Y-P (2015) Abandoned object detection via temporal consistency modeling and back-tracing verification for visual surveillance. IEEE Trans Inf Forensics Secur 10(7):1359–1370
2. Lin C-Y, Muchtar K, Yeh C-H (2014) Grabcut-based abandoned object detection. In: IEEE 16th international workshop on multimedia signal processing (MMSP), pp1–6
3. Grab cut segmentation to extract foreground, http://docs.opencv.org/3.1.0/d8/d83/tutorial_py_grabcut.html
4. Gaussian Mixture Learning Model, http://u.cs.biu.ac.il/~jkeshet/teaching/spr2017/lecture4.pdf
5. Jadhav LH, Momin BF (2016) Detection and identification of unattended/removed objects in video surveillance. In: IEEE international conference on recent trends in electronics, information & communication technology, pp 1770–1773
6. Jaouedi N, Boujnah N, Htıwıch O, Bouhlel MS (2016) Human action recognition to human behavior analysis. In: 7th international conference on sciences of electronics, technologies of information and telecommunications, pp 263–266
7. Kalman Filtering, https://en.wikipedia.org/wiki/Kalman_filter
8. Kumar S, John M (2016) Human activity recognition using optical flow based feature set. In: IEEE international carnahan conference on security technology (ICCST), pp 1–5
9. Optical flow, https://en.wikipedia.org/wiki/Optical_flow
10. Zhu S, Xia L (2015) Human action recognition based on fusion features extraction of adaptive background subtraction and optical flow model. Math Probl Eng, Article id:387464, 2015:1–11 (2015) http://dx.doi.org/10.1155/2015/387464
11. Histogram of oriented gradients feature extraction, http://mccormickml.com/2013/05/09/hog-peron-detector-tutorial/
12. Support Vector Machine (SVM), scikit-learn.org/stable/modules/svm.html
13. Gunnar Farneback Optical Flow, www.diva-portal.org/smash/get/diva2:273847/FULLTEXT01.pdf
14. Hierarchical Clustering, https://en.wikipedia.org/wiki/Hierarchical_Clustering
15. K-Means Clustering, https://en.wikipedia.org/wiki/K-Means_Clustering
16. Pets Dataset www.cvg.reading.ac.uk/PETS2006/data.html
17. ABODA Dataset imp.iis.sinica.edu.tw/ABODA/index.html
18. UMN Dataset http://mha.cs.umn.edu/
19. Lin Y, Tong Y, Cao Y, Zhou Y, Wang S (2017) Visual- attention-based background modelling for detecting infrequently moving objects. IEEE Trans Circuits Syst Video Technol 27(6):1208–1221
20. Lee D-G, Suk H-I, Park S-K, Lee S-W (2015) Motion influence map for unusual human activity detection and localization in crowded scenes. IEEE Trans Circuits Syst Video Technol 25(10):1612–1623

Scaled Conjugate Gradient Algorithm and SVM for Iris Liveness Detection

Manjusha N. Chavan, Prashant P. Patavardhan and Ashwini S. Shinde

Abstract With the extensive use of biometric systems at most of the places for authentication, there are security and privacy issues concerning with it. Most of the public places we usually visit are under reconnaissance and may keep track of face, fingerprints, and iris etc. Thus, biometric information is not secret anymore and sensitive to spoofing attacks. Therefore, not only biometric traits but also liveness detection must be deployed in an authentication mechanism which is a challenging task. Iris is the most accurate trait and increasingly in demand in applications like national security, duplicate-free voter registration list, and Aadhar program its detection must be made robust. Solution to the iris liveness detection by extracting distinctive textural features from genuine (live) and fake (print) patterns using statistical approaches GLCM and GLRLM are implemented. Popular supervised SVM algorithm and PatternNet neural network with second-order scaled conjugate gradient training algorithm are assessed. Both of these algorithms are found to be faster with PatternNet outperforms over SVM.

1 Introduction

With the advancement of technology and machine learning, biometric systems are adopted in various large-scale applications over typical password-based authentication. These systems based on physiological attributes of individual offer sophisticated and reliable authentication far better than systems which need the user to remember or carry some identity [1, 2]. Most accurate biometric trait reported among fingerprint, face, and Iris, is Iris which is increasingly in demand in applications like national security, duplicate-free voter registration list and Aadhar program in various countries as each individual's iris is unique, having complex texture pattern. In

M. N. Chavan (✉) · A. S. Shinde
E & TC Department, ADCET, Ashta, India
e-mail: manju3205@gmail.com

P. P. Patavardhan
E & C Department, GIT Belgaum, Belgaum, India

© Springer Nature Switzerland AG 2019
D. Pandian et al. (eds.), *Proceedings of the International Conference on ISMAC in Computational Vision and Bio-Engineering 2018 (ISMAC-CVB)*, Lecture Notes in Computational Vision and Biomechanics 30,
https://doi.org/10.1007/978-3-030-00665-5_80

spite of these advantages, biometric systems over traditional methods are vulnerable to spoofing attacks which are either direct attacks at sensor level or tempering with database or storage media. Therefore, there is a continuous need for improvement and research going on to deal with spoofing attacks by distinguishing between the acquisition of genuine or fake images at the sensor level, inclusive of some additional response or intrinsic features of a person. Commercial iris technology based biometric authentication systems use algorithms developed by Daugman [3, 4] are not suffice and prone to spoofing attacks like printed iris images, replay attack or use of video, use of glasses or contact lenses. Therefore, there is a need for iris liveness detection to detect any kind of attack.

For liveness detection, many researchers developed new models/algorithms based on pupil dilation and or pupil dynamics, which though reliable require additional hardware, may take longer time for processing and also needs user cooperation [5–7]. Hardware-based techniques may not be suitable for detecting fake attack using unknown material and also if the user is wearing a cosmetic lens. The periodic pattern of dots is observed in case of spoofing attack with printed iris images which have reduced energy contents at higher frequencies [8]. Most of the time printed fake images used for spoofing, exhibits attributes like low quality, poor focus, blur, and poor contrast, unlike generic images which show quality features like sharpness, high contrast. Liveness detection will be carried out by making use of these attributes [9]. In recent years research using micro-textures is gaining attention as biometric traits are rich in textural information and can provide reasonably good results. Distinctive textural features form rich iris patterns like furrows, ridges and pigment spots are extracted to discriminate between live and fake samples [5]. Proposed work is a contribution in finding out a robust solution in analyzing textural pattern.

2 Related Work

Biometric systems used for authentication in security applications need to be reliable, simple and robust. To prevent and detect intruders or attackers in various biometric systems various methods are proposed so far. The multimodal approach provides good reliability and is resilient to specific attacks however it fails to detect vulnerable attacks, which are not considered at design time and may not work well for other unknown materials used for spoofing. Software-based spoofing methods are more flexible and reliable and can provide a generalized solution for spoofing attacks of any type by incorporating some intrinsic characteristics of biometric traits used. Biometric traits are characterized by textural information by extracting features using various descriptors. Many researchers have examined the discriminative power of descriptors on publicly available datasets using linear SVM classifier, which does not require parameter tuning and prior segmentation was not carried out K-means clustering with Euclidean distance for joint quantization of features is implemented and tested for percentage of fake/genuine misclassification. SID is found to be outperforming for both types of attacks like printed iris patterns and of the cosmetic contact lens. SIFT descriptors found to be computation intensive, SID, LCPD also shows moderate computation time simpler descriptors like LBP take much less time and suitable for low power applications like mobile phone authentication [1]. Deep learning techniques are nowadays seeking the attention of researchers, which have shown great success in computer vision tasks can be efficiently used to model the distribution of real/fake classes for liveness detection. Instead of relying on image quality features [9], deep learning framework learns representational and discriminative features. Though results are promising the approach is more computationally expensive than other existing solutions. Iris spoofing is examined for only specific printed and cosmetic contact lens dataset which is available publicly.

Iris textures are stochastic in nature with small repetitiveness. Variety of textures are available so a single operator is not adequate to describe. The operator should have qualities like efficient discrimination of different types of textures, robust to pose and scale variation and to spatial nonuniformity. Also, it should work well for small sample size and should have low computational complexity. The texture-based approaches reported to perform well for constrained databases with partial occlusion, uniform illumination, and high-quality input, therefore, there is need of feature extractor invariant to transformation and occlusions. Use of key point descriptors like SIFT and SURF improve the accuracy of iris recognition, it is found that SURF requires less time for identification compared to SIFT. It can be used for highly trust-based applications so useful for security [10]. For irises which textures are not distinct due to defocusing, blurring or occlusion by extracting local texture features and use of K-means clustering assessment time and hence system performance can be improved [11]. Using a spatial pyramid model and relational measure among neighboring regions, features are extracted in regions of different shapes which achieve improved performance for iris liveness detection for benchmark datasets [12].

3   Overview of Iris Recognition System and Feature Extraction

Typical iris recognition system as shown in Fig. 1 [3, 4]. Image acquisition is mostly carried out under constrained environment wherein eye image is captured by cameras under visible or near-infrared illumination. Captured irises are preprocessed by removing noise and reflections with the use of suitable filtering mostly median filtering and enhanced for further processing. Iris segmentation is the most important step, which is to approximately localize boundaries between iris-sclera and iris-pupil region. Eyelashes and eyelids are also localized and removed as they do not contribute any significant information in recognition. Poor segmentation would result in insufficient textural information which may lead to increased rejection rates and hence affects iris recognition accuracy. Normalization is carried out after segmentation to compensate for deformations like varying iris sizes, pupil dilation, and other factors. In the normalization process, iris rings are unwrapped into a rectangular

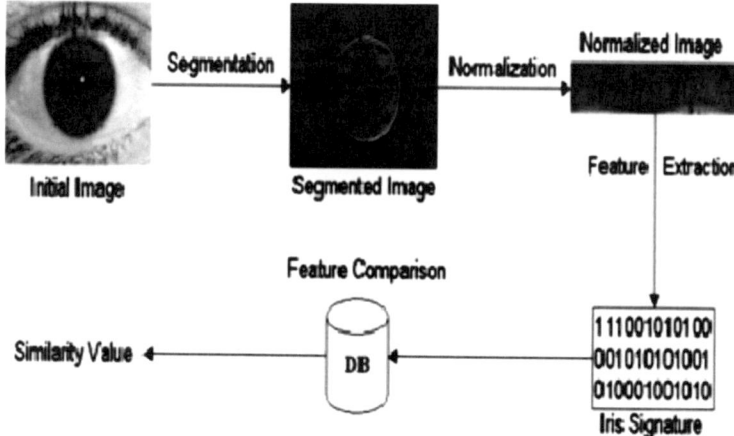

Fig. 1 Typical iris recognition system

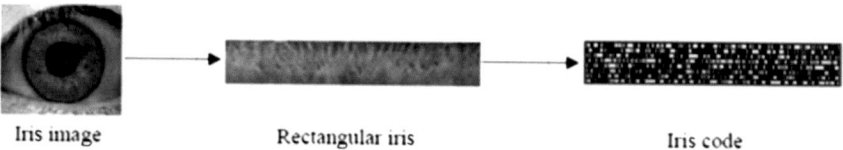

Fig. 2 Iris normalization using Daugman's rubber sheet model and generation of iris code from features

textural block of fixed size (20 × 240). Now after normalizing iris is ready for feature extraction [3, 4, 10] (Fig. 2).

Textural features of normalized iris are extracted which are unique for each individual. Extracted features are then encoded and iris code is generated for each registered iris and stored in the dataset. At verification and or matching stage enquired iris is processed through all these stages and code is compared against each iris code in the dataset and the user is either accepted as an authenticated user or rejected by the system.

In this paper, popular circular Hough transform is implemented as a segmentation algorithm and statistical textural methods gray-level co-occurrence matrix and gray-level run length matrix are used for feature extraction from normalized iris.

3.1 Proposed Work

Warsaw dataset consisting of eye images captured in NIR illumination. Fake images are created using laser printers with both low and high resolution on matt papers. Images which successfully spoofed the commercial iris recognition system are

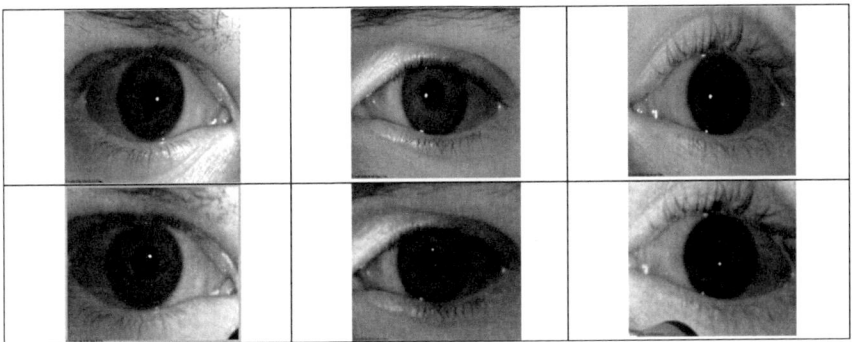

Fig. 3 Top row—Sample live iris images, bottom row—Sample print iris images

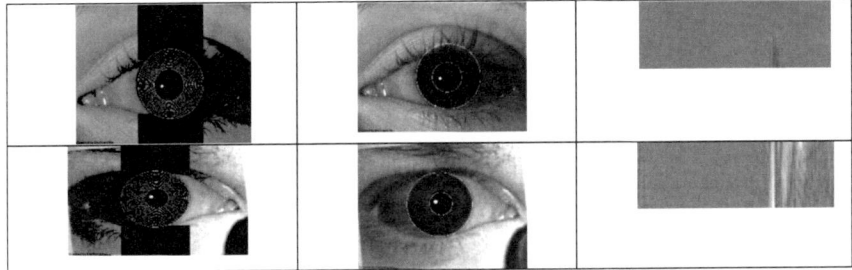

Fig. 4 Top row—Iris and pupil localization, segmentation and normalization for live iris Bottom row—Iris and pupil localization, segmentation, and normalization for print iris

preserved in the database. [12] Real and fake irises may reveal different optical characteristics under near-infrared lighting. These optical characteristics are helpful for iris liveness detection, which includes frequency distribution, Purkinje images, image quality features, and statistical texture features [13]. An optimization approach used in scaled conjugate gradient makes it effective over others as it does not contain any user-dependent parameters. It also avoids line search per learning iteration leading to the fastest algorithm among other gradient-based algorithms [14]. SVM is one of the most popular supervised machine learning algorithms used to analyze data and recognize patterns, used for classification with high performance and requires less tuning [5, 15, 16]. In view of these advantages of classification algorithms, SVM and scaled conjugate gradient (SCG) are chosen to test performance.

The proposed work is, therefore, to make use of statistical texture features like GLCM and GLRLM for liveness detection for Warsaw images. From normalized iris, 76 and 44 features are extracted using GLCM and GLRLM, respectively. 40 Real (Live) and 40 Print (Fake) iris images are tested using SVM and SCG algorithm for classifying between fake and genuine irises. Classification accuracy and confusion matrix for both classifiers are given in Results and discussion (Figs. 3 and 4).

3.1.1 Gray-Level Co-occurrence Matrix (GLCM)

Second-order texture information can be extracted using GLCM [5, 13]. It is a matrix whose rows and columns represent a number of distinct gray levels or pixel values in the image. It calculates how often a pixel with gray level (Intensity or tone) value i occurs either horizontally, vertically or diagonally to adjacent pixel j in the region of interest. GLCM represents the distance and angular spatial relationship over an image sub-region of a specific region of a specific size are transformed into the co-occurrence matrix space by a given kernel mask such as 3 3, 5 5, 7 7, and so forth. Normally, it is analyzed in four directions (e.g., Horizontal ($0°$), Vertical ($90°$), Diagonal ($45°$ and $135°$). It is denoted as pairs $P(i, j)$, P_0, P_{45}, P_{90}, and P_{135}, respectively.

In 1973, Haralick et al. defined a set of 14 textural features which can be extracted from the co-occurrence matrix, and which contain information about image textural characteristics such as Angular Second Moment, Energy, Correlation, entropy, Inverse Difference Moment (IDM), Sum Average, Sum Variance, Sum Entropy, Entropy, Difference Variance, Difference Entropy, Difference average, information measure of correlation 1 and 2 [13, 14].

Most significant and relevant parameters used for textural features are energy, entropy, contrast, IDM, and DM defined below

Energy:

$$E = \sum_{i=0}^{G-1} [P(i)]^2 \tag{1}$$

Entropy:

$$H = -\sum_{i=0}^{G-1} P(i) \log_2 P(i) \tag{2}$$

Contrast:

$$\text{CTR} = \sum_{n=0}^{G-1} n^2 \left\{ \sum_{i=1}^{G} \sum_{j=1}^{G} P(i, j) \right\}; \quad |i - j| = n \tag{3}$$

Inverse Difference Moment:

$$\text{IDM} = \sum_{i=0}^{G-1} \sum_{j=0}^{G-1} \frac{1}{1 + (i - j)^2} P(i, j) \tag{4}$$

Directional Moment:

$$DM = \sum_{i=1}^{G} \sum_{j=1}^{G} M(i, j)|i = j| \tag{5}$$

3.1.2 Gray-Level Run Length Matrix Features

Higher order statistical texture information can be computed using gray-level run length method. A set of consecutive pixels with the same gray level is called a gray level run. Gray level run length matrix describes the total number of occurrences of run length j at gray level i in a given direction θ. In order to extract texture features, GLRLM can be computed for any direction. Mostly five features are derived from the GLRLM introduced by Galloway. These features are: Short Runs Emphasis (SRE), Long Runs Emphasis (LRE), Gray-Level Non-Uniformity (GLNU), Run Length Non-Uniformity (RLNU), and Run Percentage (RP) [13]. Some of the features are described here.

Short Run Emphasis (SRE) is evaluated as

$$SRE = \frac{1}{n_r} \sum_{i=1}^{M} \sum_{j=1}^{N} \frac{P(i, j)}{j^2} \tag{6}$$

where n_r is the total number of runs, M the number of gray levels, and N the maximum run length.

Long Run Emphasis (LRE):

$$LRE = \frac{1}{n_r} \sum_{i=1}^{M} \sum_{j=1}^{N} P(i, j) \cdot j^2 \tag{7}$$

Gray-Level Non-Uniformity (GLN):

$$GLN = \frac{1}{n_r} \sum_{i=1}^{M} \left(\sum_{j=1}^{N} P(i, j) \right)^2 \tag{8}$$

Run Length Non-Uniformity (RLN):

$$RLN = \frac{1}{n_r} \sum_{j=1}^{N} \left(\sum_{i=1}^{M} P(i, j) \right)^2 \tag{9}$$

Run Percentage (RP):

$$RP = \frac{n_r}{n_p}, \tag{10}$$

where n_p is the total number of pixels in the image.

3.2   Classification Using Scaled Conjugate Gradient (SCG) Algorithm and SVM

Feature set for all iris images under test is used to train Pattern net neural network with 50 neurons in hidden layer and SCG back propagation as training algorithm, which updates weight and bias values according to the SCG method. Basic backpropagation algorithm adjusts the weights in the steepest descent direction in which the performance is decreasing most rapidly along the negative gradient but does not necessarily produce the fastest convergence [14]. In the conjugate gradient algorithms, a search is performed along conjugate directions, which produces generally faster convergence than steepest descent directions. Maximum epoch used to train the network is 1000. Performance of the network for a given set of features GLCM, GLRLM, and combined feature set is tested with 10, 15, and 20 fold cross-validation. ROC curves and confusion matrix elaborates the overall performance of classification.

In SVM, the basic idea is to select hyperplane which is a line which best separates the input data points in space by their class as wide as possible. For 2 class problem, class 0 and class 1 can be completely separated by the line. The distance between the line and closest points is referred to as margin which is calculated as the perpendicular distance between the two [5]. These points are called support vectors and play an important role in defining the line and hence classifier. The hyperplane is learned from training data using an optimization procedure that maximizes the margin. In practice, real data cannot be perfectly separated with hyperplane and hence need some relaxation, which is called soft margin classifier. A tuning parameter C is introduced that defines the amount of violation of the margin allowed. C influences the number of support vectors used, smaller the value of C, the more sensitive the algorithm to the training data and vice versa. SVM algorithm is implemented using different kernels such as linear, polynomial, and radial. RBF kernel is found to be outperforming among all kernels. Performance of SVM classifier on feature set is tested with 10-, 15-, and 20-fold cross-validation. Overall classification accuracy is evaluated.

4   Results and Discussion

For 80 normalized iris images, 40 each class, 76 and 44 features are extracted in four different orientations using GLCM and GLRLM respectively. Combined GLCM–GLRLM features are a concatenation of the two individual features vectors. These features are used to train SCG and SVM classifiers with 10-, 15-, and 20-fold cross-validation. Experimental results given in the table describes confusion matrix of different K-fold cross-validation in terms of True positive, False positive, True Negative, and False Negative. From the table, it is evident that for GLCM overall individual class accuracy is better for 10-fold whereas for GLRLM it increases and maximum for 20-fold. The same is true for the combined performance of these two. SVM performs better for GLCM than the other two. For combined features, it performs worst. SCG shows poor performance for combined feature vector with tenfold. The overall accuracy of SCG is better than SVM.

5 Conclusion

From experimentation, it can be concluded that overall accuracy for GLCM is better. SCG is able to outperform for iris liveness detection task than SVM. By increasing, data samples and features which distinguish between fake and genuine irises must be smartly chosen to improve performance. The overall accuracy of classification can be increased with accurate iris localization, preprocessing and feature selection. As some of the images in Warsaw database are of low resolution, it can be the cause of detection and hence classification performance which range from 60–90%. Future work will be in this direction also some feature reduction techniques will be employed in Table 1.

Table 1 Classification accuracy results of SCG and SVM

Total no. of iris images = 80, 40 live and 40 print images						Overall accuracy (%)	
SCG				SVM		SCG	SVM
	$T_P = 35$	$F_P = 12$	$T_P = 29$	$F_P = 15$		78.8	72.5
	$F_N = 5$	$T_N = 28$	$F_N = 11$	$T_N = 25$			
	Sensitivity = 87.5	Specificity = 70	Sensitivity = 72.5	Specificity = 62.5			
	$T_P = 29$	$F_P = 12$	$T_P = 29$	$F_P = 12$		71.3	76.25
	$F_N = 11$	$T_N = 28$	$F_N = 11$	$T_N = 28$			
	Sensitivity = 72.5	Specificity = 70	Sensitivity = 72.5	Specificity = 70			

(continued)

Table 1 (continued)

Total no. of iris images = 80, 40 live and 40 print images					Overall accuracy (%)	
AR ROC — 20 fold	$T_P = 32$	$F_P = 14$	$T_P = 30$	$F_P = 15$	72.5	68.75
	$F_N = 8$	$T_N = 26$	$F_N = 10$	$T_N = 25$		
	Sensitivity = 80	Specificity = 65	Sensitivity = 75	Specificity = 62.5		
GLRLM — AR ROC — 10 fold	$T_P = 30$	$F_P = 16$	$T_P = 14$	$F_P = 6$	67.5	63.75
	$F_N = 10$	$T_N = 24$	$F_N = 26$	$T_N = 34$		
	Sensitivity = 75	Specificity = 60	Sensitivity = 65	Specificity = 85		
AR ROC — 15 fold	$T_P = 35$	$F_P = 8$	$T_P = 16$	$F_P = 6$	83.8	61.25
	$F_N = 5$	$T_N = 32$	$F_N = 24$	$T_N = 34$		
	Sensitivity = 87.5	Specificity = 80	Sensitivity = 60	Specificity = 85		
AR ROC — 20 fold	$T_P = 36$	$F_P = 4$	$T_P = 18$	$F_P = 7$	90	60
	$F_N = 4$	$T_N = 36$	$F_N = 22$	$T_N = 33$		
	Sensitivity = 90	Specificity = 90	Sensitivity = 55	Specificity = 82.5		
GLCM+GLRLM — AR ROC — 10 fold	$T_P = 9$	$F_P = 4$	$T_P = 29$	$F_P = 19$	56.3	65
	$F_N = 31$	$T_N = 36$	$F_N = 11$	$T_N = 21$		
	Sensitivity = 22.5	Specificity = 90	Sensitivity = 72.5	Specificity = 52.5		
AR ROC — 15 fold	$T_P = 33$	$F_P = 10$	$T_P = 32$	$F_P = 19$	78.8	58.75
	$F_N = 7$	$T_N = 30$	$F_N = 8$	$T_N = 21$		
	Sensitivity = 82.5	Specificity = 75	Sensitivity = 80	Specificity = 52.5		
AR ROC — 20 fold	$T_P = 37$	$F_P = 1$	$T_P = 31$	$F_P = 20$	95	66.25
	$F_N = 3$	$T_N = 39$	$F_N = 9$	$T_N = 20$		
	Sensitivity = 92.5	Specificity = 97.5	Sensitivity = 77.5	Specificity = 50		

References

1. Gragnaniello D, Poggi G, Verdoliva L (2015) An investigation of local descriptors for biometric spoofing detection. IEEE Trans Inf Forensics Secur 10(4):849–863
2. Unar JA, Seng WC, Abbasi A (2014) A review of biometric technology along with trends and prospects. Pattern Recognit 47(8):2673–2688
3. Daugman John (1993) High confidence visual recognition of persons by a test of statistical independence. IEEE Trans Pattern Anal Mach Intell 15(11):1148–1161
4. Daugman John (2004) How iris recognition works. IEEE Trans Circuits Syst Video Technol 14(1):21–30
5. He X, An S, Shi P Statistical texture analysis-based approach for fake iris detection using support vector machines. In: Proceedings of international conference on biometrics, pp 540–546
6. Lee E, Park K, Kim J (2005) Fake iris detection by using purkinje image. In: Advances in biometrics, ser. Lecture Notes in Computer science. Springer, vol 3832, pp 397–403
7. Kanematsu M, Takano H, Nakamura K (2007) Highly reliable liveness detection method for iris recognition. In: Annual conference SICE, pp 361–364
8. Jain AK, Bolle R, Pankanti S (1999) Personal identification in a networked society. Springer, New York, USA, pp 103–121
9. Galbally J, Ortiz-Lopez J, Fierrez J, Ortega-Garcia J (2012) Iris liveness detection based on quality related features. In: Proceedings of the 5th IAPR international conference on biometrics (ICB), pp 271–276
10. Mehrotra H et al (2012) Fast segmentation and adaptive SURF descriptor for iris recognition. Mathematical and computer modeling. Elsevier. https://doi.org/10.1016/j.mcm.2012.06.034
11. He Y et al (2012) Feature extraction of iris based on texture analysis. Advances in FCCS, vol 1, AISC 159. Springer, Berlin, Heidelberg, pp 541–546
12. Hu Y et al (2015) Iris liveness detection using regional features. Pattern Recogn Lett. https://doi.org/10.1016/j.patrec.2015.10.010
13. Sun Z, Tan T (2014) Iris Anti-spoofing. In: Marcel S et al (eds) Handbook of biometric anti-spoofing, advances in computer vision and pattern recognition. Springer, London. https://doi.org/10.1007/978-1-4471-6524-8_6
14. Haralick RM, Shanmugam K, Dijnstein I (1973) Textural features for image classification. IEEE Transactions On Systems, Man, And Cybernetics, November 1973
15. Huang X, Ti C, Hou Q-Z, Tokuta A, Yang R (2013) An experimental study of pupil constriction for liveness detection. In: Proceeding of IEEE workshop applications of computer vision (WACV), pp 252–258
16. He X, Lu Y, Shi P (2009) A new fake iris detection method. In: Proceeding 3rd international conference on advanced biometrics, pp 1132–1139
17. Pacut A, Czajka A (2006) Aliveness detection for iris biometrics. In: Proceedings of 40th annual IEEE international carnahan conference security technology, pp 122–129

Long Short-Term Memory-Based Recurrent Neural Network Approach for Intrusion Detection

Nishanth Rajkumar, Austen D'Souza, Sagaya Alex and G. Jaspher W. Kathrine

Abstract Intrusion detection is very essential in the field of information security. The cornerstone of an Intrusion Detection System (IDS) is to accurately identify different attacks in a network. In this paper, a deep learning system to detect intrusions is proposed. The existing recurrent neural network (RNN-IDS) based IDS is expanded to include Long Short term memory (LSTM) and the results are compared. The binary classification performance of the RNN-IDS is tested with various learning rates and using different number of hidden nodes. The results show that by integrating LSTM with RNN-IDS, the accuracy of intrusion prediction has improved against the benchmark dataset.

1 Introduction

Intrusion detection System (IDS) are hardware or software systems which can provide the ability to monitor networks for any type of unusual activity of intrusions such as misuse of access, unlawful access or login, hacking, etc. A very capable IDS must be able to identify and distinguish internal and external attacks. On detection of an intrusion an alarm is generated to alert the administrators or any software trigger can be used to block the further attacks. But even though intrusion systems of high detection have been deduced, the increased number of false positive attacks makes the development of any IDS technology unreliable.

N. Rajkumar · A. D'Souza · S. Alex · G. J. W. Kathrine (✉)
Karunya Institute of Technology and Sciences, Coimbatore, India
e-mail: kathrine@karunya.edu

N. Rajkumar
e-mail: rajkumar.nishanth@gmail.com

A. D'Souza
e-mail: austen.dsouza@gmail.com

S. Alex
e-mail: sagayaalex25@gmail.com

© Springer Nature Switzerland AG 2019
D. Pandian et al. (eds.), *Proceedings of the International Conference on ISMAC in Computational Vision and Bio-Engineering 2018 (ISMAC-CVB)*, Lecture Notes in Computational Vision and Biomechanics 30,
https://doi.org/10.1007/978-3-030-00665-5_81

With the pervasive use of the Internet, more and more people are able to connect with each other through various means. This increase in connectivity means that there is more room for attackers to infiltrate a network that would have been secure a few years ago. The problem with identifying an attack is that we use known models to detect threats. However, the real difficulty lies in detecting an attack that is new, i.e., an attack that does not follow any well-defined pattern. This is where an IDS comes into play. An IDS can identify a security breach, that has occurred in the past, is currently ongoing or future attacks when they occur. This is done by classifying network traffic into the categories of "normal" and "anomalous". Network traffic is classified in two ways, i.e., the binary way as stated above or the five-category classification, i.e., Normal, Denial of Service (DoS), User to Root (U2R), Probe and Root to Local (R2L). The main objective of this work is to improve the correctness of the classification to effectively identify intrusive behaviour.

Techniques based on Machine Learning have been used to implement IDS but the inherent requirement of Machine Learning algorithms is that they need feature selection and belong to the shallow learning school. The notable downside of this is that these algorithms are unable to scale to the massive datasets that we would deal with in real-world scenarios. Moreover, they are not equipped for intelligent analysis and forecasting. Deep learning algorithms have the much better probability to generate models and extract much better impressions from the data.

This paper focuses on the following:

1. Implementation of RNN-IDS using LSTM. Focus is given on binary classification and the varying performance of the network based on the number of neurons and the learning rate.
2. The results are compared with RNN-IDs that does not use LSTM on the benchmark NSL-KDD dataset.

This paper aims to propose an effective intrusion detection system for networks and also to identify the framework to prevent the further intrusions in the networks. Section 2 contains the literature survey. Section 3 gives the framework for the proposed intrusion detection system.

2   Related Work

In current times, data mining and other data analysis based techniques are being used for analysing, identifying, and detecting intrusions. Nadiammai et al. [1] has proposed an IDS to detect the relevant and hidden data using data mining techniques which has led to less execution time. The proposed algorithm solves the major four issues such as data classification, need for decision-making by humans, short of data that is ready and labelled, and distributed denial of service (DoS) attack. Sabri et al. [2] have proposed an intrusion detection system which detects denial of service attack using techniques pertaining to data mining which includes clustering and classification. Dewa et al. [3] have proposed an intrusion detection system based on data mining which has achieved higher accuracy to new types of intrusion and a more robust performance is revealed when compared to traditional IDSs. Patel et al. [4] has proposed a paper where an amalgamated model is proposed to maximize the effectiveness in identifying the attacks that integrates the anomaly-based intrusion detection technique. Miller et al. [5] has done a research in which the IDS based on anomaly is used because it is more theoretically accurate in detecting the previously undocumented threats.

In recent years, deep learning is gaining more prevalence in identifying and mitigation intrusions in networks. Javaid et al. [6] have proposed a study in NIDS which helps to detect the network security breaches using deep learning approach. Chuanlong et al. [7] have proposed a paper for IDS based on recurrent neural network using deep learning approach. Shone et al. [8] has proposed a deep learning approach to intrusion detection in networks. Anomaly-based intrusion system assists in identifying both identified and unidentified intrusions. Jabez et al. [9] has proposed outlier detection for measuring anomaly in the dataset by using the neighbourhood outlier factor. Viegas et al. [10] has proposed a new evaluation scheme specific to the machine learning ID field. Umer et al. [11] have proposed taxonomy of flow-based IDS on the foundation of detecting maliciousness in flow record. Flow-based IDS is an inventive way of detecting intrusions in high speed networks.

Deep et al. [12] has proposed an intelligent system which will perform feature ranking on the origin of information gain and correlation. Dias et al. [13] has proposed a research paper for an IDS based on artificial neural network. Wang et al. [14] has proposed an effective IDS framework based on a support vector machine (SVM). Jha and Ragha [15] have proposed two ideas. First, they did the review in current statics of intrusion detection using support vector machine, and second, they proposed a novel approach to select the good feature for detecting intrusions. Abd-Eldayem [16] proposed an intrusion detection system which is based on Naïve Bayes classifier.

3   Proposed Methodology

On the basis of the literature survey it was decided to implement an IDS using Neural Networks. The neural network used was Recurrent Neural Network (RNN). The proposed framework for the IDS using Deep Learning is shown in Fig. 1. In addition to how a traditional RNN works, the model also implements Long Short-Term Memory (LSTM) to improve accuracy.

Long Short-Term Memory (LSTM) is the basic units required when constructing an RNN. Figure 1 shows the representation of LSTM. The most common construction of an LSTM cell consists of a cell, an input gate, an output gate and a forget gate. Each gate can be likened to an artificial neuron because they compute an activation of a weighted sum. The cell and the gates are connected. The cell is accountable for the "memory" of the LSTM. LSTM's are short term memory that can last over a long period of time.

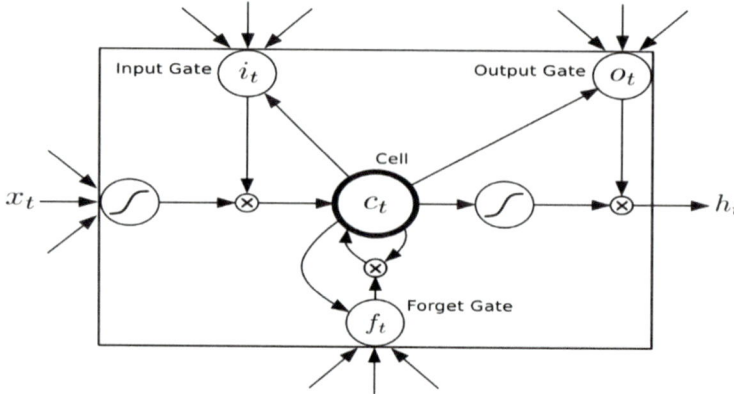

Fig. 1 Representation of LSTM

There is a debate existing over whether using LSTM improves the overall performance of a neural network. The consensus is that it works on a case by case basis. In this paper, an analysis is done to discover whether the use of LSTM is justified by augmenting its performance with the performance of the RNN. The Fig. 2 shows the architecture of the proposed and implemented Recurrent Neural Network (RNN) based IDS using Long Short-Term Memory (LSTM).

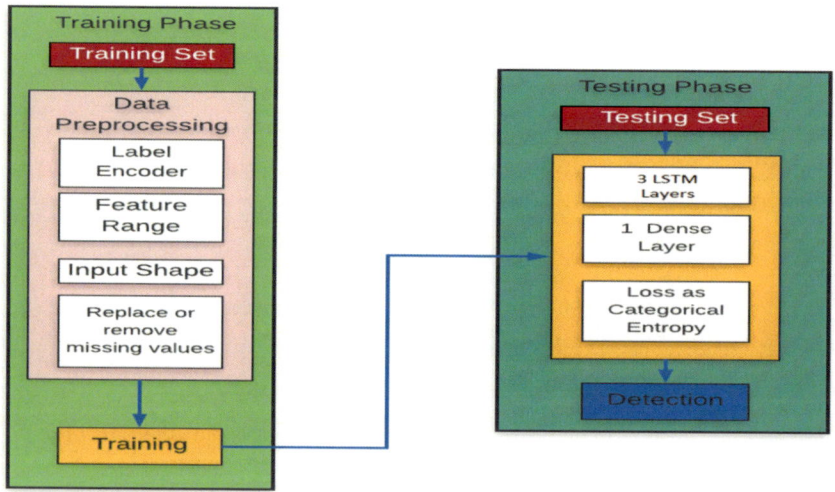

Fig. 2 Architecture of the proposed RNN-IDS using LSTM

A. Dataset

The NSL-KDD dataset was generated in the year 2009. This dataset is used in a wide range of IDS experiments and can be considered the benchmark dataset for these experiments. The NSL-KDD dataset solves the problem of the KD Cup (1999) dataset, wherein there existed many records that were inherently redundant.

Another advantage of the NSL-KDD dataset is that the number of records in the training and testing sets is better suited for the purpose of IDS testing. The KDDTest-21 dataset is a subset of the KDDTest+ dataset and possesses a greater level of difficulty for classification. The dataset contains four types of attack records. The types of attacks are Dos (Denial of Service), U2R (User to Root), R2L (Root to Local) and Probe. Table 1 shows the composition of the dataset.

Every record (traffic record) is made up of 41 features and 1 class label. The features are further categorized as Basic (1–10), Content (11–22) and Traffic (23–41); numbered according to Table 1. Another unique feature of the dataset is that the testing set has some specific types of attacks that are not present in the training set. This acts like a more realistic basis for intrusion detection development and testing.

Table 1 Selected feature of the NSL-KDD dataset

No.	Features	Types	No.	Features	Types
1	Duration	Continuous	22	is_guest_login	Symbolic
2	protocol_type	Symbolic	23	Count	Continuous
3	Service	Symbolic	24	srv_count	Continuous
4	Flag	Symbolic	25	serror_rate	Conti nuous
5	src_bytes	Continuous	26	srv_serror_rate	Continuous
6	dst_bytes	Continuous	27	rerror_rate	Continuous
7	Land	Symbolic	28	srv_rerror_rate	Continuous
8	wrong_fragment	Continuous	29	same_srv_rate	Continuous
9	Urgent	Continuous	30	diff_srv_rate	Continuous
10	Hot	Continuous	31	srv_diff_host_rate	Continuous
11	num_failed_logins	Continuous	32	dst_host_count	Continuous
12	logged_in	Symbolic	33	dst_host_srv_count	Continuous
13	num_compromised	Continuous	34	dst_host_same_srv_rate	Continuous
14	root_shell	Continuous	35	dst_host_diff_srv_rate	Continuous
15	su_attempted	Continuous	36	dst_host_same_src_port_ra	Continuous
16	num_root	Continuous	37	dst_host_srv_diff_host_rat	Conti nuous
17	num_file_creations	Continuous	38	dst_host_serror_rate	Continuous
18	num_shells	Continuous	39	dst_host_srv_serror_rate	Conti nuous
19	num_access_files	Continuous	40	dst_host_rerror_rate	Continuous
20	num_outbound_cmds	Continuous	41	dst_host_srv_rerror_rate	Continuous
21	is_host_login	Symbolic			

B. Data Preprocessing

Label Encoding
Label Encoding is done to normalize the tags such that they contain only values between 0 and n − 1 classes. Label encoding is a method of normalizing the data to assist in easier classification. The label encoder also helps in numericalization of the data. Some features are non-numerical in nature and therefore need to be converted into numerical values based on the number of values they take.

Normalization
Certain features such as duration and src_bytes have a huge difference between the maximum and minimum values. So we use the Min Max scaler to normalize the dataset, where min and max are maximum and minimum values of each feature respectively.

The formula is

$$x_i = \frac{x_i - \text{Min}}{\text{Max} - \text{Min}} \tag{1}$$

C. Evaluation metrics

An important indicator of performance is "acc" or accuracy of the lstm model on each dataset. The accuracy graph that is plotted is the accuracy of the model on each dataset for every epoch. The x axis is the epoch ranging from 1 to 100 and y axis is the accuracy ranging from 0.00 to 1.00. Another important indicator is the loss graph or "loss". Loss reviews the accuracy of a model by penalizing false classifications. The loss function that we used in this model was "mean squared error" or "MSE" since it is a classification model. MSE is a straight line between two points in Euclidean space. For classification models it is best to use mean squared error or cross entropy.

$$\text{MSE formula:} \quad \frac{1}{n} \sum_{t=1}^{n} e_t^2 \tag{2}$$

When the LSTM model is trained on a dataset it returns a variable called history. The variable history contains various metrics stated while building the model. The history variable records the loss and accuracy for each epoch and graphs are plotted based on this. We can infer three different possibilities from the loss graph. They are underfit, overfit, and goodfit models. Underfit model is when it performs well on the train dataset and poorly in the test dataset. Overfit is when it performs well on training set whereas in test set it improves up to a point and then degrades. Goodfit is when the model performs well on the train and test set.

Fig. 3 Loss graph of the
RNN-IDS on the dataset

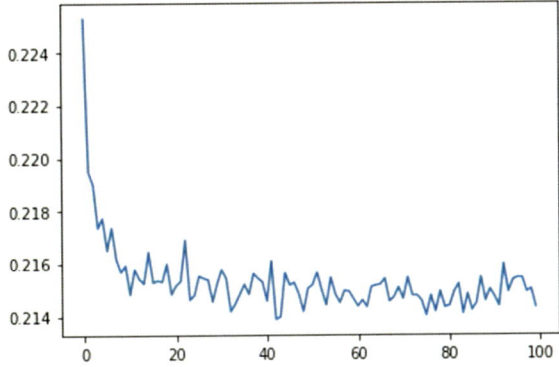

4 Results and Discussion

In our research Keras deep learning framework based on TensorFlow is used. Binary classification is used and compared with the accuracy of LSTM Recurrent Neural Network with the fully connected Recurrent Neural Network model. The experiment was carried out on a Dell Inspiron 7559, which had the following configuration: Intel Core i7 CPU @ 2.6 GHz, 8 GB of memory and uses GPU acceleration. The GPU unit is an Nvidia GeForce 960 M GTX @ 1096 MHz.

The LSTM-IDS model we used has input shape of three dimensions (batch size, timesteps, features). In our model the input shape was (50, 10, 41). The entire dataset is reshaped to this 3D input since LSTM accepts only three-dimensional input. Since it is a binary classification we have two output nodes. The number of epochs used is 100. We used various combinations of hidden nodes and learning rate. No of hidden nodes were 20, 60, 80, 100, and 120. The learning rates used were 0.01, 0.1, and 0.5. The observations for the classification accuracy are shown in Table 2. In our experiment the RNN performs best at 60 hidden nodes and the learning rate at 0.1 as seen in Fig. 4. The experiments show that we obtain 74.33% for KDDTrain+ dataset, 74.99% for KDDTest+ dataset and 75.00% for Test-21 dataset and the time taken is 3258 s. Although the 100 hidden nodes have a higher accuracy at 0.1 learning rate, the accuracy is inconsistent and the loss graph is poor for this case. Hence 100 hidden nodes at 0.1 learning rate is not considered the best performance in our research experiment. Whereas the loss graph as shown in Fig. 3 for the case of 60 hidden nodes depicts that it is continuously decreasing showing that there is very little overfitting compared to 100 hidden nodes.

In Yin Chuan-long et al., the authors have shown the results obtained by fully connected RNN model without the use of GPU on binary and multiclass classification. The results gathered are compared with that of the binary classification. The model gives an accuracy of 68.55% for binary classification on the test-21 dataset. These results are based on the same NSL-KDD dataset. The performance of the LSTM-IDS is better in Test^{-21} dataset and the time taken is also much less in binary classification.

Table 2 Accuracy and time (s) of the RNN-IDS using LSTM

Number of hidden neurons	Learning rate	KDD train+ (%)	KDD test+ (%)	KDD test^{-21} (%)	Time (s)
20	0.01	65.66	65.15	64.25	3065
	0.1	67.51	63.19	63.55	3069
	0.5	34.94	34.89	34.78	3068
60	0.01	66.64	67.90	67.30	3253
	0.1	74.33	74.99	75.00	3258
	0.5	67.11	67.22	67.18	3280
80	0.01	66.81	67.31	66.88	3427
	0.1	51.01	51.03	50.90	3435
	0.5	34.93	34.83	34.84	3420
100	0.01	66.88	67.80	67.38	3530
	0.1	75.27	75.28	75.28	3541
	0.5	34.92	3482	34.82	3550
120	0.01	66.66	66.43	66.35	3857
	0.1	34.92	34.83	34.85	3852
	0.5	34.91	34.82	34.84	3865

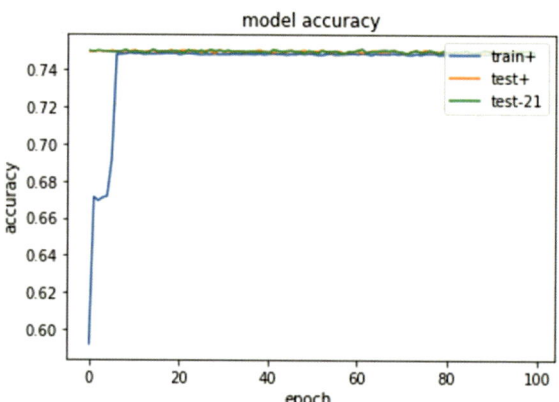

Fig. 4 Accuracy graph of the RNN-IDS on the dataset

In the observations from our research the results show that the RNN-IDS using LSTM model has a higher accuracy on the testing set than the training set as compared to RNN-IDS without LSTM. Moreover, it is seen to be consistently faster than a system that does not use LSTM. It also solves vanishing gradient problem that occurs in non LSTM models.

5 Conclusion

The RNN-IDS model using LSTM has a strong case to be used in intrusion detection, especially in binary classification. Compared to a non-LSTM-based RNN-IDS the results on the $Test^{-21}$ set of the NSL-KDD dataset is very high. This model improves the ability to detect whether an intrusion is taking place or not. In future research, time optimization to better fit real time scenarios will be undertaken. Analysis to study the effect of different combinations of timesteps on LSTM model's accuracy and time taken have to be done to enhance the optimization.

References

1. Nadiammai GV, Hemalatha M (2014) Effective approach toward intrusion detection system using data mining techniques. Egypt Inf J 15(1):37–50
2. Sabri FNM, Norita NM, Seman K (2011) Identifying false alarm rates for intrusion detection system with data mining. Int J Comput Sci Netw Secur 11:4
3. Dewa Z, Maglaras LA (2016) Data mining and intrusion detection systems data mining and intrusion detection systems. (IJACSA) Int J Adv Comput Sci Appl 7(1)
4. Patel J, Panchal K (2015) Effective intrusion detection system using data mining technique. IJSRD—Int J Sci Res Dev 3(02)
5. Miller Z, Deitrick W, Wei H (2011) Anomalous network packet detection using data stream mining. J Inf Secur 2:158–168
6. Javaid A, Niyaz Q, Sun W, Alam M (2016) A deep learning approach for network intrusion detection system. In: Proceedings of the 9th EAI international conference on bio-inspired information and communications technologies (formerly BIONETICS), pp 21–26
7. Chuan-long Y, Yue-fei Z, Jin-long F, Xin-zheng H (2017) A deep learning approach for intrusion detection using recurrent neural networks. IEEE Access, https://doi.org/10.1109/access.2017.2762418, pp 21954–21961
8. Shone N, Ngoc TN, Phai VD, Shi Q (2017) A deep learning approach to network intrusion detection. IEEE Trans Emerg Top Comput Intell 2(1):41–50
9. Jabez J, Muthukumar B (2015) Intrusion detection system (IDS): anomaly detection using outlier detection approach. In: International conference on intelligent computing, communication & convergence (ICCC-2015), Procedia Computer Science, vol 48, pp 338–346
10. Viegas EK, Santin AO, Oliveira LS (2017) Toward a reliable anomaly-based intrusion detection in real-world environments. Comput Netw 127:200–216
11. Umer MF, Sher M, Bi Y (2017) Flow-based intrusion detection: techniques and challenges. Comput Secur 70:238–254
12. Akashdeep IM, Kumar N (2017) A feature reduced intrusion detection system using ANN classifier. Expert Syst Appl 88:249–257
13. Dias LP, Cerqueira JJF, Assis KDR, Almeida RC (2017) Using artificial neural network in intrusion detection systems to computer networks. In: Proceedings of computer science and electronic engineering (CEEC)
14. Wang H, Jie G, Wang S (2017) An effective intrusion detection framework based on SVM with feature augmentation. Knowl-Based Syst 136:130–139

15. Jha J, Ragha L (2013) Intrusion detection system using support vector machine. Int J Appl Inf Syst (IJAIS) 3:25–30
16. Abd-Eldayem MM (2014) A proposed HTTP service based IDS. Egypt Inf J 15(1):13–24

Video Frame Interpolation Using Deep Convolutional Neural Network

Varghese Mathai, Arun Baby, Akhila Sabu, Jeexson Jose and Bineeth Kuriakose

Abstract Video frame interpolation fuse several low-resolution (LR) frames into one high-resolution (HR) frame. The existing methods for video frame interpolation use optical flow method to determine motion in a scene, but computation using optical flow method is difficult, which can lead to artifacts in the output video. In many applications where we use video footages, there is a similarity in the content of footages. This similarity in content recommends that using some kind of context-aware approach can do better interpolation than the different existing interpolation techniques. We propose such a context-aware approach for video interpolation, the video frame interpolation using convolutional neural networks. In this proposed method, neighboring images are given as input to an end-to-end convolutional neural network which interpolates a frame between them. A comparative analysis of video interpolation technique using proposed RGB model and HSV model using metric standards such as SSIM, PSNR, and MSE is also included in the proposed method.

1 Introduction

Interpolation can be defined as a process of approximating the intermediate values of a continuous event from different discrete samples. In the real case scenario, video

V. Mathai · A. Baby · A. Sabu · J. Jose · B. Kuriakose (✉)
Department of Computer Science and Engineering, MITS, Ernakulam, India
e-mail: bineethbinz@gmail.com

V. Mathai
e-mail: varghesemathaime@gmail.com

A. Baby
e-mail: arunnellickamuriyil@gmail.com

A. Sabu
e-mail: akhilasabu1212@gmail.com

J. Jose
e-mail: jeexsonjose1996@gmail.com

© Springer Nature Switzerland AG 2019
D. Pandian et al. (eds.), *Proceedings of the International Conference on ISMAC in Computational Vision and Bio-Engineering 2018 (ISMAC-CVB)*, Lecture Notes in Computational Vision and Biomechanics 30,
https://doi.org/10.1007/978-3-030-00665-5_82

interpolation is a technique which interpolates a frame between two neighboring frames. The dense motion between two consecutive input frames is estimated using stereo matching or optical flow algorithms and then interpolates different middle frames according to the dense correspondences estimation. This is the case happening in most of the existing frame interpolation methods. Any interpolation algorithm takes a pair of two images and produces an interpolated image. One of the main application of the interpolation algorithms is application which requires increasing frame rate [1, 2].

The most common and simple method for video frame interpolation is linear frame interpolation (LFI) [3], which simply takes a linear combination of neighboring frames to generate the target frame or the output. However, when different objects in the scene are moving over a stationary background, this technique fails. Motion compensated frame interpolation (MCFI) [4] is a popular, widely used and current industry standard algorithm for video frame interpolation, particularly in the applications concerned with upsampling of videos that is implemented in many modern HDTVs. But the problem with this techniques is that these methods are not context-aware. In many applications where we use video footages, the content of the video footages is similar. For example, consider CCTV camera footages, they have fixed background and only contains images of people moving across that background. Similarly, videos footages of football matches also mainly contain frames containing a ball and players on the green pitch. This similarity in content recommends that a context-aware approach could perform better at interpolation than the existing techniques. In this paper, we propose such a context-aware approach, video frame interpolation using deep convolutional neural network.

2 Methodology

2.1 System Description

In this method, input pairs of neighboring frames in a video are given to an end-to-end convolutional neural network which interpolates a frame between the input pairs. As a result, the frame rate of videos will be increased. For a given target frame, the corresponding inputs are the frames that are immediately before and immediately after the target frame. The input is a six-channel array composed of stacking the HSV data from the input video frames. We apply a convolutional layer on this data many times and then store the results obtained each time to ultimately arrive at the latent representation of the data. A set of filters is convolved with the image in a convolution step. In the deconvolution process, we pass the data through a deconvolutional layer, then concatenate the result with the similar sized layer from the convolution process before applying the next deconvolution. This allows the image regeneration process to have access to some details which may have been lost during the process of convolution [5]. The same process of interpolation is carried out by stacking the

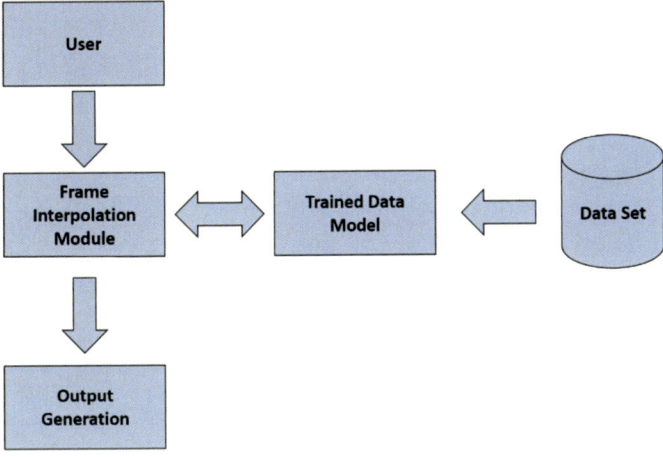

Fig. 1 Block diagram of the proposed system

RGB data from the input frames. Then, we will make a comparative analysis of both techniques using various metric standards.

Figure 1 gives the block diagram of the system which shows the working of the system. Initially, the user inputs the video to be interpolated. The neural network is trained to predict the intermediate frame based on the two input frames given. In frame interpolation, the input video is split into frames and is given as input to the trained network. Based on the previous inference, the intermediate frame is predicted. After generating frames for each set of input, all the frames are converted back to the video, and this is what happens in output generation. The RGB image data can be converted to HSV and can be used as input which increases the performance [6].

The system can be mainly divided into two parts, feed-forward neural networks and video frame interpolation module. The feed-forward neural networks is used to train the neural networks. We train the model on our training set for an epoch at a time. In frame interpolation, the first input video is converted to frames. From these frames, nth and $(n+2)$th frames are given as input to trained neural nets to generate intermediate frame. After generating all the frames, it is converted back to the video. The hardware used to enable the computation done in this study consists of a single NVIDIA 1080Ti GPU card.

Our python-based implementation of CNN for video frame interpolation enabled by the Keras and Tensorflow deep learning libraries is given in the following sections. The objective function we need to minimize is the Charbonnier error loss, given by

$$\rho(x) = \sqrt{\alpha^2 + \varepsilon^2}. \tag{1}$$

In this case, α represents the difference between the true frame f and the approximated frame f_i.

This function has various advantages over standard functions like the mean squared error loss as in this, the additional parameter ϵ lets us control the behavior of the function as the loss approaches zero. It is also possible to choose ϵ in such a way that when the loss falls below a certain threshold, the function increases quickly giving a pronounced signal that the model has converged within the acceptable given error range. For enabling gradient descent-based optimization, the Keras native Adam optimizer can be used with default parameter settings in all training cases, with a learning rate $\alpha = 0.0001$ for stochastic gradient descent (SGD) [7]. To avoid getting stuck at local optima, we can decrease the learning rate by a factor of 10 when the training loss becomes stagnant.

2.2   Data Collection and Processing

The dataset is created from a two and half hour duration movie. The entire movie is split into 15,000 frames using blender. The 9000 images are used as training dataset and 3000 frames are used as testing dataset. The videos are stored as individual frames with sequential numbering and PNG format. For a given video, there are grayscale as well as colored versions stored separately. First, a directory is chosen and all the files in that directory are enumerated. Then they are sequentially loaded into numpy arrays. We also set aside separate image sequences for validation in order to properly track the performance of our model. So, the remaining frames in the dataset can be used as validation sets.

2.3   Network Architecture

Ronnegerber et al. from Cornell University proposed U-Net in 2015 [8], which is a recursive CNN (RCNN) based network architecture that consists of a multi-layer contracting network. The network successively implements upsampling layers rather than pooling layers. This implementation increases the resolution and resultant feature space of the output data. The architecture is given in Fig. 2. The convolutional neural network architecture that we use consists of two main blocks. The first block type consists of two 3×3 convolutional layers, followed by a 2×2 max pooling layer. Similarly, the second block type contains a 2×2 upsampling layer, followed by two 3×3 convolutional layers. The entire network contains five blocks of the first type, followed by five blocks of the second type, in which the number of filters for each convolutional layer increases as the layers reach the center of the network, up to size 512. The input of the network has a shape of $6 \times 128 \times 384$, in which we concatenate the first frame f_i with the last frames f_j, on the channel axis. The final layer is again just a 1×1 convolutional layer with three filters, thus outputting a shape of $3 \times 128 \times 384$ because there is equal number of max pooling and upsampling layers.

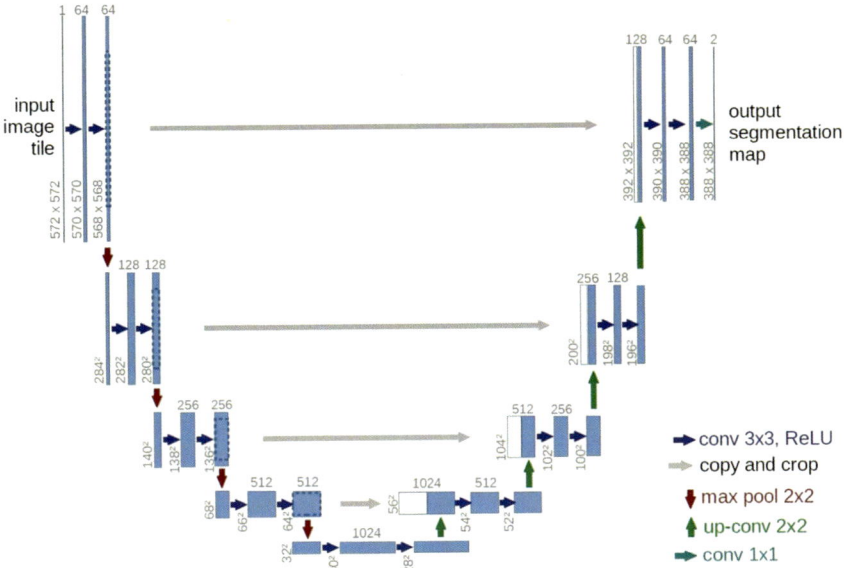

Fig. 2 UNet architecture used for the proposed system

This inherent symmetry of the network allows to increase performance. We have a merge layer before each block of the second type that concatenates the output of the opposite block in the first half of the network. This allows us to maintain feature information from the previous parts of the network, which had a thin field of view. As we use a 128 × 384 resolution image as our network input, and we have five max pooling layers, the height and width of the layer outputs reach a 4 × 12 resolution, that allows us to see a broader view of the input space as well. Along with all these, after each convolutional layer, we utilize the rectified linear unit (ReLU) nonlinearity activation function and to maintain the same input and output relative size, we also perform padding on the input in an orderly manner [9].

2.4 Network Training

The initial training can be done using the entire training set which takes over the course of some hours of time. The accuracy of the model can be increased by an increase in the loss incurred during validation, hence the weights defining the network topology are only updated when such an increase in validation loss occurs.

2.5 Metrics for Comparison

Along with analyzing the outputs from the system which implement video frame interpolation using CNN, the interpolated frames will be compared to their ground truths using various metrics such as Structural Similarity Index (SSIM), Peak Signal-to-Noise Ratio (PSNR), Mean Squared Error (MSE), and Mean Absolute Error (MAE) [10].

Mean Squared Error (MSE): The basic metric which we used to analyze the performance is the Mean Squared Error (MSE) between the generated frames and the ground truth images. Mean Squared Error gives the measure of the average of the squares of the errors or the deviations. It is basically the difference between the estimator and what is actually estimated. It is given by

$$\text{MSE}(y, \hat{y}) = \frac{\|y - \hat{y}\|_2^2}{H * W * C} \tag{2}$$

where H, W, and C are the height, width, and depth of the input frames y and $\hat{y}$.

Peak Signal-to-Noise Ratio (PSNR): The PSNR is a log-scale ratio that is actually related to the Mean Square Error between an image and its ground truth. PSNR will penalize the noise that may get introduced to video frame during interpolation. It is given by

$$\text{PSNR}(y, \hat{y}) = 10 * \log_{10} \frac{\text{max power}^2}{\text{MSE}(y, \hat{y})}$$

$$= 10 * \log_{10} \frac{255^2}{\text{MSE}(y, \hat{y})} \tag{3}$$

where max power is the maximum possible pixel value of the image. When the pixels are represented using 8 bits per sample, this is 255.

Structural Similarity Index (SSIM): The structural similarity index (SSIM) is a metric that is used for predicting the perceived quality of images and videos. It is actually used for measuring the similarity between two images. SSIM is a qualitative metric and a good supplement to PSNR or MSE. Both the PSNR or MSE are robust quantitative measurements but is inconsistent with human visual perception.

$$\text{SSIM}(x, y) = \frac{\left(2\mu_x\mu_y + c_1\right)\left(2\sigma_{xy} + c_2\right)}{\left(\mu_x^2 + \mu_y^2 + c_1\right)\left(\sigma_x^2 + \sigma_y^2 + c_2\right)} \tag{4}$$

3 Results

The input to the system is two frames of a video and the system is expected to produce a middle frame between the two input frames. The system generates the

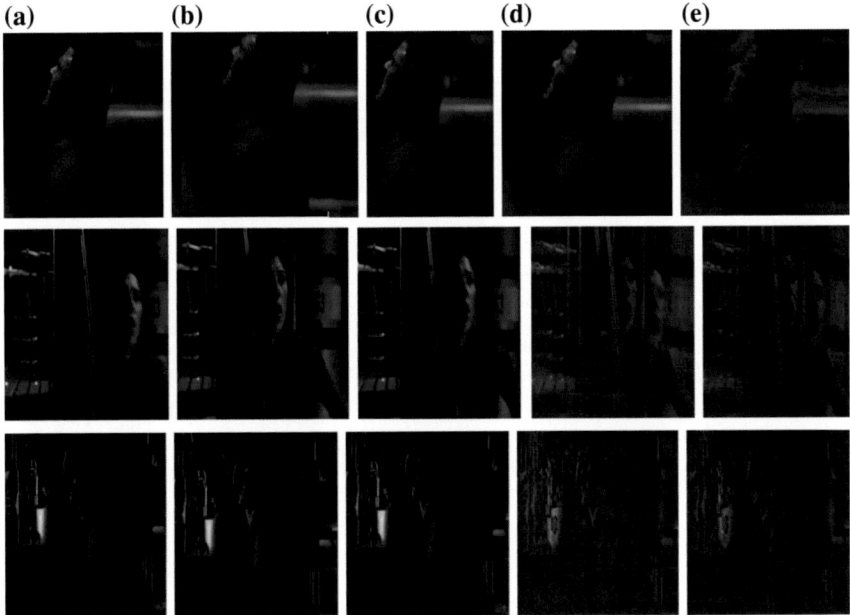

Fig. 3 Illustration of results obtained from three different movie scenes. **a** Frame 1. **b** Frame 3. **c** Ground truth. **d** Interpolated frame from RGB model. **e** Interpolated frame from HSV model

output based on the inference it made during the training phase. To estimate the effectiveness of the system, the video quality is used as the factor. Using the quality assessment metrics such as PSNR, SSIM, and the MSE, the video is compared to the ground truth. If the system does not produce output with expected accuracy, the architecture or training style has to be modified. Both RGB and HSV color models are used separately. In case of HSV model, the image was converted to HSV after it was read using OpenCV library using an inbuilt function. The HSV color space is more or less related to the way in which human beings perceive color whereas RGB color space defines the color in relation to the primary colors. The main reason behind the use of HSV version of video frame is because using the component Hue, we make the algorithm less sensitive to the variations of light. Also by using HSV instead of RGB, noise in the process output can be reduced. The result of our model will be less blurry on the moving object, thus we can expect a better SSIM. The results are shown in Fig. 3.

A comparative analysis of interpolation using RGB with the HSV is carried out using metric standards such as SSIM, PSNR, and MSE. The comparison of RGB model and HSV model using these metrics standards is given in Table 1.

The two models give the second frame by reproducing the structure successfully. But the noise is higher in the interpolated frame comparing with ground truth. The mean squared error is found to be lower for interpolation by HSV model compared to the interpolation by RGB models for all the set of frames we tested.

Table 1 Comparison of RGB model output with HSV model using quality metrics MSE, PSNR, and SSIM

Inputs	MSE		PSNR		SSIM	
	RGB	HSV	RGB	HSV	RGB	HSV
Frame set 1	436.94	343.49	21.73	22.77	0.49	0.52
Frame set 2	679.57	550.11	19.81	20.73	0.38	0.40
Frame set 3	577.81	472.21	20.51	21.39	0.37	0.41

Therefore, HSV gives a more similar interpolated frame with ground truth compared with RGB model. Noise is less in the HSV output compared to RGB output as PSNR value is higher for HSV model. Here, the results of all the testing sets will have higher value of SSIM for HSV model than the RGB model. So the HSV model will generate high-quality frame than RGB as HSV model reproduces structure better than RGB model.

4 Conclusion and Future Work

The video frame interpolation using convolutional neural network exploits a convolutional neural network that takes input pairs of neighboring images and tries to interpolate a frame between them. One of the applications for such an algorithm is in the data compression and also in upsampling process. In both cases, the generated images were compared to a ground truth using different metric standards of PSNR, SSIM, and MSE. Our output shows good results and it is clear that the network excels at predicting the interpolated location of objects in the image scenes, even if the end result is of lower quality than desired. Also, we conclude that interpolation using HSV model is slightly better than the RGB model.

This video interpolation technique can be used in HDTV's to increase frame rate of the videos. Currently, MFCI is used for frame interpolation, which has disadvantages like soap opera effect and motion blur. The proposed system uses DCNN for frame interpolation which can produce more strong and error-free results compared to all existing techniques. The proposed system can be also used in data compression and upsampling in the case where video has to be sent through a low bandwidth channel. Another advantage of the technique is that it can be also used to increase the frame rate of CCTV footages thereby making CCTV footages more reliable. It can be also used in animation domain to increase the frame rate of animation videos thereby increasing the smoothness of the animation videos. The effectiveness of the system can be further increased by using generative adversarial networks. For achieving better performance, different layer architectures can also be considered for different types of video types.

References

1. Kang K, Ouyang W, Li H, Wang X (2016) Object detection from video tubelets with convolutional neural networks. In: IEEE conference on computer vision and pattern recognition (CVPR), pp 817–825
2. Dong C, Loy CC, He K, Tang X (2016) Image super-resolution using deep convolutional networks. IEEE Trans Pattern Anal Mach Intell 38(2):295307
3. Hung K-W, Siu W-C (2011) Fast video interpolation/upsampling using linear motion model. In: 18th IEEE international conference on image processing
4. Guo D, Lu Z (2016) Motion-compensated Frame Interpolation with weighted motion estimation and hierarchical vector refinement. Neurocomputing 181
5. Ghutke RC, Naveen C, Satpute VR (2016) Temporal video frame interpolation using new cubic motion compensation technique. In: IEEE international conference on signal processing and communications (SPCOM)
6. Saravanan G, Yamuna G, Nandhini S (2016) Real time implementation of RGB to HSV/HSI/HSL and its reverse color space models. In: IEEE international conference on communication and signal processing, 6–8 April 2016
7. Kingma DP, Ba J (2014) Adam: a method for stochastic optimization. CoRR abs/1412.6980 available at http://arxiv.org/abs/1412.6980
8. Ronneberger O, Fischer P, Brox T (2015) U-Net: convolutional networks for biomedical image segmentation, CoRR abs/1505.04597, available at http://arxiv.org/abs/1505.04597
9. Song BC, Jeong SC, Choi Y (2011) Video super-resolution algorithm using bi-directional overlapped block motion compensation and on-the-fly dictionary training. IEEE Trans Circuits Syst Video Technol 21(3):274–285
10. Kotevski Z, Mitrevski P (2009) Experimental comparison of PSNR and SSIM metrics for video quality estimation. In: Davcev D, Gmez JM (eds) ICT innovations. Springer, Berlin, Heidelberg

Sentiment Analysis of Twitter Data Using Big Data Tools and Hadoop Ecosystem

Monica Malik, Sameena Naaz and Iffat Rehman Ansari

Abstract Sentiment analysis and opinion mining help in the analysis of people's views, opinions, attitudes, emotions and sentiments. In this twenty-first century, huge amount of opinionated data recorded in the digital form is available for analysis. The demand of sentiment analysis occupies the same space with the growth of social media such as Twitter, Facebook, Quora, blogs, microblogs, Instagram and other social networks. In this research work, the most popular microblogging site 'twitter' has been used for sentiment analysis. People's views, opinions, attitudes, emotions and sentiments on an outdoor game 'Lawn Tennis' have been used for the analysis. This is done by analysing people's positive, neutral and negative reviews posted on Twitter. Through this it has been analysed that how many people around the world really like this game and how popular this game is in different countries.

1 Introduction

Microblogging websites have progressed to the level of being a source of a variety of information. This can be accredited to the nature of microblogs, wherein people can present their opinions, upload real-time pictures, post comments on a discussion related to current issues or review sentiment for a product they must have used. Recently, some companies have started polling such microblogs, so that they can conclude a general sentiment for their product. A challenge faced by them is to build a technology that helps them identify and compile overall sentiment.

M. Malik · S. Naaz (✉)
Department of Computer Science and Engineering, School of Engineering Sciences and Technology, Jamia Hamdard, New Delhi 110062, India
e-mail: snaaz@jamiahamdard.ac.in

M. Malik
e-mail: monicamalik17@gmail.com

I. R. Ansari
University Women's Polytechnic, Aligarh Muslim University, Aligarh 202002, U.P., India
e-mail: iffat_rehman2002@yahoo.co.in

© Springer Nature Switzerland AG 2019
D. Pandian et al. (eds.), *Proceedings of the International Conference on ISMAC in Computational Vision and Bio-Engineering 2018 (ISMAC-CVB)*, Lecture Notes in Computational Vision and Biomechanics 30,
https://doi.org/10.1007/978-3-030-00665-5_83

Sentiment analysis is the same as opinion mining. Twitter is one of the social networking sites which is used by million number of people and it receives tweets in millions every day. And this data can be used for the business purpose or industrial purpose. Big data is datasets which are complex and voluminous and are not dealt with traditional processing application software. In this work, analyses of Twitter data is done in the Hadoop environment, and for that Cloudera has been used which provides Apache Hadoop based software. The clusters of nodes are formed for distributed processing. Tweets are in the form of comments which are the opinions, emotions or sentiments of the people. This Twitter data is gathered by using Twitter API (Application Program Interface) [1]. Later, after analyses, the system gives the required output in the form of neutral, positive or negative tweets. Further, the analysed data is represented in the form of graphs and pie charts using Tableau software for visualization.

2 Related Work

Natural level processing at various levels of granularity is used in sentiment analysis. The classification task can be carried out either at the document level [2] or at the sentence level [3, 4] or at the phrase level [5, 6]. The data on microblogging platform like Twitter has views and opinions of people on various subjects. This data is ever increasing and of diverse nature which poses a lot of difficulty in its processing. A lot of work has been done on sentiment analysis of Twitter data some of which can be found in [7–9].

One of the methods used for extracting sentiment data is distant learning and has been used in [7]. Here, the tweets finishing off with positive emojis like ':)' ':-)' are taken as positive and emojis like ':(' ':/' as negative. In this work, the models have been built using support vector machines (SVMs), Naive Bayes and MaxEnt classifiers and it has been shown that SVM performs better than the other two classifiers. They have used unigram and bigram model along with parts-of-speech (POS) features in the feature space. This paper proves that the unigram model performs better than every other model. Bigrams and POS features do not help specifically.

A similar distant learning paradigm is used by authors in reference [9] to gather data. They use an alternate method for the classification task: subjective versus objective. Twitter accounts of major newspapers like 'New York Times', 'Washington Posts' and the likes are crawled to obtain the objective data. For the data which is subjective in nature, they gather the tweets finishing with emojis as was done in [7]. In contrast to the reports obtained in [7], they report that POS and bi-grams both give better results. Both these methodologies work on the n-gram model. But the data they use for preparing and testing is gathered via search engine queries and hence may show some level of biasness.

This biasness has been eliminated in [10] which considers manually annotated data. The features used here also present better results as compared to the unigram baseline. Additionally, an alternative technique for data representation is used which provides much better results as compared to the unigram models. In this work, three type models are experimented, those three being feature based, tree kernel and unigram-based model. Feature-based model includes some features from past literature, and some newly proposed ones are used. A new tree design has been created for the tree-based model, and the unigram model is used as baseline. A unigram only fetches 20% over the chance baseline for both the tasks and is a hard baseline. Only 100 features are used in this work and it is able to provide results at par with those obtained by the unigram model that uses over 10,000 features. Both feature based and unigram model are good but not better than tree-based model when it comes to the performance.

Models can be tested in combinations, the combinations work 4% more efficiently than unigram baseline for both classification tasks. It is evident from various studies that Twitter specific features such as hashtags, emoticons etc add only a slight value to the analyser [6].

Sentiment classification on Twitter data has also been carried out in [11]. In this work, the data has been collected from three websites: Twitter Sentiment, Twendz and TweetFeel which provide real-time sentiment detection. The data taken from these sources which are noisy and biased in nature have been used to train the model. After this, tuning and testing have been carried out on 1000 manually labelled tweets each. They suggest the utilization of syntax features of tweets like retweet, link, punctuations, hashtags and exclamation marks along with features like prior polarity of words and POS of words. The outcomes here show that the features that combine prior polarity of words with their parts of speech give the best results for the classifiers considered. There is not much effect of the tweet syntax feature.

Sentiment analysis by Aue and Gamon [12] is executed on feedback data from Global Support Services survey in 2004. They have analysed the role of linguistic features like POS. They perform broad feature analysis and feature selection and show that abstract linguistic analysis features add to the precision of a classifier.

Positive and negative score is assigned to each word, which is calculated separately. This starts with one seed for calculating positive scores and one for negative score, respectively. General algorithm for label propagation is used in [13].

3   Sentiment Analysis of Twitter Data for Tennis

In this era of information where enormous amount of data is available on hand for decision-making it is very important to understand the data properly and extract the features which are important for analysis. Big data is nothing but the datasets that are not only big but also very high in velocity and variety, which is difficult to tackle with the traditional tools and techniques. And as we all know the amount of data is increasing day by day so much and for that, we need to study and provide solutions for such problem to handle and extract knowledge out of these datasets.

Also, valuable insights are needed by the decision makers from such rapidly changing data, that ranges from customer interactions to day-to-day transactions and social network data. And for that, we use the application of advanced analytics techniques which is used to provide value to the decision makers using big data analytics.

Here, in this paper, we aim to analyse opportunities provided by the application of big data analytics in various decision domains and we also analyse some of the different analytics methods and tools which can be applied to big data. We will analyse the views of people on an outdoor game 'Lawn Tennis'. What are their views about this game, how do they feel about this game and how popular this game is around the world.

This paper focuses on analysing one such micro-blog 'twitter' and would create models in order to classify 'tweets' into positive, neutral and negative sentiments.

4 Proposed Work

In this paper, twitter data is used that is 'the tweets' posted by the users on Twitter. This data has an advantage over previously used datasets is that the tweets represent a true sample of the actual tweets in terms of language use and content as in they are not filtered or modified and are collected in a flooded fashion. Only 1% of data is provided by Twitter for analysis. Two dictionary tools are used in this work.

First one is the hand-annotated dictionary for emoticons. This dictionary helps to map emoticons to their polarity and on Wikipedia, there are almost 170 emoticons listed with their emotional state. For example, we can say :) is used for happy which means positive and :(is used for sad which means negative [2–4].

Second is an acronym dictionary, which is collected from the web. This acronym dictionary has English translations of over 5000 frequently used acronyms. For example, lol is used for laughing out loud, and bff is used for best friend forever.

5 Result Analysis

Figure 1 shows the geographical representation of our datasets visualizing the popularity of tennis across the world. Contrary to the popular belief that tennis is only popular in urban parts of the world, this map shows the reality, surpassing the financial or growth index factors of a country or area of regions, and the reach of this enthralling sport. This shows the diversity in the fan base of the sport across the continents. This map has pie charts hanging over different countries representing the popular sentiment of the country for the sport.

Figure 2 is a three factors based line chart which focusses on the varying sentiments across the ten most active countries when it comes to tennis popularity. Each country's slope represents the countrymen's state of mind regarding their method of putting up

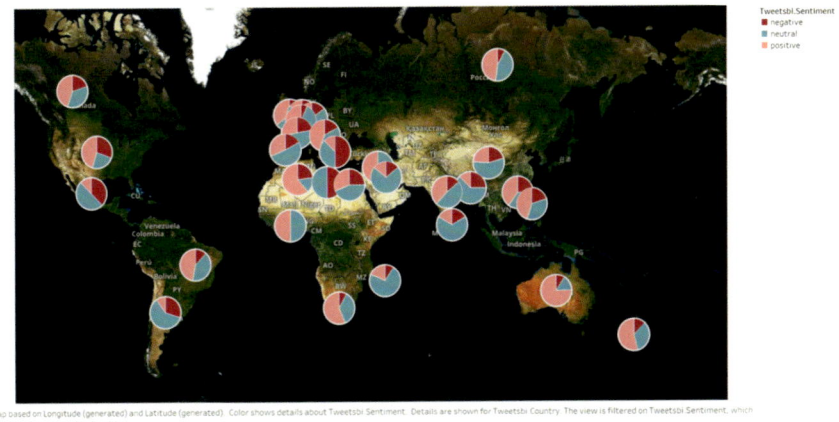

Fig. 1 Country-wise popularity of TENNIS

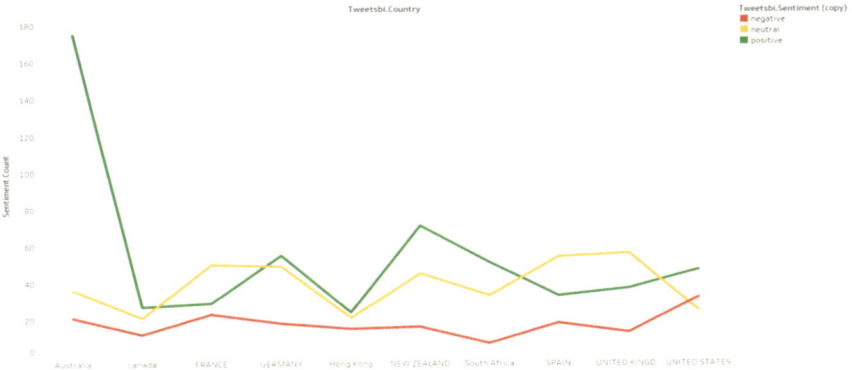

Fig. 2 Positive V/S Negative V/S Neutral slopes

opinions. For example—United States, very unbalanced nation, politically, shows the diversity in their opinions regarding the sport. There is no sheep walk, and everyone has a different set of opinions not influenced by anyone. Same goes to every other country except Australia, whose people are known for being chill and following their country's norms and traditions.

The pie chart in Fig. 3 shows the popularity of the sport across the world. From this analysis, we can draw two conclusions. First is, most of the people who use twitter as a social networking platform, use it to spread positivity and second is even if there is considerable amount of negativity around the tennis world in people's minds, they do not bother to show it to the world and they do not consider it worth their time and energy.

The ribbon chart depicted in Fig. 4 is useful for companies and industries dependent on tennis for their business. They can either shift their focus on to countries

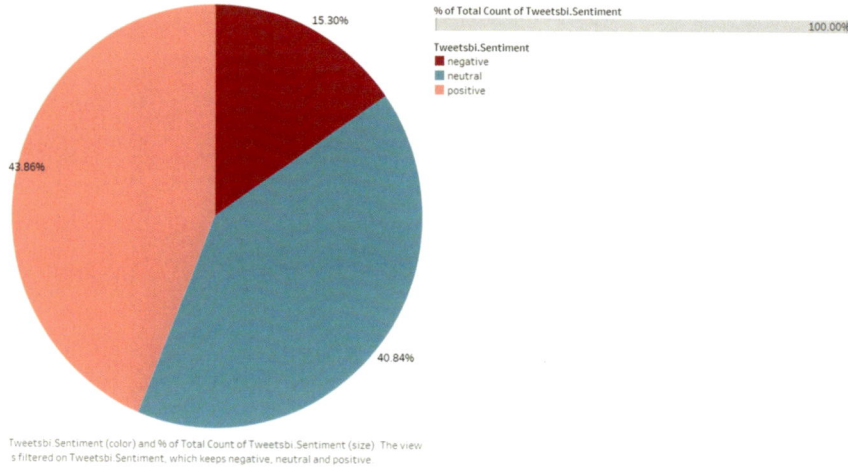

Fig. 3 Overview of the sentiments regarding 'Tennis around the World'

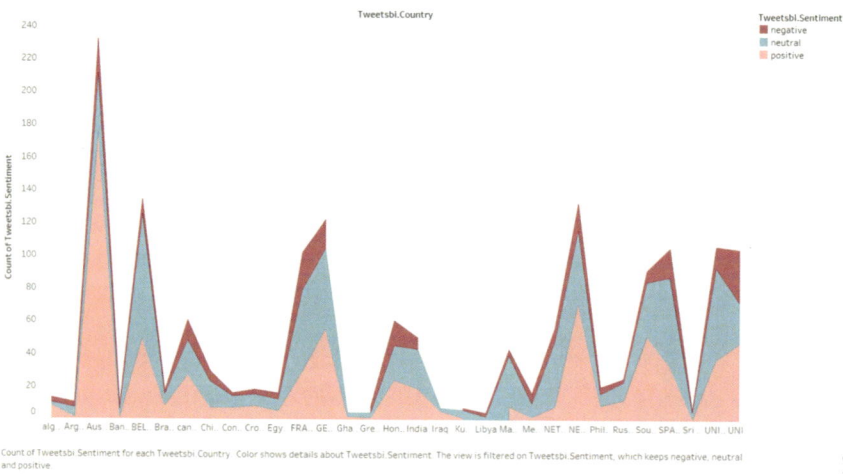

Fig. 4 Analysis of tennis popularity

that are not very developed but still have considerable amount of tennis popularity
amongst the masses where tennis industry is still in infancy. Companies can also use
this data to bank on to opportunities in rich nations where tennis popularity is already
on its peak and the masses would accept any useful tennis related product with open
hands.

6 Conclusion

Apache Hadoop is gaining a significant momentum from both industry and academia as the volume of data is growing rapidly. In this work Twitter data has been collected and analysed to see the sentiment of tennis among the people in the world. Hadoop has been used for this work due to its distributed architecture and ease of handling huge amount of data. It has been established from this work that contrary to the popular belief that tennis is only popular in urban parts of the world, it is equally popular in the rural areas. This analysis is very important for people who are involved in sports industry. The more the demand for this sport in a particular country the more their sale will be of the products related to the sport. So they can explore various parts of the world to see where their business can prosper most.

References

1. Ingle A, Kante A, Samak S, Kumari A (2015) Sentiment analysis of twitter data using hadoop. Int J Eng Res Gen Sci 3(6)
2. Pang B, Lee L (2008) Opinion mining and sentiment analysis. Now Publishers Inc., Foundations trends in information retrieval, available at http://portal.acm.org/citation.Cfm
3. Hu M, Liu B (2004) Mining and summarizing customer reviews. In: Proceedings of the tenth ACM SIGKDD international conference on knowledge discovery and data mining. ACM, pp 168–177
4. Kim SM, Hovy E (2004) Determining the sentiment of opinions. In: Proceedings of the 20th international conference on computational linguistics. Association for Computational Linguistics, p 1367
5. Wilson T, Hoffmann P, Somasundaran S, Kessler J, Wiebe J, Choi Y … Patwardhan S (2005) OpinionFinder: a system for subjectivity analysis. In:: Proceedings of hlt/emnlp on interactive demonstrations. Association for Computational Linguistics, pp 34–35
6. Bekkerman R, Gavish M (2011) High-precision phrase-based document classification on a modern scale. In: Proceedings of the 17th ACM SIGKDD international conference on Knowledge discovery and data mining. ACM, pp 231–239
7. Go A, Bhayani R, Huang L (2009) Twitter sentiment classification using distant supervision. CS224 N Project Report, Stanford 1(12)
8. Bermingham A, Smeaton AF (2010) Classifying sentiment in microblogs: is brevity an advantage? In: Proceedings of the 19th ACM international conference on Information and knowledge management. ACM, pp 1833–1836
9. Pak A, Paroubek P (2010) Twitter as a corpus for sentiment analysis and opinion mining. In: LREc, vol 10, No. 2010
10. Agarwal A, Xie B, Vovsha I, Rambow O, Passonneau R (2011) Sentiment analysis of twitter data. In: Proceedings of the workshop on languages in social media. Association for Computational Linguistics, pp 30–38
11. Barbosa L, Feng J (2010) Robust sentiment detection on twitter from biased and noisy data. In: Proceedings of the 23rd international conference on computational linguistics: posters. Association for Computational Linguistics, pp 36–44
12. Aue A, Gamon M (2005) Customizing sentiment classifiers to new domains: a case study. In: Proceedings of recent advances in natural language processing (RANLP), vol 1, No. 3.1, pp 2–1
13. Zhu X, Ghahramani Z (2002) Learning from labeled and unlabeled data with label propagation

Detection of Chemically Ripened Fruits Based on Visual Features and Non-destructive Sensor Techniques

N. R. Meghana, R. Roopalakshmi, T. E. Nischitha and Prajwal Kumar

Abstract Nowadays great concern for everyone is health; hence primary requirement for sound health is eating good quality fruits. However, most of the available fruits in the market are ripened using hazardous chemicals such as calcium carbide, which is highly hazardous to human health. In the existing literature, less focus is given towards addressing the problem of identification of artificially as well as naturally ripened fruits, due to the complex nature of problem. In order to solve this problem, a new framework is proposed in this paper, which utilizes both the image features- and sensor-based techniques to identify whether the fruit is ripened by chemicals or not. By employing pH-sensor based techniques and visual features, it is possible to detect artificially ripened fruits and save the human beings from serious health hazards. The experiments were conducted and the results indicate that the proposed technique is performing better for the identification of artificially ripened banana fruits.

1 Introduction

The primary requirement for everyone is having good health condition, so eating good quality fruits provides sound health. The fruits are sweet-tasting plant product which contains fiber, water, vitamin C, and sugars. It also contains minerals, protein, cellulose, and various photo chemicals which protect human body against various disorders. Regular consumption of fruit is associated with anti-cancer, cardiovascular disease reduction, and declines aging factor. During the natural ripening process fruits attain desirable color, quality, flavor, palatable nature, and other textural changes during natural ripening process [1]. However it is quite impossible to get naturally ripened fruits, because most of available fruits in the market are ripened

N. R. Meghana (✉) · R. Roopalakshmi · T. E. Nischitha · P. Kumar
Alvas Institute of Engineering and Technology, Mangaluru, India
e-mail: meghanagowdanr@gmail.com

R. Roopalakshmi
e-mail: drroopalakshmir@gmail.com

© Springer Nature Switzerland AG 2019
D. Pandian et al. (eds.), *Proceedings of the International Conference on ISMAC in Computational Vision and Bio-Engineering 2018 (ISMAC-CVB)*, Lecture Notes in Computational Vision and Biomechanics 30,
https://doi.org/10.1007/978-3-030-00665-5_84

using hazardous chemicals. For example, nearly 80% fruits like banana, mango, tomato, and papaya are nearly artificially ripened using different chemicals [2].

The process of fruit ripening is stimulated by applying artificial fruit ripening agents. Specifically farmers and vendors often use artificial ripening agents like calcium carbide and ethylene to control the rate of fruit ripening. Precisely, Calcium carbide is a corrosive and dangerous chemical which will be used as a ripening agent. Carbide consumption causes cancer due to phosphorous and arsenic poisoning which leads to diarrhea, burning sensation of chest and abdomen, vomiting, thirst, permanent damage of eyes, shortness of breath, weakness. Due to these reasons, as per PFA (Prevention of Food Adulteration) act in 1955, artificial ripening of fruits using calcium carbide is strictly banned.

Although calcium carbide is banned, farmers and traders are still using it for their own profit. Furthermore, due to this artificial ripening many agricultural products and fruits of Indian are rejected by few international market, which created a blackmark for all the Indian farmers as well as agricultural industry. Though India is the largest producer of bananas, its productivity has been reduced compare other countries like Guatemala, Indonesia. On the other hand, sensor-based non-destructive techniques are popularly used to detect calcium carbide in fruits. By employing sensor based technique it is possible to detect artificially ripened fruits and save the human from different hazards.

2 Related Work

Nelson et al. [1] presented a comparison of vitamin A, B, and C content of green, air, ethylene, and vine-ripened tomatoes respectively. Though this article indicates variations in the vitamin contents, but fails to detect the differences between the groups of air-ripened or ethylene-ripened tomato. Ahmad et al. [2] investigated the effect of temperature, ethylene, and their interaction towards the speed of ripening as well as quality of banana fruit. However it fails to differentiate between ethylene treated and non-ethylene treated bananas. F. J. Ramos et al. [3] estimated fruit firmness by using different strategies such as measurement of variables, acoustic responses, optical properties, analysis of impact forces, and nuclear magnetic resonance. Yet these techniques fail to predict whether fruit is artificially ripened or naturally ripened. K. de Mora et al. [4] presented an improved formulation allowing sensitive and accurate detection of less than 10 ppb arsenate in water.

In 2013, Zhao et al. [5] introduced a recognition system for artificial ripening, which extracted the image information of tomato using neural network based genetic algorithm, to identify the artificially ripened tomatoes. Even though, proposed technique achieved higher recognition rate, the complexity of the proposed algorithm is bit high. In 2013, Anmin Zhu et al. [6] proposed the concept of the neighborhood density to avoid the improper initial value and to initialize the cluster center which is based on the neighborhood data space correlation. In 2015, A. A. Bhosale et al. [7] introduced a capacitive sensing system, color index method and eco measurement to find various ripening stages of papaya. In 2016, Pratim Ray et al. [8] proposed an easy way to monitor the ripening stage of banana based on color indices and the information will be sent to the monitoring person automatically to the remote area. The information received will be processed by microcontroller to the authorize person using GSM module.

Very recently, in 2017, R. Karthika et al. [9] estimated an efficient image processing technique to detect the artificially ripened bananas. The author used threshold-based segmentation to extracted discriminatory features using Haar filter. In [10] proposed a method of Thermal Imaging Technique to detect whether fruit ripened by calcium carbide or not. This technique involves image pre-processing, segmentation, and feature extraction steps for processing of an image based on the infrared energy emitted by the fruit. Sahu et al. [11] developed an automated tool capable of identifying defects and maturity of mango fruits based on size, shape, and color features by digital image analysis. The author used MATLAB as the programming tool for classification and identification of fruits. The author failed to sort fruits based on mango quality which is essential for value addition.

To summarize, the existing systems fails to completely address the problems of identification of artificially and naturally ripened fruits. In order to overcome this proposed framework uses an image processing and pH sensor based techniques to identify whether the fruit is ripened artificially or naturally.

3 Proposed Framework

The sensor-based technique and image processing are combined to bifercate between naturally and artificially ripened bananas. The image processing involves feature extraction of banana image followed by finding the discriminatory behaviors for feature analysis. Figure 1 represents the block diagram of proposed method to classify between banana samples using image processing.

Proposed framework includes four modules: Image Enhancement, segmentation, feature extraction, classification. The image enhancement and segmentation of the image is made using K-means algorithm. Here the image is divided into discrete number of regions so that the pixels will have high contrast and similarity in each region. Followed by segmentation stage, feature extraction stage of the image is included which extracts statistical and texture features. More specifically features including variance, mean, contrast, standard deviation, energy, correlation, homogeneity, and gray-level co-occurrence (GLCM) matrix are extracted. In the classification module, SVM Classifier based on supervised learning, which is widely used for classification tasks is utilized. It consists of learning algorithms that analyze data and identify patterns, used for classification and analysis. SVM classifier distributes the data into different classes by the help of kernel function. The resultant extracted features are matched using SVM classifier which decides whether the fruit is artificially ripened or naturally ripened.

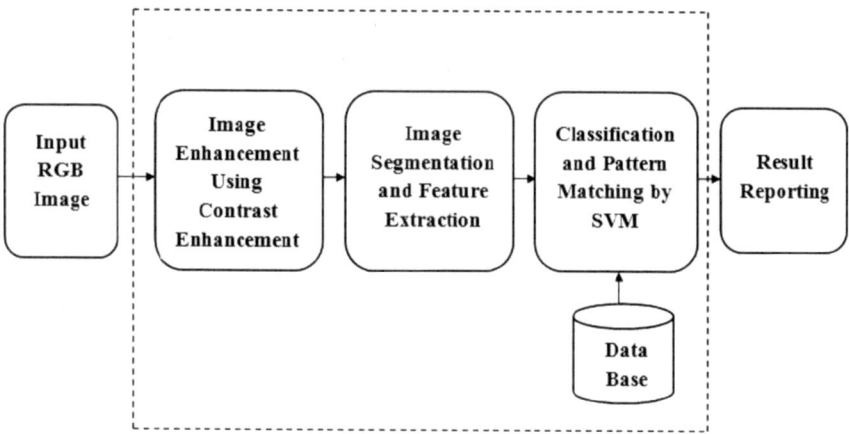

Fig. 1 Block diagram

As in case of pH sensor technique, the sample fruit is dipped into the warm water. The sensor is connected to aurdino board which in turn is connected to the PC. Based on the obtained pH value of that water it will decide whether the fruit is ripened chemically or naturally. After combining both image and sensor based technique obtained result will be displayed in the system.

Algorithm

K-means algorithm is used for the segmentation of input image. One of the unsupervised learning algorithms is K-means that solve the well-known clustering problem. The procedure is an easy and simple method to classify a given data set by a certain number of clusters. The main idea is to define k centers, one for each cluster.

4 Experimental Setup

To evaluate the proposed method pH sensor is used which is connected to the aurdino. For building electronics projects aurdino is used. It contains both physical programmable circuit board (often referred to as a microcontroller) and a software, or IDE (Integrated Development Environment) which is runed by the computer, used to write and upload computer code to the physical board. The aurdino IDE uses a simplified version of C++, making it easier to learn to program. pH is a measure of acidity of a solution the pH scale ranges from 0 to 14. pH sensor measure the level of pH in solution by measuring activity of the hydrogen ions in the solution. Below figure shows the snapshot of experimental setup of proposed framework (Fig. 2).

Experimental setup includes components like aurdino, ORP meter (pH sensor), Display, Camera. Here the sample fruit is dipped into the warm water. The sensor

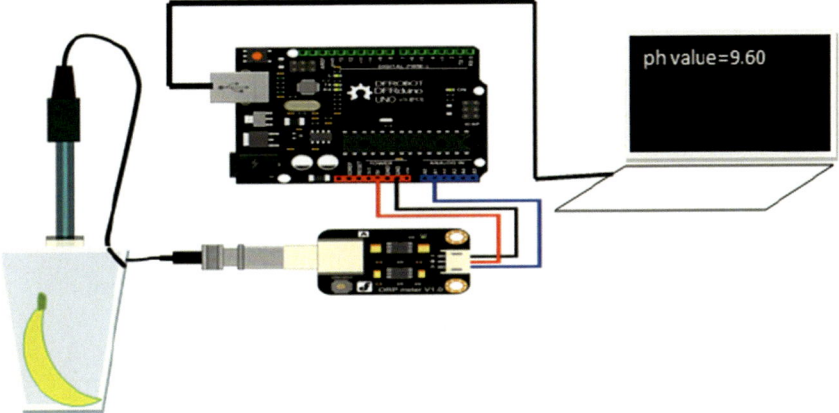

Fig. 2 Experimental setup of proposed framework

Fig. 3 Snapshot of the input image

is in contact with the sample and aurdino board which in turn is connected to the display.

5 Results and Discussion

Captured images are inputs to an image processing unit. Figure 3 shows the snapshot of input image. The important step in image processing is image segmentation. This step is to separate the image constituent areas of interest or regions. The processing and analysis of an image is based on segmented image. The segmentation based on color is used to segment the portion of banana fruit which is contaminated. Figure 4 shows the snapshot of the input image after segmentation.

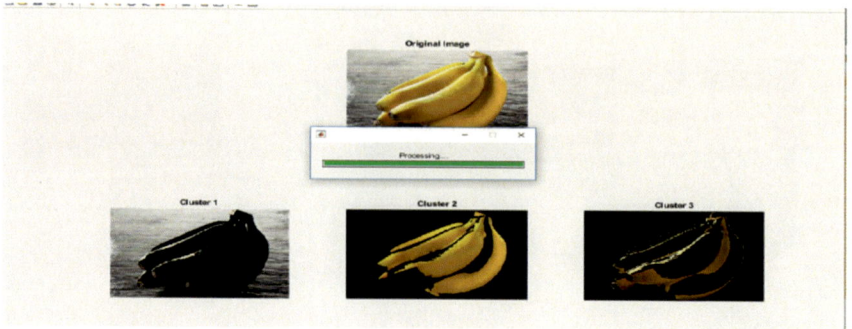

Fig. 4 Snapshot of the input image after segmentation

Fig. 5 Snapshot of the feature extraction

Image analysis is accomplished either straightly on the true image or either across individual color of plane. On the basis of intensity value of color of the image threshold is set because intensity values are sustained in homogeneous color region. Specific color range can be detected easily from whole image. The mean value of color is calculated and it is compared with each RGB pixels values of an image.

The next important step in image processing is feature extraction. The main aim of this is to extricate the feature from an image. It involves evaluating the area of an object in binary image. The input of an image may be logical or numeric. If it is numeric then nonzero pixels are appraised features like min and max, Area, Mean, Standard Deviation values of RGB are calculated for both the images and that will be stored in the database. Because all the extracted features have different values for different images. On the basis of Standard Deviation, Area, Mean, min and max values of RGB parameters of an image is compared with image stored in the database. Figure 5 shows the snapshot of feature extraction.

Figure 6 shows the snapshot of result obtained after image processing and pH sensor based technique. It concludes that the detected fruit is artificially ripened and its pH value is mismatched.

Fig. 6 Snapshot of the result

6 Conclusion

As the existing systems fail to completely address the problem of identification of artificially and naturally ripened fruits, the image processing and sensor-based techniques are proposed to identify the kind of fruit. The images which are captured will be the input to an image processing unit is used for identification of artificially and naturally ripened fruits. The processed image is compared with the database image to display whether the fruit is artificially ripened or not. By employing sensor-based technique it is possible to detect which kind of chemical component is present. So the conducted experiment indicates that the proposed method gives high accuracy compare to earlier techniques. In future, the proposed framework can be further improved for to chemically ripened fruits and value of the pH can be indicated.

References

1. Nelson M, Haber ES (1928) The Vitamin A, B, and C Content of artificially versus naturally ripened tomatoes. J Biol Chem 2:3
2. Ahmad S, Thompson AK (2001) Effect of temperature on the ripening behavior and quality of banana fruit. Int J Agric Biol 3:2
3. García-Ramos FJ, Valero C (2005) Non-destructive fruit firmness sensors. J Agric Res 3:61–73
4. de Mora K, Joshi N (2011) A pH-based biosensor for detection of arsenic in drinking water. In: Proceeding Springer, pp 1031–1039
5. Dhembare AJ (2013) Bitter truth about fruit with reference to artificial ripener. Proc Arch Appl Sci Res 5:45–54
6. Zhu A, Yang L (2013) An improved FCM algorithm for ripe fruit image segmentation. In: Proceeding of IEEE international conference on information and automation Yinchuan, pp 436–441
7. Bhosale AA, Sundaram KK (2015) Nondestructive method for ripening prediction of papaya. Proc Int Conf Interdisciplinarity Eng 19:623–630

8. Ray PP, Pradhan S, Sharma RK (2016) IoT based fruit quality measurement system. In: Proceeding of IEEE international conference on green engineering and technologies, pp 224–229
9. Karthika R, Ragadevi KVM (2017) Detection of artificially ripened fruits using image processing. Int J Adv Sci Eng Res 2(1):20–34
10. Ansari Sheeba (2017) An overview on thermal image processing. Proc Second Int Conf Res Intell Comput Eng 10:117–120
11. Sahu D, Potdar RM (2017) Defect identification and maturity detection of mango fruits using image analysis. Am J Artif Intell 1(1):5–14

Dynamic Object Indexing Technique for Distortionless Video Synopsis

G. Thirumalaiah and S. Immanuel Alex Pandian

Abstract With a development of observation cameras, the measure of caught recordings extends. Physically dissecting and recovering reconnaissance video is work concentrated and costly. It is substantially more important to create a video description and the video can be observed in a good manner. So, here we describe a novel video outline way to deal with produce consolidated video, which utilizes a protest following technique for extracting imperative items. This strategy will create video objects and a crease cutting technique to gather the first video. Finally, output results that our proposed strategy can accomplish a high buildup rate while safeguarding all the imperative objects of intrigue. Hence, in this method, we can empower clients to see the synopsis video with high impact.

1 Introduction

In the decade, an expansive number of observation cameras have been sent and used as a piece of transportation focus focuses, ATMs, and various other open or private workplaces. Due to the reducing cost of sending cameras, it is fundamentally less requesting and more affordable to surveillance a specific zone. With the change of the web, a considerable number of surveillance chronicles are transmitted via the Internet. This would require seeing dynamically to select any key events and moreover perceive any suspicious conduct for a great deal of got video by associations and security affiliations. Regardless, with the help of high skilled persons, most perception accounts are being watched.

The advantages of the synopsis are used to test and validate the video objects and their timings in the scene. And the example in this domain is dynamic video method; it can extract important properties of the video. Another method is picture

G. Thirumalaiah (✉) · S. Immanuel Alex Pandian
Department of ECE, Karunya University, Coimbatore, Tamil Nadu, India
e-mail: tiru5502@gmail.com

S. Immanuel Alex Pandian
e-mail: immans@karunya.edu

© Springer Nature Switzerland AG 2019
D. Pandian et al. (eds.), *Proceedings of the International Conference on ISMAC in Computational Vision and Bio-Engineering 2018 (ISMAC-CVB)*, Lecture Notes in Computational Vision and Biomechanics 30,
https://doi.org/10.1007/978-3-030-00665-5_85

retargeting [1, 2]; in this the selected frame is modified without changing the important movements can be preserved. This scheme of evaluation is implemented in the previously extensively. Including these methods some more schemes are utilized for video synopsis those are fast forward method [3, 4]. In that by skipping some content in the video, digested video is achieved.

So the condensed rate is very low and content loss is also possible. Another one is storyline method [5], in that important objects are extracted and present as storyline and content may be lost. Next to this content equidistant and time equidistant methods, in [6, 7] these methods, distortion is the major drawback. Hence, for solving these drawbacks and limitations, first extract the important parameters from the video background and present them in the space and take common background. Thusly, a solidified procedure video can be created. It should be seen that as opposed to various strategies, the combined video made by video once-over can express the aggregate stream of the scene. Additionally, video synopsis may change the relative arranging between items to lessen common abundance however much as could sensibly be normal. Some other authors have proposed different methods among that ribbon carving is another important aspect. In that, important aspects in the video are prompted and added to the single background scene, but the condensing ratio is very poor. In this paper, implementation of condensing ratio and content preservation is also possible.

2 Our Proposed Method

In conventional video summary methodologies, for the most part, static foundation pixels or casings are disposed of. Hence, the action of articles can be saved alongside the video being more minimal. The motivation behind video summary, which gives the watcher a reduced portrayal of the entire substance in the video, another video summation strategy. In this technique, the foundation pixels and the video objects are prepared. In this method, apllication of crease cutting to consolidate every movement. Exploiting crease cutting, key activities of the items are saved and excess developments are decreased. After these activities, a dense video can be produced. Figure 1 gives the process flow for our proposed method.

A. Object abstraction

In our object video summation technique, object abstraction is the primary step. As of late numerous protest following strategies have been presented [1]. These techniques can be straightforwardly connected in situations where a moving article may relate to an imperative occasion.

Moving shades are recognized and extracted by a simple method [8, 9] in which first the input video is converted into the frames and comparisons can be done between frames. Any value greater than 0, that current frame can be preserved and this procedure follows to last frame comparison. Figure 1 reflects complete process which is employed in this work for producing synoptic video. Figure 2 shows snapshots of tube getting and intensity flow of selected frames.

Fig. 1 Process flow

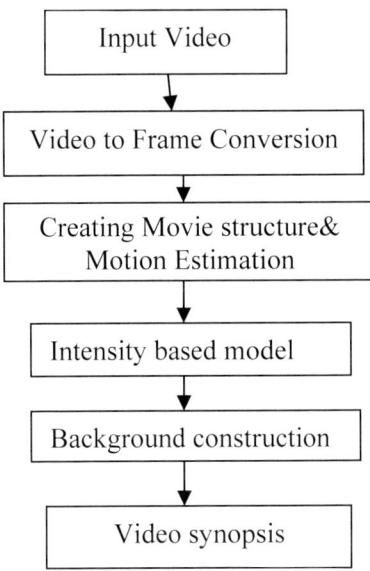

B. Object Condensing

In this segment, we use seam carving operator for extracted tube condensing. We can achieve this using spatial and temporal coherence cost functions.

(i) **Spatial Coherence cost**

In this, we are using seam carving operator which was proposed by shai avian et al. [10] by calculating the energy of the image we look forward to seams to be carved. We use forward energy idea of Rubinstein which improves the seam carving technique.

We start with the saliency map, which is used to identify the significant contents of the image. It is possible with the technique saliency region detection which is an easy and efficient one. The saliency regions can be found out by considering a range of frequencies of different pixels, i.e., using bandpass filters. The saliency map will high lighten the significant objects in the frame with perfect borders (Fig. 3).

In some cases, the edges were formed by the noise. This problem can be rectified by using canny detection. This technique uses gradients of sobel and finds the strong edges. The weak edges in the image are suppressed based on thresholding and the noise can be reduced by using Gaussian filter. The addition of saliency map and canny detected image gives the image with significant contents with strong edges so as to preserve them.

Seam carving algorithm is applied after calculating the compound energy. We are using dynamic programming for solving complex issues. This procedure requires more time and may not produce better results. So the frame undergoes quality checks before carving the seam right after seam removal computation. The similarity of

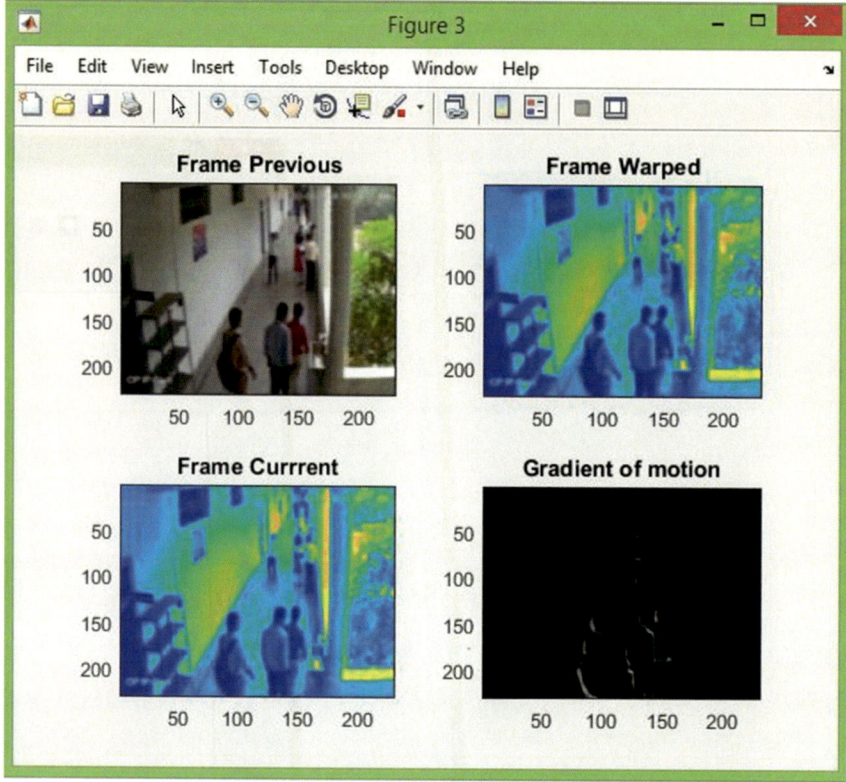

Fig. 2 Moving object detection (Testing frames from my own CCTV footage)

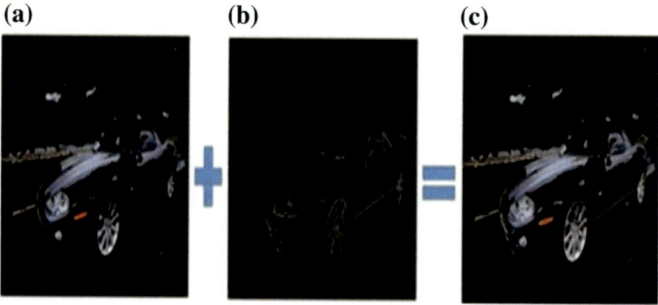

Fig. 3 **a** Saliency map **b** canny detection **c** area to be protected

processed frame and the original frame defines the cost. To measure the similarity we use bidirectional warping used specially in video retargeting. The cost may be increased due to the computation of similarity of the complete image. To reduce the cost, we calculate the window where the seam is removed and measure the similarity.

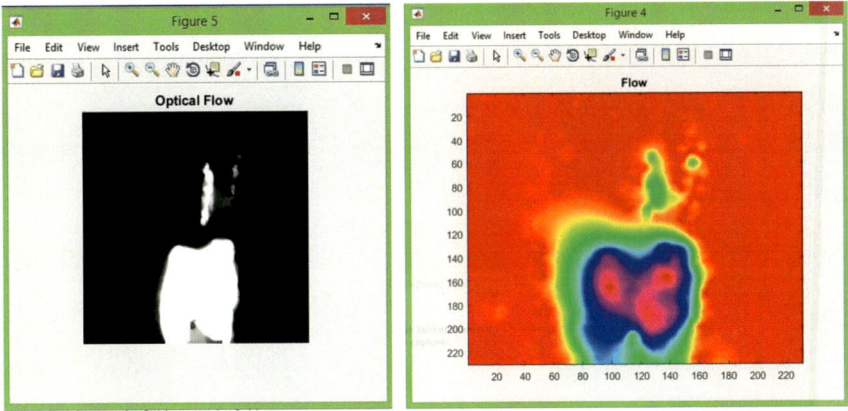

Fig. 4 Calculated seam and the cropping area for similarity check

There is no need to evaluate the similarity at the remaining positions because those are unaffected by the algorithm (Fig. 4).

(ii) **Temporal coherence cost**:

As said earlier the seam carving algorithm must be applied to individual frames because there may be chances of occurrence of artifacts if applied directly to a video. We identify the videos which are recorded from a static camera arrangement or dynamic camera arrangement.

In static camera arrangement, the recorded video background remains same throughout the video. For this, we calculated the sum of difference between the frames.

$$\sum_{i=0}^{l-1} F_i(x, y) - F_{i+1}(x, y) \qquad (1)$$

The above Eq. (1) gives the difference map that consists of the overall motion of the object in the image and the seams calculated for the first frame are directly applied to next frames. Here the cost is effective.

(iii) **Energy Optimization**:

After carved video tracks generated, subtracted objects should be added to extracted background. Normally in surveillance video, the inner scene is in stable position. Before adding objects, the inner scene is generated using optical mean filter of the video.

The process of foreground adding to background to the condensed video is evaluated as shown in Fig. (3). The framework of this process can be evaluated using Eq. (2) (Fig. 5).

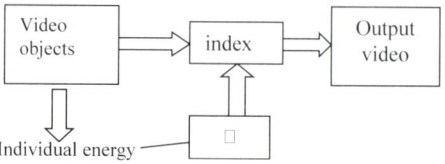

Fig. 5 Simple prototype model 66

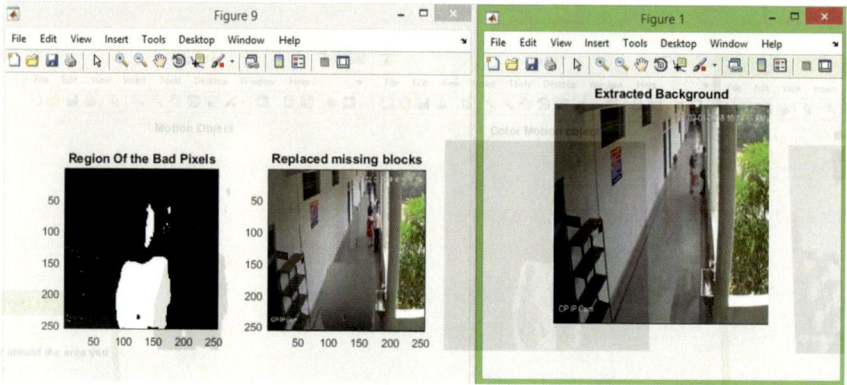

Fig. 6 Background construction of all extracted objects

$$E(M) = \alpha S_{as} + \beta \sum_i S_{in}^i + \gamma \sum_{i,j} S_{pa}^{i,j} \qquad (2)$$

In the above equation, the energy of the objects is minimized by energy minimization method, where S_{as} is the object energy and S_{in}^i is the total frame average energy. Similarly the term $S_{pa}^{i,j}$ is the total condensed frames energy. energy minimization is utilized to reduce the undesired energy objects and it can improve the efficiency of this method (Fig. 6).

By considering the huge computation of the more number of pixels and objects, optimization approaches are limited. Here, we use a greedy algorithm is used for better result. The position of each tube is defined as the term.

The evaluation of first frame number for every tube is assigned by this algorithm. Based on the assigned index, we can stitch each tube into the background and by doing this, we can generate the proper condensed video.

3 Experimental Results

To explain the accuracy and usability of the proposed scheme, five videos are used to evaluate the subjective performance: The Snooker game video, the office, Store video, and Bypass videos are used to explain the proposed method. The important

Table 1 Input video parameters

Name	Frame number	Resolution	Frame rate
Office.mp4	674	854 * 480	30 fs/s
Snooker.mp4	2400	400 * 224	20 fs/s
Store.mp4	1800	854 * 480	15 fs/s
Traffic.mp4	749	1920 * 1080	25 fs/s

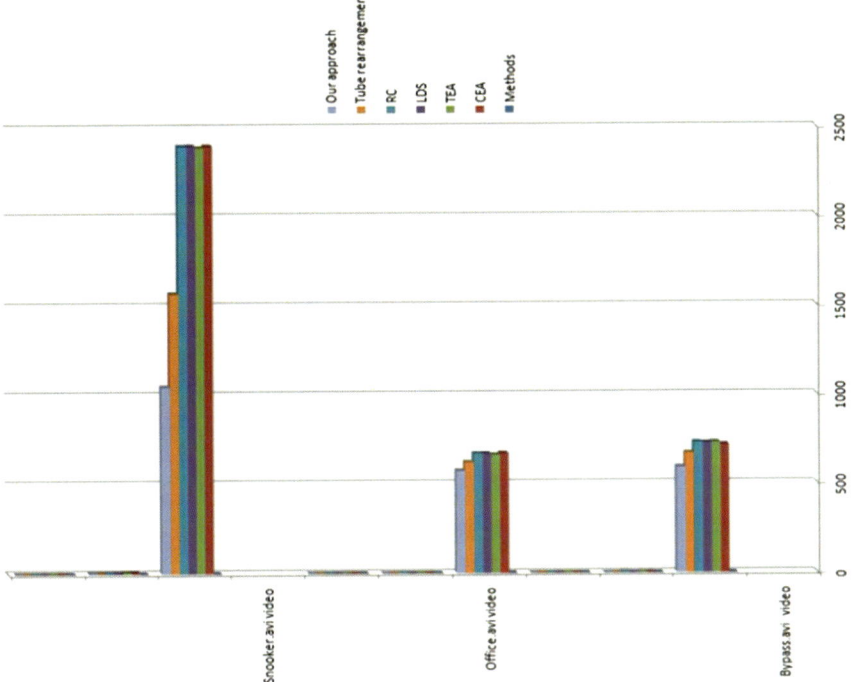

Fig. 7 Graphical view of comparative analysis

and useful parameters are listed in Table 1. All the programming codes are involved in this algorithm are related to either.mp4 or.avi formats. This is the most important and useful sole parameter of video analysis in our experimental condition in our paper. All the experiments are conducted on a laptop PC with Intel core i3, 2.5 GHz CPU, 8G RAM and under the WINDOWS 10 operating system.

As shown in Fig. 7, the performance of the moving objects tracking and extraction and video synopsis is achieved. The condensed and digested video itself is graphical representation of data which can be utilized to browse and extract the video content. So, the decrease in picture quality does not damage the summarized video.

Table 2 Comparison results for different methods

Methods	Corridor.avi video			Office.avi video			Snooker.avi video		
	Frames	Min distortion	Ratio	Frames	Min distortion	Ratio	Frames	Min distortion	Ratio
CEA [12]	720	1.5284	1.05;1	673	1.1625	7:05	2400	7.5501	4.2:1
TEA [12]	733	1.7035	2.12:1	663	1.2620	6:02	2391	7.3174	3.8:1
LDS	729	1.2888	5.13:1	669	1.0078	9:07	2400	5.0701	3.95:1
RC [13]	733	1.6258	1.14:1	670	1.1031	7:06	2399	5.0140	4.1:1
Tube rearrangement [13]	673	1.2735	2.34:1	621	1.1210	13.42:1	1572	4.7921	5.20:1
Our approach	592	1.2688	19.24:1	574	1.0732	14.84:1	1049	4.510	5.70:1

To assess the execution of the proposed approach, we have led a few analyses utilizing extraordinary observation recordings as unique video. We utilize three testing recordings from [11] for execution correlation. An evaluation and comparisons of these recordings are given in Table 1. With our proposed video summary approach, we extricate the moving objects from the video utilizing object following strategies and apply video crease cutting to the question cuts.

In this experiments results, three existing methods (content equidistant algorithm (CEA), time equidistant (TEA), noise model (LDS), and content rearrangement methods) [12, 13] are analyzed to evaluate our proposed method and its efficiency.

For producing video synopsis effectively, our proposed algorithm is employed on three different videos and parameters such as condensed frames, minimum distortion and aspect ratio are calculated and compared with the existing methods [12, 13, 2, 14]. These comparisons are shown in Table 2 and the snapshots of the tested videos are described with the help of Fig. 8.

In the last picture, a dense edge created by our technique is given, which can incorporate every one of these items. In this manner, our proposed technique can well protect every one of the items' action with a significantly more modest number of casings in the dense video.

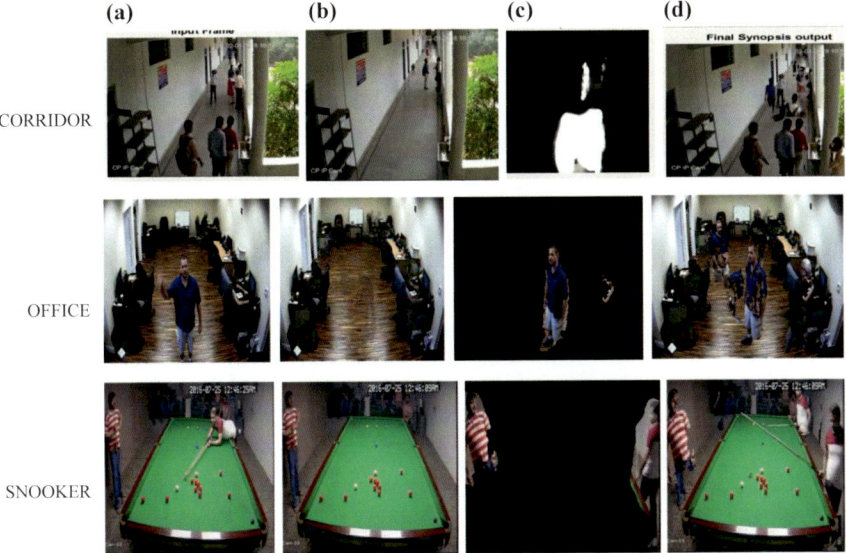

Fig. 8 Process flow of video synopsis; **a** The original video **b** Extracted background **c** Extracted foreground **d** final result video synopsis

4 Conclusion

In this paper, an effective method of video rundown is proposed. In that, the input video can be converted into frames and stored in the program. This method can utilize GUI commands from user and produce digested video. In this technique, dynamic indexing is used to track the moving objects. A facility is provided to the users for selecting keyframes in which they interested.

Acknowledgements We need to thank the accommodating remarks and recommendations from the unknown analysts. The proposed algorithm is developed by me and images which are utilized in this work are taken with the help of CCTV cameras, except office, and snooker videos. These are downloaded from the Google and open source data set.

References

1. Rubinstein M, Shamir A, Avidan S (2008) Improved seam carving for video retargeting. ACM Trans Graph 27(3):16. ACM
2. Zhong R, Hu R, Wang Z, Wang S (2014) Fast synopsis for moving objects using 925 compressed video. Signal Process Lett IEEE 21(7):834–838
3. Stauffer C, Grison WEL (1999) Adaptive background mixture models for real-time tracking. In: CVPR, pp 246–252

4. Boult TE, Micheals RJ, Gao X, Eckmann M (2001) Into the woods: visual surveillance of non-cooperative and camouflaged targets in complex outdoor settings. Proc IEEE 89:1382–1401
5. Zhang Z, Huang K, Tan T (2008) Multi-thread parsing for recognizing complex events in videos. In: Torr P, Zisserman A (eds) 10th ECCV, Part III, pp 738–751
6. Zeng W, Du J, Gao W, Huang Q (2005) Robust moving object segmentation on H.264/AVC compressed video using the block-based MRF model. Real-Time Imaging 11:290–299
7. O'callaghan D, Lew EL (1995) Method and apparatus for video on demand with fast forward, reverse and channel pause, US Patent 5,477,263, 19 Dec 1995
8. Gandhi NM, Misra R (2015) Performance comparison of parallel graph coloring algorithms on bsp model using hadoop. In: International conference on computing, networking and communications (ICNC). IEEE, pp 110–116
9. Zhong R, Hu R, Wang Z, Wang S (2014) Fast synopsis for moving objects using compressed video. IEEE Signal Process Lett 21(7):834–838
10. Li Z, Ishwar P, Konrad J (2009) Video condensation by ribbon carving. IEEE Trans Image Process 18(11):2572–2583
11. Yoo JW, Yea S, Park IK (2013) Content-driven retargeting of stereoscopic images. IEEE Signal Process Lett 20(5):519–522
12. Panagiotakis C, Ovsepian N, Michael E (2013) Video synopsis based on a sequential distortion minimization method. In: International conference on computer analysis of images and patterns
13. Ye Y, Yi-jun L, Yan-qing W (2014) An improved aco algorithm for the bin packing problem with conflicts based on graph coloring model. In: International conference on management science & engineering (ICMSE). IEEE, pp 3–9
14. Pritch Y, Rav-Acha A, Peleg S (2008) Nonchronological video synopsis and indexing. IEEE Trans Patt Anal Mach Intell 30(11):1971–1984
15. Javed O, Shafique K, Shah M (2007) Automated visual surveillance in realistic scenarios. IEEE Multimedia 14:30–39
16. Babu RV, Ramakrishnan KR, Srinivasan SH (2004) Video object segmentation: a compressed domain approach. CSVT 14:462–474
17. Oh J, Wen Q, Hwang S, Lee J (2004) Video abstraction, video data management and information retrieval, pp 321–346
18. Yeung MM, Yeo B-L (1997) Video visualization for compact presentation and 855 fast browsing of pictorial content. Circuits Syst Video Technol IEEE Trans 7(5):771–785
19. Huang C-R, Chung P-CJ, Yang D-K, Chen H-C, Huang G-J (2014) Maximum a posteriori probability estimation for online surveillance video synopsis. Circuits Syst Video Technol IEEE Trans 24(8):1417–1429
20. Feng S, Lei Z, Yi D, Li SZ (2012) Online content-aware video condensation. In: IEEE Conference 930 on computer vision and pattern recognition (CVPR). IEEE, pp 2082–2087

An Intrusion Detection and Prevention System Using AIS—An NK Cell-Based Approach

B. J. Bejoy and S. Janakiraman

Abstract The widespread use of internet in key areas has increased unauthorized attacks in the network. Intrusion detection and prevention system detects as well as prevents the attacks on confidentiality, integrity, and availability of the system. In this paper, an Artificial Immune System based intrusion detection and prevention system is designed using artificial Natural Killer (NK) cells. Random NK cells are generated and negative selection algorithm is applied to eliminate self-identifying cells. These cells detect attacks on the network. High health value cells that detect a large number of attacks are proliferated into the network. When the proliferation reaches a threshold, the NK cells are migrated into the intrusion prevention system. So NK cells in IDS are in promiscuous mode and NK cells in IPS are in inline mode. The technique yields high detection rate, better accuracy and low response time.

1 Introduction

Internet and its fabulous growth have constituted to the association of a large number of people around the world. Due to its fast evolution, all basic services now depend on the internet. This leads to a growth in attacks on basic services by malicious users. Intrusion detection and prevention system (IDPS) is used to preserve the reliability, privacy, and accessibility of the system. Any malicious activity conveyed is either notified to the administrator or some systems even have the ability to respond to an intrusion. The systems that respond to an intrusion is acknowledged as intrusion prevention system. This is done by dropping the packets or blacklisting the IP address used by the invader. Anderson [1] developed an IDS by audit trail analysis in 1980 and Denning [2] established a model that become the base of modern IDS by anomaly detection method of both user and system data.

B. J. Bejoy (✉) · S. Janakiraman
Department of Banking Technology, Pondicherry University, Pondicherry, India
e-mail: bejoybj@gmail.com

S. Janakiraman
e-mail: jana3376@yahoo.co.in

© Springer Nature Switzerland AG 2019
D. Pandian et al. (eds.), *Proceedings of the International Conference on ISMAC in Computational Vision and Bio-Engineering 2018 (ISMAC-CVB)*, Lecture Notes in Computational Vision and Biomechanics 30,
https://doi.org/10.1007/978-3-030-00665-5_86

The Human immune system (HIS) is a perfect defense mechanism that fights against diseases without any past information about these diseases. These mechanisms were impeccably similar to intrusion detection and prevention systems in the computer. The adaptation of human immune system's accomplishments to computer security emerged as a field branded as Artificial Immune System (AIS). IDPS based on AIS are generally anomaly detection models [3, 4]. In anomaly detection, a system using antigens (non-self) is fabricated using self-antigen (normal) and what the system recognizes is identified as an attack.

2 Related Work

AIS models were very effectual in designing an anomaly based IDPS system. Amalgamation of agent-based models with AIS developed a frightful combination for impostors attacking a system. AIS designs are based on four AIS Algorithms. The foremost concept is a Negative selection [4] which is commonly used to eliminate self-identifying agents. The next concept is Clonal Selection [5, 6] that is used for cloning of high fitness value agents. The third is the Danger theory [7, 8] used as a filtering technique in anomaly detection. Fourth is the Immune network [9] used mainly for communication between agents.

A dynamic real-time system was proposed by means of immune network algorithm was used for intrusion detection known as DIDAIN [10]. In this approach, three phases are used for intrusion detection. In the first phase, detectors are randomly generated and negative selection algorithm (NSA) is applied to eliminate self-identifying cell. Remaining cells are evolved into mature. During the second phase, the mature detectors are dynamically updated. When a mature detector detects another detector it is stimulated and when it is detected by another detector it is suppressed. The original intrusion detection happens in the third phase where the mature detectors are used for detecting any abnormal data. When the detector detects any data i.e. the data falls within the radius of a mature detector, the data is classified as abnormal. A multi-agent distributed intrusion detection system [11] was proposed which was used in virtual machines (VM). The system used negative selection algorithm for eliminating randomly selected detectors that identify self. During the training phase, the detectors with high fitness values were matured and migrated to the virtual machines. Clonal selection algorithm was used for cloning detectors with highest fitness value i.e. detectors that identify a large number of attacks have a high fitness value. Communication between agents takes place using immune network algorithm. The approach used static agents as well as mobile agents that travel to Virtual machines. Orchestra was used for agent communication and KVM hypervisor was used for implementing VM.

The concept of using Natural Killer Cells for Host-based intrusion detection was first devised in Fu et al. [12]. NK cells were used to detect concealed spyware in the system. If any infiltration behavior was exhibited by the program, activating signals were activated and when normal programs encounter NK cell, the inhibitory signal was activated. Low fitness NK cells were eliminated and higher fitness value cells were adapted. It even induces baits for concealed spyware to activate itself by presenting fake activities of the user. Mutation and crossover of NK cells were performed according to the fitness value of the NK cell. An Agent-based model was proposed known as ABIDS [13] that used dendritic cell algorithm (DCA). DCA is based on danger theory of HIS. The work used four types of agents. An antigen agent that sends a message to DC agent when it encounters a non-self antigen while parsing the dataset. A dendritic cell agent is distributed at the hosts that evaluate the danger value of antigens. T-Cell agents are activated when danger value exceeds a certain threshold and they inform the responding agent which response either by informing the security manager or by preventing the intrusion.

A real-time distributed network intrusion detection was proposed in Yang et al. [14] known as DAMIDAIS. Here multiple agents were used that interact with each other. Sensor agents were used to monitoring the environment. Analyzer agents were used for accumulating the information obtained from sensor agents. A manager agent is there to control the entire system. Message agents to carry the message between sensor agents. An alert agent was used to keep track of alert messages. An immune cooperation based learning (ICL) [15] was proposed based on immune cooperation mechanism. In this approach, the danger zone is not defined, but two signals were used. An antigen-specific signal from malware and an antigen-nonspecific signal from normal programs. These signals are taken together and the system is trained based on the cooperation of these two signals. A cooperative immunology based approach [16] that combines danger theory and Self/Non-Self theory for ids is presented. A sniffer module was used to capture the packets. A Non-Self Detector Module was used that compares the self-set with the captured packets and affinity was reported. A Danger detection module was used for comparing normal system profile with current profile. A vaccination Module was used that updates the system knowledge database. A result was produced by the decision module and response module was used for responses.

A Multi operational (MO) algorithm that used immune theory for fault detection was implemented [17] using Self/Non-Self theory. In this approach, random detectors were generated with variable detection radius for better coverage. For overlapped detectors, a moving detector algorithm was used. The detectors moved until the overlap measurement was less. If it does not reduce the overlap the detector radius decay was used. It was applied for fault detection in DC motor. Heavyweight agents, as well as various lightweight agents, were used for intrusion detection was proposed in Janakiraman and Vasudevan [18]. Even though the heavyweight agent increased the complexity of the system, it also increased the accuracy of the system. Ant colony optimization was used to detect intrusion in this approach. A detailed survey of AIS based IDS is presented in Bejoy and Janakiraman [19].

Even though there exists a large number of works using AIS, an effective multi-agent AIS based IDPS is still a deception and also many works in the survey lack an important parameter used for IDPS i.e. response time. These limitations motivated us to design an IDPS based on AIS using Natural Killer Cells.

3   Proposed Work

An Intrusion detection and prevention system the centered on Natural Killer (NK) Cells of AIS is projected in this paper. In the first section, we are giving an introduction to Natural Killer Cells and how artificial NK cells are used to design an IDPS. We propose an anomaly based IDPS system that can detect as well as respond to an intrusion. First, we design an IDS using NK cells and high health NK cells then migrate into the IPS thus reducing the probability of normal packets to be dropped. Random NK cells are created and the negative selection algorithm is applied to eliminate self-identifying cells. Then high health cells are cloned and based on IP threshold, effective cells are migrated into the IPS that are inline with the traffic.

3.1   Natural Killer Cells

Natural Killer (NK) Cells are lymphocytes (white blood cells) of Innate Immune System that has the capacity to detect both viral infected cells and tumors. They are cytotoxic in nature that contains special proteins that perform apoptosis or programmable cell death. NK cell activation is mainly based on a molecule known as major histocompatibility complex class 1 (MHC1) present on target cell surface. If the MHC1 class is missing NK cells perform lysis or apoptosis, hence they are called natural killers. Once considered the mainstay of innate immunity, recent studies [20] suggest that they have immunological memory also.

3.2   Artificial NK Cells

An artificial NK cell is defined as "An autonomous AIS cell that proliferates into the network and high health cells are then migrated into the Intrusion Prevention System". In this paper, we are considering NK cells with immunological memory for designing IDPS. NK cells have two type of receptors-an Activating Receptor(AR) and an Inhibitory Receptor (IR). AR is activated when an NK cell detects an antigen and an IR is activated when it encounters a normal cell.

An Artificial NK cell with immunological memory can be expressed as follows

NK (Type, AR, Health, State, CN)

Type: Types of attacks an NK cell is trained to detect. An NK cell can detect multiple numbers of attacks.

Activating Range (Ar): The range from where an NK cell can detect an attack. If the incoming MHC1 falls in the range of AR, then the NK cell detects the attack.

Health: Health value indicates the fitness value of an NK cell. When an NK cell detects an attack, its health value increases.

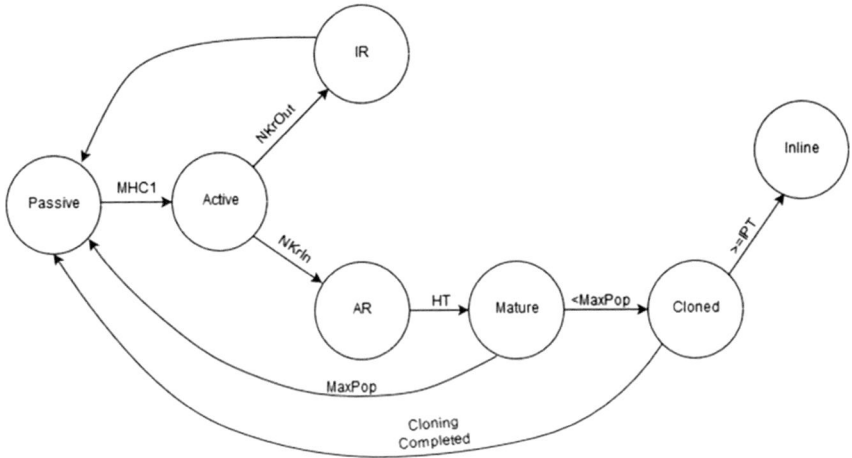

Fig. 1 State transition diagram of an NK cell

State: State indicates the states of an NK cell. There are seven states of an NK cell that are explained later. The initial state is the passive state.

CN: Number of clones indicates the total number of cloned NK cell in the network. This is the main eligibility of an NK cell to be used in the Intrusion Prevention system deployed in the network. When CN exceeds the Intrusion Prevention threshold the NK cell is migrated into the Intrusion prevention system.

An NK cell in this paper has seven states. Initially, an NK cell is in a passive state. When it encounters an MHC1 value, it changes to active state. If the MHC1 value is inside the NK cell radius (NKr), then the NK cell changes to Activating response (AR) state. If the incoming MHC1 is outside the NKr then the NK cell is in Inhibitory Response state (IR). MHC1 outside NKr means that it is a normal packet and hence no response from the NK cell. So NK cell returns back to the Passive state. On the other hand, if it's in AR means that the Nk cell detected an attack. Hence the health value is incremented. If the health value reaches a particular threshold (HT) the NK cells are in the Mature state. When NK cell is in the mature state the system checks whether the maximum population is reached (MaxPop). If MaxPop is reached the NK cell returns to the passive state. If MaxPop is not reached, NK cell is proliferated and it changes to Cloned State. When cloning is completed, the number of total clones is calculated and compared with Intrusion Prevention Threshold (IPT). If its less than the IPT, the NK cell returns back to the Passive state. If Number of clones is greater than IPT the cell is in inline state and migrated to the intrusion prevention system. After reaching IPT an NK is not proliferated hereafter. The state transition is shown in Fig. 1 and states are described in Table 1.

Table 1 States description

States	Description
Passive	Initial state
Active	When it encounters MHC1
IR	When MHC1 inside NKr
AR	When MHC1 outside NKr
Mature	When health > Health threshold
Cloned	When number < Maximum population
Inline	When number of clones >= Intru.Preven.Threshold

3.3 Working

During the initial stage, random NK cells are created based on Eq. (1).

$$\text{NKR}_i = \text{Min} \sum_{j=1}^{M} \left(C_i - R_{ij} \right)^2 \tag{1}$$

where $\text{NK}_i = (C_1, C_2 \ldots)$ and $S_j = (R_{1j}, R_{2j} \ldots)$.

S_j refers to the self -antigens and NK_i refer to NK cells and C and R are the detectors and self-antigen coordinates respectively. More than one attacks can be identified by an NK cell. This helps us to eliminate self-identifying cells and thus negative selection algorithm is used to eliminate such cells. Throughout testing phase, when an NK cell detects an attack its health value increases. When the health value reaches a threshold an NK cell is proliferated.

3.4 Working of NK Cell

MHC1 is created based on the incoming packets. MHC1 can be extracted from port number, IP address etc. MHC1 is given to NK cell. If MHC1 matches any detectors, then activating response is activated. It means the incoming MHC1 is an attack. If it does not match any detectors, then the inhibitory response is activated and MHC1 is considered normal. The working of NK cell is illustrated in Fig. 2.

3.5 NK Cell-Based IDPS Architecture

NK cell-based IDPS architecture consists of Sensors, Analyzers, IDS module and IPS module. Passive sensors are used which collect a copy of the incoming traffic and provide it to analyzers as NK cells are positioned in promiscuous mode. When

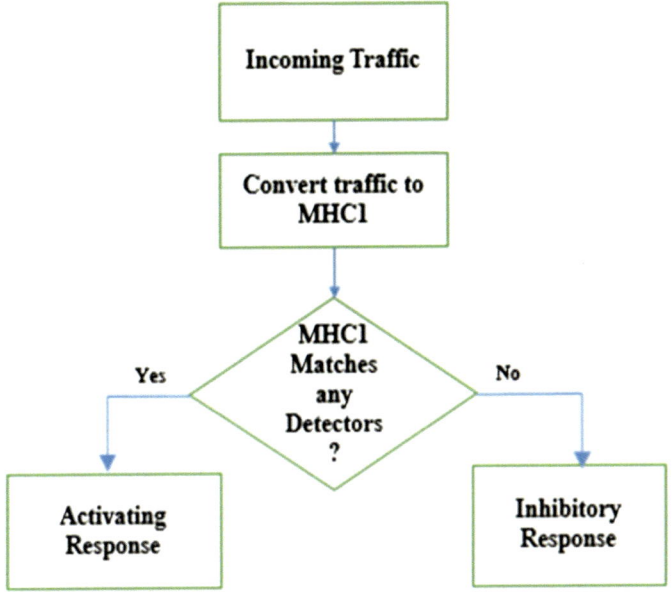

Fig. 2 Working of NK Cell

NK cells detect various attacks the health value is increased which result in the proliferation of NK cells. Analyzer extracts only useful information from the packet that is used to create an MHC1. The analyzer also accomplishes load balancing when a number of NK cells are available. All NK cells work parallelly so that the overall detection time is very quick. When the number of cloned NK cell reaches a threshold value, the NK cell migrates into the Intrusion Prevention System module that connects the NK cell to the incoming traffic in inline mode. NK cell-based IDPS is illustrated in Fig. 3.

4 Results and Discussion

We implement the work on MATLAB R2017a. 10% of NSL KDD Cup 99 dataset [21] is used as it is the only popular benchmark dataset available for IDS. The dataset is preprocessed to eliminate redundant entries and normalized before giving to the algorithm. Min–max normalization is used to make the data within a range from 0 to 1. The output was verified based on the five classes available in the dataset—Probe, DoS, U2R, R2L and Normal. The work was compared with that of [17, 18]. Four parameters were used to evaluate the work. Detection rate (DR), Accuracy (Acc), False alarm rate (FAR) and Response Time (RT) [19].

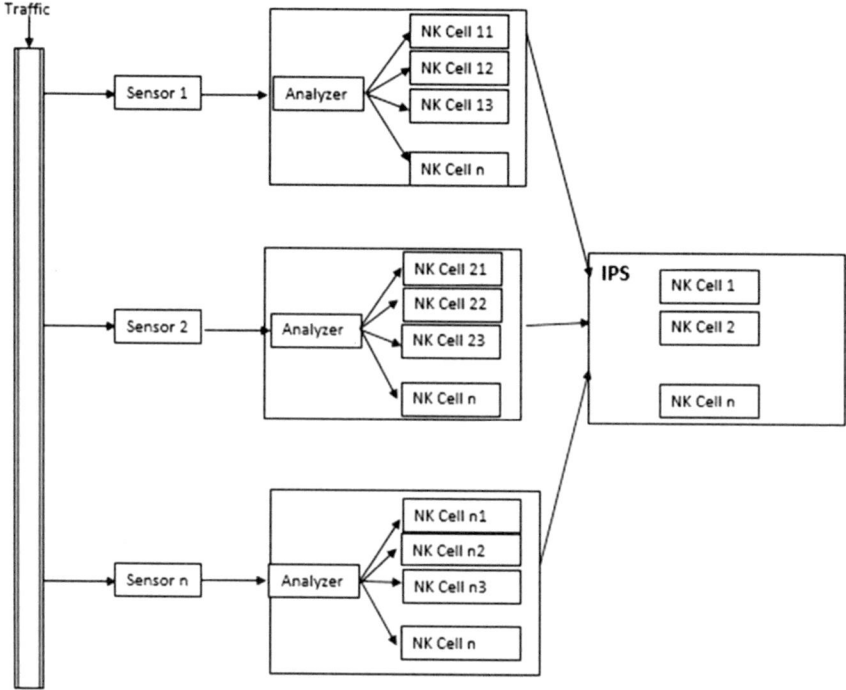

Fig. 3 NK Cell-based IDPS architecture

$$DR = TP/(TP + FN) * 100 \tag{2}$$

$$Acc = (TP + TN)/(TP + TN + FP + FN) * 100 \tag{3}$$

$$FAR = FP/(FP + TN) * 100 \tag{4}$$

$$RT = Dd + Dr \tag{5}$$

where TP-True Positive, FN-False Negative, TN-True Negative, FP-False Positive, Dd-Detection Delay and Dr-Response delay.

NK cell-based approach has an average detection rate of 96.7% whereas the other two approaches have 95.6 and 95% respectively (Fig. 4).

The average accuracy of (Fig. 5) NK cell-based approach is 99.4% whereas the other two approaches have 99.1 and 98.4% respectively.

The average FAR for NK cell-based approach is 0.4 while others have 0.7 and 1.1 respectively (Fig. 6). Low FAR of NK cell approach makes it appropriate to be used for Intrusion prevention system.

The NK cell-based approach has a quick response time than other two approaches making it an idle candidate for IPS (Fig. 7).

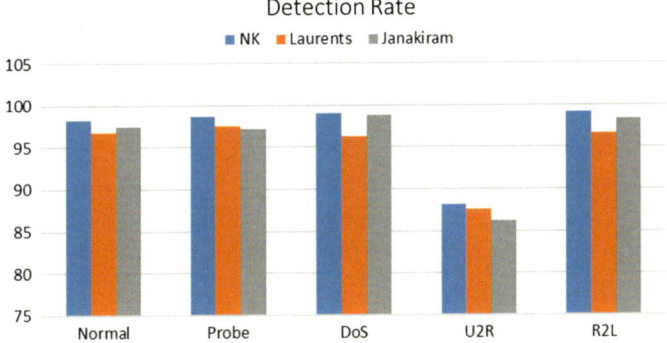

Fig. 4 Detection rate

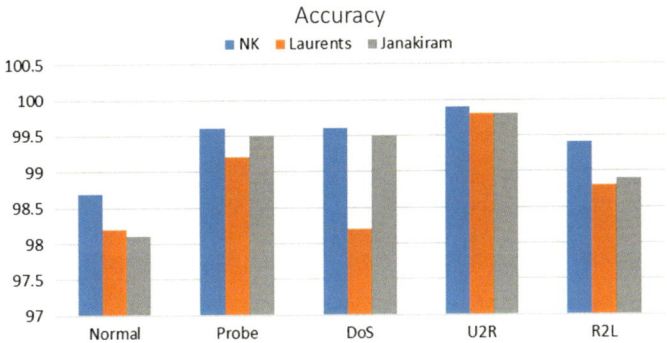

Fig. 5 Accuracy

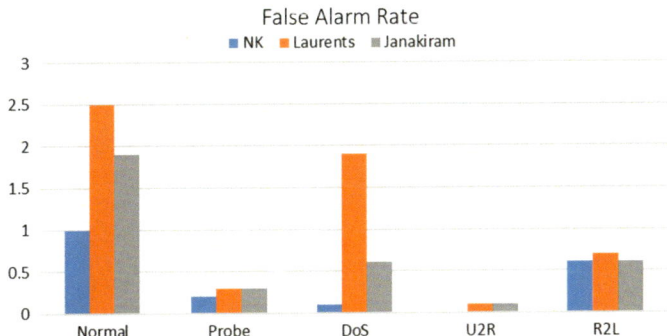

Fig. 6 False alarm rate

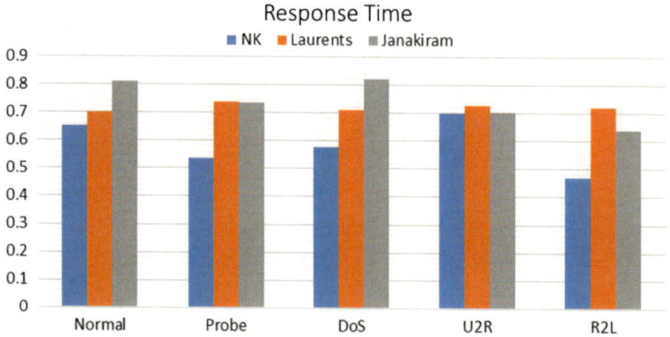

Fig. 7 Response time

5 Conclusion and Future Work

An NK cell-based Intrusion detection and prevention system is suggested. The approach has high accuracy, better detection rate, low false alarm rate and quick response time making it idle to be used in intrusion detection and prevention systems. Random detectors are generated and negative selection is used to eliminate self-identifying detectors. High health NK cells are proliferated using the clonal selection algorithm. Frequently cloned NK cells are migrated into an intrusion prevention system that can give an active response by dropping packets.

Future work includes using a Heavyweight agent that can detect new attacks as well as improve the accuracy. The system will be tested real time in a network.

References

1. Anderson JP (1980) Computer security threat monitoring and surveillance. Washing, PA, James P. Anderson Co
2. Denning DE (1986) An intrusion detection model. In: Proceedings of the seventh IEEE symposium on security and privacy
3. Forrest S, Hofmeyr SA, Somayaji A (1997) Computer Immunology. Commun ACM 40(10):88–96
4. Forrest S, Pereslon AS, Allen L, Cherukuri R (1994) Self-nonself discrimination in a computer. In: Proceedings of the 1992 IEEE symposium on security and privacy. IEEE Computer Society Press, pp 202–212
5. Burnet FM (1959) The clonal selection theory of acquired immunity. Vanderbilt University Press, USA
6. de Castro LN, Von Zuben FJ (2002) Learning and optimization using the clonal selection principle. IEEE Trans Evol Comput 6:239–251
7. Matzinger P (2002) The danger model: a renewed sense of self. Science 296(5566):301–305
8. Greensmith J, Aickelin U, Cayzer S (2005) Introducing dendritic cells as a novel Immune-Inspired algorithm for anomaly detection. Lecture Notes in Computer science, 3627. Springer, Berlin, Heidelberg

9. Jerne NK (1974) Towards a network theory of the immune system. Ann Immunol (Inst Pasteur) 125C:373–389
10. Hu X, Liu X, Li T, Yang T, Chen W, Liu Z (2015) Dynamically real-time intrusion detection algorithm with immune network. J Comput Inf Syst 11:587–594
11. Afzali Seresht N, Azmi R (2014) MAIS-IDS: a distributed intrusion detection system using multi-agent AIS approach. Eng Appl Artif Intell 35:286–298
12. Fu J, Yang H, Liang Y, Tan C (2012) Bait a trap: introducing natural killer cells to artificial immune system for spyware detection. Lecture Notes Computer Science (including Subser. Lect. Notes Artif. Intell. Lect. Notes Bioinformatics), 7597. LNCS, pp 125–138
13. Ou CM (2012) Host-based intrusion detection systems adapted from agent-based artificial immune systems. Neurocomputing 88:78–86
14. Yang J, Liu X, Li T, Liang G, Liu S (2009) Distributed agents model for intrusion detection based on AIS. Knowl-Based Syst 22:115–119
15. Zhang P, Tan Y (2015) Immune cooperation mechanism based learning framework. Neurocomputing 148:158–166
16. Sobh TS, Mostafa WM (2011) A cooperative immunological approach for detecting network anomaly. Appl Soft Comput J 11:1275–1283
17. Laurentys CA, Ronacher G, Palhares RM, Caminhas WM (2010) Design of an artificial immune system for fault detection: a negative selection approach. Expert Syst Appl 37:5507–5513
18. Janakiraman S, Vasudevan V (2009) Agent-based DIDS: a intelligent learning approach. Int J Intell Inf Process 8. Serials Publications
19. Bejoy BJ, Janakiraman S (2017) Artificial immune system based intrusion detection systems—a comprehensive review. Int J Comput Eng Technol 8(1):85–95
20. Arina A, Murillo O, Dubrot J, Azpilikueta A, Alfaro C, PÃrez-Gracia JL, Bendandi M, Palencia B, Hervás-Stubbs S, Melero I (2007) Cellular liaisons of natural killer lymphocytes in immunology and immunotherapy of cancer. Expert Opin Biol Ther 7(5):599–615
21. NSL KDD CUP 99 Dataset. http://nsl.cs.unb.ca/NSL-KDD

HORBoVF—A Novel Three-Level Image Classifier Using Histogram, ORB and Dynamic Bag of Visual Features

Vishwas Raval and Apurva Shah

Abstract Every country has its own currency in terms of coins and paper notes. Each of the currency of Individual County has its unique features, colors, denominations, and international value. Though it is easy for us to identify the denomination but for the blind people, it is a not at all possible to do this! Especially when size of currency of different denominations is same, it becomes almost impossible for them to do this and there are chances that they might got cheated by others. This paper proposes and discusses a novel three-stage image classifier algorithm for the same purpose to help the blind people in identifying denomination more accurately and to check if the currency is real or fake.

1 History of Currency

In early age of human civilization, people used to have barter systems to carry out daily routines (transactions!). Gradually, barter system became obsolete and currencies came into existence for day to day life from 2000 BC where it used to be in the form of coins. Paper-based currencies started in between 618 AD and 907 AD, in pre-modern china, during Tang dynasty. During seventh to twelfth century, the paper-based currency was introduced in the Islamic countries which later on became the base for a stable and a high valued currency called "dinar". In 1661, it was Sweden which introduced the first paper-based currency in Europe. Each country in this world has its own currency with specific denomination indicating its monetary value. US Dollars, British Pound, Japanese Yen, EURO are the examples of currencies of different countries.

V. Raval (✉) · A. Shah
CSE Department, Faculty of Technology and Engineering,
The M S University of Baroda, Vadodara, India
e-mail: vishwas.raval@gmail.com

A. Shah
e-mail: apurva.shah-cse@msubaroda.ac.in

© Springer Nature Switzerland AG 2019
D. Pandian et al. (eds.), *Proceedings of the International Conference on ISMAC in Computational Vision and Bio-Engineering 2018 (ISMAC-CVB)*, Lecture Notes in Computational Vision and Biomechanics 30,
https://doi.org/10.1007/978-3-030-00665-5_87

In India, Rupee is the main currency and printing and distribution is controlled by the Reserve Bank of India. Till 2010, "Rs." was the word which was used for Indian currency which was replaced by the symbol for ₹, designed by D. Uday Kumar, and Government of India (GoI) conferred it as an official symbol for Rupee.

Traditionally, starting from British rule, Indian currency was ranging from 1 Aana to 100 Rupees. The currency had denomination like 1, 5, 10, 20, 25, and 50 in the form of Paise. Later on, new coins of ₹1, ₹2, ₹5, ₹10 were introduced and rests of the coins have been taken back from market. The currency notes are available in a denomination value of ₹5, ₹10, ₹20, ₹50, ₹100, ₹200, ₹500, and ₹2000.

When, on November 8, 2016, Hon'ble Prime Minister of India demonetized the existing currencies of ₹500 and ₹1000 and introduced new ₹500 and ₹2000 currencies. In his speech, he told the motive behind demonetization, to curb the black money, eradiate corruption menace and prevent the terror funding and Hawala business being carried out for terrorist activities in J&K.

2 Motivation, Existing Tools, Applications and Related Work

Money is something for which people gets cheated in different ways. When it comes to blind people, it becomes nightmare for them to identify it and perform daily transactions. Talking with them, they said that they rely on others and what others say, they have to trust. In addition to this, post-demonetization, GoI introduced few new currencies of ₹200, ₹500, and ₹2,000. The dimensions of these new currencies are, in fact, smaller than regular currency notes. This has made blind people's life more miserable. Earlier, the size of ₹2,000 was the largest one, then ₹500, then ₹100 and so on. Now, the ₹500 currency is smaller in size than ₹100. So there are chances that blind people might get cheated as they do not have now any way out to identify it. Though GoI has introduced some embossing for blind people but that embossing becomes useless as the currency gets older [1]. So, the only motivation behind this work is to help the blind people in currency identification, where there are highest chances of being cheated.

In United States, all the currency denominations are of same size which makes the identification more difficult for blind people. In such situation, the Governments provide some way to help them to identify the different money denominations. The countries like Australia and Malaysia have distinct sized notes for various denominations which helps the blind people to recognize the denomination easily. In Canadian Dollars, for every denomination, there is a specific Braille mark so that the blind people can easily read those Braille marks and recognize the denomination.

Smart Saudi Currency Recognizer (SSCR) [2] is a currency recognizer for Saudi Arabia. LookTel App [3] supports the US Dollar, Australian Dollar, Bahraini Dinar, Brazilian Real, Belarusian Ruble, British Pound, Canadian Dollar, Euro, Hungarian Forint, Israeli Shekel, Indian Rupee, Japanese Yen, Kuwaiti Dinar, Mexican Peso,

New Zealand Dollar, Polish Zloty, Russian Ruble, Saudi Arabian Riyal, Singapore Dollar, and United Arab Emirates Dirham. EyeNote [4] and MoneySpeaker [5] are the other applications used for currency recognition.

Apart from these applications, many research oriented attempts, in the form of algorithms, have been carried out in order to give a robust way for currency recognition across the world since 1992. For all the work that has been carried out, the preprocessing and feature extraction are the common techniques which have been used in the initial phase. The following is the list of the classification approaches that have been used in the individual work in order to recognize the currency accurately (Table 1).

In India, the currency recognition tools are available with ATMs and Banks. But these tools are not affordable to everyone cannot afford and not handy as well. Hence, there is a real need for a system that can help the blind people in India to recognize currency properly, especially in the situations like demonetization when real currency also becomes un-useful.

3 HORBoVF—Histogram and ORB Based Bag of Visual Features Classifier

Seeing the demand of a robust, affordable and handy system to help the visually challenged people in currency identification, the authors have developed a computationally lighter computer vision algorithm. This was earlier named as *iCu₹e* [6], now *D₹ushti*, an Android App for Indian currency recognition in Indian vernacular languages. The main reason behind having an Android App is the blind people are quite acquainted in using smartphones with Talk-Back feature. The algorithm has been developed in order to provide accuracy in currency identification and fake currency recognition. At the same time, as it has been targeted for the mobile devices so care has been taken to make it faster and lighter so as to run on low memory devices too.

The overall algorithm has been divided into three stages after Image Capture, called Preprocessing, Image Classification (Identification) and Text-to-Speech Conversion. The preprocessing performs background removal in order to improve accuracy and remove outliers and irrelevant parts from the background using GrabCut [7]. The Image classification itself has been divided into three stage Filtered Classification process.

First, the captured image is converted into histogram and matched with dataset. Histogram intersection is used for finding the closest match. Here, top 10 histogram intersections are fetched. However, the histogram of a banana and a grassfield could have more intersection value and may lead to wrong identification. To filter-out any such Banana-Grassfield combination, the images with top ten histogram intersections are passed to the next layer of classification. At second layer, Oriented BRIEF (ORB) [8], alternative to SIFT and SURF, is used for feature extraction and matching. The

Table 1 List of currency recognition algorithms/work carried out for various currencies

Sr.#	Currency recognition approach	Currency	Accuracy	Year
1	Neural network [11]	USD	98.08	1993
2	Neural network, optimized mask and genetic algorithm [12]	USD	>95%	1995
3	Hybrid neural network [13]	USD & Japanese	92	1996
4	Multilayered peceptrons in NN [14]	USD		1996
5	Neural NN with Gaussian distribution [15]	USD, CA$, AU$, Krone, Franc, GBP, Mark, Pesetas		1998
6	Neural NN and axis-symmetrical mask [16]	EURO	97	2000
7	Neural NN and principal component analysis [17]	USD	95	2002
8	Back propagation NN [18]	Chinese Renminbi	96.6	2003
9	Speeded-up robust feature (SURF) [19]	USD		2007
10	Markov models [20]	USD, EURO, Dirham, Rial	95	2007
11	Ada-boost classification [21]	USD		2008
12	Artificial neural network [22]	SL Rupee		2008
13	Neural network [23]	Malaysian Ringit		2008
14	Data acquisition [24]	Chinese Renminbi	100	2008
15	Bio-inspired image processing [25]	EURO	100	2009
16	Ensemble neural network with negative correlation learning [26]	Bangladeshi Taka	98	2010
17	Local binary patterns [27]	Chinese Renminbi	100	2010
18	Wavelet transform [28]	Rials	81	2010
19	Intersection change [29]	Chinese Renminbi	97.5	2010
20	Image processing & neural network [30]	Indian Rupee		2010
21	Support vector machine [31]	Chinese Renminbi	87.097	2011
22	Speeded-up robust feature (SURF) [32]	USD	100	2012
23	Wavelet transform & neural network [33]	Dirham	99.12	2012
24	Local binary patterns and RGB space [34]	Mexican	97.5	2012

(continued)

Table 1 (continued)

Sr.#	Currency recognition approach	Currency	Accuracy	Year
25	Quaternion wavelet transform & generalized Gaussian density [35]	USD, Renminbi and EURO	99.68	2013
26	Basic feature extraction using Euler numbers [36]	Pakistani Rupee		2013
27	Number recognition [37]	Chinese Renminbi	95.92	2014
28	Instance retrieval and indexing [38]	Indian Rupee	96.7	2014
29	Radial basis Kenrel function [39]	Dirham	91.51	2015
30	Segmentation, feature extraction [40]	Bangladeshi Taka		2015
31	Region of interest (ROI), discrete wavelet transform, linear regression and SVM [41]	Indian Rupee		2015

best matched image passing through a specific threshold value is selected from the top ten images. This image is sent to the third and final stage of classification for verification purpose.

Bag of Words classification [9, 10] is one a preferred technique for image classification. Here, in the third stage, Bag of Visual Features is created for the dataset images. ORB is used to detect features and feature descriptor is computed for those images. These descriptors form a raw bag of features is created. These features are clustered using K-Means algorithm in order to create final visual vocabulary. The test image received from second stage is compared with visual vocabulary and appropriate descriptor is calculated to label the image and converted into speech using Tect2Speech. The *HORBoVF* structure is shown in Fig. 1.

The pseudo code of the algorithm is given below

1. Initialize *totalNoOfImages (N)* in the dataset, the *featureSet* with distinguishable features for all images, *imageObject*, in the dataset.
2. Initialize *featureThreshold, similarityDistanceThreshold*
3. Histogram Intersection Filtering:

   a. Generate histograms of all N images, *imageObject*, in dataset and histogram of *inputImage*
   b. For each histogram of *imageObject*, perform histogram intersection with a histogram of *inputImage*
   c. Find top K histogram intersection values and corresponding *imageObject* into *topKIntersects* and *topKImageObjects*, where $K < N$.

4. ORB based decision making:

   a. Create an ORB feature detector for *inputImage*
   b. For each *imageObject* in *topKImageObjects* with the corresponding *featureSet*, Perform feature matching:

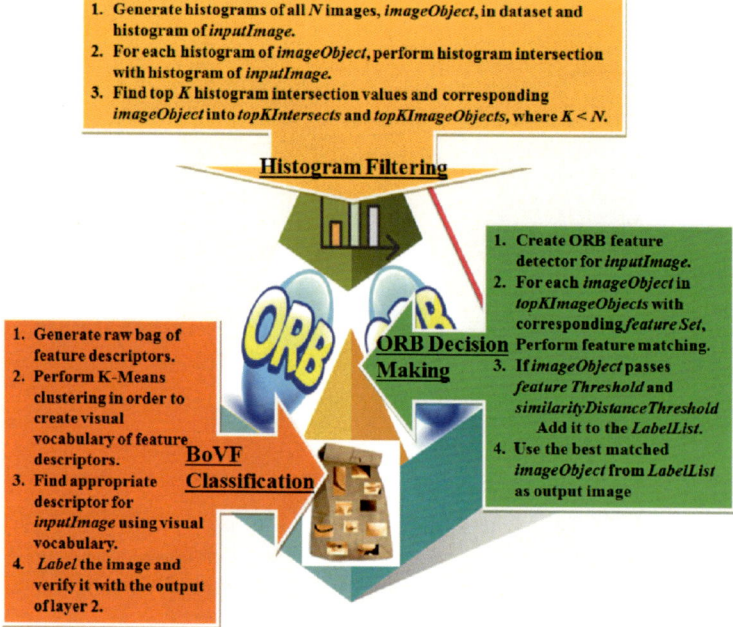

1. Generate histograms of all *N* images, *imageObject*, in dataset and histogram of *inputImage*.
2. For each histogram of *imageObject*, perform histogram intersection with histogram of *inputImage*.
3. Find top *K* histogram intersection values and corresponding *imageObject* into *topKIntersects* and *topKImageObjects*, where *K < N*.

Histogram Filtering

1. Generate raw bag of feature descriptors.
2. Perform K-Means clustering in order to create visual vocabulary of feature descriptors.
3. Find appropriate descriptor for *inputImage* using visual vocabulary.
4. *Label* the image and verify it with the output of layer 2.

BoVF Classification

1. Create ORB feature detector for *inputImage*.
2. For each *imageObject* in *topKImageObjects* with corresponding *feature Set*, Perform feature matching.
3. If *imageObject* passes *feature Threshold* and *similarityDistanceThreshold* Add it to the *LabelList*.
4. Use the best matched *imageObject* from *LabelList* as output image

ORB Decision Making

Fig. 1 The *HORBoVF* pyramid

If *imageObject* passes *featureThreshold* and *similarityDistanceThreshold* then,

Add it to the *LabelList*

c. Use the best-matched *imageObject* from *LabelList* as an output image.

5. Decision Verification using BoVF:

a. Generate raw bag of feature descriptors.
b. Perform K-Means clustering to create a visual vocabulary of feature descriptors.
c. Find appropriate descriptor for *inputImage* using visual vocabulary.
d. *Label* the image and verify it with the output of stage 4.

Note: *featureThreshold* and *similarityDistanceThreshold* are experimentally set values which may vary.

4 Testing and Conclusion

This paper proposes a three-layered image classifier for currency (image) recognition using Histogram, ORB, and Bag of Visual Features. The experiments are carried out on an Android (Kitkat) phone with 1 GB RAM and Quad-Core Max 1.6 GHz processor. Once the Visual Vocabulary is generated and loaded into the memory, which takes around 4–5 s, image classification is carried out in average 2–3 s with an average accuracy of 91.54% for $K = 24$ clusters. It has been observed also that wrong prediction occurs in case of improper capturing of image, how much portion of image covering distinguishable features is visible and mainly the distance of the object from camera.

Acknowledgements We are thankful to the Omnipotent God for making us able to do something for the society. We are thankful to our parents for bringing us on this beautiful planet. We are grateful to our department and University for providing support and resources for this work. Finally, we acknowledge the authors and researchers whose papers helped us to move ahead for this work.

References

1. http://indianexpress.com/article/india/visually-disabled-are-struggling-with-new-currency-notes-nab-demonetisation-note-ban-4621070
2. https://play.google.com/store/apps/details?id=sscr.imagemanipulations
3. http://www.looktel.com/moneyreader
4. http://www.eyenote.gov/
5. https://www.androidpit.com/app/com.hmi.moneyspeaker
6. Raval V, Shah A (2017) iCuře—an IoT application for indian currency recognition in vernacular languages for visually challenged people. In: Confluence-2017, 7th international conference on cloud computing, data science and engineering. Amity University, Noida
7. Rother C, Kolmogorov V, Blake A (2004) Grabcut: interactive foreground extraction using iterated graph cuts. ACM Trans Graph (TOG) 23(3):309–314. ACM
8. Ethan R, Vincent R, Kurt K, Gary B ORB: an efficient alternative to SIFT or SURF. In: IEEE international conference on computer vision
9. Cordelia S "Bag-of-features for image classification", Thesis, INRIA (2014)
10. Dimitri L (2006) Image classification with bags of local features. Ph.D. Thesis
11. Takeda F, Omatu S, Onami S (1993) Recognition system of US dollars using a neural network with random masks. IJCNN'93-Nagoya. In: Proceedings of 1993 IEEE international joint conference on neural networks
12. Takeda F, Omatu S (1995) A neuro-paper currency recognition method using optimized masks by genetic algorithm. In: IEEE International conference on intelligent systems, man and cybernetics for the 21st century, vol 5, pp 4367–4371
13. Takeda F, Sigeru O (1995) High speed paper currency recognition by neural networks. IEEE Trans Neural Netw 6:73–77
14. Frosini Angelo, Gori Marco, Priami Paolo (1996) A neural network-based model for paper currency recognition and verification. IEEE Trans Neural Netw 6:1482–1490
15. Masahiro T, Takeda F, Ohkouchi K, Michiyuki Y (1998) Recognition of paper currencies by hybrid neural network
16. Takeda F, Toshihiro N (2000) Multiple kinds of paper currency recognition using neural network and application for Euro currency. In: Proceedings of the IEEE international joint conference on neural networks, vol 2, pp 143–147
17. Ahmadi A, Sigeru O, Michifumi Y (2002) Implementing a reliable neuro-classifier for paper currency using PCA algorithm. In: Proceedings of the 41st SICE annual conference, vol 4, pp 2466–2468
18. Zhang E-H, Jiang B, Duan J-H, Bian Z-Z (2003) Research on paper currency recognition by neural networks. Int Conf Mach Learn Cybern 4:2193–2197

19. Chen W, Yingen X, Jiang G, Natasha G, Radek G (2007) Efficient extraction of robust image features on mobile devices. In: Proceedings of the 2007 6th IEEE and ACM international symposium on mixed and augmented reality, pp 1–2

20. Hassanpour H, Yaseri A, Ardeshiri G (2007) Feature extraction for paper currency recognition. In: 9th international symposium on signal processing and its applications, pp 1–4

21. Liu X (2008) A camera phone based currency reader for the visually impaired. In: Proceedings of the 10th international ACM SIGACCESS conference on Computers and accessibility, pp 305–306

22. Gunaratna DAKS, Kodikara ND, Premaratne HL (2008) ANN based currency recognition system using compressed gray scale and application for Sri Lankan currency notes-slcrec. In: Proceedings of World academy of science, engineering and technology, pp 235–240

23. Nurlaila H (2008) Currency recognition and converter system. Ph.D. dissertation. Universiti Malaysia Pahang

24. He J, Zhigang H, Pengcheng X, Ou J, Minfang P (2008) The design and implementation of an embedded paper currency characteristic data acquisition system. In: International conference on information and automation, pp 1021–1024

25. Parlouar R, Florian D, Marc M, Christophe J (2009) Assistive device for the blind based on object recognition: an application to identify currency bills. In: Proceedings of the 11th international ACM SIGACCESS conference on Computers and accessibility, pp 227–228

26. Debnath K, Sultan U, Shahjahan Md, Kazuyuki Mu (2010) A paper currency recognition system using negatively correlated neural network ensemble. J Multimedia 5(6):560–567

27. Guo J, Yanyun Z, Anni C (2010) A reliable method for paper currency recognition based on LBP. In: 2nd IEEE international conference on network infrastructure and digital content, pp 359–363

28. Daraee F, Saeed M (2010) Eroded money notes recognition using wavelet transform. In: 6th Iranian conference on machine vision and image processing, pp 1–5

29. Shao K, Yang G, Na W, Hong-Yan Z, Fei L, Wen-Cheng L (2010) Paper money number recognition based on intersection change. In: Third International workshop on advanced computational intelligence (IWACI), pp 533–536

30. Gopal K (2010) Image processing based feature extraction of Indian currency notes; MTech Thesis. Thapar University

31. Yeh C, Wen-Pin S, Shie-Jue L (2011) Employing multiple-kernel support vector machines for counterfeit banknote recognition. Appl Soft Comput 11(1):1439–1447

32. Hasanuzzaman F, Xiaodong Y, Yingli T (2012) Robust and effective component-based banknote recognition for the blind. IEEE Trans Syst Man Cybern (Appl Rev) 42(6):1021–1030

33. Ahangaryan F, Mohammadpour T, Kianisarkaleh A (2012) Persian banknote recognition using wavelet and neural network. In: International conference on computer science and electronics engineering (ICCSEE), pp 679–684

34. García-Lamont F, Jair C, Asdrúbal L (2012) Recognition of Mexican banknotes via their color and texture features. Expert Syst Appl 39(10):9651–9660

35. Gai S, Guowei Y, Minghua W (2013) Employing quaternion wavelet transform for banknote classification. Neurocomputing 118:171–178

36. Ali A, Mirfa M (2013) Recognition system for Pakistani paper currency. World Appl Sci J 28(12):2069–2075

37. Yu H, Yingyong Z (2014) Study on money number recognition arithmetic. Int J Multimedia Ubiquit Eng 9(11):189–196

38. Suriya S, Shushman C, Vishal K, Jawahar C (2014) Currency recognition on mobile phones. In: Proceedings of IEEE 22nd international conference on pattern recognition, pp 2661–2666

39. Sarfraz M (2015) An intelligent paper currency recognition system. In: Procedia Computer Science-65, pp 538–545

40. Saifullah S, Rahman M, Hossain Md (2015) Currency recognition using image processing. Am J Eng Res 4(11):26–32

41. Pham T, Danh Y, Seung Y, Dat T, Husan V, Kang R, Dae S, Sungsoo Y (2015) Recognizing banknote fitness with a visible light one dimensional line image sensor. Sensors 15(9):21016–21032

Neural Network Based Image Registration Using Synthetic Reference Image Rotation

S. Phandi and C. Shunmuga Velayutham

Abstract Typical image registration techniques use a set of features from a target and reference images and search in the affine transformation space using a similarity metric. Neural Networks typically have employed two choices—geometric transformations to find correlation between images and a similarity metric. In this paper, however, we have proposed and employed a simple and effective method for image registration using neural networks. The image registration has been formulated as a classification problem. By generating and learning exhaustive synthetic reference image transformations appropriate re-transformation for target image is computed for effective registration. The proposed work is tested on satellite imagery.

1 Introduction

Image Registration is the process of aligning multiple images of same scene into a single integrated image with a common co-ordinate system. In essence, this spatial alignment of images essentially involves determination of appropriate geometrical transformation that aligns one image with reference to a particular image. Registration is often carried out primarily for information fusion—multiple images offer more comprehensive information than provided by individual images. Typically, registration is most often employed in medical [1, 2] and satellite imagery [3–5] for multi-view, multi-temporal and multi-modal analyses. In both application domains, factors related to image acquisition devices, images themselves and objects in images often make registration a hard problem.

Most image registration methods, proposed in literature, typically involves *feature space* (used for matching), *search space* (of geometrical transformations), *a*

S. Phandi · C. Shunmuga Velayutham (✉)
Department of Computer Science and Engineering, Amrita School
of Engineering, Coimbatore, Amrita Vishwa Vidyapeetham, India
e-mail: cs_velayutham@cb.amrita.edu

S. Phandi
e-mail: cb.en.p2cvi16009@cb.students.amrita.edu

© Springer Nature Switzerland AG 2019
D. Pandian et al. (eds.), *Proceedings of the International Conference on ISMAC in Computational Vision and Bio-Engineering 2018 (ISMAC-CVB)*, Lecture Notes in Computational Vision and Biomechanics 30,
https://doi.org/10.1007/978-3-030-00665-5_88

search strategy and *a similarity metric*. Extrinsic methods [6], surface methods [7], moments and principal axes methods [8], correlation-based methods [9], mutual information based methods [10], and wavelet-based methods [11] are various image registration methods to cite by way of few examples are some of the methods of registration in literature. Soft computing based methods [12] are relatively recent and have acquired prominence by virtue of their capability to handle uncertainty effectively. Among softcomputing techniques, Neural Networks have been one of the popular and effective techniques employed for image registration.

Against the iterative optimization procedures, Neural Networks possess the ability to learn the input-output relationship from a data set with no prior knowledge. Typical image registration techniques use a set of features from a target and reference images and search in the affine transformation space using a similarity metric. Neural Networks, in the literature, typically have employed two choices—geometric transformations to find correlation between images and a similarity metric.

In this paper, however, we have proposed and employed a simple and effective method for image registration using neural networks. By generating and learning exhaustive synthetic reference image transformations (restricted only to rotation in this work), appropriate re-transformation for target image is computed for effective registration. The proposed work is tested on satellite imagery.

This paper is organized as follows. Section 2 presents the related works. Section 3 details the proposed neural network based image registration method. Section 4 presents experimental design, simulation results and analysis. Finally, Sect. 5 concludes the paper.

2   Related Works

In case of registration, same co-ordinate points from reference image and target image is often determined with the help of common features (like edge information, edge outline [13]) that are present in both images. In simplest form, features extracted from distorted image are trained using back propagation algorithm [13]. Target image coordinates are then fed into the trained network. In [14], Radial Basis function neural network has been used (with lesser training time) to determine geometrical transformation. DCT (Discrete Cosine Transform) co-efficient and moments are extracted from reference image and are used to train the network.

In [15], fourier transform of both reference and target images are fed into a feedforward network to estimate transformation parameter viz. translation, rotation, and magnification to register images. Feature-based image registration in [16], investigates the efficacy of Speeded Up Robust Feature (SURF), Scale Invariant Feature Transform (SIFT), Maximally Stable Extremal Regions (MSER) features for registering images. These features are invariant to zoom, noise, rotation, and illumination and are suitable for image registration. RANSAC algorithm is used for outlier removal and Recall as well as RMSE measures are used to estimate quality of transformation. In case of high resolution images, spatial information [17] is very essential for effective registration.

Texture analysis is employed to find texture boundaries for satellite image registration in [3]. GLCM (Gray-Level Co-Occurrence Matrix) is used for texture analysis of images. Pattern recognition has been used to estimate rate of changes in satellite images for registration [4]. In [5], neural network has been employed to classify pattern like buildings, farm land etc. in satellite images.

3 Proposed Work

Essentially, the problem of image registration can be reduced to finding the appropriate transformation between the target and reference images. The typical transformation model used for image registration is affine transformation since it is sufficiently general by virtue of its capability to handle translations, rotations and scaling. This affine transformation, for a general 3D case can be represented as

$$x' = A \cdot x + b,$$

where A (a 3×3 matrix) embeds the rotations and scaling, x, x' and b are three dimensional arrays and respectively represent the original position of feature vectors, the transformed positions and the translation information.

Typical image registration techniques use a set of features from a target and reference images and search in the affine transformation space using a similarity metric. Neural Networks are by no means an exception to this. This demands that image registration needs two choices viz. the geometric transformations to be considered for finding correlation between images and a suitable measure of match. The proposed method, in this paper, formulates finding appropriate affine transformation as a classification problem. By generating and learning exhaustive reference image transformations (restricted only to rotation in this work), appropriate re-transformation for target image is computed for effective registration. This alleviates the need for a similarity metric which often has a considerable impact on the registration process. Algorithm for the proposed method is explained below

1. SURF features from Reference image is taken at each degree of rotation.
2. SURF features from Target image is taken without rotation of image.
3. Train the reference image features using Patternet architecture.
4. Match the Trained reference image features with target image features.
5. Find Magnitude and Direction between matched Reference and Target image features.
6. Find angle of rotation and register the image.

The reference image is rotated a full circle, say with $1°$ each, thus obtaining 360 images. From each of those rotated images, N features are extracted. This $360 \times N$ matrix forms the feature matrix that forms input to the neural network. Each rotated image represents a single class, thus going with our example we essentially have 360 classes! So, 360×1 class labels forms the teaching signal. The feature matrix along with the class label vector forms the training set for the neural network.

The neural network has been trained on the above said training set about different classes of rotated images. After having learnt the rotation classes, features are extracted from the target image without any rotation. This feature vector (of size $1 \times N$) serves as a test vector which when administered into the neural network will get classified. Then with the features from trained reference image and target images features are matched then into an appropriate rotation class. The rotation class, thus identified, can be considered as the rotation the target image has undergone during the imaging process. Consequently, the target image has to be rotated counterclockwise $x°$ assuming that the target image has been classified under class x. After having rotated, the target image will be registered with the reference image.

4   Simulation Results

Pattern recognition network (*patternet*) which is part of MATLAB's Neural Network toolbox has been employed to learn and classify the synthetically generated reference image transformations. Patternet is essentially a feedforward that classify input as per the classes. The implementation of patternet demands that the class labels should consist of vectors of all zeros except for a 1 in location i where i is the class they are to represent. So, instead of 360×1 class label vector, we employ 360×360 matrix which is more of an implementation specific requirement.

Figure 1 shows a schematic of patternet architecture employed for the simulation experiments. As can be seen from the figure, 20 SURF features, form the input, with 30 hidden layers and 360 output classes accounting for 360 times 1° rotations.

The reference image, as detailed in the previous section, is rotated clockwise 360° and for each degree 20 strongest features were extracted using SURF feature extraction method [18]. So, the input data for patternet is essentially a 360×20 matrix and the supervisor signal for patternet is a 360×360 matrix containing class

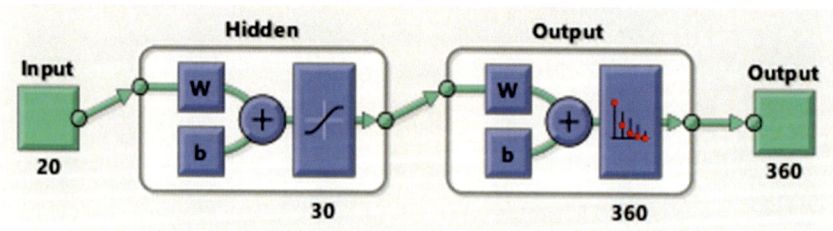

Fig. 1 Patternet architecture employed in this study

Fig. 2 Reference image

Fig. 3 Target image

labels for each of the 360 classes. The SURF features of target image (without any rotation) is also extracted to be used a target vector.

A pair of satellite images provided by National Remote Sensing Centre (NRSC), Hyderabad, India serves as the reference and target images. Figures 2 and 3 show the reference and target images used for the simulation experiments respectively.

Figure 4 shows the matched features between trained reference image and target image features. With the help of matched features, both image magnitude and direction of image is determined. We also experimented the rotation with 0.2° granularity. Table 1 shows the details of both the simulation experiments. It explained about 1° and 0.2° rotation. It took 350 iterations with gradient value of 1.84e-05 and performance value as 0.00749 with training time 57.54 min for 1° rotation. Similarly 389 iterations with gradient value of 6.42e-05 and performance value around 0.00748

Fig. 4 Image obtained from matching features at an angle of 90.20° rotation

Table 1 Simulation results

S. No.	Degree of rotation	No. of iteration	Gradient value	Performance value	Training time
1	1°	350	1.84e-05	0.00749	57:54 min
2	0.2°	389	6.42e-05	0.00778	11:26:52 h

Fig. 5 Registered image by 1° rotation

with training time 11:26:52 h for 0.2° rotation. Figures 5 and 6 respectively show the registered images after 1° and 0.2° rotation experiments. Finally Figs. 5 and 6 shows the registration of image at 1° and 0.2° rotation.

Fig. 6 Registered image by 0.2° rotation

5 Conclusion

In this paper, we have employed a simple and effective method for image registration using neural networks. The method involved generating and learning exhaustive synthetic image transformations (primarily rotation). Essentially, the reference image is rotated every 1° for 360 times and SURF features are extracted for each rotation with each rotation labelled as a class. Thus image registration is formulated as a classification problem. Once the patternet (the neural network employed) learnt the rotation classes, the target image, as is without any rotation, has been tested with the neural network to estimate the rotation the image might have undergone during the imaging process. After undoing the rotation (by rotating the image counter clockwise) the target image has been registered with the reference image. This registration method has been tested with satellite imagery. The granularity of the rotation as well as considering synthetic affine transformation will be part of the future work.

References

1. Menon HP, Narayanakutty KA (2010) Applicability of non-rigid medical image registration using moving least squares. Int J Comput Appl 1(6):79–86
2. Menon HP, Narayanakutty KA, Indulekha TS (2014) Feature point selection using structural graph matching for MLS based image registration. Int J Comput Appl Found Comput Sci 100(4):18–23
3. Balamurugan C (2013) Automated feature extraction from high resolution satellite imagery using ANN. Asian Rev Civil Eng 1:1–6
4. Gautier C, Somervile RCJ, Volfson LB (1986) Pattern Recognition of Satellite cloud imagery for improved weather prediction. National Aeronautics and Space Administration
5. Bhadran B, Nair JJ (2015) Classification of pattern on high resolution SAR images. In: Conference on computing and network communication. IEEE, pp 784–792

6. Wyawahare MV, Patil PM, Abhyankar HK (2009) Image registration techniques: an overview. Int J Signal Process 2:11–28
7. Xie Z, Pena JT, Gieseg M, Liachenko S, Dhamija S (2015) Segmentation by Surface to Image Registration. Virtual Scopics
8. Bulow H, Dooley L, Wermser D (2000) Application of principal axes for registration of NMR image sequence. Pattern Recognition Letters
9. Rao YR, Prathapani N, Nagabhooshanam E (2014) Application of normalized cross correlation to image registration. Int J Res Eng Technol 3:12–16
10. Kosinski W, Michalak P, Gut P (2012) Robust image registration based on mutual information measure. J Signal Inf Process 3:175–178
11. Mekky NE, Abou-Chadi FE-Z, Kishk S (2011) Wavelet based Image registration technique: a study of performance. Int J Comput Sci Netw Secur 11:188–196
12. Costin H, Rotariu C (2011) Medical Image Processing by using soft computing and information fusion. WSEAS international conference on mathematical methods and computational techniques and intelligent system
13. Wang S, Lei S, Chang F (2008) Image Registration based on neural network. In: International conference on information technology and application in biomedicine. IEEE, pp 74–77
14. Sarnel H, Senol Y, Sagirlibus D (2008) Accurate and robust image registration based on radial basis neural network. In: International symposium on computer and information science. IEEE
15. Abche AB, Yaacoub F, Maalouf A, Karam E (2006) Image registration based on neural network and fourier transform. In: Engineering in medicine and biology society. IEEE, pp 4803–4806
16. Krishna S, Varghese A (2015) Feature based automatic multiview image registration. Int J Comput Sci Softw Eng 4:308–314
17. Ariputhiran G, Gandhimathi S (2013) Feature extraction and classification of high resolution satellite images using GLCM and back propagation techniques. Int J Eng Comput Sci 2:525–528
18. Karami E, Prasad S, Shehata M (2015) Image matching using SIFT, SURF, BRIEF, and ORB: performance comparison for distorted image. In: Newfoundland electrical and computer engineering conference

Design and Development of Efficient Algorithms for Iris Recognition System for Different Unconstrained Environments

M. R. Prasad and T. C. Manjunath

Abstract One of the important concepts of identification of human beings in various sectors across the universe is the biometrics. In this paper, a brief report of the biometric recognition is being presented in a nutshell. This paper gives a brief conceptual view of the research work done on the topic titled, "Design and Development of Efficient Algorithms for Iris Recognition System for Different Unconstrained Environments" as the research topic chosen.

1 Introduction to the Research Work

One of the important concepts of identification of human beings in various sectors across the universe is the biometrics [1, 2]. This biometrics is defined as the art of identifying a human being by different methods. Identifying or verifying one's identity using biometrics is attracting considerable attention in this modern day automated world, one of the main reasons being the security issues in various places. Biometrics is the beautiful science of automatic identification of individuals that uses the unique physical or behavioral traits/characteristics of individuals to recognize them. Since biometrics is extremely difficult to forge and cannot be forgotten or stolen, biometric authentication offers a convenient, accurate, irreplaceable, and high secure alterna-

M. R. Prasad (✉)
Department of CSE, VTU RRC-Belagavi, Belgaum, Karnataka, India
e-mail: mrp.prasad@gmail.com

M. R. Prasad
Department of Computer Science and Engineering, JSS Academy
of Technical Education (JSSATE), Dr. Vishnuvardan Road (Kengeri-Uttarahalli Road),
Srinivapura Post, Bengaluru 560060, Karnataka, India

T. C. Manjunath
Department of ECE, Dayananda Sagar College of Engineering, Bangalore,
Karnataka, India
e-mail: dr.manjunath.phd@ieee.org

© Springer Nature Switzerland AG 2019
D. Pandian et al. (eds.), *Proceedings of the International Conference on ISMAC
in Computational Vision and Bio-Engineering 2018 (ISMAC-CVB)*, Lecture Notes
in Computational Vision and Biomechanics 30,
https://doi.org/10.1007/978-3-030-00665-5_89

tive for an individual, which makes it have more advantages over the traditional cryptography-based authentication schemes [3].

2 Brief Insight into the Literature Survey

In the recent digital era of the current central government, biometrics has been made compulsory in all the places (e.g., UID Aadhar, fingerprint, PAN, etc.). We had seen that even though there were lot of biometric methodologies, each one was suffering from one or the other drawbacks. Finally, in this context after studying the implications of each of the biometric methodologies, we arrived at the selection of the iris as the best method of biometric identification of human beings due to its large number of advantages [4]. Also, due to the current initiative taken up by the central government in the field of biometrics to be implemented in all sectors, this had further motivated us to take up the research work on the iris biometric field. This has made us identify the problem because of its vast application in each and every sector in this automated digital world [5].

Hence, in continuation, with zeal of this research work after making a thorough survey, we proposed some methodologies for the automatic recognition of biometric using iris by developing some algorithms in Matlab/LabVIEW, the problem finally, being defined as the research problem statement as "Design and Development of Efficient Algorithms for Iris Recognition System for Different Unconstrained Environments" with a brief analysis and the same was presented in the form of an exhaustive literature survey.

In majority of the work done by the various researchers in the relevant field [6–9], there were certain drawbacks, disadvantages, lacunas such as high computation to achieve good accuracy, longer execution time, large amount of storage to achieve secure authentication, and methodology failed when there was a noise like reflection in the image, thereby producing complex or extra edges of images; in fact, few works were done on increasing the performance and accuracy of the system w.r.t. high-speed computations. Some of the drawbacks were considered in our work and algorithms were developed, which was verified through effective simulation results in the Matlab and LabVIEW environments (software and hardware implementation) along with an application development incorporating the microcontroller for implementation.

3 Objective of the Research Work

The main objective of the research work is to develop image processing algorithms in Matlab/LabVIEW environments for the biometric identification of human beings through the iris part of the human eye under unconstrained environments [2, 10]. The other objective being to develop high-speed and efficient algorithms to overcome the security and recognition problems faced in many of the existing biometric authentication fields, to develop a system to work properly with all the types of human eyes and to improve the performance of the commonly used existing algorithms. The main motto of our research work is to make use of the iris images taken under unconstrained environments [11–14] and to develop algorithms to correctly identify an individual is present or not. The major scope of the proposed research work is mainly to develop some hybrid algorithms to overcome the security and recognition problems faced in many fields of the works as mentioned above, to develop an iris recognition system to provide fast identification and recognition for all human beings, to work properly with all eyes of different sizes using different types of classifiers, and finally to improve the performance of the commonly used existing algorithms.

4   Proposed Block Diagrams for the Recognition Process

The proposed overall block diagram of the developed authentication system makes uses of only two phases, viz., the iris enrolment/training phase and the iris recognition/evaluation/testing phase [15]. The iris images used in the research work were acquired from the CASIA-IrisV4 databases, which consists of all types of constrained and unconstrained image datasets. The complete iris scan recognition system could be summarized as consisting of different blocks with each block having its own functionality and all the blocks are used in our research work. They are the database (general/generated one), image acquisition/capturing, grayscale conversion, identification of ROI [16], preprocessing, resizing, boundary detection, segmentation [17–20], localization, normalization, noise removal, enhancement, feature processing, feature extraction, feature encoding, matching, classifiers, testing, decision taking, authentication, identification, recognition/matched, and the non-recognition/unmatched blocks.

The time consumption of the biometric authentication system developed was also very low, as it can identify an iris within few seconds. This compilation time includes all the times that were taken by the various processes mentioned precedingly. In fact, the proposed algorithms were worked out in the abovementioned phases. Successful simulations and experimentations were carried out, and encouraging results were achieved, thus claiming that the proposed systems are capable of fast and efficient iris identification over the existing ones under extreme conditions. In this context, it is iterated that nine contributions were developed as a part of the research work undertaken in the field of iris biometric recognition systems under unconstrained environments. Those nine contributions involved six in Matlab and three in Lab-VIEW, out of which one is a LabVIEW application of the voting process followed by the hardware implementation of the iris recognition system with a microcontroller using LabVIEW.

5 Proposed Identification Schemes in Seven Stages

The proposed identification scheme involved seven important stages in the detection process, viz.,

Stage 1   Preprocessing, segmentation, normalization, and enhancement,
Stage 2   Feature extraction,
Stage 3   Classification or matching,
Stage 4   Functional block diagram,
Stage 5   Overall working of the developed IRS,
Stage 6   Simulation results and the finally,
Stage 7   Conclusions.

The flow of our implemented research work is explained as follows.

In stage 1, the input is an eye image of a person who wants to get the authorization to the iris recognition system and the next step is preprocessing of the image considered (taken from CASIA-V4 database). In preprocessing step, the original image is enhanced and then the image iris is localized; once after localizing, the iris of the image pupil is localized by removing the occlusion present in the eye image. The various preprocessing operations are done with the help of Canny edge detection, Hough transforms, Fuzzy trapezoidal, Sobel operator method, OTSU algorithm, morphological operators, and the boundary detection methods. It has to be noted that in our work, the hybrid combination of preprocessing and segmentation concepts were used to obtain high degree of accuracy. In this content, normalization, segmentation, enhancement, and noise removal are also carried out, thus obtaining a noise-free, good resolution image that could be used further to extract its features [21, 22].

Coming to stage 2, which is the feature extraction stage, the various features of iris images are extracted using the concepts of local binary pattern features method, Gabor wavelets, log Gabor convolution wavelet method, hybrid SVD decomposition method, SFTA method, gray-level co-occurrence matrix method, fruit fly with cuckoo search algorithm, heuristic algorithm, local binary pattern features method, and the Haar wavelet methods. Once the features are extracted, it is stored in a refined database, sometimes being called the knowledge database. It has to be noted that in the work considered, the hybrid combination of different feature extraction methodologies was used to obtain high degree of accuracy. After all, the abovementioned steps in stage 2 were carried out to compare the feature vectors of the iris image with already stored feature vectors in database as templates (reference code).

So, the final step was to classify the test iris image by using different types of classifiers to classify the extracted iris into recognized ones and the unrecognized ones, which was considered in stage 3. Also in some cases, instead of classifiers, the matching techniques have been used to match the test iris with the already stored iris in the database. The different types of classifiers that were used for classification purposes are the SVM, ANNs, Multi-SVM, neural networks, and the radial basis functions. The different types of matching techniques that were used for matching purposes are the Hausdorff matching, surf matching, and the Hamming distance methods. Thus, by using these classifiers as a result if the stored feature vectors and the extracted template of test iris image are one, the same person is verified and authenticated, else it is rejected. It has to be noted that in the work considered, the hybrid combination of classifiers was used to obtain high degree of accuracy. If the match is found, then the iris is recognized, and the person is found in the database, else it is not recognized, i.e., the person is not there in the database.

In stage 4, all the block diagrams for the various contributions are proposed using hybrid combinations of the various processes used in the first three stages, followed by the brief working of the recognition process in stage 5. Codes are developed in the Matlab environment as .m files or .vi files in stage 6.

The developed .m/.vi files are run, the test query image is given as the input to the developed code, and after running the simulation, the simulation results are observed for the various contributions for the two cases, viz., matched (recognized) and unmatched (not recognized). Images which are captured under different unconstrained environments such as bad light, bad illumination, iris captured at a distance, at an angle or having parallax error, squint eyes, tilted iris, wearing glasses, eyes affected with diseases or having cataract done, etc. are given as the input to the proposed algorithms which are compared with the iris part which is already in the database; if the match occurs, then the person is verified or authenticated or matched or recognized, else it is rejected. Other type being the classifier approach used to classify the recognized irises and the unrecognized irises. All these methods proved to be more effective when the algorithms were developed and compared with the existing ones, thus establishing the supremacy of our methods over the others.

The proposed research work gives a simple and stable solution for iris recognition system for secure authentication in unconstrained environments. Although there are numerous works done on the chosen subject [23, 24], the proposed work stands unique with its ability of providing excellent results for various unconstrained cases of images considered in this work. The proposed algorithm/s focusses on the methods for rapid and accurate iris identification and authentication, especially in the field of preprocessing, feature extractions, and in the case of classification and matching of the irises. Satisfying and convincing results are obtained for all the defined objectives of the proposed algorithms which could be seen in a separate section of the simulation results presented under each contribution. The different algorithms which have been developed have provided very good results when it was implemented on CASIA-Iris Version 4 database along with better accuracy, high performance, and less error rate. At the end, also the results have been compared with some of the existing technologies and w.r.t. the work done by other authors to establish the supremacy of the proposed methods with them.

All the different proposed methods developed gave satisfying results for the recognition process, but the performance metrics, accuracy, differed among them. Performance metrics was also plotted to judge the best methodology. At the same time, in each contribution, a user-interactive automatic graphical user interface (GUI) was also developed as a part of the research work. This GUI takes the test pattern, compares with the database available, and if it is existing, immediately authenticates it well within couple of seconds, else it rejects the test pattern saying not authenticated or recognized, thus concluding the iris recognition process in the final stage presented as "conclusions" in stage 7.

6   Brief Insight into the Iris Image Databases

A brief review of the iris image databases that are being used in our research work was also explored. In fact, an image database can be defined as a collection of image data in .jpeg format, and typically all the images are associated with the activities of one or more related organizations, and thus the databases focus on the organization of images and its metadata in an efficient manner. There are different types of image databases and in our work, we have mainly concentrated on the CASIA-IrisV4 database for the analysis purposes.

Next, the implementation of the automatic biometric iris recognition system under unconstrained environments using the proposed methodologies is being presented in a nutshell. It also describes various steps that are used in the proposed methodology. In order to achieve the better accuracy, performance, and error rate than the existing methods, nine different iris recognition system techniques (software and hardware approaches) have been proposed which involve different preprocessing, feature extraction technique, and matching/classification algorithms. The entire work is presented in three stages in this research paper, viz.,

- design of the algorithms for iris recognition,
- the simulation results, and
- the development of an automated GUI for iris recognition.

7   Contributions of the Research Work

A brief introduction to the nine contributed works: The first six contributory works (software implementations in Matlab), next two contributory works (software implementations in LabVIEW) followed by the last contributory work (hardware implementations using a LabVIEW coupled with C), uses the concepts of preprocessing, edge detection, segmentation, normalization, feature extraction, matching, and classification for the recognition of an iris of any human being under the unconstrained environments.

Preprocessing (P), segmentation (S) and normalization (N) use the following:

1. PSN 1: CED—Canny Edge Detection and CHT—Circular Hough Transforms,
2. PSN 2: HT—Hough Transform and CED—Canny Edge Detection,
3. PSN 3: FTM—Fuzzy Trapezoidal and SOM—Sobel Operator Method,
4. PSN 4: OTSU—OTSU Threshold Values and BDM—Boundary Detection Method,
5. PSN 5: MO—Morphological Operators,

6. PSN 6: CED—Canny Edge Detection method and HT—Hough Transforms,
7. PSN 7: BD—Boundary Detection method,
8. PSN 8: BD—Boundary Detection method, and
9. PSN 9: MO—Morphological Operators.

Feature extraction (FE) use the following:

1. FE 1: LBP—Local Binary Pattern method,
2. FE 2: GW—Gabor Wavelets,
3. FE 3: GCW—1D Log Gabor Convolution Wavelet and HSVD—Hybrid SVD,
4. FE 4: SFTA—SFTA method,
5. FE 5: GLCM method and fruit fly with cuckoo search algorithm, heuristic algorithm,
6. FE 6: LBP—Local Binary Pattern method,
7. FE 7: 2D Gabor Wavelets and Haar Wavelets,
8. FE 8: 2D Gabor Wavelets and Haar Wavelets, and
9. FE 9: LBP—Local Binary Pattern method.

Classification (CN) use the following:

1. CN 1: m-SVM—Multi-SVM,
2. CN 2: HD—Hamming Distance,
3. CN 3: HM—Hausdorff Matching and SM—Surf Matching,
4. CN 4: KDB—Knowledge Data Base and ANN—Neural Network Algorithm,
5. CN 5: RBFNN, Neural Network, and SVM,
6. CN 6: HDM—Hamming Distance Method,
7. CN 7: ANN—Artificial Neural Network,
8. CN 8: ANN—Artificial Neural Network, and
9. CN 9: ANN—Artificial Neural Network.

Overall working of the six proposed methodologies in Matlab: The overall working of the developed methodology/flow diagram of the proposed method, i.e., the flow of research work, is as shown in Fig. 1 and is summarized as follows.

- Input is an eye image of a person who wants to get the authorization to the system and the next step is preprocessing of the inputted image considered.
- In preprocessing step, the original inputted image is enhanced and then the image iris is localized, and once after localizing, the iris of the image pupil is localized by removing the occlusion present in the eye image; different preprocessing techniques such as CED, CHT, HT, FTM, SOM, OTSU, BDM, and the MO methods are used for processing before the features are extracted [25].
- Then, the various features of an iris image are extracted using hybrid combination of LBP, GW, GCW, HSVD, SFTA, GLCM, FF, and CSA methods [26].
- After all, in the abovementioned steps, these feature vectors of an iris image are then made ready to compare with the already stored feature vectors in database as reference templates (reference code).
- So, the final step is to classify the test iris image by using hybrid combinations of classifiers that are m-SVM, HD, HM, SM, KDB, ANN, and the RBFNN methods.

- Thus, by using these classifiers as a result, if the stored feature vectors and the extracted template of test iris image are one and the same, then person is verified and recognized, else it is not recognized.

The general data flow diagram employed in this work for the nine contributory works is shown in Fig. 1.

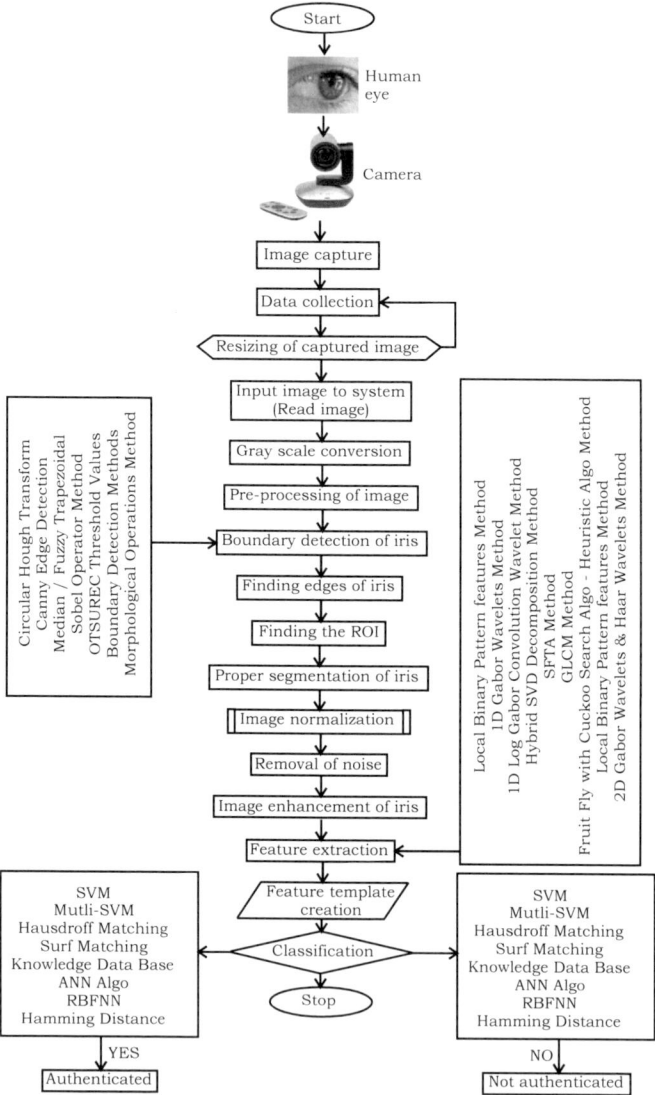

Fig. 1 General flowchart for the detection of iris in unconstrained environments for all the nine contributions

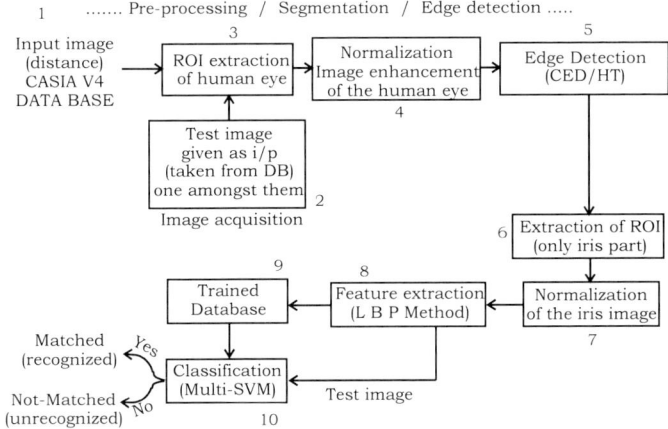

Fig. 2 Contribution 1—Iris recognition using CED/HT/LBP and Multi-SVM

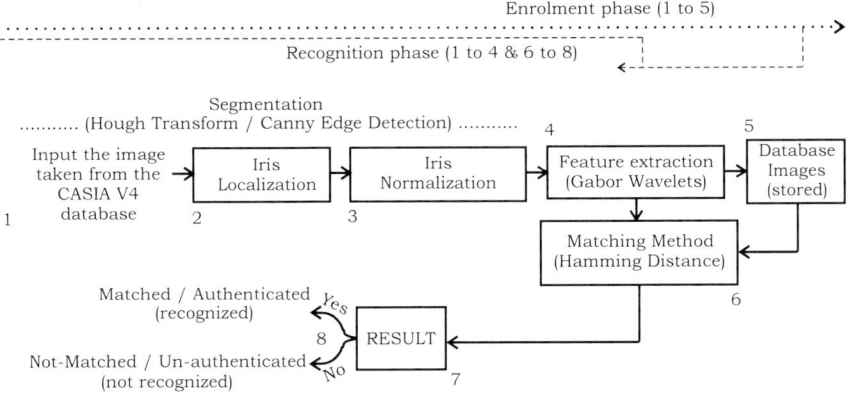

Fig. 3 Contribution 2—Iris recognition using (CED-HT)/(Gabor Wavelets)/(HD)

Contribution 1: Development of iris recognition using the concepts of preprocessing, segmentation (Canny Edge Detection/Circular Hough Transforms), feature extraction (Local Binary Pattern Features Method), and classification (Multi-SVM) of iris images with the development of an automatic GUI using Matlab. The proposed block diagram of this contributory work is shown in Fig. 2.

Contribution 2: Development of iris recognition using the concepts of preprocessing, segmentation (Hough Transform and Canny Edge Detection), feature extraction (Gabor Wavelets), and matching (Hamming Distance) of iris images with the development of an automatic GUI using Matlab. The proposed block diagram of this contributory work is shown in Fig. 3.

Contribution 3: Development of iris recognition using the concepts of preprocessing, segmentation (Fuzzy Trapezoidal and Sobel Operator Method), feature extrac-

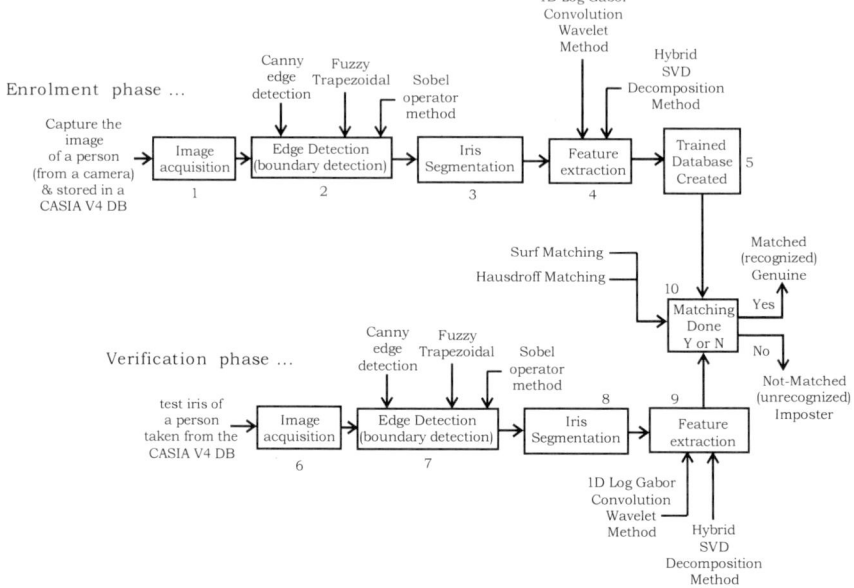

Fig. 4 Contribution 3—Iris recognition using (CED-Sobel-FTM)/(Gabor-SVD)/(Hausdorff–Surf method)

tion (1D Log Gabor Convolution Wavelet Method and Hybrid SVD Method), and Matching (Hausdorff Matching and Surf Matching) of iris images with the development of an automatic GUI using Matlab. The proposed block diagram of this contributory work is shown in Fig. 4.

Contribution 4: Development of iris recognition using the concepts of preprocessing, segmentation (OTSU Algorithm), feature extraction (SFTA Method), and Classification (Neural Network Algorithm) of iris images with the development of an automatic GUI using Matlab. The proposed block diagram of this contributory work is shown in Fig. 5.

Contribution 5: Development of iris recognition using the concepts of preprocessing, segmentation (Morphological Operators), feature extraction (GLCM Method and Fruit Fly with Cuckoo Search Algorithm, Heuristic Algorithm), and Classification (RBFNN, Neural Network, and SVM) of iris images with the development of an automatic GUI using Matlab. The proposed block diagram of this contributory work is shown in Fig. 6.

Contribution 6: Development of iris recognition using the concepts of preprocessing, segmentation (Canny Edge Detection Method and Hough Transforms), feature extraction (Local Binary Pattern Method), and matching of iris images (Hamming Distance Method) with the development of an automated GUI for iris biometric recognition using the Matlab tool [27–30]. The proposed block diagram of this contributory work is shown in Fig. 7.

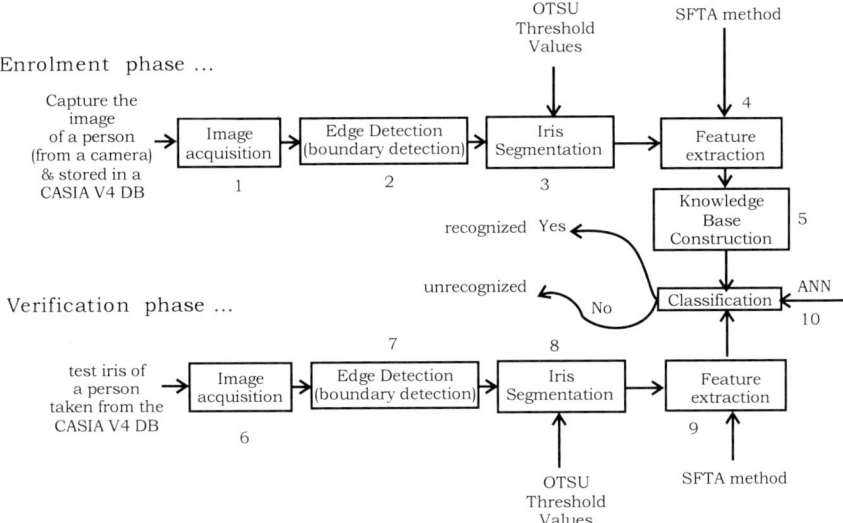

Fig. 5 Contribution 4—Iris recognition using (OTSU Algo)/(SFTA)/(ANN)

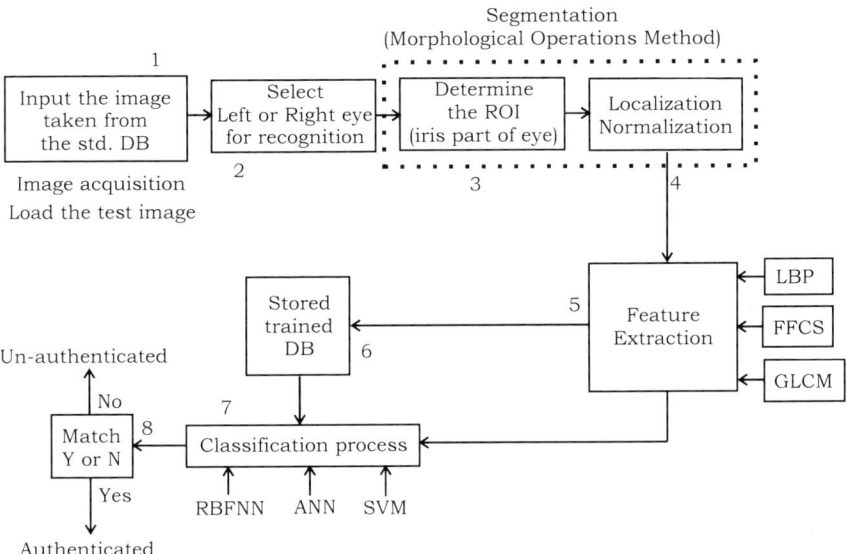

Fig. 6 Contribution 5—Iris recognition using (MO)/(GLCM-LBP-FFCS)/(RBFNN-ANN-SVM) method

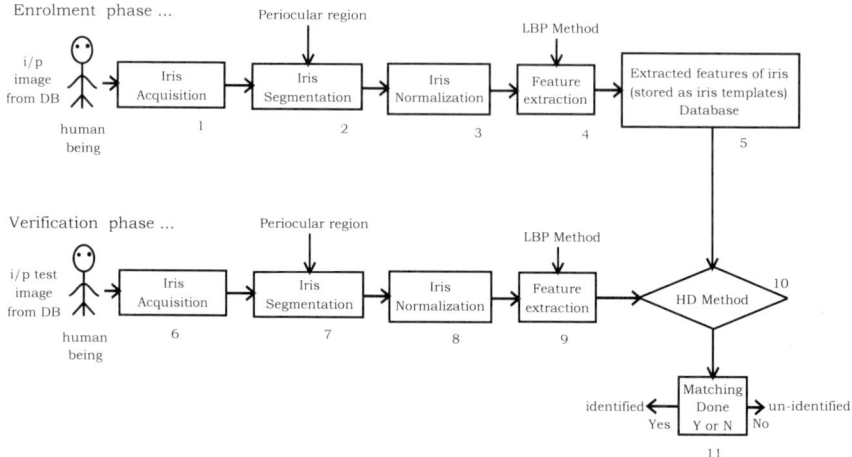

Fig. 7 Contribution 6—Iris recognition using (CED-HT)/LBP/HD method

8 Comparisons of the Contributions from C1 to C6

The six different methods which are proposed for the iris recognition are compared for their best performance, i.e., the comparisons of the different contributions from 1 to 6 are presented with their quantitative results for different cases of image processing such as segmentation, preprocessing, recognition, etc. It was found that for the entire set of images given from CASIA-V4 database, more than 99% of the images present in the database were recognized for all the six contributory works and the result of the percentage acceptance is shown in Table 1. Further, the accuracy of segmentation and recognition rate for all the contributions is shown in Table 1. Also, one of the contributions C5 has been compared with the work done by others which shows the supremacy of our method as shown in Table 2.

Table 1 Comparison rate of all the contributions

Method	Segmentation	Feature extraction	Matching/ classification	Percentage of segmentation	Percentage of recognition
1	CED/HT	LBP	SVM	99.8	99.72
2	CED	1D-GW	HDM	98.6	99.63
3	SO/Fuzzy/CED	1D-logGW/SVD	HDM/SURF	98.5	99.68
4	OTSU	SFTA	ANN	99.82	99.72
5	MO	GLCM/FFCS	RBFNN/ANN/SVM	99.7	99.85
6	CED/HT	LBP	HD	99.8	99.81

Table 2 Comparison rate of the proposed work with others

Algorithm	Correct recognition rate (%)
Daugman	99.5
Wildes	98
Boles	92.6
Li Ma	94.9
Roy	99.4
Ma	99.3
Liu	97.08
Avilla	97.89
Masek	83.92
Mayank	90
Proposed C5	99.85

9   Conclusion of the Six Proposed Methodologies (C1–C6)

The implementation results of the above six methodologies are, if the templates stored in the database is matched with the inputted image feature vector by using the hybrid methodology of classification as the classifiers, then it results into the conclusion that the person is recognized and genuine, else its imposter or not recognized (image not in the database).

10   Design and Developments Using the Contributions C7–C9

Contribution 7: Development of iris recognition using the concepts of preprocessing, segmentation (Boundary Detection Method), feature extraction (2D Gabor Wavelets and Haar Wavelets), and classification of iris images (ANN Method) with the development of an automated GUI for iris biometric recognition using the Lab-VIEW tool [31]. The proposed block diagram of this contributory work is shown in Fig. 8.

Contribution 8: Development of application of iris recognition system developed for electronic voting using LabVIEW. The proposed block diagram of this contributory work is shown in Fig. 9.

Contribution 9: Hardware implementation of the iris recognition concept using ATMEL microcontroller interfaced with LabVIEW with the development of an automated GUI. The proposed block diagram of this contributory work is shown in Fig. 10.

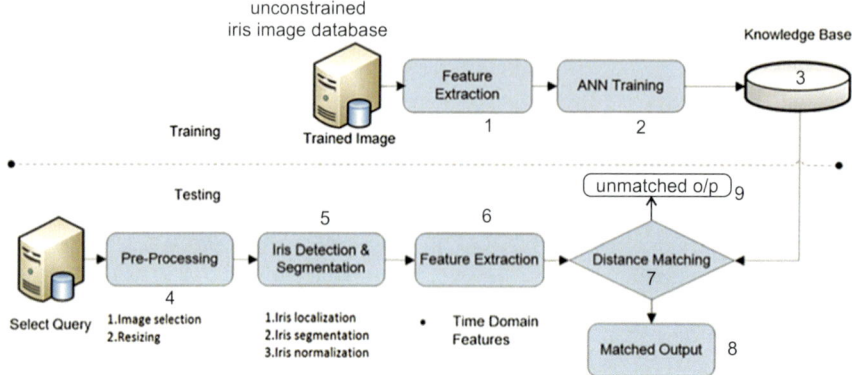

Fig. 8 Contribution 7—Proposed iris recognition BD using ANN with LabVIEW tool

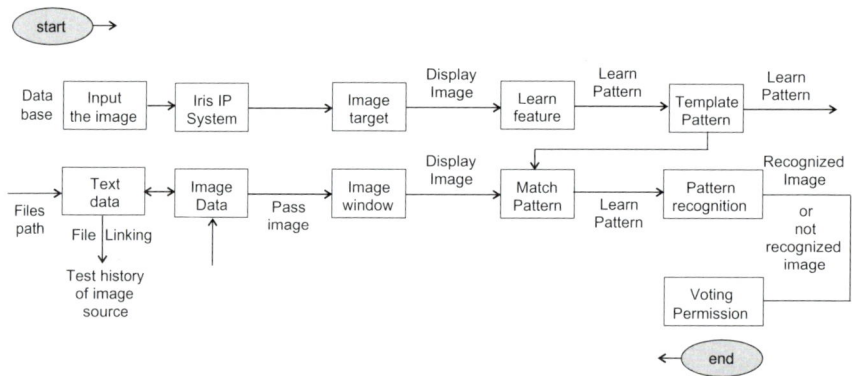

Fig. 9 Contribution 8—Data flow diagram of the IRS used for voting purposes

11 Working Principles of Contributions C7–C9

Working principle of C7: The seventh contributory work consists of developing an interactive iris recognition system for unconstrained images using the concept of preprocessing, segmentation (Boundary Detection Method), feature extraction (2D Gabor Wavelets and Haar Wavelets), and classification of iris images (ANN) with the development of an automated GUI for iris biometric recognition using the LabVIEW tool, and the iris recognition system is developed as per the block diagram shown in Fig. 8. Here, all the concepts of the various biomedical image processing such as the database generation, image acquisition/capturing, grayscale conversion, identification of ROI, preprocessing, resizing, boundary detection, segmentation, localization, normalization, noise removal, enhancement, feature processing, feature extraction, feature encoding, matching, classifiers, testing, decision taking, authentication, identification, and recognition are being used.

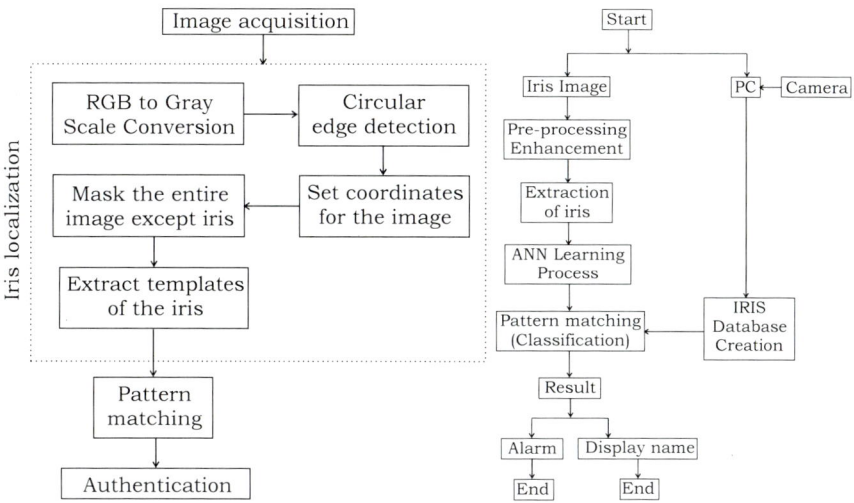

Fig. 10 Contribution 9—Algorithm for the hardware implementation using a μC

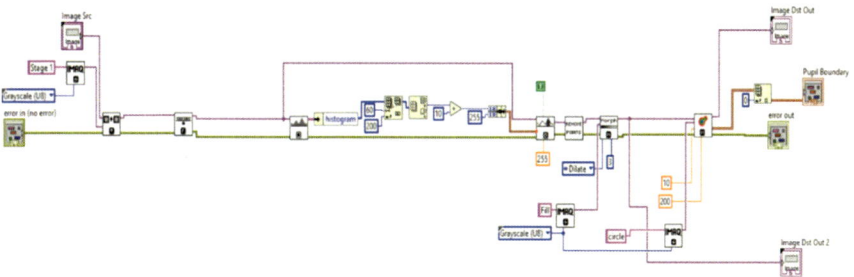

Fig. 11 Block diagram of the determination of pupil in LabVIEW (C7)

Circuit diagrams are being developed in the LabVIEW environment for each of the separate IP processes (such as pupil detection, limbic boundary detection, normalization, segmentation, resizing, feature extraction, noise removal, recognition, etc.) and then finally compiled into a hybrid model. One such diagram developed is shown in Fig. 11. When the overall hybrid model is run, the various simulation results are observed as shown in Figs. 12 and 13, respectively, thus showing the efficacy of the developed iris recognition system. If the test iris is matched with the iris in the database, then it appears as [Matched IRIS | 1], else it appears as [Matched IRIS | 0] (Fig. 14).

Working principle of C8: Eighth contributory work is an application-oriented one, which includes the development of an intuitive iris acknowledgment framework for unconstrained pictures utilizing the idea of IP and implementing it for a voting application. Circuit diagrams are being developed in the LabVIEW environment for each of the separate IP processes (such as pupil detection, limbic boundary detection,

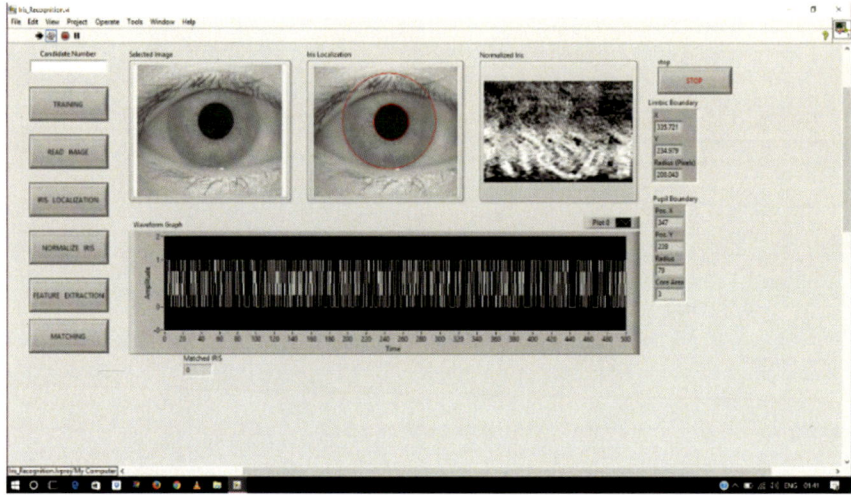

Fig. 12 GUI of the IRS application generated using LabVIEW—Matched iris (recognized—display shows 1)…(C7)

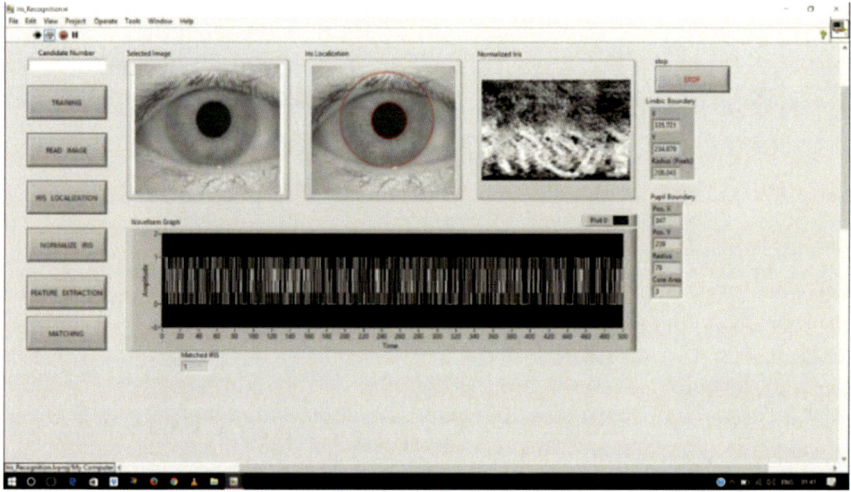

Fig. 13 GUI of the IRS application generated using LabVIEW—Not matched iris (not recognized—display shows 0)…. (C7)

normalization, segmentation, resizing, feature extraction, noise removal, recognition, etc.) for the considered voting application scenario and finally merged to form a hybrid model. To start with, all the iris images are taken from the standard available database one by one, preprocessing done, and iris parameters are being found out and saved in a DB. Next phase is called as verification or testing phase; the test iris which is to be detected to find whether its presence is there in the refined database or not is

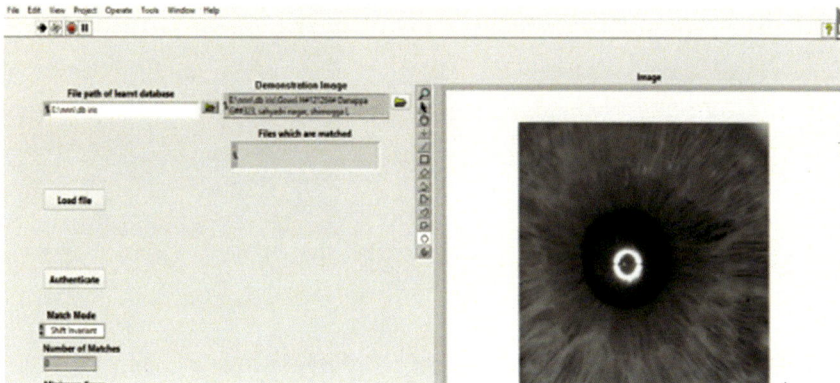

Fig. 14 Loading of the image of the human being (C8)

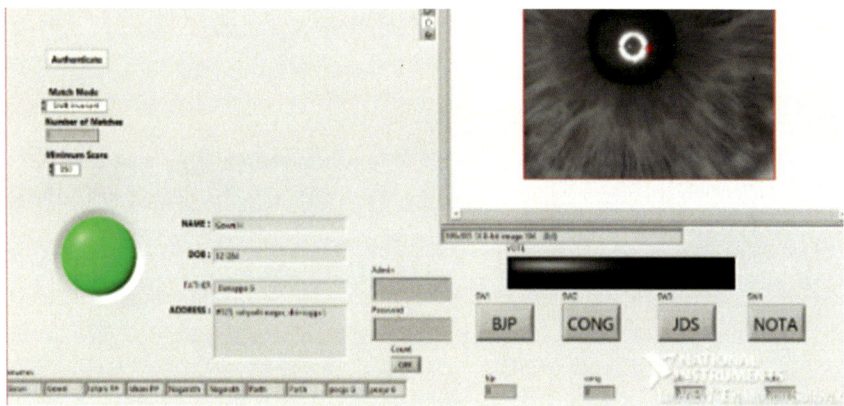

Fig. 15 Recognition success by an authorized person in the database (C8)

given as input to the proposed algorithm and all the processes as followed previously are carried out, and the matching of the test iris with its counterpart in the refined database is found out whether it is there or not using a user-friendly developed GUI.

If the recognition is successful, then the iris is authenticated (tab appears green in color), else it is not recognized (tab appears red in color). Now, once the iris is being recognized (Fig. 15), the person can do the voting by pressing the corresponding button against the party name, after which the red LED glows indicating that the person has voted (Fig. 16). For an unrecognized case, the test iris which is not there in the database is being loaded in the GUI and immediately the matching score says "0" indicating that the person is not present in the database and the tag glows red showing that the person is not there in the database and hence he cannot vote, proving that recognition is unsuccessful (Fig. 17), indicating red button.

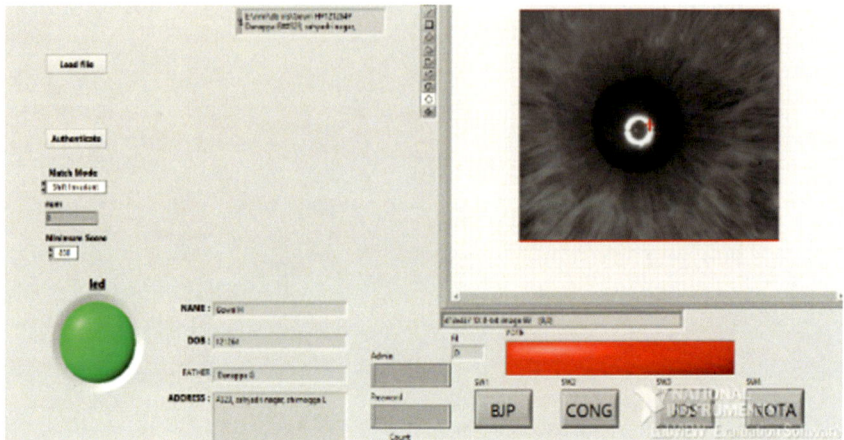

Fig. 16 Simulation results after being voted by a recognized human being (C8), shown in green color

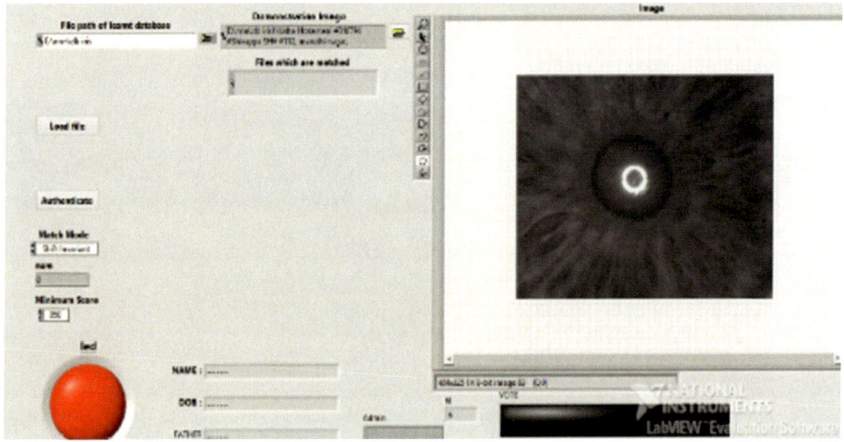

Fig. 17 Recognition unsuccessful by an unauthorized person in the database (failure)...,(C8), shown in red color

Working principle of C9: The ninth contributory work involves the hardware implementation of the iris recognition process using an Atmel microcontroller with the simulation being done in the LabVIEW environment and then dumping the code onto the μC. All the concepts of the various biomedical image processing such as the database generation, image acquisition/capturing, grayscale conversion, identification of ROI, preprocessing, resizing, boundary detection, segmentation, localization, normalization, noise removal, enhancement, feature processing, feature extraction, feature encoding, matching, classifiers, testing, decision taking, authentication, identification, and recognition are being used [32–34]. Once the code has been developed,

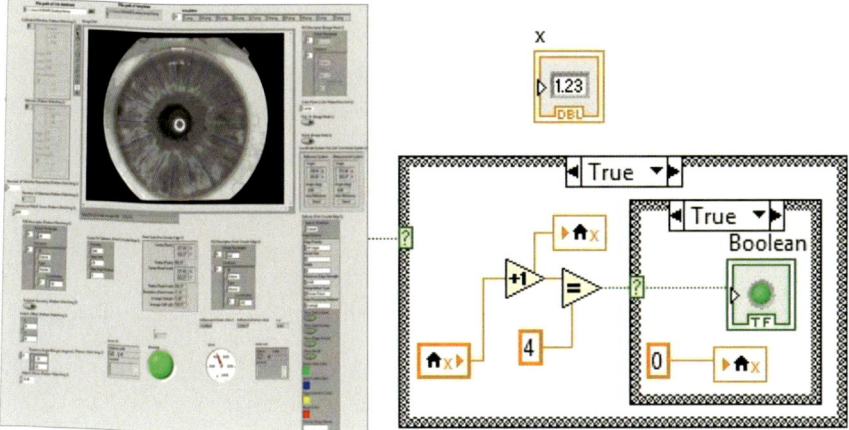

Fig. 18 VI G-code for recognition with recognized status shown in green color (C9) with the developed LabVIEW circuit diagram

Fig. 19 Hardware interfaced to μC using a laptop equipped with an NI LabVIEW tool

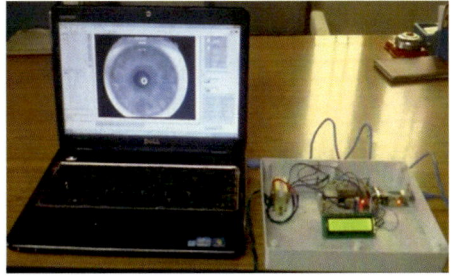

Fig. 20 Iris being recognized (authenticated) using the μC hardware kit (Contribution—9)

it is run and the various simulation results are observed as shown in Figs. 18, 19, and 20, respectively. It has to be noted that the iris images are being obtained from the standard iris image database. The LabVIEW tool is being interfaced to the μC, and the hardware results are observed on the display which is interfaced to the μC (Fig. 14).

Finally, the simulation results are validated with the hardware results justifying the same. The hardware unit of the IRS consists of the ATmega32 a Hi-Pi, LP AVR®

8-bit μC which is interfaced with a (16 × 2) Hitachi JHD 162A LCD display for displaying the iris recognition results. Also, a USB to UART converter adapter is being used for interfacing this hardware with the computer system using an SFE USB-RS232 Controller Driver. Along with this hardware, a 6-0-6 transformer, a 7805 5 V IC regulator, ST-232 driver circuits, USB cable, and power cords are being used for the hardware implementation purposes. The LCD display will display the authentication results obtained from the PC. The MAX-232 is a coordinated circuit, which gives the signal from an RS-232 serial port to the concerned signals for the use in a TTL good advanced rationale circuitry.

The MAX232 is a double drive/beneficiary and normally changes over the R_X, T_X, CTS, and the RTS signals. Figures 19 and 20 show the interfacing connections required to interface the ATmega32 with an LCD. The port A is used to get the o/p from microcontroller such that to display the test results in the display. The figure of hardware implementation (Figs. 19 and 20) is interfaced to a laptop/PC, which is fully loaded with NI LabVIEW s/w along with flowchart of the recognition process in Fig. 10. To start with, all the iris images are taken from the standard available database one by one, preprocessing done, extract iris features, train the different irises using ANN, and stored in a refined database. In the second phase, called as the verification or the testing phase, the test iris which is to be detected to find whether its presence is there in the refined database or not is given as input to the proposed algorithm and all the processes as followed previously are carried out and then, the matching of the test iris with its counterpart in the refined database is found out whether it is there or not. If the recognition is successful, then the iris is authenticated (LED turns green), else it is not recognized (LED turns red).

This work portrays a substitute strategy to distinguish people utilizing pictures of their iris using simulations as well as using hardware. Pattern recognition or template matching is carried out using the National Instruments LabVIEW tool. The developed NI LV script is finally run with the pictures which are inputted from the standard database. The GUI is developed using the NI LabVIEW and NI VISION Assistant is used for the extraction of templates characteristics. This recognizable proof framework is basic requiring just couple of parts and is sufficiently compelling to be coordinated inside security frameworks that require a personality check. Once simulations could be observed, then G-code is dumped on to the μC through the relevant ports and the recognition phenomenon is observed on the display.

12 Conclusions of Contributions C7–C9

Research was done in the area of iris recognition systems with the help of LabVIEW tool. Three different contributions (C7–C9) were carried out, with each one having one or the other advantages over the others. First one being a typical iris recognition system design using LabVIEW, second one being an application of iris recognition system using LabVIEW for a voting application, and the third one being the hardware implementation of IRS using LabVIEW. Simulations presented under the simulation results section in each contribution show the effectiveness of the methods proposed for developing a user-friendly graphical user interface for the iris recognition.

13 Overall Conclusions of the Research Work

In this section, the brief outcome, i.e., the conclusions about the research work done, followed by the scope for the future work in this exciting field of biometric authentication and identification under *unconstrained environments* is being presented.

Research was done for the efficient approach for secure biometric authentication using iris. Broad writing literature overview was being done in this emerging–developing field. In this context, a sincere attempt was being made to develop a simple and efficient method for iris recognition using simple segmentation method. The time consumption of the system is also very low, as it can identify an iris within few seconds. This time includes segmentation, feature extraction, feature selection/dimension reduction, and classification time. After the successful experiments and the encouraging results achieved, it can be claimed that the proposed system is capable of fast and efficient iris identification.

The proposed methodology developed by us has a wide range of advantages such as simple algorithm and easy to implement, less complicated, less execution time, faster authentication and recognition, iris has a fine texture because of which the extraction of features becomes easy, speed of the recognition system increases significantly, false acceptance ratio reduces up to a significant extent, performs robustly with different image qualities, highly reliable system for authentication, highly stable over lifetime, and many more. The recognition system developed is quite simple requiring only few components with the concept of the usage of hybrid methodologies and is highly effective enough to be integrated within security systems that require an identity check in any engineering application.

Research was done for the efficient approach for secure biometric recognition using iris under a large number of unconstrained environments such as tilted iris, iris captured under bad light illumination, from a distance, at an angle, wearing spectacles, occluded, etc.

In the proposed system, CASIA-V4 databases have been used for the recognition of iris in human beings and the work can be extended to other databases also in the near future by the future researchers. Moreover, average time consumption of the system could be improved by changing/improving the segmentation technique and other classifiers may also be used to evaluate the system. A critical feature of this coding approach is the achievement of commensurability among iris codes, by mapping all irises into a representation having universal format and constant length, regardless of the apparent amount of iris detail.

Any iris extraction from the acquired image needs four important parameters, viz., preprocessing, segmentation, feature extraction, and classifiers approaches. Different algorithms were used in the above four cases, each of which is mentioned below in greater detail. The various preprocessing methodologies used in our work are Hough

Transforms, Mean Median Std. Deviation and Entropy Methods, Morphological Operations Method, and the Boundary Detection Method.

The various segmentation methods that were used in our work for segmenting the iris image from the background are canny edge detection, fuzzy trapezoidal, and the SFTA—Segmentation-based Fractal Texture Analysis method, rising and falling edge method. The various feature extractions of iris images are done by using hybrid combinations of Gray-Level Co-occurrence Matrix (GLCM), LBP, FT, 2D-Gabor filter, and Fast Fourier Transform (FFT).

The various classifiers/matching concepts that were used in our work for classifying the recognized images are SVM, ANN, RBFNN, HD, and SURF which give effective results with less number of false detections (fake detection and recognition). Satisfying and convincing results are obtained for all the defined objectives of the proposed algorithms. Thus, biometric recognition system using iris which is developed by us is quite simple and requires few components and is effective enough to be integrated within security systems that require an identity check. The work can be used in security- and authentication-related applications for unconstrained environments successfully, and the errors that have occurred can be easily overcome by the use of stable equipment.

A. *Nine contributory works were carried out in this research work which was briefly explained in different sections.*
B. *The eight different algorithms provided good results when implemented on the iris CASIA Version 4 database under unconstrained environments.*
C. *One application problem was considered using LabVIEW and was successfully tackled upon with.*
D. *A hardware implementation of the iris recognition using a microcontroller was also implemented which yielded good results of iris authentication.*

Related to the methods mentioned in the exhaustive literature review, the proposed methods vary from different aspects and take into account all possible outcomes such as variations in the iris part in varying sunlight and brightness (*low light, high glares, darker regions, wearing spectacles, half eyes closed, half eye open, squint eyes, partially occluded eyes,* etc.). Positions of iris at different distances and angles [35–37], when iris part is not captured completely due to blink of eyelashes, presence of sunglasses, and other factors based on the work done to simulate and capture the iris image with different cases like angle, light incident on surface, and different other real-time scenarios are also considered.

The outcome of the research work is to show that when the designed algorithm/s developed in the Matlab/LabVIEW environment is run, the automatic recognition of the iris is done with minimum computational time in comparison with the work done by the other researchers till date taking into consideration many of the drawbacks of the fellow researchers, thus enhancing and improving the performance of the existing algorithms under unconstrained environments with the development of an automated GUI for iris recognition purposes. The nine different algorithms which were proposed provided good results when implemented on the different standard CASIS Version 4 iris databases under unconstrained environments.

14 Scope for the Future Work

At the end, the scope for future work, i.e., extensions of the research work that is being done, is being presented in this paper, which can be explored upon by the future research scholars. Iris recognition is an important and challenging task in secure authentication. Even though the proposed work gives good results with many advantages, many improvements can be done. Future work can aim to enhance the system to deal with image degradation by noise, duty, and glasses. Color feature can be also used to increase the recognition accuracy. There is scope to work on developing an algorithm based on multimodal biometrics system which will combine both the biometric characteristics derived from one or modalities such as palm print and iris which give high level of security. To conclude, a sincere attempt was being made to develop simple and efficient methods for iris recognition in "***unconstrained environments***" using a combination of several methodologies (hybrid algorithms) in the various processes of the iris biometric recognition which could be seen from the results of simulation and experiments.

References

1. Daugman J (2004) How iris recognition works. IEEE Trans Circuits Syst Video Technol 14(I):21–30
2. Daugman J (2004) Recognizing persons by their iris patterns. In: Proceedings of Advances in Biometric Person Authentication, vol 3338, pp 5–25
3. Daugman J (2007) New methods in iris recognition. IEEE Trans Syst Man Cybern Part B-Cybern 37(5):1167–1175
4. Daugman J (2003) The importance of being random: statistical principles of iris recognition. Pattern Recognit 36:279–291
5. Daugman JG, Downing CJ (2001) Epigenetic randomness, complexity & singularity of human iris patterns. Proc R Soc Lond B Biol Sci 268:1737–1740
6. Zhou Z, Du YZ et al (2009) Transforming traditional iris recognition systems to work in non-ideal situations. IEEE Trans Ind Electr 56(8):3203–3213
7. Daniel DM, Monica B (2010) Person authentication technique using human iris recognition.In: IEEE Conference Paper, pp 265–268. 978-1-4244-8460-7/10/$26.00©2010
8. Li P, Ma H (2012) Iris recognition in non-ideal imaging conditions. Pattern Recogn Lett 33:1–9
9. Karakaya M (2016) A study of how gaze angle affects the performance of iris recognition. Elsevier Sci Direct's Pattern Recog Lett 82:132–143
10. Fancourt C, Bogoni L, Hanna K, Guo Y, Wildes R, Takahashi N, Jain U (2005) Iris recognition at a distance. In: Proceeding of international conference on audio and video based biometric person authentication, vol 3546, pp 1–13
11. Moi SH, Asmuni H, Hassan R, Othman RM (2014) A unified approach for unconstrained off-angle iris recognition. In: International symposium on biometrics & security technologies (ISBAST), pp 39–44. 978-1-4799-6444-4/14/$31.00©2014. IEEE
12. Kaur N, Juneja M (2014) A novel approach for iris recognition in unconstrained environment. J Emerg Technol Web Intell Acad Publishers 6(2):243–246
13. Tsai Y-H (2014) A weighted approach to unconstrained iris recognition. World Acad Sci Eng Technol Int J Comput Inf Eng 8(1):30–33. ISSN:1307-6892
14. Shin KY, Nama GP, Jeong DS, Cho DH, Kang BJ, Park KR, Kim J (2012) New iris recognition method for noisy iris images. Elsevier Sci Direct's Pattern Recogn Lett 33:991–999

15. Chen Y (2010) A high efficient biometrics approach for unconstrained iris segmentation and recognition. Ph.D. Thesis. College of Engineering and Computing, Florida International University, Greater Miami

16. Roy K, Bhattacharya P et al (2010) Unideal iris segmentation using region-based active contour model. In: Campilho A, Kamel M (eds) ICIAR 2010, Part II. LNCS 6112, © Springer, Berlin, Heidelberg, Germany, pp 256–265

17. Jan F (2017) Segmentation and localization schemes for non-ideal iris biometric systems. Elsivier's Sci Direct Jr Signal Proc 133:192–212

18. Reddy N, Rattani A, Derakhshani R (2016) A robust scheme for iris segmentation in mobile environment. In: IEEE symposium on technologies for homeland security (HST). IEEE, Florida, USA. 978-1-5090-0770-7/16/$31.00©2016

19. Santoso A, Choirunnisa S, Prihasto B, Wang J-C (2016) Improving iris image segmentation in unconstrained environments using NMF-based approach. In: 2016 IEEE international conference on consumer electronics-Taiwan (ICCE-TW). IEEE, pp 27–29. 978-1-5090-2073-7/16/$31.00©2016

20. Yahiaoui M, Monfrini E, Dorizzi B (2016) Markov Chains for unsupervised segmentation of degraded NIR iris images for person recognition. Elsevier Sci Direct's Jr Pattern Recogn Lett 82:116–123. http://dx.doi.org/10.1016/j.patrec.2016.05.025

21. Sahmoud SA, Abuhaiba IS (2013) Efficient iris segmentation ethod in unconstrained environments. Elsivier's Science Direct Jr. of Pattern Recognition, vol 46, pp 3174–3185

22. Mahlouji1 M, Noruzi A (2012) Human iris segmentation for iris recognition in unconstrained environments. IJCSI Int J Comput Sci Issues 9(1), No. 3:149–155. ISSN (Online): 1694-0814

23. Proenca H (2010) An iris recognition approach through structural pattern analysis methods. Expert Syst 27(1):6–16

24. Raffei AFM, Hishammuddin A, Hassan R, Othman RM (2013) Feature extraction for different distances of visible reflection iris using multiscale sparse representation of local Radon transforms. Elsivier's Sci Direct Jr Patt Recogn 46:2622–2633

25. Haindl M, Krupička M (2015) Unsupervised detection of non-iris occlusions. Elsevier Sci Direct's Pattern Recogn Lett 57:60–65

26. Barpanda SS, Majhi B, PankajKumar S (2015) Region based feature extraction from non-cooperative iris images using triplet half-band filter bank. Elsevier Sci Direct's Jr Opt Laser Technol 72:6–14

27. Tan C-W et al (2013) Towards online iris & periocular recognition under relaxed imaging constraints. IEEE Trans Image Proc 22(10):3751–3765

28. Santos G, Hoyle E (2012) A fusion approach to unconstrained iris recognition. Elsevier Sci Direct's Pattern Recogn Lett 33:984–990

29. Proença H, Neves JC (2016) Visible-wave length iris/periocular imaging and recognition surveillance environments. Elsevier Sci Direct's Jr Image Vis Comput 55:22–25

30. Liu B, Lam S-K, Srikanthan T, Yuan W (2011) Utilizing dark features for iris recognition in less constrained environments. In: Fourth IEEE international symposium on parallel architectures, algorithms and programming. IEEE Comp Soc, Tianjin, China, pp 110-114. ISBN 978-0-7695-4575-2/11 $26.00©2011, https://doi.org/10.1109/paap.2011.51

31. Hajaria K, Gawandeb U, Golharc Y (2016) Neural network approach to iris recognition in noisy environment. Elsivier's Sci Direct Int Conf Inf Secur Priv (ICISP2015) Procedia Comput Sci 78:675–682, 11–12. https://doi.org/10.1016/j.procs.2016.02.116.Nagpur, India

32. Liu J, Sun Z, Tan T (2014) Distance metric learning for recognizing low-resolution iris images. Elsevier Sci Direct's Jr Neurocomp 144:484–492

33. Alvarez-Betancourt Y, Garcia-Silvente M (2016) A key points—based feature extraction method for iris recognition under variable image quality conditions. Elsevier Sci Direct's Jr Knowl-Based Syst 92:169–182

34. Hu Y, Sirlantzis K, Howells G (2015) Exploiting stable and discriminative iris weight map for iris recognition under less constrained environment. In: 7th IEEE international conference on biometrics theory, applications and systems (BTAS), Arlington, VA, USA, ISBN: INSPEC Accession Number: 15668109, https://doi.org/10.1109/btas.2015.7358759, 8–11 Sept 2015

35. Hu Y, Sirlantzis K, Howells G (2017) A novel iris weight map method for less constrained iris recognition based on bit stability and discriminability. Elsevier Sci Direct's Jr Image Vis Comput 58:168–180
36. Arya KV, Gupta A, Kumar G, Singhal P (2012) A novel approach to minimize the impact of non ideal samples in iris recognition system. In: Third IEEE international conference on computer and communication technology. IEEE Comp Soc, Allahabad, U.P., India, pp 352–356. 978-0-7695-4872-2/12 $26.00©2012 IEEE, https://doi.org/10.1109/iccct.2012.77, 23–25 Nov 2012
37. Perez C, Lazcano V, Estévez P, Held C (2009) Real-time template based face and iris detection on rotated faces. Int J Opto-mech 3:54–67. Taylor & Francis Group, UK. https://doi.org/10.1080/15599610902717801, ISSN: 1559-9612 (Print) 1559-9620 (Online)